METHODS IN MOLECULAR MEDICINE™

Lung Cancer

*Volume 2:
Diagnostic and Therapeutic
Methods and Reviews*

Edited by

Barbara Driscoll, PhD

*Children's Hospital,
Los Angeles, CA*

Humana Press ✳ Totowa, New Jersey

© 2003 Humana Press Inc.
999 Riverview Drive, Suite 208
Totowa, New Jersey 07512

www.humanapress.com

All rights reserved. No part of this book may be reproduced, stored in a retrieval system, or transmitted in any form or by any means, electronic, mechanical, photocopying, microfilming, recording, or otherwise without written permission from the Publisher. Methods in Molecular Medicine™ is a trademark of The Humana Press Inc.

All papers, comments, opinions, conclusions, or recommendations are those of the author(s), and do not necessarily reflect the views of the publisher.

This publication is printed on acid-free paper. ∞
ANSI Z39.48-1984 (American Standards Institute) Permanence of Paper for Printed Library Materials.

Cover design by Patricia F. Cleary.

Cover illustration: Figure 1B from Volume 1, Chapter 2/Clinical and Biological Relevance of Recently Defined Categories of Pulmonary Neoplasia by E. Gabrielson.

Production Editor: Mark J. Breaugh.

For additional copies, pricing for bulk purchases, and/or information about other Humana titles, contact Humana at the above address or at any of the following numbers: Tel.: 973-256-1699; Fax: 973-256-8341; E-mail: humana@humanapr.com; Website: http://humanapress.com

Photocopy Authorization Policy:
Authorization to photocopy items for internal or personal use, or the internal or personal use of specific clients, is granted by Humana Press Inc., provided that the base fee of US $10.00 per copy, plus US $00.25 per page, is paid directly to the Copyright Clearance Center at 222 Rosewood Drive, Danvers, MA 01923. For those organizations that have been granted a photocopy license from the CCC, a separate system of payment has been arranged and is acceptable to Humana Press Inc. The fee code for users of the Transactional Reporting Service is [0-89603-920-X/03 $10.00 + $00.25].

Printed in the United States of America. 10 9 8 7 6 5 4 3 2 1

Library of Congress Cataloging in Publication Data

Main entry under title: Methods in molecular medicine™.

Lung cancer / edited by Barbara Driscoll.
    p. ; cm. -- (Methods in molecular medicine ; 74-75)
  Includes bibliographical references and index.
  Contents: v. 1. Molecular pathology methods and reviews -- v. 2. Diagnostic and therapeutic methods and reviews.
   ISBN 0-89603-985-4 (v. 1 : alk. paper) -- ISBN 0-89603-920-X (v. 2 : alk. paper)
   1. Lungs--Cancer--Molecular aspects--Laboratory manuals. 2. Lungs--Cancer--Gene therapy--Laboratory manuals. 3. Molecular diagnosis--Laboratory manuals. 4. Tumor markers--Laboratory manuals. I. Driscoll, Barbara. II. Series.
   [DNLM: 1. Lung neoplasms--physiopathology. 2. Lung Neoplasms--diagnosis. 3. Lung Neoplasms--therapy. 4. Molecular Biology--methods. WF 658 L96062 2002]
RC280.L8 L76532 2002
616.99'424--dc21

                                                2002024050

Lung Cancer
Volume 2

METHODS IN MOLECULAR MEDICINE™

John M. Walker, Series Editor

79. **Drugs of Abuse:** *Neurological Reviews and Protocols*, edited by *John Q. Wang*, 2003
78. **Wound Healing:** *Methods and Protocols*, edited by *Luisa A. DiPietro and Aime L. Burns*, 2003
77. **Psychiatric Genetics:** *Methods and Reviews*, edited by *Marion Leboyer and Frank Bellivier*, 2003
76. **Viral Vectors for Gene Therapy:** *Methods and Protocols*, edited by *Curtis A. Machida*, 2003
75. **Lung Cancer:** *Volume 2, Diagnostic and Therapeutic Methods and Reviews*, edited by *Barbara Driscoll*, 2003
74. **Lung Cancer:** *Volume 1, Molecular Pathology Methods and Reviews*, edited by *Barbara Driscoll*, 2003
73. **E. coli:** *Shiga Toxin Methods and Protocols*, edited by *Dana Philpott and Frank Ebel*, 2003
72. **Malaria Methods and Protocols,** edited by *Denise L. Doolan*, 2002
71. **Hemophilus influenzae Protocols,** edited by *Mark A. Herbert, E. Richard Moxon, and Derek Hood*, 2002
70. **Cystic Fibrosis Methods and Protocols,** edited by *William R. Skach*, 2002
69. **Gene Therapy Protocols, 2nd ed.,** edited by *Jeffrey R. Morgan*, 2002
68. **Molecular Analysis of Cancer,** edited by *Jacqueline Boultwood and Carrie Fidler*, 2002
67. **Meningococcal Disease:** *Methods and Protocols*, edited by *Andrew J. Pollard and Martin C. J. Maiden*, 2001
66. **Meningococcal Vaccines:** *Methods and Protocols*, edited by *Andrew J. Pollard and Martin C. J. Maiden*, 2001
65. **Nonviral Vectors for Gene Therapy:** *Methods and Protocols*, edited by *Mark A. Findeis*, 2001
64. **Dendritic Cell Protocols,** edited by *Stephen P. Robinson and Andrew J. Stagg*, 2001
63. **Hematopoietic Stem Cell Protocols,** edited by *Christopher A. Klug and Craig T. Jordan*, 2002

62. **Parkinson's Disease:** *Methods and Protocols*, edited by *M. Maral Mouradian*, 2001
61. **Melanoma Techniques and Protocols:** *Molecular Diagnosis, Treatment, and Monitoring*, edited by *Brian J. Nickoloff*, 2001
60. **Interleukin Protocols,** edited by *Luke A. J. O'Neill and Andrew Bowie*, 2001
59. **Molecular Pathology of the Prions,** edited by *Harry F. Baker*, 2001
58. **Metastasis Research Protocols:** *Volume 2, Cell Behavior In Vitro and In Vivo*, edited by *Susan A. Brooks and Udo Schumacher*, 2001
57. **Metastasis Research Protocols:** *Volume 1, Analysis of Cells and Tissues*, edited by *Susan A. Brooks and Udo Schumacher*, 2001
56. **Human Airway Inflammation:** *Sampling Techniques and Analytical Protocols*, edited by *Duncan F. Rogers and Louise E. Donnelly*, 2001
55. **Hematologic Malignancies:** *Methods and Protocols*, edited by *Guy B. Faguet*, 2001
54. **Mycobacterium tuberculosis Protocols,** edited by *Tanya Parish and Neil G. Stoker*, 2001
53. **Renal Cancer:** *Methods and Protocols*, edited by *Jack H. Mydlo*, 2001
52. **Atherosclerosis:** *Experimental Methods and Protocols*, edited by *Angela F. Drew*, 2001
51. **Angiotensin Protocols**, edited by *Donna H. Wang*, 2001
50. **Colorectal Cancer:** *Methods and Protocols*, edited by *Steven M. Powell*, 2001
49. **Molecular Pathology Protocols**, edited by *Anthony A. Killeen*, 2001
48. **Antibiotic Resistance Methods and Protocols,** edited by *Stephen H. Gillespie*, 2001
47. **Vision Research Protocols,** edited by *P. Elizabeth Rakoczy*, 2001
46. **Angiogenesis Protocols,** edited by *J. Clifford Murray*, 2001
45. **Hepatocellular Carcinoma:** *Methods and Protocols*, edited by *Nagy A. Habib*, 2000
44. **Asthma:** *Mechanisms and Protocols*, edited by *K. Fan Chung and Ian Adcock*, 2001

Preface

This work, *Lung Cancer, Volume 2: Diagnostic and Therapeutic Methods and Reviews*, in the *Methods in Molecular Medicine* series presents an overview of the current status of those methods useful in the diagnosis and treatment of lung cancer—both as it exists in the clinic and as it is being revolutionized in the laboratory. The book is intended to serve as a resource for researchers wishing to increase their knowledge of current and cutting edge technologies, in order that their investigations into neoplasms of the lung may benefit from this enriched diversity of techniques and approaches.

Owing to the complex nature of the disease and the variety of methods available to analyze and attack it, no volume attempting to define diagnostic and therapeutic approaches to lung cancer can ever be complete. The sheer number of investigators involved in lung cancer research guarantees that some aspect will be inadvertently excluded. However, I hope that the range of techniques included herein will serve to open up new avenues of investigation for both the novice and experienced researcher.

As with all volumes in the *Methods in Molecular Medicine* series, the reader should find that each methods-based chapter provides clear instructions for the performance of various protocols, supplemented by additional technical notes that provide valuable insight. These notes should offer the reader a perspective on the skills and materials required for performance of both standard and novel techniques. In putting together this volume, one aim was to provide clinicians and laboratory investigators an appreciation of the current status of lung cancer diagnosis and treatment. However, *Lung Cancer, vol. 2* is also a look into the future, with descriptions of novel methods for molecular diagnosis as well as techniques for treatment based on gene therapies, new anticancer approaches, immune therapies, and chemoprevention. Chapters describing the challenges faced by those attempting to employ these new methods should also be of use to those determined to translate basic research from the laboratory to the clinic.

I would like to express my gratitude to the contributors who made this volume possible and for their patience during the period the volume was collated. I am grateful to Professor John Walker for his encouragement and guidance as series editor.

This volume is dedicated to the memory of Richard Mackenzie Brown, Warren Reardon, and Ching-Tuan T'Ang, three fathers, deeply missed.

***Barbara Driscoll,* PhD**

Contents

Preface v

Contents of Volume I xiii

Contributors xvii

PART I. INTRODUCTION
1 Molecular and Genetic Aspects of Lung Cancer
   William N. Rom and Kam-Meng Tchou-Wong 3

PART II. DETECTION OF ALTERED TUMOR MARKERS IN CLINICAL SAMPLES FOR DIAGNOSIS AND PROGNOSIS

   A. MOLECULAR METHODS FOR PROGNOSTIC DETECTION OF ALTERED TUMOR MARKERS: CLINICAL IMPLICATIONS

2 Molecular Alterations in Lung Cancer: *Impact on Prognosis*
   Takashi Kijima, Gautam Maulik, and Ravi Salgia 29
3 Cellular Predictive Factors of Drug Resistance
   in Non-Small Cell Lung Carcinomas
   Manfred Volm, Reet Koomägi, and Werner Rittgen 39
4 Clinical Implications of p53 Mutations in Lung Cancer
   Barbara G. Campling and Wafik S. El-Deiry 53
5 An Epidemiologic and Clinicopathologic Overview
   of AIDS-Associated Pulmonary Kaposi's Sarcoma
   David M. Aboulafia 79
6 Myeloperoxidase Promoter Region Polymorphism
   and Lung Cancer Risk
   Xifeng Wu, Matthew B. Schabath, and Margaret R. Spitz 121
7 Clinical Utility of Tumor Markers in the Management
   of Non-Small Cell Lung Cancer
   Dennis E. Niewoehner and Jeffrey B. Rubins 135

B. Molecular Methods for Prognostic Detection of Altered Lung Tumor Markers in Clinical Samples

I. DNA-Based Analysis

8 Detection of Chromosomal Aberrations in Lung Tissue and Peripheral Blood Lymphocytes Using Interphase Fluorescence *In Situ* Hybridization (FISH)
   Randa A. El-Zein and Sherif Z. Abdel-Rahman 145

9 *In Situ* Analysis of Telomerase RNA Gene Expression as a Marker for Tumor Progression
   W. Nicol Keith 163

10 A High-Throughput Methodology for Identifying Molecular Targets Overexpressed in Lung Cancers
   Tongtong Wang and Steven G. Reed 177

11 Expression Profiling of Lung Cancer Based on Suppression Subtraction Hybridization (SSH)
   Simone Petersen and Iver Petersen 189

12 Comparative Genomic Hybridization of Human Lung Cancer
   Iver Petersen 209

13 Sensitive Assays for Detection of Lung Cancer: Molecular Markers in Blood Samples
   Marcia V. Fournier, Katherine J. Martin, Edgard Graner, and Arthur B. Pardee 239

14 Fluorescent Microsatellite Analysis in Bronchial Lavage as a Potential Diagnostic Tool for Lung Cancer
   John K. Field and Triantafillos Liloglou 251

15 Southern Blotting of Genomic DNA from Lung and Its Tumors: Application to Analysis of Allele Loss on Chromosome 11p15.5
   Diana M. Pitterle and Gerold Bepler 263

16 Assessment of Insulin-Like Growth Factors and Mutagen Sensitivity as Predictors of Lung Cancer Risk
   Xifeng Wu, He Yu, Nimisha Makan, and Margaret R. Spitz 279

II. PCR-Based Analyses

17 Comparative Multiplex PCR and Allele-Specific Expression Analysis in Human Lung Cancer: *Tools to Facilitate Target Identification*
   Jim Heighway, Daniel Betticher, Teresa Knapp, and Paul Hoban 291

18 Detection of K-ras Point Mutations in Sputum from Patients with Adenocarcinoma of the Lung by Point-EXACCT
   Veerle A. M. C. Somers and Frederick B. J. M. Thunnissen 305

19 Detection of K-ras and p53 Mutations
by "Mutant-Enriched" PCR-RFLP
Marcus Schuermann *325*

20 Detection of Small Cell Lung Cancer by RT-PCR
for Neuropeptides, Neuropeptide Receptors,
or a Splice Variant of the Neuron Restrictive Silencer Factor
**Judy M. Coulson, Samreen I. Ahmed, John P. Quinn,
and Penella J. Woll** *335*

III. Immunohistochemical Analyses

21 Utilization of Thyroid Transcription Factor-1 Immunostaining
in the Diagnosis of Lung Tumors
Nelson G. Ordóñez *355*

22 Molecular Biologic Substaining of Stage I NSCLC
Through Immunohistochemistry Performed on Formalin-Fixed,
Paraffin-Embedded Tissue
**Mary-Beth Moore Joshi, Thomas A. D'Amico,
and David H. Harpole, Jr.** *369*

23 A Sensitive Immunofluorescence Assay for Detection
of p53 Protein in the Sputum
Zumei Feng, Defa Tian, and Radoslav Goldman *389*

24 SPR1: *An Early Molecular Marker for Bronchial Carcinogenesis*
Derick Lau, Linlang Guo, Andrew Chan, and Reen Wu *397*

25 Determination of Biological Parameters on Fine-Needle Aspirates
from Non-Small Cell Lung Cancer
**Cecilia Bozzetti, Annamaria Guazzi, Rita Nizzoli, Nadia Naldi,
Vittorio Franciosi, and Stefano Cascinu** *405*

IV. Imaging Analysis

26 Detection and Analysis of Lung Cancer Cells from Body Fluids
Using a Rare Event Imaging System
Stine-Kathrein Kraeft, Ravi Salgia, and Lan Bo Chen *423*

27 Reconstruction of Geno-Phenotypic Evolutionary Sequences
from Intracellular Patterns of Molecular Abnormalities
in Human Solid Tumors
**Stanley E. Shackney, Charles A. Smith, Agnese A. Pollice,
and Jan F. Silverman** *431*

PART III. NOVEL THERAPIES

A. INTRODUCTION TO CURRENT TREATMENTS

28 Surgical Treatment of Lung Cancer: *Past and Present*
Clifton F. Mountain and Kay E. Hermes *453*

29 Recent Advances and Dilemmas
    in the Radiotherapeutic Management
    of Locally Advanced Non-Small Cell Lung Cancer
    Benjamin Movsas **489**

30 Photodynamic Therapy in Lung Cancer: *A Review*
    David Ost **507**

   B. GENE THERAPIES
   *I. Overview: Advantages and Pitfalls
   of Lung-Targeted Gene Therapies*

31 Gene Therapy for Lung Cancer: *An Introduction*
    Eric B. Haura, Eduardo Sotomayor, and Scott J. Antonia **529**

32 Identifying Obstacles to Viral Gene Therapy for Lung Cancer:
      Malignant Pleural Effusions as a Paradigm
    Raj K. Batra, Raymond M. Bernal, and Sherven Sharma **545**

33 Topical Gene Therapy for Pulmonary Diseases
      with PEI-DNA Aerosol Complexes
    **Ajay Gautam, J. Clifford Waldrep, Frank M. Orson,
    Berma M. Kinsey, Bo Xu, and Charles L. Densmore** **561**

   II. Plasmid and Virus-Based Gene Therapies

34 Targeted Delivery of Expression Plasmids to the Lung
      via Macroaggregated Polyethylenimine-Albumin Conjugates
    **Frank M. Orson, Berma M. Kinsey, Balbir S. Bhogal,
    Ling Song, Charles L. Densmore, and Michael A. Barry** **575**

35 Preparation of Retroviral Vectors for Cell-Cycle-Targeted Gene
      Therapy of Lung Cancer
    Pier Paolo Claudio and Antonio Giordano **591**

36 Aerosol Gene Therapy for Metastatic Lung Cancer Using PEI-p53
      Complexes
    **Ajay Gautam, J. Clifford Waldrep, Eugenie S. Kleinerman,
    Bo Xu, Yuen-Kai Fung, Anne T'Ang,
    and Charles L. Densmore** **607**

   III. Antisense Oligonucleotide Gene Therapies

37 Clinical Development of Antisense Oligonucleotides
      as Anti-Cancer Therapeutics
    Helen X. Chen **621**

38 Antisense Oligonucleotides Targeting RIα Subunit of Protein Kinase A:
      In Vitro and In Vivo Anti-Tumor Activity and Pharmacokinetics
    Ruiwen Zhang and Hui Wang **637**

39 Use of Antisense Oligonucleotides for Therapy: *Manipulation of Apoptosis Inhibitors for Destruction of Lung Cancer Cells*
   Siân H. Leech, Robert A. Olie, and Uwe Zangemeister-Wittke 655

40 Induction of Programmed Cell Death with an Antisense Bcl-2 Oligonucleotide
   Patrick P. Koty, Wendong Lei, and Mark L. Levitt 671

   C. IMMUNE THERAPIES

41 Tumor Vaccination with Cytokine-Encapsulated Microspheres
   Nejat K. Egilmez, Yong S. Jong, Edith Mathiowitz, and Richard B. Bankert 687

42 Adoptive Immunotherapy
   Julian A. Kim and Suyu Shu 697

43 Intratumoral Therapy with Cytokine Gene-Modified Dendritic Cells in Murine Lung Cancer Models
   Sherven Sharma, Seok Chul Yang, Raj K. Batra, and Steven M. Dubinett 711

44 Cyclooxygenase 2-Dependent Regulation of Antitumor Immunity in Lung Cancer
   Sherven Sharma, Min Huang, Mariam Dohadwala, Mehis Pold, Raj K. Batra, and Steven M. Dubinett 723

   D. CHEMOPREVENTION

45 Chemoprevention of Lung Cancer
   Hanspeter Witschi 739

46 Assessing the Interaction of Particulate Delivery Systems with Lung Surfactant
   Timothy Scott Wiedmann 755

47 Chemopreventive Therapeutics: *Inhalation Therapies for Lung Cancer and Bronchial Premalignancy*
   Missak Haigentz, Jr. and Roman Perez-Soler 771

Index 781

CONTENTS OF THE COMPANION VOLUME

Lung Cancer

Volume I: Molecular Pathology Methods and Reviews

PART I. INTRODUCTION

1 Characteristic Genetic Alterations in Lung Cancer
   Ignacio I. Wistuba and Adi F. Gazdar

PART II. ETIOLOGY AND CLASSIFICATION OF LUNG TUMORS

2 Clinical and Biological Relevance of Recently Defined Categories of Pulmonary Neoplasia
   Edward Gabrielson

3 Molecular Epidemiology of Human Cancer Risk: *Gene–Environment Interactions and p53 Mutation Spectrum in Human Lung Cancer*
   Kirsi Vähäkangas

4 Neuroendocrine Phenotype of Small Cell Lung Cancer
   Judy M. Coulson, Marta Ocejo-Garcia, and Penella J. Woll

5 AIDS-Associated Pulmonary Cancers
   Kushagra Katariya and Richard J. Thurer

PART III. MOLECULAR ABNORMALITIES IN LUNG CANCER

   A. DETECTION OF ALTERATIONS IN THE CELL CYCLE

6 Abrogation of the RB-p16 Tumor Suppressor Pathway in Human Lung Cancer
   Joseph Geradts

7 Sensitive Detection of Hypermethylated $p16^{INK4a}$ Alleles in Exfoliative Tissue Material
   Marcus Schuermann and Michael Kersting

   B. DETECTION OF ALTERATIONS IN SIGNAL TRANSDUCTION PATHWAYS

8 Role of Receptor Tyrosine Kinases in Lung Cancer
   Gautam Maulik, Takashi Kijima, and Ravi Salgia

9 Relationship of EGFR Signal-Transduction Modulation by Tyrosine Kinase Inhibitors to Chemosensitivity and Programmed Cell Death in Lung Cancer Cell Lines
   Wendong Lei, Patrick P. Koty, and Mark L. Levitt

10 Detection of the Transcripts and Proteins for the Transforming Growth Factor-β Isoforms and Receptors in Mouse Lung Tumorigenesis Using *In Situ* Hybridization and Immunohistochemistry in Paraffin-Embedded Tissue Sections
    Sonia B. Jakowlew and Jennifer Mariano

11 Alterations in the Expression of Insulin-Like Growth Factors
    and Their Binding Proteins in Lung Cancer
    Carrie M. Coleman and Adda Grimberg

12 Screening of Mutations in the ras Family of Oncogenes
    by Polymerase Chain Reaction-Based Ligase Chain Reaction
    **Alfredo Martínez, Teresa A. Lehman, Rama Modali,
    and James L. Mulshine**

13 Assays for Raf-1 Kinase Phosphorylation and Activity
    in Human Small Cell Lung Cancer Cells
    Rajani K. Ravi and Barry D. Nelkin

14 γ-Glutamylcysteine Synthetase in Lung Cancer:
    Effect on Cell Viability
    Kristiina Järvinen, Ylermi Soini, and Vuokko L. Kinnula

15 Localization of Cyclooxygenase-2 Protein Expression
    in Lung Cancer Specimens by Immunohistochemical Analysis
    Sarah A. Wardlaw and Thomas H. March

16 Non-RI Protocols for L-*myc* Allelotyping and Deletion Mapping
    of Chromosome 1p in Primary Lung Cancers
    Keiko Hiyama, Ciro Mendoza, and Eiso Hiyama

    C. DETECTION OF ALTERED CELL SURFACE MARKERS

17 Intercellular Adhesion Molecules (ICAM-1, VCAM-1, and LFA-1)
    in Adenocarcinoma of Lung: *Technical Aspects*
    **Armando E. Fraire, Bruce A. Woda, Louis Savas,
    and Zhong Jiang**

18 MUC1 Expression in Lung Cancer
    Eva Szabo

19 Altered Surface Markers in Lung Cancer: *Lack of Cell-Surface
    Fas/APO-1 Expression in Pulmonary Adenocarcinoma
    May Allow Escape from Immune Surveillance*
    Yoshihiro Nambu and David G. Beer

    D. DETECTION OF ALTERATIONS IN THE APOPTOTIC PATHWAY

20 Prognostic Relevance of Angiogenic, Proliferative,
    and Apoptotic Factors in Lung Carcinomas: *A Case Review*
    Manfred Volm and Reet Koomägi

21 Morphological Assessment of Apoptosis in Human Lung Cells
    **Lori D. Dwyer-Nield, David Dinsdale, J. John Cohen,
    Margaret K. T. Squier, and Alvin M. Malkinson**

Contents of Volume I

22 [D-Arg^6, D-Trp^{7,9}, N^{me}Phe^8]-Substance P (6-11) Activates JNK and Induces Apoptosis in Small Cell Lung Cancer Cells via an Oxidant-Dependent Mechanism
   Alison MacKinnon and Tariq Sethi

23 Isolation of Cells Entering Different Programmed Cell Death Pathways Using a Discontinuous Percoll Gradient
   Patrick P. Koty, Haifan Zhang, Wendong Lei, and Mark L. Levitt

   E. DETECTION OF ALTERATIONS IN LUNG CELL-DIRECTED ANGIOGENESIS

24 Angiogenesis, Metastasis, and Lung Cancer: *An Overview*
   Amir Onn and Roy S. Herbst

25 Matrix Metalloproteinase Expression in Lung Cancer
   Melissa Lim and David M. Jablons

26 Vascular Endothelial Growth Factor Expression in Non-Small Cell Lung Cancer
   Kenneth J. O'Byrne, Jonathan Goddard, Alexandra Giatromanolaki, and Michael I. Koukourakis

27 Regulation of Angiostatin Mobilization by Tumor-Derived Matrix Metalloproteinase-2
   Marsha A. Moses and Michael S. O'Reilly

28 In Vitro and In Vivo Assays for the Proliferative and Vascular Permeabilization Activities of Vascular Endothelial Growth Factor (VEGF) and Its Receptor
   Seiji Yano, Roy S. Herbst, and Saburo Sone

   F. DETECTION OF ALTERATIONS IN DNA REPLICATION AND REPAIR

29 Detection of Telomerase Activity in Lung Cancer Tissues
   Keiko Hiyama and Eiso Hiyama

30 Analysis of Alterations in a Base-Excision Repair Gene in Lung Cancer
   Nandan Bhattacharyya and Sipra Banerjee

PART IV. LUNG CANCER MODEL SYSTEMS

   A. ANIMAL MODELS FOR LUNG CANCER

31 Induction of Lung Cancer by Passive Smoking in an Animal Model System
   Hanspeter Witschi

32 Metastatic Orthotopic Mouse Models of Lung Cancer
   Robert M. Hoffman

33 Lung-Specific Expression of Mutant p53 as a Mouse Model
for Lung Cancer
Kam-Meng Tchou-Wong and William N. Rom

34 Use of Nucleotide Excision Repair-Deficient Mice as a Model
for Chemically Induced Lung Cancer
David L. Cheo and Errol C. Friedberg

B. ANIMAL MODELS FOR TUMOR METASTASES TO THE LUNG

35 Nude Mouse Lung Metastases Models of Osteosarcoma
and Ewing's Sarcoma for Evaluating New Therapeutic Strategies
Shu-Fang Jia, Rong-Rong Zhou, and Eugenie S. Kleinerman

36 Tumor-Specific Metastasis to Lung
Using Reporter Gene-Tagged Tumor Cells
**Lloyd A. Culp, Wen-Chang Lin, Nanette Kleinman,
Julianne L. Holleran, and Carson J. Miller**

C. MODELS FOR DEVELOPMENT OF TARGETED THERAPEUTICS

37 Cultures of Surgical Material from Lung Cancers: *A Kinetic Approach*
**Bruce C. Baguley, Elaine S. Marshall,
and Timothy I. Christmas**

38 The Hollow Fiber Assay
**Leslie-Ann M. Hall, Candice M. Krauthauser,
Roseanne S. Wexler, Andrew M. Slee, and Janet S. Kerr**

Contributors

SHERIF Z. ABDEL-RAHMAN • *Department of Preventive Medicine and Community Health, The University of Texas Medical Branch, Galveston, TX*
DAVID M. ABOULAFIA • *Virginia Mason Clinic, Division of Hematology/Oncology, University of Washington, Seattle, WA*
SAMREEN I. AHMED • *Department of Medical Oncology, Weston Park Hospital, Sheffield, UK*
SCOTT J. ANTONIA • *H. Lee Moffitt Cancer Center, Tampa, FL*
RICHARD B. BANKERT • *Department of Immunology, Roswell Park Cancer Institute, Buffalo, NY*
MICHAEL A. BARRY • *Departments of Microbiology and Immunology and Molecular and Human Genetics, Baylor College of Medicine, Houston, TX*
RAJ K. BATRA • *Wadsworth Pulmonary Immunology Laboratory, Veterans Administration Greater Los Angeles Healthcare System; Division of Pulmonary and Critical Care Medicine, University of California at Los Angeles School of Medicine, Los Angeles, CA*
GEROLD BEPLER • *H. Lee Moffitt Cancer Center and Research Institute, Tampa, FL*
RAYMOND M. BERNAL • *Division of Pulmonary and Critical Care Medicine, UCLA School of Medicine, and Veterans Administration Greater Los Angeles Health Care System, Los Angeles, CA*
DANIEL BETTICHER • *Department of Clinical Research, Institute of Medical Oncology, University of Bern, Switzerland*
BALBIR S. BHOGAL • *Veterans Affairs Medical Center and Department of Internal Medicine, Baylor College of Medicine, Houston TX*
CECILIA BOZZETTI • *Divisione di Oncologia Medica, Azienda Ospedaliera-Universitaria di Parma, Parma, Italy*
BARBARA G. CAMPLING • *Laboratory of Molecular Oncology and Cell Cycle Regulation, Howard Hughes Medical Institute; Department of Medicine, University of Pennsylvania School of Medicine, Philadelphia, PA*
STEFANO CASCINU • *Divisione di Oncologia Medica, Azienda Ospedaliera-Universitaria di Parma, Parma, Italy*

ANDREW CHAN • Division of Pulmonary Medicine; Division of Hematology/ Oncology, University of California, Davis Cancer Center, Sacramento, CA

HELEN X. CHEN • Cancer Therapy Evaluation Program, National Cancer Institute, Bethesda, MD

LAN BO CHEN • Department of Cancer Biology, Dana-Farber Cancer Institute and Harvard Medical School, Boston, MA

PIER PAOLO CLAUDIO • Department of Biotechnology, Temple University, Philadelphia, PA

JUDY M. COULSON • Departments of Physiology and Human Anatomy, University of Liverpool, Liverpool, UK

THOMAS A. D'AMICO • Thoracic Oncology Program, Duke University Medical Center, Durham, NC

CHARLES L. DENSMORE • Department of Molecular Physiology and Biophysics, Baylor College of Medicine, Houston, TX

MARIAM DOHADWALA • Wadsworth Pulmonary Immunology Laboratory, Veterans Administration Greater Los Angeles Healthcare System; University of California Los Angeles School of Medicine, Los Angeles, CA

STEVEN M. DUBINETT • Wadsworth Pulmonary Immunology Laboratory, Veterans Administration Greater Los Angeles Healthcare System; Jonsson Comprehension Cancer Center and Division of Pulmonary and Critical Care Medicine, University of California Los Angeles School of Medicine, Los Angeles, CA

NEJAT K. EGILMEZ • Department of Microbiology, State University of New York, Buffalo, NY

WAFIK S. EL-DEIRY • Laboratory of Molecular Oncology and Cell Cycle Regulation, Howard Hughes Medical Institute, Department of Medicine, University of Pennsylvania School of Medicine, Philadelphia, PA

RANDA A. EL-ZEIN • Department of Epidemiology, The University of Texas M. D. Anderson Cancer Center, Houston, TX

ZUMEI FENG • Department of Oncology, Lombardi Cancer Center, Georgetown University Medical Center. Washington, DC

JOHN K. FIELD • Roy Castle International Centre for Lung Cancer Research and Molecular Genetics & Oncology Group, Clinical Dental Sciences, The University of Liverpool, Liverpool, UK

MARCIA V. FOURNIER • Adult Oncology Division, Dana-Farber Cancer Institute, Boston, MA

VITTORIO FRANCIOSI • Divisione di Oncologia Medica, Azienda Ospedaliera-Universitaria di Parma, Parma, Italy

Contributors

YUEN-KAI FUNG • *Divisions of Hematology/Oncology and Ophthalmology, Childrens Hospital Los Angeles Research Institute, and Departments of Pediatrics, Microbiology and Ophthalmology, Keck School of Medicine, University of Southern California, Los Angeles, CA*

AJAY GAUTAM • *Department of Molecular Physiology and Biophysics, Baylor College of Medicine, Houston, TX*

ANTONIO GIORDANO • *Department of Biotechnology, Temple University, Philadelphia, PA*

RADOSLAV GOLDMAN • *Department of Oncology, Lombardi Cancer Center, Georgetown University Medical Center, Washington, DC*

EDGARD GRANER • *Dana-Farber Cancer Institute, Department of Adult Oncology, Boston, MA*

ANNAMARIA GUAZZI • *Divisione di Oncologia Medica, Azienda Ospedaliera-Universitaria di Parma, Parma, Italy*

LINLANG GUO • *Division of Hematology/Oncology, University of California Davis Cancer Center, Sacramento, CA*

MISSAK HAIGENTZ, JR. • *Division of Medical Oncology, Albert Einstein College of Medicine/Montefiore Medical Center, Bronx, NY*

DAVID H. HARPOLE, JR. • *Thoracic Oncology Program, Duke University Medical Center, Durham, NC*

ERIC B. HAURA • *H. Lee Moffitt Cancer Center, Tampa, FL*

JIM HEIGHWAY • *Roy Castle International Centre for Lung Cancer Research, Liverpool, UK*

KAY E. HERMES • *Clifton F. Mountain Foundation, Houston, TX*

PAUL HOBAN • *Centre for Cell and Molecular Medicine, Keele University School of Postgraduate University, North Staffordshire Hospital, Stoke-on-Trent, UK*

MIN HUANG • *Wadsworth Pulmonary Immunology Laboratory, Veterans Administration Greater Los Angeles Healthcare System, University of California Los Angeles School of Medicine, Los Angeles, CA*

YONG S. JONG • *Department of Molecular Pharmacology, Physiology and Biotechnology, Brown University, Providence, RI*

W. NICOL KEITH • *CRC Department of Medical Oncology and CRC Beatson Laboratories, University of Glasgow, Glasgow, UK*

TAKASHI KIJIMA • *Division of Adult Oncology and Thoracic Oncology Program, Department of Medicine, Dana-Farber Cancer Institute, Brigham and Women's Hospital and Harvard Medical School, Boston, MA*

JULIAN A. KIM • *Center for Surgery Research, Cleveland Clinic, Cleveland, OH*

BERMA M. KINSEY • *Veterans Affairs Medical Center and Department of Internal Medicine, Baylor College of Medicine, Houston, TX*
EUGENIE S. KLEINERMAN • *Department of Pediatrics, The University of Texas M. D. Anderson Cancer Center, Houston, TX*
TERESA KNAPP • *Roy Castle International Centre For Lung Cancer Research, Liverpool, UK*
REET KOOMÄGI • *Department of Oncological Diagnostics and Therapy, Central Unit of Biostatistics, German Cancer Research Center, Heidelberg, Germany*
PATRICK P. KOTY • *Department of Environmental and Occupational Health, Graduate School of Public Health, University of Pittsburgh, Pittsburgh, PA*
STINE-KATHREIN KRAEFT • *Department of Cancer Biology, Dana-Farber Cancer Institute and Harvard Medical School, Boston, MA*
DERICK LAU • *Division of Pulmonary Medicine; Division of Hematology/ Oncology, University of California, Davis Cancer Center, Sacramento, CA*
SIÂN H. LEECH • *School of Biological Sciences, The University of Manchester, Manchester, UK*
WENDONG LEI • *Cancer Institute and Hospital, Chinese Academy of Medical Sciences and Peking Union Medical College, Beijing, China*
MARK L. LEVITT • *Department of Environmental and Occupational Health, Graduate School of Public Health, University of Pittsburgh, Pittsburgh, PA; Sheba Medical Center, Institute of Oncology, Tel Hashomer, Israel*
TRIANTAFILLOS LILOGLOU • *Roy Castle International Centre for Lung Cancer Research and Molecular Genetics & Oncology Group, Clinical Dental Sciences, The University of Liverpool, UK*
NIMISHA MAKAN • *Department of Epidemiology, The University of Texas M. D. Anderson Cancer Center, Houston, TX*
KATHERINE J. MARTIN • *Department of Cancer Biology, Dana-Farber Cancer Institute, Boston, MA*
EDITH MATHIOWITZ • *Department of Molecular Pharmacology, Physiology and Biotechnology, Brown University, Providence, RI*
GAUTAM MAULIK • *Division of Adult Oncology and Thoracic Oncology Program, Department of Medicine, Dana-Farber Cancer Institute, Brigham and Women's Hospital and Harvard Medical School, Boston, MA*
MARY-BETH MOORE JOSHI • *Thoracic Oncology Program, Duke University Medical Center, Durham, NC*
CLIFTON F. MOUNTAIN • *Division of Cardiothoracic Surgery, The University of California at San Diego, San Diego, CA*

BENJAMIN MOVSAS • *Department of Radiation Oncology, Fox Chase Cancer Center, Philadelphia, PA*
NADIA NALDI • *Divisione di Oncologia Medica, Azienda Ospedaliera-Universitaria di Parma, Parma, Italy*
DENNIS E. NIEWOEHNER • *Pulmonary Section, Veterans Affairs Medical Center, University of Minnesota, Minneapolis, MN*
RITA NIZZOLI • *Divisione di Oncologia Medica, Azienda Ospedaliera-Universitaria di Parma, Parma, Italy*
ROBERT A. OLIE • *Division of Medical Oncology, Department of Internal Medicine, University Hospital Zurich, Zurich, Switzerland*
NELSON G. ORDÓÑEZ • *Department of Pathology, The University of Texas MD Anderson Cancer Center, Houston, TX*
FRANK M. ORSON • *The Center for AIDS Research, Veterans Affairs Medical Center; Departments of Internal Medicine and Microbiology and Immunology, Baylor College of Medicine, Houston, TX*
DAVID OST • *North Shore University Hospital and New York University School of Medicine, Manhasset, NY*
ARTHUR B. PARDEE • *Department of Cancer Biology, Dana-Farber Cancer Institute, Boston, MA*
ROMAN PEREZ-SOLER • *Division of Medical Oncology, Albert Einstein College of Medicine/Montefiore Medical Center, Bronx, NY*
IVER PETERSEN • *Institute of Pathology, Medical School Charite, Humboldt University, Berlin, Germany*
SIMONE PETERSEN • *Institute of Pathology, Medical School Charite, Humboldt University, Berlin, Germany*
DIANA M. PITTERLE • *Department of Surgery, University of Wisconsin-Madison Medical School, Madison, WI*
MEHIS POLD • *Wadsworth Pulmonary Immunology Laboratory, Veterans Administration Greater Los Angeles Healthcare System, University of California Los Angeles School of Medicine, Los Angeles, CA*
AGNESE A. POLLICE • *Department of Human Oncology, Laboratory of Cancer Cell Biology and Genetics, MCP/Hahnemann University, Allegheny General Hospital, Pittsburgh, PA*
JOHN P. QUINN • *Departments of Physiology & Human Anatomy and Cell Biology, University of Liverpool, Liverpool, UK*
STEVEN G. REED • *Department of Pathobiology, University of Washington; Department of Tumor Antigen Discovery, Corixa Corporation, Seattle WA*

WERNER RITTGEN • *Department of Oncological Diagnostics and Therapy, Central Unit of Biostatistics, German Cancer Research Center, Heidelberg, Germany*
WILLIAM N. ROM • *Bellevue Chest Service, NYU Medical Center and Division of Pulmonary and Critical Care Medicine, Department of Medicine, NYU School of Medicine, New York, NY*
JEFFREY B. RUBINS • *Pulmonary Section, Veterans Affairs Medical Center, University of Minnesota, Minneapolis, MN*
RAVI SALGIA • *Division of Adult Oncology and Thoracic Oncology Program, Department of Medicine, Dana-Farber Cancer Institute, Brigham and Women's Hospital and Harvard Medical School, Boston, MA*
MATTHEW B. SCHABATH • *Department of Epidemiology, The University of Texas M. D. Anderson Cancer Center, Houston, TX*
MARCUS SCHUERMANN • *Department of Hematology, Oncology and Immunology, Phillipps-University of Marburg, Marburg, Germany*
STANLEY E. SHACKNEY • *Laboratory of Cancer Cell Biology and Genetics, Department of Human Oncology and Human Genetics, MCP/Hahnemann University, Allegheny General Hospital, Pittsburgh, PA*
SHERVEN SHARMA • *Wadsworth Pulmonary Immunology Laboratory, Veterans Administration Greater Los Angeles Healthcare System, University of California Los Angeles School of Medicine, Los Angeles, CA*
SUYU SHU • *Center for Surgery Research, Cleveland Clinic, Cleveland, OH*
JAN F. SILVERMAN • *Department of Pathology and Laboratory Medicine, MCP/Hahnemann University, Allegheny General Hospital, Pittsburgh, PA*
CHARLES A. SMITH • *Laboratory of Cancer Cell Biology and Genetics, Department of Human Oncology, MCP/Hahnemann University, Allegheny General Hospital, Pittsburgh, PA*
VEERLE A. M. C. SOMERS • *Department of Pathology, Maastricht University Hospital, Maastricht, The Netherlands*
LING SONG • *Department of Internal Medicine, Veterans Affairs Medical Center, Baylor College of Medicine, Houston, TX*
EDUARDO SOTOMAYOR • *H. Lee Moffitt Cancer Center & Research Institute, University of South Florida, Tampa, FL*
MARGARET R. SPITZ • *Department of Epidemiology, The University of Texas MD Anderson Cancer Center, Houston, TX*
ANNE T'ANG • *Division of Hematology/Oncology and Department of Pathology, Childrens Hospital Los Angeles Research Institute; Department of Pathology, Keck School of Medicine, University of Southern California, Los Angeles, CA*

KAM-MENG TCHOU-WONG • *Division of Pulmonary and Critical Care Medicine, Department of Medicine, NYU School of Medicine, New York, NY*
FREDERICK B. J. M. THUNNISSEN • *Department of Pathology, Canisius Wilhelmina Hospital, Nijmegen, The Netherlands*
DEFA TIAN • *Department of Oncology, Lombardi Cancer Center, Georgetown University Medical Center, Washington, DC*
MANFRED VOLM • *Department of Oncological Diagnostics and Therapy, Central Unit of Biostatistics, German Cancer Research Center, Heidelberg, Germany*
J. CLIFFORD WALDREP • *Department of Molecular Physiology and Biophysics, Baylor College of Medicine, Houston, TX*
HUI WANG • *Division of Clinical Pharmacology, Department of Pharmacology and Toxicology, University of Alabama at Birmingham, Birmingham, AL*
TONGTONG WANG • *Department of Tumor Antigen Discovery, Corixa Corporation, Seattle, WA*
TIMOTHY SCOTT WIEDMANN • *Department of Pharmaceutics, University of Minnesota, Minneapolis, MN*
HANSPETER WITSCHI • *CHE and Department of Molecular Biosciences, School of Veterinary Medicine, University of California, Davis, CA*
PENELLA J. WOLL • *FRCP Department of Clinical Oncology, University of Nottingham, Nottingham, UK*
REEN WU • *Division of Pulmonary Medicine and Division of Hematology/ Oncology, University of California, Davis Cancer Center, Sacramento, CA*
XIFENG WU • *Department of Epidemiology, The University of Texas M. D. Anderson Cancer Center, Houston, TX*
BO XU • *Department of Molecular Physiology and Biophysics, Baylor College of Medicine, Houston, TX*
SEOK CHUL YANG • *Wadsworth Pulmonary Immunology Laboratory, Veterans Administration Greater Los Angeles Healthcare System, Los Angeles, CA*
HE YU • *Section of Cancer Prevention and Control, Feist-Weiller Cancer Center, Louisiana State University Medical Center, Shreveport, LA*
UWE ZANGEMEISTER-WITTKE • *Division of Medical Oncology, Department of Internal Medicine, University Hospital Zurich, Zurich, Switzerland*
RUIWEN ZHANG • *Division of Clinical Pharmacology, Department of Pharmacology and Toxicology, Comprehensive Cancer Center, and Gene Therapy Center, University of Alabama at Birmingham, Birmingham, AL*

I

INTRODUCTION

1

Molecular and Genetic Aspects of Lung Cancer

William N. Rom and Kam-Meng Tchou-Wong

1. Introduction

Lung cancer is the leading cause of cancer death among men and women in the United States with 170,000 deaths per year. This exceeds the sum of the next three leading causes of death due to cancer: breast, colon, and prostate. There are over 1 million deaths worldwide due to lung cancer, making it truly an epidemic. Fewer than 15% achieve a 5-yr survival. The vast majority (85%) present with advanced disease, although stage I patients may have a 5-yr survival approaching 70% *(1)*. 80% of the lung cancers are non-small cell lung cancer (NSCLC; adenocarcinomas, squamous cell, bronchoalveolar and large cell carcinomas) and 20% are small cell lung cancer (SCLC). Cigarette smoking constitutes 80% of the attributable risk and asbestos, radon, other occupational and environmental exposures and genetic factors contribute to the rest. The purpose of this state of the art review is to introduce the molecular genetics of lung cancer for the clinician in this rapidly progressing field. Many of the basic science concepts to follow already are being studied in clinical trials of new chemotherapeutic agents or gene therapy.

2. Diagnosis (Clinical and Molecular Approaches)

James Alexander Miller, the first Director of the Bellevue Chest Service, reviewed primary carcinoma of the lung in 1930 *(2)*. He presented 32 cases from Bellevue Hospital, and noted that the disease appeared to be due to urban dust and bronchial irritation but did not explicitly indict tobacco or cigarette smoking. In 1939, Ochsner and DeBakey presented a case series of seven lung cancers treated surgically by pneumonectomy and discussed the possibility that smoking caused lung cancer by irritating the bronchial mucosa *(3)*.

Lung cancer can progress significantly before symptoms are manifest although the common symptoms of expectoration and cough increase in frequency over time in clinical cases. Dyspnea, wheeze, heaviness in the chest, chest pain, and hoarseness are not particularly helpful, but hemoptysis increases 12-fold at time of diagnosis compared to matched controls and loss of weight increases threefold *(4)*. Among helpful clinical signs is digital clubbing which recently was observed in 29% of 111 consecutive patients with lung cancer *(5)*. Clubbing was more common in NSCLC than SCLC, and among women than men. Paraneoplastic conditions may give rise to symptoms and signs including syndrome of inappropriate antidiuretic hormone, ectopic adrenocorticotrophic hormone, Eaton-Lambert syndrome, neurologic syndromes, hypercalcemia, deep vein thrombosis, marantic endocarditis, disseminated intravascular coagulation, and hypertrophic osteoarthropathy. The staging of lung cancer has recently been reviewed by Mountain *(6)*. Evaluation for metastases must include a clinical and laboratory examination and if abnormal followed by CT scan of the head and abdomen and a radionuclide bone scan *(7)*.

Appropriately stratified case-control studies that take cigarette smoking into account typically report that lung cancer cases have an odds ratio for having a first-order relative with a history of lung cancer of approx 1.7 to 5.3 *(8,9)*. Chronic obstructive lung disease and pulmonary fibrosis are clinical risk factors for lung cancer.

Low-dose spiral computed tomography (CT) chest scan has tremendous promise in detecting stage I lung cancer compared to the chest X-ray. Henschke and colleagues screened 1000 persons aged 60 or over with at least 10 pack years' smoking finding noncalcified nodules in 23% *(10)*. Among those with positive CT, 28 were recommended for biopsy and 27 of these were malignant. Pathological and clinical staging classified 23 of the 27 as stage I and potentially curable. In the whole study population, malignant disease was detected four times more frequently on low-dose CT than on chest radiography.

Although sputum cytology is regarded as having too low a sensitivity to be useful in screening for lung cancer, it can be useful for detecting dysplasia. Kennedy and colleagues reported that 26% of a high-risk cohort (FEV_1 <70% predicted, FEV_1/FVC <70% predicted, 40 pack years of smoking) had moderate to severe dysplasia and should be a target group for research programs focusing on lung cancer prevention, early detection, and exploratory biomarker studies *(11)*. Tockman and colleagues have used a monoclonal antibody (MAb) to hnRNP (Ribonucleoprotein) A2/B1 as a cancer antigen that can be detected in sputum specimens for up to 2 yr before the tumor is detectable radiographically *(12)*. He and his colleagues reported hnRNP overexpression with a sensitivity of 91% and specificity of 88% on archived sputum of smokers who went

Molecular and Genetic Aspects of Lung Cancer

on to develop lung cancer *(13)*. They performed two prospective studies on sputum detection with overexpression of hnRNP A2/B1: first, 32 of 40 surgically treated primary lung cancer patients with recurrence over 12 mo were identified, and second, the test detected 69 of 94 high-risk Chinese tin miners with primary lung cancer. Computer-assisted cytometry techniques may detect early nuclear morphological changes on sputum samples *(14)*.

Autofluorescence bronchoscopy using the laser-induced fluorescence emission system has been optimistically demonstrated to increase the dysplasia detection rate over that obtained by white light bronchoscopy from approx 40–80% *(15,16)*. Considerable operator skill is required to detect brownish red discoloration on tertiary carinas and to distinguish these sites from the background greenish discoloration *(17)*.

3. Cigarette Smoking and Molecular Damage to the Lung

The World Health Organization (WHO) estimates that 47% of men and 12% of women worldwide aged 15 and over are smokers *(18)*. Although smoking rates have decreased in industrialized countries since 1975, there has been a corresponding 50% increase in developing countries.

Case control studies reported an association between lung cancer and smoking in 1950 with a risk ratio of approx 10, which were quickly followed by cohort studies in the United States and United Kingdom. The cohort studies enrolled healthy people who recorded their smoking habits and were then followed up to determine the variation in mortality with the amount smoked. All showed that the mortality from lung cancer increased approximately in proportion to the amount smoked *(19,20)*. The American Cancer Society enrolled one million citizens prospectively in 1982 and found that the lung cancer mortality rate ratio for smokers vs nonsmokers after nine yr follow-up was 23.9 for men and 14 for women *(21)*. Sir Richard Doll established a cohort of 34,000 British doctors in 1951 that has been followed for over 40 years with cigarette smoking habits recorded periodically *(22)*. The mortality rate ratio for lung cancer in smokers vs nonsmokers was 14.9 and this dropped to 4.1 in ex-smokers. The lung cancer mortality rate ratio increased from 7.5 among current smokers smoking 1–14 cigarettes per day to 25.4 for those smoking 25 or more cigarettes per day. The loss of expectation of life for all cigarette smokers in the British doctor's study was 8.0 yr. It has been known since 1981 that passive smoke also increases risk for lung cancer when Hirayama and Trichopoulos et al. independently reported an increased risk of lung cancer in nonsmokers if their spouses smoked *(23,24)*. Ex-smokers have a progressive reduction in risk approaching 90% with most of the reduction occurring five or more years after quitting.

There are substantial racial differences for the incidence of lung cancer with African- Americans having a 1.8-fold higher risk than Caucasians *(25)*, and Hispanics and Asian/Pacific Islander groups having a reduced incidence compared to Caucasians. Interestingly, women are at a higher risk than men for a given level of smoking with a relative risk of 1.7. Lung cancers from women have significantly greater polycyclic aromatic DNA adducts per pack year than men *(26)*. As tar and nicotine per cigarette have dropped by more than two-thirds from 38 mg to 12 mg and 2.3 mg to 1.2 mg, respectively, there has been a concomitant change in the histologic type of lung cancer *(27)*. While SCLC has persisted at about 20% in most series, adenocarcinoma has increased to 45% with declines in squamous cell and large cell carcinoma. Thun and colleagues have suggested that these changes are due to cigarette design, e.g., the smoke in filter-tip cigarettes is inhaled more deeply than earlier, unfiltered cigarettes (more toxic), and deeper inhalation transports tobacco-specific carcinogens more distally toward the bronchoalveolar junction where adenocarcinomas often arise *(28)*. In addition, blended reconstituted tobacco includes more stems than leaves, which release higher concentrations of nitrosamines.

Pershagen and colleagues demonstrated that residential exposure to radon gas increases lung cancer risk in relation to cumulative and time-weighted exposure *(29)*. The excess relative risk of lung cancer was 3.4% per 27 pCi/L, which is in the range reported for underground miners at 2–10% per 27 pCi/L. Selikoff assembled a cohort of 17,800 asbestos insulators in the United States and Canada in 1967 and followed them prospectively to assess lung cancer and mesothelioma risk *(30)*. Compared to nonsmoking controls who had no exposure to asbestos, asbestos workers who had a history of smoking had a 53-fold increased mortality ratio from lung cancer. This was greater than the sum of the increases for lung cancer from asbestos exposure alone (5-fold) or cigarette smoking alone (11-fold). Other exposures for increased risk for lung cancer include silica, metal mining and smelting (chromium, cadmium, nickel, and arsenic), bischloromethyl ether, coke ovens (polycyclic aromatic hydrocarbons), and ionizing radiation. Diet may also influence lung cancer risk with a high-fat diet similar to that consumed in the United States enhancing risk posed by tobacco-smoke carcinogens.

Tobacco smoke is complex, with over 4000 compounds identified that are suspended in an aerosol of over 10^{10} particles per milliliter of mainstream smoke. Among the more than 60 carcinogens in tobacco and cigarette smoke, the two major classes are polycyclic aromatic hydrocarbons and nitrosamines. Mainstream smoke contains 20–40 ng of benzo(a)pyrene per cigarette and 0.08–0.77 mg of the nitrosamine NNK per cigarette. The total amount of NNK required to produce lung cancer in rats is similar to the total amount of this compound to which a smoker would be exposed in a lifetime of smoking (31).

Metabolism of inhaled carcinogens was recently reviewed by Spivack and colleagues (32). Since most tobacco-derived organic carcinogens are water-insoluble, they require oxidation and conjugation for excretion in aqueous environments. The aryl hydrocarbon receptor binds incoming aromatic hydrocarbons and members of the cytochrome P450 family activate polycyclic aromatics whereas members of the glutathione-S-transferase family inactivate these carcinogens. Combined phenotypes such as CYPIAI plus GSTMI null can accelerate carcinogen activation and impair inactivation leading to increased risk for lung cancer (32). DNA repair capacity as measured in a host-cell reactivation assay with plasmids damaged by exposure to benzo(a)pyrene diol epoxide was significantly lower in lung cancer cases (3.3%) than in controls (5.1%) (33). After adjustment for age, gender, ethnicity, and smoking status, the cases were five times more likely than controls to have reduced DNA repair capacity.

4. Molecular Abnormalities in Lung Cancer: A Disease of the Cell Cycle

Approximately 50 tumor-suppressor genes and over 100 oncogenes have now been described. Since tumor-suppressor genes, telomeres, and oncogenes are intimately involved in the regulation of cell growth and division, cancer can be considered a disease of deregulation of the cell cycle. Oncogenes result from gain-of-function mutations in their normal cellular counterpart protooncogenes and act in a dominant fashion.

The classical cell-cycle model, consisting of a DNA synthesis (S) phase, a mitosis (M) phase, and two gap (G_1 and G_2) phases, has now been elucidated in molecular detail (34–36; see **Fig. 1**). Critical components of the cycle include the cyclins, cyclin-dependent kinases (Cdk), and the retinoblastoma (Rb), p53, and E2F proteins. Each Cdk is regulated by a cyclin subunit, which is required for catalytic activity and substrate specificity. A first crucial step in the cell cycle occurs late in the G_1 phase at the restriction point, when a cell commits to completing the cycle. Competence factors such as platelet-derived growth factor (PDGF) and progression factors such as insulin-like growth factor-1 (IGF-1) can interact at this point to stimulate cell proliferation. Both growth factors can be made by lung tumor cells to enhance tumor growth in an autocrine fashion, usually in the late stage of tumorigenesis. Engagement of growth factors with their respective receptors leads to receptor dimerization, phosphorylation, and transmission of growth signals to the nucleus. Growth-promoting signals transduced from the cell surface to the nucleus cause a rapid

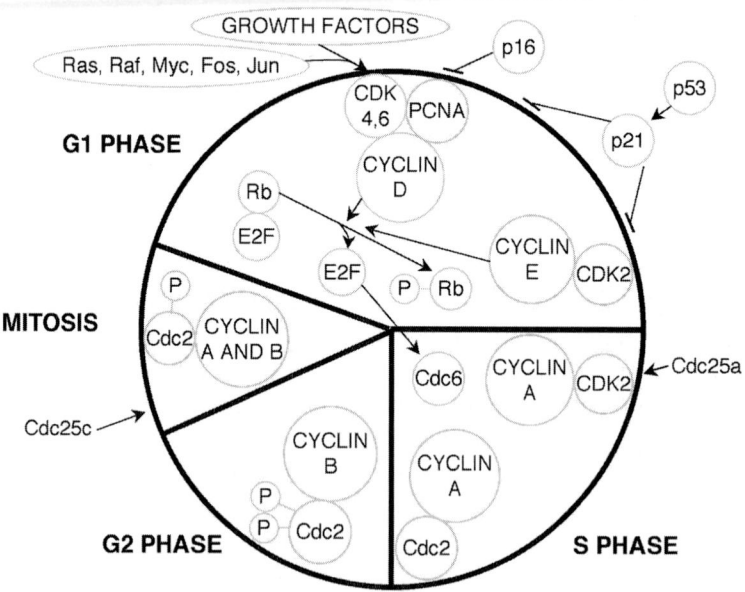

Fig. 1. Cell-cycle regulators implicated in lung cancer. (Adapted from **ref. 36**.)

and transient elevation in the D-type cyclins (early G_1). Cyclin D_1 complexes with Cdk4/6 and phosphorylates the Retinoblastoma (Rb) protein (*see* **Fig. 2**; *36*). Cyclin D_1 overexpression is a common molecular abnormality in lung cancer *(37)*. Hyperphosphorylation of Rb in G_1 releases the transcription factor E2F, which activates S-phase genes, including thymidine kinase, c-*myc*, dihydrofolate reductase, Cdc6, and DNA polymerase-α *(38)*.

Two families of Cdk inhibitors are crucial in G_1 progression (*see* **Fig. 3**). The INK4 family on chromosome 9p21 encodes four genes (INK4a, b, c, and d) whose products bind cyclin D-Cdk4/6 dimers to inactivate the kinase function. Members of the Kip1 family (p21, p27, p57) bind the cyclin D-Cdk 4/6, cyclin E-Cdk2, and cyclin A-Cdk2 complexes *(39)*. The cyclin E-Cdk2 complex mediates progression out of G_1, and cyclin A expression increases dramatically with the onset of S phase. Cyclin A-Cdk2 function appears to be required for DNA replication and the G_2/M transition. Loss of p53 function leads to reduced levels of p21 and hyperactivity of both cyclin D-Cdk and cyclin E-Cdk complexes, hyperphosphorylation of the Rb gene, and elevated levels of E2F *(40)*. Inactivation of the tumor-suppressor gene Rb produces the same effect, resulting in increased levels of free E2F in the cell. Cooperation between the Rb and p53 pathways likely determines whether p53 induces G_1 arrest or apoptosis

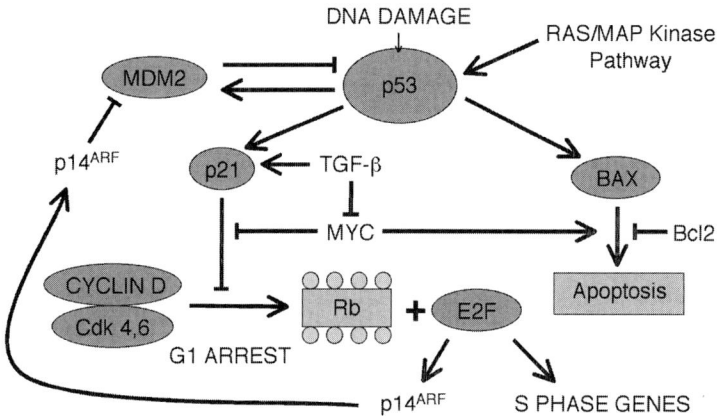

Fig. 2. p53 and Rb pathways in molecular carcinogenesis.

in response to DNA damage, with the loss of Rb tilting the balance toward apoptosis *(35)*. Preventing p53-dependent apoptosis is a key to carcinogenicity, and lung cancers that have wild-type p53 usually have increased expression of the MDM2 gene product, which binds to the p53 transactivation domain and targets p53 for ubiquitin-mediated degradation *(41)*. Overexpression of MDM2 overcomes wild-type p53-mediated suppression of transformed cell growth (*see* **Fig. 2**).

Because E2F is a transcription factor that activates S-phase genes, E2F may be critically important for replication of DNA in the cell cycle. DNA replication occurs at multiple chromosomal sites called origins of DNA replication and is controlled, in part, by origin recognition complex (ORC) proteins *(42)*. The ORC proteins are bound to Cdc6 which controls initiation of DNA replication *(42)*. A prereplication complex is formed when the Cdc6/ORC interaction directs the loading of minichromosome maintenance (MCM) proteins onto chromatin; the MCM proteins are on chromatin in G_1, much less so in S, and not at all in G_2/M. Human Cdc6 mRNA and protein are not detectable in serum-deprived human diploid fibroblasts, but increase prior to the G_1/S transition as the cells are stimulated with serum *(43)*. This transition is regulated by E2F proteins, as revealed by a functional analysis of the Cdc6 promoter showing E2F binding sites and stimulation of the Cdc6 gene by exogenous E2F *(44)*. Immunodepletion with anti-Cdc6 antibodies prevents initiation of DNA replication *(44)*. In lung cancer, E2F is free and may upregulate Cdc6 leading to a deregulated cell cycle with abnormal cellular proliferation. Cdc6 may be a marker for cell-cycle deregulation and a target for detection or therapeutics.

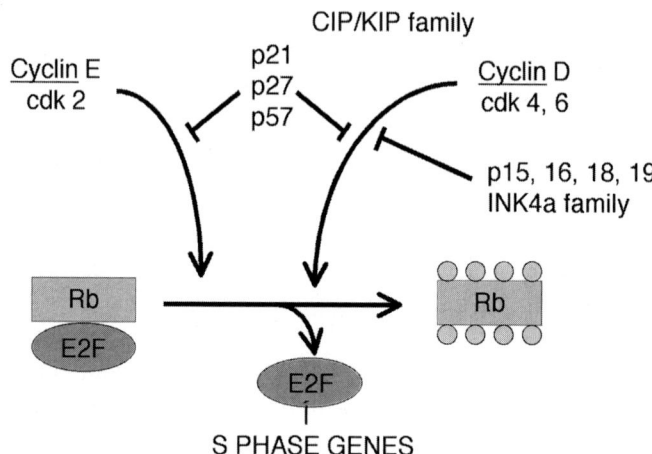

Fig. 3. Sites where p21 and p16 work as checkpoint inhibitors in the cell cycle.

4.1. Role of p53 as the Guardian of the Genome and Protector of the Lung from Environmental Carcinogens

The p53 tumor-suppressor gene is the most commonly mutated gene in cancer *(45)* and is mutated in 50% (NSCLC) to 70% (SCLC) of lung cancer. Mutations in p53 commonly reflect exposures to environmental carcinogens, e.g., cigarette smoke and lung cancer or aflatoxin and liver cancer in Southeast Asia. The p53 protein has been aptly referred to as the "guardian of the genome" because the p53 gene is induced by DNA damaging agents and subsequently either delays cell-cycle progression, or steers the damaged cell headlong into programmed cell death *(46)*. The p53 protein is a nuclear transcription factor that binds to the p21 promoter inducing its expression and inhibiting cell-cycle progression at the G_1/S cell-cycle checkpoint *(39)*. Mutant p53 cannot activate p21, and the cell cycle proceeds unabated; thus the term "tumor suppressor." Alternatively, p53 may induce *bax*, a gene promoting apoptosis *(47)*. Most missense mutations in the p53 gene occur in the DNA binding domain consequently inactivating its transactivation function *(48)*. Mutations of p53 greatly enhance the half-life of the protein, allowing for frequent immunohistochemical detection of mutant p53, e.g., in the severely dysplastic bronchial epithelium or in the tumor tissue. For tumor-suppressor genes, phenotypic expression requires that both alleles be lost through mutations, large deletions, or other recombinant mechanisms *(49)*. In lung cancer cell lines Calu-1 (both p53 alleles are deleted) and A549 (containing wild-type p53), growth arrest can be induced after in vitro treatment with phorbol ester *(50)*, which activates a protein kinase C (PKC) signaling cascade. The induction of p21 expression by phorbol ester temporally coincides with growth arrest in G_2/M.

p53 is located on chromosome 17p and is composed of 393 amino acids. The transactivation domain is at the N-terminus followed by the sequence specific DNA binding domain and oligomerization domain at the C-terminus. p53 mutations in lung cancer are clustered in the middle of the gene at codons 157, 245, 248, and 273 *(51)*. The apparent significance of these mutational sites became clear when the tobacco-smoke carcinogen, benzo(a)pyrene, was shown to induce benzo(a)pyrene diol-epoxide (BPDE) adducts at CpG sites in codons 157, 248, and 273 in vitro in bronchial epithelial cells *(52)*. These codons contain CpG islands, and the presence of 5-methyl cytosine greatly enhances BPDE binding to guanine *(53,54)*. The p53 mutations seen in lung cancer are guanine to thymine transversions that occur at the CpG sites where BPDE-DNA adducts are formed in vitro *(54)*. Interestingly, these mutations occur on the nontranscribed DNA strand, which is repaired relatively inefficiently. Codon 157 mutations appear to be unique to lung cancer, whereas codon 248 and 273 mutations occur at hot spots in other cancers, e.g., colon, liver, and prostate. Nonsmokers who develop lung cancer have a completely different, almost random grouping of p53 mutations.

p21 has been shown to inhibit DNA replication in vitro by a second mechanism dependent on proliferating cell nuclear antigen (PCNA) *(55)*. Another molecule stimulated by p53 is the growth arrest and DNA damage gene (Gadd 45), which binds PCNA, inhibits growth, and directs DNA nucleotide excision repair *(56)*. Inactivation of wild-type p53 function can occur through complex formation with viral oncogene products such as the large T antigen of SV40, the E1b-55 kDa protein of adenovirus type 5, and the E6 gene product of the human papilloma virus types 16 and 18 *(57)*. Mutant p53 can derepress the insulin-like growth factor-1 receptor (IGF-1R) promoter allowing for high-level expression in cancer cell lines and enhancing growth-promoting signals *(58)*. Stable expression of a dominant-negative mutant of IGF-1R in the lung cancer cell line A549 enhances sensitivity to apoptosis-inducing agents and suppresses tumor formation in nude mice by promoting glandular differentiation in vivo *(59)*. Wild-type p53 when introduced into a variety of cancer cell lines reduces colony formation in agar and carcinogenicity in animal models.

4.2. The p16 Tumor-Suppressor Pathway

The p16 protein from chromosome 9p21 binds to Cdk4 (hence *in*hibitor of *k*inase 4, or INK4) and inhibits phosphorylation of Rb (*see* **Fig. 2**; *60*).

Disruption of p16 function results in inappropriate hyperphosphorylation and, therefore, inactivation of Rb. Overexpression of the E2F transcription factor upregulates p16 expression and inhibits cyclin D-dependent kinase activity, suggesting the presence of a feedback loop. Inactivation of p16 may occur by homozygous or hemizygous deletion *(61,62)*, inactivation of the remaining p16 allele by point mutation *(63)*, or by gene silencing through methylation of CpG islands surrounding the first exon of p16 *(64)*. Methylation of CpG sequences in the p16 gene provides a way of suppressing expression of p16 in the absence of any mutation in the DNA and has been referred to as epigenetic regulation *(64)*. p16 may be silenced by DNA methylation in early stages of NSCLC, whereas homozygous deletions and/or mutations may occur more frequently in later stages of NSCLC development. Alterations in both the p16/pRb and p53 pathways lead to enhanced proliferation of NSCLC cell lines, and correlate with significantly shorter 5-yr survival, suggesting an aggressive tumor phenotype *(65)*. These genetic lesions can be mutually exclusive within any given tumor, consistent with the concept that they constitute equivalent steps in a single critical pathway *(66)*. There is a reciprocal relationship between Rb mutations and p16 expression, whereas Rb is less frequently mutated in NSCLC than in SCLC, p16 expression is commonly absent *(67)*. Functional Rb protein was absent in 90% of SCLC, and 15–30% of NSCLC primary lesions and tumor cell lines studied *(68)*. Kelley and colleagues *(69)* found 18/77 (23%) of NSCLC to have p16 homozygously deleted compared to one percent of SCLC, and coincident loss of p16 and functional Rb protein was rarely observed. Immunohistochemistry showed strong p16 nuclear staining in Rb-negative NSCLC, which correlated with increased proliferative activity, especially in NSCLC with p53 mutations. Thus, there is an interesting inverse relation between p16 and Rb in lung cancer: in SCLC, Rb is mutated and p16 is intact, whereas in NSCLC, p16 expression is disrupted and Rb is usually intact. A deregulated Rb pathway may correlate with overexpression of p53 and decreased MDM2, suggesting synergism in the deregulation of these pathways *(70)*.

The INK4a locus at 9p21 gives rise to two RNA transcripts: each transcript has a distinct 5′ exon, E1a or E1b, which is spliced into common exons E2 and E3. p16 arises from the E1a-containing transcript while p14^{ARF} (*a*lternate *r*eading *f*rame) contains the E1b transcript *(66)*. The p14^{ARF} protein is not a cdk inhibitor and has no sequence homology to p15 or p16, but can induce cell-cycle arrest, both in G_1 and G_2 *(44)*. E2F and c-myc recently have been shown to directly activate p14^{ARF} *(71,72)*, and p14^{ARF} binds to the MDM2-p53 complex preventing p53 degradation *(73,74)*. p14^{ARF} complexes with MDM2 and p53, which is localized in the nucleolus, and nuclear export of MDM2 and p53 is blocked *(75)*. This provides a link of the E2F-Rb pathway to prolongation of activation of p53 and cell-cycle arrest, allowing for the repair of damaged DNA. This constitutes a further fail-safe mechanism to protect against aberrant cell growth. Loss of nuclear staining for p14^{ARF} occurs in over 70% of SCLC and 25% of NSCLC *(76)*. SCLC may have a greater propensity for cell proliferation through the loss of both the p14^{ARF} fail-safe mechanism and p53.

4.3. Transforming Growth Factor-β Induces p15

Transforming growth factor-β (TGF-β) is a key cytokine mediating inflammation in the lung; accumulation of matrix proteins in fibrosis; deactivation of macrophage immune response; and inhibition of growth of most epithelial, endothelial, myeloid, and lymphoid cells. Cancer cell lines may express integrins such as $α_vβ_1$ that bind latency associated peptide (LAP) that covalently binds inactive TGF-β; integrin binding on the surface of lung cancer cells may contribute to the release of active TGF-β. Because of its role in growth control, TGF-β is implicated in many cancer networks and is one of the strongest checkpoint inhibitor at G_1/S. TGF-β influences the cell cycle, inducing p15 selectively as a checkpoint control and causing cells to cease proliferation and arrest in G_1 *(77)*. The Rb protein is a transcriptional activator of TGF-$β_1$ and TGF-$β_2$ *(78)*. TGF-β treatment causes the accumulation of Rb in the underphosphorylated state, and expression of Rb-inactivating carcinogens prevents TGF-β-induced cell-cycle arrest. Withdrawal from the cell cycle may also induce differentiation, and TGF-β is a key molecule that may contribute to this process. TGF-β has also been shown to induce p21 and to repress c-*myc*, although these mechanisms have not been demonstrated in lung cancer cell lines or in vivo *(79)*. TGF-β inhibition of Cdk 4/6 and Cdk2 can also occur via increased tyrosine phosphorylation by repression of the tyrosine phosphatase cdc25A *(80)*; this has been found in cell lines deficient in p15. However, no effect on cdc25A was noted in the A549 lung adenocarcinoma cell line. The G1/S arrest caused by TGF-β, p16, and contact inhibition is mediated by the Rb-E2F complex *(81)*.

5. Role of Activated Oncogenic *ras* in the Genesis of Lung Cancer

Activation of the K-*ras* oncogene by point mutations in codon 12 occurs in 50% of lung adenocarcinomas *(82)*, and PCR techniques can identify these mutations in bronchoalveolar lavage (BAL) cells from patients suspected of having lung cancer *(83)*. For example, in 52 patients with confirmed lung cancer, BAL cells contained K-*ras* codon-12 mutations in 14/25 adenocarcinomas, 1/3 bronchoalveolar carcinomas, 1/5 large cell carcinomas, and 0/14 squamous cell carcinomas. Tissue analysis matched the BAL-cell mutation in 35 cases, and no mutation was found in 30 patients with diagnoses other than NSCLC. K-*ras* codon-12 point mutations in lung cancer may predict significantly poorer survival and shorter duration of disease-free survival *(84)*. An antisense K-*ras* construct in a retrovirus has been shown to inhibit ras protein expression in a lung cancer cell line with mutant *ras*; colony formation in soft agar and tumorigenicity in nude mice were dramatically reduced in NSCLC cells expressing antisense K-*ras (85)*.

The three 21-kD *ras* proteins (H-Ras, N-Ras, K-Ras) are members of a superfamily of proteins that in the active state bind to GTP and in the inactive state bind to GDP. Through the intrinsic *ras* GTPase activity, *ras* returns to the quiescent state after interacting with its substrate c-Raf1 *(86)*. The signal is subsequently transmitted by a cascade of kinases, resulting in the activation of MAP kinases (ERK1 and ERK2), which translocate to the nucleus and activate transcription factors. Most *ras* mutants are defective in GTPase activity and thus are locked into the growth stimulatory GTP-bound form. ras mutations usually occur by point mutations at codons 12, 13, or 61 *(87)* and in lung cancer most *ras* mutations occur at codon 12.

The *ras*-MAP kinase pathway is involved in establishing basal and induced levels of p53 *(88)*. The mechanism of the *myc-ras* collaboration relates to activation of cyclin E-Cdk activity, loss of p27 inhibition, and induction of S phase *(89)*. ras also positively regulates the synthesis of cyclin D1 *(90)* and stabilizes the short-lived *myc* protein *(91)*. p16 can block the *ras* plus *myc*-induced transformation *(92)*. An intact Rb protein is essential for ras signaling effects on the cell cycle. In Rb-deficient cells, inactivation of *ras* with a MAb fails to cause G_1 arrest and the cells proliferate, demonstrating that multiple genetic lesions further enhance cell proliferation *(90)*. ras activates the serine/threonine kinase *Raf*, which induces S-phase genes, but excess *Ras/Raf* can induce p21 *(93)*. Recently, *Rho* has been shown to suppress the expression of p21 and overcome the cell-cycle block *(93)*. It will be interesting to examine the levels of expression of *Rho* in lung adenocarcinomas.

The discovery of p14^{ARF} has provided further insights into how the oncogenes c-*myc* and *ras* promote carcinogenesis. p14^{ARF} is essential for the p53-dependent arrest provoked by *ras (94)*, and a loss of either of p14^{ARF} or p53 would contribute to *ras* transformation. p14^{ARF} is also upregulated by c-*myc (72)*. For c-*myc* overexpression to succeed in cell transformation and proliferation, p53-induced apoptosis must by blocked. Analogous to *ras*, loss of p14^{ARF} or p53, which are common genetic lesions in lung cancer, would enable an amplified c-*myc* unfettered opportunity for cell proliferation and transformation. p14^{ARF} appears to bridge a gap between oncogenic signals and p53 whereby p14^{ARF}-induced activation would be critical to move the compromised cell toward apoptosis. Mice with targeted deletions of p14^{ARF} are prone to develop cancers at an early age and methylation of INK4a or mutations or deletions of exon 2, which disrupt p16^{INK4a} and p14^{ARF} are common in human lung cancer *(81,95)*.

6. Oncogenic Pathways: c-Myc in Lung Cancer

The c-*Myc* proto-oncogene belongs to a family of related genes (c-*Myc*, N-*Myc*, L-*Myc*) that are amplified in a subset of SCLC and, less commonly, in NSCLC. The product of *c-Myc* is a transcription factor that forms a heterodimer with Max that activates genes involved in growth control and apoptosis. Myc-Max dimers activate the promoter of cdc25A, which activates Cdk2 and Cdk4, growth-factor-responsive stimulators of G_1/S progression *(96)*. Cdc25A and cdc25B can cooperate with activated *ras* to transform primary rodent fibroblasts *(97)*. The Mad family of proteins bind Max and antagonize the *c-Myc* transactivation function *(98)*. The Mad proteins contain a Sin 3 interaction domain that complexes with histone deacetylase, which exerts transcriptional repression.

A novel growth enhancing effect of c-*Myc* is to repress growth arrest genes, e.g., *gas1*, which activates a transactivation-independent p53-mediated growth arrest function *(99)*, gadd 45 *(100)*, and p21. The growth arrest gene, *gas1*, is activated in G_0 growth-arrested cells, and its expression keeps cells in G_0 arrest *(101)*. The activity of *gas1* in G_0 arrest is dependent on the presence of wild-type p53 *(101)*.

c-*Myc* is a positive regulator of G_1-specific cyclin dependent kinases, particularly of cyclin E/CDK2 complexes. We have observed that *c-Myc* protein is overexpressed in tumor samples compared to non-neoplastic lung tissue, and that the *c-Myc* antagonist Mxi1 is abundantly expressed in nonmalignant lung samples but barely detectable in tumors (Lee, T. C. and Rom, W. N., unpublished observations). These results are consistent with active cell cycling in lung cancer tissue. *c-Myc* upregulates and prevents inhibition of cyclin E/Cdk2 activity by causing inactivation of the CDK inhibitor p27 and probably p21 and p57 by transcriptional and/or post-translational mechanisms. The cell-cycle deregulation seen in NSCLC may be explained, at least in part, by *c-Myc* overexpression, which leads to enhanced cyclin E/Cdk2 activity and Rb phosphorylation/inactivation, and entry into S phase. The most common abnormality involving c-*Myc* and its other family members in lung cancer is gene amplification or gene overexpression without amplification. Overexpression of a *c-Myc* family gene, with or without amplification, occurs in 80–90% of SCLCs *(102)*. Only one *c-Myc* gene family member is amplified in any one given tumor. In contrast to SCLC, amplification of the *c-Myc* gene occurs only in about 10% of NSCLCs. However, c-Myc overexpression in the absence of gene amplification occurs in over 50% of NSCLC specimens *(103,104)*.

7. Chromosomal Abnormalities: Preneoplastic Changes in Bronchial Epithelial Cells

Field cancerization is a concept that applies to lung cancer to describe the frequent occurrence of multiple primary tumors *(105)* or metachronous second primary lung cancer. Auerbach dissected airways of cigarette smokers and observed widespread and dispersed metaplasia *(106)*. He and Saccomanno *(107)* suggested a progressive pathway to bronchial carcinogenesis in smoking uranium miners whereby dysplasia progressed to carcinoma-in-situ over a period of 10–15 yr. Dysplastic lesions followed progressively have a risk for developing into invasive cancer; approx 25% progress over 36 mo for lung, and similar incidences occur for bladder, breast, and cervical carcinomas *(108)*.

Franklin and colleagues *(105)* recently observed widely dispersed p53 mutations in dysplastic respiratory epithelium dissected from a lifelong smoker who had died suddenly from coronary artery disease. Seven out of ten microdissected dysplastic lesions from both lungs had an identical G→T transversion of codon 245 in exon 7, which is a "hot spot" for mutation in cancer. Widely dispersed loss of heterozygosity (LOH) has also been reported in the respiratory epithelium for chromosome 3p *(109)*. It is likely that multiple clones with varying genetic mutations develop concurrently.

7.1. Chromosomal Abnormalities: Telomeres and Telomerases in Lung Cancer

The telomere-telomerase hypothesis states that continued shortening of telomere length, which occurs in normal cells eventually results in the induction of cellular senescence, and that activation of telomerase results in unlimited replicative potential. This hypothesis is based on observations that most normal human somatic cells do not have detectable telomerase activity, whereas most human tumors have shortened telomeres and demonstrate telomerase activity.

Telomeres are repetitive noncoding DNA (TTAGGG)n nucleoprotein structures that protect the ends of linear chromosomes. Maintenance of telomere length and function depends on a specialized reverse transcriptase known as telomerase, which consists of two components: the telomerase reverse transcriptase (TERT) component, and the telomerase RNA (TR) component *(110)*.

Telomerase activity is very low or undetectable in most human somatic tissues and primary cells. Telomeres shorten with each cell division in vivo and in vitro. A critical telomeric length, known as the Hayflick limit *(111)*, is reached in human primary cells, which limits replicative capacity and induces cellular senescence. This telomeric length checkpoint response is mediated by the Rb and p53 tumor-suppressor pathways. Primary cells deficient in Rb or p53 demonstrate continued growth beyond the Hayflick limit, and suffer from marked telomere shortening, genetic instability, and massive cell death—a phenomenon known as crisis. Telomere dysfunction activates a p53-dependent checkpoint *(112)*. The loss of telomere function and p53 deficiency as seen in mice doubly null for mTR and p53 cooperate to initiate the process of cellular transformation *(112)*. Thus, potential cancer cells must overcome two telomeric tumor-suppression mechanisms: replicative senescence and crisis. Ectopic expression of human TERT in normal human primary cells results in maintenance of telomeric length and unlimited growth *(113)*. Telomere shortening in the absence of telomerase activity, therefore, is a critical signal for entry into senescence, and that activation of telomerase blocks this process. Immortalization of some epithelial cells, however, requires not only TERT expression but also a defective RB/p16 pathway *(114)*. In mice doubly null for the telomerase RNA (mTR) and the INK4a tumor-suppressor genes, the loss of telomere function, and the inability to activate telomerase reduced the cancer incidence by greater than 50% in vivo *(115)*. Reintroduction of mTR into cells significantly restored the oncogenic potential, demonstrating that telomerase activation is a cooperating event in the malignant transformation of cells containing very shortened telomeres *(115)*.

Telomerase is expressed in most human cancers, including lung cancers. Telomerase activity in 136 primary lung cancer resection specimens and 68 adjacent nonmalignant tissues were evaluated using a polymerase chain reaction (PCR)-based telomeric repeat amplification protocol (TRAP assay) *(116)*. Telomerase activity was detected in 80% (109 of 136) of primary lung cancer samples vs 4% (3 of 68) normal adjacent tissue samples. Eleven of the 136 surgically resected specimens (from 11 patients) were primary SCLCs, which demonstrated very high levels of telomerase activity whereas the other 125 specimens (primary NSCLCs from 125 patients) had a wider range of telomerase activity. A high telomerase activity in primary NSCLC was found to be associated with increased cell proliferation rates and advanced pathologic stage *(117)*.

Telomerase activity was also detected in lung cancer cells obtained from bronchial washings from 82% (18 of 22) lung cancer patients *(118)*, whereas cytologic examination detected malignant cells in only 41% (9 of 22). Telomerase activity was detectable regardless of the location of the tumor (central vs peripheral). In a similar study of 37 primary lung cancer patients diagnosed histologically, there were 24 positive cytologies and 29 positive for telomerase activity *(119)*. A positive diagnostic outcome increased to 32 when both cytology and telomerase activity were considered. Thus, assaying for telomerase activity with the TRAP assay in addition to cytologic examination increases the sensitivity of cytology alone in making the diagnosis of lung cancer in bronchial washings.

Reactivation of telomerase expression is necessary for the continuous proliferation of cancerous cells to reach immortality and its deregulation may occur in preneoplastic bronchial epithelial dysplasias. Fresh and archival tissue samples from 40 patients (34 invasive lung cancers, 5 carcinoma *in situ* (CIS) without invasion, and 1 without lung carcinoma), were studied using the TRAP assay and *in situ* hybridization for hTR *(120)*. Telomerase positivity was present in basal epithelial cells of normal bronchial epithelium (7 of 27, 26%) and in peripheral lung samples (14 of 60, 23%; epithelium of small bronchi and bronchioles) *(120)*. Telomerase activity was detected in a much higher proportion of abnormal bronchial epithelial samples: hyperplasia (20 of 28, 71%), metaplasia (4 of 5, 80%), dysplasia (9 of 11, 82%), and CIS (11 of 11, 100%). Whereas normal cells demonstrate shortening of telomere length with each cell division, tumor cells show no net loss of telomere length, suggesting that telomere stability may be a requirement for bronchial epithelial cells to escape replicative senescence.

8. Summary: Cell-Cycle Networking

Insights into cell-cycle networking have grown exponentially in the past several years, leading to the concept that lung cancer is a disorder of the cell cycle. Although many of these findings are applicable to the lung, lung cancer may be unusual in that the progenitor cells give rise to squamous carcinoma, adenocarcinoma, small cell carcinoma or other cell types. The lung is also the target organ for many environmental toxicants; consequently extrapolating from in vitro studies to the lung requires studies of various lung cells directly. It is clear that mutations of cell-cycle genes occur in a sequential manner in the lung eventually leading to clonal cell expansion. After 8–12 mutations, a malignant clone proliferates into a CIS lesion where the apoptotic pathway to destroy wayward cells has been sabotaged. Important to the progression from a colony of cells to a growing tumor are induction of genes that stimulate endothelial cell incursion to form capillaries, and nearby stromal cell activation to release metalloproteinases with the capability to digest matrix proteins and allowing for tumor cell invasion. Central to these concepts is a central hypoxic region in the tumor mass, which leads to induction of transcription factors, e.g., hypoxia inducing factor (HIF-1) to activate genes such as vascular endothelial growth factor (VEGF) necessary for capillary formation *(121)*. At this juncture, the orchestration of the cancer phenotype is well underway, albeit clinically undetectable. Treatment strategies to cure lung cancer will have to focus on these early genetic lesions to enhance their repair, or to trigger the apoptotic pathway to eliminate wayward cells. The lung would be an excellent target for a strategy that involves inhalation of such a chemopreventive or protective agent.

References

1. Nesbitt, J. C., Putnam, J. B. Jr., Walsh, G. L., Roth, J. A., and Mountain, C. F. (1995) Survival in early-stage lung cancer. *Ann. Thorac. Surg.* **60**, 466–472.
2. Miller, J. A. and Jones, O. R. (1930) Primary carcinoma of the lung. *Am. Rev. Tuberc.* **21**, 1–56.
3. Ochsner, M. and DeBakey, M. (1939) Symposium on cancer. Primary pulmonary malignancy. Treatment by total pneumonectomy; analyses of 79 collected cases and presentation of 7 personal cases. *Surg. Gynecol. Obstet.* **68**, 435–451.
4. Weiss, W., Seidman, H., and Boucot, K. R. (1978) The Philadelphia pulmonary neoplasm research project. Symptoms in occult lung cancer. *Chest* **73**, 57–61.
5. Sridhar, K. S., Lobo, C. F., and Altman, R. D. (1998) Digital clubbing and lung cancer. *Chest* **114**, 1535–1537.
6. Mountain, C. F. (1997) Revisions in the International System for staging lung cancer. *Chest* **111**, 1710–1717.
7. ATS/ERS Statement. (1997) Pretreatment evaluation of non-small-cell lung cancer. *Am. J. Respir. Crit. Care Med.* **156**, 320–332.
8. Shaw, G. L., Falk, R. T., Pickle, L. W., Mason, T. J., and Buffler, D. A. (1991) Lung cancer risk associated with cancer in relatives. *J. Clin. Epidemiol.* **44**, 429–437.
9. Saraceno, J. and Spivack, S. D. (1999) Strategies for early detection of lung cancer. *Clin. Pulm. Med.* **6**, 66–72.
10. Henschke, C. I., McCauley, D. I., Yankelevitz, D. F., Naidich, D. P., McGuinness, G., Miettinen, O. S., et al. (1999) Early lung cancer action project: overall design and findings from baseline screening. *Lancet* **354**, 99–105.
11. Kennedy, T. C., Proudfoot, S. P., Franklin, W. A., et al. (1996) Cytopathological analysis of sputum in patients with airflow obstruction and significant smoking histories. *Cancer Res.* **56**, 4673–4678.
12. Tockman, M. S. Mulshine, J. L., Piantadosi, S., et al. (1997) Prospective detection of preclinical lung cancer: results from two studies of heterogeneous nuclear ribonucleoprotein A2/B1 overexpression. *Clin. Cancer Res.* **3**, 2237–2246.
13. Tockman, M. S., Gupta, P. K., Myers, J. D., et al. (1988) Sensitive and specific monoclonal antibody recognition of human lung cancer antigen on preserved sputum cells: a new approach to early lung cancer detection. *J. Clin. Oncol.* **6**, 1685–1693.
14. Payne, P. W., Sebo, T. J., Doudkine, A., et al. (1997) Sputum screening by quantitative microscopy: a re-examination of a portion of the NCI Cooperative Early Lung Cancer Study. *Mayo Clin. Proc.* **72**, 697–704.
15. Lam, S., MacAulay, C., Hung, J., LeRiche, J., Profio, A. E., and Palcic, B. (1993) Detection of dysplasia and carcinoma in situ with a lung imaging fluorescence endoscopy (LIFE) device. *J. Thoracic Cardiovasc. Surg.* **105**, 1035–1040.
16. Lam, S., Kennedy, T., Unger, M., et al. (1998) Localization of bronchial intraepithelial neoplastic lesions by fluorescence bronchoscopy. *Chest* **113**, 696–702.

17. Kurie, J. M., Lee, J. S., Morice, R. C., Walsh, G. L., Khuri, F. R., Broxson, A., et al. (1998) Autofluorescence bronchoscopy in the detection of squamous metaplasia and dysplasia in current and former smokers. *J. Natl. Cancer Inst.* **90,** 991–995.
18. Collisharv, N. E. and Lopez, A. D. (1996) The tobacco epidemic: a global public health emergency. Tobacco Alert, World Health Organization, Geneva, Switzerland.
19. Doll, R. and Hill, A. B. (1954) The mortality of doctors in relation to their smoking habits. A Preliminary report. *BMJ* **1,** 1451–1455.
20. Hammond, E. C. and Horn, D. (1954) The relationship between human smoking habits and death rates: a follow-up study of 187 766 men. *JAMA* **155,** 1316–1328.
21. Thun, M. J., Day-Lalley, C. A., Calle, E. E., Flanders, W. D., and Heath, C. A. (1995) Excess mortality among cigarette smokers: changes in a 20-year interval. *Am. J. Public Health* **85,** 1223–1230.
22. Doll, R., Peto, R., Wheatley, K., Gray, R., and Sutherland, I. (1994) Mortality in relation to smoking: 40 years' observations on male British doctors. *BMJ* **309,** 901–911.
23. Hirayama, T. (1981) Non-smoking wives of heavy smokers have a higher risk of lung cancer: a study from Japan. *BMJ* **282,** 183–185.
24. Trichopoulos, D., Kalandida, A., Sparros, L., and MacMahon, B. (1981) Lung cancer and passive smoking. *Int. J. Cancer* **23,** 803–807.
25. Harris, R. E., Zang, E. A., Anderson, J. I., and Wynder, E. L. (1993) Race and sex differences in lung cancer risk associated with cigarette smoking. *Int. J. Epidemiol.* **22,** 592–599.
26. Ryberg, D., Hewer, A., Phillips, D. H., and Haugen, A. (1994) Different susceptibility to smoking induced DNA damage among male and female lung cancer patients. *Cancer Res.* **54,** 5801–5803.
27. Fielding, J. E. (1985) Smoking: health effects and control. *N. Engl. J. Med.* **313,** 491–498, 555–562.
28. Thun, M. J., Lally, C. A., Flannery, J. T., Calle, E. E., Flanders, W. D., and Heath, C. W. (1997) Cigarette smoking and changes in the histopathology of lung cancer. *J. Natl. Cancer Inst.* **89,** 1580–1586.
29. Pershagen, G., Akerblom, G., Axelson, O., et al. (1994) Residential radon exposure and lung cancer in Sweden. *N. Engl. J. Med.* **330,** 159–164.
30. Hammond, E. C., Selikoff, I. J., and Seidman, H. (1979) Asbestos exposure, cigarette smoking, and death rates. *Ann. NY Acad. Sci.* **330,** 473–490.
31. Hecht, S. S. (1998) Cigarette smoking and cancer, in *Environmental and Occupational Medicine*, 3rd ed. (Rom, W. N., ed.), Lippincott-Raven, Philadelphia, pp. 1479–1500.
32. Spivack, S. D., Fasco, M. J., Walker, V. E., and Kaminsky, L. S. (1997) The molecular epidemiology of lung cancer. *Crit. Rev. Toxicol.* **27,** 319–365.
33. Wei, Q., Cheng, L., Hong, W. K., and Spitz, M. R. (1996) Reduced DNA repair capacity in lung cancer patients. *Cancer Res.* **56,** 4103–4107.

34. Wuerin, J. and Nurse, P. (1996) Regulating S phase: CDKs, licensing and proteolysis. *Cell* **85,** 785–787.
35. Sherr, C. (1996) Cancer cell cycles. *Science* **274,** 1672–1677.
36. Hunter, T. and Pines, J. (1994) Cyclins and Cancer II: Cyclin D and CDK inhibitors come of age. *Cell* **79,** 573–582.
37. Schaur, E. I., Siriwardana, S., Langan, T. A., and Sciafani, R. A. (1994) Cyclin D1 overexpression *vs.* retinoblastoma inactivation: implications for growth control evasion in non-small cell and small cell lung cancer. *Proc. Natl. Acad. Sci. USA* **91,** 7827–7831.
38. Weintraub, S. J. (1996) Inactivation of tumor suppressor proteins in lung cancer. *Am. J. Respir. Cell Mol. Biol.* **15,** 150–155.
39. El-Deiry, W. S., Tokino, T., Velculescu, V. E., et al. (1993) WAF1, a potential mediator of p53 tumor suppression. *Cell* **75,** 817–825.
40. Nevins, J. R. (1992) E2F: a link between the Rb tumor suppressor protein and viral oncoproteins. *Science* **258,** 424–429.
41. Barak, Y., Juven, T., Haffner, R., and Oren, M. (1993) Mdm2 expression is induced by wild type p53 activity. *EMBO J.* **12,** 461–468.
42. Stillman, B. (1996) Cell cycle control of DNA replication. *Science* **274,** 1659–1663.
43. Sanders Williams, R., Shohet, R. V., and Stillman, B. (1997) A human protein related to yeast Cdc6p. *Proc. Natl. Acad. Sci. USA* **94,** 142–147.
44. Yan, Z., DeGregori, J., Shohet, R., et al. (1998) Cdc6 is regulated by E2F and is essential for DNA replication in mammalian cells. *Proc. Natl. Acad. Sci. USA* **95,** 3603–3608.
45. Levine, A. J., Momand, J., and Finlay, C. A. (1991) The p53 tumour suppressor gene. *Nature* **351,** 453–456.
46. Greenblatt, M. S., Bennett, W. P., Hollstein, M., and Harris, C. C. (1994) Mutations in the p53 tumor suppressor gene: clues to cancer etiology and molecular pathogenesis. *Cancer Res.* **54,** 4855–4878.
47. Prives, C. (1998) Signaling to p53: breaking the MDM2-p53 circuit. *Cell* **95,** 1437–1443.
48. Weinberg, R, A. (1991) Tumor suppressor genes. *Science* **254,** 1138–1146.
49. Knudson, A. G. (1993) Antioncogenes and human cancer. *Proc. Natl. Acad. Sci. USA* **90,** 10914–10921.
50. Tchou, W. W., Rom, W. N., and Tchou-Wong, K.-M. (1996) Novel form of p21 WAF1/C1P1/SC11 protein in phorbol ester-induced G2/M arrest. *J. Biol. Chem.* **271,** 29556–29560.
51. Ramet, M., Casten, K., Jarvinen, K., et al. (1996) p53 protein expression is correlated with benzo[a]pyrene-DNA adducts in carcinoma cell lines. *Carcinogenesis* **16,** 2117–2124.
52. Denissenko, M. F., Pao, A., Tang, M., and Pfeifer, G. P. (1996) Preferential formation of benzo[a] pyrene adducts at lung cancer mutational hotspots in p53. *Science* **274,** 430–432.

53. Denissenko, M. F., Chen, J. Y., Tang, M. S., and Pfeifer, G. P. (1997) Cytosine methylation determines hot spots of DNA damage in the human p53 gene. *Proc. Natl. Acad. Sci. USA* **94,** 3893–3898.
54. Chen, J. X., Zheng, Y., West, M., and Tang, M. (1998) Carcinogens preferentially bind at methylated CpG in the p53 mutational hot spots. *Cancer Res.* **58,** 2070–2075.
55. Waga, S., Hannon, G. J., Beach, D., and Stillman, B. (1994) The p21 inhibitor of cyclin-dependent kinases controls DNA replication by interaction with PCNA. *Nature* **369,** 574–578.
56. Smith, M. L., Chen, I.-T., Zhan, Q., et al. (1994) Interaction of the p53-regulated protein Gadd45 with proliferating cell nuclear antigen. *Science* **266,** 1376–1380.
57. Zambetti, G. and Levine, A. J. (1993) A comparison of the biological activities of wild-type and mutant p53. *FASEB J.* **7,** 855–865.
58. Werner, H., Karnieli, E., Rauscher, III F. J., and LeRoith, D. (1996) Wild-type and mutant p53 differentially regulate transcription of the insulin-like growth factor I receptor gene. *Proc. Natl. Acad. Sci. USA* **93,** 8318–8323.
59. Jiang, Y., Rom, W. N,, Yie, T. A., Chi, C., and Tchou-Wong, K. M. (1999) Induction of tumor suppression and glandular differentiation of A549 lung carcinoma cells by dominant negative IGF-I receptor. *Oncogene* **18,** 6071–6077.
60. Kamb, A., Gruis, N. A., Weaver-Feldhaus, J., et al. (1994) A cell cycle regulator potentially involved in genesis of many tumor types. *Science* **264,** 436–440.
61. Cairns, P., Polascik, T. J., Eby, Y., et al. (1995) Frequency of homozygous deletion at p16/CDKN2 in primary human tumors. *Nat. Genet.* **11,** 210–212.
62. Okamoto, A., Hussain, S. P., Hagiwara, K., et al. (1995) Mutations in the p16INK4/MTS1/CDKN2, p15INK4B/MTS2, and p18 genes in primary and metastatic lung cancer. *Cancer Res.* **55,** 1448–1451.
63. Rusin, M. R., Okamoto, A., Chorazy, M., et al. (1996) Intragenic mutations of the $p16^{INK4}$, $p15^{INK4B}$ and p18 genes in primary non-small-cell lung cancers. *Int. J. Cancer* **65,** 734–739.
64. Belinsky, S. A., Nikula, K. J., Palmisano, W. A., Michels, R., Saccomanno, G., Gabrielson, E., Baylin, S. B., and Herman, J. G. (1998) Aberrant methylation of $p16^{INK4a}$ is an early event in lung cancer and a potential biomarker for early diagnosis. *Proc. Natl. Acad. Sci. USA* **95,** 11891–11896.
65. Michalides, R. J. A. M. (1998) Deregulation of the cell cycle in lung cancer, in *Lung Biology in Health and Disease* (Brambilla, C. and Brambilla, E., eds.), Marcell Dekker Inc., New York, pp. 211–225.
66. Haber, D. A. (1997) Splicing into senescence: the curious case p16 and $p19^{ARF}$. *Cell* **91,** 555–558.
67. Harbour, J. W., Lai, S. L., Whang-Peng, J., et al. (1988) Abnormalities in structure and expression of the human retinoblastoma gene in SCLC. *Science* **241,** 353–357.
68. Shapiro, G. I., Edwards, C. D., Kobzik, L., et al. (1995) Reciprocal Rb inactivation and $p16^{INK4}$ expression in primary lung cancers and cell lines. *Cancer Res.* **55,** 505–509.

69. Kelley, M. J., Nakagawa, K., Steinberg, S. M., Mulshine, J. L., Kamb, A., and Johnson, B. E. (1995) Differential inactivation of CDKN2 and Rb protein in non-small-cell and small-cell lung cancer cell lines. *J. Natl. Cancer Inst.* **87**, 756–761.
70. Gorgoulis, V. G., Zacharatos, P., Kotsinas, A., et al. (1998) Alterations of the p16-pRb pathway and the chromosome locus 9p21-22 in non-small-cell lung carcinomas. *Am. J. Path.* **153**, 1749–1765.
71. Bates, S., Phillips, A., Clarke, P., et al. (1998) p14^{ARF} links the tumour suppressors RB and p53. *Nature* **395**, 124–125.
72. Zindy, F., Eischen, C. M., Randle, D. H., Kamijo, T., Cleveland, J. L., Sherr, C. J., and Roussel, M. F. (1998) Myc signaling via the ARF tumor suppressor regulates p53-dependent apoptosis and immortalization. *Genes Dev.* **12**, 2424–2433.
73. Pomerantz, J., Schreiber-Angus, N., Liegeois, N. J., et al. (1998) The *INK4a* tumor suppressor gene product, p19^{ARF}, interacts with MDM2 and neutralizes MDM2's inhibition of p53. *Cell* **92**, 713–723.
74. Kamijo, T., Weber, J. D., Zambetti, G., Zindy, F., Roussel, M. F., and Sherr, C. J. (1998) Functional and physical interactions of the ARF tumor suppressor with p53 and Mdm2. *Proc. Natl. Acad. Sci. USA* **95**, 8292–8297.
75. Zhang, Y. and Xiong, Y. (1999) Mutations in human ARF exon 2 disrupt its nucleolar localization and impair its ability to block nuclear export of MDM2 and p53. *Mol. Cell* **3**, 579–591.
76. Gazzeri, S., Della Valle, V., Chaussade, L., Brambilla, C., Larsen, C. J., and Brambilla, E. (1998) The human p19^{INK4a} gene is frequently lost in small cell lung cancer. *Cancer Res.* **58**, 3926–3931.
77. Hannon, G. J. and Beach, D. (1994) p15^{INK4B} is a potential effector of TGF-β-induced cell cycle arrest. *Nature* **371**, 257–261.
78. Kim, S. J., Wagner, S., Lin, F., O'Reilly, M. A., Robbins, P. D., and Green, M. R. (1992) Retinoblastoma gene product activates expression of the human TGF-β2 gene through transscription factor ATF-2. *Nature* **358**, 331–334.
79. Li, C.-Y., Suardet, L., and Little, J. R. (1995) Potential role of WAF1/Cip/p21 as a mediator of TGF-β cytoinhibitory effect. *J. Biol. Chem.* **270**, 4971–4974.
80. Iavarone, A. and Massague, J. (1997) Repression of the CDK activator Cdc 25A and cell-cycle arrest by cytokine TGF-β in cells lacking the CDK inhibitor p15. *Nature* **387**, 417–421.
81. Zhang, H. S., Postigo, A. A., and Dean, D. C. (1999) Active transcriptional repression by the Rb-E2F complex mediates G1 arrest triggered by p16^{INK4a}, TGFβ, and contact inhibition. *Cell* **97**, 53–61.
82. Mills, N. E., Fishman, C. L., Rom, W. N., et al. (1995) Increased prevalence of K-ras oncogene mutations in lung adenocarcinoma. *Cancer Res.* **55**, 1444–1447.
83. Mills, N. E., Fishman, C. L., Scholes, J., Anderson, S. E., Rom, W. N., and Jacobson, D. R. (1995) Detection of K-ras oncogene mutations in bronchoalveolar lavage fluid as a diagnostic test for lung cancer. *J. Natl. Cancer Inst.* **87**, 1056–1060.

84. Slebos, R. J. C., Kibbelaar, R. E., Dalesio, O., et al. (1990) K-*ras* oncogene activation as a prognostic marker in adenocarcinoma of the lung. *N. Engl. J. Med.* **9,** 561–565.
85. Zhang, Y., Mukhopadhyay, T., Donehower, L. A., Georges, R. N., and Roth, J. A. (1993) Retroviral vector-mediated transduction of K-*ras* antisense RNA into human lung cancer cells inhibits expression of the malignant phenotype. *Hum. Gene Ther.* **4,** 451–460.
86. Moodie, S. A., Willumsen, B. M., Weber, M. J., and Wolfman, A. (1993) Complexes of ras-GTP with raf-1 and mitogen-activated protein kinase kinase. *Science* **260,** 1658–1661.
87. Li, Z. H., Zheng, J., Weiss, L. M., and Shibata, D. (1994) K-*ras* and p53 mutations occur very early in adenocarcinoma of the lung. *Am. J. Pathol.* **144,** 303–309.
88. Agarwal, M. L., Taylor, W. R., Chernov, M. V., Chernova, O. B., and Stark, G. R. (1998) The p53 network. *J. Biol. Chem.* **273,** 1–4.
89. Leone, G., DeGregori, J., Sears, R., Jakoi, L., and Nevins, J. R. (1997) Myc and ras collaborate in inducing accumulation of active cyclin E/Cdk2 and E2F. *Nature* **387,** 422–426.
90. Peeper, D. S., Upton, T. M., Ladha, M., et al. (1997) Ras signaling linked to the cell-cycle machinery by the retinoblastoma protein. *Nature* **386,** 177–181.
91. Sears, R., Leone, G., DeGregori, J., and Nevins, J. R. (1999) Ras enhances myc protein stability. *Mol. Cell* **3,** 169–179.
92. Serrano, M., Gomez-Lahoz, E., DePinho, R. A., Beach, D., and Bar Sagi, D. (1995) Inhibition of ras-induced proliferation and cellular transformation by p16^{ink4}. *Science* **267,** 249–252.
93. Olson, M. F., Peterson, H. F., and Marshall, C. F. (1998) Signals from Ras and Rho GTPases interact to regulate expression of p21$^{WAF1/C1P1}$. *Nature* **394,** 295–299.
94. Palmero, I., Pantoja, C., and Serrano, M. (1998) p19^{ARF} links the tumour suppressor p53 to ras. *Nature* **395,** 125–126.
95. Kamijo, T., Bodner, S., van de Kamp, E., Randle, D. H., and Sherr, C. J. (1999) Tumor spectrum in *ARF*-deficient mice. *Cancer Res.* **59,** 2217–2222.
96. Galaktionov, K., Chen, X., and Beach, D. (1996) Cdc25 cell-cycle phosphatase as a target of *c-myc*. *Nature* **382,** 511–517.
97. Hunter, T. (1997) Oncoprotein networks. *Cell* **88,** 333–346.
98. Lee, T. C. and Ziff, E. B. (1999) Mxi1 is a repressor of the c-myc promoter and reverses activation by USF. *J. Biol. Chem.* **274,** 595–606.
99. Lee, T. C., Li, L., Philipson, L., and Ziff, E. B. (1997) Myc represses transcription of the growth arrest gene *gas*1. *Proc. Natl. Acad. Sci. USA* **94,** 12886–12891.
100. Marhin, W. W., Chen, S., Facchini, L. M., Fornace, A. J. Jr., and Penn, L. Z. (1997) Myc represses the growth arrest gene gadd45. *Oncogene* **14,** 2825–2834.
101. Del Sal, G., Ruaro, E. M., Utrera, R., Cole, C. N., Levine, A. J., and Schneider, C. (1995) Gas1-induced growth suppression requires a transactivation-independent p53 function. *Mol. Cell. Biol.* **15,** 7152–7160.

102. Viallet, J. and Minna, J. (1990) Dominant oncogenes and tumor suppressor genes in the pathogenesis of human lung cancer. *Am. J. Respir. Cell Mol. Biol.* **2,** 225–232.
103. Broers, J. L. V., Viallet, J., Jensen, S. M., Pass, H., Travis, W. D., Minna, J. D., and Linnoila, R. I. (1993) Expression of c-myc in progenitor cells of the bronchopulmonary epithelium and in a large number of non-small cell lung cancers. *Am. J. Respir. Cell Mol. Biol.* **9,** 33–43.
104. Gazzeri, S., Brambilla, E., Caron de Fromentel, C., Bouyer, V., Moro, D., Perron, P., et al. (1994) p53 genetic abnormalities and myc activation in human lung carcinoma. *Int. J. Cancer* **58,** 24–32.
105. Franklin, W. A., Gazdar, A. F., Haney, J., et al. (1997) Widely dispersed p53 mutation in respiratory epithelium: a novel mechanism for field carcinogenesis. *J. Clin. Invest.* **100,** 2133–2137.
106. Auerbach, O. C., Hammond, C., and Garfinkel, L. (1979) Changes in bronchial epithelium in relation to cigarette smoking, 1955–1960 vs 1970–1977. *N. Engl. J. Med.* **300,** 381–386.
107. Saccomanno, G., Archer, V. E., Auerbach, O., Saunders, R. P., and Brennan, L. U. (1974) Development of carcinoma of the lung as reflected in exfoliated cells. *Cancer* **33,** 256–270.
108. Risse, E. K. J., Vooijs, G. P., and van't Hof, M. A. (1988) Diagnostic significance of 'severe dysplasia' in sputum cytology. *Acta Cytol.* **32,** 629–634.
109. Keith, R. L., Varella-Garcia, M., Gemmill, R. M., et al. (1998) Morphologic and genetic abnormalities in the bronchial epithelium of high risk smokers. *Am. J. Respir. Crit. Care Med.* **157,** 692A.
110. van Steensel, B. and de Lange, T. (1998) Control of telomere length by the human telomeric protein TRF1. *Nature* **385,** 740–743.
111. Hayflick, L. and Moorhead, P. S. (1961) The serial cultivation of human diploid cells strains. *Exp. Cell Res.* **25,** 585–621.
112. Chin, L., Artandi, S. E., Shen, Q., Tam, A., Lee, S. L., Gottlieb, G. J., et al. (1999) p53 deficiency rescues the adverse effects of telomere loss and cooperates with telomere dysfunction to accelerate carcinogenesis. *Cell* **97,** 527–538.
113. Bodnar, A. G., Ouellette, M., Frolkis, M., Holt, S. E., Chiu, C. P., Morin, G. B., et al. (1998) Extension of life-span by introduction of telomerase into normal human cells. *Science* **279,** 349–352.
114. Kiyono, T., Foster, S. A., Koop, J. I., McDougall, J. K., Galloway, D. A., and Klingelhutz, A. J. (1998) Both Rb/p16^{INK4a} inactivation and telomerase activity are required to immortalize human epithelial cells. *Nature* **396,** 84–88.
115. Greenberg, R. A., Chin, L., Femino, A., Lee, K. H., Gottlieb, G. J., Singer, R. H., et al. (1999) Short dysfunctional telomere impair tumorigenesis in the INK4a$^{\Delta 2/3}$ cancer-prone mouse. *Cell* **97,** 515–525.
116. Hiyama, K., Hiyama, E., Ishioka, S., Yamakido, M., Inai, K., Gazdar, A. F., et al. (1995) Telomerase activity in small-cell and non-small-cell lung cancers. *J. Natl. Cancer Inst.* **87,** 895–902.

117. Albanell, J., Lonardo, F., Rusch, V., Engelhardt, M., Langenfeld, J., Han, W., et al. (1997) High telomerase activity in primary lung cancers: association with increased cell proliferation rates and advanced pathologic stage. *J. Natl. Cancer Inst.* **89,** 1609–1615.
118. Yahata, N., Ohyashiki, K., Ohyashiki, J. H., et al. (1998) Telomerase activity in lung cancer cells obtained from bronchial washings. *J. Natl. Cancer Inst.* **90,** 684–690.
119. Arai, T., Yasuda, Y., Takaya, T., Ito, Y., Hayakawa, K., Toshima, S., et al. (1998) Application of telomerase activity for screening of primary lung cancer in broncho-alveolar lavage fluid. *Oncol. Reports* **5,** 405–408.
120. Yashima, K., Litzky, L. A., Kaiser, L., et al. (1997) Telomerase expression in respiratory epithelium during the multistage pathogenesis of lung carcinomas. *Cancer Res.* **57,** 2373–2377.
121. Folkman, J. (1995) Angiogenesis in cancer, vascular, rheumatoid, and other disease. *Nature Med.* **1,** 27–31.

II

DETECTION OF ALTERED TUMOR MARKERS IN CLINICAL SAMPLES FOR DIAGNOSIS AND PROGNOSIS

A. MOLECULAR METHODS FOR PROGNOSTIC DETECTION OF ALTERED TUMOR MARKERS: CLINICAL IMPLICATIONS

2

Molecular Alterations in Lung Cancer

Impact on Prognosis

Takashi Kijima, Gautam Maulik, and Ravi Salgia

1. Introduction

Lung cancer is a devastating illness, and very few of the advances in chemotherapeutics have enhanced survival over the past decade. In order to make an impact on this disease, we must understand the molecular abnormalities to target better therapeutics. In this chapter, we will describe the better understanding at the molecular level that has been gained for lung cancer and its prognosis. Over the last decade, advances in molecular biology have provided important information about potentially significant determinants of prognosis in lung cancer *(1,2)*. These molecular biomarkers are typically expressed in neoplastic lung tissue. Molecular abnormalities include chromosomal aberrations, telomerase expression, expression of oncogenes, and loss of tumor suppressor genes. Based on the understanding of the abnormalities at the cellular and molecular levels, newer therapies can be pursued.

2. Molecular Biological Abnormalities in Lung Cancer
2.1. Model of Lung Cancer Development

For any tumor to become cancerous, various mutations/alterations occur in the cell, which render it neoplastic. To understand the mechanisms for transformation, we will briefly describe several pathways that can be activated or suppressed in the pathogenesis of lung cancer. In terms of development of a lung cancer cell, the cell is surrounded by other cells and extracellular matrix (ECM). The ECM and various molecules in the circulation can communicate with the cell through receptors such as growth factor receptors, adhesion molecules (such as cadherins), and integrins. Once a transducing signal is achieved in the cell, different pathways may be activated, which can lead to dramatic changes that could affect the cytoskeleton, morphology, or migration of the cell. Eventually, the signal is transmitted to the nucleus, and certain genes can be activated or suppressed, which results in malignant transformation.

When a cell becomes neoplastic, many changes occur inside and outside the cellular environment. The cell not only changes its internal homeostatic mechanisms, but also changes the extracellular environment. Eventually, the cancerous population multiplies and, thereafter, may metastasize and colonize in different sites. For a carcinoma to metastasize, many events occur such as: invasion with changes in tumor cell adhesion, proteinase production, and locomotion; intra-vasation and extra-vasation from the circulatory system; colonization in the distant site; and angiogenesis in the new site of implant. For the purposes of this review, we will restrict ourselves to chromosomal abnormalities, activation of oncogenes, and loss of tumor-suppressor genes in lung cancer. We will also emphasize the importance of prognostication with the various genes that become abnormal in lung cancer.

2.2. Chromosomal Abnormalities, Telomerase Activation, and Implications

Using both actual tumor specimens and cell lines, various chromosomal and oncogene abnormalities have been identified in lung cancer. There have been a number of reports on the various chromosomal abnormalities (including loss of complete chromosomes or portions thereof) that can occur in lung cancer. For example, in non-small cell lung cancer (NSCLC), chromosomal aberrations have been described on 3p, 8p, 9p, 11p, 15p, and 17p with deletions of chromosomes 7, 11, 13, or 19. Also, in small cell lung cancer (SCLC), chromosomal abnormalities have been described on 1p, 3p, 5q, 6q, 8q, 13q, or 17p *(3)*.

One of the most consistent chromosomal abnormalities in lung cancer has been the loss of the short arm of chromosome 3 (3p[14-25]) *(4)*. The loss of alleles at 3p is observed in >90% of SCLC tumors and approx 50% of NSCLC tumors *(5)*. Various groups are trying to clone out the tumor-suppressor genes, which may be involved in the loss of 3p regions. As an example, the FHIT gene (for fragile histidine triad) has been localized to 3p14.2 and about 80% of SCLC tumors shown abnormalities of this gene *(6)*. The protein product of the FHIT gene is involved in the metabolism of diadenosine tetraphosphate into ATP and AMP. Loss of FHIT gene results in the accumulation of diadenosine tetraphosphate and could lead to the stimulation of DNA synthesis and proliferation. The FHIT gene may represent one of several potential tumor suppressor genes located on chromosome 3p involved in the pathogenesis of SCLC.

Other genetic losses have, although not consistently, been identified in lung cancer. In NSCLC, these include genetic loss at chromosome 8p(21.3-22) and may affect in 50% of multiple samples *(7)*. Genetic loss at 9p(21-22) could potentially involve the p16 (MTS1/p16^{INK4A}) and p15 (MTS2/p15^{INK4B}) tumor suppressor genes, which are involved in cell-cycle regulation at the G1 checkpoint by inhibiting cyclin-dependent kinase CDK4 and may be affected in 67% of tumor samples *(8,9)*. Genetic loss at 11p (p13 and p15) may involve the Wilms' tumor suppressor gene at region p13 and can be affected in 20% to 46% of tumor samples *(10)*. SCLC exhibits infrequent loss of 9p, but more losses than NSCLC of 3p, 5q, 13q, and 17p *(5)*.

Telomeres, which are genetic elements at the ends of linear eukaryotic chromosomes from degradation, illegitimate recombination, or cellular senescence. Long telomeres are present in germ cells and most cancer cells via the telomerase enzyme, and this probably maintains the ability of the cells to divide indefinitely *(11,12)*. Telomerase activity has been directly correlated with malignant and metastatic phenotype of a wide array of solid tumors. In one study, 80% of tumor tissue from lung cancer had telomerase activity *(13)*. Telomerase activity was measured in bronchial washings with 18 of 22 patients with lung cancer being positive, whereas only 1 of 19 without cancer *(14)*. Because telomerase activation is essential for long-term growth of many malignancies, inhibition of this enzyme would be an attractive target for therapy *(11)*.

Compared to the chronology of colon cancer development, it is quite difficult to arrive at the chronology of events for a normal cell to develop from a preoplastic lesion to a frank neoplasia in lung cancer. It is possible that multiple synchronous molecular abnormalities occur in response to toxins such as cigarette smoke, which transform the normal lung cell into a cancerous cell.

2.3. Oncogenes and Tumor-Suppressor Genes

Various oncogene expressions have been investigated in NSCLC and SCLC. There are two forms of oncogenes: dominant oncogenes, and tumor-suppressor genes. Dominant oncogenes, such as RAS, MYC, HER-2/NEU, and BCL-2, exert their effect by overtaking the normal cellular growth function; tumor-suppressor genes that exert their effect in controlling cellular growth. Once suppressor genes such as p53, RB, p16^{INK4A}, p15^{INK4B}, and genes on chromosome 3p are deleted or mutated, normal control mechanisms are not available. None of the genes have been implicated in the etiology of lung cancer 100% of the time *(1)*.

2.3.1. Oncogenes

2.3.1.1. RAS Genes

The RAS-dominant oncogenes play an important role in signal transduction and cellular proliferation. The RAS proteins have a molecular weight of 21 kDa and consist of K-RAS, H-RAS, and N-RAS. The RAS proteins are active when bound to guanosine triphophate (GTP) and are inactivated by GTPase-activating protein (GAP) by hydrolyzing GTP to guanosine diphosphate (GDP). These proteins acquire transforming potential secondary to a point mutation at codon 12, 13, or 61 in the encoding gene. Mutations at or near the GTP-binding domain of RAS protein prevents the inactivation of GTP, thereby resulting in continuous RAS activity.

Activation of the K-RAS oncogene is an adverse prognostic factor in resectable adenocarcinoma of the lung. As an example, a point mutation in the K-RAS oncogene was noted in 29% of 69 resected specimens in one series *(15)*. Tumors associated with the mutation tended to be smaller but more poorly differentiated. Death during a median follow-up of 3 yr was more common in patients with the K-RAS mutation than in those without mutation (63% vs. 32%). There were no significant associations between K-RAS mutations and tumor size, stage, nodal status, or tumor differentiation. Several other reports also demonstrated a decrease in survival associated with K-RAS mutations in patients with respectable NSCLC *(16,17)*, but such findings have not been universal *(18,19)*.

2.3.1.2. MYC Genes

The MYC-dominant oncogenes, c-MYC and N-MYC, and L-MYC, encode for nuclear DNA binding proteins, which are involved in transcriptional regulation. The general mechanism of activation of MYC genes in lung cancer is gene amplification with resulting overexpression *(20,21)*. The frequency of abnormal expression of MYC genes is low in NSCLC (10%) and variable in SCLC (10–40%). Studies have shown that amplification of c-MYC genes adversely affect survival in SCLC *(22–24)*. In cell lines established after progression from cytotoxic chemotherapy, 44% of the cell lines had MYC amplification *(25)*. A total of 80–90% of SCLC also show overexpression of MYC RNA as compared with normal lung tissue *(26)*. c-MYC amplification is rarely seen in NSCLC *(27)*. In a study of Japanese patients with NSCLC *(28)*, the restriction fragment-length polymorphism (RFLP) of the L-MYC gene may be a marker for metastatic potential. However, there may be geographical differences, since no RFLP changes of the L-MYC gene were detected in NSCLC tumor samples from Australian, Norwegian, and North American patients *(2,10)*.

2.3.1.3. HER2/NEU GENE

c-erbB-1 proto-oncogene encodes the epidermal growth factor receptor (EGFR) and has been a classic model for signal-transduction events in normal and transformed cells. A related proto-oncogene, c-erbB-2 (also known as HER2/NEU), encodes for a protein product of molecular weight 185 kDa (p185^{neu}), and is a growth-factor receptor. The frequency of normal expression of c-erbB-2 in NSCLC is approx 25%, and it has not been reported to be abnormal in SCLC. Overexpression of the c-erbB-2 has been shown to be associated with an adverse prognosis in adenocarcinoma of the lung *(29)*. An antibody against the p185^{neu} has been shown to inhibit proliferation of NSCLC cell lines *(30)*.

2.3.1.4. BCL-2 GENE

The BCL-2 proto-oncogene product inhibits programmed cell death, termed apoptosis. BCL-2 overexpressing cells have expansion of cell populations secondary to lack of apoptosis. The expression of BCL-2 in NSCLC has been evaluated *(31)*. Of the various tumors evaluated, BCL-2 protein was abnormally expressed (where the pattern of expression in lung cancer cells differs from that in adjacent normal tissue) in 20 of 80 squamous cell carcinomas and in 5 of 42 adenocarcinomas. Basal cells in adjacent normal epithelium were positively stained for BCL-2; however, the more differentiated columnar cells were negative. In a group of patients with squamous cell carcinomas, 5-yr survival was better for patients with BCL-2- positive tumors (78% vs 48%, $p < .05$) evaluated *(31)*. The explanation for better prognosis for BCL-2-positive tumors has not yet been determined; however, there is a possibility that the increased expression of BCL-2 involved in the pathogenesis of some cases of lung cancer confers a survival advantage through its anti-apoptotic effects *(32)*.

2.3.2. Tumor-Suppressor Genes

2.3.2.1. P53 GENE

The p53 family of genes includes p53, p73, and p63. The original p53 gene, located at the chromosome 17p13.1 encodes a nuclear protein that acts as a transcription factor and blocks the progression of cells through the cell-cycle late in the G_1 phase. The most common genetic changes associated with cancer involve mutations of the p53 gene. p53 gene mutations cause a loss of tumor-suppression function, promoting cellular proliferation. Some p53 mutant proteins also have transforming properties, and can bind and inactivate available wild-type (normal) p53. The Li-Fraumeni cancer syndrome that is typified by multiple tumors at an early age of onset is characterized by inherited forms of p53 mutations. p53 genetic mutations can involve deletions, point mutations, and overexpression. In lung cancer, the prevalent type of point mutation is a GC to TA transversion and related to adducts of benzo(a)pyrene from cigarette smoking *(12)*. Frequency of mutations may be up to 50% in NSCLC and 80% in SCLC *(33)*. Abnormal expression of p53 has been shown to correlate with both better and worse prognosis and further work is needed to refine these observations *(34,35)*. In contrast to p53, the p73 gene (located on chromosome 1p36) is not dramatically altered in lung cancer, as observed by analysis of 17 lung cancer cell lines, only three of which exhibit mutations that affect the amino acid sequence *(36)*. Finally, the p63 (located on chromosome 3p27-28) mutation was detected in the DNA-binding domain in one squamous cell carcinoma cell line of the lung but generally appears to be rare *(37)*.

2.3.2.2. RB Gene

The RB gene, located on chromosome 13q14.11, encodes for a nuclear protein which was determined to be abnormal in patients with retinoblastoma. Knudson predicted the tumor-suppressor nature of this gene by studying inheritance patterns of familial retinoblastoma *(38)*. The protein encoded for by RB is a 105 kDa phosphoprotein, important in regulating the cell cycle during G_0/G_1 phase. A deletion of the RB gene can be found in >90% of SCLC, and the abnormal expression of the tumor-suppressor gene RB may be an adverse prognosticator in SCLC *(39,40)*.

In NSCLC, there is absence of normal RB mRNA in 10% of cell lines, and absence of normal RB protein in up to 30% of tumors *(41,42)*. Most of the RB-positive lung cancer cell lines that have been tested are also negative for $p16^{INK4A}$, a kinase inhibitor of CDK4, and thus a strong inhibitor of RB phosphorylation *(9)*. This would indicate that the pathogenesis of some lung cancers could occur by either the mutational disruption of RB protein or by the absence of the $p16^{INK4A}$ inhibitor that functions to keep RB hypophosphorylated and therefore active *(43)*. In addition, there may be a correlation between increased abnormal RB protein expression and stage in NSCLC. For example, in one study, abnormal expression was positive at a ratio of 20% for stages I and II and 60% for stages III and IV *(42)*.

2.3.2.3. $p16^{INK4A}$ and $p15^{INK4B}$ Genes

Certain lung cancer cells have a characteristic deletion of chromosome 9p21, thus implicating one or more tumor suppressor genes in this region as being important. From genetic analysis, $p16^{INK4A}$ (hereafter designated p16) and $p15^{INK4B}$ (designated p15) have been identified to map within 30kb of each other in this region. p16 was originally identified as a binding protein to CDK4 in a yeast two-hybrid screen, and contains four ankyrin repeats *(44)*. p15 was thereafter cloned from a low stringency screen using p16 probe on a human keratinocyte cDNA library, and has a very high homology with p16. In some cells stimulated by TGF-β, p15 induction is a 30-fold greater than p16 *(45)*. Both p16 and p15 encode for proteins that inhibit CDK4-regulated cell cycle control, thereby preventing progression from G_1 to S phase. The expression of p16 has been evaluated in lung cancer, and one study reported that 18 of 27 primary NSCLC contained no detectable protein *(46)*. The p16-negative tumors were invariably positive for RB protein expression.

3. Summary

There are multiple molecular abnormalities that can occur in lung cancer. Based on the aberrancies described previously, many investigators and drug companies are designing novel therapies. The molecular markers can also be used as prognostic variables for future clinical trials and therapeutic interventions. It will not be an easy task to make an impact on lung cancer using these methods, since multiple pathways are abnormal in its pathogenesis. It is hoped that with the advent of novel and directed therapeutics, we may soon show some impact on the survival of this devastating disease.

References

1. Salgia, R. and Skarin, A. T. (1998) Molecular abnormalities in lung cancer. *J. Clin. Oncol.* **16,** 1207–1217.
2. Mountain, C. F. (1995) New prognostic factors in lung cancer. Biologic prophets of cancer cell aggression. *Chest* **108,** 246–254.
3. Devereux, T. R., Taylor, J. A., and Barrett, J. C. (1996) Molecular mechanisms of lung cancer. Interaction of environmental and genetic factors. Giles F. Filley Lecture. *Chest* **109,** 14S–19S.
4. Hibi, K., Takahashi, T., Yamakawa, K., Ueda, R., Sekido, Y., Ariyoshi, Y., et al. (1992) Three distinct regions involved in 3p deletion in human lung cancer. *Oncogene* **7,** 445–449.
5. Otterson, G., Lin, A., and Kay, F. (1992) Genetic etiology of lung cancer. *Oncology (Huntingt).* **6,** 97–104, 107; discussion 108, 111–112.
6. Sozzi, G., Veronese, M. L., Negrini, M., Baffa, R., Cotticelli, M. G., Inoue, H., et al. (1996) The FHIT gene 3p14.2 is abnormal in lung cancer. *Cell* **85,** 17–26.
7. Ohata, H., Emi, M., Fujiwara, Y., Higashino, K., Nakagawa, K., Futagami, R., et al. (1993) Deletion mapping of the short arm of chromosome 8 in non-small cell lung carcinoma. *Genes Chromosomes Cancer* **7,** 85–88.
8. Merlo, A., Gabrielson, E., Askin, F., and Sidransky, D. (1994) Frequent loss of chromosome 9 in human primary non-small cell lung cancer. *Cancer Res.* **54,** 640–642.
9. Shapiro, G. I. and Rollins, B. J. (1996) p16^{INK4A} as a human tumor suppressor. *Biochim. Biophys. Acta.* **1242,** 165–169.

10. Fong, K. M., Zimmerman, P. V., and Smith, P. J. (1995) Lung pathology: the molecular genetics of non-small cell lung cancer. *Pathology* **27**, 295–301.
11. Meyerson, M. (2000) Role of telomerase in normal and cancer cells. *J. Clin. Oncol.* **18**, 2626–2634.
12. Carbone, D. P. (1997) The biology of lung cancer. *Semin. Oncol.* **24**, 388–401.
13. Hiyama, K., Ishioka, S., Shirotani, Y., Inai, K., Hiyama, E., Murakami, I., et al. (1995) Alterations in telomeric repeat length in lung cancer are associated with loss of heterozygosity in p53 and Rb. *Oncogene* **10**, 937–944.
14. Yahata, N., Ohyashiki, K., Ohyashiki, J. H., Iwama, H., Hayashi, S., Ando, K., et al. (1998) Telomerase activity in lung cancer cells obtained from bronchial washings. *J. Natl. Cancer Inst.* **90**, 684–690.
15. Slebos, R. J., Kibbelaar, R. E., Dalesio, O., Kooistra, A., Stam, J., Meijer, C. J., et al. (1990) K-ras oncogene activation as a prognostic marker in adenocarcinoma of the lung. *N. Engl. J. Med.* **323**, 561–565.
16. Rosell, R., Li, S., Skacel, Z., Mate, J. L., Maestre, J., Canela, M., et al. (1993) Prognostic impact of mutated K-ras gene in surgically resected non-small cell lung cancer patients. *Oncogene* **8**, 2407–2412.
17. Mitsudomi, T., Steinberg, S. M., Oie, H. K., Mulshine, J. L., Phelps, R., Viallet, J., et al. (1991) ras gene mutations in non-small cell lung cancers are associated with shortened survival irrespective of treatment intent. *Cancer Res.* **51**, 4999–5002.
18. Graziano, S. L., Gamble, G. P., Newman, N. B., Abbott, L. Z., Rooney, M., Mookherjee, S., et al. (1999) Prognostic significance of K-ras codon 12 mutations in patients with resected stage I and II non-small-cell lung cancer. *J. Clin. Oncol.* **17**, 668–675.
19. Greatens, T. M., Niehans, G. A., Rubins, J. B., Jessurun, J., Kratzke, R. A., Maddaus, M. A., and Niewoehner, D. E. (1998) Do molecular markers predict survival in non-small-cell lung cancer? *Am. J. Respir. Crit. Care Med.* **157**, 1093–1097.
20. Paget, S. (1989) The distribution of secondary growths in cancer of the breast. *Cancer Metastasis Rev.* **8**, 98–101.
21. Prins, J., De Vries, E. G., and Mulder, N. H. (1993) The myc family of oncogenes and their presence and importance in small-cell lung carcinoma and other tumour types. *Anticancer Res.* **13**, 1373–1385.
22. Brennan, J., O'Connor, T., Makuch, R. W., Simmons, A. M., Russell, E., Linnoila, R. I., et al. (1991) myc family DNA amplification in 107 tumors and tumor cell lines from patients with small cell lung cancer treated with different combination chemotherapy regimens. *Cancer Res.* **51**, 1708–1712.
23. Noguchi, M., Hirohashi, S., Hara, F., Kojima, A., Shimosato, Y., Shinkai, T., and Tsuchiya, R. (1990) Heterogenous amplification of myc family oncogenes in small cell lung carcinoma. *Cancer* **66**, 2053–2058.
24. Funa, K., Steinholtz, L., Nou, E., and Bergh, J. (1987) Increased expression of N-myc in human small cell lung cancer biopsies predicts lack of response to chemotherapy and poor prognosis. *Am. J. Clin. Pathol.* **88**, 216–220.

25. Johnson, B. E., Ihde, D. C., Makuch, R. W., Gazdar, A. F., Carney, D. N., Oie, H., et al. (1987) myc family oncogene amplification in tumor cell lines established from small cell lung cancer patients and its relationship to clinical status and course. *J. Clin. Invest.* **79,** 1629–1634.
26. Takahashi, T., Obata, Y., Sekido, Y., Hida, T., Ueda, R., Watanabe, H., et al. (1989) Expression and amplification of myc gene family in small cell lung cancer and its relation to biological characteristics. *Cancer Res.* **49,** 2683–2688.
27. Slebos, R. J., Evers, S. G., Wagenaar, S. S., and Rodenhuis, S. (1989) Cellular protoonocogenes are infrequently amplified in untreated non-small cell lung cancer. *Br. J. Cancer* **59,** 76–80.
28. Kawashima, K., Nomura, S., Hirai, H., Fukushi, S., Karube, T., Takeuchi, K., et al. (1992) Correlation of L-myc RFLP with metastasis, prognosis and multiple cancer in lung-cancer patients. *Int. J. Cancer* **50,** 557–561.
29. Kern, J. A., Schwartz, D. A., Nordberg, J. E., Weiner, D. B., Greene, M. I., Torney, L., and Robinson, R. A. (1990) p185^{neu} expression in human lung adenocarcinomas predicts shortened survival. *Cancer Res.* **50,** 5184–5187.
30. Kern, J. A. and Filderman, A. E. (1993) Oncogenes and growth factors in human lung cancer. *Clin. Chest Med.* **14,** 31–41.
31. Pezzella, F., Turley, H., Kuzu, I., Tungekar, M. F., Dunnill, M. S., Pierce, C. B., et al. (1993) bcl-2 protein in non-small-cell lung carcinoma. *N. Engl. J. Med.* **329,** 690–694.
32. Uren, A. G. and Vaux, D. L. (1996) Molecular and clinical aspects of apoptosis. *Pharmacol. Ther.* **72,** 37–50.
33. Sidransky, D. and Hollstein, M. (1996) Clinical implications of the p53 gene. *Annu. Rev. Med.* **47,** 285–301.
34. Mitsudomi, T., Lam, S., Shirakusa, T., and Gazdar, A. F. (1993) Detection and sequencing of p53 gene mutations in bronchial biopsy samples in patients with lung cancer. *Chest* **104,** 362–365.
35. Lee, J. S., Yoon, A., Kalapurakal, S. K., Ro, J. Y., Lee, J. J., Tu, N., et al. (1995) Expression of p53 oncoprotein in non-small-cell lung cancer: a favorable prognostic factor. *J. Clin. Oncol.* **13,** 1893–1903.
36. Yoshikawa, H., Nagashima, M., Khan, M. A., McMenamin, M. G., Hagiwara, K., and Harris, C. C. (1999) Mutational analysis of p73 and p53 in human cancer cell lines. *Oncogene* **18,** 3415–3421.
37. Osada, M., Ohba, M., Kawahara, C., Ishioka, C., Kanamaru, R., Katoh, I., et al. (1998) Cloning and functional analysis of human p51, which structurally and functionally resembles p53. *Nat. Med.* **4,** 839–843.
38. Knudson, A. G., Jr. (1989) The ninth Gordon Hamilton-Fairley memorial lecture. Hereditary cancers: clues to mechanisms of carcinogenesis. *Br. J. Cancer* **59,** 661–666.
39. Yokota, J., Akiyama, T., Fung, Y. K., Benedict, W. F., Namba, Y., Hanaoka, M., et al. (1988) Altered expression of the retinoblastoma (RB) gene in small-cell carcinoma of the lung. *Oncogene* **3,** 471–475.

40. Harbour, J. W., Lai, S. L., Whang-Peng, J., Gazdar, A. F., Minna, J. D., and Kaye, F. J. (1988) Abnormalities in structure and expression of the human retinoblastoma gene in SCLC. *Science* **241,** 353–357.
41. Shimizu, E., Coxon, A., Otterson, G. A., Steinberg, S. M., Kratzke, R. A., Kim, Y. W., et al. (1994) RB protein status and clinical correlation from 171 cell lines representing lung cancer, extrapulmonary small cell carcinoma, and mesothelioma. *Oncogene* **9,** 2441–2448.
42. Xu, H. J., Hu, S. X., Cagle, P. T., Moore, G. E., and Benedict, W. F. (1991) Absence of retinoblastoma protein expression in primary non-small cell lung carcinomas. *Cancer Res.* **51,** 2735–2739.
43. Kratzke, R. A., Greatens, T. M., Rubins, J. B., Maddaus, M. A., Niewoehner, D. E., Niehans, G. A., and Geradts, J. (1996) Rb and p16^{INK4A} expression in resected non-small cell lung tumors. *Cancer Res.* **56,** 3415–3420.
44. Serrano, M., Hannon, G. J., and Beach, D. (1993) A new regulatory motif in cell-cycle control causing specific inhibition of cyclin D/CDK4. *Nature* **366,** 704–707.
45. Hannon, G. J. and Beach, D. (1994) p15^{INK4B} is a potential effector of TGF-β-induced cell cycle arrest. *Nature* **371,** 257–261.
46. Shapiro, G. I., Edwards, C. D., Kobzik, L., Godleski, J., Richards, W., Sugarbaker, D. J., and Rollins, B. J. (1995) Reciprocal Rb inactivation and p16^{INK4A} expression in primary lung cancers and cell lines. *Cancer Res.* **55,** 505–509.

3

Cellular Predictive Factors of Drug Resistance in Non-Small Cell Lung Carcinomas

Manfred Volm, Reet Koomägi, and Werner Rittgen

1. Introduction

Resistance to chemotherapy is a major source of failure in cancer treatment. A large number of cancers are intrinsically resistant to cytostatic agents; others initially respond to treatment, but subsequently develop a resistance to the drugs being used. Of the human solid malignancies, small cell lung carcinomas (SCLCs) are among the most chemosensitive and radiosensitive tumors, with most of them responding to initial therapy. However, the majority not only will recur but will remain largely refractory to further treatment. Most non-small cell lung carcinomas (NSCLCs) are primarily resistant to therapy; it is rare to obtain a complete response.

1.1. Underlying Causes of Multidrug Resistance

1.1.1. MDR Family Glycoproteins

During the past several decades, our knowledge of the molecular genetics and biology of lung cancer has greatly increased. One type of resistance that has now been well-characterized by examining these parameters is the so-called multidrug resistance (MDR) *(1)*. Investigations conducted with tumor cell lines and tumor specimens obtained from cancer patients have shown that cross-resistance between different drugs that are structurally and functionally dissimilar is a common phenomenon. The MDR phenomenon is associated with the overexpression of a 170 kDa membrane-associated glycoprotein (P-170) that decreases intracellular drug accumulation. The expression of P-170 results in the resistance to a variety of anticancer drugs such as vinva alkaloids, anthracyclines, epipodophylotoxins, and taxol. But the overexpression of P-170 alone does not completely explain multidrug resistance. During the past few years, two new proteins have been identified. These proteins, which are distant members of the MDR family, are called multidrug resistance-associated protein (MRP) and lung resistance-related protein (LRP).

1.1.2. DNA Replication and Repair Enzymes

In addition to P-170, MDR and LRP other mechanisms for resistance to multiple drugs have been described *(2)*. These frequently include the alteration of topoisomerase II. This is a ubiquitous nuclear enzyme that is essential for replication and transcription. The enzyme is the target of many clinically important anti-neoplastic drugs, such as the anthracyclines, amsacrines, and epipodophylotoxins. These drugs stabilize the cleavable complex formed between topoisomerase and DNA that is visible as DNA strand breaks and DNA/protein cross-links. O6-methylguanine-DNA-methyltransferase (MGMT) is a ubiquitous DNA repair protein that transfers alkyl groups of guanine from the O6-position to an internal cysteine residue and prevents the binding of alkylating drugs to DNA. It forms S-methylcysteine on MGMT and guanine with the substrate DNA. Additionally, thymidylate synthase, which plays a central role in DNA biosynthesis, is the target of many chemotherapeutic agents, such as 5-fluorouracil and methotrexate. Tumor cells that are resistant to cisplatin and doxorubicin have increased levels of this enzyme.

1.1.3. Glutathione-S-transferases

Additionally, glutathione-S-transferases play an important role in the detoxification of cytostatic compounds. They mediate the conjugation of cytostatic agents with reduced glutathione. Melphalan, cyclophosphamide, chlorambucil, and doxorubicin are substrates for these proteins.

1.1.4. Metallothonein

Another important protein for drug resistance is metallothionein, which binds heavy metal ions such as zinc, cooper, cadmium and platinum and inactivates metal-containing anticancer agents.

1.1.5. Dihydrofolate Reductase

Dihydrofolate reductase is the primary target for the action of antifolate drugs in cancer chemotherapy. The increased production of dihydrofolate reductase has been identified as an important mechanism for acquiring methotrexate resistance.

1.1.6. Heat-Shock Proteins

Heat-shock proteins are a family of proteins that protect cells from toxic external stimuli. Cells that overexpress heat shock proteins are resistant to doxorubicin, colchicine, and vincristine.

1.2. The Influence of Proliferative, Apoptotic, and Angiogenic Factors

The present study examines the interrelationships between the expression of P-glycoprotein, lung resistance-related protein (LRP), topoisomerase II, glutathione-S-transferase-π, metallothionein, O6-methyl-guanine-DNA-methyltransferase, thymidylate synthase, dihydrofolate reductase, heat-shock proteins, and the multidrug resistance of NSCLCs. However, the aforementioned proteins alone do not completely explain the resistance of human cancer. Therefore, we investigated several other resistance-related proteins in our study.

1.2.1. Proliferative Factors

Cancer chemotherapy is most successful when used on rapidly growing malignant cells *(3)*. Experimental and clinical data show that tumors with a low rate of proliferation are less responsive to treatment than tumors with a high rate of proliferation. On average, NSCLCs exhibit a lower labeling index and a longer doubling time than SCLC. This partly accounts for the resistance of NSCLC to cytostatic drugs. Protein complexes that are composed of cyclins and cyclin-dependant kinases (cdks) are important factors in cellular proliferation *(4)*. Cyclin A achieves its maximum level during the S and G_2 phases and together with cdk2 regulates the transition to mitosis. Cyclin D reaches its peak during the G_1 phase and regulates the transition from this phase to the S phase. The proliferating cell nuclear antigen (PCNA) is essential for cellular DNA synthesis. In the present study, we determined the expression of cyclin A, cyclin D, cdk2, and PCNA and compared the expression of these factors with the resistance of the tumors. Additionally, we investigated the effect exhibited by the proportion of S phase cells found in tumors on the resistance to doxorubicin.

Some indications exist that proto-oncogenes may be involved in tumor resistance *(7)*. Elevated gene transcripts of c-*fos*, c-*myc*, and c-H-*ras* have been found in cisplatin-resistant tumors. In several types of tumors, it has been confirmed that bcl-2-negative tumors are more often sensitive to anticancer drugs than bcl-2-positive tumors. P53 may also play an important role in resistant tumors. Our study evaluated the proto-oncogene products, Fos, Jun,

Myc, Erb-B1, Erb-B2, N-Ras, K-Ras, and H-Ras, and their relationships to drug resistance.

1.2.2. Apoptotic Factors

Apoptosis also plays an important role in the response of tumors to cytostatic agents *(5)*. This process involves the death-inducing ligand receptor systems and the cleavage of caspases. Cytostatic agents have resulted in the induction of Fas ligand and the upregulation of Fas. Therefore, this study analyzes the relationships between Fas ligand, Fas, caspase 3 and the response of NSCLC to drugs. Additionally, we evaluated the relationship between the response obtained and the apoptotic index as measured by the TUNEL reaction.

1.2.3. Angiogenic Factors

Angiogenesis, the development and formation of new blood vessels, plays an important role in a variety of processes including resistance. Solid tumors with few blood vessels contain a fraction of hypoxic cells that are relatively resistant to radiotherapy and certain cytostatic drugs *(6)*. The vascular endothelial growth factor (VEGF) and the fibroblast growth factor (FGF) are molecules that directly exert an angiogenic effect and might also influence the therapeutic response of tumors. Expression of platelet-derived endothelial cell growth factor (PD-ECGF) is elevated in several types of tumors and plays a role in tumor vascularization. PD-ECGF catalyzes the reversible phosphorylation of thymidine, deoxyuridine and their analogs to their respective bases and 2-deoxyribose-1-phosphate. Therefore, PD-ECGF may play a role in the activation of drugs such as 5-fluorouracil (5-FU). It has been suggested that tissue factor (TF), a physiological initiator of blood coagulation, may also be involved in tumor growth and angiogenesis. In the present study, we analyzed the relationships between expressions of VEGF, basic fibroblast growth factor (bFGF), PD-ECGF, TF and the drug response. Additionally, we investigated the relationship between microvessel density and the resistance of NSCLC.

2. Results

2.1. Patient Population Statistics

The main objective of this study was to evaluate which cellular factors are most predictive of the resistance exhibited by NSCLC and whether or not a combination of factors can improve the prognostic information. Ninety-four patients with previously untreated NSCLC were admitted into this study. The morphological classification of the carcinomas was conducted according to WHO specifications. Of the carcinomas, 48 were squamous carcinomas, 34 were adenocarcinomas, and 12 were large cell carcinomas. Sixteen patients had stage I, 12 patients stage II, and 66 patients had stage III tumors (according to the guidelines of the American Joint Committee on Cancer). Most of the patients were only treated with surgery with a small group of patients undergoing a combination of surgery and radiotherapy or chemotherapy.

2.2. Methods for Evaluating Drug Resistance

2.2.1. In Vitro Postdrug Treatment Proliferative Assay

2.2.1.1. METHOD

To determine resistance, we used a short-term in vitro test that measures changes in the rate at which radioactive nucleic acid precursors are incorporated into tumor cells after the addition of cytostatic agents. After the removal of fatty and necrotic tissue, fresh tumor material is cut into small pieces. The tissue pieces, suspended in Hanks balanced salt solution (HBSS), are pipetted several times with a sharp-edged glass pipet and the resultant cell suspension filtered through gauze. The cells are sedimented and then resuspended in culture medium at a defined cell density (5×10^5 cells/mL). After preincubation at 37°C in a shaking water bath, the chemotherapeutic agents are added. The drugs are tested over a concentration range that extends over four powers of ten. Following incubation for 2 h, radioactive nucleotide precursors are added and the incubation is continued for an additional hour. 100-µL aliquots are pipetted from each test tube onto round filter discs and dried in a stream of warm air. The unincorporated radioactivity is extracted with ice-cold trichloroacetic acid (TCA). The filters were then washed in ethanol/ether and the incorporated radioactivity determined by liquid scintillation counting. Uptake values for the individual concentrations are then expressed as percentages of the controls.

2.2.1.2. ANALYSIS OF RESULTS

Tumors were defined as being sensitive or resistant based on prior clinical studies *(8,9)*. In a clinical pilot study, patients with lung carcinomas underwent chemotherapy and the effect of the treatment was compared with the results obtained by the in vitro short-term test (*see* **Fig. 1**). The data indicated that carcinomas that exhibited only a weak in vitro response also failed to respond to clinical chemotherapy. Tumors that were strongly inhibited in vitro generally showed some degree of clinical remission. When the patients were subdivided into two groups based on the test results, the groups exhibited different survival curves (*see* **Fig. 2**). Patients with tumors that were insensitive to the test died sooner than those with carcinomas that yielded an obvious in vitro response. Results of this pilot study encouraged our group to start a controlled clinical trial that focused mainly on lung but also on other types of cancer. This trial

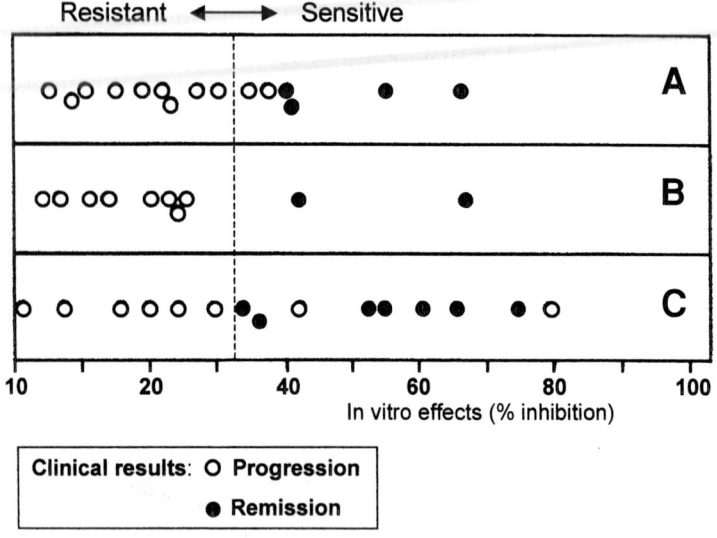

Fig. 1. A comparison of the in vitro short-term test results and the clinical results in patients with lung cancer. (**A–C**) different chemotherapy protocols.

employed defined standard chemotherapy schedules to compare the clinical response with the in vitro test results. In this study, 55 of 57 tumors that proved resistant in the test were clinically progressive and 40 of 58 tumors that were sensitive in the test demonstrated a clinical response. Therefore, we believe that the in vitro short-term test is suited to determine the greatest number of cellular predictive factors in NSCLC.

2.2.2. Comparison of Multiple Methods for Analyzing Multidrug Resistance

Many methods are available to determine resistance-relevant factors. **Figure 3** shows the results of one sensitive (S) and two multidrug-resistant cell lines (R1, R2) obtained by different methods. The resistance was detected by the in vitro short-term test (**A**). Flow cytometry measured the lower accumulation of cytostatic agents (**B**). Southern blotting detected the MDR1 gene amplification in resistant cells (**C**) and Northern blotting detected the MDR1-gene expression (**D**). Western blotting (**E**), immunofluorescence (**F**) and immunohistochemistry (**G**) were used to detect the increase of P-glycoprotein in resistant cells.

2.2.3. Advantages of Immunohistochemistry

The immunohistochemical method holds particular promise for clinical applications, since the procedure is practical and easy to perform. We used immunohistochemistry to determine which resistance-related proteins are expressed in NSCLC and which of the 30 investigated factors (resistance

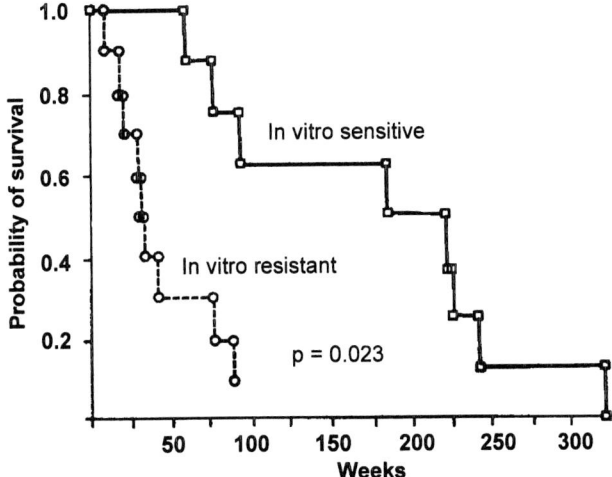

Fig. 2. Kaplan-Meier curves for patients with stage III adenocarcinomas of the lung treated with surgery and chemotherapy. The patients were separated into either resistant or sensitive groups according to the in vitro short-term test results. The median survival times for the groups were 31 and 185 wk, respectively.

proteins, proliferative, apoptotic, angiogenenic factors, proto-oncogenes) most accurately predicts the resistance of NSCLC. Significant correlations were detected between the data obtained by the in vitro short-term test and expressions of the resistance proteins.

A significant correlation was obtained with P-glycoprotein (P-170, $p = 0.00004$), glutathione-S-transferase-π (GST-π, $p = 0.0002$), metallothionein (MT, $p = 0.0008$), thymidylate synthase (TS, $p = 0.002$), O6-methylguanine-DNA-methyltransferase (O6, $p = 0.008$) and lung resistance-related protein (LRP, $p = 0.03$). A weak correlation existed between the heat-shock proteins (HSP70, $p = 0.05$) and no correlation was observed for the expression of topoisomerase II (TOPO II) (*see* **Table 1**). **Figure 4** (left) shows that the resistance increases with the level of P-glycoprotein.

It is well known that rapidly growing tumors usually respond to treatment and that tumors with a low rate of proliferation very often show no response. Therefore, we used immunohistochemistry to determine the proliferation markers PCNA, cyclin A, cyclin D1 and cdk2, and flow cytometry to determine the S phases. **Table 1** shows that only a weak relationship exists between the expressions of cdk2 ($p = 0.04$) and PCNA ($p = 0.05$) and the drug response. We were also unable to detect a significant relationship between the drug response and the proportion of S phases.

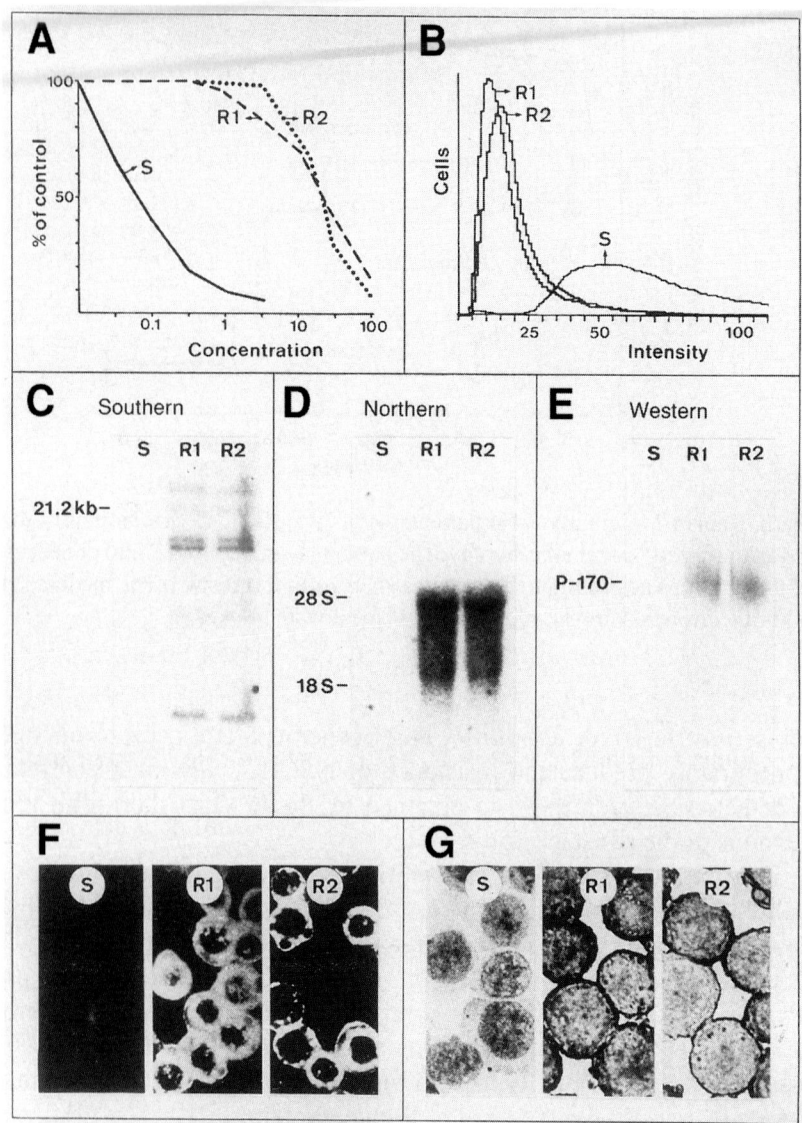

Fig. 3. Different methods for detecting multidrug resistance. S, sensitive (parental) cell line; R1, R2, multidrug-resistant cell lines. The resistance was developed in response to doxorubicin (R1) and daunomycin (R2). **(A)** Results of the in vitro short term test; **(B)** Flow cytometry (accumulation of cytostatic agents); **(C)** Southern Blotting (MDR1); **(D)** Northern Blotting; **(E)** Western Blotting (P-170 glycoprotein); **(F)** Immunofluorescence; **(G)** Immunohistochemistry.

Table 1
Correlation Between the In Vitro Short-Term Test and Resistance-Related Cellular Factors as Measured by Immunohistochemistry and Flow Cytometry (p Values)

Res-proteins		Proliferation		Apoptosis		Angiogenesis		Oncogene	
Factor	p value	Factor	p value	Factor	p value	Factor	p value	Factor	p value
P170	0.00004	Cdk2	0.04	Fas	0.007	PD-ECGF	0.0006	Erb-B2	0.04
GST	0.0002	PCNA	0.05	FasL	0.16	VEGF	0.004	Fos	0.12
MT	0.0008	Cyclin A	0.09	Casp3	0.19	FGF	0.007	Myc	0.14
TS	0.002	S-phase	0.25	AI	0.70	MVD	0.06	Jun	0.16
O6	0.008	Cyclin D	0.27			TF	0.1	N-Ras	0.21
LRP	0.03							K-Ras	0.36
HSP	0.05							H-Ras	0.43
Topo	0.17							Erb-B1	0.46

Apoptosis is regulated by a variety of pro-apoptotic and anti-apoptotic factors. In this investigation, we measured the apoptotic indices (AI), the expressions of the Fas receptor (Fas), Fas ligand (FasL), and caspase 3. We could only detect a significant relationship to the drug response for the Fas receptor ($p = 0.007$).

Indications exist that proto-oncogenes may play a role in tumor resistance. Therefore, we examined the expressions of the proteins Fos, Jun, ErbB-1, ErbB-2, Myc, and Ras and compared these results with data obtained by the in vitro short-term test. Only ErbB-2 correlated with the data of the in vitro short-term test ($p = 0.04$).

2.2.4. Microvessel Density (MVD) Assay

Angiogenesis was measured by MVD and analysis of expression of several angiogenic factors. Platelet-derived endothelial growth factor (PD-ECGF, $p = 0.0006$), vascular endothelial growth factor (VEGF, $p = 0.004$) and fibroblast growth factor (FGF, $p = 0.007$) exhibit significant inverse correlations to the resistance of non-small cell lung cancer (**Table 1**; **Fig. 4** right). In contrast, MVD and tissue factor (TF) do not exhibit a significant relationship.

2.3. Analysis of Results

2.3.1. Contribution of Individual Factors to Resistance

To summarize, this analysis shows that the resistance proteins are the most important factors associated with the resistance of NSCLC. Angiogenic and apoptotic factors are of secondary importance. In contrast, the predictive value

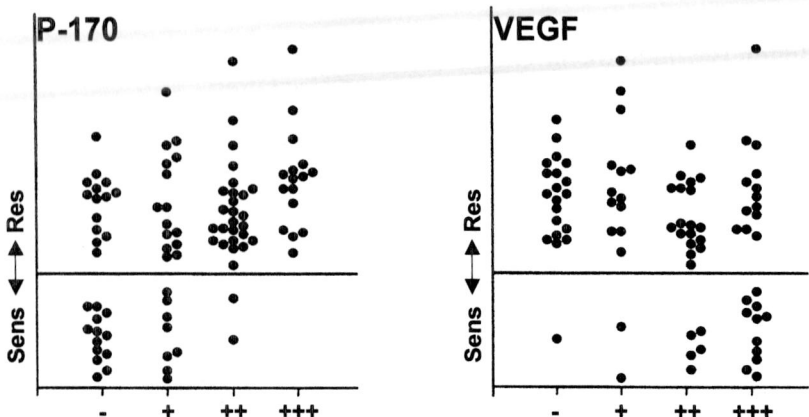

Fig. 4. Relationship between the drug response (doxorubicin) as determined by the in vitro short-term test (ordinate) and the immunohistochemical reaction of P-170 and VEGF. The intensity of immunostaining (negative, weak, moderate, high) is specified as –, +, ++, +++. P-170 and VEGF exhibit an inverse reaction to the drug response.

of the proliferative factors and proto-oncogenes is marginal at best. With respect to the resistance, an inverse relationship exists between angiogenic or apoptotic factors and the resistance proteins.

Determining the sensitivity and specificity of all the parameters and then using the sum of the sensitivity and specificity as a measure of the accuracy of the diagnostic test, reveals that the maximum accuracy (150.5%) is attained by P-glycoprotein (P170). 76.6% of the tumors could be correctly diagnosed with thymidylate synthase (TS). **Table 2** presents the six best factors.

2.3.2. Contribution of Combined Factors to Resistance

To determine whether a combination of factors could yield improved information, the sensitivity and specificity of all pairs of factors were evaluated. In this manner, the maximum accuracy of the diagnostic test increased to 168.3%. The best three pairs were P170/VEGF (*see* **Fig. 5, left**), P170/FGF, and VEGF/GST-π. Using the combination of VEGF/GST-π, 85.4% of the tumors could be correctly diagnosed. The best three triplets were VEGF/FGF/P170, VEGF/P170/TS, and VEGF P170/MT (*see* **Fig. 5, right**). The maximum accuracy achieved was 174.7% by the combination of VEGF/FGF/P170 with 89.47% of the tumors diagnosed correctly. **Table 2** presents the six best pairs and triplets.

3. Discussion

Although the statistical probability of therapeutic success is known from appropriate studies, the clinical response of the individual patient still remains

Table 2
Sensitivity and Specificity Analysis of the Most Significant Resistance Factors

Factor1	Factor2	Factor 3	Sp (%)	Se (%)	Se+Sp (%)	CD (%)
P170			59.15	91.30	**150.46**	67.02
GST-π			73.24	69.57	142.80	72.34
TS			81.69	60.87	142.56	**76.60**
VEGF			79.37	57.89	137.26	74.39
FGF			42.62	90.00	132.62	54.32
MT			71.43	60.87	132.30	68.82
VEGF	P170		84.13	84.21	**168.34**	84.15
FGF	P170		78.69	85.00	163.69	80.25
VEGF	GST-π		92.06	63.16	155.22	**85.37**
P170	TS		54.93	100.00	154.93	65.96
P170	GST-		56.34	95.65	151.99	65.96
VEGF	FGF		63.79	83.33	147.13	68.42
VEGF	FGF	P170	91.38	83.33	**174.71**	**89.47**
VEGF	P170	TS	77.78	89.47	167.25	80.49
VEGF	P170	MT	70.97	94.74	165.70	76.54
FGF	P170	TS	70.49	95.00	165.49	76.54
VEGF	GST-π	MT	82.26	78.95	161.21	81.48
FGF	P170	GST-π	70.49	90.00	160.49	75.31

Se, Sensitivity; Sp, Specificity; CD, Correctly diagnosed.

uncertain. Therefore, a number of test systems to detect tumor resistance against cytostatic agents have been developed over the past several years. The following approaches to predictive testing are being investigated: measuring the cellular damage in a monolayer or organ culture, measuring the inhibition of radioactive precursor incorporation, measuring clonogenic cell survival and human tumor xenografts. However, an investigation of the clinical correlation between the test results and the clinical data reveals that although tests of this kind can determine which anticancer drugs will not be clinically useful, none can satisfactorily predict which drug will prove most effective. Nevertheless, the in vitro short-term test exhibited good agreement between the test results and the clinical response. There was also good agreement between the test results and the survival time.

Advances in molecular and cell biology have opened new avenues for the characterization of drug-resistant tumors. Interest is especially focused on protein and RNA expression. Protein detection by immunohistochemistry and

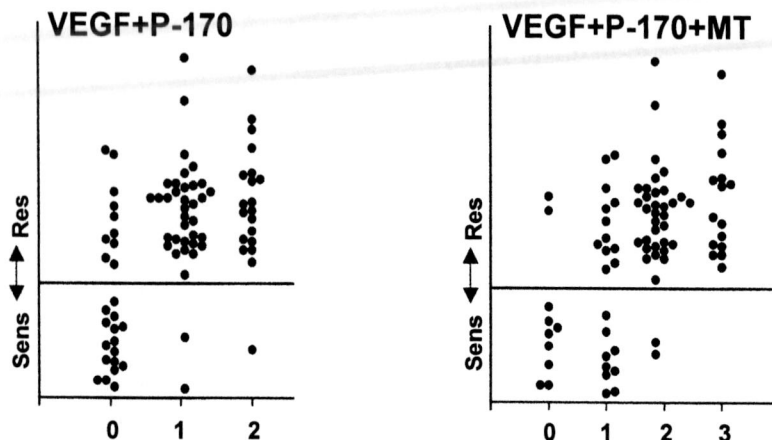

Fig. 5. Relationship between the drug resistance as measured by the in vitro short-term test and the number of resistance-related factors as measured by immunohistochemistry. 0, no single factor indicates resistance; 1, only one factor; 2, both (two) factors; and 3, three factors indicate resistance.

RNA detection by Northern blotting or by the reverse transcription-polymerase chain reaction (RT-PCR) may be better suited for clinical applications than assays such as Western blotting or RNAse protection.

Evidence exists that a great diversity of drug-resistance mechanisms function in clinically relevant drug resistance. The greater the number of resistance-related proteins in a given tumor, the greater can be the accuracy of determining the resistance *(10)*. The systematic investigation of combinations of cellular factors in NSCLC clearly yields improved predictive information. By using three cellular factors, the responsiveness and resistance exhibited by NSCLC could be correctly diagnosed in about 90% of the cases.

Many genes have been implicated in resistance and some must still be discovered. The recently developed microarray technology facilitates the simultaneous analyses of thousands of genes in a single experiment. Corresponding analyses of proteins can be conducted with peptide arrays. RNA and protein expression levels are not closely related and quantitative mRNA data only inadequately predict the quantity of a protein. Furthermore, RNA does not detect functional protein modifications such as phosphorylation and glycosylation. Both technologies are complementary in molecular screening. With such novel approaches, scientists and clinicians can generate a predictive test for lung cancer with the highest possible degree of accuracy. Furthermore, individual cytostatic treatment schedules for each patient can be generated.

References

1. Bradley, G., Juranka, P. F., and Ling, V. (1988) Mechanisms of multidrug resistance. *Biochim. Biophys. Acta* **948,** 87–128.
2. Volm, M. and Mattern, J. (1996) Resistance mechanisms and their regulation in lung cancer. *Crit. Rev. Oncogenesis* **7,** 227–244.
3. Valeriote, F. and van Putten, L. (1975) Proliferation-dependant cytotoxicity of anticancer agents. A review. *Cancer Res.* **3,** 2619–2630.
4. Cordon-Cardo, C. (1995) Mutation of cell cycle regulators. Biological and clinical implications for human neoplasia (review). *Am. J. Pathol.* **147,** 545–560.
5. Hickman, J. A. (1996) Apoptosis and chemotherapy resistance. *Eur. J. Cancer* **6,** 921–926.
6. Giaccia, A. J. (1996) Hypoxic stress proteins: Survival of the fittest. *Sem. Rad. Oncol.* **6,** 46–58.
7. Scanlon, K. J., Jiao, L., Wang, W., Tone, T., Rossi, J. J., and Kashani-Sabet, M (1991) Ribozyme-mediated cleavage of c-fos mRNA reduces gene expression of DNA synthesis enzymes and metallothionein. *Proc. Natl. Acad. Sci. USA* **88,** 10591–10595.
8. Volm, M., Wayss, K., Kaufmann, M., and Mattern, J. (1979). Pretherapeutic detection of tumour resistance and the results of tumor chemotherapy. *Eur. J. Cancer* **15,** 938–993.
9. Group for sensitivity testing of tumors (KSST) (1981) In vitro short-term test to determine the resistance of human tumors to chemotherapy. *Cancer* **48,** 2127–2135.
10. Volm, M. and Rittgen, W. (2000) Cellular predictive factors for the drug response of lung cancer. *Anticancer Res.* **20,** 3449–3458.

4

Clinical Implications of p53 Mutations in Lung Cancer

Barbara G. Campling and Wafik S. El-Deiry

1. Introduction

The p53 tumor-suppressor gene plays a crucial role in the cellular response to stress (reviewed by Vogelstein et al. *[1]*). Under normal conditions, p53 is rapidly degraded and thus is not present in detectable levels within the cell. A variety of cellular stresses, including DNA damage and oncogene activation, result in stabilization and activation of p53, causing the protein to accumulate within the nucleus. Stabilized p53 then transcriptionally activates the expression of a variety of proteins that are involved in cell-cycle regulation and apoptosis. The damaged cells may then undergo cell-cycle arrest, allowing them to repair the genetic damage. Alternatively, if the damage is irreparable, the p53 protein initiates a cascade of events that culminate in programmed cell death (apoptosis) *(2)*.

The absence or inactivation of p53 by a variety of mechanisms allows genomic damage to persist, eventually resulting in the uncontrolled cell growth that is characteristic of cancer. The p53 can become inactivated through allelic deletion and mutation, as well as through a variety of mechanisms that affect the p53 signaling pathway in various tumor types. For example, a negative regulator of p53, the MDM2 protein, is frequently amplified or overexpressed in soft tissue sarcomas *(3)*. The p53 protein may also be degraded by the human papilloma virus (HPV) E6 protein in high-risk HPV infections that are common in patients with cervical cancer *(4)*. Another common mechanism of p53 disruption involves deletion of the ARF (alternative reading frame) gene at the INK4a locus on chromosome 9p *(5)*. This disrupts the signaling pathway leading to p53 stabilization in response to inappropriate growth signals or oncogenes. More uncommon defects in the p53 pathway include mutations in the ATM gene (ataxia telangiectasia mutated) in patients with the cancer-prone ataxia-telangiectasia syndrome *(6)*, or mutation in the CHK2 gene in a subset of patients with Li-Fraumeni Syndrome who do not harbor germline p53 mutations *(7)*. Inactivation of p53 by mutation or deletion is the most common genetic abnormality in a variety of malignancies, including lung cancer *(8,9)*. More than 10,000 individual mutations in human tumors and cell lines have now been reported *(10)*.

Lung cancers (bronchogenic carcinomas) arise from bronchial epithelial cells, and can be broadly classified into two major groups, small cell lung cancer (SCLC) and non-small cell lung cancer (NSCLC). NSCLC consists of a variety of histologic types including adenocarcinoma, squamous cell and large cell anaplastic carcinoma. SCLC can be distinguished from NSCLC by a number of clinical and biological features. For example, SCLC tends to grow and disseminate more rapidly, but is more responsive to chemotherapy than NSCLC.

The molecular abnormalities found in SCLC and NSCLC have some common as well as some distinguishing features (reviewed by Sekido et al. *[11]*). Lung cancers have numerous accumulated genetic abnormalities, including activation of protooncogenes and inactivation of tumor-suppressor genes *(11)*. One of the earliest clues suggesting the inactivation of tumor-suppressor genes in lung cancer was the detection of loss of genetic material on the short arm of chromosome 3 *(12)*. Loss of heterozygosity (LOH) in this region, often at multiple sites, is found in more than 90% of SCLC and more than 80% of NSCLC, and is one of the earliest alterations detectable during malignant transformation of bronchial epithelium *(11)*. This has led to a search for tumor-suppressor genes on chromosome 3p, which may be important in the pathogenesis of lung cancer *(13)*. However, it is now clear that not one, but multiple tumor-suppressor genes are inactivated in lung carcinomas. For example, inactivation of the retinoblastoma tumor suppressor gene (Rb) on chromosome 13q14 occurs in approx 90% of SCLC but only 15–30% of NSCLC. On the other hand the p16 tumor-suppressor gene on chromosome 9p21 is inactivated in 30–70% of NSCLC, but only 1–10% of SCLC *(11)*.

One of the most consistently detected alterations in lung cancer cells is LOH at chromosome 17p, the location of the p53 gene, accompanied by mutation of the remaining p53 allele. Despite the accumulated genetic damage in lung cancer cells, the reintroduction of normal p53 into cells lacking p53 function is sufficient to suppress tumor growth. For example, Takahashi et al. introduced wild-type p53 into lung cancer cells with either homozygous deletion or mutation of p53, and this resulted in suppression of tumor growth both in vitro and in vivo *(14)*.

Germline mutations of the p53 gene cause an inherited cancer predisposition syndrome, initially described by Li and Fraumeni *(15,16)*. These patients develop a variety of cancers including sarcomas, brain tumors, and breast cancer. Although lung cancers have been reported in patients with this familial syndrome, these tumors are not usually considered part of the spectrum of malignancies commonly associated with the Li-Fraumeni syndrome *(17)*. Lung cancers are one of the most common tumors in the general population, and it is uncertain whether the incidence is increased in patients with germline p53 mutations. Furthermore, it is not clear whether patients with this syndrome who develop lung cancer are also smokers. However, it should be noted that the median age of onset of lung cancer in Li-Fraumeni syndrome patients is considerably younger (around 50 yr) than in the general population (68 yr) *(18)*.

2. Detection of p53 Mutations

A number of methods have been used to detect alterations in the p53 gene. The "gold standard" involves sequencing the entire p53 gene. Because most mutations involve exons 5–8, most investigators have confined their search to these regions. Many groups have initially used the techniques of single-strand conformational polymorphism (SSCP) or denaturing gradient gel electrophoresis (DGGE) of polymerase chain reaction (PCR)-amplified segments of the gene to screen for p53 mutations. These methods are based on the fact that point mutations in the gene cause the subsequently PCR-amplified DNA to migrate differentially during gel electrophoresis. The abnormally migrating bands can then be sequenced, thus avoiding sequencing of the entire gene. Although this method is sensitive for specific alterations, ultimately it is not possible to exclude p53 mutations unless the entire p53 cDNA is sequenced. Numerous clinical studies of p53 alterations have utilized immunohistochemistry. Many p53 mutations will result in abnormal accumulation of the protein, which can be detected in tissue sections with p53-specific antibodies. Wild-type p53 has a very short half-life and is not usually detectable. Immunohistochemistry can be readily applied in routine histopathology laboratories, but the correlation between p53 mutations detected by sequencing and abnormal p53 protein accumulation is imperfect, as discussed later in this chapter. A functional yeast-based p53 assay has also been employed and may be used in conjunction with sequencing to characterize the activity of altered p53 *(19)*. More recently, oligonucleotide microarrays have been used to screen for p53 mutations in clinical samples *(20)*. These arrays contain a series of oligonucleotides coding for wild-type p53 sequence as well as sequences of the most common p53 mutations. Although expensive, large numbers of samples can be easily screened using oligonucleotide microarrays. However, in a series of 100 NSCLC samples, screening using a p53 GeneChip did not identify all of the mutations detected by sequencing, and did not identify any frameshift mutations *(20)*.

3. p53 Mutations in Lung Cancer

The frequent detection of LOH in lung cancer cell lines and tumor samples at the location of the p53 gene on chromosome 17p13, suggested that this gene was likely to be involved in the pathogenesis of lung cancer. An early study, utilizing a collection of 30 lung cancer cell lines indicated that p53 was frequently mutated or inactivated in all major histologic types of lung cancer *(21)*. The types of abnormalities included homozygous deletions, DNA rearrangements, abnormally sized mRNAs as well as point mutations *(21)*. Another early study of 51 resected NSCLC tumors detected p53 mutations in 23 samples (45%) *(22)*. Mutations were found in tumors both with and without allele loss at 17p13, and were most often located within the DNA binding domain of p53 *(22)*.

The highest frequency of p53 alterations is found in SCLC specimens *(23–26)*. For example, Takahashi et al. studied tumor samples from 15 SCLC patients and found mutations in 11 of these (73.3%) *(23)*. They were able to establish continuous tumor cell lines from 9 of these patients, and all of these cell lines had p53 mutations. Other studies of SCLC samples have found p53 mutations in 23/25 cases (85%) *(25)*, 5/9 cases (56%) *(24)*, and 16/20 cases (80%) *(26)*. Combining these results gives a frequency of 55/69 cases (80%). It is unclear from these studies whether one or both p53 alleles are mutated or whether such mutations result in a dominant negative or "gain of function" form of p53.

Among NSCLC tumor samples, the frequency of p53 mutations is highest in squamous cell carcinomas and lowest in adenocarcinomas. For example, in a study of 115 surgical samples of NSCLC, Kishimoto et al. found p53 alterations in 8 of 14 large cell carcinomas (57%), 24 of 58 adenocarcinomas (41%), 25 of 37 squamous cell carcinomas (68%) and 3 of 6 adenosquamous carcinomas (50%), with an overall frequency of 60/115 tumors (52%) *(27)*. A number of these studies have also examined normal tissues from lung cancer patients, both SCLC and NSCLC, and found no evidence of germline mutations, providing convincing evidence that the p53 mutations in these lung tumors are somatically acquired *(22,24,26,28,29)*.

3.1. p53 Mutations and Bronchial Carcinogenesis

As noted, the location of the p53 gene on the short arm of chromosome 17 is one of the most frequent sites of allelic loss in lung cancer (in addition to LOH on chromosome 3p, 5q, 9p, and 13q) *(30)*. At what stage during the malignant transformation of bronchial epithelium do these p53 abnormalities occur? Molecular studies of microdissected areas of dysplasia within the bronchi of smokers with and without lung cancer have facilitated studies of the timing and frequency of p53 alterations during the process of multistage bronchial carcinogenesis.

Bennett et al. found that nuclear p53 accumulation (an indirect indication of p53 mutation) was not detectable in normal bronchial mucosa, whereas it was found in 6.7% of squamous metaplasias, 29.5% of mild dysplasias, 59.7% of severe dysplasias, 58.5% of carcinomas *in situ*, 67.5% of microinvasive carcinomas, and 79.5% of invasive squamous cell carcinomas *(31)*. Similar results were reported by Walker et al. *(32)*. This group also found that p53 accumulation could be detected in normal appearing bronchial mucosa of lung cancer patients *(32)*. Another group was able to detect p53 mutations in sputum specimens obtained as part of a screening study, up to 1 yr prior to the diagnosis of lung cancer *(33)*.

Thus, it appears that abnormalities of p53 may be detected in the earliest recognizable premalignant lesions of the bronchi. p53 mutations are an early but not obligatory step in bronchial carcinogenesis. The earliest detectable molecular lesions in bronchial premalignancy are loss of genetic material on chromosome 3p *(34)* and on chromosome 9p *(35)*. Damage to the p53 gene appears to occur after allelic loss on chromosome 3 *(35)*, whereas mutations in the K-ras oncogene occur late in the course of bronchial carcinogenesis, and are not usually detectable in premalignant lesions *(36)*. In studies in which areas of premalignant change were compared to areas of carcinoma in the same patient, the allele which was lost was always identical (allele-specific mutations) *(34,35)*.

Mao et al. examined bronchial biopsy specimens from 54 current and former smokers (without lung cancer) for LOH at loci on chromosome 3p14 (location of the FHIT gene), 9p21 (location of the p16 gene), and 17p13 (location of the p53 gene) *(37)*. Among informative cases, DNA losses were detected in 75% of subjects at 3p14, 57% of subjects at 9p21, and 18% of subjects at 17p13 *(37)*. Overall, the frequency of LOH was 82% in current smokers and was still remarkably high at 62% in former smokers. Another similar study compared molecular abnormalities at specific chromosomal loci (including 17p13) in bronchial biopsy specimens from current smokers, former smokers, and those who never smoked *(38)*. No abnormalities were detected in the group who never smoked. However, about half of the histologically normal samples from smokers showed evidence of LOH. The frequency of abnormalities was even higher in dysplastic lesions. There was no statistically significant difference in the frequency of LOH in the group of current smokers compared to the former smokers. This high rate of persistent genetic damage following smoking cessation may partially explain the fact that more than half of lung cancer cases currently diagnosed in the United States occur in former smokers *(39,40)*.

3.2. Tobacco Specific Carcinogens and p53

The spectrum of p53 mutations found in lung cancer as well as in other smoking-associated malignancies is quite different from that found in other tumor types. For example, the most frequently detected mutations in lung cancers are G to T transversions, whereas G to A transitions are more common in other tumors such as breast and colon cancers *(41)*. There are a number of mutational hotspots along the p53 gene in lung cancer, including codons 157, 248, and 273, all located within the central domain which is required for DNA binding *(41)*.

A number of studies in lung cancer have shown a clear-cut dose-response relationship between p53 mutations and smoking, with a higher frequency of mutations in tumors from patients with a higher cumulative tobacco exposure *(42,43)*. Furthermore, studies in those uncommon cases of lung cancer that occur in nonsmokers have detected a much lower frequency of p53 mutations (8–26%) compared to the more common tumors obtained from current or ex-smokers *(43–45)*. As well, the characteristic G to T transversions are unusual in lung tumors from nonsmokers *(44,45)*. p53 mutations occur even more frequently in the lung cancers of smokers who are also drinkers compared to smokers who do not drink *(46)*. The same observation has also been made in head and neck cancer *(47)*. It is possible that alcohol may enhance the mutagenic properties of tobacco by affecting the metabolism of tobacco-specific carcinogens *(48)*.

Tobacco smoke contains many carcinogens, including benzo[a]pyrene, one of the most potent carcinogens known. After metabolic activation to benzo[a]pyrene diol epoxide (BPDE), it binds avidly to DNA. Denissenko et al. have mapped the distribution of BPDE adducts along the p53 gene of HeLa cells as well as normal bronchial epithelial cells, and found selective adduct formation at codons 157, 248, and 273 *(49)*. Not surprisingly, these sites correspond to major mutational hotspots in human lung cancers. As an extension of this work, the same group examined DNA adduct formation in p53 caused by several other polycyclic aromatic hydrocarbons found in tobacco smoke *(50)*. These toxins preferentially bound to methylated CpG sequences within the p53 gene, and the codons most frequently affected were those most commonly mutated in lung cancers. Thus, it appears that tobacco-specific carcinogens leave a unique and indelible signature on the p53 gene.

3.3. Clinical Prognostic Studies of p53 Mutations in Lung Cancer

Despite a multitude of clinical studies that have specifically examined the prognostic significance of p53 alterations in lung cancer, the effect of these mutations on survival is still unclear. Most of the clinical correlative studies have been surgical series utilizing tumor tissues obtained from NSCLC patients (summarized in **Table 1**). Very few clinical studies have been performed in SCLC, presumably because of the difficulty in obtaining sufficient tumor tissue from large numbers of these patients. Furthermore, because approx 80% of SCLC patients have p53 mutations, it would be difficult to detect a prognostic impact in the small number of patients who lack mutations.

The majority of studies, using either immunohistochemistry or molecular analysis of mutations have suggested that patients with alterations in p53 have a worse outlook than those without (as shown in **Table 1**). Nevertheless some carefully conducted prospective studies have shown the opposite association *(51–53)*. Still others have found no correlation between p53 status and survival. One study even found that those patients with intermediate levels of p53 had a worse outcome than those with either very high or undetectable levels *(54)*.

How does one account for these disparate results? Most of the clinical studies have utilized immunohistochemistry, probably because this technique is available in routine pathology laboratories. Positive immunostaining for p53 is presumptive evidence for gene mutation. More than 90% of missense mutations in p53 will result in abnormal accumulation, as a result of increased stability of the mutant protein *(55,56)*. However, not all p53 alterations will be detected using immunohistochemistry. For example, homozygous deletions, frameshifts, or nonsense mutations do not cause p53 accumulation *(56,57)*. Conversely, overexpression of p53 may not necessarily be an indication of gene mutation. It is possible that ongoing DNA damage or oncogene activation within the tumor may result in stabilization of genetically normal p53 protein.

Thus, it is not surprising that studies which have compared immunostaining to analysis of gene mutations have shown poor concordance *(58–62)*. For example, Carbone et al. examined 85 NSCLC tumors using both immunohistochemistry and DNA sequencing *(58)*. Sixty-four percent of the tumors had p53 overexpression by immunostaining and 51% had mutant p53 sequences, but the concordance between the two types of studies was only 67%. Another study examined 30 cases of stage I adenocarcinoma of the lung using both

Table 1
Prognostic Significance of p53 Alterations in NSCLC[a]

Author/ year/ref.	Number of cases	Stage	Method of detection	Frequency of alterations	Correlation with survival (p-value)	Comments
Quinlan (1992) *(69)*	114	I, II	IHC	43%	Negative ($p < 0.001$)	Median survival 16 vs 38 mo
McLaren (1992) *(65)*	125	NS	IHC	54%	None	
Mitsudomi (1993) *(70)*	120	I–IV	PCR-SSCP	43%	Negative (p 0.01)	Adverse prognostic effect only in stage III and IV disease
Horio (1993) *(71)*	71	I–IIIA	PCR-SSCP	49%	Negative ($p = 0.014$)	p53 mutation an independent adverse prognostic factor
Brambilla (1993) *(66)*	70	I–IV	IHC	41%	None	
Ebina (1994) *(72)*	123	I–IV	IHC	39%	Negative ($p = 0.027$)	Survival analysis in 63 of the patients
Morkve (1993) *(54)*	112	I–II	Flow cytometry	76.7%	Negative (not significant)	Patients with intermediate levels of expression did worse
Carbone (1994) *(58)*	85	I–III	IHC SSCP	IHC-64% SSCP-51%	IHC-negative ($p = 0.05$) SSCP-none	Concordance of 67% between IHC and SSCP results
Ebina (1994) *(72)*	123	I–IV	IHC	39%	Negative ($p = 0.0011$)	Survival analysis in 63 patients treated with curative intent
Isobe (1994) *(59)*	30	I	IHC PCR-DGGE	IHC-37% DGGE-37%	Negative ($p = 0.003$)	Concordance of 73% between IHC and DGGE results
Passlick (1994) *(51)*	73	I–IV	IHC	45.2%	Positive ($p = 0.004$)	Improved survival only in patients with early stage disease

(continued)

Table 1 *(continued)*

Author/ year/ref.	Number of cases	Stage	Method of detection	Frequency of alterations	Correlation with survival (p-value)	Comments
Volm (1994) *(52)*	209	I–IV	IHC	51%	Positive ($p=0.002$)	5-Yr survival 38% for p53 positive vs 20% for p53 negative patients
Fontanini (1995) *(73)*	101	I–III	IHC	50%	None	Median p53 level used as cut off
Harpole (1995) *(74)*	271	I	IHC	38%	Negative ($p=0.001$)	p53 accumulation an independent adverse prognostic factor
Lee (1995) *(53)*	156	I–IIIB	IHC	31–66% (depending on cut-off)	Positive	Survival difference only in lymph node-positive patients
Fujino (1995) *(75)*	96	I–IV	IHC	58%	Negative ($p=0.024$)	
Top (1995) *(57)*	54	Not stated	PCR-SSCP IHC	69%	Positive ($p=0.06$)	No survival difference when patients with mutated K-ras excluded
Langendijk (1995) *(76)*	65	III	IHC	57%	None	p53-negative patients more chemosensitive
Ohsaki (1996) *(77)*	99	I–IV	IHC	44%	Negative (not significant)	
Pappot (1996) *(78)*	214	I–IIIB	ELISA	NS	None	
Nishio (1996) *(60)*	208	I–IIIB	IHC	46%	None	Borderline significant negative prognostic factor only in adenocarcinoma
Fontanini (1996) *(79)*	70	I–III	IHC	50%	Negative ($p=0.008$)	Used median p53 level as cut off
Xu (1996) *(80)*	119	I–II	IHC	45%	Negative ($p=0.01$)	p53 positivity not an independent adverse prognostic factor

(continued)

Table 1 *(continued)*

Author/ year/ref.	Number of cases	Stage	Method of detection	Frequency of alterations	Correlation with survival (p-value)	Comments
Pastorino (1996) *(81)*	557	I	IHC	43%	None	
Dalquen (1996) *(82)*	247	I–III	IHC	47.8%	Negative (borderline significance, $p = 0.056$)	In stage I patients p53 overexpression was an independent indicator of poor prognosis ($p = 0.033$)
Komiya (1997) *(83)*	137	I–IIIA	IHC	42%	None	
Kawasaki (1997) *(67)*	111	I–IV	IHC	55%	Negative for stage III and IV ($p = 0.02$)	Not a significant prognostic factor for stage I and II
Fukuyama (1997) *(84)*	159	I–IV	PCR-SSCP	35.8%	Negative for stage I and II ($p < 0.05$)	Not a significant prognostic factor for stage III and IV
Vega (1997) *(61)*	81	I–IV	IHC, PCR-SSCP	IHC: 46.9% PCR-SSCP: 21%	IHC: none PCR-SSCP: negative ($p = 0.04$)	Worst prognosis was for mutations in exon 5
Giatromanolaki (1998) *(85)*	120	I–II	IHC	29–60%	None	Three different antibodies gave different results
Huang (1998) *(63)*	204	I–III	PCR-SSCP	36.8%	Overall no survival difference; exon 7 and 8 mutations had worse prognosis	Higher mutation frequency in squamous compared to adenocarcinoma
Kwiatkowski (1998) *(86)*	242	I	IHC	44%	Negative ($p = 0.018$)	p53 accumulation an independent adverse prognostic factor

(continued)

Table 1 (continued)

Author/ year/ref.	Number of cases	Stage	Method of detection	Frequency of alterations	Correlation with survival (p-value)	Comments
D'Amico (1999) *(87)*	408	I	IHC	43%	Negative ($p = 0.001$)	p53 accumulation an independent adverse prognostic factor
Kandioler-Eckersberger (1999) *(62)*	24	IIIA and B	IHC PCR and sequencing	33%	Negative ($p = 0.027$)	
Geradts (1999) *(88)*	103	I–III	IHC	48.5%	None	
Skaug (2000) *(64)*	148	I–IIIB	SSCP	54%	Negative ($p = 0.022$)	Worse prognosis for mutations in exon 8
Moldvay (2000) *(89)*	227	I–IV	IHC	60%	None overall	p53 accumulation an independent negative prognostic factor in stage I and II adenocarcinoma

*a*IHC, immunohistochemistry; PCR, polymerase chain reaction; SSCP, single strand conformational polymorphism; DGGE, denaturing gradient gel electrophoresis; ELISA, enzyme-linked immunosorbent assay; NS, not stated; negative, worse prognosis for tumors with p53 alteration; positive, better prognosis for tumors with p53 alterations; none, no prognostic significance of p53 alterations.

molecular analysis and immunohistochemistry *(59)*. Thirty seven percent of cases had mutated p53 sequences and 37% had p53 accumulation but the concordance between the two techniques was only 73%. Vega et al. studied 81 cases of NSCLC and detected p53 protein accumulation by immunostaining in 47% of these *(61)*. However, only 45% of p53 immunopositive tumors had detectable p53 mutations. Even among the immunohistochemical studies, it is difficult to make comparisons, because different staining techniques have been used, with different antibodies, with or without antigen retrieval *(60)*, and with different cut-off levels for positivity.

Most of the studies of p53 mutations in lung cancer have examined exons 5–8 because the majority of mutations occur within this part of the gene. There have been some attempts to determine whether mutations in specific exons of p53 are of prognostic importance *(61,63,64)*. However, the small sample sizes make it difficult to draw any definite conclusions. Vega et al. found that patients with tumors harboring mutations in exon 5 had a shorter survival than those with mutations in other exons *(61)*. Another group found that exon 7 mutations carried the poorest prognosis whereas there was no survival difference for patients with mutations in exon 5 *(63)*. Yet another study found that mutations in exon 8 had the worst outlook *(64)*.

There have been very few studies of the prognostic importance of p53 alterations in patients with SCLC, and overall the number of patients is too small to make any definite conclusions *(26,65–68)*. Kawasaki et al. *(67)* examined 64 transbronchial biopsy specimens from SCLC patients by immunostaining with a monoclonal antibody (MAb) specific for p53. 58% of these specimens were positive for p53, but there was no correlation with response to chemotherapy or survival. Another study of 65 SCLC patients found no significant difference in overall survival in patients with or without p53 mutations *(68)*.

Thus, most of the evidence indicates that alterations in p53 are associated with a poor prognosis, at least in NSCLC patients. Nevertheless, there is no definitive evidence yet that the knowledge of p53 status could play a role in the management of individual patients with lung cancer. Furthermore, it remains to be seen whether the identification of patients with poor prognostic features can lead to improved therapeutic outcome in lung cancer patients.

3.4. Influence of p53 Status on Response to Therapy

Most of the aforementioned studies of the prognostic importance of p53 mutations in lung cancer were not designed to assess the effect of these mutations on response to chemotherapy or radiation. Chemotherapy is the major modality of treatment of SCLC and plays a significant role in the treatment of locally advanced NSCLC as well as the palliative therapy of metastatic NSCLC. SCLC usually responds dramatically to chemotherapy and radiotherapy, but frequently recurs and is resistant to further treatment *(90)*. Although NSCLC tumors often respond to chemotherapy and radiation, the responses are usually not as dramatic as for SCLC *(91)*.

Radiation and most chemotherapeutic agents exert their cytotoxic effects by causing cancer cells to undergo apoptosis. Because normal p53 function is often required for apoptosis to occur, it is expected that alterations of p53 might confer resistance to these forms of treatment. The data implicating p53 mutations as a cause of resistance to drug and radiation therapy are compelling. Studies of murine cells bearing homozygous mutations of p53 have shown that these cells are highly resistant to the cytotoxic effects of radiation as well as certain chemotherapeutic agents including 5-flurouracil (5-FU), etoposide, and doxorubicin, compared to cells expressing wild-type p53 *(92-95)*. Furthermore, transfer of wild-type p53 into a human NSCLC cell line with a homozygous deletion of p53 resulted in marked enhancement of sensitivity to the DNA damaging chemotherapeutic drug cisplatin *(96)*.

Most of the studies utilizing collections of cell lines of different histologic types have confirmed the important role of p53 in chemosensitivity. For example, Fan et al. *(97)* examined a panel of 8 lymphoma and lymphoblastoid cell lines, 4 with mutant and 4 with wild-type p53. The cells with wild-type p53 tended to be more sensitive to a variety of DNA damaging agents, including γ-irradiation, nitrogen mustard, cisplatin, and etoposide. The largest of these studies used the NCI drug screening panel of 60 cell lines *(98)*. This collection of cell lines from diverse tumor types was analyzed for p53 mutations, and the results were correlated with previously obtained drug-sensitivity data. Cell lines with mutant p53 tended to be more resistant to the majority of clinically useful chemotherapeutic drugs, including DNA damaging agents, antimetabolites, and topoisomerase I and II inhibitors *(98)*. Surprisingly, our group did not detect an association between p53 mutations and chemoresistance in three human lung cancer cell lines *(99)*. As well, targeted degradation of p53 did not appear to increase the resistance of a lung cancer line bearing wild-type p53 *(99)*. Another study of 20 SCLC cell lines was unable to detect a correlation between chemoresistance and p53 mutations, presumably because the majority of the cell lines (18/20) had mutated p53 *(100)*.

A notable exception to the correlation between p53 alterations and chemoresistance in the NCI panel of cell lines were the antimitotic drugs, including vinca alkaloids and taxanes *(98)*. The cytotoxicity of these agents appeared to be independent of p53 status *(98)*. On the other hand, Wahl et al. found that human fibroblasts that were depleted of functional p53 were actually more sensitive to paclitaxel than comparable cells with intact p53 *(101)*. However, our group found the opposite effect in ovarian teratocarcinoma cells *(102)*. Thus it appears that, at least in some cell types, paclitaxel is able to induce apoptosis by a mechanism that is independent of p53.

Does p53 status influence the response to therapy in lung cancer patients? In other tumor types, the evidence is conflicting. For example, ovarian cancer patients with mutant p53 tumors were less likely to respond to cisplatin chemotherapy than those with intact p53 *(103)*. On the other hand, there was no significant evidence that p53 status influenced the outcome of adjuvant chemotherapy in high risk stage I breast cancer patients *(104)*.

Most of the studies in lung cancer have examined only NSCLC patients *(62,105–107)*. Although the frequency of p53 mutations is higher in SCLC than in NSCLC, SCLC tumors tend to be more chemo- and radio-responsive than NSCLC tumors. Kawasaki et al. examined transbronchial biopsy specimens from both NSCLC as well as SCLC patients. Positive immunostaining for p53 correlated significantly with resistance to chemotherapy in NSCLC, whereas there was no correlation with drug response in the SCLC patients *(67)*. Other clinical studies in NSCLC patients have found a correlation between p53 alterations and treatment resistance. In a study of locally advanced NSCLC patients who received high-dose cisplatin prior to surgery, Rusch et al. found a statistically significant correlation between positive immunostaining for p53 and lack of pathological response to cisplatin *(105)*. Another group studied patients with recurrent NSCLC who were treated with radiation *(106)*. Thirteen of the 34 patients had p53 mutations, and these patients had an overall response rate of 15.4% compared to 61.9% for patients with tumors with wild-type p53 ($p = 0.013$) *(106)*. In a study of 24 patients with locally advanced NSCLC, who were treated with cisplatin and ifosfamide prior to surgery there was a clear correlation between chemoresistance and p53 gene mutations, but not with p53 accumulation as detected by immunohistochemistry *(62)*. On the other hand, in 25 patients with metastatic NSCLC treated with paclitaxel, response rates were 75% for patients with tumors with p53 mutations, and 47% for patients without p53 mutations, although this result was not statistically significant *(107)*.

3.5. p53 Antibodies in Lung Cancer Patients

In the course of investigating the murine immune response to tumor antigens, DeLeo et al. immunized mice with murine sarcoma cells. The antisera that they produced identified a 53 kDa antigen, which was present in a variety of murine tumors, but was not present in normal tissues *(108)*. They speculated that this antigen was associated with malignant transformation, and termed it p53. Subsequently, Crawford et al. identified antibodies to human p53 in 14/155 sera from breast cancer patients *(109)*. None of 164 control sera were positive. Thus, it has been known for some time—long before the role of this protein as a tumor suppressor was clearly understood—that cancer patients may develop antibodies to p53.

The development of antibodies to p53 in a variety of human tumors has recently been reviewed *(110)*. It appears that the presence of these antibodies is almost exclusively associated with a diagnosis of cancer. In some cases a humoral immune response to p53 may precede the diagnosis of malignancy. For example, anti-p53 antibodies have been identified in stored sera of vinyl chloride-exposed workers who subsequently developed angiosarcoma of the liver *(110)*. These antibodies have also been detected in sera from heavy smokers prior to the diagnosis of lung cancer *(111–113)*.

Antibodies to p53 in cancer patients react with both normal and mutant p53 *(110)*. With few exceptions, they occur only in patients with those mutations that result in p53 accumulation: namely missense mutations within the DNA binding domain *(114–115)*. Even among those patients with tumors with abnormal accumulation of p53, only a subset will develop antibodies *(110,114,116)*. It has been suggested that only those mutations in exons 5 and 6 will result in the formation of antibodies (116). Tumors with these particular mutations have been shown to form complexes with a 70 kDa heat-shock protein, which may be involved in antigen presentation *(116)*. However, another group found that tumors with mutations in exon 7 or 8 were more likely to be associated with an immune response to p53 *(114)*. Detailed epitope-mapping studies have shown that these autoantibodies react with immunodominant epitopes within the amino or carboxyl terminus of the p53 protein, and not within the DNA binding domain where most of the mutations occur *(112,117)*.

Tumors with the highest frequency of p53 mutations, including breast, ovarian, head and neck, lung, and colon cancer are most likely to elicit autoantibodies to p53 *(110,117,118)*. Clinical correlative studies in a variety of tumor types have demonstrated an association with poor survival *(110)*. In lung cancer, the frequency of these antibodies ranges from 8–27% *(112,114,115,117–126)*. Studies examining the prognostic impact of p53 antibodies in lung cancer have been contradictory, as was the case for studies of the prognostic implications of p53 mutations. Some have shown an association with poor prognosis *(124,126)*, whereas others have shown an improved outcome for p53 antibody-positive patients *(121)*, and still others have shown no correlation with survival *(114,120,122)*. Most of the clinical studies have examined predominantly NSCLC patients. It is unclear from any of these studies whether the prognosis of those patients with p53 mutations who develop antibodies is any different from those who do not.

Is it likely that measurement of p53 antibodies will be of any value in the screening, diagnosis and management of lung cancer? As a diagnostic test, the sensitivity is quite poor, because the prevalence of these antibodies is low, even among patients known to have p53 mutations in their tumors. On the other hand, the specificity is very high. It is clear that p53 does not elicit a humoral immune response under normal circumstances. Thus, the presence of p53 antibodies in a person without a diagnosis of cancer should prompt a search for a primary tumor.

4. Summary

The process of bronchial carcinogenesis is characterized by accumulated genetic abnormalities which ultimately lead to malignant transformation of bronchial epithelial cells, followed by invasion and metastasis. One of the most common and consistent of these genetic lesions is inactivation of the p53 tumor suppressor gene by mutation or deletion. The frequency of p53 alterations in lung cancer is highest in those subtypes of bronchial carcinomas that are most consistently associated with smoking, especially SCLC and squamous cell carcinomas. The frequency is lower in adenocarcinomas, in which the association with smoking, although present, is not as strong. The frequency of p53 abnormalities is higher in patients with greater cumulative tobacco exposure. Tobacco-specific carcinogens, in particular BPDE, cause a unique spectrum of p53 mutations, quite distinct from those found in cancers that are not associated with smoking. This characteristic genetic "signature" may persist even decades following smoking cessation.

The prognostic significance of p53 mutations in lung cancer is not entirely clear despite the multitude of clinical studies that have been carried out. Nevertheless, the majority of clinical studies suggest that lung cancers with p53 alterations carry a worse prognosis. Furthermore, those tumors with mutant p53 may be relatively more resistant to chemotherapy and radiation.

An understanding of the role of p53 in human lung cancer may lead to more rational targeted approaches for treating this disease. For example, the observation that the introduction of wild-type p53 into lung cancer cells with mutant or deleted p53 may reverse the malignant phenotype despite the presence of multiple other genetic abnormalities *(14)* suggests that replacement of this gene may be an effective clinical strategy. Preclinical and early clinical studies indicate that this is a promising approach, but clearly more effective means of gene delivery to the tumor cells are required *(127–129)*, as discussed elsewhere in this volume.

References

1. Vogelstein, B., Lane, D., and Levine, A. J.(2000) Surfing the p53 network. *Nature* **408**, 307–310.
2. Raff, M. (1998) Cell suicide for beginners. *Nature* **396**, 119–122.
3. Leach, F. S., Tokino, T., Meltzer, P., Burrell, M., Oliner, J. D., Smith, S., et al. (1993) p53 mutation and MDM2 amplification in human soft tissue sarcomas. *Cancer Res.* **53**, 2231–2234.
4. Scheffner, M., Werness, B. A., Huibregtse, J. M., Levine, A. J., and Howley, P. M. (1990) The E6 oncoprotein encoded by human papillomavirus types 16 and 18 promotes the degradation of p53. *Cell* **63**, 1129–1136.
5. Sherr, C. J. (1998) Tumor surveillance via the ARF-p53 pathway [Review]. *Genes Dev.* **12**, 2984–2992.
6. Rotman, G. and Shiloh, Y. (1999) ATM: a mediator of multiple responses to genotoxic stress. *Oncogene* **18**, 6135–6144.
7. Bell, D. W., Varley, J. M., Szydlo, T. E., Kang, D. H., Wahrer, D. C. R., Shannon, K. E., et al. (1999) Heterozygous germ line *hCHK2* mutations in Li-Fraumeni syndrome. *Science* **286**, 2528–2531.
8. Levine, A. J., Momand, J., and Finlay, C. A. (1991) The p53 tumor suppressor gene. *Nature* **351**, 453–456.

9. Harris, C. C. and Hollstein, M. (1993) Clinical implications of the p53 tumor suppressor gene. *N. Eng. J. Med.* **329,** 1318–1327.
10. Hernandez-Boussard, T. M., Rodriguez-Tome, P., Montesano, R., and Hainaut, P. (1999) IARC p53 mutations database: a relational database to compile and analyze p53 mutations in human tumors and cell lines. *Hum. Mutat.* **14,** 1–8.
11. Sekido, Y., Fong, K. M., and Minna, J. D. (2001) Molecular biology of lung cancer, in *Cancer: Principles and Practice of Oncology* (DeVita, V. T., Hellman, S., and Rosenberg, S. A., eds.), Lippincott Williams & Wilkins, Philadelphia, pp. 917–925.
12. Whang-Peng, J., Kao-Shan, C. S., Lee, E. C., Bunn, P. A., Carney, D. N., Gazdar, A. F., and Minna, J. D. (1982) Specific chromosome defect associated with human small-cell lung cancer: deletion 3p(14-23). *Science* **215,** 181–182.
13. Wistuba, I. I., Behrens, C., Virmani, A. K., Mele, G., Milchgrub, S., Girard, L., et al. (2000) High resolution chromosome 3p allelotyping of human lung cancer and preneoplastic/preinvasive bronchial epithelium reveals multiple, discontinuous sites of 3p allele loss and three regions of frequent breakpoints. *Cancer Res.* **60,** 1949–1960.
14. Takahashi, T., Carbone, D., Nau, M. M., Hida, T., Linnoila, I., Ueda, R., and Minna, J. D. (1992) Wild-type but not mutant p53 suppresses the growth of human lung cancer cells bearing multiple genetic lesions. *Cancer Res.* **52,** 2340–2343.
15. Li, F. P. and Fraumeni, J. F., Jr. (1969) Soft-tissue sarcomas, breast cancer, and other neoplasms. A familial syndrome? *Ann. Intern. Med.* **71,** 747–752.
16. Malkin, D., Li, F. P., Strong, L. C., and Fraumeni, J. F. (1990) Germline p53 mutations in a familial syndrome of breast cancer, sarcomas, and other neoplasms. *Science* **250,** 1233–1238.
17. Kleihues, P., Schauble, B., zur Hausen, A., Estève, J., and Ohgaki, H. (1997) Tumors associated with p53 germline mutations: a synopsis of 91 families. *Am. J. Pathol.* **150,** 1–13.
18. Nichols, K. E., Malkin, D., Garber, J. E., Fraumeni, J. F., Jr., and Li, F. P. (2001) Germ-line p53 mutations predispose to a wide spectrum of early-onset cancers. *Cancer Epidemiol. Biomarkers Prev.* **10,** 83–87.
19. Ishioka, C., Frebourg, T., Yan, Y. X., Vidal, M., Friend, S. H., Schmidt, S., and Iggo, R. (1993) Screening patients for heterozygous p53 mutations using a functional assay in yeast. *Nat. Genet.* **5,** 124–129.
20. Ahrendt, S. A., Halachmi, S., Chow, J. T., Halachmi, N., Yang, S. C., Wehage, S., et al. (1999) Rapid p53 sequence analysis in primary lung cancer using an oligonucleotide probe array. *Proc. Natl. Acad. Sci. USA* **96,** 7382–7387.
21. Takahashi, T., Nau, M. M., Chiba, I. B., Rosenberg, R. K., Vinocour, M., Levitt, M., et al. (1989) p53: a frequent target for genetic abnormalities in lung cancer. *Science* **246,** 491–494.
22. Chiba, I., Takahashi, T., Nau, M. M., D'Amico, D., Curiel, D. T., Mitsudomi, T., et al. (1990) Mutations in the p53 gene are frequent in primary, resected non-small cell lung cancer. *Oncogene* **5,** 1603–1610.

23. Takahashi, T., Suzuki, H., Hida, T. S., Ariyoshi, Y., and Ueda, R. (1991) The p53 gene is very frequently mutated in small-cell lung cancer with a distinct nucleotide substitution pattern. *Oncogene* **6,** 1775–1778.
24. Hensel, C. H., Xiang, R. H., Sakaguchi, A. Y., and Naylor, S. L. (1991) Use of the single strand conformation polymorphism technique and PCR to detect p53 gene mutations in small cell lung cancer. *Oncogene* **6,** 1067–1071.
25. Samesima, Y., Matsuno, Y., Hirohashi, S., Shimosato, Y., Mizoguchi, H., Sugimura, T., et al. (1992) Alterations of the p53 gene are common and critical events for the maintenance of malignant phenotypes in small-cell lung carcinoma. *Oncogene* **7,** 451–457.
26. D'Amico, D., Carbone, D., Mitsudomi, T., Nau, M., Fedorko, J., Russell, E., et al. (1992) High frequency of somatically acquired p53 mutations in small-cell lung cancer cell lines and tumors. *Oncogene* **7,** 339–346.
27. Kishimoto, Y., Murakami, Y., Shiraishi, M., Hayashi, K., and Sekiya, T. (1992) Aberrations of the p53 tumor suppressor gene in human non-small cell carcinomas of the lung. *Cancer Res.* **52,** 4799–4804.
28. Mitsudomi, T., Steinberg, S. M., Nau, M. M., Carbone, D., D'Amico, D., Bodner, S., et al. (1992) p53 gene mutations in non-small-cell lung cancer cell lines and their correlation with the presence of ras mutations and clinical features. *Oncogene* **7,** 171–180.
29. Shipman, R., Schraml, P., Colombi, M., Raefle, G., Dalquen, P., and Ludwig, C. (1996) Frequent TP53 gene alterations (mutation, allelic loss, nuclear accumulation) in primary non-small cell lung cancer. *Eur. J. Cancer* **32A,** 335–341.
30. Girard, L., Zochbauer-Muller, S., Virmani, A. K., Gazdar, A. F., and Minna, J. D. (2000) Genome-wide allelotyping of lung cancer identifies new regions of allelic loss, differences between small cell lung cancer and non-small cell lung cancer and loci clustering. *Cancer Res.* **60,** 4894–4906.
31. Bennett, W. P., Colby, T. V., Travis, W. D., Borkowski, A., Jones, R. T., Lane, D. P., et al. (1993) p53 protein accumulates frequently in early bronchial neoplasia. *Cancer Res.* **53,** 4817–4822.
32. Walker, C., Robertson, L. J., Myskow, M. W., Dixon, G. R., and Pendleton, N. (1994) p53 expression in normal and dysplastic bronchial epithelium and in lung carcinomas. *Br. J. Cancer* **70,** 297–303.
33. Mao, L., Hruban, R. H., Boyle, J. O., Tockman, M., and Sidransky, D. (1994) Detection of oncogene mutations in sputum precedes diagnosis of lung cancer. *Cancer Res.* **54,** 1634–1637.
34. Hung, J., Kishimoto, Y., Sugio, K., Virmani, A., McIntire, D. D., Minna, J. D., and Gazdar, A. F. (1995) Allele-specific chromosome 3p deletions occur at an early stage in the pathogenesis of lung carcinoma [published erratum appears in JAMA 1995]. *J. Am. Med. Assoc.* **273,** 558–563.
35. Kishimoto, Y., Sugio, K., Hung, J. Y., Virmani, A. K., McIntire, D. D., Minna, J. D., and Gazdar, A. F. (1995) Allele-specific loss in chromosome 9p loci in preneoplastic lesions accompanying non-small-cell lung cancer. *J. Natl. Cancer Inst.* **87,** 1224–1229.

36. Sugio, K., Kishimoto, Y., Virmani, A. K., Hung, J. Y., and Gazdar, A. F. (1994) K-ras mutations are a relatively late event in the pathogenesis of lung carcinomas. *Cancer Res.* **54,** 5811–5815.
37. Mao, L., Lee, J. S., Kurie, J. M., Fan, Y. H., Lippman, S. M., Lee, J. J., et al. (1997) Clonal genetic alterations in the lungs of current and former smokers. *J. Natl. Cancer Inst.* **89,** 857–862.
38. Wistuba, I. I., Lam, S., Behrens, C., Virmani, A. K., Fong, K. M., LeRiche, J., et al. (1997) Molecular damage in the bronchial epithelium of current and former smokers. *J. Natl. Cancer Inst.* **89,** 1366–1373.
39. Strauss, G., DeCamp, M., Dibiccaro, E., Richards, W., Harpole, D., Healey, E., and Sugarbaker, D. (1995) Lung cancer diagnosis is being made with increasing frequency in former cigarette smokers. *Proc. Am. Soc. Clin. Oncol.* **14,** 362-Abst 1106.
40. Tong, L., Spitz, M. R., Fueger, J. J., and Amos, C. I. (1996) Lung carcinoma in former smokers. *Cancer* **78,** 1004–1010.
41. Hernandez-Boussard, T. M. and Hainaut, P. (1998) A specific spectrum of p53 mutations in lung cancer from smokers: review of mutations compiled in the IARC p53 database. *Environ. Health Perspect.* **106,** 385–391.
42. Suzuki, H., Takahashi, T., Kuroishi, T., Suyama, M., Ariyoshi, Y., and Ueda, R. (1992) p53 mutations in non-small cell lung cancer in Japan: association between mutations and smoking. *Cancer Res.* **52,** 734–736.
43. Kondo, K., Tsuzuki, H., Sasa, M., Sumitomo, M., Uyama, T., and Monden, Y. (1996) A dose-response relationship between the frequency of p53 mutations and tobacco consumption in lung cancer patients. *J. Surg. Oncol.* **61,** 20–26.
44. Takagi, Y., Osada, H., Kuroishi, T., Mitsudomi, T., Kondo, M., Niimi, T., et al. (1998) p53 mutations in non-small-cell lung cancers occuring in individuals without a past history of active smoking. *Br. J. Cancer* **77,** 1568–1572.
45. Husgafvel-Pursiainen, K., Boffetta, P., Kannio, A., Nyberg, F., Pershagen, G., Mukeria, A., et al. (2000) p53 mutations and exposure to environmental tobacco smoke in a multicenter study on lung cancer. *Cancer Res.* **60,** 2906–2911.
46. Ahrendt, S. A., Chow, J. T., Yang, S. C., Wu, L., Zhang, M.-J., Jen, J., and Sidransky, D. (2000) Alcohol consumption and cigarette smoking increase the frequency of p53 mutation in non-small cell lung cancer. *Cancer Res.* **60,** 3155–3159.
47. Brennan, J. A., Boyle, J. O., Koch, W. M., Goodman, S. N., Hruban, R. H., Eby, Y. J., et al. (1995) Association between cigarette smoking and mutation of the p53 gene in squamous-cell carcinoma of the head and neck. *N. Eng. J. Med.* **332,** 712–717.
48. Anderson, L. M., Chabra, S. K., Nerurkar, P. V., Souliotis, V. L., and Kyrtopoulos, S. A. (1995) Alcohol-related cancer risk: a toxicologic hypothesis. *Alcohol* **12,** 97–104.
49. Denissenko, M. F., Pao, A., Tang, M., and Pfeifer, G. P. (1996) Preferential formation of benzo[a]pyrene adducts at lung cancer mutational hotspots in p53. *Science* **274,** 430–432.

50. Smith, L. E., Denissenko, M. F., Bennett, W. P., Li, H., Amin, S., Tang, M., and Pfeifer, G. P. (2000) Targeting of lung cancer mutational hotspots by polycyclic aromatic hydrocarbons. *J. Natl. Cancer Inst.* **92,** 803–811.
51. Passlick, B., Izbicki, J. R., Reithmüller, G., and Pantel, K. (1994) p53 in non-small cell lung cancer. *J. Natl. Cancer Inst.* **86,** 801–802.
52. Volm, M. and Mattern, J. (1994) Immunohistochemical detection of p53 in non-small-cell lung cancer (letter). *J. Natl. Cancer Inst.* **86,** 1249.
53. Lee, J. S., Yoon, A., Kalapurakal, S. K., Ro, J. Y., Lee, J. J., Tu, N., et al. (1995) Expression of p53 oncoprotein in non-small-cell lung cancer: a favorable prognostic factor. *J. Clin. Oncol.* **13,** 1893–1903.
54. Morkve, O., Halvorsen, O. J., Skjaerven, R., Stangeland, L., Gulsvik, A., and Laerum, O. D. (1993) Prognostic significance of p53 protein expression and DNA ploidy in surgically treated non-small cell lung carcinomas. *Anticancer Res.* **13,** 571–578.
55. Iggo, R., Gatter, K., Bartek, J., Lane, D., and Harris, A. L. (1990) Increased expression of mutant forms of p53 oncogene in primary lung cancer. *Lancet* **335,** 675–679.
56. Bodner, S. M., Minna, J. D., Jensen, S. M., D'Amico, D., Carbone, D., Mitsudomi, T., et al. (1992) Expression of mutant p53 proteins in lung cancer correlates with the class of p53 gene mutation. *Oncogene* **7,** 743–749.
57. Top, B., Mooi, W. J., Klauer, S. G., Boerrigter, L., Wisman, P., Elbers, M., et al. (1995) Comparative analysis of p53 gene mutations and protein accumulation in human non-small cell lung cancer. *Int. J. Cancer* **64,** 83–91.
58. Carbone, D. P., Mitsudomi, T., Chiba, I., Piantadosi, S., Rusch, V., Nowak, J. A., et al. (1994) p53 immunostaining positivity is associated with reduced survival and is imperfectly correlated with gene mutations in resected non-small cell lung cancer. *Chest* **106,** 377S–381S.
59. Isobe, T., Hiyama, K., Yoshida, Y., Fujiwara, Y., and Yamakido, M. (1994) Prognostic significance of p53 and *ras* gene abnormalities in lung adenocarcinoma patients with stage I disease after curative resection. *Jap. J. Cancer Res.* **85,** 1240–1246.
60. Nishio, M., Koshikawa, T., Kuroishi, T., Suyama, M., Uchida, K., Takagi, Y., et al. (1996) Prognostic significance of abnormal p53 accumulation in primary, resected non-small-cell lung cancers. *J. Clin. Oncol.* **14,** 497–502.
61. Vega, F. J., Iniesta, P., Caldes, T., Sanchez, A., Lopez, J. A., de Juan, C., et al. (1997) p53 exon 5 mutations as a prognostic indicator of shortened survival in non-small-cell lung cancer. *Br. J. Cancer* **76,** 44–51.
62. Kandioler-Eckersberger, D., Kappel, S., Mittlbock, M., Dekan, G., Ludwig, C., Janschek, E., et al. (1999) The TP53 genotype but not immunohistochemical result is predictive of response to cisplatin-based neoadjuvant therapy in stage III non-small cell lung cancer. *J. Thorac. Cardiovasc. Surg.* **117,** 744–750.
63. Huang, C., Taki, T., Adachi, M., Konishi, T., Higashiyama, M., and Miyake, M. (1998) Mutations in exon 7 and 8 of p53 as poor prognostic factors in patients with non-small cell lung cancer. *Oncogene* **16,** 2469–2477.

64. Skaug, V., Ryberg, D., Kure, E. H., Arab, M. O., Stangeland, L., Myking, A. O., and Haugen, A. (2000) p53 mutations in defined structural and functional domains are related to poor clinical outcome in non-small cell lung cancer patients. *Clin. Cancer Res.* **6,** 1031–1037.
65. McLaren, R., Kuzu, I., Dunnill, M., Harris, A., Lane, D., and Gatter, K. C. (1992) The relationship of p53 immunostaining to survival in carcinoma of the lung. *Br. J. Cancer* **66,** 735–738.
66. Brambilla, E., Gazzeri, S., Moro, D., de Fromental, C. C., Gouyer, V., Jacrot, M., and Brambilla, C. (1993) The relationship of p53 immunostaining correlates with a poor prognosis in human lung cancer. *Am. J. Pathol.* **143,** 199–220.
67. Kawasaki, M., Nakanishi, Y., Kuwano, K., Yatsunami, J., Takayama, K., and Hara, N. (1997) The utility of p53 immunostaining of transbronchial biopsy specimens of lung cancer: p53 overexpression predicts poor prognosis and chemoresistance in advanced non-small cell lung cancer. *Clin. Cancer Res.* **3,** 1195–1200.
68. Tseng, J. E., Rodriguez, M., Ro, J., Liu, D., Hong, W. K., and Mao, L. (1999) Gender differences in p53 mutational status in small cell lung cancer. *Cancer Res.* **59,** 5666–5670.
69. Quinlan, D. C., Davidson, A. G., Summers, C. L., Warden, H. E., and Doshi, H. M. (1992) Accumulation of p53 protein correlates with poor prognosis in human lung cancer. *Cancer Res.* **52,** 4828–4831.
70. Mitsudomi, T., Oyama, T., Kusano, T., Osaki, T., Nakanishi, R., and Shirakusa, T. (1993) Mutations of the p53 gene as a predictor of poor prognosis in patients with non-small-cell lung cancer. *J. Natl. Cancer Inst.* **85,** 2018–2023.
71. Horio, Y., Takahashi, T., Kuroishi, T., Hibi, K., Suyama, M., Niimi, T., et al. (1993) Prognostic significance of p53 mutations and 3p deletions in primary resected non-small cell lung cancer. *Cancer Res.* **53,** 1–4.
72. Ebina, M., Steinberg, S. M., Mulshine, J. L., and Linnoila, R. I. (1994) Relationship of p53 overexpression and up-regulation of proliferating cell nuclear antigen with the clinical course of non-small cell lung cancer. *Cancer Res.* **54,** 2496–2503.
73. Fontanini, G., Vignati, S., Bigini, D., Mussi, A., Lucchi, M., Angeletti, C. A., et al. (1995) Bcl-2 protein: a prognostic factor inversely correlated to p53 in non-small-cell lung cancer. *Br. J. Cancer* **71,** 1003–1007.
74. Harpole, D. H., Herndon, J. E., Wolfe, W. G., Iglehart, J. D., and Marks, J. R. (1995) A prognostic model of recurrence and death in stage I non-small cell lung cancer utilizing presentation, histopathology, and oncoprotein expression. *Cancer Res.* **55,** 51–56.
75. Fujino, M., Akita, H. D., Harada, M., Hiroumi, H., Kinoshita, I., Akie, K., and Kawakami, Y. (1995) Prognostic significance of p53 and ras p21 expression in non small cell lung cancer. *Cancer* **76,** 2457–2463.
76. Langendijk, J. A., Thunnissen, F. B., Lamers, R. J., de Jong, J. M. A., ten Velde, G. P. M., and Wouters, E. F. M. (1995) The prognostic signficance of accumulation of p53 protein in stage III non-small cell lung cancer treated with radiotherapy. *Radiother. Oncol.* **36,** 218–224.

77. Ohsaki, Y., Toyoshima, E., Fujiuchi, S., Matsui, H., Hirata, S., Miyokawa, N., et al. (1996) bcl-2 and p53 protein expression in non-small cell lung cancers: Correlation with survival time. *Clin. Cancer Res.* **2,** 915–920.
78. Pappot, H., Francis, D., Brünner, N., Grondahl-Hansen, J., and Osterlind, K. (1996) p53 protein in non-small cell lung cancer as quantitated by enzyme-linked immunosorbent assay: Relation to prognosis. *Clin. Cancer Res.* **2,** 155–160.
79. Fontanini, G., Vignati, S., Bigini, D., Mussi, A., Lucchi, M., Chine, S., et al. (1996) Recurrence and death in non-small cell lung carcinomas: A prognostic model using pathological parameters, microvessel count, and gene protein products. *Clin. Cancer Res.* **2,** 1067–1075.
80. Xu, H.-J., Cagle, P. T., Hu, S.-X., Li, J., and Benedict, W. F. (1996) Altered retinoblastoma and p53 protein status in non-small cell carcinoma of the lung: Potential synergistic effects on prognosis. *Clin. Cancer Res.* **2,** 1169–1176.
81. Pastorino, U., Andreola, S., Tagliabue, E., Pezzella, F., Incabone, M., Sozzi, G., et al. (1997) Immunocytochemical markers in stage I lung cancer: Relevance to prognosis. *J. Clin. Oncol.* **15,** 2858–2865.
82. Dalquen, P., Sauter, G., Torhorst, J., Schultheiss, E., Jordan, P., Lehmann, S., et al. (1996) Nuclear p53 overexpression is an independent prognostic parameter in node-negative non-small cell lung carcinoma. *J. Pathol.* **178,** 53–58.
83. Komiya, T., Hosono, Y., Hirashima, T., Masuda, N., Yasumitsu, T., Nakagawa, K., et al. (1997) p21 expression as a predictor for favorable prognosis in squamous cell carcinoma of the lung. *Clin. Cancer Res.* **3,** 1831–1835.
84. Fukuyama, Y., Mitsudomi, T., Sugio, K., Ishida, T., Akazawa, K., and Sugimachi, K. (1997) K-ras and p53 mutations are an independent unfavourable prognostic indicator in patients with non-small-cell lung cancer. *Br. J. Cancer* **75,** 1125–1130.
85. Giatromanolaki, A., Koukourakis, M. I., Kakolyris, S., Turley, H., O'Byrne, K., Scott, P. A. E., et al. (1998) Vascular endothelial growth factor, wild-type p53, and angiogenesis in early operable non-small cell lung cancer. *Clin. Cancer Res.* **4,** 3015–3024.
86. Kwiatkowski, D. J., Harpole, D. H., Godleski, J., Herndon, J. E. II., Shieh, D.-B., Richards, W., et al. (1998) Molecular pathologic substaging in 244 Stage I non-small-cell lung cancer patients: Clinical implications. *J. Clin. Oncol.* **16,** 2468–2477.
87. D'Amico, T. A., Massey, M., Herndon, J. E. II., Moore, M.-B., and Harpole, D. H. (1999) A biologic risk model for stage I lung cancer: Immunohistochemical analysis of 408 patients with the use of ten molecular markers. *J. Thorac. Cardiovasc. Surg.* **117,** 736–743.
88. Geradts, J., Fong, K. M., Zimmerman, P. V., Maynard, R., and Minna, J. D. (1999) Correlation of abnormal RB, p16[ink4a], and p53 expression with 3p loss of heterozygosity, other genetic abnormalities, and clinical features in 103 primary non-small cell lung cancers. *Clin. Cancer Res.* **5,** 791–800.
89. Moldvay, J., Scheid, P., Wild, P., Nabil, K., Siat, J., Borrelly, J., et al. (2000) Predictive survival markers in patients with surgically resected non-small cell lung carcinoma. *Clin. Cancer Res.* **6,** 1125–1134.

90. Murren, J., Glatstein, E., and Pass, H. I. (2001) Small cell lung cancer, in *Cancer: Principles & Practice of Oncology* (DeVita, V.T., Hellman, S., and Rosenberg, S.A., eds.), Lippincott Williams & Wilkins, Philadelphia, pp. 983–1018.
91. Ginsberg, R. J., Vokes, E. E., and Rosenzweig, K. (2001) Non-small cell lung cancer, in *Cancer: Principles & Practice of Oncology* (DeVita, V. T., Hellman, S., and Rosenberg, S. A., eds.), Lippincott Williams & Wilkins, Philadelphia, pp. 925–983.
92. Lowe, S. W., Schmitt, E. M., Smith, S. W., Osborne, B. A., and Jacks, T. (1993) p53 is required for radiation-induced apoptosis in mouse thymocytes. *Nature* **362**, 847–849.
93. Clarke, A. R., Purdie, C. A., Harrison, D. J., Morris, R. G., Bird, C. C., Hooper, M. L., and Wyllie, A. H. (1993) Thymocyte apoptosis induced by p53-dependent and independent pathways. *Nature* **362**, 849.
94. Lee, J. M. and Bernstein, A. (1993) p53 mutations increase resistance to ionizing radiation. *Proc. Natl. Acad. Sci. USA* **90**, 5742–5746.
95. Lowe, S. W., Ruley, H. E., Jacks, T., and Housman, D. E. (1993) p53-dependent apoptosis modulates the cytotoxicity of anticancer agents. *Cell* **74**, 957–967.
96. Fujiwara, T., Grimm, E. A., Mukhopadhyay, T., Zhang, W.-W., Owen-Schaub, L. B., and Roth, J. A. (1994) Induction of chemosensitivity in human lung cancer cells in vivo by adenovirus-mediated transfer of the wild-type p53 gene. *Cancer Res.* **54**, 2287–2291.
97. Fan, S., El-Deiry, W. S., Bae, I., Freeman, J., Jondle, D., Bhatia, K., et al. (1994) p53 gene mutations are associated with decreased sensitivity of human lymphoma cells to DNA damaging agents. *Cancer Res.* **54**, 5824–5830.
98. O'Connor, P. M., Jackman, J., Bae, I., Myers, T. G., Fan, S., Mutoh, M., et al. (1997) Characterization of the p53 tumor suppressor pathway in cell lines of the National Cancer Institute anticancer drug screen and correlation with the growth-inhibitory potency of 123 anticancer agents. *Cancer Res.* **57**, 4285–4300.
99. Wu, G. S. and El-Deiry, W. S. (1996) Apoptotic death of tumor cells correlates with chemosensitivity, independent of p53 or bcl-2. *Clin. Cancer Res.* **2**, 623–633.
100. Tsai, C.-M., Chang, K.-T., Wu, L.-H., Chen, J.-Y., Mitsudomi, T., Chen, M.-H., and Perng, R.-P. (1996) Correlations between intrinsic chemoresistance and HER-2/neu gene expression, p53 gene mutations, and cell proliferation characteristics in non-small cell lung cancer cell lines. *Cancer Res.* **56**, 206–209.
101. Wahl, A. F., Donaldson, K. L., Fairchild, C., Lee, F. Y., Foster, S. A., Demers, G. W., and Galloway, D. A. (1996) Loss of normal p53 function confers sensitization to Taxol by increasing G2-M arrest and apoptosis. *Nature Med.* **2**, 72–79.
102. Wu, G. S. and El-Deiry, W. S. (1996) p53 and chemosensitivity. *Nature Med.* **2**, 255–256.
103. Righetti, S. C., Della Torre, G., Pilotti, S., Menard, S., Ottone, F., Colnaghi, M. I., et al. (1996) A comparative study of p53 gene mutations, protein accumulation, and response to cisplatin-based chemotherapy in advanced ovarian carcinoma. *Cancer Res.* **56**, 689–693.

104. Elledge, R. M., Gray, R., Mansour, E., Yu, Y., Clark, G. M., Ravdin, P., et al. (1995) Accumulation of p53 protein as a possible predictor of response to adjuvant combination chemotherapy with cyclophosphamide, methotrexate, fluoruracil, and prednisone for breast cancer. *J. Natl. Cancer Inst.* **87,** 1254–1256.
105. Rusch, V., Klimstra, D., Venkatraman, E., Oliver, J., Martini, N., Gralla, R., et al. (1995) Aberrant p53 expression predicts clinical resistance to cisplatin-based chemotherapy in locally advanced non-small cell lung cancer. *Cancer Res.* **55,** 5038–5042.
106. Matsuzoe, D., Hideshima, T., Kimura, A., Inada, K., Watanabe, K., Akita, Y., et al. (1999) p53 mutations predict non-small cell lung carcinoma response to radiotherapy. *Cancer Lett.* **135,** 189.
107. King, T. C., Akerley, W., Fan, A. C., Moore, T., Mangray, S., Chen, M. H., and Safran, H. (2000) p53 mutations do not predict response to paclitaxel in metastatic nonsmall cell lung carcinoma. *Cancer* **89,** 769–773.
108. De Leo, A. B., Jay, G., Appela, E., Dubois, G. C., Law, L. W., and Old, L. J. (1979) Detection of a transformation-related antigen in chemically induced sarcomas and other transformed cells of the mouse. *Proc. Natl. Acad. Sci. USA* **76,** 2420–2424.
109. Crawford, L. V., Pim, D. C., and Bulbrook, R. D. (1982) Detection of antibodies against the cellular protein p53 in sera from patients with breast cancer. *Int. J. Cancer* **30,** 403–408.
110. Soussi, T. (2000) p53 antibodies in the sera of patients with various types of cancer: a review. *Cancer Res.* **60,** 1777–1788.
111. Lubin, R., Zalcman, G., Bouchei, L., Bouchet, L., Tredaniel, J., Legros, Y., et al. (1995) Serum p53 antibodies as early markers of lung cancer. *Nature Med.* **11,** 701–702.
112. Schlichtholz, B., Trédaniel, J., Lubin, R., Zalcman, G., Hirsch, A., and Soussi, T. (1994) Analysis of p53 antibodies in sera of patients with lung carcinoma define immunodominant regions in the p53 protein. *Br. J. Cancer* **69,** 809–816.
113. Trivers, G. E., De Benedetti, V. M. G., Cawley, H. L., Caron, G., Harrington, A. M., Bennet, W. P., et al. (1996) Anti-p53 antibodies in sera from patients with chronic obstructive pulmonary disease can predate a diagnosis of cancer. *Clin. Cancer Res.* **2,** 1767–1775.
114. Winter, S. F., Minna, J. D., Johnson, B. E., Takahashi, T., Gazdar, A. F., and Carbone, D. P. (1992) Development of antibodies against p53 in lung cancer patients appears to be dependent on the type of p53 mutation. *Cancer Res.* **52,** 4168–4168.
115. Wild, C. P., Ridanpaa, M., Anttila, S., Lubin, R., Soussi, T., Husgafvel-Pursiainen, K., and Vainio, H. (1995) p53 antibodies in the sera of lung cancer patients: comparison with p53 mutation in the tumour tissue. *Int. J. Cancer* **64,** 176–181.
116. Davidoff, A. M., Inglehart, J. D., and Marks, J. R. (1992) Immune response to p53 is dependent upon p53/HSP70 complexes in breast cancers. *Proc. Natl. Acad. Sci. USA* **89,** 3439–3442.

117. Lubin, R., Schlichtholz, B., Bengoufa, D., Zalcman, G., Tredaniel, J., Hirsch, A., et al. (1993) Analysis of p53 antibodies in patients with various cancers define B-cell epitopes of human p53: distribution on primary structure and exposure on protein surface. *Cancer Res.* **53,** 5872–5876.
118. Angelopoulou, K., Diamandis, E. P., Sutherland, D. J. A., Kellen, J. A., and Bunting, P. S. (1994) Prevalence of serum antibodies against the p53 tumor suppressor gene protein in various cancers. *Int. J. Cancer* **58,** 480–487.
119. Komiya, T., Hirashima, T., Takada, M., Masuda, N., Yasumitsu, T., Nakagawa, K., et al. (1997) Prognostic significance of serum p53 antibodies in squamous cell carcinoma of the lung. *Anticancer Res.* **17,** 3721–3724.
120. Rosenfeld, M. R., Malats, N., Schramm, L., Graus, F., Cardenal, F., Vinolas, N., et al. (1997) Serum anti-p53 antibodies and prognosis of patients with small-cell lung cancer. *J. Natl. Cancer Inst.* **89,** 381–385.
121. Bergqvist, M., Brattstrom, D., Larsson, A., Holmertz, J., Rosenberg, L., Wagenius, G., and Brodin, O. (1998) p53 auto-antibodies in non-small cell lung cancer patients can predict increased life expectancy after radiotherapy. *Anticancer Res.* **18,** 1999–2002.
122. Mitsudomi, T., Suzuki, S., Yatabe, Y., Nishio, M., Kuwabara, M., Gotoh, K., et al. (1998) Clinical implications of p53 autoantibodies in the sera of patients with non-small cell lung cancer. *J. Natl. Cancer Inst.* **90,** 1563.
123. Iizasa, T., Fujisawa, T., Saitoh, Y., Hiroshima, K., and Ohwada, H. (1998) Serum anti-p53 autoantibodies in primary resected non-small-cell lung carcinoma. *Cancer Immunol. Immunother.* **46,** 345–349.
124. Lai, C.-L., Tsai, C.-M., Tsai, T.-T., Kuo, B. I.-T., Chang, K.-T., Fu, H.-T., et al. (1998) Presence of serum anti-p53 antibodies is associated with pleural effusion and poor prognosis in lung cancer patients. *Clin. Cancer Res.* **4,** 3025–3030.
125. Segawa, Y., Kageyama, M., Suzuki, S., Jinno, K., Fujimoto, N., Hotta, K., and Eguchi, K. (1998) Measurement and evaluation of serum anti-p53 antibody levels in patients with lung cancer at its initial presentation: a prospective study. *Br. J. Cancer* **78,** 667.
126. Laudanski, J., Burzykowski, T., Niklinska, W., Chyczewski, K., Furman, M., and Niklinski, J. (1998) Prognostic value of serum p53 antibodies in patients with resected non-small cell lung cancer. *Lung Cancer* **22,** 191–200.
127. Roth, J. A., Nguyen, D., Lawrence, D. D., Kemp, B. L., Carrasco, C. H., Ferson, D. Z., et al. (1996) Retrovirus-mediated wildtype p53 gene transfer to tumors of patients with lung cancer. *Nature Med.* **2,** 985–991.
128. Swisher, S. G., Roth, J. A., Nemunaitis, J., Lawrence, D. D., Kemp, B. L., Carrasco, C. H., et al. (1999) Adenovirus-mediated p53 gene transfer in advanced non-small-cell lung cancer. *J. Natl. Cancer Inst.* **91,** 763–771.
129. Zou, Y., Zong, G., Ling, Y.-H., Hao, M. M., Lozano, G., Hong, W. K., and Perez-Soler, R. (1998) Effective treatment of early endobronchial cancer with regional administration of liposome-p53 complexes. *J. Natl. Cancer Inst.* **90,** 1130–1137.

5

An Epidemiologic and Clinicopathologic Overview of AIDS-Associated Pulmonary Kaposi's Sarcoma

David M. Aboulafia

1. Introduction

Before the first clinical descriptions of the acquired immunodeficiency syndrome (AIDS), Kaposi's sarcoma (KS) was a rare tumor among Western populations, occurring in only 0.02–0.06% per 100,000 people *(1)*. By June and July of 1981, however, reports from California and New York described large numbers of homosexual men who were afflicted with pigmented skin lesions of KS, either as an initial manifestation of a compromised immune system or following opportunistic infections such as oral candidiasis and *Pneumocystis carinii* pneumonia (PCP) *(2–4)*. Since then, approximately 15–25% of human immunodeficiency virus (HIV)-infected men in the United States have been diagnosed with KS *(5)*. Typically, these tumors involve skin and lymph nodes and, less frequently, visceral organs *(6)*.

The natural history of AIDS-related KS has changed with the widespread use of highly active antiretroviral therapy (HAART). Recent declines in morbidity and mortality due to AIDS have been attributed to the use of these three-drug or four-drug combination antiretroviral regimens, which generally include nucleoside analog reverse transcriptase inhibitors and either protease inhibitors or non-nucleoside reverse transcriptase inhibitors.

This chapter will briefly analyze the changing epidemiology of KS in the HAART era. The pathology and pathogenesis of KS will also be discussed, followed by a discussion on the radiographic and clinical features of pulmonary KS. Pulmonary KS may be difficult to distinguish from other infections or neoplastic conditions, yet the distinction is an important one, for without treatment, patients with this condition in the pre-HAART era had a median survival of only a few months *(7)*. I will also review new developments in the treatment of HIV-associated KS, including the potential to modulate the natural history of this tumor with HAART.

2. Epidemiology of Kaposi's Sarcoma

Moritz Kaposi, a Hungarian dermatologist, described the first cases of KS in 1872 *(8)*. He reviewed the clinical course of 5 men with aggressive "idiopathic multiple pigmented sarcomas of the skin." One patient died of gastrointestinal bleeding 15 mo after the initial appearance of the skin lesions, and an autopsy showed visceral lesions in the lung and the gastrointestinal tract. Classical KS, as it is now called, was later characterized as a slowly progressive disease involving the cutaneous surfaces of the lower extremities. The condition has been found to be more common among elderly men from Eastern Europe (Jewish) or Mediterranean countries.

As early as 1971 KS accounted for roughly 3–9% of all reported malignancies in Uganda *(9)*. Four clinically distinct forms of endemic African KS have been described (*see* **Table 1**). Benign nodular endemic African KS most commonly appears as papules or nodules on the extremities of men in their 40s. Aggressive endemic African KS is also seen more commonly among male patients, but differs from classical KS in that it may affect a younger population and is more likely to spread to visceral organs and lymphatics *(10)*. In addition to the usual number of patients with typical endemic African KS, an increasing number of patients with an aggressive (florid) variant that respond poorly to conventional treatment has been reported *(10)*. Lymphadenopathic African KS may also affect African children in particular. In Eastern and Southern Africa, KS accounts for 25–50% of soft tissue sarcomas in children and 2–10% of all cancers in children *(11)*. It generally grows rapidly and contributes to death within 1–2 yr.

While HIV is also endemic to equatorial Africa, African endemic KS is not related to HIV infection and is a clinical entity distinct from AIDS-KS in Africa. KS in HIV seronegative and HIV seropositive patients is now the most common tumor in central Africa, accounting for 50% of tumors reported in some countries *(12,13)*.

Iatrogenic or transplant-related KS affects patients receiving chronic immunosuppressive therapy such as azathioprine, cyclosporine, or corticosteroids to prevent organ rejection or a variety of other medical conditions *(14)*. It tends to be aggressive, involving lymph nodes, mucosa, and visceral organs in about half of patients, sometimes in the absence of skin lesion *(15)*. This form of KS predominates in men, although less dramatically so than does classical KS. The Cincinnati Transplant Tumor Registry contains data collected during the

Table 1
Epidemiology and Clinical Characteristics of Kaposi's Sarcoma Variants

Type	Epidemiology	Occurrence	Lesions	Distribution	Lymph nodes	Visceral	Behavior
Classical	Mediterranean or Ashkenazic descent; 40–70 yr of age; male:female ratio 10–15:1	0.2% of cancers in US	Some patches, mostly plaques and nodules (usually rounded)	Usually localized to lower extremities; disseminated lesions late in course of disease	Rare	Sometimes	Indolent; gradual increase in number of lesions often associated with lymphedema; visceral lesions occur late, often discovered at autopsy; survival: 10–15 yr
Endemic African	Blacks in equatorial Africa; middle age and children; male:female ratio, 17:1 (adults) and 3:1 (children) Men in 40s	9% of all malignant tumors in equatorial Africa					
1. Benign nodular			Papules and nodules	Multiple usually localized to the extremities	Rare	Rare	Indolent; resembles classical type disease; survival: 8–10 yr

(continued)

Table 1 (continued)

Type	Epidemiology	Occurrence	Lesions	Distribution	Lymph nodes	Visceral	Behavior
2. Aggressive	Younger population, usually men		Large exophytic nodules and fungating tumor	Usually localized to the extremities	Rare	Sometimes	Progressive development of multiple lesions with invasion and destruction of underlying subcutaneous tissue and bone; survival: 5–8 yr
3. Florid			Nodules	Widely disseminated	Sometimes	Sometimes	Rapidly progressive; locally aggressive and invasive, early visceral involvement; survival: 3 to 5 yr
4. Lymph-adenopathic	Children	manifests lesions	Rarely	Minimal	Always	Frequent	Rapidly progressive; survival: 2 to 3 yr
Iatrogenic/transplant immuno-suppression	Immuno-suppressive therapy; any age male:female ratio, 2.3:1	400% greater incidence than in the population at large	Patches, plaques, and nodules	Usually localized to the extremities; rarely disseminated	Rare	Sometimes	Variable; tumor may regress after immunosuppressive therapy is discontinued
Epidemic 1. HIV-associated	Homosexual males and intravenous drug users; 20–50 yr of age; male:female ratio, 106:1	15–35% of AIDS patients in early years of the epidemic	Patches, plaques, nodules; often fusiform and irregular	Multifocal; widely disseminated often symmetric; frequent oral lesions	Frequent	Frequent	Rapidly progressive; survival: 2 mo to 5 yr (median, 18 mo) with visceral disease; cutaneous disease may be indolent or progress gradually
2. HIV-negative	Homosexuals	Currently unknown	Small nodules, patches or	Multifocal; often on extremities	Rare	Rare	Indolent; appears to be more benign than classical type

past 30 years on almost 11,000 recipients of solid-organ transplants, many of whom later developed various malignancies *(15)*. These data indicate that, while KS constitutes a negligible percentage of all cancers among the general population, it constitutes approx 6% of all cancers in solid-organ transplant recipients, appearing a median of 12 mo (average 21 mo) after transplantation.

In male patients of European, Semitic, or African ancestry, the rate of KS after renal transplantation is 500-fold greater than in other populations who undergo this procedure *(14,15)*. KS constitutes approx 80% of all tumors among Saudi Arabian solid-organ transplant recipients *(16)*.

Given the link between iatrogenic or transplant-related KS and immune status, the use of new targeted immunomodulating therapies, which may cause less systemic immunosuppression, may contribute to a reduced incidence of this disease. However, as the number of solid-organ transplants continues to increase, the absolute number of transplant-related KS also may increase *(17)*.

Until recently, the rate of epidemic KS was 100,000-fold greater among patients infected with HIV than that of the general population. This risk appeared most concentrated in men who acquired HIV infection through unprotected sex with men *(18)*. Specifically, KS is 20 times more prevalent in men who have sex with men than among heterosexual HIV-infected hemophiliacs. The disparity between the sexes is reflected by a male-to-female ratio of greater than 20:1 among HIV transmission groups *(19)*.

Reasons for the higher incidence of KS among certain HIV transmission groups has baffled researchers since the beginning of the AIDS epidemic. Investigators have speculated on the role that different sexual practices, exposure to various viruses, hormonal milieu, and class II human lymphocyte antigen-DR 5 antigens might have in promoting KS growth *(19,20)*. More recently, a newly identified herpes virus, termed human herpesvirus-8 or KS-associated herpesvirus (KSHV), has been noted in KS tissues *(21)*. In contrast to other viruses previously linked to KS pathogenesis (including Epstein-Barr virus [EBV], cytomegalovirus [CMV], and human papilloma virus [HPV]), KSHV has been consistently detected in all forms of KS *(22–24)*. In addition, KSHV DNA is present in the lymphoid system, peripheral blood mononuclear cells (PBMCs), saliva, and semen of patients with KS *(25–27)*. In HIV-infected individuals the presence of antibodies to this virus is predictive of KS development *(28,29)*.

In the United States, up to 40% of homosexual men who received an AIDS diagnosis in the early 1980s presented with KS at the time of their initial AIDS diagnosis. Ten years later, the percentage of people with HIV infection who had KS as their initial AIDS-defining illness had declined to approx 10–20% *(19)*. Factors that antedated the HAART era and that may have contributed to this phenomenon include expansion of the AIDS case definition to encompass conditions that may be diagnosed earlier than KS, a decrease in identification or reporting of relatively minor KS lesions, and a decline in exposure to environmental factors associated with KS. In particular, the adoption of "safer" sex practices, which theoretically have reduced the rate of transmission of the putative KS infectious agent KSHV, is considered to be an important modifier of incidence *(19,20)*.

The decline in KS incidence has been even more marked during the HAART era, which began in late 1995 and early 1996. In a prospective study of 6,704 men who have sex with men, Buchbinder and colleagues evaluated the relationship between the rate of new cases of AIDS in general and the incidence of AIDS-defining malignancies *(30)*. Index of AIDS diagnoses per 100,000 patient years fell dramatically from 17.6/100 PY in 1993 to 1.7/100 PY in 1996. Likewise, the risk of death declined dramatically in the same period. A significant decrease in the incidence of KS was reported from 3/100 PY (1993–1997) to 0/100 PY in 1996 ($p = 0.06$).

Data collected from a large multistate observational cohort study, the Adult/Adolescent Spectrum of HIV Disease (ASD) project, indicates that the incidence of KS declined 8.8% per yr between 1990 (observed incidence, 4.1/100 PY) and 1998 (observed incidence, 0.7/100PY) *(31)*. The ASD data analysis shows that the use of antiretroviral therapy is associated with reduced risk for the development of AIDS-KS ranging from a 13% reduction with monotherapy or dual therapy to a 59% reduction with triple therapy (*see* **Table 2**). Improvements in HAART have resulted in prolongation of the duration of HIV infection prior to the development of profound immunosuppression, one of the pathogenic factors in the development of KS *(32,33)*.

3. Incidence of Pulmonary KS

A broad range of pulmonary diseases complicate the course of HIV-infected patients. These include opportunistic and bacterial infections, neoplasms, and non-HIV-associated pulmonary disorders *(34,35)*. The prevalence of pulmonary KS in patients with AIDS is unclear because radiographic changes due to KS have often been attributed to other pulmonary processes *(36–39)*. Definitive antemortem diagnosis necessitates lung biopsy, but this must be counterbalanced by other factors such as the clinician's differential diagnosis and the patient's willingness to undergo an invasive procedure that may be associated with such iatrogenic complications as pain, hemorrhage, or pneumothorax. When postmortem examinations have been conducted on HIV-infected patients already known to have cutaneous disease, the prevalence of pulmonary involvement has ranged from 47–75% *(40–43)*. In other clinical studies, pulmonary KS has been diagnosed by fiberoptic bronchoscopy in 8–14% of patients with AIDS and respiratory symptoms, in 6–32% of patients with AIDS who had cutaneous

Table 2
Effect of Antiretroviral Therapy on the Risk of Development of AIDS-KS[31]

Antiretroviral therapy	Number of patients	Relative risk (95% CI)
Mono or dual therapy	21,080	0.87 (0.78–0.97)
Triple therapy	802	0.41 (0.35–0.98)

CI, Confidence Interval.

KS, and in 21–49% of HIV-infected persons with known mucocutaneous KS and respiratory symptoms *(6,7,43–48)*.

Pulmonary KS without mucocutaneous involvement is widely regarded a rare event *(49)*. The incidence of pulmonary KS without mucocutaneous involvement is unknown but has likely been underestimated. Physicians rarely consider this diagnosis unless mucocutaneous disease is also present. The frequency of isolated pulmonary KS has ranged from 0–15.3% *(39,42,43,47,50)*. As with the incidence of KS in general, the actual frequency of pulmonary KS in North America and Europe will continue to decline in the short term. Nevertheless, with a prevalence of 20,000 cases in the United States alone, KS will remain a clinical problem in the foreseeable future *(50,51)*. Factors that may ultimately contribute to the durability of KS control will likely be interrelated to the changing clinical expression and natural history of AIDS and the long-term effectiveness of HAART in the backdrop of emerging trends in HIV-viral resistance *(52)*. Without an effective HIV vaccine strategy, third-world countries will likely continue to document large case loads of pulmonary KS *(24,53)*.

4. KS Pathology

Although the clinical expression of disease may vary, KS histopathology does not differ among the various risk groups *(54)*. Furthermore, the histopathology of KS involving the lymph nodes and viscera is similar to that affecting the skin. KS is an angioproliferative tumor that is characterized by slit-like neovascular processes and the presence of proliferating endothelial cells, fibroblasts, infiltrating leukocytes, and a population of spindle-shaped tumor cells *(55)*. In the very early stages, cutaneous KS is characterized by inflammatory cell infiltration, extravasation of red blood cells, endothelial cell activation, and angiogenesis. Later, typical spindle-shaped cells appear that represent a heterogeneous population dominated by activated endothelial cells mixed with macrophages and dendritic cells (*see* **Fig. 1**) *(56,57)*. Mononuclear cell infiltrates are seen, especially in younger lesions *(58)*. In advancing lesions,

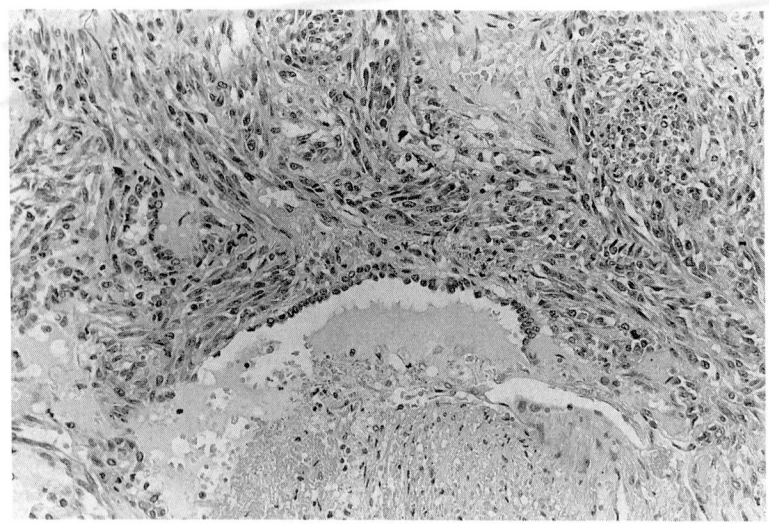

Fig. 1. Spindle tumor cells with slit-like small vascular spaces replace smooth muscle around this bronchiole.

spindle cells tend to become the predominant cell type, although angiogenesis remains always a prominent feature (see **Fig. 2**) *(59)*.

Pulmonary KS can have classic, telangiectatic, and inflammatory microscopic forms, all of which have been reported before the era of AIDS *(60)*. The histopathological differential diagnosis includes both inflammatory conditions and other vascular proliferations such as capillary hemangiomatosis, pulmonary epithelioid hemangioendothelioma (also called intravascular sclerosing bronchoalveolar tumor), primary or metastatic angiosarcoma, and pulmonary artery sarcoma *(40,61)*.

Microscopically, pulmonary KS lesions may be subtle, particularly when focal *(62)*. Attention must be focused on the areas of expected lymphatic routes. In the more solid regions, spindle cells are in loose fascicles, which may interdigitate. Slit-like spaces often do not have identifiable endothelial cells or lining tumor cells, but do possess extravasated red blood cells. The smooth muscle of the bronchioles (see **Figs. 1** and **2B**) and pulmonary arteries (see **Figs. 2A** and **3**) may be infiltrated by tumor, which may mimic granulation tissue. Fairly extensive acute intra-alveolar hemorrhage may also be present, especially at time of death. Nodular forms of the disease possess abundant spindle cells and vascular clefts. Mitoses are not prominent, but tumor nuclei are elongated, moderately dark, and not greatly enlarged, although they rarely

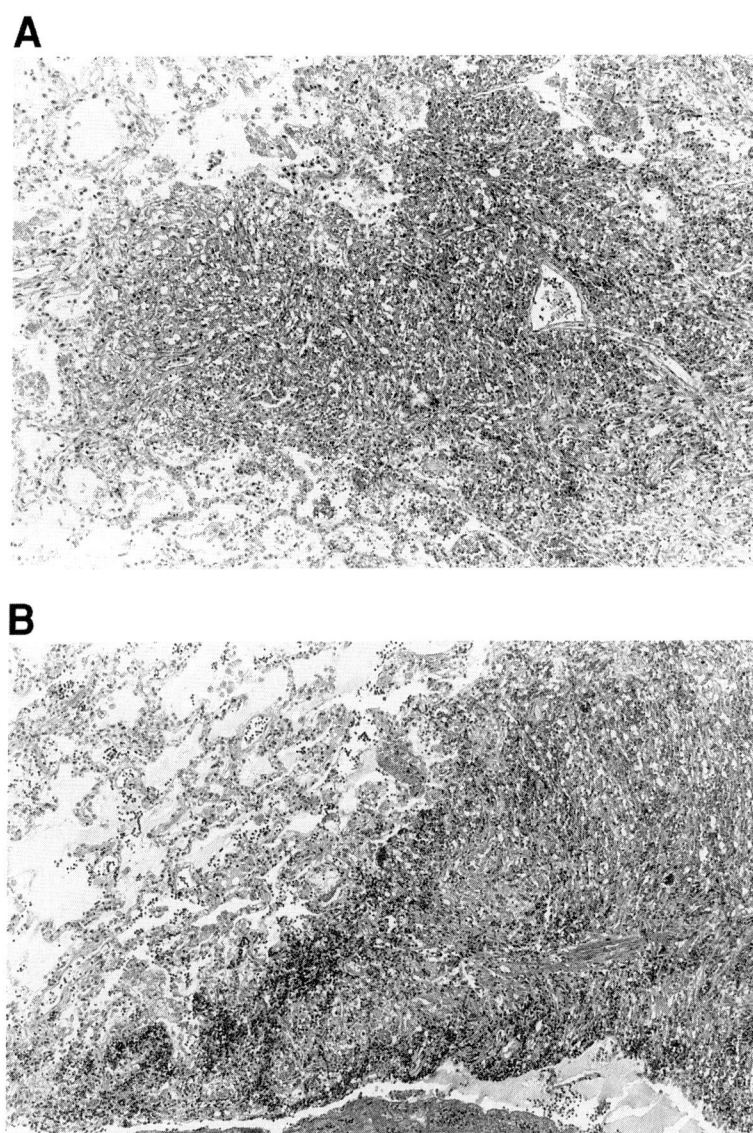

Fig. 2. Kaposi's sarcoma following lymphangitics. (**A**) Around vessels. (**B**) In bronchi and in each location, it replaces the smooth muscle with spindle tumor cells and small vascular spaces, often filled with blood.

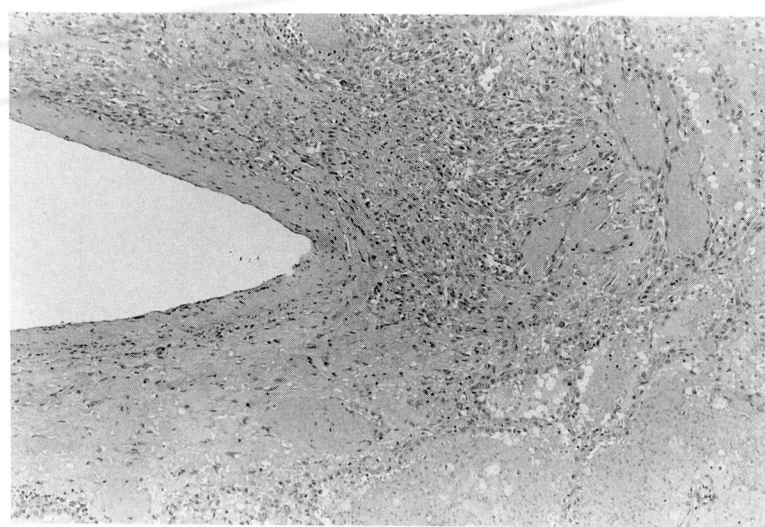

Fig. 3. Kaposi's sarcoma is seen here filling lymphatic spaces halfway around this pulmonary artery. Small vascular lakes are at right and represent the telangiectatic variant of tumor.

possess very anaplastic features. Lung necrosis is rarely caused by KS, but rather the result of coexistent lung infection *(49)*.

Grossly, the surface of the lungs may show flat to slightly raised disk-shaped red to violaceous plaques confined to the visceral pleura (*see* **Fig. 4**). The most striking parenchymal changes include lymphangitic thickening of tumor causing a red to red-blue discoloration about interlobular septa and bronchopulmonary rays (*see* **Fig. 5**) *(40)*. Bronchoscopically, similar bright to dark red to violaceous lesions may be seen in the mucous membranes of the transbronchial tree. Occasional red to purple to gray nodules of various sizes may coalesce to form larger tumor densities. The nodes may be involved with spongy red to gray material replacing the usual translucent tan architecture. Pleural effusions typically are exudative and often contain reactive mesothelial cells without evidence of neoplasm. At autopsy, effusions are rarely an isolated event; in addition to KS (which infrequently involves the pleura), there is also a mixture of other disease events such as opportunistic infections and respiratory distress syndrome *(62,63)*.

5. KS Pathogenesis

A detailed understanding of the factors that contribute to KS initiation and growth occurred only after laboratory techniques were perfected to sustain the

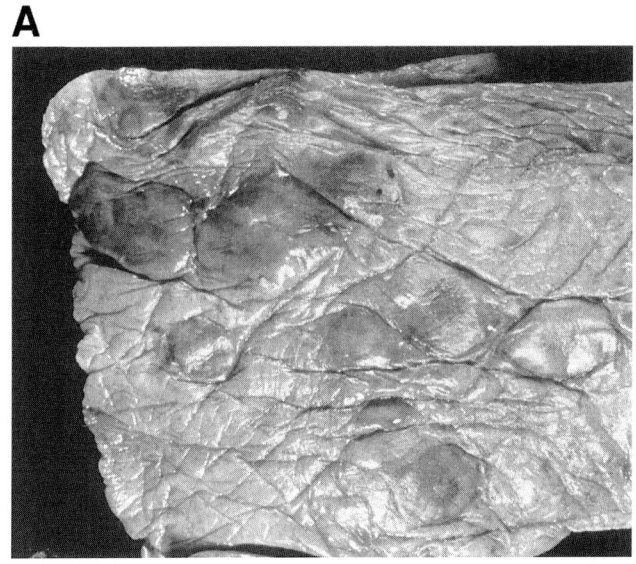

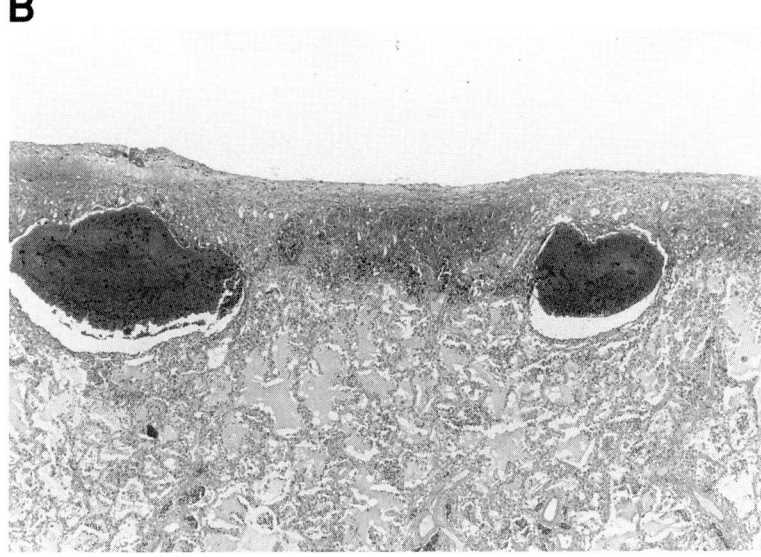

Fig. 4. Pleural involvement. (**A**) Grossly vascular lakes appear on pleura as dome-shaped smooth purple mounds. (**B**) Microscopically these often are telangiectatic dilatations of tumor in pleura.

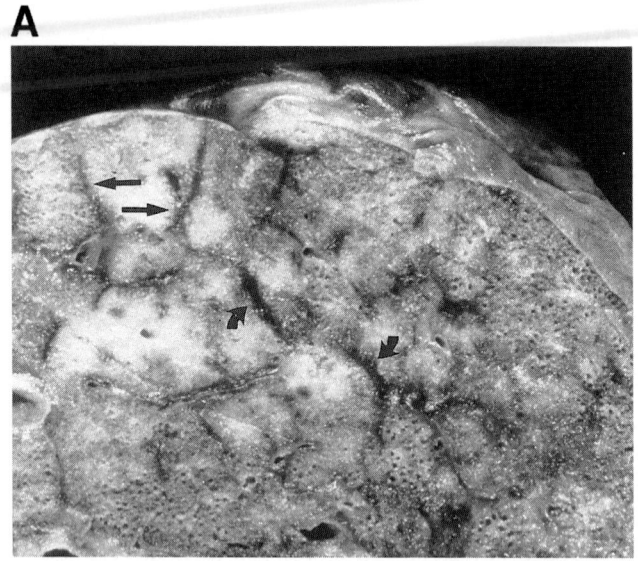

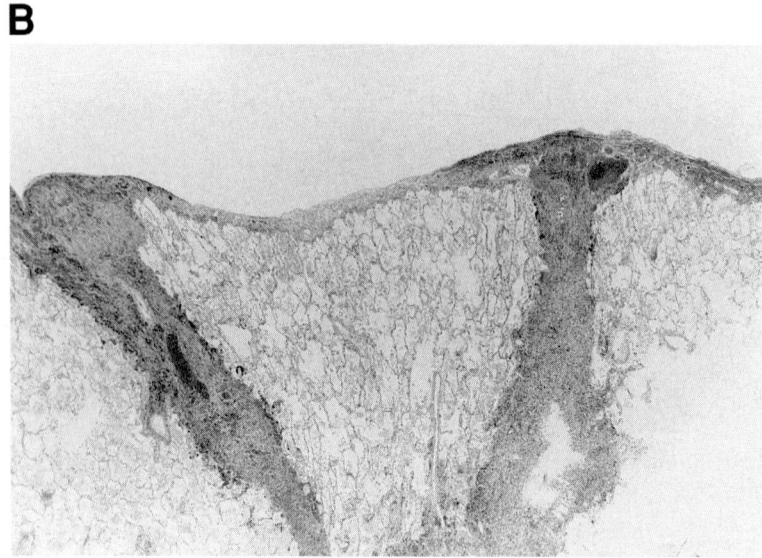

Fig. 5. Interlobular septal involvement. (**A**) Kaposi's sarcoma is seen following lymphatics, grossly darkening interlobular septa (arrows) and microscopically doing same (**5B** low power, **5C** medium power) and again note blood-filled spaces.

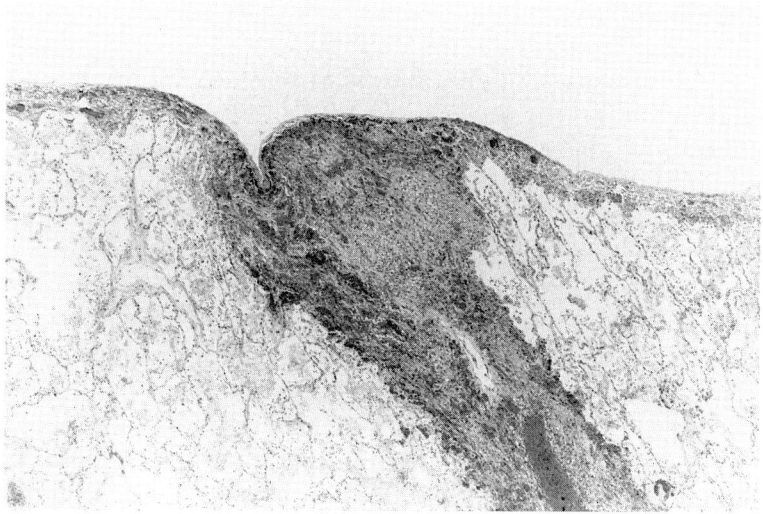

Fig. 5C.

growth of a large number of KS cells in culture *(64)*. Cultured spindle cells from KS lesions are especially responsive to a variety of growth factors that transform culture media and promote normal endothelial cells to acquire the features of the KS phenotype *(65)*. These factors, which include Oncostatin M, gamma interferon IFN-γ, interleukins (IL) -1, -6, and -8, tumor necrosis factor (TNF), granulocyte macrophage-colony stimulating factor (GM-CSF), platelet derived growth factor, basic fibroblast growth factor (bFGF), and vascular endothelial growth factor (VEGF) are also found in the inflammatory cell infiltrates of KS lesions *(66–68)*. The fact that this inflammatory infiltrate can promote KS growth was elegantly illustrated in experiments whereby nude mice were inoculated with KS cells and the media used to support growth of the cells and subsequently developed cutaneous KS-like lesions *(69)*.

Early in its growth, KS is better characterized as a hyperplastic proliferative disease than as a true cancer *(69,70)*. BFGF and VEGF mediate spindle cell growth, angiogenesis, and edema of KS. The abnormal cytokine milieu of HIV infection and the mitogenic effects of HIV-*tat* protein may further act synergistically to stimulate KS spindle cell growth through autocrine and paracrine loops, leading to an increased frequency and aggressiveness of KS *(69,70–74)*. However, the recognition that KS lesions are clonal and that

KS-like cells can be detected in the peripheral blood of patients with AIDS-associated KS provides evidence for the malignant potential of KS spindle cell proliferation *(75–77)*.

How KSHV stimulates KS growth remains uncertain, but a number of clues are emerging. The genetic sequence of KSHV largely has been determined and portions of its genome are analogous to DNA sequences believed to have oncogenic potential *(78)*. The bcl-2 family of proteins modulate apoptosis and KSHV DNA codes for a functional bcl-2 homolog *(79)*. Dysregulation of bcl-2 may contribute to neoplastic cell expansion via an anti-apoptotic effect that enhances cell survival rather than by accelerating rates of cellular proliferation. The KSHV-encoded G-protein coupled receptor (GPCR) may also act as a viral oncogene. In conjunction with VEGF, it appears capable of inducing angiogenesis in transformed mouse kidney cells containing the KSHV-GPCR gene *(80)*. In nude mice, the introduction of KSHV-GPCR transformed kidney cells results in tumor formation. KSHV may also code for proteins that mimic human cytokine and cytokine response pathways (including IL-6 and macrophage inhibitory protein-1) or stimulate supporting cells to produce angiogenesis factors *(81,82)*.

Multiple factors contribute to the creation of an inflammatory-angiogenic environment. According to one model *(83)*. circulating KS progenitors and cells latently infected with KSHV migrate to inflammatory sites. Exposure to various inflammatory cytokines results in dedifferentiation of these latently infected cells into KS-like spindle cells and induces KSHV reactivation. Reactivation of KSHV leads to the expression of pathogenic early genes, including viral IL-6, which can activate VEGF and induce angiogenesis. Viral lytic replication in the same cells activates inflammation, which may also stimulate angiogenesis. The HIV-1 Tat protein enhances this inflammatory-angiogenic state by increasing the angiogenic activities of VEGF, bFGF, and IFN-γ, and by increasing the expression of matrix metalloproteinases (MMPs) *(84)*.

6. Natural History of Pulmonary KS

Prior to HAART, patients with pulmonary KS faced an uncertain future. It was assumed that they had a shorter life-expectancy than other AIDS patients, yet important factors such as co-morbid illness and immune profile were not consistently taken into account *(85–87)*. Gill and colleagues reported a median CD4+ count of 94 cells/µL in 20 patients with pulmonary KS *(48)*. Gruden and associates reported the median CD4+ count was 34/µL; 84% of patients had counts <100 cells/µL, 96% had counts <150 cells/µL *(88)*. In the largest series describing the clinical presentation of pulmonary KS, investigators from San Francisco reported a median CD4+ count of 19 cells/µL among 168 patients

(50). These and other studies imply that pulmonary KS is confined primarily to patients with marked immune impairment, but in each of these analyses patients had radiographically and bronchoscopically advanced disease and may have had early KS at a time when CD4+ cell counts were at a higher level. As immune impairment worsens, so too may the clinical expression of KS *(89)*.

Several studies suggest that patients with KS as their first AIDS-defining illness may actually have a longer survival than patients without KS *(90,91)*. Early in the AIDS epidemic, the median CD4+ count was higher at diagnosis of KS than at diagnosis of other AIDS-defining opportunistic infections. This factor may have accounted for the observed survival differences. However, most analyses indicate that patients with KS have a diminished survival. In a retrospective analysis of 241 HIV-infected homosexuals with KS compared to 241 HIV-infected controls, the KS group had a greater number of opportunistic infections (5.99 vs 3.88 infections) and an increased risk of death (odds ratio, 1:28), even when adjusting for age, previous opportunistic infection, baseline CD4+ cell count, and antiretroviral therapy *(92)*.

The implementation of a uniform staging system for classifying patients with AIDS-related KS has been difficult. This tumor, unlike other cancers, is largely affected by the underlying HIV infection, which influences growth as well as overall outcome. Comparative assessment of the efficacy of different treatment regimens was historically compromised by the lack of established criteria for classifying extent of disease, tumor stage, and response to treatment. In 1989, the National Institute of Health (NIH)-sponsored AIDS Clinical Trials Group (ACTG) developed a system for classifying AIDS-related KS in order to categorize patients more effectively for clinical trial participation and subsequent evaluation *(93)*. Stratifying patients into good or poor-risk groups, this three-tiered staging system characterized disease severity according to the TIS system: clinical extent of tumor (T), immunologic status (I), and evaluation of HIV-related systemic illness (S). More recently, Krown et al. conducted a prospective validation of the original TIS staging classification developed by the ACTG in order to reflect its impact on patient survival *(94)*. The analysis demonstrated that patients with KS confined to the skin and/or lymph nodes who possess minimal oral KS lived significantly longer than patients with visceral KS, bulky oral KS, or tumor-associated edema (27 vs 15 mo; $p < 0.001$). A change in CD4+ count from 200 to 150 cells/µL or lower was the only modification needed to distinguish between the good and poor immune system categories. Patients with a CD4+ count >150 cells/µL had a median survival of 39 months where those with <150 cells/µL survived a median of only 12 mo ($p < 0.001$). **Table 3** illustrates the recommended staging classification according to these criteria.

Table 3
Revised ACTG Staging Classification for Kaposi's Sarcoma (94)

Staging	Good risk (0) (All of the following)	Poor risk (1) (Any of the following)
Tumor (T)	Confined to skin and/or lymph nodes and/or nodular oral disease confined to the palate	Tumor-associated edema or ulceration Extensive oral KS Gastrointestinal KS KS in other non-nodal viscera
Immune system (I)	CD4 cells >150/mL	CD4 cells <150/mL
Systemic illness (S)	No history of opportunistic infection or thrush No "B" symptoms (unexplained fever, night sweats, >10% involuntary weight loss, or diarrhea) persisting more than 2 wk Performance status 70 (Karnofsky)	History of opportunistic infections and/or thrush "B" symptoms present Performance status <70 Other HIV-related illness (e.g., neurologic disease, lymphoma)

The revised CD4+ cut-off of 150 cells/mL is lower than the original proposal of 200 cells/mL. Example of staging: a patient with KS restricted to the skin, CD4+ count of 10 cells/mL, and a history of *Pneumocystis carinii* pneumonia would be $T_0I_1S_1$.

Surprisingly, among patients with extensive KS who were enrolled in ACTG KS studies, pulmonary involvement was not associated with diminished survival. The 160 T_1 subgroup of patients without lung involvement possessed a 16-mo median survival, whereas the 25 T_1 patients with established lung involvement had a 14-mo median survival ($p = 0.59$). These results are different than the usual 4- to 8-mo survivals that have been reported in pulmonary KS clinical trials (see below). Poor prognostic factors for survival among patients with thoracic KS in non-ACTG clinical trials have included: 1) absence of cutaneous KS; 2) prior opportunistic infections; 3) CD4+ <100 cells/mm^3; 4) leukopenia and anemia; and 5) large pleural effusions *(48,95)*.

Krown and colleagues were cautious not to trumpet the relatively good prognosis they noted among patients receiving treatment for pulmonary KS. Their analysis contained only a small number of patients with documented pulmonary disease, which limited their ability to detect survival differences if such differences existed. The data may also have been skewed either because of selection bias or under-ascertainment of pulmonary KS. Patients with the most severe pulmonary KS may have been excluded from participation in trials for a number of reasons (e.g., low performance status, prior treatment, inadequate pulmonary function making it difficult to enroll into clinical trials, or a perceived need to treat immediately with noninvestigational agents before trial entry evaluations could be completed) *(94)*. Some patients with early or subtle forms of pulmonary KS may have gone undetected, even after undergoing radiographic and scintigraphic scans and bronchoscopy.

7. Clinical Manifestations of AIDS-Associated KS

The clinical features of epidemic KS differ markedly from those seen in classical and endemic African forms (*see* **Table 1**) *(96,97)*. AIDS-KS tends to be multicentric, often involving mucous membranes along the entire gastrointestinal tract and occurring in atypical locations. Patients may have small, innocuous-looking skin blemishes that are easily overlooked. Alternatively, large and complex skin lesions may be scattered over the body, manifested as red, purple, or brown patches, plaques, or nodules. In some patients, only a few skin lesions are apparent, and the lesions may remain unchanged for several years; in others, lesions appear rapidly, particularly during a period of heightened immunosuppression or illness. In most patients, new lesions appear gradually during a period of several weeks to months.

KS often involves the head and neck, including the tip of the nose and the retroauricular and periorbital areas. KS involving the eyelid or conjunctiva can interfere with vision *(98)*. Careful examination of the oropharynx may uncover clinically silent hard and soft palate lesions that, if allowed to grow, can interfere with eating and, rarely, can cause airway obstruction. Lymphatic involvement may produce debilitating and cosmetically unacceptable edema, particularly in the periorbital areas, genitalia, and lower extremities. Edema may be complicated by skin breakdown and cellulitis. Foot lesions are common.

Lung KS is often insidious in onset. As with most pulmonary processes in patients with AIDS, the clinical presentation of intrathoracic KS is nonspecific and may be indistinguishable from pneumonia *(99,100)*. Dyspnea and cough are the most common presenting symptoms, and >50% of patients with respiratory symptoms have previously treated cutaneous lesions *(36,43–50,85,99–103)*. Fever and night sweats may be present and usually suggest a concomitant infection. Hemoptysis may also occur, although its rate varies from series to series. Among 30 AIDS patients with symptomatic pulmonary KS, 100% had dyspnea, 80% cough, 47% chest pain, 30% hemoptysis, and fever >38.5°C was seen initially in 4 of 30 (13%) *(95)*. In one report of 7 women with AIDS, pulmonary KS caused diffuse lung disease and was usually mistaken clinically for pulmonary infection *(104)*. In medically underserved or economically disadvantaged areas such as rural Africa, hemoptysis may be a more frequent event, perhaps reflecting the severity of disease and its late presentation and fundamental differences in host-virus interactions *(53,105,106)*. When hemoptysis does occur, its origin is not always obvious. Meduri and colleagues were able to identify the site of bleeding via bronchoscopy in only a minority of patients with pulmonary KS *(36)*.

Hoarseness and stridor are relatively infrequent problems in patients with pulmonary KS. When present, these symptoms usually connote tumor involvement of the trachea or larynx. However, life-threatening narrowing of the tracheobronchial airways due to KS growth is fortunately rare and was not seen in one series of 168 patients with bronchoscopically diagnosed KS *(50)*. In the absence of upper respiratory tract KS or pleural effusion, physical examination of the thorax is usually normal *(44)*.

Some investigators believe that wheezing, hemoptysis, pleuritic chest pain, and stridor are more likely due to KS than to opportunistic infections, and that pleural effusions (which in the setting of KS typically develop over 1–2 wk) and pulmonary nodules are more likely due to an infectious etiology *(35,36,46)*. Others contend hemoptysis presages a worse prognosis *(106,107)*. In contrast, Huang and colleagues found that fever ≥38.5°C and elevated serum LDH concentration were the only clinical or laboratory parameters that distinguished patients with isolated pulmonary KS from those with superimposed opportunistic infection *(50)*.

Three important points emerge when reviewing the intrathoracic manifestations of HIV-related KS: 1) respiratory involvement nearly always follows rather than precedes the development of mucocutaneous lesions; 2) the frequency of pulmonary involvement that can be documented at autopsy (50%) is higher than that detected clinically (33%); and 3) approx two-thirds of patients with known KS who present with new pulmonary findings actually have a coexisting, usually treatable, opportunistic infection *(88,108)*. The latter observation punctuates the importance of evaluating each patient with KS and pulmonary complaints not only for intrathoracic spread of the malignancy, but for respiratory tract infections as well.

When assessing a patient with AIDS and respiratory symptoms, I wish to emphasize that although most patients with pulmonary KS will have cutaneous, mucocutaneous, or endobronchial involvement, this is not invariably true *(95,106,109)*. Among the first descriptions of AIDS-associated pulmonary KS was the case of a male homosexual with fever, weight loss, diarrhea, no skin lesions, and a chest roentgenogram that revealed bilateral nodular interstitial infiltrates *(110)*. Isolated pulmonary KS may present in a variety of fashions ranging from a slow-growing and asymptomatic peripheral nodule without accompanying adenopathy or pleural effusions *(35,111)*. to a rapidly progres-

Table 4
Chest Radiographic Features of Pulmonary Kaposi's Sarcoma *(114,115)*

Parenchyma	Reticulonodular infiltrates due to tumor nodules.
	Diffuse interstitial infiltrates or linear/septal angiomatous infiltration.
	Focal air space consolidation or collapse. Parenchymal lesions may appear normal, particularly in the early stages of disease.
Pleura	Pleural effusions on one or both sides of the chest that may vary considerably in size.
Lymphadenopathy	In advanced pulmonary Kaposi's sarcoma, 10–20% of patients have enlarged hilar or mediastinal nodes.

sive interstitial infiltrate culminating in acute respiratory failure *(112)*. It may also be the sole cause of persistent fever in a patient with AIDS *(113)*.

9. Radiologic Findings of Pulmonary KS

Table 4 summarizes the classic chest radiographic features of pulmonary KS *(114,115)*. In 5–20% of cases, chest roentgenograms may be normal, but more commonly they reveal thickening along bronchovascular bundles, often emanating from a perihilar origin. As KS grows, a reticulonodular infiltrate appears, mainly in the lower lobes (**Figs. 6A,C**) *(88,116)*. With continued growth, nodules become irregular and confluent and this, along with interstitial infiltrates, leads to dense air-space consolidation *(43)*. Hilar or mediastinal lymphadenopathy is evident in 10–16% of patients with pulmonary KS *(88,114)*. Pleural effusions are present in as many as 50% of cases and are most often large, bilateral, and associated with parenchymal lesions.

A staging system for pulmonary KS exists based on radiographic findings *(88)*. Among 76 patients with pulmonary KS, coarsened bronchovascular bundles tended to coalesce, small nodules and effusions became larger, and changes in previously abnormal lung sounds became more prominent as tumors grew. Although degree of radiographic abnormalities usually correlated with findings on endoscopy, there was often substantial variation among patients with regard to tumor growth. Occasionally, parenchymal disease was present on radiographs even when no detectable tracheobronchial lesions were noted on endoscopy.

The striking radiographic characteristics of intrathoracic KS provide important clues to the radiologist. In a study of 102 HIV-infected patients with

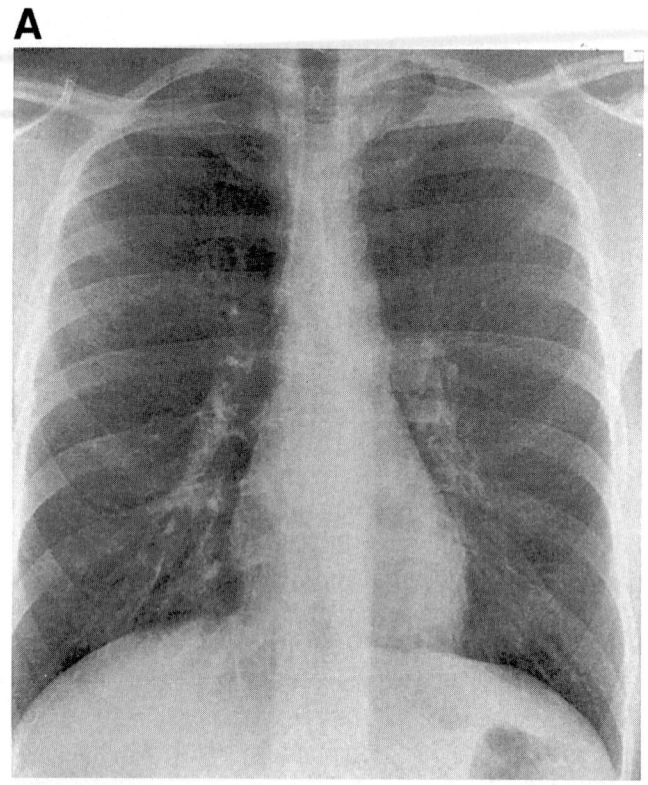
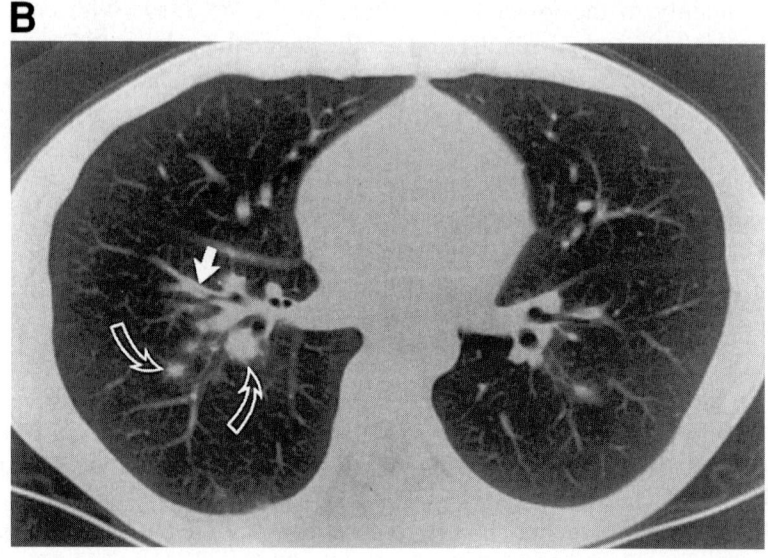

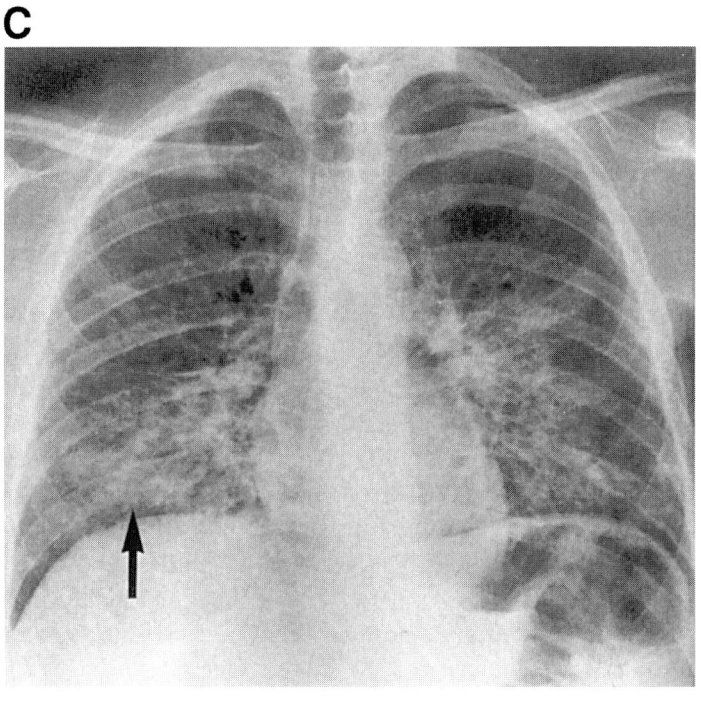

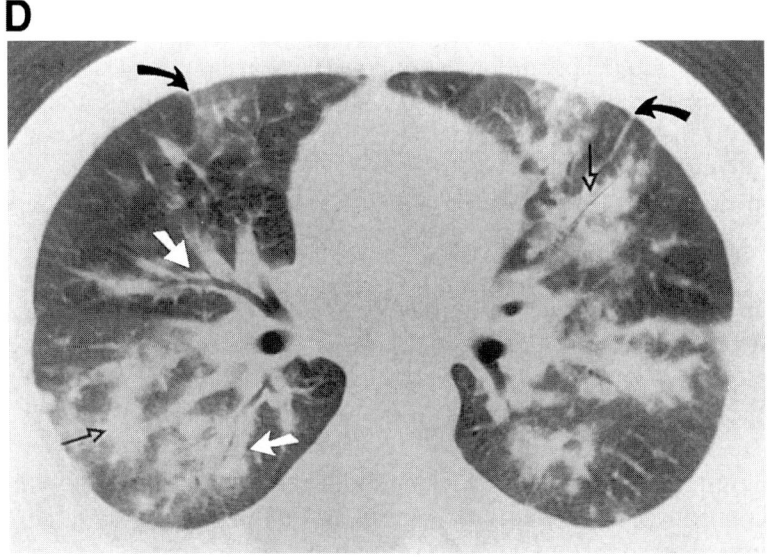

Fig. 6. Early and progressive pulmonary Kaposi's sarcoma (KS) in AIDS. (**A**) Chest radiograph reveals subtle bronchial wall thickening of the infrahilar regions. (**B**) CT image through the infrahilar regions demonstrates peribronchovascular thickening characteristic of KS (arrow), and irregular thickening of the interface between the vessel and lung (curved arrows). (**C**) Chest radiograph demonstrates advanced pulmonary KS with more nodular thickening of the bronchovascular perihilar mid- and lower lung. Confluence of nodules in the right lower lung leads to airspace consolidation (arrow).

documented thoracic disease and 20 patients without thoracic complications, radiologists achieved a 90% accuracy score when asked to interpret the radiologic findings while not knowing the clinical diagnosis *(116)*.

Computerized tomographic (CT) scans of the chest will often identify bronchial wall thickening and spiculated lesions in patients with pulmonary KS (**Figs. 6B,D**). Among 53 patients with pulmonary KS, such abnormalities occurred in 66% and 79% of cases, respectively *(117)*. CT scans also show the distinct pattern of pulmonary KS following bronchovascular pathways, whereby poorly marginated nodular infiltrates radiate out from both pulmonary hilum along the bronchovascular structures into surrounding interlobular septa *(118)*. CT scans detect enlarged lymph nodes (15–53% of cases) more readily than chest roentgenograms *(117–119)*. and provide greater detail regarding the presence of intrapulmonary chest disease. Eight of 15 (53%) patients with pulmonary KS also had sternal, rib, thoracic spine, and/or subcutaneous tissue involvement incidentally detected by CT scanning *(119)*. Other radiographic findings such as pericardial effusions and septal lines are infrequent; ground-glass opacities are very unusual and suggest alveolar hemorrhage *(120)*.

CT scans offer advantages over conventional chest roentgenograms not only in terms of identifying KS, but by providing greater detail regarding other lung diseases *(121)*. This difference is, however, usually modest and in most patients a chest roentgenogram provides adequate information. CT scan is less valuable for the detection of endobronchial tumors. Only the largest of lesions are easily identified by this modality.

For patients with significant dyspnea, the technique of spiral CT scanning is particularly useful because of its shorter imaging time and higher resolution images *(115)*. Although not extensively evaluated in the setting of HIV infection, one study did compare this modality to gallium scanning in patients with AIDS and pulmonary symptoms and normal or equivocal chest radiographs. High resolution CT scanning was most helpful in guiding the method of biopsy and directing the bronchoscopist to the diseased lung segment that would maximize diagnostic yield *(122)*.

Gallium-thallium radionuclide imaging is also used in patients with AIDS to distinguish KS from other processes *(123)*. KS is thallium-avid, but unlike other infectious or neoplastic AIDS-related complications, does not take up gallium. The combination of focal KS pulmonary involvement and concurrent infection can occasionally lead to a negative thallium and a positive gallium pattern scan *(124)*. Like gallium, indium-111 labeled polyclonal immunoglobulin also localizes to infection but does not accumulate in KS or lymphoma *(125)*. Radiologists who have experience in interpreting these scans can sometimes provide useful information when evaluating a patient with an uncertain diagnosis. In practice, only a few medical centers in the United States use these studies to establish the diagnosis of pulmonary KS. This is due to the high cost of such tests and the accuracy of CT scan and bronchoscopy in the identification of characteristic KS lesions.

The magnetic resonance imaging (MRI) features of pulmonary KS have not been extensively detailed. In one study of patients with established pulmonary KS, characteristic findings of MRI included an increased signal intensity on T1-weighted images, markedly reduced signal intensity on T2-weighted images, and strong lesional contrast enhancement after administration of gadolinium *(126)*. This pattern of signal abnormalities, particularly when seen in a peribronchial vascular distribution, was most suggestive of KS. Further analyses will better define how specific this finding is for pulmonary KS and what role MRI may have in evaluating pulmonary disease in patients with HIV infection.

Positron emission tomography with 18-fluorodeoxyglucose (FDG-PET) is a functional imaging technique used to localize malignant lesions by detecting increased cellular metabolic activity. It has not been well-studied in the setting of HIV infection. In current practice, FDG-PET has high sensitivity and intermediate specificity for non-AIDS-related malignancies *(127)*.

10. Clinical Evaluation of Cutaneous and Pulmonary KS

Initial evaluation of a patient with KS includes a physical examination, with particular attention given to the skin and rectal and oral cavities. Clinically suspected AIDS-related KS should be confirmed by biopsy and histologic examination of a skin lesion, lymph node, or other tumor-involved tissue. Biopsies are important for excluding other diseases that may mimic the appearance of KS, including bacillary angiomatosis, cellulitis, vasculitis, or other angiopathic lesions *(128,129)*. Bacillary angiomatosis *Bartonella* organisms can be identified by Warthin-Starry silver staining. A chest roentgenogram and routine blood tests, including CBC, serum albumin, cholesterol, CD4+ T-lymphocyte cell count, and HIV viral load, may help stratify patients into good or poor prognosis groups. Additional studies (such as upper and lower gastrointestinal tract evaluation and CT scans) are sometimes necessary to exclude other conditions but are rarely necessary as part of a routine staging evaluation. Symptomatic gastrointestinal involvement is best evaluated with endoscopy because barium contrast studies often produce false-negative findings *(130)*.

Pulmonary KS can involve the tracheobronchial tree, the pulmonary parenchyma, and, infrequently, the pleura. It is the most common endobronchial lesion associated with HIV and has a characteristic red or purple macular or papular appearance often located at airway bifurcations *(131)*. If parenchymal pulmonary lesions are not seen on chest radiographs, it is rare for bronchoscopy to identify KS lesions below the level of the carina *(132)*. In a retrospective analysis of 76 patients with bronchoscopically proven KS, a correlation was noted between the extent of endobronchial and radiographically documented parenchymal disease *(109)*. A strong correlation between endobronchial lesions and underlying parenchymal KS has also been observed in a small number of patients at autopsy *(36,107,133)*. The presence of characteristic tracheobronchial KS is thus sufficient to make a presumptive diagnosis of pulmonary KS *(50)*. When confronted with an atypical endobronchial lesion or in rare cases in which there are no cutaneous manifestations of KS, bronchial biopsy may be indicated, although one series reported a 30% incidence of significant hemorrhage *(36)*. This risk may be greatest when lesions located in the central airway are sampled *(60,134)*.

Although direct inspection of lesions at bronchoscopy is the most sensitive technique available for establishing a diagnosis of pulmonary KS, only 45–73% of cases have readily identifiable endobronchial lesions *(44,45)*. Parenchymal lesions may occur in the absence of tracheobronchial lesions and they too may be missed at bronchoscopy *(130)*. Reasons why bronchoscopic evaluation of the airways may not provide a diagnosis of pulmonary KS include the following: 1) the bronchoscope may not be advanced far enough to detect distal airway disease; 2) KS may not involve the bronchi or adjacent tissue; 3) the lesions may not extend into the submucosal space where they can be visualized; and 4) the interstitial involvement of KS may be microscopic *(95,108,117)*.

Both endobronchial and transbronchial biopsy have a diagnostic yield of only 26–60% because of the patchy submucous nature of the lesions *(47,49,63)*. Histological identification is difficult because the lesions are composed of spindle cells and blood vessels, some of which may appear normal. Because of the paucity of malignant features, biopsies may be misinterpreted as reactive fibrous tissue *(135)*.

Alveolar hemorrhage (detected in bronchoalveolar lavage fluid) is sometimes associated with pulmonary KS. This is a nonspecific finding and is associated with a variety of infective and noninfective HIV-related pulmonary diseases *(131,136,137)*. In the absence of pulmonary infection, studies on bronchial alveolar lavage from AIDS patients with pulmonary KS have shown a high frequency of alveolar hemorrhage *(49)*. Hemorrhagic or chylous pleural effusions are suggestive of KS *(120)*. Biochemical analyses are usually exudative on the basis of protein and LDH, but transudates may also be recovered. In a recent study of the radiographic features of pulmonary KS, pleural involvement occurred only in the presence of parenchymal abnormalities *(132)*. Thoracentesis rarely leads to a diagnosis of KS, but is important to perform in order to exclude the possibility of malignancy or pyogenic, mycobacterial, or fungal infection *(115,133)*.

AIDS-Associated Pulmonary KS

Open lung biopsy has a diagnostic yield of approx 50%, but it is rarely performed due to pain and other potential complications associated with this procedure *(47–49,133)*. A successful thorascopic biopsy obviates the need for diagnostic thoracotomy, although results in the setting of HIV-infection have not been extensively published *(138)*.

An investigational approach to aid in the diagnosis of pulmonary KS involves polymerase chain reaction (PCR)-based localization studies for KSHV. Among 9 patients with KS, normal skin and lung did not reveal KSHV infection, but diseased tissue showed KSHV-specific infection of endothelial cells and KS tumor cells, as well as the epithelioid pneumocytes *(139)*. KSHV has also been detected in the bronchoalveolar lavage-recovered cells from 7 of 12 patients with endobronchial KS and in only 2 of 39 samples from HIV-infected patients without KS *(140)*. The presence of KSHV DNA sequences has also recently been identified in KS pleural effusions *(114)*.

11. Treatment of Pulmonary KS
11.1. Radiation Therapy

Electron-beam radiation therapy is the most common modality employed for the treatment of localized KS. Electron-beam therapy is highly effective in relieving facial and eyelid edema. It has also been used to shrink inguinal lymphadenopathy and plantar lesions but is usually not a first-line treatment for control of lower extremity edema or oral lesions because of the potential to worsen lymphedema and skin compliance when applied to the legs and to cause mucositis when used to treat oral lesions *(142,143)*. Other complications of radiotherapy include hair loss, hyperpigmentation, and fibrosis *(144)*.

External beam radiation therapy was first reported for treatment of HIV-associated pulmonary KS in 1985 *(145)*. Since then, several case series have described patients with advanced pulmonary KS not responsive to systemic chemotherapy who were treated with radiation therapy. These patients had limited functional reserve and achieved only short-term palliation before succumbing to pulmonary embarrassment and superimposed opportunistic infections. Among 11 patients with pulmonary KS who were treated with radiotherapy, 2 died during therapy; the remaining 9 had significant relief of symptoms until death *(146)*. Similarly, among 4 patients with advanced disease who were treated at Walter Reed Hospital, pulmonary symptoms were transiently alleviated, but median survival was only 1.5 mo *(147)*. Meyer and colleagues treated 25 patients with whole lung irradiation for symptomatic pulmonary KS. Treatment was given 4 d/wk. Eighteen of the patients who presented with dyspnea reported improvements, although there were no long-term survivors *(148)*.

Among 25 men with pulmonary KS who failed to respond to chemotherapy, their median survival was <1 mo. For those who had not yet received chemotherapy, roughly two-thirds survived >3 mo *(148)*. It is not known if treatment of pulmonary KS at an earlier stage of disease would offer patients longer term palliation.

Thoracic irradiation has been poorly tolerated among HIV-infected patients with lung cancer *(149)*. Fortunately, KS is a very radiosensitive tumor. With the moderate doses that patients with pulmonary KS typically receive, side effects are tolerable and radiation pneumonitis does not occur. However, for patients with oropharyngeal KS, care must be taken when radiating the oral mucosa. In this setting, significant mucositis, thrush, or reactivation of oral herpes simplex virus infection may supervene *(150)*. The use of antifungals and antivirals may reduce this risk. Amifostine has been used to diminish radiation-induced mucositis in the setting of head and neck and anal-rectal cancers. Its role in the management of HIV-associated tumors is investigational.

11.2. Systemic Chemotherapy

A number of cytotoxic chemotherapeutic agents have systemic activity in AIDS-KS, including bleomycin *(151)*, vinca alkaloids *(152,153)*, etoposide *(154)*, and anthracyclines *(155)*. For patients with extensive or advanced disease, combination regimens containing bleomycin and vincristine *(156)*; or doxorubicin, bleomycin, and vincristine/vinblastine (ABV) *(154,157–160)*, have produced response rates as high as 60–88% but with appreciable toxicity. Liposomal preparations of doxorubicin and daunorubicin are also frequently employed for KS treatment due in part to their lower toxicity profile *(161–163)*. In randomized multicenter trials, each of these liposomal agents has been found to be superior to conventional chemotherapy (bleomycin and vincristine with or without nonliposomal doxorubicin) in terms of response rate and toxicity profiles *(164–166)*. At a dose of 20 mg/m^2 every 3 wk for liposomal doxorubicin, and 40 mg/m^2 every 2 wk for liposomal daunorubicin, side effects including alopecia, neuropathy, and cardiomyopathy are rare. Contrary to initial assumptions *(167)*, these agents may also offer a cost-effective alternative to ABV or BV *(168)*. For advanced KS unresponsive to first and second line options, infusions of low-to-moderate dose paclitaxel are associated with response rates of 53–65% *(169,170)*. An NIH sponsored AIDS-Malignancy Consortium (AMC) study is currently randomizing patients with KS to receive either liposomal doxorubicin or paclitaxel. The results of this study will likely determine whether the potential side effects of paclitaxel, such as leukopenia, hair loss, and peripheral neuropathy, are sufficiently burdensome to limit its use only for patients with advanced and refractory disease.

Chemotherapy drugs that are active in treating mucocutaneous KS are also useful for the treatment of pulmonary KS *(171)*. The highest response rates have been achieved when these agents are used in combination. Using various combination chemotherapy regimens Garay and colleagues and Meduri and associates achieved median survivals that ranged between 3.8 and 6.0 mo in patients with pulmonary KS *(36,47)*. A recurring theme in these and other early trials of pulmonary KS was that the tumor was a direct cause of or contributed to death in the majority of patients.

ABV has been the regimen most often employed for the treatment of pulmonary KS *(172)*. The benefits of such chemotherapy are usually short-lived, and even with combination therapy, time until relapse is usually brief. In a representative study, Gill and colleagues treated 20 patients with ABV or BV *(48)*. Twelve (60%) showed a favorable response to therapy. The median survival for responders was 12 mo vs 6 mo for the nonresponders (range 1–17+ mo). With ABV and other comparable chemotherapy regimens, responding patients may achieve dramatic improvements in pulmonary symptoms even before radiographic changes are appreciated.

Chemotherapy treatment can lead to significant myelosuppression, alopecia, mucositis, GI symptoms, cardiac dysfunction, and aggravate peripheral neuropathy. The limitations of chemotherapy were illustrated among 30 patients with respiratory symptoms and bronchoscopically confirmed KS who received ABV every 4 wk *(95)*. Sixty-four percent had radiographic responses and improvements in respiratory symptomatology. Yet, the median survival in this study was only 6.5 mo (range 1–23 mo) after the first course of chemotherapy, 9 mo after the diagnosis of pulmonary KS, and 11.4 mo after the onset of respiratory symptoms. Roughly half of patients died from progressive KS and respiratory failure. Neutropenia and infection were common complications.

In an effort to improve clinical responses, oncologists have used more myelosuppressive chemotherapies coupled with granulocyte-colony stimulating factor (G-CSF) support. Unfortunately, such strategies have proven dangerous to patients and have not improved outcomes. For example, Sloand and colleagues used a complex chemotherapy regimen consisting of ABV, actinomycin-D, and dacarbazine with concurrent antiretroviral therapy and G-CSF *(173)*. Fifteen of 18 (83%) patients had improvements in their pulmonary infiltrates and resolution of dyspnea and cough. Responses were rapid and skin activity mirrored pulmonary disease. However, in this group with a median CD4+ count of 73 cells/μL, the median survival was only 9 mo and maintenance therapy with etoposide was not useful.

Among the first descriptions of liposomal daunorubicin for treatment of AIDS-related pulmonary KS was the case of a patient with severe pulmonary symptoms and advanced immunosuppression *(174)*. After receiving chemotherapy, he obtained a radiographic partial response and survived an additional 12 mo before succumbing to further HIV-related complications. More recently, in a Phase II study, 11 patients with a median CD4+ count of 52 cells/μL received liposomal daunorubicin (40 mg/m^2 every 2 wk) and concurrent antiretroviral therapy *(175)*. Toxicities included mild nausea, fatigue, and modest granulocytopenia. After 3 mo of therapy, only 1 patient had obvious tumor progression. In another study, liposomal anthracyclines were associated with a median survival of 11 mo compared to 4 mo for patients who received other forms of chemotherapy *(176)*. In an open-label Phase II clinical trial of liposomal daunorubicin at a dose of 60 mg/m^2 every 2 wk, the vast majority of patients had significant improvements in pulmonary symptoms and achieved a median survival of 7.1 mo *(177)*.

Oncologists experienced in HIV-care now consider liposomal anthracyclines as mainstay treatment for patients with advanced/visceral KS. Combining liposomal anthracyclines with BV does not appear to offer additional antitumor benefits for patients with pulmonary or cutaneous KS *(178)*. Experience with Taxol for pulmonary KS is limited, but in one study all 5 patients with pulmonary involvement responded *(170)*.

11.3. HAART

Patients who respond to HAART (typically consisting of two nucleoside analogs and a protease inhibitor) achieve decreased plasma levels of HIV, decreased incidence of opportunistic infections, increased circulating CD4+ T-cells, and decreased short-term mortality *(179)*. Reports further suggest that conditions previously deemed intractable, such as cytomegalovirus retinitis, progressive multifocal leukoencephalopathy, azole-resistant mucocutaneous oral candidiasis, and intestinal cryptosporidiosis and microsporidiosis may stabilize or even diminish after increases in CD4+ cell counts or significant reductions in HIV plasma viral RNA loads *(180–182)*. The effect that potent combination antiretroviral therapy will have in altering the clinical course of AIDS-related neoplasms is uncertain and necessitates long-term study *(183)*.

Recently, I described a patient with symptomatic pulmonary KS, a CD4+ count <50 cells/μL, and an elevated HIV viral load *(184)*. He declined recommendations to receive chemotherapy, but reluctantly agreed to begin HAART. Over the next month, his pulmonary symptoms resolved and over the ensuing year his chest roentgenogram gradually improved. His clinical and radiographic improvements corresponded to suppression of his HIV viral load to nondetectable levels and a progressive rise in his CD4+ cell count.

Support for the assertion that effective suppression of HIV replication has important clinical implications comes from a preliminary report involving 13 patients with AIDS-related KS *(185)*. Before initiation of HAART, these patients received one or more systemic therapies for severe KS for a median of 8 mo. After taking HAART, their median HIV viral load fell from 43,000 copies/mL to nondetectable levels and their median CD4+ count increased from 79–180 cells/µL. In no instance did KS progress, even though they discontinued chemotherapy for a median of 10 wk (range 0–41). Similar results were reported among 8 patients in Italy with KSHV antibodies and documented KS. The initiation of HAART led to tumor shrinkage and a decline in KSHV viral loads *(186)*. In London, effective suppression of HIV viral RNA levels among patients with previously treated KS considerably prolonged time to treatment failure *(187)*. A report from Chanan-Khan and colleagues described 6 patients with pulmonary KS who required ongoing chemotherapy *(188)*. Shortly after beginning HAART, HIV viral levels fell to undetectable levels and the patients were successfully weaned off maintenance chemotherapy without progression of their pulmonary disease. With a median follow-up of 78 wk off chemotherapy (range 40–100 wk), all 6 patients continued to do well. We now know that HAART alone is sometimes an effective maintenance therapy for some patients with pulmonary KS.

The mechanisms by which HAART influences growth of KS lesions remain incompletely understood. The antiviral drugs that are typically utilized in HIV drug regimens appear to have little if any intrinsic inhibitory activity against KSHV. Rather, by downregulating HIV expression while promoting some component of immune reconstitution it appears that HAART may play an active role in regression of KS tumors *(189,190)*.

Although preliminary clinical reports concerning the impact of HAART on modifying both the incidence and clinical expression of KS are encouraging, enthusiasm must be tempered by several mitigating factors: our uncertain knowledge of the length of time that HAART can effectively suppress viral replication; the inability of antiviral therapies to restore severely impaired immunity; and our imperfect knowledge of how best to maintain patient compliance in a setting in which medications must be taken multiple times a day, have unpleasant side effects, and interact unpredictably with numerous other medications *(191–194)*.

For individuals with pulmonary KS, the goals of therapy need to be clearly delineated. Despite dramatic advancements in the care of HIV-infected patients, neither KS nor AIDS are yet considered curable conditions. Options need to account for the patient's level of activity, co-morbid illness, immune reserves, and the input of other members of the multidisciplinary team.

Acknowledgment

I wish to thank David Dail, MD, for providing photomicrographs and accompanying legends and Ms. Arleen Sierra for manuscript preparation.

References

1. Hymes, K. D., Greene, J. B., Marcus, A., et al. (1981) Kaposi's sarcoma in homosexual men: a report of eight cases. *Lancet* **2,** 598–608.
2. Centers for Disease Control (1981) Pneumocystis carinii pneumonia—Los Angeles. *MMWR Morb. Mortal. Wkly. Rep.* **30,** 250–252.
3. Centers for Disease Control (1981) Kaposi's sarcoma and Pneumocystis pneumonia among homosexual men—New York City, California. *MMWR Morb. Mortal. Wkly. Rep.* **30,** 305–308.
4. Friedman-Kein, A. E. (1981) Disseminated Kaposi's sarcoma syndrome in young homosexual men. *J. Am. Acad. Dermatol.* **5,** 468–471.
5. Goedert, J. J. (2000) The epidemiology of acquired immune deficiency syndrome malignancies. *Semin. Oncol.* **27,** 390–401.
6. Mitsuyasu, R. T. (2000) AIDS-related Kaposi's sarcoma: Current treatment options, future trends. *Oncology* **14,** 868–878.
7. Aboulafia, D. M. (2001) The epidemiologic, pathologic and clinical features of AIDS-associated pulmonary Kaposi's sarcoma. *Chest* **17,** 1128–1145.
8. Kaposi, M. (1872) Idiopathisches multiples pigment-sarkom der haut. *Arch. Dermatol. Syph.* **4,** 265–273.
9. Taylor, J. F., Templeton, A. C., Vogel, C. L., et al. (1971) Kaposi's sarcoma in Uganda: a clinicopathologic study. *Int. J. Cancer* **8,** 122–135.
10. Bayley, A. C. (1984) Aggressive Kaposi's sarcoma in Zambia, 1983. *Lancet* **1,** 1318–1320.
11. Atnal, V. H., Patil, P. S., Chintu, C., and Elem, B. (1995) Influence of HIV epidemic on the incidence of Kaposi's sarcoma in Zambian children. *J. Acquir. Immune Defic. Syndr. Hum. Retrovirol.* **8,** 96–100.
12. Stein, M. E., Spencer, D., Ruff, P., Lakier, R., MacPhail, P., and Bezwoda, W. R. (1994) Endemic African Kaposi's sarcoma: Clinical and therapeutic implications: 10-year experience in the Johannesburg Hospital (1980–1990). *Oncology* **51,** 63–69.
13. Antman, K. and Chang, Y. (2000) Kaposi's sarcoma. *N. Engl. J. Med.* **342,** 1027–1038.
14. Penn, I. (1994) Depressed immunity and the development of cancer. *Cancer Detect. Prev.* **18,** 241–252.
15. Penn, I. (1995) Sarcomas in organ allograft recipients. *Transplantation* **60,** 1485–1491.
16. Al-Sulaiman, M. H. and Al-Kadar, A. A. (1994) Kaposi's sarcoma in renal transplant recipients. *Transplant Sci.* **4,** 48.
17. Friedman-Kien, A. E. and Mitsuyasu, R. T. (1999) Redefining clinical unmet needs on the treatment of Kaposi's sarcoma. Oxford Institute for Continuing Education, OCC North America, Inc., Newton, PA, pp. 1–41.

18. Beral, V., Peterman, T. A., Berkelman, R. C., and Jaffe, H. W. (1990) Kaposi's sarcoma among persons with AIDS: A sexually transmitted infection? *Lancet* **335**, 123–128.
19. Biggar, R. J. and Rabkin, C. S. (1996) The epidemiology of AIDS-related neoplasms. *Hematol. Oncol. Clin. North Am.* **10**, 997–1010.
20. Beral, V., Bull, D., Darby, S., et al. (1992) Risk of Kaposi's sarcoma and sexual practices associated with fecal contact in homosexual or bisexual men with AIDS. *Lancet* **339**, 632–635.
21. Chang, Y., Cesarman, E., Pessin, M. S., et al. (1994) Identification of herpesvirus-like DNA sequences in AIDS-associated Kaposi's sarcoma. *Science* **266**, 1865–1869.
22. Moore, P. S. and Chang, Y. (1995) Detection of herpesvirus-like sequences in Kaposi's sarcoma in patients with and without HIV infection. *N. Engl. J. Med.* **332**, 1181–1185.
23. Huang, Y. Q., Li, J. J., Kaplan, M. H., et al. (1995) Human herpesvirus-like nucleic acid in various forms of Kaposi's sarcoma. *Lancet* **345**, 759–761.
24. Sitas, F., Carrara, H., Beral, V., et al. (1999) Antibodies against human herpesvirus 8 in black South African patients with cancer. *N. Engl. J. Med.* **340**, 1863–1871.
25. Humphrey, R. W., O'Brien, T. R., Newcomb, F. M., et al. (1996) Kaposi's sarcoma (KS)-associated herpesvirus-like DNA sequences in peripheral blood mononuclear cells: association with KS and persistence in patients receiving anti-herpesvirus drugs. *Blood* **88**, 297–301.
26. Bigoni, B., Dolcetti, R., de Lellis, L., et al. (1996) Human herpesvirus 8 is present in the lymphoid system of healthy persons and can reactivate in the course of AIDS. *J. Infect. Dis.* **173**, 542–549.
27. Li, J. J., Huang, Y. Q., and Friedman-Kien, A. (1995) Detection of DNA sequences of KSHV in blood, semen, KS tumor, and uninvolved skin of AIDS-KS patients. *AIDS Res. Hum. Retroviruses* **11(Suppl. 1):5**, 98.
28. Moore, P. S., Kingsley, L. A., Holmberg, S. D., et al. (1996) Kaposi's sarcoma-associated herpesvirus infection prior to the onset of Kaposi's sarcoma. *AIDS* **10**, 175–180.
29. Ressa, G., Dorrucci, M., Pezzotti, P., Andreoni, M., and Ensoli, B. (1998) HHV-8 seropositivity and risk of developing Kaposi's sarcoma (KS) and other AIDS-related diseases among individuals with known dates of HIV seroconversion. 12th World AIDS Conference (abstract no. 13310) Geneva, Switzerland.
30. Buchbinder, S. P., Vittinghoff, E., Colfax, G., and Holmberg, S. (1998) Declines in AIDS incidence associated with highly active anti-retroviral therapy (HAART) are not reflected in KS and lymphoma incidence (abstract S7). 2nd National AIDS Malignancy Conference, Bethesda, MD.
31. Jones, J., Hanson, D. L., Dworkin, J. W., Ward, H. W., Jaffe, H. W., and the Adult/Adolescent Spectrum Disease Project Group. (2000) Incidence and trends in Kaposi's sarcoma in the era of effective antiretroviral therapy. *J. AIDS* **24**, 270–274.
32. Centers for Disease Control and Prevention. (1998) *HIV/AIDS Surveil. Rep.* **10(no.1)**, 1–40.

33. Palella, F. J., Delaney, K. M., Moorman, A. C., et al. (1998) Declining morbidity and mortality among patients with advanced human immunodeficiency virus infection. *N. Engl. J. Med.* **338,** 853–860.
34. Wallace, J. M., Hansen, N. I., LaVarge, L., et al. (1997) Respiratory Disease Trends in the Pulmonary Complications of HIV Infection Study Cohort. *Am. J. Respir. Crit. Care Med.* **155,** 72–80.
35. Jasmer, R. M., Edinburgh, K. J., Thompson, A., et al. (2000) Clinical and radiographic predictors of the etiology of pulmonary nodules in HIV infected patients. *Chest* **117,** 1023–1030.
36. Meduri, G. U., Stover, D. E., Lee, M., et al. (1986) Pulmonary Kaposi's sarcoma in the acquired immune deficiency syndrome: clinical, radiographic, and pathologic manifestations. *Am. J. Med.* **81,** 11–18.
37. Pothoff, G., Pohler, E., Diekmann, M., Krueger, G., and Hilger, H. H. (1990) Difficulties in the diagnosis of AIDS-associated intrapulmonary Kaposi's sarcoma. *Pneumologie* **44,** 74–77.
38. Kang, E. Y., Staples, C. A., McGuinness, G., Primack, S. L., and Müller, N. L. (1996) Detection and differential diagnosis of pulmonary infections and tumors in patients with AIDS: value of chest radiography versus CT. *AJR* **166,** 15–19.
39. Stover, D. E., White, D. A., Romano, P. A., Gellene, R. A., and Robeson, W. A. (1985) Spectrum of pulmonary diseases associated with acquired immune deficiency syndrome. *Am. J. Med.* **78,** 429–437.
40. Dail, D. H. (1994) Pulmonary Kaposi's Sarcoma, in: Introduction to AIDS Pathology in Pulmonary Pathology, 2nd ed. (Dail, D. H. and Hammer, S. P., eds.), Springer-Verlag, New York, NY, pp. 121–153.
41. Welch, K., Finkbeiner, W., and Alpers, C. E. (1984) Autopsy findings in the acquired immune deficiency syndrome. *JAMA* **252,** 1152–1159.
42. Lemlich, G., Schwam, L., and Lebwhol, M. (1987) Kaposi's sarcoma in acquired immunodeficiency syndrome. Postmortem findings in twenty-four cases. *J. Am. Acad. Dermatol.* **16,** 319–325.
43. Miller, R. F., Tomlinson, N. C., Cottrill, C. P., et al. (1992) Bronchopulmonary Kaposi's sarcoma in patients with AIDS. *Thorax* **47,** 721–725.
44. Mitchell, D. M., McCarthy, M., Fleming, J., and Moss, F. M. (1992) Bronchopulmonary Kaposi's sarcoma in patients with AIDS. *Thorax* **47,** 726–729.
45. Zibrak, J. D., Silvestri, R. C., Costello, P., et al. (1986) Bronchoscopic and radiologic features of Kaposi's sarcoma involving the respiratory system. *Chest* **90,** 476–479.
46. Irwin, D. H. and Kaplan, L. D. (1993) Pulmonary manifestations of acquired immunodeficiency syndrome—associated malignancies. *Semin. Respir. Infect.* **8,** 139–148.
47. Garay, S. M., Belenko, M., Fazzini, E., and Schinella, R. (1987) Pulmonary manifestations of Kaposi's sarcoma. *Chest* **91,** 39–43.
48. Gill, P. S., Akil, B., Colletti, P., et al. (1989) Pulmonary Kaposi's sarcoma: clinical findings and results of therapy. *Am. J. Med.* **87,** 57–61.

49. Fouret, P. J., Touboul, J. L., Maynaud, C. M., Akoun, G. M., and Roland, J. (1987) Pulmonary Kaposi's sarcoma in patients with acquired immunodeficiency syndrome: a clinicopathological study. *Thorax* **42**, 262–268.
50. Huang, L., Schnapp, L. M., Gruden, J. F., Hopewell, P. C., and Stansell, J. D. (1996) Presentation of AIDS-related pulmonary Kaposi's sarcoma diagnosed by bronchoscopy. *Am. J. Resp. Crit. Care Med.* **153**, 1385–1390.
51. Northfelt, D. W. (1994) AIDS-related Kaposi's sarcoma: still a problem, still an opportunity. *J. Clin. Oncol.* **12**, 1109–1110.
52. Young, B., Johnson, S., Bahktiari, M., et al. (1998) Resistance mutations in protease and reverse transcriptase genes of HIV-1 isolates from patients failing combination antiretroviral therapy. *J. Infect. Dis.* **178**, 1497–1501.
53. Pozniak, A. L., Latif, A. S., Neill, P., Houston, S., Chen, K., and Robertson, V. (1992) Pulmonary Kaposi's sarcoma in Africa. *Thorax* **47**, 730–733.
54. McNutt, N. S., Fletcher, V., and Conant, M. A. (1983) Early lesions of Kaposi's sarcoma in homosexual men. An ultrastructural comparison with other vascular proliferations in skin. *Am. J. Pathol.* **111**, 62–77.
55. Nickoloff, B. J. and Griffiths, C. E. (1989) The spindle-shaped cells in cutaneous Kaposi's sarcoma. Histologic simulators include factor XIIIa dermal dendrocytes. *Am. J. Pathol.* **135**, 793–800.
56. Zhang, Y. M., Bachmann, S., Hemmer, C., et al. (1994) Vascular origin of Kaposi's sarcoma. Expression of leukocyte adhesion molecule-1, thrombomodulin, and tissue factor. *Am. J. Pathol.* **144**, 51–59.
57. Kaaya, E. E., Parravicini, C., Ordonez, C., et al. (1995) Heterogeneity of spindle cells in Kaposi's sarcoma: comparison of cells and lesions and in culture. *J. AIDS Hum. Retrov.* **10**, 295–305.
58. Safai, B., Johnson, K., Myskowski, P., and Al, E. (1985) The natural history of Kaposi's sarcoma in the acquired immune deficiency syndrome. *Ann. Intern. Med.* **103**, 744–750.
59. Fiorelli, V., Gendelman, R., Sirianni, M. C., et al. (1998) Gamma-interferon produced by CD8+ T cells infiltrating Kaposi's sarcoma induces spindle cells with angiogenic phenotype and synergy with human immunodeficiency virus-1 *Tat* protein: an immune response to human herpesvirus-8 infection? *Blood* **91**, 956–967.
60. Purdy, J. J., Colby, T. V., Yousem, S. A., and Battifora, H. (1986) Pulmonary Kaposi's sarcoma. Premortem histologic diagnosis. *Am. J. Surg. Pathol.* **10**, 301–311.
61. Case records of the Massachusetts General Hospital. (1993) Weekly clinicopathologic exercises. Cases 1-1990. A 29-year-old man with a positive test for HIV and reticulonodular pulmonary infiltrate. *N. Engl. J. Med.* **322**, 43–51.
62. Nash, G. and Fligiel, S. (1984) Kaposi's sarcoma presenting as pulmonary disease in the acquired immunodeficiency syndrome: diagnosis by lung biopsy. *Hum. Pathol.* **15**, 999–1001.
63. Case records of the Massachusetts General Hospital. (1993) Weekly clinicopathological exercises. Case 34-1993. A 36-year-old man with AIDS and respiratory failure. *N. Engl. J. Med.* **329**, 645–653.

64. Nakamura, S., Salahuddin, S. Z., Biberfeld, P., et al. (1988) Kaposi's sarcoma cells: long-term culture with growth factor from retrovirus-infected CD4+ T cells. *Science* **242,** 426–430.
65. Gallo, R. C. (1998) The enigmas of Kaposi's Sarcoma. *Science* **282,** 1837–1839.
66. Purdy, J. J., Colby, T. V., Yousem, S. A., and Battifora, H. (1986) Pulmonary Kaposi's sarcoma. Premortem histologic diagnosis. *Am. J. Surg. Pathol.* **10,** 301–311.
67. Ensoli, B. and Gallo, R. C. (1995) AIDS-associated Kaposi's sarcoma: a new perspective of its pathogenesis and treatment. *Proc. Assoc. Am. Phys.* **107,** 8–18.
68. Kelly, G. D., Ensoli, B., Gunthel, C. J., and Offerman, M. K. (1998) Purified Tat induces inflammatory response genes in Kaposi's sarcoma. *AIDS* **12,** 1753–1761.
69. Salahuddin, S. Z., Nakamura, S., Biberfeld, P., et al. (1988) Angiogenic properties of Kaposi's sarcoma-derived cells after long-term culture in vitro. *Science* **242,** 430–433.
70. Jung-Chung, L. (1998) Kaposi's sarcoma and HHV-8: Implications for therapy. *Infect. Med.* **15,** 337–343.
71. Barillari, G., Gendelman, R., Gallo, R. C., and Ensoli, B. (1993) The *Tat* protein of human immunodeficiency virus type I, a growth factor for AIDS Kaposi's sarcoma and cytokine-activated vascular cells, induces adhesion of the same cell types by using integrin receptors recognizing the RGD amino acid sequence. *Proc. Natl. Acad. Sci. USA* **90,** 7941–7945.
72. Ensoli, G., Nakamura, S., Salahuddin, S. Z., et al. (1989) AIDS—Kaposi's sarcoma-derived cells express cytokines with autocrine and paracrine growth effects. *Science* **243,** 223–226.
73. Cornali, E., Zeitz, C., Benelli, R., et al. (1996) Vascular endothelial growth factor regulates angiogenesis and vascular permeability in Kaposi's sarcoma. *Am. J. Pathol.* **149,** 1851–1869.
74. Ensoli, B., Gendelman, R., Markham, P., et al. (1994) Synergy between basic fibroblast growth factor and HIV-1 *Tat* protein in induction of Kaposi's sarcoma. *Nature* **371,** 674–680.
75. Rabkin, C. S., Janz, S., Lash, A., et al. (1997) Monoclonal origin of multicentric Kaposi's sarcoma lesions. *N. Engl. J. Med.* **336,** 988–993.
76. Vecini, S., Sirianni, M. C., Vincenzi, L., et al. (1997) Kaposi's sarcoma cells express the macrophage-associated antigen mannose receptor and develop in peripheral blood cultures of Kaposi's sarcoma patients. *Am. J. Pathol.* **150,** 929–937.
77. Barkin, C., Janz, S., Pack, S., et al. (1998) Genetic abnormalities in Kaposi's sarcoma tumors. 12th World AIDS Conference; (abstract no. 22284) Geneva, Switzerland.
78. Moore, P. S., Gao, S. J., Dominguez, G., et al. (1996) Primary characterization of a herpesvirus agent associated with Kaposi's sarcoma. *J. Virol.* **70,** 549–558.
79. Sarid, R., Sato, T., Bohensky, R. A., Russo, J. J., and Chang, Y. (1997) Kaposi's sarcoma-associated herpesvirus encodes a functional bcl-2 homologue. *Nat. Med.* **3,** 293–298.

80. Bais, C., Santomasso, B., Coso, O., et al. (1998) G-protein-coupled receptors of Kaposi's sarcoma-associated herpesvirus is a viral oncogene and angiogenesis activator. *Nature* **391,** 86–89.
81. Nicholas, J., Ruvolo, V. R., Burns, W. H., et al. (1997) Kaposi's sarcoma-associated human herpesvirus-8 encodes homologues of macrophage inflammatory protein-1 and interleukin-6. *Nat. Med.* **3,** 287–292.
82. Sirianni, M., Vincenzi, L., Fiorelli, V., et al. (1998) Gamma-interferon production in peripheral blood mononuclear cells and tumor infiltrating lymphocytes from Kaposi's sarcoma patients: correlation with the presence of human herpesvirus-8 in peripheral blood mononuclear cells and lesional macrophages. *Blood* **91,** 968–976.
83. Mesri, E. A. (1999) Inflammatory reactivation and angiogenicity of Kaposi's sarcoma-associated herpesvirus/HHV8: A missing link in the pathogenesis of acquired immune deficiency syndrome-associated Kaposi's sarcoma. *Blood* **93,** 4031–4033.
84. Dezube, B. J. (2000) The role of human immunodeficiency virus-1 in the pathogenesis of acquired immunodeficiency syndrome-related Kaposi's sarcoma: The importance of an inflammatory and angiogenic milieu. *Semin. Oncol.* **27,** 420–422.
85. Lemp, G. F., Payne, S. F., Neal, D., Temelso, T., and Rutherford, G. M. (1990) Survival trends for patients with AIDS. *JAMA* **263,** 402–406.
86. Payne, S. F., Lemp, G. F., and Rutherford, G. W. (1990) Survival following diagnosis of oral Kaposi's sarcoma for AIDS patients in San Francisco. *J. Acquir. Immune Defic. Syndr.* **3,** S14–S17.
87. Miles, S. A., Wang, H., Elashoff, R., and Mitsuyasu, R. T. (1994) Improved survival for patients with AIDS-related Kaposi's sarcoma. *J. Clin. Oncol.* **12,** 1910–1916.
88. Gruden, J. F., Huang, L., Webb, W. R., Gamsu, G., Hopewell, P. C., and Sides, D. M. (1995) AIDS-related Kaposi's sarcoma of the lung: radiographic findings and staging system with bronchoscopic correlation. *Radiology* **195,** 545–552.
89. Fife, K. and Bower, M. (1996) Recent insights into the pathogenesis of Kaposi's sarcoma. *Br. J. Cancer* **73,** 1317–1322.
90. Mocroft, A. J., Lundgren, J. D., d'Armino Monforte, A., et al. (1997) Survival of AIDS patients according to the type of AIDS-defining event. The AIDS in Europe Study Group. *Int. J. Epidemiol.* **26,** 400–407.
91. Zwhalen, M., Neunschwander, B. E., Vlahor, D., and Brookmeyer, R. (1998) Using registry data to estimate trends in survival after AIDS diagnosis in Switzerland, diagnosed between 1988 and 1992. *12th World AIDS Conference* (abstract no. 60253) Geneva, Switzerland.
92. Brodt, H. R., Kamps, B. S., Helm, E. B., Schofer, H., and Mitrou, P. (1998) Kaposi's sarcoma in HIV infection: impact on opportunistic infections and survival. *AIDS* **12,** 1475–1481.
93. Krown, S. E., Metroka. C., and Wernz, J. C. (1989) Kaposi's sarcoma in the acquired immunodeficiency syndrome: A proposal for uniform evaluation, response, and staging criteria. *J. Clin. Oncol.* **7,** 1201–1207.

94. Krown, S. E., Testa, M. A., and Huang, J. (1997) AIDS-related Kaposi's sarcoma: prospective validation of the AIDS Clinical Trials Group staging classification: AIDS Clinical Trials Group Oncology Committee. *J. Clin. Oncol.* **15,** 3085–3092.
95. Cadranel, J. C., Kammoun, S., Chevret, S., et al. (1994) Results of chemotherapy in 30 AIDS patients with symptomatic pulmonary Kaposi's sarcoma. *Thorax* **49,** 958–960.
96. Aboulafia, D. M. (1994) Human immunodeficiency virus-associated neoplasms. Epidemiology, pathogenesis and review of current therapy. *Cancer Practice* **2,** 297–306.
97. Tappero, J. W., Conant, M. A., Wolfe, S. F., et al. (1993) Kaposi's sarcoma: epidemiology, pathogenesis, histology, clinical spectrum, staging criteria and therapy. *J. Am. Acad. Dermatol.* **28,** 371–395.
98. Shuler, J., Nolland, G., Miles, S., Miller, B., and Grossman, I. (1989) Kaposi's sarcoma of the conjunctiva and eyelids associated with acquired immunodeficiency syndrome. *Arch. Ophthalmol.* **107,** 858–862.
99. Couderc, L. J., Detruchis, P., Mathern, S., Katlama, C., Caubarrere, I., and Clauvel, J. P. (1990) Pulmonary Kaposi's sarcoma (KS) in patients with HIV infection. *Int. Conf. AIDS* **6(2),** 376(abstract no. 2091).
100. Pozniak, A., Latif, A., McLead, D., Ndemera, B., and Neill, P. (1990) Clinical, radiological, and bronchoscopic features of pulmonary Kaposi's sarcoma. *Int. Conf. AIDS* **6(2),** 378(abstract no. 2099).
101. Tupule, A., Yung, R. C., Wernz, J., et al. (1998) Phase II trial of liposomal daunorubicin in the treatment of AIDS-related pulmonary Kaposi's sarcoma. *J. Clin. Oncol.* **16,** 3369–3374.
102. Meyer, J. L. (1993) Whole-lung irradiation for Kaposi's sarcoma. *Am. J. Clin. Oncol.* **16,** 372–376.
103. Aboulafia, D. M., Olson, M., and Bundow, D. (1999) Clinical characteristics of advanced pulmonary KS (PKS): a difficult disease to treat. Third National AIDS Malignancy Conference. *J. Acquir. Immune Defic. Syndr.* **21,** A23(abstract 53).
104. Haramati, L. B. and Wong, J. (2000) Intrathoracic Kaposi's sarcoma in women with AIDS. *Chest* **117,** 410–414.
105. Kanter, A., Spencer, D. C., and Steinberg, M. H. (1998) Incidence and characteristics of HIV-related conditions in distinct South Africa populations. 12th World AIDS Conference, (abstract no. 13381) Geneva, Switzerland.
106. Pozniak, A. L., Latif, A. S., Neill, P., Houston, S., Chen, K., and Robertson, V. (1992) Pulmonary Kaposi's sarcoma in Africa. *Thorax* **47,** 730–733.
107. Kaplan, L. D., Hopewell, P. C., Jaffe, H., Goodman, P. C., Bottles, K., and Volderding, P. A. (1988) Kaposi's sarcoma involving the lung in patients with acquired immunodeficiency syndrome. *J. Acquir. Immune Defic. Syndr.* **1,** 23–30.
108. Stansel, J. D. and Murray, J. F. (1996) Pulmonary complications of human immunodeficiency virus infection, in *Textbook of Respiratory Medicine*, 2nd ed. (Murray, J. F., Nadel, J. A., and Sanders, W. B., eds), Lippincott Williams and Wilkins, Philadelphia, PA, pp. 2333–2367.

109. Jeyapalan, M. and Steffenson, S. (1997) Diagnosis of pulmonary Kaposi's sarcoma in AIDS patients. *AIDS Patient Care STDs* **11**, 9–12.
110. Kornfield, H. and Axelrod, J. L. (1983) Pulmonary presentation of Kaposi's sarcoma in a homosexual patient. *Am. Rev. Respir. Dis.* **127**, 248–249.
111. Roux, F. J., Bancal, C., Dombret, M. C., et al. (1994) Pulmonary Kaposi's sarcoma revealed by a solitary nodule in a patient with acquired immunodeficiency syndrome. *Am. J. Resp. Crit. Care Med.* **149**, 1041–1043.
112. Sadaghdar, H. and Eden, E. (1991) Pulmonary Kaposi's sarcoma presenting as fulminant respiratory failure. *Chest* **100**, 858–860.
113. Bach, M. C., Bagwell, S. P., and Fanning, J. P. (1988) Primary pulmonary Kaposi's sarcoma in the acquired immunodeficiency syndrome: a cause of persistent pyrexia. *Am. J. Med.* **85**, 274–275.
114. McGuinnes, G. (1997) Changing trends in the pulmonary manifestation of AIDS. *Radiol. Clin. North Am.* **35**, 1066–1082.
115. Denton, A. S., Miller, R. F., and Spittle, M. F. (1995) Management of pulmonary Kaposi's sarcoma: new perspectives. *Br. J. Hosp. Med.* **53**, 344–350.
116. Hartman, T. E., Primack, S. L., Müller, N. L., and Staples, C. A. (1994) Diagnosis of thoracic complications in AIDS: accuracy of CT. *AJR Am. J. Roentgenol.* **162**, 547–553.
117. Khalil, A. M., Carette, M. R., Cadranel, J. L., Mayaud, C. M., and Bigot, J. M. (1995) Intrathoracic Kaposi's sarcoma. CT findings. *Chest* **108**, 1622–1626/*Emr. Respir. J.* **7**, 1285–1289.
118. Sivit, C. J., Schwartz, A. M., and Rockoff, S. D. (1987) Kaposi's sarcoma of the lung in AIDS: radiologic-pathologic analyses. *AJR Roentgenol.* **145**, 25–28.
119. Wolff, S. D., Kuhlman, J. E., and Fishman, E. K. Thoracic Kaposi's Sarcoma in AIDS: CT findings. *J. Comput. Assist. Tomogr.* **17**, 60–62.
120. Cadranel, J., Naccache, J.-M., Wislez, M., and Mayaud, C. (1999) Pulmonary malignancies in the immunocompromised patient. *Respiration* **66**, 289–309.
121. Kang, E. Y., Staples, C. A., McGuinness, G., Primack, S. L., and Müller, N. L. (1996) Detection and differential diagnosis of pulmonary infections and tumors in patients with AIDS: value of chest radiography versus CT. *AJR Am. J. Roentgenol.* **1**, 15–19.
122. Kirshenbaum, K. J., Burke, R., Fanapour, F., et al. (1998) Pulmonary high-resolution computed tomography versus gallium scintigraphy: Diagnostic utility in the diagnosis of patients with AIDS who have chest symptoms and normal or equivocal chest radiographs. *J. Thorac. Imaging* **13**, 52–57.
123. Lee, V. W., Fuller, J. D., O'Brien, M. J., Parker, D. R., Cooley, T. P., and Liebman, H. A. (1991) Pulmonary Kaposi's sarcoma in patients with AIDS: scintigraphic diagnosis with sequential thallium and gallium scanning. *Radiology* **180**, 409–412.
124. Delval Gomez, M. A., Castro Beiras, J. M., Gallardo, F. G., and Verdejo, J. (1994) Thallium and gallium scintigraphy in pulmonary Kaposi's sarcoma in an HIV-positive patient. *Clin. Nucl. Med.* **19**, 467–468.

125. Buscombe, J. R., Oyen, W. J., Grant, A., et al. (1993) Indium-111-labeled polyclonal human immunoglobulin: identifying focal infection in patients positive for human immunodeficiency virus. *J. Nucl. Med.* **34,** 1621–1625.
126. Khalil, A. M., Carette, M. F., Cadranel, J. L., Mayaud Ch, M., Akoun, G. M., and Bigot, J. M. (1994) Magnetic resonance imaging findings in pulmonary Kaposi's sarcoma: a series of 10 cases. *Eur. Respir. J.* **7,** 1285–1289.
127. Gould, M., MacLean, C. C., Kushner, W. G., Rydzak, C. E., and Owens, D. K. (2001) Accuracy of positron emission tomography for diagnosis of pulmonary nodules and mass lesions: a meta-analysis. *JAMA* **285,** 914–924.
128. Caldwell, B. D., Kushner, D., and Young, B. (1996) Kaposi's sarcoma versus bacillary angiomatosis. *J. Am. Podiatr. Med. Assoc.* **86,** 260–262.
129. Kauffman, C. L., Holman, R. P., and Herbert, C. R. (2000) Tender nodules and hyperpigmented plaques in a man with AIDS. *AIDS Reader* **10(1),** 41–44.
130. Levine, A. M., Gill, P., and Salahuddin, S. (1992) Neoplastic complications of HIV infection, in *AIDS and Other Manifestations of HIV Infection,* 2nd ed. (Wormser, G., ed.), Raven Press, New York, NY, pp. 443–454.
131. Hamm, P. G., Judson, M. A., and Aranda, C. P. (1987) Diagnosis of pulmonary Kaposi's sarcoma with fiberoptic bronchoscopy and endobronchial biopsy. A report of five cases. *Cancer* **59,** 807–810.
132. Judson, M. A. and Sahn, S. A. (1994) Endobronchial lesions in HIV-infected individuals. *Chest* **105,** 1314–1323.
133. Afessa, B., Green, W., Chiao, J., and Frederick, W. (1998) Pulmonary complications of HIV infection: autopsy findings. *Chest* **113,** 1225–1229.
134. Heitzman, E. R. (1990) Pulmonary neoplastic and lymphoproliferative disease in AIDS: a review. *Radiology* **177,** 347–351.
135. Zibrak, J. D., Silvestri, R. C., Costello. P., et al. (1986) Bronchoscopic and radiologic features of Kaposi's sarcoma involving the respiratory system. *Chest* **90,** 476–479.
136. Broderick, P. A. and Krinsley, J. S. (1990) Pulmonary Kaposi's disease diagnosis by transbronchial biopsy. *Conn. Med.* **54,** 555–557.
137. Hughes-Davies, L., Kocjan, G., Spittle, M. F., and Miller, R. F. (1992) Occult alveolar hemorrhage in bronchopulmonary Kaposi's sarcoma. *J. Clin. Pathol.* **45,** 536–537.
138. Hill, A. D. K., Darzi, A., Menzies-Gow, N., and Riordan, J. F. (1993) Thorascopic biopsy in the diagnosis of pulmonary Kaposi's sarcoma. *J. Laparoendosc. Surg.* **3,** 571–572.
139. Foreman, K. E., Bacon, P. E., Hsi, E. D., and Nickoloff, B. J. (1997) In situ polymerase chain reaction-based localization studies support role of human herpesvirus-8 as the cause of two AIDS-related neoplasms: Kaposi's sarcoma and body cavity lymphoma. *J. Clin. Invest.* **99,** 2971–2978.
140. Benfield, T. L., Dodt, K. K., and Lundgren, J. D. (1997) Human herpes virus-8 DNA in bronchoalveolar lavage samples from patients with AIDS-associated pulmonary Kaposi's sarcoma. *Scand. J. Infect. Dis.* **29,** 13–16.

141. Beck, J. M. (1998) Pleural disease in patients with acquired immune deficiency syndrome. *Clin. Chest Med.* **19**, 341–349.
142. Tappero, J. W., Conant, M. A., Wolfe, S. F., et al. (1993) Kaposi's sarcoma. Epidemiology, pathogenesis, histology, clinical spectrum, staging criteria, and therapy. *J. Am. Acad. Dermatol.* **28**, 371–395.
143. Chak, L. Y., Gill, P. S., Levine. L. M., et al. (1998) Radiation therapy for acquired immunodeficiency syndrome-related KS. *J. Clin. Oncol.* **6**, 863–867.
144. Tappero, J. W., Grekin, R. C., Zanelli, G. A., et al. (1992) Pulsed-dye laser therapy for cutaneous Kaposi's sarcoma associated with acquired immunodeficiency syndrome. *J. Am. Acad. Dermatol.* **28**, 117–122.
145. Nobler, M. P. (1985) Pulmonary irradiation for Kaposi's sarcoma. *Am. J. Clin. Oncol.* **8**, 441–444.
146. Berson, A. M., Quivey, J. M., Harris, J. W., and Wara, W. M. (1990) Radiation therapy for AIDS-related Kaposi's sarcoma. *Int. J. Radiat. Oncol. Biol. Phys.* **19**, 569–575.
147. Doyle, M., Johnstone, P. A., and Watkins, E. B. (1993) Role of radiation therapy in management of pulmonary Kaposi's sarcoma. *South. Med. J.* **86**, 285–288.
148. Meyer, J. L. (1993) Whole-lung irradiation for Kaposi's sarcoma. *Am. J. Clin. Oncol.* **16**, 372–376.
149. Leigh, B. R. and Lau, D. H. (1998) Severe esophageal toxicity after thoracic radiation therapy for lung cancer associated with the human immunodeficiency virus: a case report and review of the literature. *Am. J. Clin. Oncol.* **21**, 479–481.
150. Swift, P. S. (1996) The role of radiation therapy in the management of HIV-related Kaposi's sarcoma. *Hematol. Oncol. Clin. North Am.* **10**, 1069–1080.
151. Caumes, E., Guermonprez, G., Katlama, C., and Gentilini, M. (1992) AIDS-associated mucocutaneous Kaposi's sarcoma treated with bleomycin. *AIDS* **6**, 1483–1487.
152. Volberding, P. A., Abrams, D. I., Conant, M., Kaslow, K., Vranizan, K., and Ziegler, J. (1985) Vinblastine therapy for Kaposi's sarcoma in the acquired immunodeficiency syndrome. *Ann. Intern. Med.* **103**, 335–338.
153. Mintzer, D. M., Real, F. X., Jovino, L., and Krown, S. E. (1985) Treatment of Kaposi's sarcoma and thrombocytopenia with vincristine in patients with the acquired immunodeficiency syndrome. *Ann. Intern. Med.* **102**, 200–202.
154. Laubenstein, L. J., Krigel, R. L., Odajnk, C. M., et al. (1984) Treatment of epidemic Kaposi's sarcoma with etoposide or a combination of doxorubicin, bleomycin and vinblastine. *J. Clin. Oncol.* **2**, 1115–1120.
155. Shepherd, F. A., Burkes, R. L., Paul, K. E., and Goss, P. E. (1991) A phase II study of 4^1-epirubicin in the treatment of poor-risk Kaposi's sarcoma and AIDS. *AIDS* **5**, 305–309.
156. Gompels, M. M., Hill, A., Jenkins, P., et al. (1992) Kaposi's sarcoma in HIV infection treated with vincristine and bleomycin. *AIDS* **6**, 1175–1180.
157. Gill, P. S., Miles, S. A., Mitsuyasu, R. T., et al. (1994) Phase I AIDS Clinical Trial Group (075) study of adriamycin, bleomycin, and vincristine chemotherapy

with zidovudine in the treatment of AIDS-related Kaposi's sarcoma. *AIDS* **8**, 1695–1699.
158. Gill, P. S., Wernz, J., Scadden, D. T., et al. (1996) Randomized phase III trial of liposomal daunorubicin versus doxorubicin, bleomycin, and vincristine in AIDS-related Kaposi's sarcoma. *J. Clin. Oncol.* **14**, 2353–2364.
159. Gill, P. S., Rarick, M., McCutchan, J. A, et al. (1991) Systemic treatment of AIDS-related Kaposi's sarcoma: results of a randomized trial. *Am. J. Med.* **90**, 427–433.
160. Gill, P. S., Mitsuyasu, R. T., Montgomery, T., et al. (1997) AIDS Clinical Trials Group Study 094: a phase I/II trial of ABV chemotherapy with zidovudine and recombinant human GM-CSF in AIDS-related Kaposi's sarcoma. *Cancer J. Sci. Am.* **3**, 278–283.
161. Northfelt, D. W. (1997) Liposomal anthracycline chemotherapy in the treatment of AIDS-related Kaposi's sarcoma. *Oncology* **11(Suppl. 11)**, 21–32.
162. Bogner, J. R., Kronawitter, U., Rolinski, B., Truebenbach, K., and Goebel, F. D. (1994) Liposomal doxorubicin in the treatment of advanced AIDS-related Kaposi's sarcoma. *J. Acquir. Immune Defic. Syndr. Hum. Retrovirol.* **7**, 463–468.
163. Gill, P. S., Espina, B. M., Muggia, F., et al. (1995) Phase I/II clinical and pharmacokinetic evaluation of liposomal daunorubicin. *J. Clin. Oncol.* **13**, 996–1003.
164. Stewart, S., Jablonowski, H., Goebel, F. D., et al. (1994) Randomized comparative trial of pegylated liposomal doxorubicin versus bleomycin and vincristine in the treatment of AIDS-related Kaposi's sarcoma. International Pegylated Liposomal Doxorubicin Study Group. *J. AIDS* **7**, 463–468.
165. Gill, P. S., Wernz, J., Scadden, D. T., et al. (1996) Randomized phase III trial of liposomal daunorubicin versus doxorubicin, bleomycin, and vincristine in AIDS-related Kaposi's sarcoma. *J. Clin. Oncol.* **14**, 2353–2364.
166. Northfelt, D. W., Dezube, B. J., Thommes, J. A., et al. (1998) Peglyted-liposomal doxorubicin versus doxorubicin, bleomycin, and vincristine in the treatment of AIDS-related Kaposi's sarcoma: results of a randomized phase III clinical trial. *J. Clin. Oncol.* **16**, 2445–2451.
167. Saville, M. W. (1998) Epidemic Kaposi's sarcoma (editorial). *Oncology* **12**, 1390–1391.
168. Bennett, C. L., Golub, R. M., Stinson, T. J., et al. (1998) Cost-effectiveness analysis comparing liposomal anthracyclines in the treatment of AIDS-related Kaposi's sarcoma. *J. Acquir. Immune Defic. Syndr. Hum. Retrovirol.* **18**, 460–465.
169. Gill, P. S., Tupule, A., Espina, B. M., et al. (1999) Paclitaxel is safe and effective in the treatment of advanced AIDS-related Kaposi's sarcoma. *J. Clin. Oncol.* **17**, 1876–1883.
170. Welles, L., Saville, M. W., Lietzau, J., et al. (1998) Phase II trial with dose titration of paclitaxel for the therapy of human immunodeficiency virus-associated Kaposi's sarcoma. *J. Clin. Oncol.* **16**, 1112–1121.
171. Nasti, G., Errante, D., Santarossa, S., Vaccher, E., and Tirelli, U. (1999) A risk and benefit assessment of treatment for AIDS-related Kaposi's sarcoma. *Drug Safety* **20**, 403–425.

172. Von Roenn, J., Krown, S. E., Benson, C. A., et al. (1998) Management of AIDS-associated Kaposi's sarcoma: A multidisciplinary perspective. **12(Suppl. 3)**, 9–24.
173. Sloand, E., Kumar, P. N., and Pierce, P. F. (1993) Chemotherapy for patients with pulmonary Kaposi's sarcoma: benefit of Filagrastin (G-CSF) in supporting dose administration. *South Med. J.* **86**, 1219–1224.
174. Schurmann, D., Dormann, A., Grunewald, T., and Ruf, B. (1994) Successful treatment of AIDS-related pulmonary Kaposi's sarcoma with liposomal daunorubicin. *Eur. Respir. J.* **7**, 824–825.
175. Saint-Marc, T., Jeanblanc, F., Makhloufi, D., and Touraine, J. L. (1996) Phase II clinical trial of liposomal daunorubicin in the treatment of pulmonary Kaposi's sarcoma. *Int. Conf. AIDS* **11(1)**, 303 (abstract no. Tu. B. 2219).
176. Grunaug, M., Bogner, J. R., Loch, O., and Goebel, F. D. (1998) Liposomal doxorubicin in pulmonary Kaposi's sarcoma: improved survival as compared to patients without liposomal doxorubicin. *Int. Conf. AIDS Eur. J. Med. Res.* **3**, 13–19.
177. Tupule, A., Yung, R. C., Wernz, J., et al. (1998) Phase II trial of liposomal daunorubicin in the treatment of AIDS-related pulmonary Kaposi's sarcoma. *J. Clin. Oncol.* **16**, 3369–3374.
178. Mitsuyasu, R., von Roen, J., Krown, S., Kaplan, L., and Testa, M. (1997) Comparison study of liposomal doxorubicin (DOX) alone or with bleomycin and vincristine (DBV) for treatment of advanced AIDS-associated Kaposi's sarcoma (AIDS-KS): AIDS clinical trials group (ACTG) protocol 286. *Proc. ASCO* **16**, 55a (abstract no. 191).
179. Palella, F. J., Jr., Delaney, K. M., and Moorman, A. C. (1998) Declining morbidity and mortality among patients with advanced human immunodeficiency virus infections. *N. Engl. J. Med.* **338**, 853–860.
180. Whitcup, S. M., Fortin, E., Nussenblatt, R. B., Polis, M. A., Muccioli, C., and Belfort, R. J. (1997) Therapeutic effect of combination antiretroviral therapy on cytomegalovirus retinitis [letter]. *JAMA* **277**, 1519–1520.
181. Power, C., Nath, A., Aoki, F. Y., and Bigio, M. D. (1997) Remission of progressive multifocal leukoencephalopathy following splenectomy and antiretroviral therapy in a patient with HIV infection [letter]. *N. Engl. J. Med.* **336**, 661–662.
182. Call, A., Foudraine, N., Reiss, P., et al. (1997) Resolution of antibiotic-resistant cryptosporidiosis (C) and microsporidiosis (M) with potent combination antiretroviral therapy [abstract]. *Fourth Conf. Retroviruses Opportun. Infect.* **4**, 191.
183. Peters, P. S. (1997) What does HIV teach us about cancer [editorial]. *Clin. Oncol. (R. Coll. Radiol.)* **9**, 355–356.
184. Aboulafia, D. M. (1998) Regression of acquired immunodeficiency syndrome-related pulmonary Kaposi's sarcoma after highly active antiretroviral therapy. *Mayo Clin. Proc.* **73**, 439–443.
185. Volm, M. D. and Wenz, J. (1997) Patients with advanced AIDS-related Kaposi's sarcoma (EKS) no longer require systemic therapy after introduction of effective antiretroviral therapy [abstract]. *Proc. Ann. Meet. Am. Soc. Clin. Oncol.* **16**, A162.

186. Santambrogio, S., Ridolfo, A. L., Galli, T. M., and Corbellino, P. M. (1998) Effect of highly active antiretroviral treatment (HAART) in patients with AIDS-associated KS. *12th World AIDS Conference* (abstract no. 22275), Geneva, Switzerland.
187. Bower, M., Fox, P., Fife, K., Gill, J., Nelson, M., and Gazzard, B. (1999) HAART prolongs time to treatment failure (TTF) in Kaposi's sarcoma (KS). Third National AIDS Malignancy Conference. *J. Acquir. Immune Defic. Syndr.* **21,** A24 (abstract 58).
188. Chanan-Khan, Goldman, M., Volm, C., et al. (1999) Patients with AIDS-related pulmonary Kaposi's sarcoma (PKS) treated with highly active antiretroviral therapy (HAART) alone after induction with systemic chemotherapy. Third National AIDS Malignancy Conference. *J. Acquir. Immune Defic. Syndr.* **21,** A24 (abstract 59).
189. Ganem, D. (1999) KSHV/HHV8 infection and the pathogenesis of AIDS-related neoplasms. An overview. Third National AIDS Malignancy Conference. *J. Acquir. Immune Defic. Syndr.* **21,** A8 (abstract S7).
190. Lederman, M. M., Connick, E., Landay, A., et al. (1998) Immunologic responses associated with 12 weeks of combination antiretroviral therapy consisting of zidovudine, lamivudine, and ritonavir: Results of AIDS Clinical Trials Group Protocol 315. *J. Infect. Dis.* **78,** 70–79.
191. Kempf, D. J., Rode, R. A., Xu, Y., et al. (1998) The duration of viral suppression during protease inhibitor therapy for HIV-1 infection is predicted by plasma HIV-1 RNA at the nadir. *AIDS* **12,** F9–F14.
192. Young, B., Johnson, S., Bahktiari, M., et al. (1998) Resistance mutations in protease and reverse transcriptase genes of HIV-1 isolates from patients failing combination antiretroviral therapy. *J. Infect. Dis.* **178,** 1497–1501.
193. Furtado, M. R., Callaway, D. S., Phair, J. P., et al. (1999) Resistance of HIV-1 transcription in peripheral-blood mononuclear cells in patients receiving potent antiretroviral therapy. *N. Engl. J. Med.* **340,** 1614–1622.
194. Desrosiers, R. C. (1999) Strategies used by human immunodeficiency virus that allow persistent viral replication. *Nature Med.* **5,** 723–725.

6

Myeloperoxidase Promoter Region Polymorphism and Lung Cancer Risk

Xifeng Wu, Matthew B. Schabath, and Margaret R. Spitz

1. Introduction

Globally, lung cancer is the leading cause of cancer death accounting for nearly one million deaths each year *(1)*. In the United States, lung cancer accounts for approx 13% of all incident cases and 28% of all cancer deaths *(2)*. It has been estimated that 90% of male lung cancer deaths and over 75% of female lung cancer deaths in the U.S. are caused by cigarette smoking *(3,4)*. Thus, elucidating the mechanisms of tobacco-induced lung cancer could lead to new strategies for decreasing lung cancer risk, for identifying susceptible individuals, and for developing innovative techniques for early detection *(4)*. One such mechanism that has been intensively investigated is the structure, function, and end-point effects of genetic polymorphisms in human metabolic genes.

Investigating polymorphic metabolic genes utilizing a molecular epidemiologic study design provides an approach for the identification of subpopulations that are more susceptible to chemically induced cancer *(5)*. The enzymatic biotransformation of xenobiotic exposures is controlled by genetically variant mechanisms. Specifically, the biologically available dose of carcinogens may be modulated by genetic variations in cellular mechanisms responsible for procarcinogen metabolism. In genetically susceptible individuals, a lower dose of an exposure could result in the same health endpoint as observed with higher exposures in nonsusceptible individuals *(6)*. Determining these interindividual differences in susceptibility may be important in assessing the human health risk from specific exogenous compounds.

Although this chapter only discusses the promoter region polymorphism associated with the myeloperoxidase (*MPO*) gene, there are numerous other polymorphic metabolic enzymes including the cytochrome P450 family (e.g., 1A1, 1A2, 2A, 2D6, etc.), glutathione-S-transferases (GST), N-acetyltransferases (NAT), Ah receptors, and mircosomal epoxide hydrolase. Moreover, genetic variability exists in genes involved in DNA repair capacity (e.g., *XRCC1*), cell-cycle check points (e.g., *cyclin D1*), and cancer-invasion and cancer-progression genes (e.g., *MMP*s and *VEGF*). Thus, the molecular epidemiologist interested in genetic susceptibility has a multitude of pathways to investigate in studying genetic susceptibility to carcinogenesis.

2. Molecular Basis of Myeloperoxidase
2.1. MPO and Carcinogenesis

Myeloperoxidase, first isolated in 1941 *(7)*, is an iron-containing heme protein localized in the azurophilic granules of neutrophil granulocytes and in the lysosomes of monocytes *(8)*. MPO is the most abundant protein in neutrophils constituting approx 5% of their dry weight *(9)*. The *MPO* gene, localized to the long arm of chromosome 17, consists of 11 introns and 12 exons accounting for approx 14 kb *(10,11)*. The single mRNA is translated into a protein approx 150 kD *(12)*. MPO enzyme synthesis is restricted to late myeloblasts and promyelocytes in the bone marrow *(13)*.

MPO catalyzes the reaction between hydrogen peroxide and the chloride ion and generates hypochlorous acid and other reactive oxygen species (ROS) *(14,15)*. The reactive by-products produced by MPO can react with most biological molecules to produce secondary free radicals *(16)* and have been associated with DNA damage *(17)*, DNA crosslinking *(18)*, and carcinogenesis *(19)*. MPO also metabolizes DNA-damaging intermediates; a variety of tobacco smoke mutagens; and environmental pollutants, including aromatic amines *(20)*, the promutagenic derivatives of polycyclic aromatic hydrocarbons (PAHs) *(18,21,22)*, and heterocyclic amines (HCAs) *(23)*. MPO has been reported to bioactivate specific procarcinogens, including benzo[a]pyrene intermediates (BPDE), 4-aminobiphenyl, and the arylamines *(18,20,24)*. Initially, exposure to these xenobiotic compounds induces pulmonary inflammation that results in the immune mediated recruitment of neutrophils containing the MPO enzyme *(25)*. As part of the immunological response to the pulmonary insult, neutrophils release MPO in the localized microenvironment *(26,27)* to oxidize and metabolize the xenobiotic. This reaction is characterized by a substantial increase in oxygen consumption and a subsequent NADPH-dependent production of superoxide and other free radicals *(23)*. Other potential mechanisms for carcinogenesis include the MPO-mediated binding of specific carcinogens

2.2. MPO Polymorphism

Austin et al. *(29)* examined the regulatory region of the *MPO* gene to determine if mutations might be responsible for differences in gene expression among different classes and cases within a class of acute leukemia. They noted a G→A nucleotide transition in the Alu region located 463 bases upstream from the transcription start site. The single nucleotide base transition is located in a HRE (hormone response element) and has been shown to negate the binding region for the SP1-transcription factor (TF) *(30)*. Piedrafita et al. *(30)* demonstrated that individuals with the A-allele genotype had overall weaker transcriptional activity of the *MPO* gene due to the reduced binding of the SP1-TF. The loss of the SP1-TF binding results in a marked decrease in mRNA expression. This decrease in MPO expression results in less enzyme available for carcinogenic activation of carcinogens, enzyme-mediated DNA damage, and free radical production. Several molecular epidemiologic studies *(31–36)*, that are reviewed below, have tested the hypothesis that individuals with one or more copies of the A-allele are at a decreased risk for lung cancer.

Because MPO is present at high levels in circulating monocytes and neutrophils, its role in other pathological processes has also been studied. Other studies demonstrated that the *MPO* polymorphisms are associated with numerous diseases including acute promyelocytic leukemia (APL) *(14)*, multiple sclerosis (MS) *(16)*, cystic fibrosis (CF) *(37)*, Alzheimer's disease (AD) *(38)*, and atherosclerosis *(39)*. Reynolds et al. *(14)* reported the higher expressing G/G genotype is shown to be overrepresented in acute promyelocytic leukemia-M3 (79%) and -M4 (82%), suggesting that higher levels of MPO are associated with an increased risk for these subclasses of leukemias. The wildtype genotype has also been noted to be over-represented (74%) in woman with early onset MS, suggesting that increased levels of MPO may accelerate neural damage *(16)*. Witko-Sarsat et al. *(37)* examined linkage between the CF genetic autosomal recessive disorder and disturbance in neutrophil function and concluded that a modification of intracellular pH and/or ionic concentrations may be related to the altered MPO enzymatic activity observed in CF neutrophils. The higher expressing G/G genotype is also associated with increased incidence of AD in females, and decreased incidence in males ($p = 0.006$) suggesting that the MPO polymorphism is a gender-specific risk factor for AD *(38)*. Daugherty et al. *(39)* explored the potential role of MPO in the development of atherosclerosis by determining whether the enzyme was present in surgically excised human vascular tissue. Their findings identified

MPO as a component of human vascular lesions, which suggests MPO may contribute to atherogenesis by catalyzing oxidative reactions in the vascular wall. The current body of literature suggests that MPO is an important factor in a myriad of diseases including carcinogenesis of the lung. The next section of this review focuses on the *MPO* polymorphism as a susceptibility factor for lung cancer.

3. Molecular Epidemiology

As of December 2001, there have been five published reports *(31–35)* and one study in press *(36)* on the *MPO* polymorphism and lung cancer risk (*see* **Table 1**). The study in press *(36)* was confirmatory of a smaller report that was previously published *(34)*. Thus, for clarity, this review will only consider the more recent report's results. Four of the five studies demonstrated consistent evidence of an association between the *MPO* genotypes and lung cancer risk, especially among Caucasians. The protective effects associated with the A/A genotype (compared to the G/G genotype) varied from 40–70%, while the protective effects of the A-allele genotypes (G/A + A/A) ranged from 17–42%. Among Caucasians, the A-allele frequency in controls varied from 20.3–25.8%, and in lung cancer cases from 15.6–20.5%. Although all five studies reported data on Caucasians, two of the studies also examined other ethnic groups. London et al. *(31)* noted that the A-allele was slightly less frequent in African-American controls (29.9%) as compared to cases (31.2%). Le Marchand et al. *(33)* noted A-allele frequencies of 13.1% in Hawaiian controls (lowest of any control group reported) and 16.9% in Japanese controls. The Japanese lung cancer cases had the lowest A-allele frequency (11.6%) of any case or control group currently reported.

London et al. *(31)* first reported a 70% protective effect (OR = 0.30; 95% CI 0.10–0.93) associated with A/A genotypes in 182 Caucasian lung cancer cases. A smaller protective effect that was statistically nonsignificant (OR = 0.61; 95% CI 0.26–1.41) was also noted in 157 African-American cases. All subjects from this study were recruited from Los Angeles County, CA. Incident cases were recruited from 35 different hospitals and controls were recruited from driver's license lists and Medicare beneficiaries. This study did not observe a modified risk of lung cancer for the G/A genotype. When we calculated the univariate odds ratios for the A-allele genotypes for both ethnic groups, a small protective effect that is statistically nonsignificant is observed among Caucasians (OR = 0.83; 95% C.I. 0.57–1.18) and no effect in African-Americans (OR = 1.19; 95% CI 0.79–1.78).

Another study *(33)* observing supportive findings examined 323 lung cancer patients identified through the Hawaiian Tumor Registry and controls randomly recruited from Oahu residents via the State of Hawaii Department of Health.

Author/yr/ref.	Ethnicity and cases/controls (n)	MPO genotypes G/G (N)	G/A (N)	A/A (N)	A-Allele frequency	OR (95% CI) A-Allele genotypes vs G/G genotype	A/A Genotype vs G/G genotype	Comments
London et al. (1997) (*31*)	Caucasian							Frequency matched on age, gender, and ethnicity
	Cases (n = 182)	119	59	4	18.4%	0.83 (0.58–1.19)[a]		
	Controls (n = 459)	280	143	36	23.4%			
	African-American							
	Cases (n = 157)	71	74	12	31.2%	1.19 (0.79–1.78)[a]	0.61 (0.26–1.41)	
	Controls (n = 244)	121	100	23	29.9%			
Le Marchand et al. (2000) (*33*)	Caucasian							1:1 matched on age (±2 yr), gender, and ethnicity
	Cases (n = 135)	90	38	7	19.3%	0.67 (0.42–1.07)[a]	0.60 (0.20–2.00)	
	Controls (n = 171)	98	58	15	25.7%			
	Japanese							
	Cases (n = 108)	84	23	1	11.6%	0.69 (0.39–1.20)[a]	0.10 (0.00–1.50)	
	Controls (n = 163)	115	41	7	16.9%			
	Hawaiian							
	Cases (n = 80)	60	16	4	15.0%	1.23 (0.62–2.44)[a]	1.50 (0.20–1.60)	
	Controls (n = 103)	81	17	5	13.1%			
Cascorbi et al. (2000) (*32*)	Caucasian							1:1 Matched controls on age and gender
	Cases (n = 196)	141	49	6	15.6%	0.47 (0.28–0.79)	1.24 (0.34–4.52)[a]	
	Controls (n = 196)	117	75	4	21.2%			
Misra et al. (2001) (*35*)	Caucasian							Used incidence density sampling for 1:1 matching on age (±5 yr), intervention group, study clinic, and date of blood draw (± 45 d)
	Cases (n = 315)	191	108	16	22.2%	1.13 (0.80–1.61)	0.72 (0.32–1.65)	
	Controls (n = 311)	206	84	21	20.3%			
Schabath et al. (2002) (*36*)	Caucasian							Frequency matched on age (± 5 yr), gender, smoking status, and ethnicity
	Cases (n = 375)	235	126	14	20.5%	0.66 (0.49–0.90)	0.59 (0.27–1.30)	
	Controls (n = 378)	202	157	19	25.8%			

[a]Crude (univariate) odds ratio.

The authors tested the association of the *MPO* polymorphisms in 3 ethnic groups of Caucasian, Japanese, or native Hawaiian ancestry. Notable differences in the allele frequencies were observed between the three ethnic groups. The A/A genotype yielded protective effects that were not statistically significant when the three ethnic groups were combined (OR = 0.5, 95% CI 0.2–1.3), in Caucasians (OR = 0.6, 95% CI 0.2–2.0), and in Japanese (OR = 0.1; 95% CI 0.0–1.5). An elevated odds ratio was noted in Hawaiians with the A/A genotype (OR = 0.5; 95% 0.2–1.6). When we combined the A-allele genotypes, similar results are observed in Caucasians (OR = 0.67; 95% CI 0.42–1.07) and Hawaiians (OR = 1.23; 95% CI 0.62–2.44), while a modulation of the overall protective effects are observed in Japanese (OR = 0.69 95% CI 0.39–1.20).

In a study from Berlin, Germany *(32)* consisting of 196 lung cancer, 245 laryngeal cancer, and 255 pharyngeal cancer patients, the A-allele genotypes were associated with protective effects for both lung cancer (OR = 0.47; 95% CI 0.28–0.79), and laryngeal cancer (OR = 0.66; 95% CI 0.44–1.01) (but not for pharyngeal cancer [OR = 0.75; 95% CI 0.51–1.12]). The A/A genotype was associated with an elevated odds ratio among the lung cancer patients (OR = 1.25; 95% CI 0.34–4.52). Only participants of German ancestry were included in this study.

In Schabath et al. *(36)* protective effects for lung cancer were observed in 375 Caucasian cases with the A/A genotype (OR = 0.59; 95% CI 0.27–1.30), the G/A genotype (OR = 0.67; 95% CI 0.49–0.92), and for the A-allele genotypes (OR = 0.66; 95% CI 0.49–0.90). Lung cancer cases were hospital based while controls were identified and recruited from a multi-specialty managed care organization in the Houston, TX area.

The consistency of these findings, that together account for over 1200 cases and over 1700 controls, suggests an important role for MPO in lung cancer etiology, possibly through activation of carcinogens and/or production of free radicals in or near susceptible target cells *(23)*. However, from a nested case-control sample consisting of only male smokers participating in the Alpha-Tocopherol, Beta-Carotene Cancer Prevention (ATBC) Study, Misra et al. *(35)* found no evidence of an association with the *MPO* genotypes among 315 lung cancer cases and 311 controls. A small protective effect that was statistically nonsignificant was observed in individuals with the A/A genotype (OR = 0.72; 95% CI 0.32–1.65), but no protection was observed in the G/A genotype (OR = 1.21; 95% CI 0.84–1.75) or A-allele genotypes groups combined (OR = 1.13; 95% CI 0.80–1.61).

3.1. Gender Effects

Variations in metabolic activation of carcinogens could also account for gender differences in overall lung cancer risk. Gender-specific differences with other metabolic Phase I enzymes have been demonstrated in animal models *(41–44)*. There is plausible evidence as well that a gender-specific role exists for the *MPO* polymorphism. Reynolds et al. *(38)* reported an increased incidence of AD in women with the G/G genotype, while Nagra et al. *(16)* found that this same genotype was over-represented in early-onset MS in women. It has also been suggested that gender-associated hormones directly or indirectly result in differential MPO gene expression in MS and AD *(38)*.

The current published data have yet to demonstrate a consistent gender effect associated with the *MPO* genotypes and lung cancer risk. Schabath et al. *(36)* noted a statistically significant reduced risk of lung cancer for men (OR = 0.55), but not women (OR = 0.81) (*see* **Table 2**). Additionally, an age effect (discussed below) was only evident in men. Conversely, Misra et al. *(35)* reported a statistically significant elevated odds ratio among a subset of older men (age 64–69) with the A-allele genotypes (OR = 2.92; 95% 1.33–6.43). However, their nested case-control study population consisted exclusively of male smokers selected from the ATBC clinical trial. The distinctiveness of examining only chronic male smokers could limit their ability to discern any differences. In the Berlin study, Cascorbi et al. *(32)* did not observe a gender effect among 150 men with lung cancer as compared to a limited sample size of 46 women with lung cancer. No gender-specific effect was observed by Le Marchand et al. *(33)*.

3.2. Age Effects

Advanced age is a risk factor for numerous chronic illnesses, including lung cancer. Therefore, it could be hypothesized that the protective effects associated with the A-allele genotypes diminish as age increases. Cascorbi et al. *(32)* reported no appreciable difference in the odds ratios for younger or older individuals when stratified on the median age. Schabath et al. *(36)* reported protective effects associated with the variant allele genotypes that decreased with increasing age for men (*see* **Table 2**), but not women. As mentioned previously, Misra et al. *(35)* reported a statistically significant elevated odds ratio among older men with the A-allele genotypes. The authors did report a statistically significant age interaction term that modified the association between the *MPO* genotypes and lung cancer risk. A decline in mediated immune responses to pulmonary insults, reduced DNA repair capacity, and cumulative xenobiotic exposures may contribute to the diminished protective effects in older individuals. These cumulative factors in older aged individuals may counteract the protective effects associated with the *MPO* variant allele genotypes. Additionally, susceptibility factors may play a more important role in early onset cancer as demonstrated by the protective effects in younger individuals as compared to older patients.

Table 2
Summary of Point Estimates Associated with the A-allele Genotypes by Selected Host Characteristics

Variable(s)			Cases (n)	Controls (n)	Univariate OR (95% CI)	Multivariate OR (95% CI)[a]
Gender						
	Men		195	207	0.52 (0.35–0.78)	0.55 (0.36–0.84)
	Women		180	171	0.91 (0.59–1.39)	0.81 (0.55–1.26)
Age (Men only)						
	32–49		15	30	0.29 (0.07–1.16)	0.38 (0.08–1.86)
	50–59		52	57	0.46 (0.21–1.01)	0.48 (0.20–1.15)
	60–69		83	76	0.53 (0.28–1.02)	0.59 (0.30–1.14)
	+70		45	44	0.67 (0.29–1.54)	0.65 (0.27–1.58)
Gene-environmental interactions						
	Pack-Yr					
	<30		62	77	0.47 (0.24–0.94)	0.39 (0.19–0.82)
	30–64.5		137	163	0.66 (0.41–1.05)	0.66 (0.41–1.06)
	≥64.6		113	78	0.75 (0.42–1.36)	0.76 (0.42–1.36)
Asbestos	Genotype					
–	G/G		155	156	1.0	1.0[b]
+	G/G		80	46	1.75 (1.14–2.68)	1.72 (1.09–2.66)
+	G/A + A/A		49	55	0.89 (0.57–1.39)	0.89 (0.56–1.44)
–	G/A + A/A		91	121	0.76 (0.53–1.08)	0.73 (0.49–1.06)
Asbestos	Genotype	Pack-yr				
–	G/G	<45	51	74	1.0	1.0[b]
–	G/A + A/A	<45	30	52	0.84 (0.4–1.48)	0.84 (0.46–1.54)
+	G/G	<45	25	14	2.59 (1.22–5.46)	3.13 (1.42–6.93)
+	G/A + A/A	<45	17	27	0.91 (0.46–1.84)	0.92 (0.44–1.93)
–	G/G	≥45	74	52	2.06 (1.25–3.42)	1.89 (1.11–3.25)
–	G/A + A/A	≥45	41	44	1.35 (0.78–2.35)	1.13 (0.63–2.03)
+	G/G	≥45	49	28	2.54 (1.42–4.55)	2.19 (1.16–4.11)
+	G/A + A/A	≥45	25	27	1.34 (0.70–2.56)	1.18 (0.58–2.38)

[a]Adjusted by age, gender, and smoking, where appropriate.
[b]Reference Strata.
Adapted from ref. *36,45*.

3.3. Gene-Environmental Interactions

The MPO-mediated conversion of specific procarcinogens may only occur if there is a biological available dose. Thus, it is reasonable to hypothesize that the protective effects are only observed in smokers, since a substrate is necessary to facilitate the metabolic bioconversions of procarcinogens. It is also plausible that in heavier smokers, the concentration of procarcinogens may overwhelm any protective effects associated with the A-allele genotypes.

London et al. *(31)* reported no appreciable change in the point estimate for never smokers vs all study subjects for either Caucasians or African-Americans. Similarly, Le Marchand et al. *(33)*, Cascorbi et al. *(32)*, and Misra et al. *(35)* observed no difference in lung cancer risk by cigarette pack-years. Misra et al. *(35)* noted a statistically significant interaction term for smoking duration which modified the association between the *MPO* genotypes and lung cancer risk. In Schabath et al. *(36)*, a 37% protective effect (OR = 0.63; 95% CI 0.45–0.87) in ever smokers with the A-allele genotypes was evident as opposed to no effect in never smokers (OR = 1.14; 95% CI 0.42–3.11) (*see* **Table 2**). Additionally, there was an incremental decrease in the protective effects as cigarette pack-years increased. Thus, lightest smokers were provided the greatest protection (OR = 0.39; 95% CI 0.19–0.82). In the study from Berlin, Germany, no effect by pack-years was observed.

Published reports have demonstrated an association between lung cancer and occupationally related exposures. There have not been published reports investigating a possible link between the *MPO* genotypes, occupational exposures, and lung cancer risk. Schabath et al. *(45)* recently performed an analysis of asbestos-related occupations and exposures to determine if lung cancer risk is modulated by the *MPO* genotypes among individuals exposed to asbestos and/or tobacco smoke. Asbestos exposure was selected *a priori* for the occupational analysis in this analysis since considerable evidence exists for markedly elevated lung cancer risk associated with exposure to asbestos.

There was a statistically significantly elevated estimate of risk (OR = 1.45; 95% CI 1.04–2.02) associated with a summary measure of asbestos exposure after controlling for age, sex, and smoking status. Initially, two separate stratified analyses were performed to examine the joint effects between 1) the *MPO* genotypes and asbestos exposure, and 2) the *MPO* genotypes and pack-years of cigarette smoking (*see* **Table 2**). Both analyses yielded quite similar results. Individuals with the A-allele genotypes vs the G/G genotype were provided a greater protection among those with the same level of environmental exposure (cigarette smoking or asbestos). When comparing individuals with the A-allele genotypes, individuals with the low-risk exposure profile (not exposed to asbestos or light cigarette smokers) were provided greater protection vs those in the high-risk exposure profile (exposed to asbestos or heavier cigarette smokers). Asbestos exposure, cigarette smoking, and the *MPO* genotypes were also analyzed simultaneously in one model. Among individuals with the same exposure profile for asbestos exposure and cigarette smoking, the A-allele genotypes consistently provided a reduced risk for lung cancer. This analysis provides reasonable epidemiologic evidence that lung cancer risks are modified by the A-allele genotypes among individuals exposed to asbestos as well as tobacco smoke *(45)*.

4. Summary

Historically, myeloperoxidase activity and subsequent production of hypochlorous acid has been associated with the killing of host-invading microorganisms (bacteria, viruses, and fungi). Currently, there is a wealth of evidence that the *MPO* polymorphism and enzyme activity is associated with a wide range of pathological and biological processes, including lung cancer carcinogenesis. Although the molecular epidemiology reports reviewed in this chapter are not in complete agreement on all aspects of their findings, it is evident that the *MPO* polymorphism contributes to the modulation of overall lung cancer risk. Four of the five molecular epidemiologic studies reviewed in this chapter utilized similar case-control study designs with quite different sources of populations. These studies all demonstrate that the *MPO* variant genotype modulates overall lung cancer risk. However, these studies are not in agreement regarding age and gender effects, and gene-environmental interactions. The nested case control study designed utilized by Misra et al. *(35)* is a valid and sound approach, but their results found no evidence of an association. Certainly, heterogeneity in study populations can contribute to the variability between these studies. Additionally, epidemiologic issues such as case-control matching and sources of control populations may contribute to the conflicting findings. Thus, the publication of inconsistent or null studies as well as other positive findings is certainly encouraged to elucidate the range of effects associated with the *MPO* polymorphism and lung cancer risk.

Acknowledgments

This work was supported by Grants CA55769, CA68437, and CA74880 from the National Cancer Institute. Matthew B. Schabath was also supported, in part, by a cancer prevention fellowship supported by the National Cancer Institute grant R25 CA57730, Robert M. Chamberlain, PhD, Principal Investigator.

References

1. Parkin, D. M., Pisani, P., and Ferlay, J. (1999) Global cancer statistics. *CA A. Cancer J. Clin.* **49,** 33–64.
2. Cancer Facts and Figures (2001) American Cancer Society, Atlanta, GA.
3. Shopland, D. R. (1995) Tobacco use and its contribution to early cancer mortality with a special emphasis on cigarette smoking. *Environ. Health Perspect.* **8,** 131–142.
4. Hecht, S. S. (1999) Tobacco smoke carcinogens and lung cancer. *J. Natl. Cancer Inst.* **91,** 1194–1210.

5. Vineis, P., Malats, N., and Boffetta, P. (1999) Why study metabolic susceptibility to cancer? in *Metabolic Polymorphisms and Susceptibility to Cancer* (Vineis, P., Malats, N., Lang, M., d'Errico,, A., Caporaso, N., Cuzick, J., and Boffetta, P., eds.), IARC Scientific Publications, Lyon, France, pp. 1–3.
6. Garte, S. (1998) The role of ethnicity in cancer susceptibility gene polymorphisms: the example of CYP1A1. *Carcinogenesis* **19,** 1329–1332.
7. Agner, K. (1941) Verdoperoxidase. A ferment isolated from leukocytes. *Acta Physiol. Scand.* **2(Suppl. 8)** 1–62.
8. Lanza, F. (1998) Clinical manifestation of myeloperoxidase deficiency. *J. Mol. Med.* **76,** 676–681.
9. Weil, S. C., Rosner, G. L., Reid, M. S., Chisholm, R. L., Farber, N. M., Spitznagel, J. K., and Swanson, M. S. (1987) cDNA cloning of human myeloperoxidase: decrease in myeloperoxidase mRNA upon induction of HL-60 cells. *Proc. Natl. Acad. Sci. USA* **84,** 2057–2061.
10. Morishita, K., Kubota, N., Asano, S., Kaziro, Y., and Nagata, S. (1987) Molecular cloning and characterization of cDNA for human myeloperoxidase. *J. Biol. Chem.* **262,** 3844–3851.
11. Morishita, K., Tsuchiya, M., Asano, S., Kaziro, Y., and Nagata, S. (1987) Chromosomal gene structure of human myeloperoxidase and regulation of its expression by granulocyte colony-stimulating factor. *J. Biol. Chem.* **262,** 15208–15213.
12. Podrez, E. A., Abu-Soud, H. M., and Hazen, S. L. (2000) Myeloperoxidase-generated oxidants and atherosclerosis. *Free Radic. Biol. Med.* **28,** 1717–1725.
13. Petrides, P. E. (1998) Molecular genetics of peroxidase deficiency. *J. Mol. Med.* **76,** 688–698.
14. Reynolds, W. F., Chang, E., Douer, D., Ball, E. D., and Kanda, V. (1997) An allelic association implicates myeloperoxidase in the etiology of acute promyelocytic leukemia. *Blood* **90,** 2730–2737.
15. Kizaki, M., Miller, C. W., Selsted, M. E., and Koeffler, H. P. (1994) Myeloperoxidase (MPO) gene mutation in hereditary MPO deficiency. *Blood* **83,** 1935–1940,
16. Nagra, R. M., Becher, B., Tourtellotte, W. W., Antel, J. P., Gold, D., Paladino, T., et al. (1997) Immunohistochemical and genetic evidence of myeloperoxidase involvement in multiple sclerosis. *J. Neuroimmunol.* **78,** 97–107.
17. Pero, R. W., Sheng, Y., Olsson, A., Bryngelsson, C., and Lund-Pero, M. (1996) Hypochlorous acid/N-chloramines are naturally produced DNA repair inhibitors. *Carcinogenesis* **17,** 13–18.
18. Petruska, J. M., Mosebrook, D. R., Jakab, G. J., and Trush, M. A. (1992) Myeloperoxidase-enhanced formation of (+–)-trans-7,8-dihydroxy-7, 8-dihydrobenzo[a]pyrene-DNA adducts in lung tissue in vitro: a role of pulmonary inflammation in the bioactivation of a procarcinogen. *Carcinogenesis* **13,** 1075–1081.
19. Josephy, P. D. and Coomber, B. L. (1998) The 1996 Veylien Henderson Award of the Society of Toxicology of Canada. Current concepts: neutrophils and the activation of carcinogens in the breast and other organs. *Can. J. Physiol. Pharmacol.* **76,** 693–700.

20. Tsuruta, Y., Subrahmanyam, V. V., Marshall, W., and O'Brien, P. J. (1985) Peroxidase-mediated irreversible binding of arylamine carcinogens to DNA in intact polymorphonuclear leukocytes activated by a tumor promoter. *Chem-Biol. Interact.* **53,** 25–35.
21. Trush, M. A., Seed, J. L., and Kensler, T. W. (1985) Oxidant-dependent metabolic activation of polycyclic aromatic hydrocarbons by phorbol ester-stimulated human polymorphonuclear leukocytes: possible link between inflammation and cancer. *Proc. Natl. Acad. Sci. USA* **82,** 5194–5198.
22. Mallet, W. G., Mosebrook, D. R., and Trush, M. A. (1991) Activation of (+–)-trans-7,8-dihydroxy-7,8-dihydrobenzo[a]pyrene to diolepoxides by human polymorphonuclear leukocytes or myeloperoxidase. *Carcinogenesis* **12,** 521–524.
23. Williams, J. A. (2001) Single nucleotide polymorphisms, metabolic activation and environmental carcinogenesis: why molecular epidemiologists should think about enzyme expression. *Carcinogenesis* **22,** 209–214.
24. Culp, S. J., Roberts, D. W., Talaska, G., Lang, N. P., Fu, P. P., Lay, J. O., Jr., et al. (1997) Immunochemical, 32P-postlabeling, and GC/MS detection of 4-aminobiphenyl-DNA adducts in human peripheral lung in relation to metabolic activation pathways involving pulmonary N-oxidation, conjugation, and peroxidation. *Mutat. Res.* **378,** 97–112.
25. Hunninghake, G. W. and Crystal, R. G. (1983) Cigarette smoking and lung destruction. Accumulation of neutrophils in the lungs of cigarette smokers. *Am. Rev. Respir. Dis.* **128,** 833–838.
26. Schmekel, B. and Venge, P. (1991) The distribution of myeloperoxidase, eosinophil cationic protein, albumin and urea in sequential bronchoalveolar lavage. *Eur. Respir. J.* **4,** 517–523.
27. Schmekel, B., Karlsson, S. E., Linden, M., Sundstrom, C., Tegner, H., and Venge, P. (1990) Myeloperoxidase in human lung lavage. I. A marker of local neutrophil activity. *Inflammation* **14,** 447–454.
28. Shen, Z., Wu, W., and Hazen, S. L. (2000) Activated leukocytes oxidatively damage DNA, RNA, and the nucleotide pool through halide-dependent formation of hydroxyl radical. *Biochemistry* **39,** 5474–5482.
29. Austin, G. E., Lam, L., Zaki, S. R., Chan, W. C., Hodge, T., Hou, J., et al. (1993) Sequence comparison of putative regulatory DNA of the 5′ flanking region of the myeloperoxidase gene in normal and leukemic bone marrow cells. *Leukemia* **7,** 1445–1450.
30. Piedrafita, F. J., Molander, R. B., Vansant, G., Orlova, E. A., Pfahl, M., and Reynolds, W. F. (1996) An Alu element in the myeloperoxidase promoter contains a composite SP1-thyroid hormone-retinoic acid response element. *J. Biol. Chem.* **271,** 14412–14420.
31. London, S. J., Lehman, T. A., and Taylor, J. A. (1997) Myeloperoxidase genetic polymorphism and lung cancer risk. *Cancer Res.* **57,** 5001–5003.
32. Cascorbi, I., Henning, S., Brockmoller, J., Gephart, J., Meisel, C., Muller, J. M., Loddenkemper, R., and Roots, I. (2000) Substantially reduced risk of cancer of

the aerodigestive tract in subjects with variant—463A of the myeloperoxidase gene. *Cancer Res.* **60,** 644–649.
33. Le Marchand, L., Seifried, A., Lum, A., and Wilkens, L. R. (2000) Association of the myeloperoxidase -463G→a polymorphism with lung cancer risk. *Cancer Epidemiol. Biomarkers Prev.* **9,** 181–184.
34. Schabath, M. B., Spitz, M. R., Zhang, X., Delclos, G. L., and Wu, X. (2000) Genetic variants of myeloperoxidase and lung cancer risk. *Carcinogenesis* **21,** 1163–1166.
35. Misra, R. R., Tangrea, J. A., Virtamo, J., Ratnasinghe, D., Andersen, M. R., Barrett, M., et al. (2001) Variation in the promoter region of the myeloperoxidase gene is not directly related to lung cancer risk among male smokers in Finland. *Cancer Lett.* **164,** 161–167.
36. Schabath, M. B., Spitz, M. R., Hong, W. K., Delclos, G. L., Reynolds, W. F., Gunn, G. B., et al. (2002) A genetic polymorphism associated with the myeloperoxidase (MPO) gene and reduced risk of lung cancer. *Lung Cancer*, in press.
37. Witko-Sarsat, V., Allen, R. C., Paulais, M., Nguyen, A. T., Bessou, G., Lenoir, G., and Descamps-Latscha, B. (1996) Disturbed myeloperoxidase-dependent activity of neutrophils in cystic fibrosis homozygotes and heterozygotes, and its correction by amiloride. *J. Immunol.* **157,** 2728–2735.
38. Reynolds, W. F., Rhees, J., Maciejewski, D., Paladino, T., Sieburg, H., Maki, R. A., and Masliah, E. (1999) Myeloperoxidase polymorphism is associated with gender specific risk for Alzheimer's disease. *Exp. Neurol.* **155,** 31–41.
39. Daugherty, A., Dunn, J. L., Rateri, D. L., and Heinecke, J. W. (1994) Myeloperoxidase, a catalyst for lipoprotein oxidation, is expressed in human atherosclerotic lesions. *J. Clin. Invest.* **94,** 437–444.
40. Pampori, N. A. and Shapiro, B. H. (1999) Gender differences in the responsiveness of the sex-dependent isoforms of hepatic P450 to the feminine plasma growth hormone profile. *Endocrinology* **140,** 1245–1254.
41. Watanabe, M., Tanaka, M., Tateishi, T., Nakura, H., Kumai, T., and Kobayashi, S. (1997) Effects of the estrous cycle and the gender differences on hepatic drug-metabolising enzyme activities. *Pharmacol. Res.* **35,** 477–480.
42. Sharma, M. C., Agrawal, A. K., Sharma, M. R., and Shapiro, B. H. (1998) Interactions of gender, growth hormone, and phenobarbital induction on murine Cyp2b expression. *Biochem. Pharmacol.* **56,** 1251–1258.
43. Wong, S. G., Kobus, S. M., McNamee, J. P., and Marks, G. S. (1998) Gender differences in N-alkyl protoporphyrin IX production in rats after the administration of porphyrinogenic xenobiotics. *Drug Metab. Dispos.* **26,** 739–744.
44. Lee, R. P. and Forkert, P. G. (1995) Pulmonary CYP2E1 bioactivates 1, 1-dichloroethylene in male and female mice. *J. Pharmacol. Exp. Ther.* **273,** 561–567.
45. Schabeth, M. B., Spitz, M. R., Delclos, G. L., Gunn, G. B., Whitehead, L. W., and Wu, X. (2002) An association between asbestos exposure, cigarette smoking, myeloperoxidase genotypes, and lung cancer risk. *Am. J. Ind. Med.,* in press.

7

Clinical Utility of Tumor Markers in the Management of Non-Small Cell Lung Cancer

Dennis E. Niewoehner and Jeffrey B. Rubins

1. Introduction

Non-small cell lung cancer (NSCLC), a leading cause of cancer-related death for men and women worldwide, exhibits a highly variable clinical course. Death may occur within a few weeks of diagnosis at one extreme, whereas other cases have apparently benign outcomes for periods of up to 20 years with no treatment (*1*). The current tumor-node-metastasis (TNM) staging system for NSCLC, which is based primarily on imaging studies and which serves as the principal guide for therapy, has significant limitations. For example, five-yr survival after "curative" resection for apparently localized, stage I NSCLC is only 46%, with death usually owing to recurrent or metastatic cancer (*2*). The reasons for this variability in clinical outcomes are largely unknown.

The past two decades have witnessed rapid advances in our understanding of the cellular mechanisms governing carcinogenesis in NSCLC. Available evidence indicates that multiple gene mutations are central to this process (*3*). It is also clear that a specific histological type of tumor may exhibit highly variable patterns of gene mutations and gene expressions. It is reasonable to suspect that specific molecular markers, or combinations of such markers, might help explain the unpredictable clinical behavior of individual tumors.

2. Molecular Markers in NSCLC

The identification of reliable molecular markers in NSCLC could serve several purposes. At the more basic level this information could yield new insights about pathogenesis, which in turn might lead to novel therapeutic modalities. At the clinical level it might improve staging as a guide to therapy.

For example, molecular markers might identify patients with early-stage NSCLC who are at high risk for disease recurrence or surgical failure and who might benefit from adjuvant or nonsurgical therapies.

Tumor markers are used clinically in the staging and surveillance of breast, colon, and prostate cancers. There have been numerous efforts to identify comparable markers in NSCLC. From a MEDLINE search we found several hundred articles addressing this topic. Most of these articles report a positive outcome in the sense that one or more molecular markers were found to be statistically associated with some measure of clinical outcome. An incomplete list of tumor markers with putative prognostic value includes K-*ras* *(4)* and *p53* gene mutations *(5)*, as well as expression of the *p53* protein *(6)*, the *bcl*-2 protein *(7)*, the c-*erb*B-2 protein *(8)*, the epidermal growth factor receptor (EGFR) or c-*erb*B-1 protein *(9)*, an antigen contained in the H/Le^y/Le^b complex *(10)*, and a variant isoform of CD44 *(11)*.

Despite these many studies, tumor markers are not presently recommended in the evaluation of NSCLC, because of continuing uncertainty about their prognostic value. Several methodologic problems limit the utility of tumor markers in current clinical practice.

One fundamental limitation is the practical problem of obtaining sufficient tissue for detailed laboratory studies. Most clinical studies of molecular tumor markers in NSCLC have been limited to patients undergoing surgical exploration where sizable tumor samples can be obtained. Since only a minority of patients with NSCLC undergoes surgery, results are heavily biased towards patients of lesser stage disease and better functional status. This may not be a problem if the sole purpose for identifying prognostic tumor markers is to assist therapeutic decisions in early stage NSCLC. However, the inclusion of tumors from all stages of the disease might yield a broader and clearer picture of their biological importance.

Study design is another limitation that has produced the seemingly inconsistent and in some cases frankly contradictory results in the literature. Even a cursory review of this subject reveals many examples. One relatively large study ($n = 271$) reported that overexpression of *p53* protein adversely affected survival *(12)*, whereas another somewhat smaller study ($n = 156$) reported that the same variable conferred a better prognosis *(13)*. Similarly contradictory results can be found for other molecular markers, including expression of the bcl-2 oncogenic protein *(14,15)* and expression of an antigen in the H/Le^y/Le^b complex detected by the MIA-15-5 antibody *(10,16)*.

3. Meta-analyses

Many of these apparent inconsistencies can probably be attributed to the limited statistical power of the many published studies that include only modest numbers of cases. Quantitative systematic review, meta-analysis, may improve the precision of the overall estimate by pooling much larger numbers of outcomes for comparison. Meta-analyses may also provide insights about other apparent disparities among different studies.

Huncharek served as the lead author for two separate study groups that performed meta-analyses on the associations of the K-*ras* mutation and of the *p53* mutation with survival in NSCLC *(17,18)*. Employing accepted methods and including all studies published through 1997 that met pre-established quality criteria, the authors combined 881 cases from eight studies for their analyses of the K-*ras* mutation and 829 cases from eight studies for their analyses of the *p53* mutation. Summary estimates indicated that a mutation at either site conferred an unfavorable prognosis; the relative risk for death at 2 yr in the presence of a K-*ras* mutation was 2.35 (95% CI, 1.61–3.22) and in the presence of a *p53* mutation was 1.52 (95% CI, 1.07–2.16).

These systematic reviews suggest that both K-*ras* and *p53* mutations do have some prognostic value. However, the authors themselves raised serious questions about the accuracy of their pooled estimates. The statistical test for homogeneity indicated that the range of values from individual studies included in both meta-analyses was so large that it was unlikely that the patients were all drawn from the same population. Results from individual studies varied by more than an order of magnitude in both cases. The heterogeneity among these studies might have multiple sources. Most of the studies that were included in both meta-analyses did not specify the method of case selection, so that arbitrary case selection might have introduced a bias. Different studies were not well-matched according to anatomical stage or histological subtype. Methods of quantifying tissue markers, particularly those determined by immunoblots or immunohistochemistry, require subjective interpretation; yet information was rarely provided about intra- and inter-observer reproducibility. Similarly, genetic methods to amplify and identify mutations varied between studies with respect to the protocols used and specific genetic mutations reported. The accuracy of these methods may vary between studies, depending on the method and degree of preservation of the tissue analyzed. Mutation rates at specific codons within the *p53* locus varied inexplicably from study to study and these differences might be biologically important. In addition, true differences may exist among the North American, European, and Asian populations from which cases in the various studies were drawn.

4. Publication Bias

Neither of the meta-analyses addressed the possibility that their results might have been influenced by publication bias. Publication bias is a well-recognized phenomenon in which studies with statistically significant results are more likely to be published, to be published sooner, or to be published in more prestigious journals than those showing null effects *(19,20)*. Redundant publication of positive results is the same bias in another form.

Because tumor-specimen banks are widely available and because the methodology is not complex, studies of tumor markers in NSCLC can be done by small laboratories with limited resources. It requires no stretch of the imagination to suspect that results from "negative" studies are less likely to be submitted for publication and, if submitted, less likely to be accepted for publication than are studies with "positive" results. The failure to report all "negative" outcomes could lead to inappropriately optimistic conclusions about the prognostic value of a specific tumor marker.

We have no reliable way of judging whether publication bias has substantially influenced our perceptions about the predictive value of tumor markers in NSCLC. Funnel plots, a statistical method used to detect publication bias, are not very helpful when the meta-analyses include only a small number of studies *(21)*. Extensive surveys of all investigators who might have conducted relevant studies would be extremely laborious, and in practice it has proven difficult to obtain such information, even when these efforts have focused exclusively on large trials *(22)*.

5. Subset Analyses

Another pervasive methodological problem among published reports is that of multiple statistical comparisons. By convention we attribute an outcome with a probability of less than 0.05 as not being due to chance, a conclusion that will be wrong less than once in every 20 times. Thus, with multiple comparisons, the chance of finding at least one "significant" result increases substantially. With five comparisons, the probability of finding a "significant" association is about 23%, and with 10 comparisons it is about 40%.

We found many examples where the purported associations between NSCLC tumor markers and survival were based largely on subgroup analyses. It is apparent that many investigators do not appreciate the full extent to which subgroup analyses may subvert the scientific process *(23)*. Consider the hypothetical but not atypical example in which investigators studied three putative tumor markers simultaneously. We now have three comparisons and a 14% likelihood of finding at least one "positive" result. Encouraged by apparent trends with each of the three markers, the investigators perform a subset analysis after first stratifying each of the three markers by three different anatomical tumor stages. We now have nine separate comparisons and a 37% likelihood of finding at least one "positive" result. Finally, results are divided according to histological subtype (adenocarcinoma and squamous cell carcinoma), so that we now have up to 18 separate comparisons and a 60%

likelihood of finding at least one "positive" result. Thus, multiple subgroup analyses may produce spurious statistical associations, accounting for some of the inconsistent and contradictory results found in the literature.

6. Future Studies

Of the studies published to date on NSCLC, few have simultaneously examined as many as five individual tumor markers. High-throughput technologies now provide the capability of quickly analyzing thousands of gene expressions in the same tumor *(24)*. These sophisticated methods will ultimately provide a better understanding of cancer pathogenesis and will also afford many new opportunities to correlate genetic abnormalities with biological behavior. However, the task of properly analyzing such studies will be daunting, because the number of measured variables may well exceed the sample size by orders of magnitude. Conventional statistical methods may not be equal to the task. Alternative forms of analysis may be required so as to identify distinctive profiles that are useful for clinical staging purposes *(25)*. Expertise in information theory may be an essential adjunct to the design and analysis of such trials.

It is highly desirable that studies of tumor markers be extended to include all stages of NSCLC. For this purpose methods must be developed to allow assay of tumor markers from small tissue samples that can be obtained from biopsies and needle aspirates. These samples frequently contain a high proportion of normal cells, so reliable methods for separating tumor cells must be found. There has been some encouraging progress in this area *(26)*.

Few published studies relating to NSCLC tumor markers contain more than 150 cases, and one suspects that the sizes of most studies were determined more by case availability than by conscious study design. A major problem with inadequate sample size is the danger that the study will fail to detect a clinically meaningful difference in survival. Under many scenarios there is a 5% chance that a study with a sample size of only 100 cases will fail to detect an absolute survival difference as large as 20%. About 500 cases are required to detect a survival difference of less than 10%. Few investigators can acquire numbers of this magnitude except through multicenter studies. Broad accessibility to tumor tissue from large studies would be an invaluable aid for independent investigators and would obviate many of the problems associated with small sample sizes and nonstandardized collection of samples. Ideally, there would be at least two such groups of tissue samples, one to test the hypothesis and a second to confirm any positive result from the first. Finally, establishment of a national registry of all clinical studies of tumor markers, published or not, would allow comparisons of pooled data from individual studies without the confounding effects of publication bias.

The ongoing identification of new biological markers of cancer pathogenesis holds promise for a better understanding of the molecular biology and genomics of NSCLC. In order to translate these discoveries into better clinical management of lung cancer, we must bring together disciplines of molecular biology and clinical epidemiology to carefully design studies that can answer relevant questions regarding the diagnosis and treatment of NSCLC.

References

1. Marcus, P. M., Bergstralh, E. J., Fagerstrom, R. M., Williams, D. E., Fontana, R., Taylor, W. F., and Prorok, P. C. (2000) Lung cancer mortality in the Mayo Lung Project: impact of extended follow-up. *J. Natl. Cancer Inst.* **92,** 1308–1316.
2. Mountain, C. F. (1997) Revisions in the international system for staging lung cancer. *Chest* **111,** 1710–1717.
3. Vogelstein, B. and Kinzler, K. W. (1993) The multi-step nature of cancer. *Trends Genet,* **9,** 138–141.
4. Slebos, R. J. C., Kibbelaar, R. E., Dalesio, O., Kooistra, A., Stam, J., Meijer, C. J. L. M., et al. (1990) K-*ras* oncogene activation as a prognostic marker in adenocarcinoma of the lung. *N. Engl. J. Med.* **323,** 561–565.
5. Horio, Y., Takahashi, T., Kuroishi, T., Hibi, K., Suyama, M., Niimi, T., et al. (1993) Prognostic significance of *p53* mutations and 3p deletions in primary resected non-small cell lung cancer. *Cancer Res.* **53,** 1–4.
6. Quinlan, D. C., Davidson, A. G., Summers, C. L., Warden, H. E., and Doshi, H. M. (1992) Accumulation of *p53* protein correlates with a poor prognosis in human lung cancer. *Cancer Res.* **52,** 4828–4831.
7. Pezzella, F., Turley, H., Kuzu, I., Tungekar, M. F., Dunnill, M. S., Pierce, C. B., et al. (1993) *bcl-2* protein in non-small-cell lung carcinoma. *N. Engl. J. Med.* **329,** 690–694.
8. Kern, J. A., Schwartz, D. A., Nordberg, J. E., Weiner, D. B., Greene, M. I., Torney, L., and Robinson, R. A. (1990) p185^{neu} expression in human lung adenocarcinoma predicts shortened survival. *Cancer Res.* **50,** 5184–5189.
9. Dazzi, H., Hasleton, P. S., Thatcher, N., Barnes, D. M., Wilkes, S., Swindell, R., and Lawson, R. A. M. (1989) Expression of epidermal growth factor receptor (EGF-R) in non-small cell lung cancer. Use of archival tissue and correlation of EGF-R with histology, tumour size, node status and survival. *Br. J. Cancer* **59,** 746–749.
10. Miyake, M., Taki, T., Hitomi, S., and Hakomori, S.-I. (1992) Correlation of expression of H/Le^y/Le^b antigens with survival in patients with carcinoma of the lung. *N. Engl. J. Med.* **327,** 14–18.
11. Pirinen, R., Hirvikoski, P., Bohm, J., Kellokoski, J., Moisio, K., Johansson, R., et al. (2000) Reduced expression of CD44v3 variant isoform is associated with unfavorable outcome in on-small cell lung carcinoma. *Human Pathol.* **31,** 1088–1095.
12. Harpole, D. H., Jr, Herndon, J. E., II, Wolfe, W. G., Iglehart, J. D., and Marks, J. R. (1995) A prognostic model of recurrences and death in stage I non-small cell

lung cancer utilizing presentation, histopathology, and oncoprotein expression. *Cancer Res.* **55,** 51–56.
13. Lee, J. S., Yoon, A., Kalapurakal, S. K., Ro, J. Y., Lee, J. J., Tu, N., et al. (1995) Expression of *p53* oncoprotein in non-small-cell lung cancer: a favorable prognostic factor. *J. Clin. Oncol.* **13,** 1893–1903.
14. Irie, K., Ishida, H., Furukawa, T., Koyanagi, K., and Miyamoto, Y. (1996) Clinicopathological study on primary lung cancer—immunohistochemical expression of p53 suppressor gene and bcl-2 oncogene in relation to prognosis. *Japn. J. Clin. Pathol.* **44,** 32–41.
15. D'Amico, T. A., Aloia, T. A, Moore, M.-B. H., Herndon, J. E., Brooks, K. R., Lau, C. L., and Harpole, D. H., Jr. (2000) Molecular biologic substaging of Sate I lung cancer according to gender and histology. *Ann. Thorac. Surg.* **69,** 882–886.
16. Greatens, T. M., Niehans, G. A., Rubins, J. B., Jessurun, J., Kratzke, R. A., Maddaus, M. A., and Niewoehner, D. E. (1998) Do molecular markers predict survival in non-small cell lung cancer? *Am. J. Respir. Crit. Care Med.* **157,** 1093–1097.
17. Huncharek, M., Muscat, J., and Geschwind, J.-F. (1999) K-*ras* oncogene mutation as a prognostic marker in non-small cell lung cancer: a combined analysis of 881 cases. *Carcinogenesis* **20,** 1507–1510.
18. Huncharek, M., Kupelnick, B., Geschwind, J. F., and Caubet, J. F. (2000) Prognostic significance of p53 mutations in non-small lung cancer: a meta-analysis of 829 cases from eight published studies. *Cancer Lett.* **153,** 219–226.
19. Dickersin, K. (1990) The existence of publication bias and risk factors for its occurrence. *JAMA* **263,** 1385–1389.
20. Misakian, A. L. and Bero, L. A. (1998) Publication bias and research on passive smoking: Comparison of published and unpublished studies. *JAMA* **280,** 250–253.
21. Egger, M., Smith, G. D., Schneider, M., and Minder, C. (1997) Bias in meta-analysis detected by a simple graphical method. *BMJ* **315,** 629–634.
22. Drummond, R. (1999) Fair conduct and fair reporting of clinical trials. *JAMA* **282,** 1766–1768.
23. Counsell, C. E., Clarke, M. J., Slattery, J., and Sandercock, P. A. G. (1994) The miracle of DICE therapy for acute stroke: fact or fictional product of subgroup analysis. *BMJ* **309,** 1677–1681.
24. Wang, T., Hopkins, D., Schmidt, C., Silva, S., Houghton, R., Takita, H., et al. (2000) Identification of genes differentially over-expressed in lung squamous cell carcinoma using combination of cDNA subtraction and microarray analysis. *Oncogene* **19,** 1519–1528.
25. Alizadeh, A. A., Eisen, M. B., Davis, R. E., Ma, C., Lossos, I. S., Rosenwald, A., et al. (2000) Distinct types of diffuse large B-cell Iymphoma identified by gene expression profiling. *Nature* **403,** 503–511.
26. Maitra, A., Wistuba, I. I., Vimani, A. K., Sakaguchi, M., Park, I., Stucky, A., et al. (1999) Enrichment of epithelial cells for molecular studies. *Nat. Med.* **5,** 459–463.

II

DETECTION OF ALTERED TUMOR MARKERS IN CLINICAL SAMPLES FOR DIAGNOSIS AND PROGNOSIS

B. MOLECULAR METHODS FOR PROGNOSTIC DETECTION OF ALTERED LUNG TUMOR MARKERS IN CLINICAL SAMPLES

I. DNA-Based Analysis

8

Detection of Chromosomal Aberrations in Lung Tissue and Peripheral Blood Lymphocytes Using Interphase Fluorescence *In Situ* Hybridization (FISH)

Randa A. El-Zein and Sherif Z. Abdel-Rahman

1. Introduction

Many cancers arise from gene-environment interactions, where susceptible individuals develop cancer after exposure to toxic or mutagenic environmental agents *(1)*. Genetic instability, whether constitutional or induced, has long been suspected to predispose to carcinogenesis. Cytogenetic assays are classical methods for detecting genetic instability, in which chromosome aberrations are used as biomarkers of the effect of exposure to genotoxic agents. The conceptual basis of this approach revolves around the assumption that the extent of genetic damage reflects critical events in the carcinogenic process, such as an impaired ability to remove damaged DNA or failure to correctly rejoin DNA breaks *(2)*. In a prospective study, Sorsa et al. *(3)* reported that subjects with a high level of chromosomal aberrations appeared to be at an elevated risk for cancer. Bonassi et al. *(4)* and Hagmar et al. *(5)*, in two independent prospective studies, reported a significant increase in the mortality ratio for all cancers in subjects who had earlier shown elevated levels of chromosomal aberrations in their lymphocytes. Recently, the data from these two studies were pooled and the results indicate that the frequency of chromosome aberrations in peripheral blood lymphocytes is a relevant biomarker for cancer risk in humans, reflecting early biological effects of exposure to genotoxic carcinogens and/or individual susceptibility to cancer *(6,7)*.

While conventional cytogenetic assays reveal structural chromosomal abnormalities caused by breakage and rejoining of chromosome material, the relatively new technique of fluorescence *in situ* hybridization (FISH) allows the rapid detection of aberrations that are more transmissible and persistent, such as inversions and translocations between selectively painted and non-painted chromosomes *(8)*. In addition, FISH has been demonstrated to be more sensitive than flow cytometry for detecting aneuploidy *(9,10)*. The FISH assay therefore offers a powerful tool for many clinical and research applications in the delineation of complex structural chromosomal abnormalities, including chromosome enumeration using alpha-satellite probes, marker identification, gene mapping, and chromosome painting *(11)*.

1.1. Principle of the FISH Approach

FISH is a technique that takes advantage of the ability of double-stranded DNA to re-anneal after denaturation. During hybridization, a DNA, RNA, or synthetic probe (which, generally, is a relatively small piece of DNA that is used to find another piece of DNA called the target) is annealed to a complementary target sequence of denatured double-stranded DNA, to create a stable double-stranded hybrid molecule. The target can be genes, sections of chromosomes, or whole chromosomes. The hybrid molecule can then be directly or indirectly labeled. In the direct labeling method, a fluorescent-tagged nucleotide is incorporated into the DNA sequence of the probe. This process uses either nick translation, polymerase chain reaction (PCR), or random DNA priming techniques. The products can then be visualized using fluorescence microscopy immediately after hybridization. In the indirect labeling method, on the other hand, the probes may be tagged with a hapten, such as a biotin or digoxigenin. These haptens can then be detected with avidin or antidigoxigenin antibodies with conjugated fluorochromes. This results in the probe being tagged with multiple fluorochrome molecules at each nucleotide with a hapten molecule and results in a process of indirect detection. Visualization of the products requires a number of additional steps following hybridization.

1.2. Advantages of Using the FISH Approach in Cytogenetic Studies

In contrast to conventional cytogenetic techniques, which usually require metaphase cells, the FISH technique allows information to be obtained rapidly from both metaphase and interphase cells, depending on the study design and the availability of the tissues under investigation. Interphase FISH is very useful when working with tissue samples that do not divide readily in culture (such as buccal smears, cells from bronchio-alveolar lavage, bladder washes, and exfoliated cells). Interphase FISH is less time consuming and does not require highly trained personnel. Specific DNA probes allow the rapid determination of the numerical abnormalities and specific structural abnormalities in a large number of cells *(12–14)*. Another advantage of interphase FISH is the large number of cells that can be analyzed, 1000 or more, compared to the usual 25–100 scorable metaphase spreads. The detection of aneuploidy in interphase FISH is accomplished by counting the number of labeled regions, which represent the chromosome of interest within the interphase nucleus. Taken together, several studies have determined that the FISH cytogenetic approach is more sensitive and convenient than classical cytogenetics *(15–20)*.

1.3. Application of the FISH approach in Lung Cancer Studies

In studies using the traditional chromosome aberration assay, cigarette smokers do not consistently show elevated chromosome aberration frequencies. A number of reports do indicate increased chromosome aberrations in lymphocytes of heavy smokers compared to nonsmokers *(21–24)*, but others fail to support this observation *(25–27)*. Using the sensitive FISH cytogenetic assay, studies of human lymphocytes from individuals exposed to cigarette smoke carcinogens and a variety of other environmental clastogens, such as benzene and certain pesticides, have consistently shown an elevated frequency of breakage in the heterochromatin regions in several chromosomes, including 1q12 and 9q12, *(14,28,29)*. There is increasing evidence that the human heterochromatin regions, particularly those of chromosomes 1, 9, 16, and Y, are frequently involved in stable chromosome rearrangements *(30,31)*. Smith and Grosovsky *(32)* and Grosovsky et al. *(33)* reported that breakage affecting the centromeric and pericentromeric heterochromatin regions of human chromosomes could lead to mutations, chromosomal rearrangements and increased genomic instability. In a previous study, we applied the FISH tandem probe assay to elucidate the frequency of chromosome breakage among cigarette smokers who developed lung cancer. Our findings indicated that smokers had a significantly higher aberration frequency ($p < 0.05$) than the corresponding nonsmoker controls *(29)*.

Although a defined histological sequence of events has been identified for lung cancer, the evolutionary sequence of genetic alterations that take place during the progression of lung cancer is yet to be determined. In a follow-up study, we extended our investigation to determine whether the frequency of chromosome aberrations in peripheral blood lymphocytes from lung cancer patients reflects specific clinical variables of the disease, such as the histological type, grade, and stage of the tumors. Our results indicate a significant linear increase in the level of breaks in the patients' lymphocyte chromatin with respect to the grade of their lung carcinoma. In patients with poorly differentiated tumors, lymphocytes had a significantly higher level of chromosome breaks ($p<0.05$), compared to patients with well-differentiated tumors *(34)*. Taken together, these results indicate that chromosome aberration frequencies, as determined by FISH, have the potential for use as a biomarker for identifying individuals with aggressive types of lung cancer and, perhaps, as an indicator of the possible prognostic outcome of the disease *(29,34,35)*.

In the following sections, we provide an outline for performing the FISH assay in cytogenetic studies using peripheral blood lymphocytes or tissue sections as examples. A large variety of FISH probes can be used, depending on the study design and the purpose of the study. We also describe a method for direct and indirect fluorescence labeling. It should be mentioned that, while the principle of the method remains the same, fine adjustment of the techniques is required when using different probes from different manufacturers or different types of cells (such as cells from buccal swabs, bronchio-alveolar lavage, or tissue sections).

2. Material
2.1. General (see Note 1)

The general laboratory equipment and supplies necessary to successfully perform the FISH assay include:

1. Epi-fluorescence microscope with appropriate filters.
2. Coplin jars, cover slips, standard microscope slides.
3. Humidified chamber (constructed using a slide box with a lid and placing several water dampened paper towels at the bottom.
4. Cell culture CO_2 incubator set at 37°C.
5. Adjustable water bath (preferably two) that can be set at 37°C and 74°C.
6. Micropipettors and disposable tips.
7. Microcentrifuge tubes (1.5–2.0 mL).
8. Slide warmer or hot plate.
9. Silanized slides or Plus slides (Fisher Scientific, Pittsburgh, PA).
10. DAPI Counterstain (Vector Laboratories, Burlingame, CA).
11. NP-40 (Tegritol) (Vysis, Downers Grove, IL).
12. Tween-20 (Sigma, St. Louis, MO).
13. Propidium iodide (Sigma).
14. 12 *N* Hydrochloric acid (HCl).
15. 1 *N* Sodium hydroxide (NaOH).
16. Formamide (Roche, Indianapolis, IN).
17. Purified water.
18. Sealant or rubber cement for sealing cover slips (Vysis).

19. Antifade solution (Vysis).
20. Hybridization buffer (Vysis).

2.2. Chromosome Preparations from Peripheral Blood

1. Whole blood samples.
2. RPMI 1640 medium supplemented with 15% heat-inactivated fetal bovine serum-FBS, 2 m*M* L-glutamine, 100 U/mL penicillin, 100 µg/mL streptomycin (Life Technologies Inc., Rockville, MD) and 2% of reagent grade (9 mg/mL) phytohemagglutinin (PHA, Murex Diagnostics, Norcross, GA).
3. 15-mL culture tubes.
4. Colcemid (Life Technologies Inc.).
5. Hypotonic solution: 0.075 *M* KCl.
6. Fixative solution: 3 to 1 ratio of methanol to acetic acid.

2.3. Fixed Tissue Sample Preparation for FISH Cytogenetic Assay

1. Silanized microscope slides (Fisher Scientific, Pittsburgh, PA).
2. Xylene.
3. Paraffin pretreatment kits (Vysis).
4. 2X SSC.
5. 100% Ethanol.

2.4. Preparations of Fresh Tissue Samples for FISH Cytogenetic Assay

1. Silanized microscope slides (Fisher Scientific).
2. Methanol.

2.5. Slide Preparation for the FISH Assay

1. 2X SSC.
2. 100% Ethanol.
3. Silanized microscope slides (Fisher Scientific).

2.6. Slide Denaturation

1. Denaturation solution: 70% formamide/2X SSC. Mix thoroughly 49 mL formamide, 7 mL 20X SSC, pH 5.3, and 14 mL purified H_2O in a glass Coplin jar wrapped in aluminum foil to minimize light exposure. With solution at ambient temperature, adjust the pH to 7.0 using HCl or NaOH. Between uses, store covered at 2–8°C. Discard after 7 d.

2.7. Indirect Labeling

1. 20X Sodium chloride, sodium citrate solution: 20X SSC. Solution: 87.65 gm sodium chloride (NaCl) + 44.1 gm sodium citrate (Na citrate) in 400 mL purified H_2O. With solution at ambient temperature, measure pH using a pH meter with a glass electrode and adjust to pH 5.3 with HCl. Add purified H_2O to bring the final

volume to 500 mL. Store at ambient temperature (18°C–25°C. Discard stock solution after 6 mo, or sooner if solution appears cloudy or contaminated.
2. 2X SSC solution: Mix thoroughly 100 mL of 20X SSC with 850 mL purified H_2O. Add purified H_2O to bring final volume to one liter. Adjust the pH to 7.0 ± 0.2 with NaOH. Store at ambient temperature. Discard stock solution after 6 mo, or as soon as the solution appears cloudy or contaminated.
3. Posthybridization wash solution (satellite): 55% formamide/2X SSC. Mix thoroughly 115 mL formamide, 21 mL 20X SSC, pH 5.3, and 84 mL purified H_2O. With solution at ambient temperature, adjust pH to 7.0 ± 0.2. Pour equal volumes of the prepared solution into each of three glass Coplin jars wrapped in aluminum foil to minimize light exposure. Store covered at 2–8°C. Discard after 7 d.
4. Ethanol solutions: 70, 85, 100%. Prepare dilutions of 70, 85, and 100% ethanol with purified H_2O. Between uses, store covered at ambient temperature. Discard stock solutions after 6 mo.
5. Phosphate-buffered saline (PBS)/0.1% NP-40 wash solution: Mix thoroughly one phosphate buffer saline (PBS) sachet (Sigma) in 1 L purified H_2O. Add 1 mL NP-40 (or Tween 20). Measure pH with the solution at ambient temperature, and adjust to pH 7.0 ± 0.2 with NaOH. Store at ambient temperature. Discard stock solution after 6 mo, or as soon as the solution appears cloudy or contaminated (*see* **Note 2**).

2.8. Direct Labeling

1. 2X SSC/0.1% NP-40 wash solution: Mix thoroughly 100 mL 20X SSC, pH 5.3, with 850 mL purified H_2O. Add 1 mL NP-40. Add purified H_2O to bring final volume to 1 L. Measure pH with solution at ambient temperature, and adjust to pH 7.0 ± 0.2 with NaOH. Store at ambient temperature. Discard stock solution after 6 mo, or as soon as the solution appears cloudy or contaminated.
2. Formamide wash solution: 50% formamide/2X SSC. Mix thoroughly 105 mL formamide, 21 mL 20X SSC, pH 5.3, and 84 mL purified H_2O. With solution at ambient temperature, adjust pH at 7.0 ± 0.2. Pour equal volumes of the prepared solution into each of three glass Coplin jars wrapped in aluminum foil to minimize light exposure. Store covered at 2–8°C. Discard after 7 d.

3. Methods
3.1. Chromosome Preparations from Peripheral Blood

Blood cultures are prepared following the standard procedures of Hsu et al. *(1)*.

1. Cultures are set up with 0.5 ml whole blood and 4.5 mL RPMI 1640 medium containing 15% heat-inactivated FBS, 2 m*M* L-glutamine, 100 U/mL penicillin, 100 µg/mL streptomycin, and 2% of reagent grade (9 mg/mL) phytohemagglutinin (PHA) in a 15 mL culture tube.

2. Cultures are then incubated at 37°C in a CO_2 incubator for 72 h, with the cap on the tubes slightly loose.
3. Colcemid is added to each culture tube 1.5 h before harvest (final concentration 0.1 µg/mL).
4. Cells are harvested by centrifugation at 400*g* for 10 min.
5. The cell pellets are then treated with 5 mL warm hypotonic solution (0.075 *M* KCl) for 15–20 min in a 37°C water bath.
6. 1 mL of fresh fixative solution, composed of a 3 to 1 ratio of methanol to acetic acid, is then added to each tube and mixed well by inverting.
7. Cells are centrifuged at 400*g* for 10 min. The supernatant is then discarded, and the pellets are resuspended in 5 mL of cold fixative, and mixed well to break-up the pellet completely.
8. The tubes are centrifuged again at 400*g* for 10 min. The pellet is washed twice with fixative solution and stored at –20°C until processed for the FISH assay.
9. For the cytogenetic FISH assay: The cell pellet is spun down at 400*g* for 10 min. The supernatant is discarded and the cells are washed once with fresh fixative and resuspended in an appropriate amount of fixative depending on the cell density (about 0.5–1 mL).
10. Clean slides are dipped in 100% ice cold methanol and, using a Pasteur pipet, a few drops of the fixed cell suspension are dropped onto the slide and the slides are left to air dry after blotting-up the excess fixative by patting the side of the slide on filter paper.

3.2. Fixed Tissue Sample Preparation for FISH Cytogenetic Assay

1. Tissues from paraffin blocks should be cut into 4 micron-thick sections and placed on a silanized (plus) slide. This is done by floating the paraffin-embedded section in warm distilled water then scooping the tissue section with the silanized slide and lifting the slide with the attached tissue up and out of the water.
2. The slide is allowed to air dry and it is then baked in an oven set at 65°C for 4–16 h. The slides can then be stored indefinitely at room temperature, but long storage is not recommended as this can change the pretreatment requirements.
3. Slides are then de-paraffinized by placing them in a jar containing 40 mL xylene. The slides should soak in xylene at room temperature for 10 min.
4. Slides are then transferred to a Coplin jar containing 100% ethanol and soaked for 5 minutes at room temperature.
5. The last step is repeated one more time, then the slides are removed form the ethanol and allowed to air dry.
6. Slides are then processed for protein digestion, using the paraffin pretreatment kit from Vysis. Pretreatment solution is placed in a glass Coplin jar and pre-warmed to 45°C.
7. The slides are incubated in the prewarmed solution for 15–20 min.
8. The slides are rinsed by dipping them in fresh 2XSSC for 5 s, then processed for protein digestion.

9. Prewarm the protein digestion solution for 45°C and incubate the slides for 15–20 min, maintaining the temperature of the digestion solution at 45°C.
10. The slides are rinsed by dipping in 2X SSC for 5 s.
11. The slides are then dehydrated in 70, 80, and 95% ethanol at room temperature for 1 min each and allowed to air dry.

3.3. Touch Preparations of Fresh Tissue Samples for FISH Cytogenetic Assay

1. Slice through fresh tissue, exposing a fresh surface.
2. Gently touch the tissue surface onto a silanized glass slide several times without overlapping the touch sites. Immediately place the slide in a Coplin jar containing cold methanol; fix for 20 min at 4°C.
3. Immediately transfer the wet slide to a Coplin jar containing fresh 3:1 methanol/acetic acid fixative and fix for 20 min at room temperature.
4. Dehydrate the slide in 70, 80, 95% ethanol, 2 min each at room temperature and allow the slide to air dry (*see* **Note 3**).

3.4. Slide Preparation for the FISH Assay (see Notes 4 and 5)

1. Prepare metaphase chromosome spreads or interphase nuclei on a glass microscope slide according to standard cytogenetic procedures *(1)*, as described earlier. If adequate material is available, prepare 2–3 extra slides for each specimen. Slides may be stored for up to 3 wk in a slide box at room temperature.
2. Aging the slides: Prewarm to 37°C 40 mL of 2X SSC, pH 7.0, in a Coplin jar placed in a water bath (*see* **Note 6**).
3. Following the aging step, dehydrate the slides in 70, 85, and 100% ethanol at room temperature for 2 min each. Allow the slides to air dry (*see* **Note 7**).

3.5. Slide Denaturation

The following steps should be performed in a dark room under yellow light:

1. Prewarm 40 mL of denaturation solution in a glass Coplin jar in a water bath (*see* **Notes 8** and **9**).
2. Immerse slides in the pre-warmed denaturation solution for 3 min.
3. Dehydrate slides in an ice-cold (–20°C) series of 70, 85, and 100% ethanol for 2 min each. Cold ethanol quickly stops denaturation while dehydrating the slides.
4. Allow the ethanol to evaporate rapidly by placing the slides on a hotplate prewarmed to 50°C.
5. The denatured slides should be labeled for FISH on the same day

3.6. Indirect Labeling FISH Assay

3.6.1. Probe Preparation

There are a number of satellite probes commercially available such as: (a) alpha satellite probes or centromere probes, useful in determining aneuploidy

Detection of Chromosomal Aberrations

of specific chromosomes in both metaphase and interphase cells; (b) classical satellite probes targeting the heterochromatic region of chromosome 1,9,16 and Y, also used in both metaphase and interphase cells; and (c) telomere probes targeting the terminal ends of the chromosomes.

1. Combine 1.5 µL of probe (or 1.5 µL of each probe for dual color) with 30 µL hybridization buffer in a microcentrifuge tube and vortex gently to mix.
2. Denature the probe solution by heating in a 72°C ± 2°C water bath for 5 min, then centrifuge for only 2–3 s to collect the contents in the bottom of the tube. Place the probe in a 2–8°C ice bath until ready to use for hybridization.

3.6.1.1. TOTAL CHROMOSOME PROBES

Also known as whole chromosome painting probes, these hybridize to the entire chromosome or to chromosome arms. These probes are useful in studying marker chromosomes, translocations, and aneuploidy in metaphase cells.

1. Aliquot 10 µL probe into a 0.5 mL microcentrifuge tube.
2. Denature probe at 72°C ± 2°C for 10 min. Briefly vortex and centrifuge for 2–3 s to collect contents in the bottom of the tube.
3. Place in a 37°C water bath to pre-anneal for 0.5–2 h. This incubation blocks hybridization of repetitive sequences. Briefly vortex and centrifuge for 2–3 s to collect contents in the bottom of the tube.

3.6.2. Probe Hybridization of Satellite DNA Probes

1. Place 30 µL of satellite probe solution on each slide and cover with a 22 × 50 mm glass cover slip. Alternatively, if the whole slide is not needed, you may use 10 µL of probe solution and use a 22 × 22 mm cover slip (*see* **Note 10**).

3.6.2.1. CHROMOSOME PAINTING PROBES

1. Place 10 µL of probe on each slide and cover with a 22 × 22 mm glass cover slip. If the use of a volume other than 10 µL is desired, larger cover slips should be used to cover the desired area of hybridization.
2. Seal the perimeter of the glass cover slip to the slide with a thick layer of Sealant or rubber cement. Incubate slides at 37°C in a prewarmed humidified chamber (for 0.5–16 h depending on probe that is used (*see* **Notes 11–13**).

3.6.3. Posthybridization

1. Pre-warm 40 mL 2X SSC wash solution in a glass Coplin jar to 72°C ± 2°C in a water bath. Verify the temperature of the solution by placing a clean thermometer directly into the Coplin jar.
2. Carefully remove the cover slip sealant or rubber cement with a sharp forceps. Slide the cover slip to the edge and gently lift the cover slip off of the slide.

3. For satellite DNA probes, wash the slides with intermittent agitation for 5 min each in a series of 3 glass Coplin jars labeled I, II, and III, containing 40 mL posthybridization wash solution (55% formamide) at 43°C. For Chromosome painting probes, wash slides with intermittent agitation for 5 min each in a series of 3 glass Coplin jars labeled I, II, and III containing 40 mL posthybridization wash solution (*see* **Note 14**) at 37°C.
4. Wash slides in 40 mL 2X SSC at 37°C for 5 min with intermittent agitation.

3.6.4. Pre-Detection Washes

The purpose of these washes is to provide the proper pH for the detection reagents and to remove any unbound reagents from the previous incubation.

1. After the last 2X SSC wash, pass the slides in a series through 3 glass Coplin jars containing 40 mL of 0.1% NP-40 PBS for 2 min each. Blot the slides by patting the side of the slide on filter paper between transfers (*see* **Note 15**).

3.6.5. Detection and Counterstaining Procedure (see **Note 16**)

1. Remove the slides from the last 0.1% NP-40 PBS wash and blot excess fluid from the edge. Do not allow the slide surface to dry; this may cause nonspecific binding of the detection reagent and high background fluorescence.
2. Apply the appropriate detection reagent. For example, to detect a digoxigenin-rhodamine labeled probe use first an antidigoxigenin-rhodamine detection reagent (reagent 1). Add 45 µL of reagent 1 to each slide and place a cover slip over the solution to prevent it from drying up. Incubate slides at 37°C for 30 min in a prewarmed humidified chamber in the dark.
3. When the incubation is over, remove the cover slips. Wash the slides for two minutes in a series of each of three Coplin jars containing 1% NP-40 PBS.
4. Place 45 µL of rabbit anti-sheep antibody (detection reagent 2) on each of the slides. Cover with a cover slip and incubate slides at 37°C for 30 min in a prewarmed humidified chamber in the dark.
5. When the incubation is over, remove the cover slips. Wash slides for 2 min in a series of each of three Coplin jars with 1% NP-40 PBS.
6. Place 45 µL of rhodamine anti-rabbit (detection reagent 3) on each of the slides. Cover with a cover slip and incubate the slides at 37°C for 30 min in a prewarmed humidified chamber in the dark.
7. When the incubation is over, remove the cover slips. Wash the slides for 2 min in a series of each of three Coplin jars with 1X PBS.
8. Remove the slides from the PBS and blot lightly to allow the extra fluid to drain. Counterstain by adding 25 µL of DAPI to each slide. Cover slip the slides and place them flat in a slide box or a foil wrapped plastic box lined with an absorbent pad. This box may be placed in a –20°C freezer until the slides are scored.

3.7. Direct Labeling FISH Assay

3.7.1. Preparing the Specimen Target (see **Note 17**)

1. Place Coplin jars containing 70% formamide in a 73 ± 1°C water bath approx 30 min prior to use to bring the solution to the required temperature. Measure the temperature of the solution inside the Coplin jars not just the water bath temperature.
2. Mark the area for hybridization on the back of the specimen slide.
3. Ensure that the temperature of the denaturation solution is 73 ± 1°C.
4. Immerse the slides in the denaturation solution for 5 min.
5. Dehydrate the slides for 2 min in 70% ethanol, followed by 2 min in 85% ethanol, and 2 minutes in 100% ethanol. Keep the slides in 100% ethanol until you are ready to dry all slides and apply the probe mixture.

3.7.2. Preparing the Probe Mixture

1. Add the following solutions to a microcentrifuge tube at ambient temperature: 7 μL hybridization buffer (70%), 1 μL the probe (10%), 2 μL purified H_2O (20%) (see **Note 18**).
2. Centrifuge the tube for 5–10 s.
3. Vortex and then centrifuge again for 5–10 s.
4. Place the tube in a 73 ± 1°C water bath to denature for 5 min.
5. Remove the tube from the water bath.
6. Place the tube on a 45–50°C slide warmer until ready to apply the probe to target DNA.

3.7.3. Hybridizing the Probe to the Specimen Target

1. Remove the slides from the 100% ethanol (see **Note 19**).
2. Carefully dry the slides by touching the bottom edge of the slides to a blotter and wiping the underside of the slides dry with a paper towel.
3. Place slides on a 45–50°C slide warmer to evaporate any remaining ethanol (do not leave for longer than 2 min on the hot plate).
4. Apply 10 μL probe mixture to one target area and immediately apply a cover slip. Repeat for additional target areas.
5. Seal cover slip with rubber cement or sealant.
6. Place slides in a prewarmed humidified box and place the box in a 42°C incubator. (Note that the temperatures may differ depending on the type of probe used and the manufacturer's recommendations). Hybridize for 30 min to overnight depending on the target size.

3.7.4. Posthybridization Washing of the Slide

The formamide wash procedure takes approx 25–30 min for a group of four slides.

1. Prepare a series of 3 Coplin jars labeled I, II, and III containing 50% formamide. Place the jars in a 46 ± 1°C water bath at least 30 min prior to use.
2. Prepare a Coplin jar containing 2X SSC and another containing 2X SSC/0.1% NP-40. Place both jars in a 46 ± 1°C water bath at least 30 min prior to use.
3. Remove the cover slip from the slides and immediately immerse the slides into the first of series of 3 Coplin jars containing 50% formamide/2X SSC for 5 min (*see* **Note 20**).
4. Remove the slides form the first jar and place them into the second Coplin jar containing 50% formamide/2X SSC for 5 min.
5. Remove the slides from the second jar and place them into the third Coplin jar containing 50% formamide/2X SSC for 5 min.
6. Immerse the slides in a jar containing 2X SSC for 5–10 min.
7. Immerse the slides in a jar containing 2X SSC/0.1% NP-40 solution for 5 min.

3.7.5. Detection and Counterstaining Procedure

1. Air-dry the slides in the dark or under indirect yellow light.
2. Apply 10 µL of appropriate counterstain to the target area of the slide (*see* **Note 21**).
3. Apply a cover slip and store hybridized slides at –20°C in a slide box.
4. Use an epifluorescence microscope equipped with filters appropriate for the probes used in order to view the slides.

3.8. Fluorescence Microscopy and Photography

1. Following complete staining, slides are visualized and the target sites are identified and scored under a suitable fluorescent microscope (*see* **Notes 22–24**).
2. Objectives used for fluorescence microscopy are usually 10X dry, 40X dry, and 63X oil or 100X oil-immersion objectives. The lower power objectives are used to scan the slides for metaphase spreads or interphase nuclei, while the high power objectives are used for visualization of fluorescein signal and for photography.
3. The filters needed to visualize FISH slides are chosen based on the fluorochrome(s) used. The filters allow certain wavelengths of light through, while blocking others. Use the counterstain filter to scan the slide for cells and the probe signal filter for visualization of both the counterstain and the probe signal (*see* **Note 25**).

4. Notes

1. For good results, and as a general rule, always ensure that reagents are made fresh and used at the temperatures described. For best results, the ambient temperature for the experiments should be ~24°C and the relative humidity ~50%.
2. A solution of 2X SSC/0.005% Tween 20 can be substituted for the 2X SSC/0.1% NP-40 in the washes (prepared as follows: 10 mL 20X SSC + 90 mL water + 50 µL Tween 20).

Detection of Chromosomal Aberrations

3. Protein digestion is usually not necessary with touch preparations.
4. The protocol described here uses metaphase cells prepared from peripheral blood lymphocytes as an example. The same approach can be used with cells obtained from other sources (e.g., interphase cells, cells from tissue sections, touch preparations, and others). It should be mentioned that, regardless of the sample source, slide quality is one of the most important factors for successful FISH. A phase contrast microscope should be used for evaluation of the slides before hybridization. Under phase contrast, cells should appear dark gray in color.
5. If cells have too much cytoplasm, a cytoplasmic cleaning is recommended. One of the following methods can be used: Suspend the slides over a water bath set at 37°C for a few minutes, then spot your sample onto the humidified slides. Double fix the cells, where after spotting the sample and watching it spread out on the slide, add an extra drop of fresh fix to the slide.
6. Treatment of slides in 2X SSC renders the chromosomes less sensitive to over-denaturation. Place prepared slides in the Coplin jar and incubate for 30 min. This treatment is not necessary if the slides are more than 2 wk old.
7. Slide aging and dehydration can be performed up to 2 wk before hybridization. Store the pretreated slides at room temperature in a slide box.
8. Care should be taken to insure that the actual temperature of the denaturation solution is 74°C ± 2°C. Verify the temperature of the solution by placing a clean thermometer directly into the Coplin jar. For all procedures, always measure the temperatures of the solutions inside the Coplin jar and not in the water bath.
9. Time and temperature are very important for the maintenance of chromosome morphology. Slides that are over-denatured may not hybridize or counterstain well. For best results do not process more than 4 slides at a time.
10. Pipetting larger volumes (3 µL of probe with 60 µL of hybridization buffer) provides better accuracy than handling small amounts. The probe can be stored in a hybridization buffer at −18°C to −25°C for the designated life of the probe and the hybridization buffer. Denatured probes can be stored at −18°C to −25°C and used as is, or they may be re-denatured.
11. The 30-min hybridization is only suitable for satellite or repetitive DNA probes. All other probes should be hybridized for 16 h.
12. The hybridization temperature should not drop below 37°C at any time, to minimize cross-hybridization to alpha satellite sequences at the centromere of other chromosomes. The hybridization temperature also should not be above 37°C to insure maximum signal intensity.
13. In general, adequate hybridization depends on a number of factors affecting the stringency such as temperature, formamide, and salt concentration. The higher the temperature and formamide concentration, the more stringent the conditions. The lower the salt concentration the higher the stringency and thus the adequacy of the hybridization.
14. Post-hybridization wash for the chromosome painting probe hybridization should use 50%, rather than 55%, formamide.

15. It is important to ensure that the slides did not dry out completely once the 2X SSC wash is done, as this will affect the detection.
16. The detection reagent to be used depends on the type of probe used for labeling. For example: if the probe is labeled with digoxigenin-FITC, then the detection reagent will be fluorescein-labeled anti-digoxigenin and the counterstain should be propidium iodide (PI). If the probe is labeled with digoxigenin-rhodamine then the detection reagent to be used is rhodamine-labeled anti-digoxigenin and the counterstain is DAPI. For biotin-FITC labeled probes, the detection reagent to be used is fluorescein-labeled avidin and the counterstain is PI. For dual color detection use anti-digoxigenin-rhodamine + FITC avidin and DAPI as a counterstain. These are only examples, and the reader is referred to additional information regarding detection and counterstaining, which can be found in the manufacturer's recommendations accompanying the probes that are purchased.
17. Slides that are over-denatured may not hybridize or counterstain well. For best results do not process more than 4 slides at a time.
18. If the whole slide is to be used, increase the volume to a total of 30 µL, keeping the same ratios of 70% hybridization buffer, 10% probe and 20% water.
19. For best results the slides should be ready when the probe is denatured. The probe can be applied immediately to target DNA.
20. Wash a maximum of four slides simultaneously. Start timing when the slides are immersed.
21. For spectrum green probes use DAPI or PI counterstain while for spectrum orange probes use DAPI
22. There are two types of fluorescence microscopes available, differing in the way the light contacts the specimen. Transmitted light microscopes illuminate from the bottom of the specimen, whereas incident or epi-illuminated microscopes reflect light onto the specimen from above. Of the two types, incident illumination is better for FISH analysis. Of the various types of bulbs available, a high-pressure mercury lamp is optimal for fluorescence detection. A 100 Watt bulb will produce a stronger signal and is essential for dual color analysis with translocation probes or any other unique sequence probes where smaller signals are expected. Microscope bulb age and alignment can affect the apparent strength of the signal. If the microscope is not well aligned or the bulb has been used for many hours (~150–200 h depending on the manufacturer's recommendations) the cell nuclei may appear dark and the signal will be weak.
23. Slides with excessive background (i.e., an increased level of non-specific labeling) may be washed with higher stringency. Soak off the cover slip in 2X SSC/0.1% NP-40. Then rinse the slides in fresh 1X SSC for 5 min. For direct labeling, repeat the detection procedure. For indirect labeling repeat counterstaining.
24. If signal is weaker than desired, decrease the concentration of counterstain by de-staining as follows: Remove the cover slip and immerse the in 2X SSC/0.1% NP-40 at room temperature for 5 min. Apply 10 µL of antifade. Add a cover slip and view with a fluorescent microscope. If signal is still inadequate, the hybridization should be repeated.

25. For single probe FISH a dual bandpass filter can be used to view both the probe and the counterstain. For example, when visualizing FITC signals and PI counterstain, use a PI filter to scan and a FITC/PI filter to visualize both the probe signal and the counterstain. For two probe (dual) FISH, a triple pass filter is used to visualize the two probes separately and the counterstain. For example, when visualizing FITC and Rhodamine signals and DAPI counterstain, use a DAPI filter to scan, and a triple-pass filter (FITC/Texas Red/DAPI) to visualize the probe signals and the counterstain simultaneously.

Acknowledgment

Gratitude is expressed to Dr. Marinel Ammenheuser for her helpful comments.

References

1. Hsu, T. C., Johnston, D. A., Cherry, L. M., Ramikisson, D., Schantz, S. P., Jessup, J. M., et al. (1989) Sensitivity of genotoxic effects of bleomycin in humans: possible relationship to environmental carcinogenesis. *Int. J. Cancer* **43**, 403–409.
2. Heim, S. and Mitelman, F. (1987) Cancer cytogenetics. Alan R. Liss, New York, NY.
3. Sorsa, M., Wilbourn, J., and Vainio, H. (1992) Human cytogenetic damage as a predictor of cancer risk, in *Mechanisms of Carcinogenesis in Risk Identification* (Vainio, H., Magee, P. N., McGregor, D. B., and McMicheal, A. J., eds.), IARC Scientific Publication No. 116. International Agency for Research on Cancer, Lyon, pp. 543–554.
4. Bonassi, S. and Abbondandolo, A. (1995) Are chromosome aberrations in circulating lymphocytes predictive of future cancer onset in humans? *Cancer Genet. Cytogenet.* **79**, 133–135.
5. Hagmar, L., Brogger, A., Hansteen, I. L., Heim, S., Hogstedt, B., Knudsen, L., et al. (1994) Cancer risk in humans predicted by increased levels of chromosomal aberrations in lymphocytes: Nordic study group on the health risk of chromosome damage. *Cancer Res.* **54(11)**, 2919–2922.
6. Hagmar, L., Bonassi, S., Stromberg, U., Brogger, A., Knudsen, L. E., Norppa, H., et al. (1998) Chromosomal aberrations in lymphocytes predict human cancer: a report from the European Study Group on Cytogenetic Biomarkers and Health (ESCH). *Cancer Res.* **58(18)**, 4117–4121.
7. Bonassi, S., Hagmar, L., Stromberg, U., Montagud, A. H., Tinnerberg, H., Forni, A., et al. (2000) Chromosomal aberrations in lymphocytes predict human cancer independently of exposure to carcinogens. European Study Group on Cytogenetic Biomarkers and Health. *Cancer Res.* **60(6)**, 1619–1625.
8. Murg, M. N., Schuler, M., and Eastmond, D. A. (1999) Persistence of chromosomal alterations affecting the 1ce-q12 region in a human lymphoblastoid cell line exposed to dieposybutane and mitomycin C. *Mutation Res.* **446**, 193–203
9. Visakorpi, T., Hyytinen, E., Kalloiniemi, A., Isola, J., and Kalloiniemi, O. P. (1994) Sensitive detection of chromosome copy number aberrations in prostate cancer by fluorescence in situ hybridization. *Am. J. Pathol.* **145**, 624–630.

10. Takahashi, S., Qian, J., Brown, J. A., Alcaraz, A., Bostwick, D. G., Liber, M. M., and Jenkins, R. B. (1994) Potential markers of prostate cancer aggressiveness detected by fluorescence in situ hybridization in needle biopsies. *Cancer Res.* **54,** 3574–3579.
11. Chang, S. S. and Mar, F. F. (1997) Emerging molecular cytogenetic technologies. *Cytobios.* **90(360),** 7–22.
12. Pinkel, D., Straume, T., and Gray, J. W. (1986) Cytogenetic analysis using quantitative, high-sensitivity, fluorescence hybridization. *Proc. Natl. Acad. Sci. USA* **83(9),** 2934–2938.
13. Eastmond, D. A. and Pinkel, D. (1990) Detection of aneuploidy and aneuploidy-inducing agents in human lymphocytes using fluorescence in situ hybridization with chromosome-specific DNA probes. *Mutat. Res.* **234(5),** 303–318.
14. Eastmond, D. A., Rupa, D. S., and Hasegawa, L. S. (1994) Detection of hyperdiploidy and chromosome breakage in interphase human lymphocytes following exposure to the benzene metabolite hydroquinone using multicolor fluorescence in situ hybridization with DNA probes. *Mutat. Res.* **32,** 9–20.
15. Poddighe, P. J., Moesker, O., Smeets, D., Awwad, B. H., Ramaekers, F. C, and Hopman, A. H. (1991) Interphase cytogenetics of hematological cancer: comparison of classical karyotyping and in situ hybridization using a panel of eleven chromosome specific DNA probes. *Cancer Res.* **51(7),** 1959–1967.
16. Kadam, P., Umerani, A., Srivastava, A., Masterson, M., Lampkin, B., and Raza, A. (1991) Combination of classical and interphase cytogenetics to investigate the biology of myeloid disorders: detection of masked monosomy 7 in AML. *Leuk. Res.* **17(4),** 365–374.
17. Kibbelaar, R. E., Kok, F., Dreef, E. J., Kleiverda, J. K., Cornelisse, C. J., Raap, A. K., and Kluin, P. M. (1993) Statistical methods in interphase cytogenetics: an experimental approach. *Cytometry* **14(7),** 716–724.
18. Abdel-Rahman, S. Z., Salama, S. A., Au, W. W., and Hamada, F. A. (2000) Role of polymorphic CYP2E1 and CYP2D6 genes in NNK-induced chromosome aberrations in cultured human lymphocytes. *Pharmacogenetics* **10(3),** 239–249.
19. Abdel-Rahman, S. Z. and El-Zein, R. A. (2000) The 399Gln polymorphism in the DNA repair gene XRCC1 modulates the genotoxic response induced in human lymphocytes by the tobacco-specific nitrosamine NNK. *Cancer Lett.* **159(1),** 63–71.
20. El-Zein, R., Bondy, M. L., Wang, L. E., de Andrade, M., Sigurdson, A. J., Bruner, J. M., et al. (2001) Risk assessment for developing gliomas: a comparison of two cytogenetic approaches. *Mutat. Res.* **490(1),** 35–44.
21. Obe, G. and Herha, J. (1987) Chromosomal aberrations in heavy smokers. *Hum. Genet.* **41,** 259–263.
22. Vijayalaxmi and Evans, H. J. (1982) In vivo and in vitro effects of cigarette smoke on chromosomal damage and sister chromatid exchange in human peripheral blood lymphocytes. *Mutat. Res.* **92,** 321–332.
23. Littlefield, L. G. and Joiner, A. A. (1986) Analysis of chromosome aberrations in lymphocytes of long-term heavy smokers. *Mutat. Res.* **170,** 145–150.

24. Sinues, B., Izquierdo, M., and Perez, V. (1990) Chromosome aberrations and urinary thioethers in smokers. *Mutat. Res.* **240,** 289–293.
25. Tawn, E. J. and Cartmell, C. L. (1989) The effect of smoking on the frequencies of asymmetrical and symmetrical chromosome exchanges in human lymphocytes. *Mutat. Res.* **224,** 151–156.
26. van Diemen, P. C., Maasadam, D., Darroudi, F., and Natarajan, A. T. (1995) Influence of smoking habits on the frequencies of structural and numerical chromosomal aberrations in human peripheral blood lymphocytes using the fluorescence in situ hybridization (FISH) technique. *Mutagenesis* **10,** 487–495.
27. Brinkworth, M. H., Yardley-Jones, A., Edwards, A. J., Hughes, J. A., and Anderson, D. (1992) A comparison of smokers and nonsmokers with respect to oncogene products and cytogenetic parameters. *J. Occup. Med.* **34,** 1181–1188.
28. Rupa, D. S., Hasegawa, L. S., and Eastmond, D. A. (1995) Detection of chromosomal breakage in the 1cen-1q12 region of interphase human lymphocytes using multicolor fluorescence in situ hybridization with tandem DNA probes. *Cancer Res.* **55,** 640–645.
29. Conforti-Froes, N., El-Zein, R., Abdel-Rahman, S. Z., Zwischenberger, J. B., and Au, W. W. (1997) Predisposing genes and increased chromosome aberrations in lung cancer cigarette smokers. *Mutat. Res.* **379,** 53–59.
30. Larizza, L., Doneda, L., Ginelli, E., and Fossati, E. (1988) C-heterochromatin variation and transposition in tumor progression, in *Cancer Metastasis: Biological and Biochemical Mechanisms and Clinical Aspects* (Prodi, G., et al., eds.), Plenum Publishing Corp., New York, pp. 309–318.
31. Doneda, L., Ginelli, E., Agresti, A., and Larizza, L. (1989) In situ hybridization analysis of interstitial C-heterochromatin in marker chromosomes of two human melanomas. *Cancer Res.* **49,** 433–438.
32. Smith, L. E. and Grosovsky, A. J. (1993) Genetic instability on chromosome 16 in a human B lymphoblastoid cell line. *Somat. Cell. Mol. Genet.* **19,** 515–527.
33. Grosovsky, A. J., Parks, K., Giver, C. R., and Nelson, S. L. (1996) Clonal analysis of delayed karyotypic abnormalities and gene mutations in radiation-induced genetic instability. *Mol. Cell. Biol.* **16,** 6252–6262.
34. El-Zein, R., Abdel-Rahman, S. Z., Conforti-Froes, N., Alpard, S. K., and Zwischenberger, J. B. (2000) Chromosome aberrations as a predictor of clinical outcome for smoking associated lung cancer. *Cancer Lett.* **158(1),** 65–71.
35. Abdel-Rahman, S. Z., El-Zein, R. A., Zwischenberger, J. B., and Au, W. W. (1998) Association of the NAT1*10 genotype with increased chromosome aberrations and higher lung cancer risk in cigarette smokers. *Mutat. Res.* **398(1–2),** 43–54.

9

In Situ Analysis of Telomerase RNA Gene Expression as a Marker for Tumor Progression

W. Nicol Keith

1. Introduction

Lung cancer is common in men and women, has a very poor prognosis, and is therefore a major cause of premature mortality. As such, any prospects for improved therapy are of great significance *(1–4)*. The promise of telomerase as a therapeutic target is now close to realisation with extremely encouraging preclinical studies aimed at the RNA component (hTERC) of telomerase *(5–11)*. The rational integration of telomerase therapeutics into clinical trials will therefore require tumors to be well characterized for hTERC expression *(12–14)*. The regulation of telomerase activity is likely to be a complex issue including the transcriptional activity of the telomerase RNA component gene hTERC and the telomerase catalytic component gene (hTERT), and the interaction of telomerase with other telomere associated proteins *(10,15–20; see* **Fig. 1**). The use of telomerase as a diagnostic marker and target for cancer therapy relies on the development of reliable assays and technologies to detect telomeres and telomerase expression *(14,21–26; see* **Table 1**). Molecular techniques can be roughly broken down into two groups, lysate analysis and *in situ* analysis *(14,23,27)*. With lysate methods, tumor biopsies are homogenized and the spatial relationships between tumor cells are destroyed (Southern-blot analysis and polymerase chain reaction [PCR]). This leads to a loss of information on heterogeneity and small subpopulations and presents an averaging of changes. However, quantitation can be simpler and more accurate than *in situ* approaches. In comparison, *in situ* techniques such as RNA *in situ* hybridization allow visualization of gene expression in individual cells within their histological context *(12–14,23,28,29)*. This is an important issue in

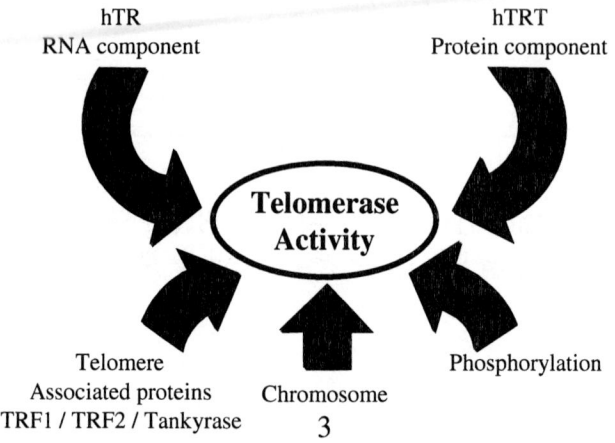

Fig. 1. Regulation of telomerase activity.

**Table 1
Methods for the Analysis of Telomeres and Telomerase**

Telomerase enzyme activity	TRAP assay
Telomerase component gene expression	Northern-blot analysis
	Nuclease protection assays
	RT-PCR
	in situ hybridization
Telomere length	Southern-blot analysis
	in situ hybridization
	flow cytometry

examining the role of telomerase in the development of immortal clones of cancer cells from a telomerase negative normal tissue. Also, for telomerase and telomerase component genes to be useful biomarkers for disease or as a therapeutic targets, differential expression is required between normal and cancerous tissue *(12–14)*. However, in both normal and cancerous tissue, admixture of cell types may confound interpretation of many assays. The *in situ* approach described here is ideally placed to solve this problem *(12–14,23)*.

Basic telomerase enzyme activity requires the expression of two genes, the telomerase catalytic component gene hTERT and the telomerase RNA component, hTERC *(10,15,30)*. This chapter will describe the *in situ* detection of hTERC gene expression *(12–14)*, however, GenBank accession numbers

Table 2
Sequences of Telomerase and Telomerase Associated Genes

Gene	Species	Accession number	Sequence length
Telomerase RNA component	Human	AF047386	1765
	Human	U85256	598
	Human	U86046	545
	Mouse	AF047387	4044
	Mouse	U33831	590
Telomerase catalytic subunit	Human	AF015950	4015
	Human	AF018167	4027
	Mouse	AF073311	3369
	Mouse	AF051911	3426
Telomerase-associated protein, TP-1	Human	U86136	8665
TRF1	Human	U40705	2686
	Mouse	U65586	1628
TRF2	Human	AF002999	2907
	Mouse	AF003000	2119
Tankyrase	Human	AF082556	4134

for a selection of telomerase and telomerase associated gene sequences are given in **Table 2** *(1,3–7)*, and can be accessed through the internet at, http://www.ncbi.nlm.nih.gov/Web/Genbank/index.html. Thus, probes to any of the genes mentioned in Table 2, can be synthesized and developed in a similar fashion to hTERC *(14,23; see* **Fig. 2**).

The principle of *in situ* hybridization is based on the specific binding of a labeled nucleic acid probe to a complementary sequence in a tissue sample, followed by visualization of the probe. This enables both detection and localization of the target sequence. A number of prerequisites for the success of this procedure include the retention of the nucleic acid sequences in the sample and its accessibility to the probe. Specimens suitable for in situ hybridization (ISH), include cells from culture and tissue from samples from whole or biopsied organs. The major steps in RNA *in situ* hybridization are shown in **Fig. 3**.

2. Materials
2.1. Linearization of Plasmid DNA

1. RNA labeling kit (Amersham, RPN 3100).
2. Diethylpyrocarbonate (DEPC) (Sigma).

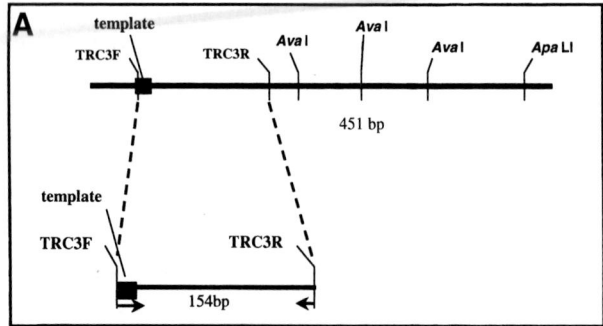

B Sequence of insert

<u>ctaaccctaactgagaagggcgta</u>ggcgccgtgcttttgct
ccccgcgcgctgtttttctcgctgactttcagcgggcggaaa
agcctcggcctgccgccttccaccgttcattctagagcaaa
caaaa<u>aatgtcagctgctggcccgttcgcc</u>

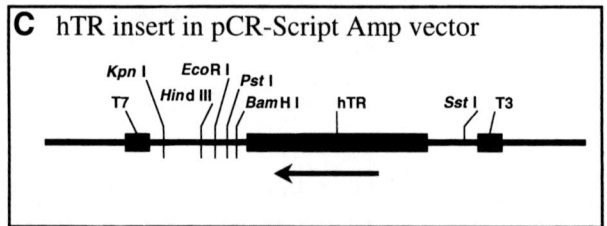

Fig. 2. Development of probe suitable for RNA in situ hybridization. (**A**) Diagramatic representation of the sequence for the hTERC gene. The 154bp insert used to generate the riboprobe for hTERC is amplified with the PCR primers TRC3F, (CTAACCCTA ACTGAGAAGGGCGTA), and TRC3R, (GGCGAACGGGCCAGCAGCTGACATT). The sequence of this region of the hTERC gene is shown in (**B**), and the sequences corresponding to the PCR primers are underlined. In order to generate riboprobes, the 154bp insert was ligated into the Stratagene vector, pCR-Script Amp. This construct is named pCRhTERC1 and the region of the vector around the cloning site is shown in (**C**). The orientation of the insert is shown by an arrow. Antisense probe is synthesized by cutting the plasmid with SstI and using T7 polymerase. Sense probe is synthesized by cutting the plasmid with KpnI and using T3 polymerase. This construct is available from the authors (contact WNK).

3. DEPC-treated distilled water (dH$_2$0): 1% DEPC, autoclave.
4. Phenol-chloroform isoamyl alcohol, pH 8.0 (Sigma).
5. 3 M Na Acetate, pH 8.0.
6. Absolute alcohol, analytical grade.
7. 1% agarose, electrophoresis grade (Gibco-BRL).

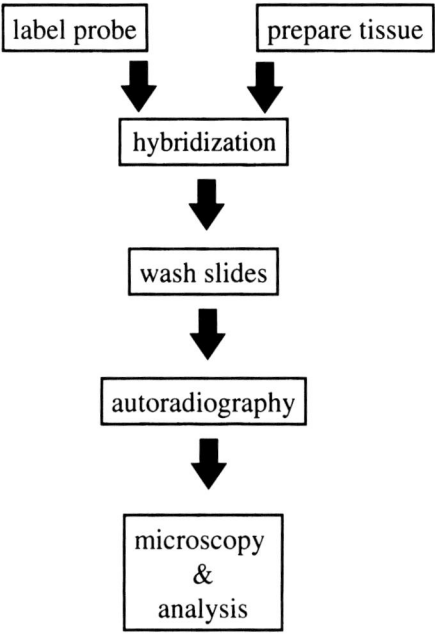

Fig. 3. Overview of RNA *in situ* hybridization.

2.2. Probe Labeling

1. 0.2 *M* Dithiothreitol (DTT) (Sigma).
2. ^{35}S UTP (Amersham SJ 603).
3. G50 Sephadex columns (Pharmacia Biotech).
4. Column buffer: 0.3 *M* Na acetate, 1 m*M* EDTA, 1% SDS, autoclave.
5. Phenol, pH 5.0 (Sigma).
6. Chloroform isoamyl alcohol (Biogene).
7. 50 m*M* DTT: aliquoted and stored at –20°C.
8. Scintillation fluid: Eoscint A (National Diagnostics).

2.3. In Situ *Hybridization*

1. Histoclear (Fisher).
2. 0.85% NaCl-DEPC: 0.85% NaCl, 1% DEPC, autoclave.
3. 1X PBS: Phosphate-buffered saline tablets (Unipath); 10 tablets per liter, 1% DEPC, autoclave.
4. 0.5 *M* EDTA: dissolve in DEPC-dH$_2$O, pH 7.5.
5. Proteinase K Buffer: 1 *M* Tris-HCl, 0.5 *M* EDTA, 1% DEPC, pH 7.5, autoclave.
6. Proteinase K stock solution (Sigma): 20 mg/mL in DEPC H$_2$O. Aliquot and store at –20°C.
7. Formalin.

8. 0.1 M Triethanolamine: 1% DEPC, autoclave.
9. Acetic anhydride (Sigma).
10. 1 M DTT: aliquoted and stored at –20°C.
11. 60% hybridmix: 6 mL formamide (Fluka), 2 mL 50% dextran sulphate-DEPC, 1 mL 20X SSC, 100 µL 1 M Tris HCl, 200 µL 50X Denhardts solution, 100 µL 10% SDS, 400 µL tRNA (10 mg/mL) (Sigma), 200 µL salmon DNA (10 mg/mL) (Sigma). Store at –20°C in 400 µL aliquots.

2.4. Membrane Washing

1. 20X SSC (Gibco/BRL); dilute for 5X SSC, 2X SSC, and 0.1X SSC.
2. 50% formamide, 2X SSC.
3. β-mercaptoethanol (Sigma).
4. RNase buffer: 0.5 M NaCl, 1 M Tris-HCl, pH 7.5, 0.5 M EDTA, pH to 7.5.
5. RNase A stock solution (Sigma): 10 mg/mL in DEPC-H_2O, store at –20°C in 400 µL aliquots.
6. Gelatin: 0.2 g in 200 mL dH_2O. Microwave for 2 min, filter, and cool.

2.5. Autoradiography

1. Emulsion for autoradiography (Amersham Hypercoat Emulsion LM-1, RPN 40).
2. Silica Gel (Fisher).
3. 20% Phenisol (Ilford).
4. 1% acetic acid (Fisher).
5. 30% sodium thiosulphate (Sigma).
6. Hematoxylin.
7. DEPX mounting solution (BDH).

3. Methods
3.1. Probe Preparation
3.1.1. Linearization of Plasmid DNA

1. Take 10–20 µg of DNA.
2. Add 10 U of restriction enzyme per µg of DNA and set up digestion as recommended by suppliers of the enzyme.
3. Leave reaction at 37°C for 3 h or overnight.

3.1.2. Phenol Chloroform Extraction

1. Add 400 µL of phenol-chloroform-isoamyl alcohol pH 8.0 and vortex.
2. Spin for 3 min at 15,000g at room temperature.
3. Keep the supernatant and add 10 µL of 3 M NaAc (pH 8.0).
4. Add 250 µL 100% ethanol (stored at –20°C).
5. Add 1 µL glycogen to help precipitate the DNA.
6. Place on dry ice for 1 h.
7. Spin for 15 min at 15,000g.

8. Remove supernatant and keep the pellet.
9. Wash pellet with 400 μL 70% ethanol (stored at –20°C).
10. Spin for 10 min at 15,000g.
11. Remove remaining 70% ethanol and air dry the pellet.
12. Resuspend pellet in 10–20 μL of DEPC-H$_2$O, depending on the volume of DNA used, aiming for a final concentration of 1 μg/μL.
13. Run 0.5 μL of this suspension on a 1% Agarose Gel.

3.2. RNA Labeling

3.2.1. Incorporation of Radioactive Nucleotides

Use Amersham kit, RPN3100 according to pack insert with reference to the method below.

1. Add 4 μL of 5X transcription buffer.
2. Add 1 μL of 0.2 M DTT.
3. Add 1 μL of HPR1.
4. Add 0.5 μL of ATP, CTP, and GTP.
5. Add 1 μL of linearized DNA template.
6. Add 9.5 μL ^{35}S UTP.
7. Add 2 μL of RNA polymerase.
8. Mix and place at 37°C for 1.5 h.

3.2.2. DNase Extraction of DNA Template

1. Add 10 U DNase I.
2. Add 1 μL RNase inhibitor.
3. Mix and place at 37°C for 10 min.

3.2.3. Removal of Unincorporated Nucleotides.

1. Equilibrate G50 sephadex column with 2-mL column buffer.
2. Add probe to the column.
3. Add 400 μL of column buffer and allow to run through.
4. Add further 400 μL of column buffer and collect in an eppendorf.

3.2.4. Phenol-Chloroform Extraction

1. Add 400 μL phenol, pH 5.0, vortex, and spin for 3 min at 15,000g.
2. Retain the supernatant and to this add 400 μL chloroform-isoamyl alcohol, vortex, and spin for 3 min at 15,000g.
3. Retain the supernatant and remove 1 μL for counting the incorporation.
4. Add 2.5 vol 100% ethanol (stored at –20°C).
5. Add yeast tRNA or glycogen to facilitate precipitation of pellet.
6. Place on dry ice for 30 min.
7. Spin for 15 min at 15,000g.
8. Remove alcohol and leave pellet undisturbed.

9. Wash pellet with 70% ethanol and spin at 15,000g for 10 min.
10. Air dry pellet and resuspend in 50 mM DTT; calculating the volume of 50 mM DTT as follows:
    a. Add the 1 µL of supernatant from **step 3** to 2–3 mL of scintillation fluid.
    b. Volume of DTT = (CMP × 400) ÷ 6 × 10^5, where 400 is the volume after phenol/chloroform extraction and 6 × 10^5 is the required total CPM in the DTT solution.
    c. This is the volume of 50 mM DTT in which the probe should be re-suspended.

3.3. In Situ Hybridization (see Notes 1–3)

3.3.1. Pretreatment of Paraffin Sections

1. Dewax with Histoclear; 2 × 10 min.
2. Rehydrate through an ethanol series; 100, 90, 70, 50, and 30%; for 10 s each.
3. Rinse in 0.85% NaCl and 1X PBS solutions; 5 min each.
4. Proteinase K digest, 400 µL of Proteinase K stock in 200 mL of Proteinase K buffer; 7.5 min.
5. Rinse in 1X PBS; 3 min.
6. Postfix in Formalin. Alternatively use 4% paraformaldehyde.
7. Rinse in DEPC treated water; 1 min.
8. Acetylate in 0.1 M triethanolamine with 500 µL of acetic anhydride; 10 min. (Stirring throughout under a fume hood).
9. Rinse in 1X PBS and 0.85% NaCl; 5 min each.
10. Dehydrate through the ethanol series; 30, 50, 70, 90, and 100%; 10 s each.
11. Air dry.

3.3.2. Preparation of Probe and Hybridization (see Notes 4–9)

1. For 20 paraffin sections, take 16 µL of 1 M DTT, 344 µL of 60% hybridmix, and 40 µL of probe. Vortex and spin briefly.
2. Denature the probe at 80°C for 3 min. Cool on ice.
3. Apply 20 µL of probe to each section and cover with a glass cover slip.
4. Hybridize at 52°C overnight in humidified chamber (*see* **Note 10**).

3.3.3. Posthybridization Wash (see Note 11)

1. Preheat solutions to required temperature.
2. Wash sections in 5X SSC with 250 µL of β-mercaptoethanol; 30 min at 50°C.
3. Wash in 50% formamide, 2X SSC with 1.4 mL of β-mercaptoethanol; 20 min at 65°C.
4. Wash in RNase buffer; 2X 10 minutes at 37°C.
5. Wash in RNase buffer with 400 µL of RNase A solution; 30 min at 37°C.
6. Repeat **step 4**; 15 min at 37°C.
7. Repeat **step 3**.
8. Wash in 5X SSC, 0.1X SSC; 15 min each at 50°C.

9. Dehydrate in ethanol series; 50, 70, 100%; 1 min each.
10. Air-dry.
11. Dip in gelatin solution; 1 min, then air dry.

3.3.4. Autoradiography (see **Note 12**)

1. Under Kodak Wratten II (or equivalent) safe light conditions, melt emulsion (Amersham LM-1) in dipping vessel immersed in water bath at 46°C.
2. Dip each slide into the emulsion and air dry.
3. Once dry place slides in a light tight box with some silica gel (wrapped in tissue paper) and store at 4°C for 10 d.

3.3.5. Development

1. Prepare solutions for development process ensuring temperature is about 20°C.
2. Under safelight conditions develop the slides in 20% Phenisol for 2.5 min.
3. Stop development in 1% acetic acid for 30 s.
4. Rinse in dH_2O for 30 s.
5. Fix in 30% sodium thiosulphate for 5 min.
6. Rinse in several changes of running water for 20 min.
7. Immerse in haematoxylin for 45 s.
8. Rinse in running water for 2 min.
9. Dehydrate in 50, 70, and 100% ethanol for 1 min each.
10. Dewax in histoclear for 10 min.
11. Mount coverslips with DEPX mounting fluid.

3.3.6. Analysis

1. Examine sections using light microscopy under light and dark field illumination.
2. Score with reference to positive and negative controls.

4. Notes

1. All solutions involved in the preparation of probe and up to the post hybridization wash steps are required to be free from RNases. Solutions should be treated with DEPC and autoclaved for 4 h at 160°C. This removes the majority of RNases but by their ubiquitous nature this is not a substitute for care in handling the solutions, glassware, and pipets. A dedicated set of pipets for use only with RNA is worthwhile and regular treatment of the pipet with DEPC-water overnight or with a proprietary anti-RNase solution such as RNaze-Zap (Ambion) may be worthwhile. All glassware should be autoclaved wrapped in aluminum foil prior to use. Plastic eppendorfs may be treated with DEPC-water or RNaze-Zap prior to autoclaving.
2. The objective is to preserve the architecture and morphology of the tissue and retain the RNA products. Rapid processing of the tissue sample either by freezing or fixing in formalin enables RNA to be preserved. Cross-linked fixatives such as 4%paraformaldehyde, 4% formaldehyde are the fixatives of choice for the

retention and/or accessibility of cellular RNA. The length of fixation will depend on the specimen size. Longer fixation will result in better tissue morphology, but reduced access to probe may be a consequence. Paraffin wax is the embedding medium of choice. It allows sectioning down to 1 µm in thickness and is easily removed prior to hybridization. As the sections will be processed through a number of solutions during processing, coated slides are recommended to the specimen on the slide. Frozen samples should be frozen down to –70°C and following cryo-sectioning, placed on a coated slide, and fixed.

3. Preparation of the tissue prior to hybridization attempts to increase the access of the probe to the target RNA sequence and reduce nonspecific background binding. The specimen is subject to protease treatment to increase the accessibility of the target nucleic acid to the probe, especially if the probe is greater than 100 base pairs. It is important to postfix the specimen in formaldehyde to prevent disintegration of the tissue. Nonspecific binding to amino groups is reduced by acetylation with acetic anhydride. During tissue preparation, great care must be taken to protect the specimen from RNAse. All glassware must be treated to remove any contamination, all solutions treated with DEPC and gloves worn throughout. Handling for the sections should be kept to a minimum.

4. These probes usually 20–30 base pairs long can be used as probes. They have the advantage of being fairly easy to generate in large quantities without the need for cloning, their shorter length enables tissue penetration and are fairly stable with no self hybridization. However their short length permits fewer labeled nucleotides to be in corporate per probe and hence reduce the sensitivity of the technique. A number of different oligonucleotides targeting different sequences can be used together to increase the signal generated. The hybrids formed by oligonucleotide probes, due to their short length, are also less stable than those formed by RNA probes.

5. This requires use of a DNA template of the target sequence, and generation of sense and antisense RNA probes, with radioactive nucleotides incorporated, are possible. Single stranded RNA probes are ideal if high sensitivity is required, probes of 200–1000 kb have been used, but probes of 150–200 base pairs are probably optimal, as tissue penetration can become reduced with longer probe size. Limited alkaline hydrolysis can be used to reduce probe size as required. The RNA/RNA or RNA/DNA hybrids are more stable than their Oligonucleotide or DNA counterparts, rendering them the most popular probes.

6. Either radioactive or nonradioactive labeling can be used for probes, with autoradiography or immunocytochemistry being the method of detection, respectively. Radioactivity is sensitive with good resolution, with some isotopes offering better resolution but requiring longer exposures, e.g., I^{125}, and others providing results after shorter exposures but reducing the resolution of the resultant image, due to the wider scatter of the higher-energy emission particles, e.g., P^{32}. S^{35} offers a compromise with exposure times of 1–2 wk being adequate and providing high resolution images. Nonradioactive probes offer the advantage

of easier working practice, and digoxigenin-labeled nucleotides can be used to generate probes.
7. It is our practice to check the size of a new probe by agarose gel, with nonradioactive nucleotides only. Once the probe is considered appropriate, a Northern blot is performed to check its specificity.
8. A count of between 3×10^5 and 6×10^5 is usually found to be satisfactory. In addition prior to hybridization the count of the probe should be rechecked. However, the half life of ^{35}Sulphur is 90 d; the probe sequence is vulnerable to radiolysis and thus probe performance is optimal only within about 5–7 d of radiolabeling.
9. We use commercially available (Ambion) DNA templates for Actin and GAPDH, to generate RNA probes for use as positive controls as they are housekeeping genes and ubiquitously expressed. Sense Probes are commonly used as negative controls and are superior to the omission of a probe as a control. We use well characterized tumor samples with a range of RNA expression as positve specimens duirng each run of slides. Commercially available tumor samples of a variety of tissue are also available for the same.
10. The hybridization temperature can be critical for some probe/target sequences. Formamide in the hybridization buffer, as a helix de-stabilizer, reduces the melting point of the hybrids and enables reduction of the temperature of hybridization. The lower temperature helps preserves tissue architecture. We find 52°C as the optimal temperature. Dextran sulphate in the hybridization buffer, by volume exclusion, increases the concentration of the probe and reduces hybridization times. Although the hybridization reaction is almost complete after 5–6 h, we find it most convenient to leave the reaction overnight. The sodium ion concentration in the buffer serves to stabilize the hybrids.
11. The main objective is to remove unbound and nonspecifically bound probe by selection of temperature, salt concentration, and formamide concentration. The use of RNAse enables the digestion of single stranded RNA, unbound to target but does not affect the bound RNA-RNA complexes.
12. This enables the detection of the bound probe by radioactivity sensitive emulsion. The emulsion is melted and thin layer used to cover the slide by dipping. The choice of emulsion will depend on the anticipated signal intensity and the method of visualization of tissue, i.e., light or electron microscopy. The slides are dipped into a dipping vessel (Amersham) filled with premelted emulsion (This takes about 10 min). Prior to dipping a clean slide is used to slowly and gently mix the emulsion. Too rapid agitation will result in the inclusion of air bubbles, which will compromise the quality of the autoradiography. Each slide is dipped for 5 s, removed and allowed to drain for 5 s and then this is repeated. The back of the slide is wiped gently and the slide placed within a light tight box on a slide drying rack for about 60–90 min. Forced drying is not recommended. The temperature of the emulsion at the time of dipping is critical to the thickness of the emulsion layer. A thin coating of emulsion will increases the resolution

of the resultant hybridization. The slides should be dried slowly and in humid conditions to prevent the emulsion/gelatin cracking and possibly acting as a signal for the formation of silver grains. We find leaving the slides in alight tight box for 2–3 h facilitates slow drying and then when the slides are "tacky" they can be stored in small light tight containers for the exposure time. Avoid touching the emulsion layer at all times and ensure that slides are not in contact with each other, and spaced well apart. After an appropriate length of exposure, the slides are developed. Try exposure of the slides for 2–3 wk. The exact time may need to be modified depending on your own results. Extending exposure times can increase the sensitivity of the experiment, however there will be a resultant loss of resolution and the increase in nonspecific background signal may render the experiment difficult to evaluate. Development of the slides must be carried out in the dark, and on completion rinsed under cold running water. Warm water will melt the emulsion and the silver deposits rinsed off. Although it is safe to switch on the light after the slides have been fixed we would recommend rinsing in cold water for a few minutes in darkness prior to switching on the light, as exposure of the fixer to light can cause a yellow/brown discoloration of the slide. Ensure that all reagents used in the development process are at the same temperature, and the working temperature is below 20°C to preserve the gelatin layer. The sections need to be counterstained with Hematoxyline to visualize the nuclei. Always ensure that the hematoxylin is freshly filtered.

References

1. Sekido, Y., Fong, K. M., and Minna, J. D. (1998) Progress in understanding the molecular pathogenesis of human lung cancer. *Biochim. Biophys. Acta* **1378,** F21–F59.
2. Boral, A. L., Dessain, S., and Chabner, B. A. (1998) Clinical evaluation of biologically targeted drugs: obstacles and opportunities. *Cancer Chemother. Pharmacol.* **42,** S3–S21.
3. Wang, D. G., Johnston, C. F., Sloan, J. M., and Buchanan, K. D. (1998) Expression of Bcl-2 in lung neuroendocrine tumours: comparison with p53. *J. Pathol.* **184,** 247–251.
4. Brambilla, E., Negoescu, A., Gazzeri, S., Lantuejoul, S., Moro, D., Brambilla, C., and Coll, J. L. (1996) Apoptosis-related factors p53, Bcl2, and Bax in neuroendocrine lung tumors. *Am. J. Pathol.* **149,** 1941–1952.
5. Wan, M. S., Fell, P. L., and Akhtar, S. (1998) Synthetic 2′-O-methyl-modified hammerhead ribozymes targeted to the RNA component of telomerase as sequence-specific inhibitors of telomerase activity. *Antisense Nucleic Acid Drug Dev.* **8,** 309–317.
6. Kondo, Y., Kondo, S., Tanaka, Y., Haqqi, T., Barna, B. P., and Cowell, J. K. (1998) Inhibition of telomerase increases the susceptibility of human malignant glioblastoma cells to cisplatin-induced apoptosis. *Oncogene* **16,** 2243–2248.
7. Kondo, S., Tanaka, Y., Kondo, Y., Hitomi, M., Barnett, G. H., Ishizaka, Y., et al. (1998) Antisense telomerase treatment: induction of two distinct pathways, apoptosis and differentiation. *FASEB J.* **12,** 801–811.

8. Kondo, S., Kondo, Y., Li, G., Silverman, R. H., and Cowell, J. K. (1998) Targeted therapy of human malignant glioma in a mouse model by 2-5A antisense directed against telomerase RNA. *Oncogene* **16**, 3323–3330.
9. Glukhov, A. I., Zimnik, O. V., Gordeev, S. A., and Severin, S. E. (1998) Inhibition of telomerase activity of melanoma cells in vitro by antisense oligonucleotides. *Biochem. Biophys. Res. Commun.* **248**, 368–371.
10. Feng, J., Funk, W. D., Wang, S. S., Weinrich, S. L., Avilion, A. A., Chiu, C. P., et al. (1995) The RNA component of human telomerase. *Science* **269**, 1236–1241.
11. White, L. K., Wright, W. E., and Shay, J. W. (2001) Telomerase inhibitors. *Trends Biotechnol.* **19**, 114–120.
12. Wisman, G. B., De Jong, S., Meersma, G. J., Helder, M. N., Hollema, H., de Vries, E. G., et al. (2000) Telomerase in (pre)neoplastic cervical disease. *Hum. Pathol.* **31**, 1304–1312.
13. Sarvesvaran, J., Going, J. J., Milroy, R., Kaye, S. B., and Keith, W. N. (1999) Is small cell lung cancer the perfect target for anti-telomerase treatment? *Carcinogenesis* **20**, 1649–1651.
14. Soder, A. I., Going, J. J., Kaye, S. B., and Keith, W. N. (1998) Tumour specific regulation of telomerase RNA gene expression visualized by in situ hybridization. *Oncogene* **16**, 979–983.
15. Nakamura, T. M., Morin, G. B., Chapman, K. B., Weinrich, S. L., Andrews, W. H., Lingner, J., et al. (1997) Telomerase catalytic subunit homologs from fission yeast and human (see comments). *Science* **277**, 955–959.
16. Nowak, R. and Chrapusta, S. J. (1998) Regulation of telomerase activity in normal and malignant human cells. *Cancer J. Sci. Am.* **4**, 148–154.
17. Nugent, C. I. and Lundblad, V. (1998) The telomerase reverse transcriptase: components and regulation. *Genes Dev.* **12**, 1073–1085.
18. Smith, S., Giriat, I., Schmitt, A., and de Lange, T. (1998) Tankyrase, a poly(ADP-ribose) polymerase at human telomeres (see comments). *Science* **282**, 1484–1487.
19. Broccoli, D., Smogorzewska, A., Chong, L., and de Lange, T. (1997) Human telomeres contain two distinct Myb-related proteins, TRF1 and TRF2. *Nat. Genet.* **17**, 231–235.
20. Zhao, J. Q., Hoare, S. F., McFarlane, R., Muir, S., Parkinson, E. K., Black, D. M., and Keith, W. N. (1998) Cloning and characterization of human and mouse telomerase RNA gene promoter sequences. *Oncogene* **16**, 1345–1350.
21. Sharma, S., Raymond, E., Soda, H., Sun, D., Hilsenbeck, S. G., Sharma, A., et al. (1997) Preclinical and clinical strategies for development of telomerase and telomere inhibitors. *Ann. Oncol.* **8**, 1063–1074.
22. Shay, J. W. (1998) Telomerase in cancer: diagnostic, prognostic, and therapeutic implications. *Cancer J. Sci. Am.* **4(Suppl. 1)**, S26–S34.
23. Soder, A. I., Hoare, S. F., Muir, S., Going, J. J., Parkinson, E. K., and Keith, W. N. (1997) Amplification, increased dosage and in situ expression of the telomerase RNA gene in human cancer. *Oncogene* **14**, 1013–1021.
24. Urquidi, V., Tarin, D., and Goodison, S. (1998) Telomerase in cancer: clinical applications (In Process Citation). *Ann. Med.* **30**, 419–430.

25. Rufer, N., Dragowska, W., Thornbury, G., Roosnek, E., and Lansdorp, P. M. (1998) Telomere length dynamics in human lymphocyte subpopulations measured by flow cytometry (see comments). *Nat. Biotechnol.* **16,** 743–747.
26. Hultdin, M., Gronlund, E., Norrback, K., Eriksson-Lindstrom, E., Just, T., and Roos, G. (1998) Telomere analysis by fluorescence in situ hybridization and flow cytometry. *Nucleic Acids Res.* **26,** 3651–3656.
27. Murphy, D. S., Hoare, S. F., Going, J. J., Mallon, E. E., George, W. D., Kaye, S. B., et al. (1995) Characterization of extensive genetic alterations in ductal carcinoma in situ by fluorescence in situ hybridization and molecular analysis. *J. Natl. Cancer Inst.* **87,** 1694–1704.
28. Yashima, K., Maitra, A., Rogers, B. B., Timmons, C. F., Rathi, A., Pinar, H., et al. (1998) Expression of the RNA component of telomerase during human development and differentiation. *Cell Growth Differ.* **9,** 805–813.
29. Ogoshi, M., Le, T., Shay, J. W., and Taylor, R. S. (1998) In situ hybridization analysis of the expression of human telomerase RNA in normal and pathologic conditions of the skin. *J. Invest. Dermatol.* **110,** 818–823.
30. Weinrich, S. L., Pruzan, R., Ma, L., Ouellette, M., Tesmer, V. M., Holt, S. E., et al. (1997) Reconstitution of human telomerase with the template RNA component hTR and the catalytic protein subunit hTRT. *Nat. Genet.* **17,** 498–502.

10

A High-Throughput Methodology for Identifying Molecular Targets Overexpressed in Lung Cancers

Tongtong Wang and Steven G. Reed

1. Introduction

As discussed in previous chapters, lung cancer is one of most deadly diseases and conventional treatments for lung cancer patients are largely ineffective. Presented with at least four major histological types of lung cancers, dependable tools for early detection and diagnosis of each type of lung cancers are urgently needed. This chapter focuses on a high-throughput methodology for identifying new molecular targets in lung cancers and these targets will potentially provide diagnostic and therapeutic values for lung cancer patients *(1)*. Previous studies have reported several lung cancer markers including carcinoembryonic antigen (CEA), urokinase plasminogen activator, squamous cell carcinoma antigen (SCC), cytokeratin 19 fragment (CYFRA 21.1) *(2–4)*, PGP 9.5 *(5)*, and RACS1 *(6)*; however, there is room for much improvement in this area.

cDNA microarray technology allows the tissue expression profiles of thousands of genes to be compared simultaneously *(7–10)*. However, genes with abundant messages tend to dominate the corresponding cDNAs to be arrayed, limiting the representation of genes expressed at lower levels. To increase the potential for identifying tumor markers with both abundant and less abundant messages expressed in lung tumors, we combined cDNA subtractive methodology with cDNA microarray technology. Here we use lung squamous cell carcinoma as a model tumor target, and the methodology can be applied to other types of lung cancers.

The first step is to use subtractive methodology to enrich for genes that are differentially overexpressed in lung cancers. The subtractive methodology was based on Hara et al. *(11)* with modifications. The inclusion of this step has several advantages. First, it will eliminate over 90% of cDNAs that are not differentially expressed in lung squamous cell carcinoma. The smaller number of cDNAs makes downstream analysis more flexible and less labor-intensive. Secondly, by analyzing a sample of cDNA clones generated from the subtracted library, the quality of the clones to be analyzed by microarray analysis is well under control. Lastly, successive subtractions can eliminate abundantly overexpressed genes, allowing recovery of genes that are differentially expressed at lower level.

The second step is to use cDNA microarray technology. This glass chip-based analysis enable investigators to fabricate many identical chips, so that expression of genes in tumors and various normal tissues can be examined simultaneously. It is not intent of this chapter to discuss cDNA microarray technology in full detail, rather to emphasize it as a tool. Investigators who are interested in setting up their own microarray technology platform should consult other resources (*see* **Note 1**). We chose to outsource the technology using custom cDNA microarray services offered by Incyte Pharmaceuticals, Inc. (Palo Alto, CA) including chip fabrication, probe labeling, and hybridization. In our case, 23 chips containing 2,002 cDNAs from subtracted lung squamous cell carcinoma libraries were fabricated and each chip was hybridized with a pair of cDNA probes, fluorescence-labeled with Cy3 and Cy5 dyes, respectively. One probe is synthesized from a tumor RNA (i.e., lung squamous cell carcinoma) and the other from a normal tissue RNA (i.e., normal lung). Thus, we have an advantage of being able to control the quality and source of RNA from various tumors and normal tissues. After hybridization, fluorescence signals for both Cy3 and Cy5 were scanned, normalized (*see* **Subheading 3.5.**), and recorded. Through GEMTools Software (Incyte), cDNA clones showed favorable expression in lung tumors were identified for further analysis.

2. Materials
2.1. Construction of cDNA Libraries

1. Poly A^+ RNA from tester (lung squamous cell carcinoma) and driver (normal tissues).
2. Superscript Plasmid System for cDNA Synthesis and Plasmid Cloning Kit (GIBCO/BRL Life Technologies). Store at –20°C.
3. pcDNA 3.1+ vector and *Bst*XI/*Eco*RI adaptors (Invitrogen). Store at –20°C.
4. ElectroMax DH 10B cells (GIBCO/BRL Life Technologies), store at –80°C.
5. Qiagen plasmid purification kits.
6. Standard Molecular Biology Grade reagents such as RNase free H_2O, ethanol, 3 M Na Acetate, pH 5.0, and 25:24:1 phenol/chloroform/isomyl alcohol.

2.2. Construction of Lung Squamous Cell Carcinoma-Specific Subtracted cDNA Libraries

1. *Bam*HI, *Xho*I, *Spe*I, and *Not*I restriction enzymes.
2. Photoprobe long-arm biotin (Vector Laboratories): 1 mg/mL in H_2O. Store at −20°C.
3. n-Butanol (water-saturated). Store at room temperature.
4. Streptavidin (GIBCO/BRL Lifetechnologies): 2 mg/mL in HES. Store at 4°C.
5. 2X hybridization buffer: 1.5 M NaCl, 10 mM EDTA, 50 mM HEPES, pH 7.5, 0.2% sodium dodecyl sulfate (SDS). Store at room temperature.
6. HES buffer: 150 mM NaCl, 10 mM HEPES, pH 7.5, 1 mM EDTA.
7. HE buffer: 10 mM HEPES, pH 7.5, 1 mM EDTA.
8. 1 M Tris-HCl, pH 9.5.
9. Yeast tRNA 5 mg/mL (−20°C).
10. pBCSK(+) (Stratagen).
11. (Optional) Chroma Spin-400 columns (Clontech). Store at room temperature.

2.3. cDNA Target Amplification

1. 10 mM dNTPs, M13 reverse and M13 forward primers at 100 µM, Elongase Amplification System (Gibco-BRL Life Technologies). Stored at −20°C.
2. PCR machine in a 96-well format
3. Millipore Multiscreen-PCR 96-well Filtration System (Cat. no. MANU 03010).

2.4. Probe Synthesis

1. 200 ng to 1 µg poly A^+ RNA prepared from each tumor or normal tissue. Aliquots of poly A RNA are analyzed by gel electrophoresis prior to probe synthesis.
2. cDNA probes directly labeled using Cy3-dCTP or Cy5-dCTP (Incyte).

2.5. Reference Genes and Internal Controls

1. Reference Genes such as beta-actin, phospholipase A2, MHC-class I gene, 23KD basic protein, ribosomal protein L31, alpha-tubulin, transferrin reductase, ribosomal protein S9, ribosomal protein S28, and ribosomal protein S7 (*see* **Note 2**).
2. cDNAs that are known to be differentially expressed in lung squamous cell carcinoma, such as keratin isoform 6 and ADH 7 (*see* **Note 3**).

3. Methods

3.1. Construction of a Tester Specific cDNA Library Using Lung Squamous Cell Carcinoma

1. Digest 1–5 µg of pcDNA 3.1 + with *Bst*XI and *Eco*RI to completion and gel purify the fragment according to standard protocol (*see* **Note 4**).
2. Use 3–5 µg of poly A^+ RNA extracted from lung squamous cell carcinoma for cDNA synthesis and library construction using a Superscript Plasmid System for

cDNA Synthesis and Plasmid Cloning Kit with following modifications. Briefly, *Bst*XI/*Eco*RI adaptors were used and cDNA was cloned into pcDNA3.1+ that was digested with *Bst*XI and *Eco*RI.
3. Set up ligation using 200 ng of pcDNA3.1+ vector with 50% of cDNA recovered. Incubate the ligation mixture at 16°C overnight.
4. Using ElectroMax DH 10B cells for transformation and plate out the entire transformations onto 20 large LB+Amp plates.
5. Determine total number of independent colonies (should be at least 1×10^6) and analyze 24–36 randomly picked clones for estimating the percentage of inserts (should be at least 90%) as well as the average insert size (should be at least 1.2 kb).
6. Collect and pool all the transformants. Freeze down aliquots of the cells as glycerol stocks and use rest of library to make 2–4 large scales of plasmid DNA preparations.

3.2. Construction of Driver cDNA Libraries Using Normal Tissues (see Note 5)

The composition of driver cDNA libraries will affect the outcome of subtraction. Either a single driver cDNA library can be made from a pool of normal tissues or several driver cDNA libraries can be generated from different normal tissues. The same procedure should be followed as for tester cDNA library construction.

3.3. Lung Squamous Cell Carcinoma Specific Subtracted cDNA Libraries

3.3.1. Preparation of Biotinylated Driver DNA

1. Digest 80–160 µg driver DNA with *Bam*HI and *Xho*I, followed by phenol-choloroform extraction and ethanol precipitation.
2. Dissolve the driver DNA in 80–160 µL H_2O and denature it completely by boiling the DNA for 5 min followed by quick chill on ice.
3. Add equal volume (80–160 µL) photoprobe biotin (1 mg/mL in H_2O) and irradiate for 20 min with a sunlamp. The reaction is carried out on ice and the distance of the sunlamp is about 10–20 cm from the tube.
4. Add one-half volume (40–80 µL) of photoprobe biotin and irradiate for another 10 min.
5. Stop biotinylation by adding 20–40 µL 1 M Tris, pH 9.5.
6. Remove free biotin by extracting the driver DNA with water-saturated n-butanol, repeat 5 more times.
7. Ethanol precipitate the biotinylated driver DNA and resuspend the driver DNA at 2 mg/mL in H_2O.

3.3.2. Preparation of Tester DNA and Vector

1. Digest 1–5 μg of pBCSK + with *Not*I and *Spe*I to completion and gel purify the plasmid DNA according to standard protocol.
2. Digest 10–20 μg of tester DNA with *Not*I and *Spe*I followed by phenol-chloroform extraction.
3. (Optional) Enrich larger tester DNA fragments by passing through a size fractionation column, such as Chroma spin-400 columns. Expect to lose up to 50% of tester DNA.
4. Ethanol precipitate the tester DNA and reconstitute the tester DNA at 1 mg/mL in H_2O.

3.3.3. Hybridization and Subtraction

1. Mix 5 μL tester DNA (5 μg) with 15 μL driver DNA (30 μg) and boil for 5 min (*see* **Note 6**). Add equal volume of 2X hybridization buffer (20 μL) and top with a drop of mineral oil. Boil again for 3 min and proceed for hybridization at 68°C overnight.
2. Add 260 μL HE buffer to the hybridizing mixture. Incubate at 55°C for 5 min.
3. Transfer the aqueous layer to a new tube and add 10 μL streptavidin. Vortex. Incubate at RT for 10–20 min, vortex occasionally. Extract with 300 μL phenol/chloroform to remove biotinylated DNA including driver DNA and tester DNA hybridized to the driver DNA.
4. Transfer the aqueous layer to a new tube and add 10 μL streptavidin. Incubate at room temperature for 3–5 min and repeat the extraction.
5. Repeat **step 4** two more times.
6. Add 1 μL tRNA and extract the DNA with chloroform followed by ethanol precipitation.
7. Resuspend the subtracted DNA in 10 μL H_2O, and mix with 10 μL (20 μg) driver DNA. Add equal volume of 2X hybridization buffer and top with a drop of mineral oil. Boil for 3 min and proceed for hybridization at 68°C for 2–3 h.
8. Repeat **steps 2–6** to complete the second-round subtraction.
9. Ligate the entire subtracted cDNA with 200 ng pBCSK+ vector overnight prior to transformation of ElectroMax DH 10 B cells. Plate out cells on 5–10 large LB+chloroamphenocol plates.

3.3.4. Characterization of Subtracted Lung Squamous Cell Carcinoma

1. Expect to recover 500–20,000 transformants. Randomly pick 96 transformants for plasmid purification and sequence analysis. The redundancy and the complexity of the subtracted cDNA library can be estimated based on the frequency of each unique cDNA recovered.
2. For rest of transformants, make a glycerol stock for each individual clone (several hundreds to several thousands depending on the complexity of the library) by

inoculating 100 μL LB+ chlroamphenicol culture overnight in 96-well plates. Add 100 μL 50% glycerol and store the plates at –80°C.

3.3.5. Generation of Additional Subtracted Lung Squamous Cell Carcinoma Libraries

To eliminate genes that are abundantly overexpressed in lung squamous cell carcinoma, such as keratin 6 isoform, as well as to recover genes that are differentially expressed at lower level, it is necessary to generate additional subtracted libraries by including abundantly overexpressed genes in the driver DNA (*see* **Note 7**).

1. Digest pool of subtracted plasmid DNA of selected tester-specific genes (~200 ng each) with noncompatible enzymes to *Not*I and *Spe*I to release the inserts and the pBCSK+ vector backbone. This will help to reduce the background for subsequent cloning.
2. Mix with normal driver DNA prior to biotinylation, and repeat the subtraction highlighted in **Subheadings 3.3.1.–3.3.4.**
3. If necessary, **steps 1** and **2** may be repeated by including more tester-specific genes in the drivers. For example, the first subtracted lung squamous cell carcinoma library repeatedly recovered abundantly overexpressed genes such as keratin 6 isoform and 58KD type II keratin. Thus, a second subtracted library was constructed by including an additional driver DNA comprised of a pool of five genes that are highly enriched in the first subtracted library. As a result, the second subtracted lung squamous cell carcinoma library contains ~20,000 independent clones. Sequence analysis of 227 clones revealed 114 unique clones and 57 of them are novel with no GenBank hits, suggesting that the complexity of the second library is improved compared with the first subtracted library. Similarly, by including 20 more gene-specific DNA in the driver, the third subtracted library yielded ~700 independent clones with little redundancy with previous subtracted libraries. These results suggest that by adding abundant tester-specific genes in the driver DNA, we were able to enrich lung squamous cell carcinoma specific genes that are expressed at lower level, and thus the complexity as well as the quality of subsequent subtracted cDNA libraries were enhanced.

3.4. Preparation of cDNA for Microarray Analysis

1. Using 0.5 μL glycerol stock from **step 2** of **Subheading 3.3.4.** as DNA template.
2. Set up PCR reaction plate (96-well format) with the final volume being 100 μL. Prepare mater mix as follows per 100 reactions (*see* **Note 8**): 7570 μL H_2O, 200 μL 10 mM dNTPs, 40 μL each 100 μM M13 forward and reverse primers, 1000 μL each 5X buffer A and 5X buffer B from Elongase Kit, 100 μL Elongase mix.
3. Purify the entire PCR products using Millipore 96-well PCR purification plate, but elute DNA in 50–60 μL H_2O. DNA is now ready for chip fabrication.

3.5. Posthybridization Analysis

Since each glass chip is hybridized with a pair of probes, labeled with Cy3 and Cy5, the expression profile reflected by Cy3 and Cy5 signals should be normalized for balance coefficient for each pair of probes (Cy3/Cy5). This is because the solubility and half-life of Cy3 and Cy5 Cyanine dyes are different, and dyes may not be incorporated into the probes with equal efficiency. Normalization can be achieved in two ways. Ideally, the Cy3 and Cy5 signals can be normalized by hybridizing the same cDNA elements with a pair of probes labeled with Cy3 and Cy5, and reversibly Cy5 and Cy3, respectively. However, these will double the amount of work and the amount of RNA used for probe synthesis, which can be very limited. Alternatively, Cy3 and Cy5 signals can be normalized using the entire cDNA elements as references. Therefore, the balanced coefficiency for each pair of probes is a ratio of average fluorescence intensity between Cy3 and Cy5 signals (*see* **Note 9**). **Table 1** summarizes the balanced coefficiency (Cy3/Cy5) for each of 23 probe pairs that we used for hybridizing with lung squamous cell carcinoma chip and the differential expressions of four reference genes (*see* **Subheading 2.5.**) after normalization. As expected, these genes are ubiquitously expressed in tumors and normal tissues with balanced Cy3 and Cy5 signals, suggesting well-controlled microarray experiments.

3.6. Identification of LSCC Specific Molecular Markers

The expression profiles of 2,002 cDNA elements on the chip are illustrated using a lung squamous tumor probe paired with a normal kidney probe (*see* **Fig. 1**, probe pair 11) *(1)*. A large number of clones (within the triangle areas) were found to be overexpressed in these lung squamous tumors, suggesting subtraction is an efficient step for enriching genes differentially overexpressed in lung squamous cell carcinoma. Lung squamous cell carcinoma specific genes were scored based on their expression in over 50% of lung squamous tumors, as compared with undetectable or limited expression in normal tissues. Upon sequencing analysis, we identified 17 genes differentially overexpressed in lung squamous cell carcinoma. **Figure 2** illustrates differential expression of four novel lung squamous cell carcinoma genes *(1)*. Of these 17 genes, seven were recovered multiple times independently through the microarray analysis, demonstrating that the combination of cDNA subtraction and microarray analysis is a reliable, reproducible, and high throughput tool for identifying new molecular targets in lung cancers.

4. Notes

1. Incyte Pharmaceuticals Inc. (Palo Alto, CA), TeleChem International Inc. (Sunnyvale, CA), Agilent Technologies, Inc., and Affimetrix Inc. (Santa Clara,

Table 1
Poly A+ Probes and Reference Guides Used for Microarray Analyses

Probes	Cy3	Cy5	Balanced coefficiency	Beta-actin	Ribosomal protein L31	Ribosomal protein 59	Ribosomal protein S28
1	Normal Lung (8009N)	Normal Skin (S5)	0.38	1.8	1.1	1.5	-1.2
2	Lung Squamous Tumor (96A)	Normal Lung (CT-2)	1.5	-1.8	-1.8	-2.9	-2.8
3	Lung Pleural Effusion (86–52)	Normal Lymph Nodes (CT6)	1.06	-1.2	-1.3	-3.2	-1
4	Colon Tumor (S18)	Normal Colon (11)	0.23	1	-1.4	-1.4	-1.2
5	Lung Squamous Tumor (9688T)	Normal Liver (CT1)	1.95	2.5	-1.5	-2.1	-4.3
6	Lung Squamous Tumor (Pooled)	Normal Pancreas (S2)	2.1	1.1	-4.4	-7.4	-1.5
7	Bronchioloalveolar Adenocarc.	Normal Breast (S73)	0.35	-2.4	-1.8	-1	-1.8
8	Lung Squamous Tumor (9681T)	Normal Heart (CT5)	1.51	-1.8	-1.4	1	-4.3
9	Lung Pleural Effusion (86–52)	Normal Bone Marrow (CT4)	0.84	-1.4	1.1	-2.1	1
10	Bronchioloalveolar Adenocarc.	Normal Large Intestine (S55)	0.48	-2.9	-1.9	-1.3	-1.3
11	Lung Squamous Tumor (9688T)	Normal Kidney (CT9)	1.35	3.3	1.1	-2.1	1
12	Lung Squamous Tumor (Pooled)	Normal Stomach (S6)	1.47	-1.9	-2.8	-1.8	-1.7
13	Lung Squamous Tumor (Pooled)	Normal Lung (NL873)	2.71	-2.2	-3.7	-2.5	-2
14	Lung Squamous Tumor (96A)	Resting PBMC (S39)	1.24	-3.5	-4.6	-6.1	-2.8
15	Lung Squamous Tumor (96A)	Normal Brain (CT2)	1.76	-1.2	1.1	1.2	-3.6
16	Lung Adenocarcinoma (9680T)	Normal Small Intestine (CT10)	0.35	-1.6	1	1.1	-2
17	Lung Adenocarcinoma (8009T)	Normal Bladder (S9)	4.77	-2.2	1.6	2.5	1.7
18	Lung Squamous Tumor (9681T)	Normal Salivary Gland (CT8)	4.99	1.1	-1.3	1.1	-1.9
19	Lung Adenocarcinoma (8009T)	Matched Normal Lung (8009N)	3.65	-1.2	3	-1.1	1.2
20	Lung Squamous Tumor (9681T)	Matched Normal Lung (9681N)	3.83	-2	-1.1	-1.8	-1.6
21	Lung Squamous Tumor (96A)	Normal Lung (CT-1)	48*		-1.1	1.2	-1.3
22	Lung Squamous Tuor (9681T)	Normal Lung (CT-2)	4.45	-1.1	-2.5	-1.4	-2.2
23	Lung Squamous Tumor (9688T)	Matched Normal Lung (9688N)	2.98	-1.9	-1.4	-2.1	-1.9

200 ng poly A+ RNA was used to generate each probe. The RNA was reverse transcribed and labeled with Cy3-dCTP or Cy5-dCTP. Balanced coefficient for each pair of probes is calculated by the average fluorescence ratio of Cy3 to Cy5 using cDNAs on the entire chip as references. Tumors listed above are primary tumors except the lung pleural effusion (86–52) is metastatic lung adenocarcinoma. '-' denotes Cy5 signal is higher than Cy3 signal. *, The balanced co-efficiency for this pair of probes is out of linear range.

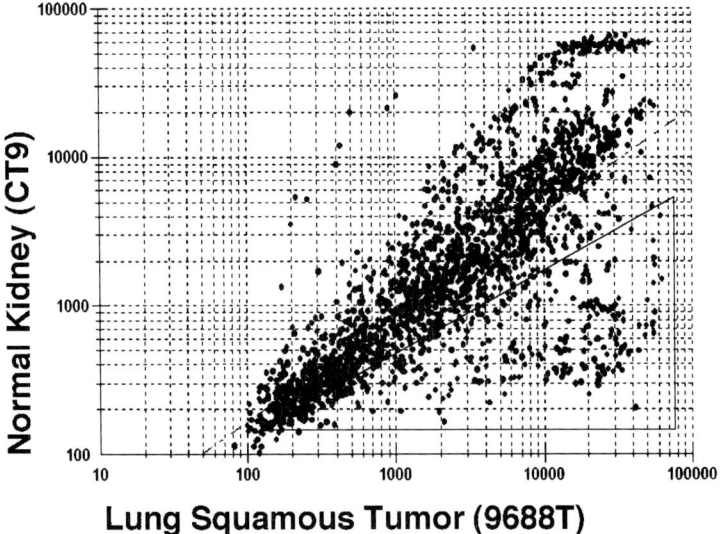

Fig. 1. The scatter plot of fluorescence intensities of 2,002 cDNA clones hybridized with a lung squamous tumor probe (9688T) and a normal kidney probe (CT9). The dots within the triangle box represent the cDNA clones that are differentially overexpressed in the lung squamous cell carcinoma.

CA) provide custom microarray services and/or microarray tools, kits, and reagents. Investigators who are interested in setting up the microarray technology platform should consult resources such as **refs. 12–15**.
2. Ubiquitously expressed genes should be included as part of internal controls for probe quality as they are generally expressed in both cancerous and normal tissues. Since expression of each of these housekeeping genes is somewhat fluctuated, it is better to use a group of internal reference genes.
3. Keratin isoform 6 and ADH7 genes were recovered multiple times during analysis of the first and second subtracted lung squamous cell lung carcinoma libraries, respectively, suggesting that keratin isoform 6 is more abundant than ADH7 even though both are differentially expressed in lung squamous cell carcinoma.
4. pcDNA3.1+ vector was chosen because it is a mammalian expression vector and the cDNA clones can be directly transduced into mammalian cells.
5. We have chosen normal tissues of lung, liver, heart, brain, and skin, etc. to be part of driver's DNA. Since there are many infiltrating T-lymphocytes particularly in late stages of tumor samples, it is also helpful to include immune tissues such as PBMC or spleen as part of driver DNA.
6. The ratio of tester to driver DNA can be varied for optimal hybridization and subtraction.
7. Recovery of highly redundant cDNA clones such as keratin 6 isoform in subtracted lung squamous cell carcinoma is an indication of successful subtrac-

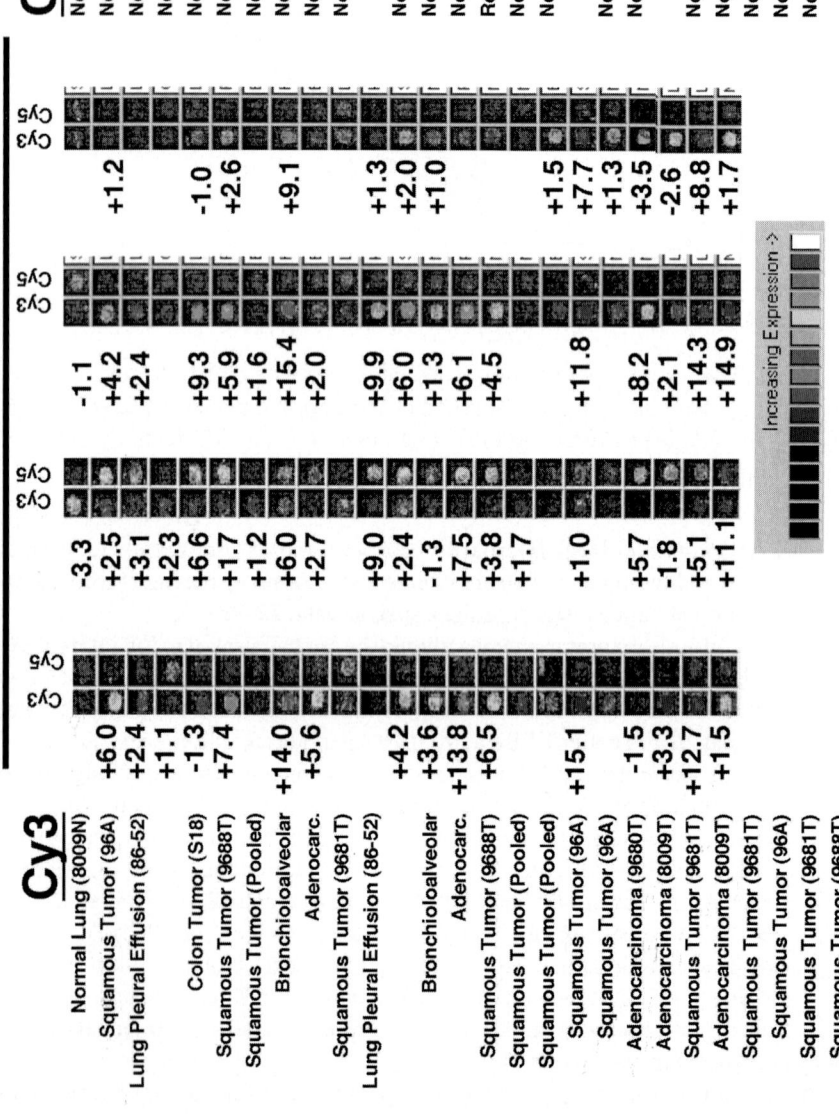

tion. This is because keratinization is one of the major characteristics of all squamous cell carcinomas. However, too many redundant clones in the subtracted library affect the complexity of the library, and these redundant genes usually represent genes that are differentially expressed in lung squamous cell carcinoma at a higher level.

8. Elongase is optimal at 68°C, and 2.5 min extension time was performed in an attempt to amplify larger subtracted cDNA fragments.
9. Since multiple probe pairs were used in this study and each hybridization is an independent experiment with a result of competitive hybridization of two probes, the fluorescence signal cannot be compared between probe pairs or chips quantitatively. In another word, quantification is only applied to cDNA elements within the chip using the same pair of probes.

References

1. Wang, T., Hopkins, D., Schmidt, C., Silva, S., Houghton, R., Takita, H., et al. (2000) Identification of genes differentially over-expressed in lung squamous cell carcinoma using combination of cDNA subtraction and microarray analysis. *Oncogene* **19,** 1519–1528.
2. Pastor, A., Menendez, R., Cremades, M. J., Pastor, V., Llopis, R., and Aznar, J. (1997) Diagnostic value of SCC, CEA and CYFRA 21.1 in lung cancer: a Bayesian analysis. *Euro. Respir. J.* **10,** 603–609.
3. Morita, S., Sato, A., Hayakawa, H., Ihara, H., Urano, T., Takada, Y. et al. (1998) Cancer cells overexpress mRNA of urokinase-type plasminogen activator, its receptor and inhibitors in human non-small-cell lung cancer tissue: analysis by Northern blotting and in situ hybridization. *Intl. J. Cancer* **78,** 286–292.
4. Brechot, J. M., Chevret, S., Nataf, J., Le Gall, C., Fretault, J., Rochemaure, J., et al. (1997) Diagnostic and prognostic value of Cyfra 21-1 compared with other tumour markers in patients with non-small cell lung cancer: a prospective study of 116 patients. *Euro. J. Cancer* **33,** 385–391.
5. Hibi, K., Westra, W. H., Borges, M., Goodman, S., Sidransky, D., and Jen, J. (1999) PGP9.5 as a candidate tumor marker for non-small-cell lung cancer. *Am. J. Pathol.* **155,** 711–715.
6. Iwasaki, T., Nakashima, M., Watanabe, T., Yamamoto, S., Inoue, Y., Yamanaka, H., et al. (2000) Expression and prognostic significance in lung cancer of human tumor-associated antigen RCAS1. *Intl. J. Cancer* **89,** 488–493.

Fig. 2. *(opposite page)* Differential expression of four novel genes, L514S, L519S, L530S, and L531S in lung squamous cell carcinoma by microarray analysis. Illustrated here are color images of hybridization intensities (white being the strongest and dark being the weakest) as well as the fold of balanced differential expression between Cy3 and Cy5 probes. The probe pairs are from 1 to 23 (also refer to **Table 1**). "+" and "–" indicate the expression being stronger in Cy3 and Cy5 channel, respectively.

7. Schena, M., Shalon, D., Davis, R. W., and Brown, P. O. (1995) Quantitative monitoring of gene expression patterns with a complementary DNA microarray. *Science* **270,** 467–470.
8. DeRisi, J., Penland, L., Brown, P. O., Bittner, M. L., Meltzer, P. S., Ray, M., et al. (1996) Use of a cDNA microarray to analyse gene expression patterns in human cancer. *Nature Genet.* **14,** 457–460.
9. DeRisi, J. L., Iyer, V. R., and Brown, P. O. (1997) Exploring the metabolic and genetic control of gene expression on a genomic scale. *Science* **278,** 680–686.
10. Iyer, V. R., Eisen, M. B., Ross, D. T., Schuler, G., Moore, T., Lee, J. C., et al. (1999) The transcriptional program in the response of human fibroblasts to serum. *Science* **283,** 83–87.
11. Hara, T., Harada, N., Mitsui, H., Miura, T., Ishizaka, T., and Miyajima, A. (1994) Characterization of cell phenotype by a novel cDNA library subtraction system: expression of CD8α in a mast cell-derived interleukin-4-dependent cell line. *Blood* **84,** 189–199.
12. Schena, M. (ed.) (2000) *Microarray Biochip Technology*. Eaton Publishing, Natick, MA.
13. Martinsky, T. and Haje, P. (2000) Microarray tools, kits, reagents, and services, in *Microarray Biochip Technology* (Schena, M., ed.), Eaton Publishing, Natick, MA, pp. 201–220.
14. Schena, M., ed. (2000) *DNA Microarrays A Practical Approach*. Oxford University Press, New York, NY.
15. Brown, P. O. http://cmgm.stanford.edu/pbrown

11

Expression Profiling of Lung Cancer Based on Suppression Subtraction Hybridization (SSH)

Simone Petersen and Iver Petersen

1. Introduction

Gene amplifications and deletions frequently contribute to tumorigenesis. The characterization of DNA copy-number changes by Comparative Genomic Hybridization (CGH) has shown that these changes are not occurring randomly. As summarized in the previous chapter, recurring patterns of chromosomal abnormalities are specific for certain histo-types of tumors and are associated with the biological behavior of the tumor, e.g., metastasis formation.

The concept of oncogenes and tumor-suppressor genes was based on the traditional finding that mutations occurred in specific genes during the process of tumorigenesis. However, a larger number of genes, fitting into neither of these two categories, are probably not mutated, but have disregulated levels of expression. This can result from changes in the promoter region (e.g., methylation), differential regulation within a cancer-associated gene cascade, or simply gene-dosage effects resulting from changes in chromosomal ploidy.

One approach to identify new tumor-associated genes is to define differences in gene expression patterns between normal and neoplastic cell populations. Techniques developed explicitly for this purpose are differential display *(1)*, representational difference analysis of cDNA *(2)*, suppression subtractive hybridization *(3)*, serial analysis of gene expression *(4)*, and cDNA microarray-based expression analysis *(5)*. In the following we will present our procedure to analyze differential gene expression of lung cancer by Suppression Subtractive Hybridization (*see* **Fig. 1**).

Subtractive hybridization enables the comparison of two populations at the RNA level and isolation of genes that are preferentially expressed in only one

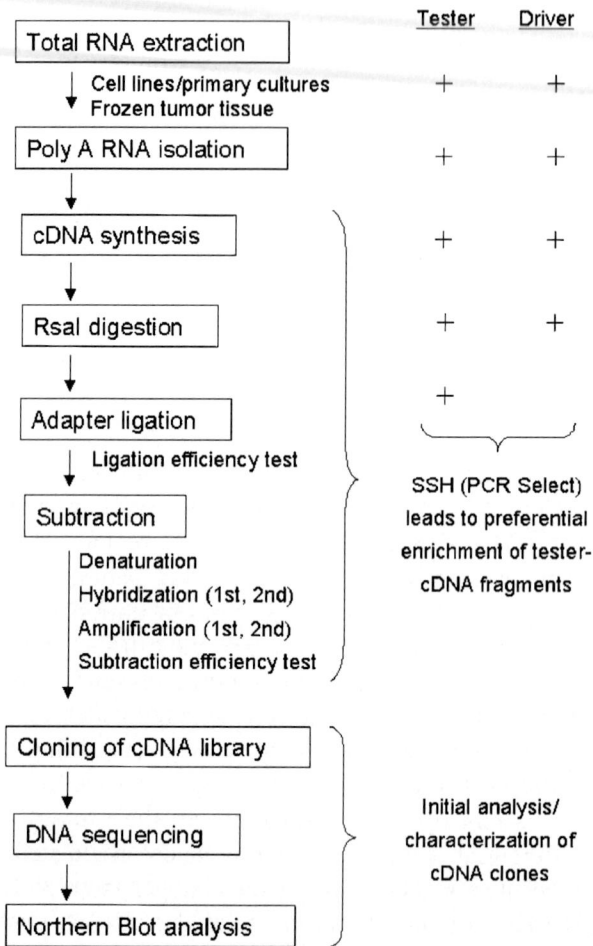

Fig. 1. Flow chart of our methodology in lung cancer expression profiling.

of the populations. It includes the reverse transcription of mRNA into cDNA. One cDNA, usually from the cells of interest, is referred as "tester," while the reference cDNA is referred to as the "driver." Tester and excess driver cDNA are allowed to anneal to each other, and the double-stranded hybrid sequences are then removed. They contain cDNA fragments abundant in both populations. The remaining single-stranded, un-annealed cDNAs represent genes that are expressed in the tester but either not at all or at significantly lower levels in the driver population.

Traditional subtractive hybridization methods have been described. They require several rounds of hybridization and are restricted to abundantly expressed messages *(6–9)*. We present here PCR-Select™ cDNA subtraction (Clontech), a modified method combining subtractive hybridization and suppression polymerase chain reaction (PCR), which allows enrichment of target molecules while suppressing background *(3,10)*.

1.1. SSH in Lung Cancer

To identify genes associated with lung cancer, we performed suppression subtractive hybridization using cDNA synthesized from normal human bronchial epithelial cells as well as small airway epithelial cells, an adenocarcinoma cell line (D51), a squamous carcinoma cell line (H2170), and a cell line (H526) derived from a metastatic SCLC. Six cDNA libraries were cloned representing up- and downregulated genes, respectively. A total of 2,452 clones were grown and plasmid-DNA was isolated. We determined the nucleotide sequence and identified more than 900 individual sequences. They encode for signal transduction and cell cycle-associated molecules, nuclear proteins (transcription factors, DNA processing enzymes), molecules involved in RNA and protein processing, protein transport and protein degrading. Furthermore, we found metabolic enzymes, transporters, ion channels, cytoskeletal components, glycoproteins, epidermal differentiation complex-associated proteins, tumor-associated antigens, extracellular proteins, and a number of unknown genes as well as cDNA fragments with not yet characterized function. More than 50% of the identified genes could be confirmed as differentially expressed by Northern-blot analysis *(11,12)*.

There are only a few studies of gene expression profiles in lung carcinomas, but these have been limited to either small cell lung cancer (SCLC) or non-small cell lung cancer (NSCLC) *(13,14)*. Our data showed that the observed expression differences are not unique to individual cell lines, since additional 28 lung cancer cell lines examined by Northern-blot analysis displayed similar patterns. The pair-wise correlation coefficient analysis between individual NSCLC and SCLC as well as bronchial epithelial cells and carcinoma cells was statistically significant. This fact, together with expression data from Anbazhagan and co-workers, which showed that SCLC are rather distinct from pulmonary carcinoids, suggest that both NSCLC and SCLC derive from a common epithelial precursor. This is supported by clinico-pathological data and genetic-based data (for review, *see* **ref. 15**) and underlines the need for a new classification of lung cancer that includes genetic composition and gene expression.

Overall, several genes found to be differentially expressed in lung carcinomas have also been described in squamous cell carcinomas of the head and neck and breast carcinomas *(16–18)*. This is suggesting that malignancies of epithelial cells have common cancer-related expression profiles. Moreover, a significant proportion of genes belong to the Ras-pathway or are p53-regulated genes *(19,20)*. This is not surprising since mutations of Ras and p53 are known to play a major role in lung cancer tumorigenesis (for review, *see* **ref. *21***). The utility of the applied method to assess functional changes occurring during development and progression of lung cancer is supported by both, the types of genes identified as differentially expressed, as well as their biological function.

Interestingly, a number of identified genes cluster within regions that are frequently affected by chromosomal imbalances. For example, the gene programmed cell death 4 (PDCD4) maps to chromosome 10q24, and could be a target of the deletions frequently observed in advanced lung carcinomas *(22)*, which is underlining the usefulness for the further identification of candidate genes within these regions.

In summary, the analysis of gene-expression profiles in lung cancer has provided detailed sequence information on transcriptional changes associated with lung tumorigenesis, many of them described the first time. The specific set of genes recovered by cDNA subtraction provides the groundwork and starting point for identifying transcripts with diagnostic and prognostic value with respect to both primary tumors and metastases. In addition, studying the function of these genes and their biological pathways may lead to the development of new therapeutic options.

2. Reagents

2.1. RNA Extraction

1. DEPC-Water: Dilute diethylpyrocarbonate (DEPC) 1:1000 in ddH$_2$O. Incubate covered overnight under the fumehood, autoclave, and store at 4°C.
2. GT-Buffer: Combine 100 g guanidine isothyocianate, 1.54 g sodium citrate, 1.01 g sarkosil, 116.8 mL DEPC water. Autoclave and add β-mercaptoethanol to 2% before using.
3. 10X MEN (1 L): Combine 50 m*M* Sodium acetate and 10 m*M* EDTA. Add DEPC water to a final volume of 900 mL. Adjust the pH to 7.0 using acetic acid or sodium hydroxide. Autoclave. Add 41.852 g RNase-free 3-Morpholinopropane sulfonic acid (MOPS, 200 m*M* final concentration) and bring volume up to 1 L with DEPC water.
4. Water saturated Phenol: Melt Phenol in a water bath at 50°C. Add an equal volume of sterile H$_2$O, shake, wait until the two phases separate. Take off upper phase, add one volume of H$_2$O and repeat. Add 0.1% hydroxyquinolin, store at 4°C in the dark.
5. RNA gel (50 mL): Dissolve 0.6 g of agarose in 45 mL of water. Add 6 mL of 10X MEN and 9.6 ml of 37% formaldehyde. Mix with stirrer and cast gel under the fumehood.

6. RNA loading buffer (1 mL): Combine 500 µL 100% glycerol, 497 µL DEPC H$_2$O, 2 µL 0.5 M EDTA, 2 µL of a 1% ethidium bromide solution. Add 0.4% Bromophenol blue.

2.2. polyA+ RNA Isolation

1. mRNA Separator kit (Clontech).
2. 2 M Potassium acetate, pH 5.0.
3. 80% Ethanol.
4. 95% Ethanol.

2.3. Subtraction Hybridization

1. AMV reverse transcriptase.
2. Phenol/Chloroform/Isoamyl alcohol (25:24:1).
3. Chloroform/Isoamyl alcohol (24:1).
4. 4 M Ammonium acetate.
5. 50X TAE.
6. Agarose.
7. Rsa I and 10X RsaI restriction buffer.
8. T4 DNA ligase (400 U/µL) and 5X ligation buffer.
9. G3DPH internal control primer.
10. β-actin internal control primer.
11. PCR purification kit (Qiagen).
12. PCR 2.1 vector (Invitrogen).
13. 0.5 M β-Mercaptoethanol.
14. Competent bacteria (One Shot TOP10F' chemically competent *Escherichia coli*, Invitrogen).
15. LB agar plates with ampicillin (50 µg/µL).
16. 5-Bromo-4-chloro-3-indolyl-β-D-galactopyranoside (X-gal).
17. DMF.
18. 0.5 mM Isopropyl-β-D-thiogalactopyranoside (IPTG).
19. QIAGEN mini spin plasmid isolation kit.

2.4. Analysis of cDNA Libraries

1. Long ranger solution: Combine 105 g urea, 30 mL Long Ranger (FMC, 50%), 110 mL sterile water, and 30 mL 10X TBE.
2. Sequencing gel: Long ranger solution 45 mL, 300 µL 10% ammonium persulfate, 30 µL TEMED.
3. Thermo Sequenase fluorescent labeled primer cycle sequencing kit with 7-deaza-dGTP (Amersham).
4. Hybond-N (Amersham).
5. 20X SSC Buffer: Combine 3 M Sodium chloride and 0.3 M Tri-Sodium citrate. Add ddH$_2$O, adjust the pH to 7.0 and bring volume up to 1 L.
6. QIAquick PCR purification kit (QIAGEN).

7. Hybridization solution: ExpressHyb Solution (Clontech).
8. MegaPrime Labeling kit (Amersham).
9. Klenow DNA polymerase.
10. ^{32}P-αdCTP.
11. Nucleotide Removal kit (QIAGEN).
12. 20X SSPE Buffer: 3.6 M sodium chloride, 0.2 M sodium dihydrogen phosphate and 0.02 M EDTA. Add ddH$_2$O, adjust the pH to 7.7 and bring volume up to 1 L. Autoclave.
13. 20% SDS.

3. Methods
3.1. RNA Extraction (see Notes 1,2)
3.1.1. Phenol-Chloroform-Extraction (23)

1. Aspirate media from growing cells cultured in either four 15 cm^2 tissue-culture dishes or four T-175 flasks and wash with 10 mL cold 1X PBS.
2. Add β-Mercaptoethanol to the GT-buffer (36 μL per 5 mL) and lyse the cells using 2 mL of buffer per plate/flask. Complete cell lysis will take about 1 min.
3. Scrape the surface of the plate in order to collect all of the cell lysate, collect with a pipet and combine the lysates from two plates/flasks in a Sarstedt-centrifuge tube (total volume 4 mL).
4. Add 400 μL of 2 M sodium acetate, pH 4.0, and vortex.
5. Add 5 mL of water saturated phenol and 1 mL chloroform-isoamyl alcohol (24:1), gently mixing after each step until the solution becomes homogeneous.
6. Incubate on ice for 7 min.
7. Centrifuge at 9800g for 20 min (15°C) (Beckman Centrifuge).
8. Transfer upper aqueous phase to a new tube (careful not to disturb protein interphase), add 1 volume (4 mL) isopropanol, mix gently and incubate at –20°C for 30 min.
9. Centrifuge at 14,600g for 20 min (4°C).
10. Discard supernatant without disturbing RNA pellet, wash once with 1 mL 80% Ethanol.
11. Centrifuge at 14,600g for 20 min (4°C).
12. Aspirate supernatant, air dry the pellet and resuspend in 500 μL GT-buffer, combine lysates of two Sarstedt tubes into one 2 mL Eppendorf tube.
13. Add 1 mL isopropanol and incubate overnight at –20°C.
14. Centrifuge tubes at 16,000g for 20 min (4°C).
15. Discard supernatant and wash pellet with 1 mL 70% ethanol.
16. Centrifuge tubes at 16,000g for 20 min (4°C), discard supernatant, air dry the pellet and resuspend in 200 μL DEPC water.
17. Determine the concentration (Abs$_{260}$) and if necessary, dilute with DEPC water to a final RNA concentration between 1000 μg/mL–2000 μg/mL.
18. Check RNA quality on a gel by combining 2 μg RNA, 1 μL 10X MEN, 10 μL formamide, 3.5 μL 37% formaldehyde and 2 μL RNA loading buffer.

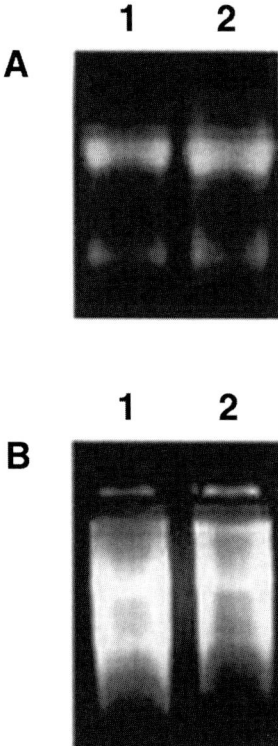

Fig. 2. **(A)** Total RNA prepared by Phenol-Chloroform extraction. Ten µg RNA separated on a 1% RNA agarose gel, lane 1; HBEC, lane 2, H526. The typical high molecular weight bands representing mainly ribosomal RNA are visible. **(B)** PolyA⁺ RNA prepared using mRNA separator method. About 1 µg RNA is loaded after precipitation on a 1% RNA agarose gel, lane 1: HBEC, lane 2: H526. The RNA should be a smear and there should be reduced amount of ribosomal bands.

19. Denature at 56°C for 10 min, chill on ice, spin down and separate by electrophoresis at 60 V for at least 1 h as shown in **Fig. 2A**.
20. Freeze RNA at –80°C until further use.

3.1.2. TRIzol Extraction (see **Note 3**)

1. Transfer 1–2 pieces (about 3 × 3 × 3 mm) of liquid nitrogen shock frozen tissue into a Sarstedt tube containing 4 mL TRIzol reagent and disperse with a homogenizer (Miccra D-8, ARTmoderneLabortechnik) for 30–60 s increasing the speed slowly up to 20,000 rpm. The homogenizer should be washed in another tube containing 4 mL TRIzol to collect the remaining tissue and both fractions should be combined. Before starting with the next probe, the homogenizer should be cleaned by running several times in sterile water.

2. Incubate the homogenates for 5–10 min at RT and centrifuge at 12,100g for 10 min (4°C). If there is a fatty phase, remove it.
3. Add 0.2 mL Chloroform per 1 mL TRIzol and mix carefully. Incubate for 5 min at RT.
4. Centrifuge at 12,100g for 15 min (4°C) and transfer upper aqueous phase to a new tube.
5. Precipitate RNA by adding 0.5 mL Isopropanol per 1 mL TRIzol and mix carefully.
6. Incubate for 10 min at RT.
7. Centrifuge at 12,100g for 10 min (4°C). Remove supernatant and wash pellet with 2 mL 80% ethanol (diluted in DEPC-H_2O).
8. Centrifuge at 6,800g for 5 min (4°C), remove supernatant and air dry the pellet for 10–20 min. Dissolve pellet in 200–300 µL DEPC-H_2O.

3.2. PolyA+ RNA isolation (see Notes 4,5)

PolyA$^+$ RNA purification protocol using the mRNA separator (Clontech):

1. Denature 1–1.5 mg total RNA (concentration 1 mg/mL in DEPC-H_2O) for 8 min at 80°C and keep afterwards at RT.
2. In the meantime, prepare the column by shaking the Oligo-dT beads in the tube, remove the caps from both ends and resuspend the beads by pipetting up and down until a homogeneous suspension is obtained.
3. Place the column containing the beads in a 50 mL Falcon tube and centrifuge in a swinging bucket rotor for at 1500 rpm for 2 min (20°C).
4. Add 2 mL High-Salt-Buffer to the beads and resuspend by pipetting up and down.
5. Centrifuge at 500g for 2 min (20°C). Oligo-dT resin is now ready for mRNA binding.
6. Add 1/5th volume Sample Buffer to the denatured RNA, vortex and mix carefully with the Oligo-dT beads by pipetting. Place the tube in a fresh 50-mL tube and allow RNA to bind to the particles for 10 min at RT.
7. Centrifuge at 500g for 2 min (4°C).
8. Wash with 900 µL High-Salt-Buffer and centrifuge at 500g for 2 min (4°C).
9. Wash with 1800 µL Low-Salt-Buffer, centrifuge again.
10. Place a 1.5-mL Eppendorf tube in the 50 mL-tube column above. Elute with 400 µL (prewarmed to 65°C) Elution-Buffer.
11. Centrifuge at 1,500 rpm 500g for 2 min (20°C).
12. Repeat to elute a second aliquot.
13. Determine concentration by diluting 20 µL of each elution fraction with 80 µL Elution-Buffer. The Abs_{260}/Abs_{280} ratio should be close to 2.0. Most of the polyA$^+$ RNA is usually in the first fraction.
14. The purity of the polyA$^+$ RNA can be increased by repeating the procedure. Therefore the beads in the column should be washed with 2 mL High-Salt-Buffer

and centrifuged as above. Denature eluent as before for the total RNA (**step 1**) and repeat **steps 6–13**.
15. Precipitate the eluted RNA with 1/10th volume 2 M Potassium acetate, pH 5.0, (approx 38 µL) and 1050 µL ethanol. Since the expected pellet is very small, it is useful to add pellet paint (Novagen Pellet Paint™ Co-Precipitant). Keep O/N at –20°C.
16. Centrifuge at 16,000g for 30 min (4°C).
17. Wash the pellet with 80% ethanol and centrifuge at 16,000g for 5 min (4°C).
18. Carefully take off supernatant and air dry the pellet for a few minutes and resuspend in DEPC-H$_2$O to a final concentration of 1 µg/µL. Check the quality by agarose electrophoresis. The RNA should be a smear and there should be reduced amount of ribosomal bands (*see* **Fig. 2B**).

3.3. Suppression Subtractive Hybridization (see Note 6)

3.3.1. cDNA-Synthesis

Synthesize double-stranded cDNA from tester as well as driver polyA$^+$ RNA.

1. Pipet 2 µg polyA$^+$ RNA into a 0.5 mL Eppendorf-tube and fill up the volume to 4 µL with sterile H$_2$O. Add 1 µL cDNA synthesis primer (10 µM) and denature for 2 min at 70°C. Place immediately on ice for 2 min and spin tube briefly.
2. Add: 2 µL 5X First-strand-Buffer, 1 µL dNTP mix (10 mM), 1 µL sterile H$_2$O, 1 µL AMV reverse transcriptase (20 U/µL).
3. Vortex carefully and incubate for 1.5 h at 42°C.
4. Place tubes on ice and immediately proceed with second strand cDNA synthesis.
5. Add: 48.8 µL sterile H$_2$O, 16.0 µL 5X Second-strand-Buffer, 1.6 µL dNTP mix (10 mM), 4.0 µL 20X Second-strand enzyme cocktail.
6. Mix, centrifuge briefly, and incubate for 2 h at 16°C. Add 2 µL (6 U) of T4 DNA polymerase, mix well, and incubate for further 30 min at 16°C.
7. Add 4 µL of 20X EDTA/glycogen to stop the reaction.
8. Add 100 µL phenol/chloroform/isoamyl alcohol (25:24:1), vortex, and centrifuge at 16,000g for 10 min (RT).
9. Carefully transfer upper phase in a new tube and add 100 µL chloroform/isoamyl alcohol (24:1), vortex, and centrifuge again.
10. Carefully transfer upper aqueous phase to a new tube and precipitate with 40 µL 4 M Ammonium acetate and 300 µL 95% ethanol.
11. Vortex and centrifuge 16,000g for 20 min (RT).
12. Wash the cDNA pellet with 80% ethanol, centrifuge for 10 min, and air dry the pellet for 10 min.
13. Resuspend the pellet in 50 µL sterile H$_2$O and check 3 µL on a 1.5% TAE-agarose gel.
14. Keep one 6 µL-aliquot at –20°C until after RsaI-digest (**step 4** below).

3.3.2. RsaI Digest

This step must be performed with both tester and driver ds-cDNA to get the short blunt-ended cDNA fragments necessary for the subtraction and adapter ligation.

1. Add to 43.5 µL ds-cDNA: 10X RsaI restriction buffer, 5.0 µL, RsaI (10 U/µL), 1.5 µL.
2. Vortex and centrifuge briefly.
3. Incubate for 4 h at 37°C in a water bath.
4. Run 3 µL of digested cDNA together with 6 µL of undigested cDNA on a 1.5% TAE-agarose gel. The undigested cDNA should give a smear of fragments ranging from 0.5–12 kb. The average fragment size of the digested cDNA should be significantly shorter, ranging from 0.1–2 kb (*see* **Fig. 3** and **Note 7**).
5. Stop the digestion reaction with 2.5 µL 20X EDTA/glycogen.
6. Purify by phenol/chloroform extraction as described, using half of the aforementioned volumes (**steps 8–12, Subheading 3.3.1.**).
7. Resuspend the air-dried cDNA pellet in 5.5 µL sterile H_2O and store at –20°C.

3.3.3. Adapter Ligation

For subtraction in the reverse direction, the samples defined as "tester" and "driver" can be switched and the entire procedure carried out again. Tester cDNA must be ligated to adapter1 (Tester-1), adapter2R (Tester-2). Tester-C (*see* below), containing a mixture of both adapter ligation reactions, is used as a ligation and subtraction control.

1. Dilute 1 µL of RsaI-digested cDNA with 5 µL H_2O.
2. Prepare enough Master Mix for 3 reactions and keep on ice: 9 µL sterile H_2O, 6 µL 5X Ligation buffer, 3 µL T4 DNA ligase (400 U/µL).
3. Pipet the following reagents:

Tube	Tester-1	Tester-2
diluted tester cDNA	2 µL	2 µL
Adaptor1 (10 µM)	2 µL	–
Adaptor2R (10 µM)	–	2 µL
Master Mix	6 µL	6 µL
Final volume	10 µL	10 µL

4. In a third tube combine 2 µL of Tester-1 and 2 µL of Tester-2 to make Tester-C.
5. Briefly centrifuge the tubes and incubate O/N at 16°C.
6. Stop the ligation reactions by adding 1 µL 20X EDTA/glycogen mix per tube and incubate at 72°C for 5 min.
7. Briefly centrifuge and store at –20°C.
8. Perform the ligation efficiency test using 1 µL of adapter-ligated probes Tester-1 and Tester-2 as recommended in the manufacturer protocol. Briefly, this consists

Suppression Subtraction Hybridization

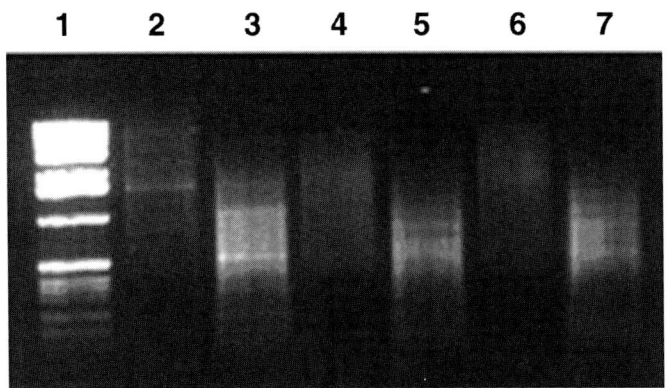

Fig. 3. Undigested and RsaI digested cDNA separated on a 1.5% TAE-agarose gel. Lane 1: 5 µL Marker X (Roche); lanes 2, 4 and 6: 5 µL of ds cDNA (control, D51 and SAEC, respectively); lanes 3, 5, and 7: 8 µL of RsaI digested cDNA. The average fragment size of the digested cDNA should be significantly shorter, ranging from 0.1–2 kb.

of two PCR reactions. The first contains of one gene-specific primer and an adapter-specific primer and the second contains a pair of gene-specific primers.
9. Analyze 5 µL on a 1.5% agarose gel. If the intensity of the first PCR product (gene-specific primer and adapter-specific primer) is less than 25% of the second (using two gene-specific primers), the ligation efficiency was poor and should be repeated. If you do not get distinct bands but a smear or several bands it is an indication of incomplete cDNA digestion.

3.3.4. Subtraction

1. Prewarm 4X Hybridization buffer to RT for at least 20 min.
2. Pipet the following first hybridization reactions:

	Tube 1	Tube 2
Rsa-digested driver cDNA	1.5 µL	1.5 µL
Tester-1	1.5 µL	–
Tester-2	–	1.5 µL
4X Hybridization buffer	1.0 µL	1.0 µL
Final volume	4.0 µL	4.0 µL

3. Overlay the probes with one drop of mineral oil, centrifuge briefly and denature for 1.5 min at 98°C, hybridize for 6–8 h at 68°C.
4. For the second hybridization, prepare 1 µL driver cDNA, 1 µL 4X Hybridization buffer and 2 µL sterile H_2O and mix.
5. Transfer 1 µL of this mixture in a new tube, overlay with one drop of mineral oil and denature for 1.5 min at 98°C and keep at RT.

6. Set pipet to a volume of approx 20 µL, pipet the hybridization mixture from tube 2, aspirate some air, take up 1 µL freshly denatured driver cDNA (**step 5**). The pipet tip should now contain two samples separated by an air bubble.
7. Transfer the contents of the pipet into tube 1 and mix by pipetting up and down.
8. Incubate O/N at 68°C.
9. Stop the reaction by adding 200 µL Dilution buffer and mix by pipetting up and down.
10. Incubate at 68°C for 7 min.
11. Keep samples at –20°C until the next step.

3.3.5. Amplification (1st PCR)

1. Dilute 1 µL of unsubtracted control cDNA (Tester-C) with 1 mL sterile H_2O.
2. Pipet 1 µL of diluted Tester-C and 1 µL of subtracted cDNA sample (**step 11**) into separate PCR tubes.
3. Make enough Master Mix for 3 reactions (for 1X Tester-C and 1xsubtracted cDNA sample):

sterile H_2O	58.5 µL
10X PCR reaction buffer	7.5 µL
dNTP Mix (10 µM)	1.5 µL
PCR primer 1 (10 µM)	3.0 µL
50X Advantage Polymerase Mix	1.5 µL
Final volume	72.0 µL

4. Vortex, spin down and add 24 µL of the mix to each of the 1 µL aliquots.
5. Run the following program in a PCR machine: 5 min at 75°C, 27 cycles (30 s at 94°C, 30 s at 66°C, 1.5 min at 72°C).
6. Run 8 µL of the PCR products on a 2% TAE-agarose gel. The subtracted samples should be enriched and show bands ranging from 0.2–2 kb.

3.3.6. Nested PCR (2nd PCR)

1. Dilute 3 µL from each of the above PCR reactions with 27 µL H_2O.
2. Aliquot 1 µL per sample into new PCR tubes.
   Prepare the following Master Mix:

Sterile H_2O	55.5 µL
10X PCR reaction buffer	7.5 µL
Nested PCR primer 1 (10 µM)	3.0 µL
Nested PCR primer 2R (10 µM)	3.0 µL
dNTP Mix (10 mM)	1.5 µL
50X Advantage Polymerase Mix	1.5 µL
Final volume	72.0 µL

3. Add 24 µL Master Mix to each sample, vortex, centrifuge and run the following PCR reaction: 12 Cycles (30 s at 94°C, 30 s at 68°C, 1.5 min at 72°C).
4. Run 8 µL of the PCR products on a 2% TAE-agarose gel. The subtracted samples should be even more enriched now *and* the bands should become more prominent than after the first PCR reaction (*see* **Fig. 4**).

Suppression Subtraction Hybridization

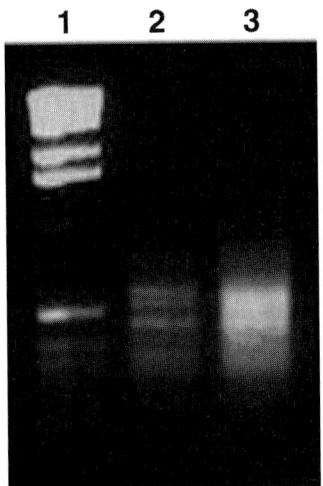

Fig. 4. Optimized nested PCR products analyzed on a 2% TAE agarose gel. Lane 1: 5 µL Marker X; lane 2: 6 µL of subtracted HBEC cDNA; lane 3: 6 µL of unsubtracted HBEC cDNA. The subtraction was performed from H526 (HBEC = tester, H526 = driver cDNA). For amplification, 27 cycles of First PCR were followed by 10 cycles of Nested PCR. The subtracted samples should be even more enriched now and the bands should become more prominent than after the first PCR reaction.

3.3.7. Subtraction-Efficiency-Test

To analyze the efficiency of the suppression subtractive hybridization, a quantitative PCR reaction must be done in which samples are taken at different cycles during the reactions and compared to one another.

1. Dilute 1 µL of the nested PCR products above 1:10.
2. Perform PCR reactions using primers specific for at least two different housekeeping genes, G3DPH, β-actin (see **Note 8**).
3. Prepare four replicate tubes for each set of PCR primers to be used.
4. Take one tube out of the PCR machine after 13 cycles, the next at 18 cycles, the next at 23 cycles, and the last at 28 cycles. Keep the first three tubes at 4°C until the 28-cycle reaction is complete.
5. Analyze five µL per reaction on a 1.5% agarose gel (see **Note 9**).

3.3.8. Cloning of the cDNA Libraries

The PCR products can now be cloned as cDNA libraries based on T/A cloning. The optimal library should be determined by ligating different amounts of PCR product as well as different transformation reactions and plating densities.

1. Repeat the nested PCR with the optimized number of PCR cycles and purify the PCR reaction (e.g. with the QIAGEN PCR purification kit) eluting in 50 µL sterile H$_2$O (see **Note 10**).
2. Pipet different ligation reactions:

Tubes	1	2	3
purified nested PCR products	1.5 µL	2.0 µL	3.0 µL
sterile H$_2$O	13.5 µL	13.0 µL	12.0 µL
10X Ligation Buffer	2.0 µL	2.0 µL	2.0 µL
PCR 2.1 vector (0.025 µg/µL, Invitrogen)	2.0 µL	2.0 µL	2.0 µL
T4-Ligase	1.0 µL	1.0 µL	1.0 µL

    Incubate at 16°C O/N and keep at –20°C.
3. For transformation, combine 2 µL 0.5 M β-Mercaptoethanol, competent bacteria (e.g., one shot TOP10F′ chemically competent *E. coli*, Invitrogen) and either 2 µL or 8 µL of each ligation reaction. Incubate for 30 min on ice. Heat shock for 30 s in a 42°C water bath and place immediately on ice for 2 min. Add 250 µL of SOC media (prewarmed to 37°C) and shake for 1 h at 37°C. Plate 50 µL or 250 µL of each transformation reaction on 12 × 12 cm plates containing 40 mL LB agar with ampicillin (50 µg/µL), X-gal (5-Bromo-4-chloro-3-indolyl-β-D-galactopyranoside dissolved in DMF, 64 µg/µL) and Isopropyl-β-D-thiogalactopyranoside (0.5 mM). Incubate at 37°C for a minimum of 18 h (see **Note 11**).

3.4. Analysis of cDNA Libraries

To assess the quality of the libraries with respect to redundancy and specificity, a minimum of 20 randomly picked cDNA transformants should be sequenced and analyzed for differential expression by Northern blot.

3.4.1. Plasmid Preparation

1. Pick individual white colonies and grow in LB media plus ampicillin (50 µg/µL) at 37°C shaking O/N.
2. Centrifuge bacteria suspension at 1000g for 10 min (RT).
3. Isolate plasmid DNA (e.g., using QIAGEN mini spin). The DNA can be used for sequencing as well as probe preparation for Northern-blot hybridization.

3.4.2. DNA Sequencing

The following protocol is for sequencing using a LiCOR sequencer (MWG) but the method can be modified for other systems.

1. Prepare Long Ranger solution, mix reagents and cast gel.
2. Pipet the following master mix for each probe:

Sterile water	4 µL
Primer (M13 forward, M13 reverse), 1 pmol/µL	4 µL
Plasmid DNA (approx 100 ng/µL)	4 µL

3. Label 4 tubes (A, C, G, T) for each probe and pipet 1 µL of the Thermo Sequenase nucleotide mix into its corresponding tube.
4. Add 3 µL of Master Mix to each nucleotide tube, overlay with one drop of wax (FMC), vortex and spin down.
5. Run the following PCR reaction: 5 min at 95°C, 30 cycles (30 s at 95°C, 15 s at 50°C, 1 min at 70°C).
6. Add 4 µL of stop buffer, denature 5 min at 70°C and load 1 µL of sample per lane.

3.4.3. Northern Blot Analysis

To confirm that the libraries are enriched for differentially expressed genes, their expression should be analyzed. If enough RNA material is available this should be done by Northern blot hybridization. The number of differentially expressed genes should account for at least about 50% of the isolated clones, otherwise the SSH procedure should be repeated.

1. Run 10 µg total RNA per sample as described (*see* **Subheading 2.1.**).
2. Wash gel for 5 min in water. Soak Hybond-N in 20X SSC for 5 min.
3. Transfer the RNA by capillary blotting: Place a container with 20X SSC and strips of Whatman paper in contact with it.
4. Place the inverted gel (e.g., wells facing down) on the Whatman paper sheets and eliminate air bubbles by rolling over it with a pipet.
5. Place soaked nylon membrane on the gel, again eliminating air bubbles. Limit membrane manipulation to a minimum and always handle with forceps.
6. Layer 2 sheets of 20X SSC soaked Whatman paper on the membrane, eliminating bubbles, and cover with 4 additional sheets of dry Whatman paper.
7. Place paper towels on top of the Whatman paper, cut towels to the size of gel and use stack 8–10 cm high to ensure sufficient capillary flow for a successful transfer.
8. Place a glass plate and weight (about 1 kg) on top of the paper towels. Transfer overnight.
9. Wash membrane briefly with 2X SCC, dry membrane on Whatman paper and crosslink membrane for 3 min in a UV crosslinker or under UV light (trans-illuminator).
10. Wrap nylon membrane in saran wrap, and store at 4°C.
11. Prepare the hybridization probe by nested PCR using plasmid DNA in the following PCR reaction:

plasmid DNA (100 ng/µL)	3.0 µL
2.5 m*M* dNTPs	2.0 µL
10X PCR buffer	5.0 µL
Nested primer 1 (10 pmol/µL)	1.0 µL
Nested primer 2 (10 pmol/µL)	1.0 µL
Taq DNA Polymerase	0.4 µL
Sterile water	39.6 µL
Final volume	50.0 µL

12. Run the following PCR: 2 min at 94°C, 45 cycles (30 s at 94°C, 15 s at 68°C, 1 min at 72°C), 10 min at 72°C.
13. Purify the PCR product (e.g., using QIAquick PCR purification kit, QIAGEN).
14. Prehybridize blots with hybridization solution for at least 30 min.
15. In the mean time label probe (MegaPrime Labeling kit, Amersham):
16. Pipet 25 ng of purified nested PCR product and add sterile water up to 28 µL. Add 5 µL of primer solution, vortex, spin down and denature at 95°C for 5 min.
17. Chill on ice, spin down and add the following reagents:
    labeling buffer   10 µL
    Klenow DNA polymerase   2 µL
    ^{32}P-αdCTP   5 µL
18. Mix by pipetting up and down and incubate at 37°C for 30 min.
19. Purify labeled probe from unincorporated nucleotides (e.g., Nucleotide Removal kit, QIAGEN; Microspin G-50, Amersham).
20. Denature probe for 5 min at 99°C, chill on ice and centrifuge to collect vapor.
21. Add probe to the hybridization solution avoiding direct contact with the membrane. Mix gently and hybridize O/N at 58°C.
22. Wash membranes after hybridization with increasing stringency (*see* **Note 12**). Use the following scheme: Wash twice with 2X SSPE/0.1% SDS for 10 min (RT).
23. Wash with 1X SSPE/0.1% SDS 10 min (RT).
24. Wash with 0.1X SSPE/0.1% SDS for 15 min at 58°C.
25. Seal blots into plastic bag and expose to film overnight at –80°C (*see* **Fig. 5**) (*see* **Note 13**).

4. Notes

1. The quality and purity of the RNA is important for the subsequent cloning of cDNA libraries and should be done very thoroughly. For the assessment of differences in expression level, it is recommended that fixed-resin based column extraction methods should not be used since the filters can retain certain RNA species, thereby randomly interfering with the abundance of different transcripts.
2. When manipulating RNA always wear gloves and work with RNase-free materials and solutions. All steps should be carried out on ice to avoid RNA degradation.
3. TRIzol extraction is a convenient alternative to the classical phenol-chloroform extraction and is recommended especially when using tumor tissue other than cell lines.
4. Since messenger RNA represents only 1.5–3% of the total cellular RNA, depending on the cell type, it is necessary to enrich for these transcripts. Therefore different systems are commercially available that use oligo-dT-Cellulose particles to bind polyA$^+$ RNA species. Again, fixed-resin based column purification kits should not be used. The mRNA separator from Clontech, when enough RNA is available, and the FastTrack 2.0 mRNA isolation kit (Invitrogen) for lower

Suppression Subtraction Hybridization

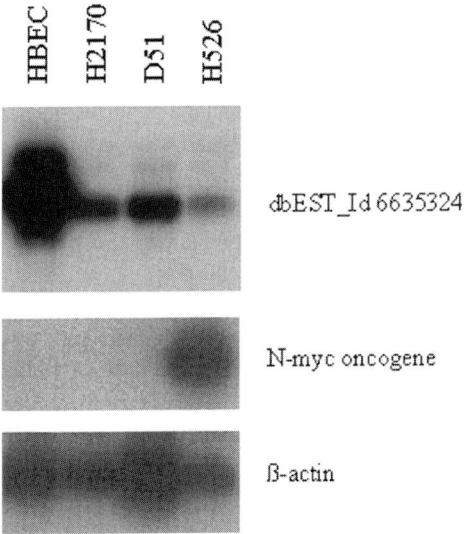

Fig. 5. Examples of Northern blot hybridizations using cloned cDNA fragments as probes. Ten µg of total RNA from bronchial epithelial cells (HBEC) and three lung cancer cell lines were electrophoresed, blotted, and hybridized with $^{32}P\alpha$-dCTP labeled cDNA probes.

amounts of RNA worked well in our hands. The method described in **Subheading 3.2.** is modified from the mRNA separator instructions from Clontech. Reagents and items described (Oligo dT beads, High-Salt-Buffer, Low-Salt-Buffer, Elution-Buffer) were obtained from this source.

5. Oligo-dT beads have only a certain life span and can dry out. This can dramatically influence the efficiency and yield of polyA$^+$ RNA isolated.
6. The method described in **Subheading 3.3.1.** is modified from the PCR-Select™ cDNA subtraction kit (Clontech). Reagents and items described (5X first-strand-Buffer, dNTP mix (10 mM), 5X second-strand-Buffer, 20X second-strand enzyme cocktail, 20X EDTA/glycogen, 4X hybridization buffer, PCR primers, 50X Advantage Polymerase Mix) were obtained from this source.
7. Undigested cDNA should give a smear of fragments ranging from 0.5–12 kb. The average fragment size of the digested cDNA should be significantly shorter, ranging from 0.1–2 kb. If this is not the case, more enzyme should be added and the digestion continued (it may also be necessary to use RsaI enzyme from a different supplier). If further incubation does not result in complete digestion, the cDNA synthesis might not have been successful.
8. Ideally, for the subtraction efficiency test, expression profiles of specific genes already known to be differentially expressed in the tester and driver cDNA populations should be determined.

9. Expression of the housekeeping genes should appear 5–15 cycles later in the subtracted samples than in the control samples, indicating the quality of the suppression. If the suppression was not efficient, the library will harbor several abundant housekeeping and ribosomal gene sequences. In this case, the nested PCR should be repeated using fewer cycles.
10. Do not freeze PCR products prior to cloning. This will decrease cloning efficiency.
11. Following PCR ligation and transformation, the most optimal plate should contain more than 400 white colonies and fewer blue ones. Choose the best plate for analysis of the cDNA libraries.
12. Every probe reacts differently, so the remaining radioactivity on the membrane in between the washes should be measured until specific radioactivity is reached.
13. Once hybridized, nylon membranes should always be kept sealed at 4°C so that they do not dry out. For re-hybridization, the membrane can be washed with boiling 0.1% SDS solution for 5 min while shaking.

References

1. Liang, P., Averboukh, L., Keyomarsi, K., Sager, R., and Pardee, A. B. (1992) Differential display and cloning of messenger RNAs from human breast cancer versus mammary epithelial cells. *Cancer Res.* **52,** 6966–6968.
2. Hubank, M. and Schatz, D. (1984) Identifying differences in mRNA expression by representational difference analysis of cDNA. *Nucleic Acids Res.* **22,** 5640–5648.
3. Diatchenko, L., Lau, Y. F., Campbell, A. P., Chenchik, A., Moqadam, F., Huang, B., et al. (1996) Suppression subtractive hybridization: a method for generating differentially regulated or tissue-specific cDNA probes and libraries. *Proc. Natl. Acad. Sci. USA* **93,** 6025–6030.
4. Zhang, L., Zhou, W., Velculescu, V. E., Kern, S. E., Hruban, R. H., Hamilton, S. R., et al. (1997) Gene expression profiles in normal and cancer cells. *Science* **276,** 1268–1272.
5. Schena, M., Shalon, D., Heller, R., Chai, A., Brown, P. O., and Davis, R. W. (1996) Parallel human genome analysis: Microarray-based expression monitoring of 1000 genes. *Proc. Natl. Acad. Sci. USA* **93,** 10614–10619.
6. Duguid, J. R. and Dinauer, M. C. (1990) Library subtraction of in vitro cDNA libraries to identify differentially expressed genes in scrapie infection. *Nucleic Acids Res.* **18,** 2789–2792.
7. Hara, E., Kato, T., Nakada, S., Sekiya, S., and Oda, K. (1991) Subtractive cDNA cloning using oligo(dT)$_{30}$-latex and PCR: isolation of cDNA clones specific to undifferentiated human embryonal carcinoma cells. *Nucleic Acids Res.* **19,** 7097–7104.
8. Hedrick, S. M., Cohen, D. I., Neilson, E. A., and Davis, M. M. (1984) Isolation of cDNA clones encoding T cell-specific membrane-associated proteins. *Nature* **308,** 149–153.
9. Sargent, T. D. and Dawid, I. B. (1983) Differential gene expression in the gastrula of Xenopus laevis. *Science* **222,** 135–139.

10. Gurskaya, N. G., Diatchenko, L., Chenchik, A., Siebert, P. D., Khaspekov, G. L., Lukyanov, K. A., et al. (1996) Equalizing cDNA subtraction based on selective suppression of polymerase chain reaction: Cloning of Jurkat cell transcripts induced by phytohemaglutinin and phorbol 12-myristate 13-acetate. *Anal. Biochem.* **240**, 90–97.
11. Petersen, S., Heckert, C., Rudolf, J., Schlüns, K., Tchernitsa, O. I., Schäfer, R., et al. (2000) Gene expression profiling of advanced lung cancer. *Int. J. Cancer* **86**, 512–517.
12. Petersen, S., Pietas, A., Chen, Y., Schlüns, K., Pacyna-Gengelbach, M., Deutschmann, N., et al. (2002) Gene expression profiles in Non-small cell lung cancer and Small cell lung cancer. Submitted.
13. Anbazhagan, R., Tihan, T., Bornman, D. M., Johnston, J. C., Saltz, J. H., Weigering, A., et al. (1999) Classification of small cell lung cancer and pulmonary carcinoid by gene expression profiles. *Cancer Res.* **59**, 5119–5122.
14. Wang, T., Hopkins, D., Schmidt, C., Silva, S., Houghton, R., Takita, H., et al. (2000) Identification of genes differentially over-expressed in lung squamous cell carcinoma using combination of cDNA subtraction and microarray analysis. *Oncogene* **19**, 1519–1528.
15. Petersen, I. and Petersen, S. (2001) Towards a genetic classification of human lung cancer. *Anal. Cell. Pathol.* **22**, 111–121.
16. Fung, L. F., Lo, A. K. F., Yuen, P. W., Liu, Y., Wang, X. H., and Tsao, S. W. (2000) Differential gene expression in nasopharyngeal carcinoma cells. *Life Sci.* **67**, 923–936.
17. Zhang, M., Martin, K. L., Sheng, S., and Sager, R. (1998) Expression genetics: a different approach to cancer diagnosis and prognosis. *Trends Biotechnol.* **16**, 66–71.
18. Nacht, M., Ferguson, A. T., Zhang, W., Petroziello, J. M., Cook, B. P., Gao, Y. H., et al. (1999) Combining serial analysis of gene expression and array technologies to identify genes differentially expressed in breast cancer. *Cancer Res.* **59**, 5464–5470.
19. Zuber, J., Tchernitsa, O. I., Hinzmann, B., Schmitz, A. C., Grips, M., Hellriegel, M., et al. (2000) A genome-wide survey of RAS transformation targets. *Nature Genet.* **24**, 144–152.
20. Zhao, R., Gish, K., Murphy, M., Yin, Y., Notterman, D., Hoffman, W. H., et al. (2000) Analysis of p53-regulated gene expression patterns using oligonucleotide arrays. *Genes Devel.* **14**, 981–993.
21. Sekido, Y., Fong, K. M., and Minna, J. D. (1998) Progress in understanding the molecular pathogenesis of human lung cancer. *Biochem. Biophys. Acta* **1378**, F21–F59.
22. Petersen, S., Rudolf, J., Bockmühl, U., Gellert, K., Wolf, G., Dietel, M., and Petersen, I. (1998) Distinct regions of allelic imbalance on chromosome 10q22-q26 in squamous cell carcinomas of the lung. *Oncogene* **17**, 449–454.
23. Chomzcynski, P. and Sacchi, N. (1987) Single-step method of RNA isolation by acid guanidium thiocyanate-phenol-chloroform extraction. *Anal. Biochem.* **162**, 156–159.

12

Comparative Genomic Hybridization of Human Lung Cancer

Iver Petersen

1. Introduction

Lung cancer is a highly aggressive neoplasm, a characteristic that is reflected by the multitude of genetic aberrations detectable on the chromosomal and molecular level. In order to understand these seemingly chaotic chromosomal alterations, we have performed Comparative Genomic Hybridization (CGH) on a large collective of human lung carcinomas. CGH was one of the first screening methods used to analyze tumor genomes for genetic imbalances *(1,2)*, and has rapidly become a popular tool for a comprehensive molecular cytogenetic analysis of solid tumors. The method is based on the hybridization of differentially labeled tumor and reference DNA on normal chromosome metaphase spreads. Both genomes compete for complementary binding sites on the chromosomal DNA. In the case of an amplification in tumor DNA, more DNA will hybridize to the corresponding band or chromosomal arm, whereas deletions will allow the binding of the normal DNA to competitively prevail. Detection of each genome is facilitated by the use of distinct fluorochrome labels. DNA imbalances are determined by comparing the intensity of the fluorescence signals along individual chromosomes. DNA gains are potentially associated with the activation of proto-oncogenes and DNA losses with the inactivation of tumor-suppressor genes.

The method involves several preparative steps, including DNA extraction, labeling, and hybridization, followed by fluorescence microscopy and digital-image analysis of fluorescence images representing the three DNA complements, i.e., the chromosomes, tumor, and normal DNA (*see* **Fig. 1**).

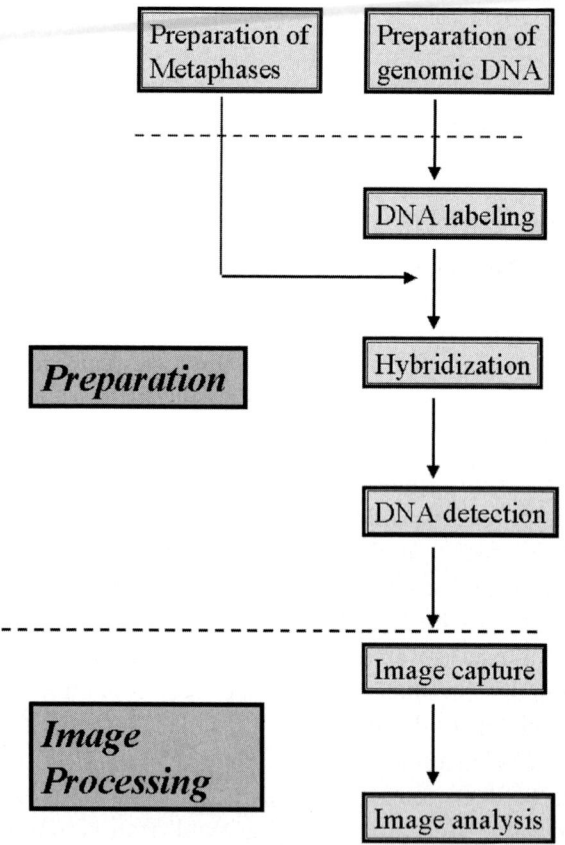

Fig. 1. Scheme of the methodological steps in CGH analysis.

1.1. Methodological Considerations

As a method of tumor analysis, CGH has many of the advantages of a DNA-based technique. Most importantly, it can be applied to archival tumor specimens that have been formalin-fixed and paraffin-embedded. In addition, optimized amplification protocols are available that enable the technique to be applied to small tumor biopsies and minute specimens following microdissection *(3–6)*.

However, there are limitations to the method and the potential to produce artifacts does exist. Several of these problems are linked to the use of normal metaphase spreads as the DNA matrix to which the test and reference genomes are hybridized. The complex structure of the chromosome hybridization targets may influence the efficiency of DNA binding and thus the fluorescence signals.

In addition, because specific chromosomal regions might harbor differences in the binding kinetics of the tumor and normal DNA, owing to distinct hapten labeling, artifactual hybridization may appear in certain regions *(7,8)*. Because of these potential pitfalls, it is important to evaluate only high-quality hybridizations that exhibit a strong and specific fluorescence signal. The use of normal chromosomes also causes the limited resolution in the assessment of genetic imbalances, which is restricted to chromosomal subregions *(7,9)*. Finally, karyotyping is quite a laborious technique that has so far escaped automation, which is the main reason for the time-consuming nature of CGH.

Although it is important to note that CGH is unable to detect balanced chromosomal alterations like translocations and inversions, in contrast to leukemias, lymphomas, and sarcomas, these do not seem to play a major role in carcinomas. Thus, the DNA gains and losses detected by CGH reflect those chromosomal alterations that are most characteristic and biologically relevant in cancers of the lung.

The application of CGH has increased reflected by the large number of recent publications describing its use. Methodologically, it has inspired tumor analysis by microarrays, using competitive hybridization of differentially labeled RNA/cDNA and the calculation of ratio values for the assessment of gene expression *(10)*.

Furthermore, the resolution and the potential for automation has been increased by applying DNA matrices other than chromosomes as hybridization targets *(11–14)*. Despite its limitations, conventional CGH has been used to reveal specific patterns of chromosomal changes in many tumor types, including lung cancer.

1.2. CGH Analysis in Lung Cancer

In recent years we have analyzed a collective of lung carcinomas comprised of over 250 tumor specimens by CGH, using custom-made CGH software *(15)*. The data has either been published *(16–23)*, or is available at our CGH online tumor database at http://amba.charite.de/cgh. Besides primary tumors, metastases and tumor cell lines were also analyzed. Tumor DNA was mainly obtained from frozen tissue derived from surgical resections at the Department of Surgery of the Charité Hospital at the Humboldt-University Berlin. Additionally, snap-frozen tumor specimens of primary and metastatic lesions were collected at post mortem examinations.

The typical findings in small cell lung cancer (SCLC) were deletions on chromosomes 3p, 4, 5q, 10q, 13q, 17p and DNA gains on 3q, 5p, 6p, 8q, and 17q (**Fig. 2B**). Interestingly, deletions are more frequent than DNA gains, suggesting that the inactivation of tumor-suppressor genes is as important as gain of function mutations of proto-oncogenes. SCLC usually harbors

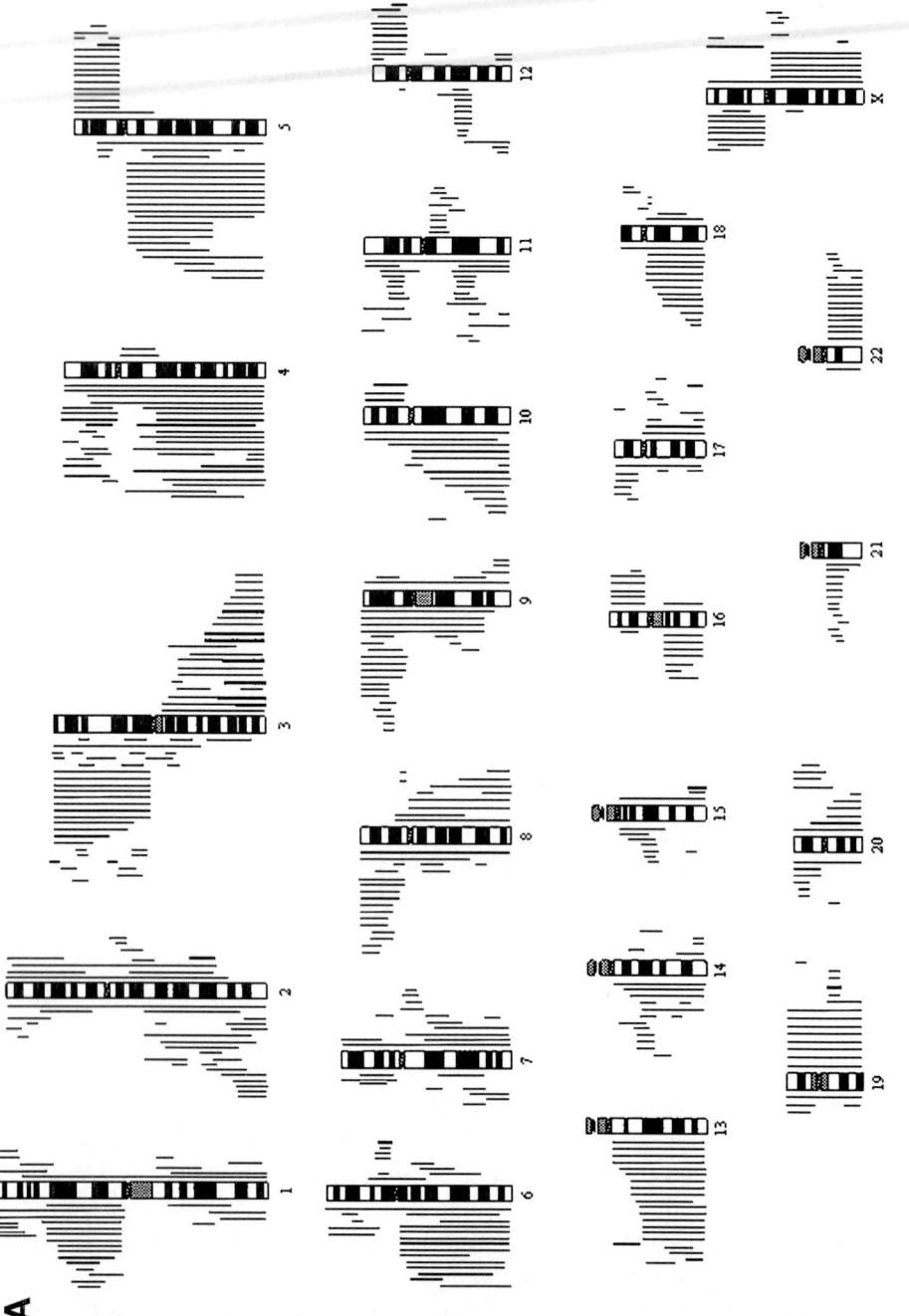

Fig. 2A.

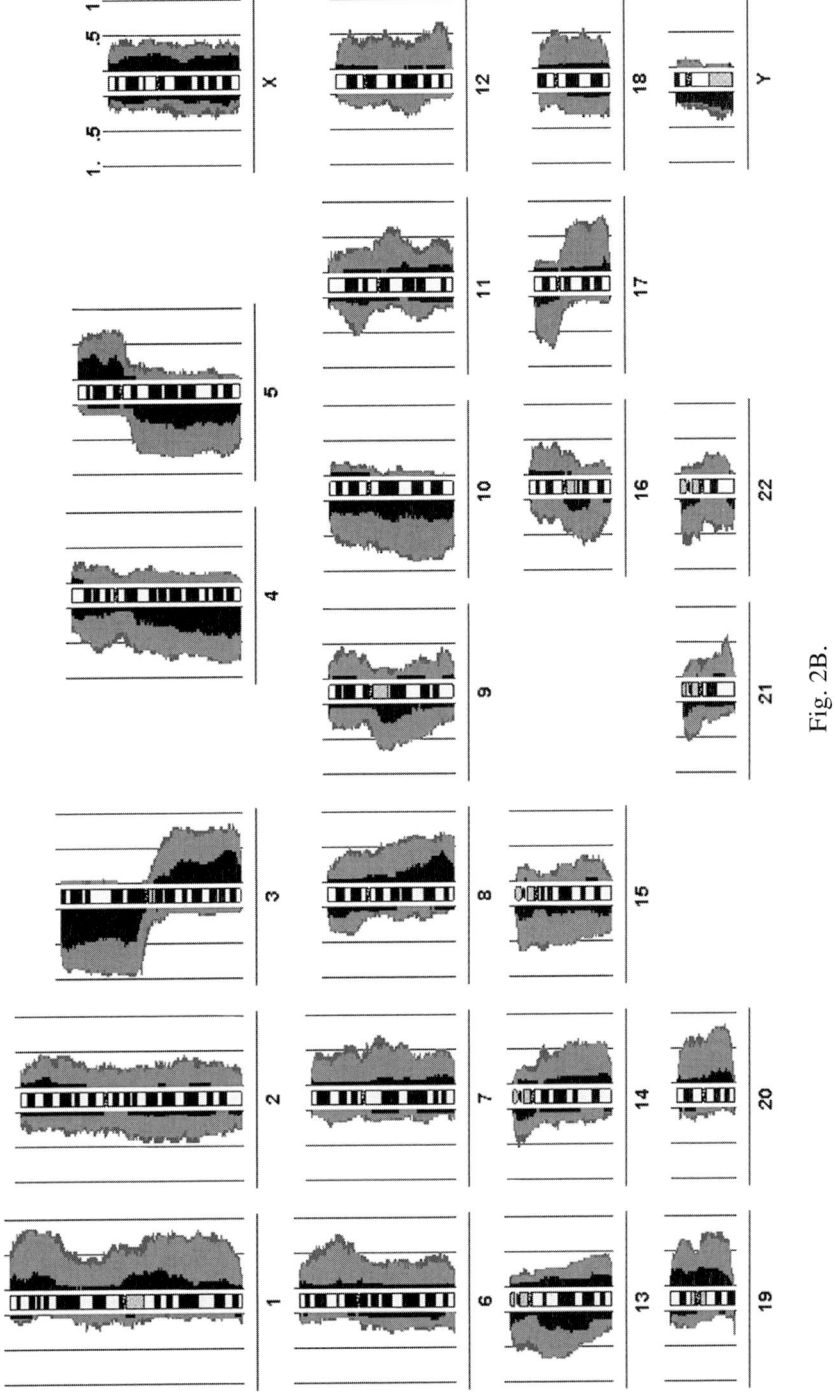

Fig. 2B.

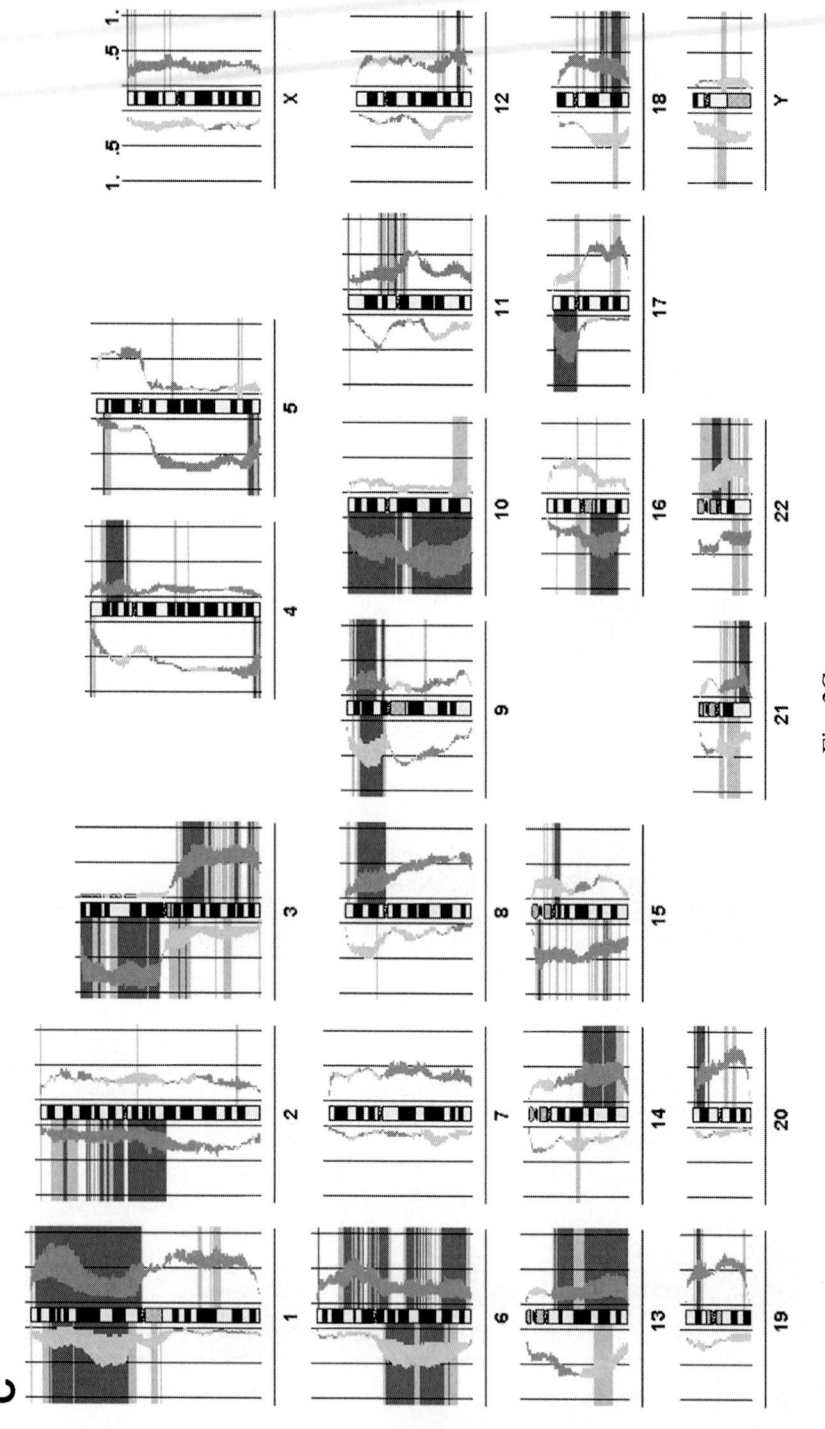

Fig. 2C.

large deletions affecting entire chromosome arms or whole chromosomes, as exemplified by the deletions on 3p, 10 and 17p. The typical pattern for chromosome 3 is the 3q isochromosome *(17,18,23)*.

Non-small cell lung cancer (NSCLC) alterations both overlapped those of SCLC, but also exhibited some distinct patterns (**Fig. 7A,C**). Like SCLC, NSCLC also carried a high incidence of deletions on chromosomes 3p, 4, 5, and 13q. In addition, DNA gains occurred at a high frequency on chromosomes 1p, 6q, 9p, 18q, and 21q. For specific chromosomes the pattern of alteration also differed. For example, chromosome 3p is particularly affected by interstitial deletions in NSCLC, whereas in SCLC the entire chromosome arm or large regions of it are affected. For chromosome 13q, SCLC showed deletions of the proximal arm, including the Rb gene locus at 13q14, whereas in NSCLC, typically the distal arm of chromosome 13 is lost. The aforementioned differences are best visualized by the difference histogram (**Fig. 2C**). It clearly indicates that the deletions of the entire chromosome 3p, 10, 4p16, 15q, 16q, and 17p, as well as the overrepresentations on chromosomes 1, 3q, 6, 13, and 17q24-q25, are significantly associated with SCLC. In contrast, the deletions of chromosome 1p, 6q, 9p, 13q22-q32, 18q, 21q, and the gain of chromosome 22q is significantly more frequent in NSCLC.

Fig. 2. *(see pages 212, 213, & 214)* (**A**) Line representation of the CGH result in 25 lung squamous cell carcinomas, using fixed ratio thresholds. Lines on the left side of the chromosome ideogram represent DNA losses, whereas lines on the right side represent DNA gains. Solid bars, rather than lines, are used to indicate high copy amplifications. (**B**) Histogram of 30 SCLCs. Chromosomal imbalances are shown as incidence curves along each chromosome. Areas on the left side of the chromosome ideogram correspond to loss of genetic material, those on the right side to DNA gains. The frequency of the alterations can be determined from the 50% and 100% incidence lines depicted parallel to the chromosome ideograms. Symbol key: Light gray curve, DNA changes with 99% significance. Dark gray rim, additional changes with 95% significance. Black area close to the chromosome ideogram, proportion of pronounced DNA imbalances that may likely represent high copy amplifications or multi-copy deletions. (**C**) Difference histogram of 30 SCLC and 110 NSCLC. Symbol key: Light gray histogram curve, percentage of changes that are exclusively present in NSCLC. Dark gray histogram curve, majority of changes in SCLC. White areas beneath the colored parts of each histogram, percentage of changes that are present in both tumor-subgroups. Grey horizontal lines, statistically significant differences. Light gray lines: regions with 95% significance. Dark gray lines, 99% significance according to the χ^2-test. In this example, the deletions on chromosomes 10 and 17p are significantly more frequent in SCLC, whereas deletions on 6q and 9p are typically observed in NSCLC.

Our studies on primary and metastatic tumors revealed a high concordance between the CGH patterns of corresponding tumors, which may be useful in establishing a clonal relationship *(18,20–22)*. However, lung cancer also carries a high chromosomal instability, which is reflected by its considerable morphological heterogeneity. In a study analyzing primary and metastatic SCC, most chromosomal regions harbored more alterations in the latter tumor group, consistent with the paradigm of tumor genetics, which postulates that tumor progression and metastasis formation is characterized by an accumulation of genetic defects. Specifically, the deletions at 3p12-p14, 4p15-p16, and 10q, as well as the gain on chromosome 1q22-q25, were associated with the metastatic phenotype *(20)*. In later study of 42 brain metastases, we additionally observed a peak in the histogram for the gain at 17q24-q25, suggesting that the amplification of a gene at these chromosomal regions might mediate tumor dissemination into the nervous system *(21)*. Interestingly, this alteration was also associated with the SCLC phenotype, providing possible genetic basis for the fact that SCLC is a highly metastatic tumor type, which often spreads into the central nervous system (CNS).

The analysis of different types of NSCLC indicated chromosomal imbalances that were associated with tumor differentiation. Adenocarcinomas, for instance, are typically characterized by overrepresentations of chromosome 1q. This again provides a genetic correlate to the fact that adenocarcinoma carry a higher risk for hematogenous metastasis formation than lung SCC *(16,22)*.

In summary, CGH has increased our genetic insight in cancer pathology. Together with other new screening methods, in particular cDNA microarray analysis, it is hoped that it will yield a refined method for lung tumor classification.

2. Materials

2.1. Metaphase Chromosome Preparation

2.1.1. Cell Preparation

1. RPMI 1640 medium.
2. Fetal calf serum (FCS), 20%.
3. Colcemid (10 µg/mL Boehringer Mannheim).
4. Cell culture flask, e.g., Falcon 250 mL, prepare up to 10 flasks (1 flask will yield about 50 slides).
5. Phythemagglutinin, PHA-L (Seromed, M 5030), 1.5 mL/5 mL blood.
6. CO_2 cell-culture incubator.
7. 50 mL Nunc/Falcon tubes.
8. 15 mL Nunc/Falcon tubes.
9. KCl solution: 0.0752 M, 0.055%.

10. Fixative: methanol/acetic acid 3:1.
11. Glass microscopy slides.
12. 5 mL peripheral blood (anticoagulated by heparin).

2.1.2. Slide Preparation

1. 70% acetic acid.

2.2. DNA Preparation

2.2.1 DNA Extraction from Frozen Tissue Using Cryotome Tissue Dissection

2.2.1.1. TISSUE DISSECTION (SEE NOTE 1)

1. Isotonic NaCl: 0.15 M.
2. Eppendorf tubes (2 mL; Safe Lock) and microscopic slides labeled with the case number.
3. Digestion buffer: 50 mL Tris-HCl, pH 8.5, 1 mM EDTA, 0.5% Tween 20.
4. Long Pasteur pipets with melted tips, one pipet per case.
5. H&E stain.

2.2.1.2. PROTEINASE K DIGESTION

1. Proteinase K: 500 µL aliquots of a 20 mg/mL stock solution, keep aliquots at −20°C.

2.2.1.3. PHENOL-CHLOROFORM EXTRACTION

1. Phenol/Chloroform/Isoamylalcohol.
2. 3 M NaCl.
3. Isopropanol.
4. 100% EtOH.
5. 70% EtOH.
6. Purified H_2O (Acqua ad iniectabile, Braun Melsungen).

2.2.2. DNA Preparation from Paraffin-Embedded Material

1. Sterile scalpels.
2. Xylol.

2.3. Nick Translation

1. DNA for labeling (concentration c > 150 ng/µL).
2. Modified nucleotides: *Bio*tin-16-dUTP, *Dig*oxigenin-11-dUTP, conc. 1 nmol/µL (Boehringer Mannheim).
3. dNTPs (unlabeled): dATP, dCTP, dGTP, 0.5 mM each, dTTP 0.1 mM.
4. NT reaction buffer 10X: 0.5 M Tris-HCl, pH 8.0, 50 mM $MgCl_2$, 0.5 mg/mL bovine serum albumin (BSA).

5. β-mercaptoethanol (β-ME): 0.1 M.
6. DNase: stock solution 3 mg/mL, working solution 1:2000 diluted in aqua bidest.
7. Pol: Kornberg DNA-polymerase 5 U/μL (Boehringer Mannheim).
8. EDTA: 0.5 M, pH 8.0.
9. Sodium dodecyl sulfate (SDS) (20%).

2.4. Hybridization

1. Labeled tumor and normal-DNA (*see* Nick translation protocol; **Subheading 3.3.**)
2. Salmon sperm DNA, 10 mg/mL (Promega).
3. Human Cot1 DNA, 1 mg/mL (GibcoBRL, Life Technologies).
4. 3 M sodium acetate.
5. 100% ethanol, 90% ethanol, 70% ethanol.
6. Formamide (FA), (Merck).
7. Master mix: 20% dextransulfate/4X SSC.
8. SSC (stock solution, 20X).
9. 70% FA/2X SSC for chromosome denaturation (120 μL per slide).
10. Cover slips, small (18 × 18 mm^2) and large (60 × 24 mm^2) (Menzel).
11. Rubber cement (Fixogum, Marabu, D-71732 Tamm, FRG).

2.5. DNA Detection

1. Formamide (FA)/2X SSC (1:1).
2. 0.1X SSC.
3. 4X SSC/0.1% Tween 20.
4. 3% BSA in 4X SSC/0.1% Tween 20.
5. DAPI (4.6-diamidino-2phenylindole dihydrochloride): stock solution 0.2 mg/mL, dissolve 1:5000 in aqua bidest, keep solution in use in a light-protected cuvet.
6. Anti-Dig-Rhodamine, 200 μg/mL (Boehringer Mannheim).
7. Fluorescein-Avidin-dcs, 1 mg/mL (Vector Laboratories).
8. Fluorochrome solution: 10 μL anti-Dig-Rhodamine + 5 μL Fluorescein-Avidin in 1000 μL 3% BSA/4X SSC/0.1% Tween.
9. DABCO (1,4-diazabicyclo(2.2.2)octane): 2.3% (w/v) in Glycerol/Tris, e.g., dissolve 23 mg DABCO in 10 mL Glycerol/Tris, (9 vol Glycerol plus 1 vol Tris-HCl, 0.2 M, pH 8.0).
10. 24 × 60 mm^2 cover slips (e.g., Menzel)

3. Methods

3.1. Metaphase Chromosome Preparation

Normal metaphase chromosomes are used as the DNA matrix to which both genomes are hybridized. The are prepared from peripheral blood lymphocytes after stimulating proliferation and metaphase arrest by colcemid. Metaphase spreads are also commercially available (for example, from Vysis). But these are not necessarily superior to conventionally prepared spreads *(8)*. For control purposes new metaphase batches should be tested by normal vs normal hybridizations.

3.1.1. Cell Preparation

1. Incubate culture (amounts per 5 mL blood: 40 mL RPMI 1640 Medium, 10 mL FCS, 5 mL peripheral blood, 1.5 mL phythemagglutinin) for 72 h in CO_2 cell-culture incubator, mix flask 1–2 times per day.
2. Add 400 µL Colcemid about 45 min before harvesting.
3. Make 2 aliquots and transfer cell into 50-mL Falcon tubes.
4. Incubate in cell culture incubator or 37°C water bath for additional 45 min.
5. Centrifuge for 10 min at 1000 rpm (230g).
6. Remove supernatant, e.g., with a cell-culture pipettor until 5 mL remain.
7. Gently add 40 mL KCl solution at 37°C, the first 5 mL drop by drop (hypotonic treatment).
8. Incubate for 25 min in a 37°C water bath.
9. Centrifuge 10 min at 1000 rpm (230g).
10. Remove supernatant, leaving about 5 mL, and resuspend pellet.
11. Add 2 mL fixative, mix well.
12. Add fixative until volume reaches 40 mL, with constant mixing.
13. Repeat **steps 9–12** until the pellet is white (at least 4 times).
14. After removal and resuspension of the pellet, transfer cells into a 15-mL Falcon tube.
15. Repeat **steps 9–12**, but adding just 10 mL fixative.
16. Remove fixative, leaving 2 mL final volume.
17. Resuspend pellet and apply suspension onto slides.

3.1.2. Slide Preparation

1. Cool slides to –20°C (e.g., put about 10 slides in a cuvet in the –20°C freezer and keep the cuvet on ice while preparing the metaphase slides).
2. Take one slide and moisten by breathing from very close.
3. Either drop 50–100 µL of the suspension on the slide or apply the same volume to an inclined slide (the fast draining and drying of the fluid is usually an indication for good spreading).
4. Let the suspension begin to dry (the fluid film will start to retract).
5. Put the slide briefly in a cuvet containing 70% acetic acid (*see* **Note 2**).
6. Air-dry the chromosome slide, check for chromosome spreading and cytoplasm debris using a phase contrast lab microscope, and adjust volume of fixative so that the density of nuclei/metaphases is appropriate. If the conditions are favorable, prepare a batch of metaphases spreads. Keep fixative with lymphocytes at –20°C until the preparation of new slides. Add new fixative and wash cells before the preparation of new metaphase spreads.
7. Store slides in a box at room temperature for up to about 1–2 mo. Metaphase spreads may be kept longer at –80°C or in 70% ethanol at 4°C.

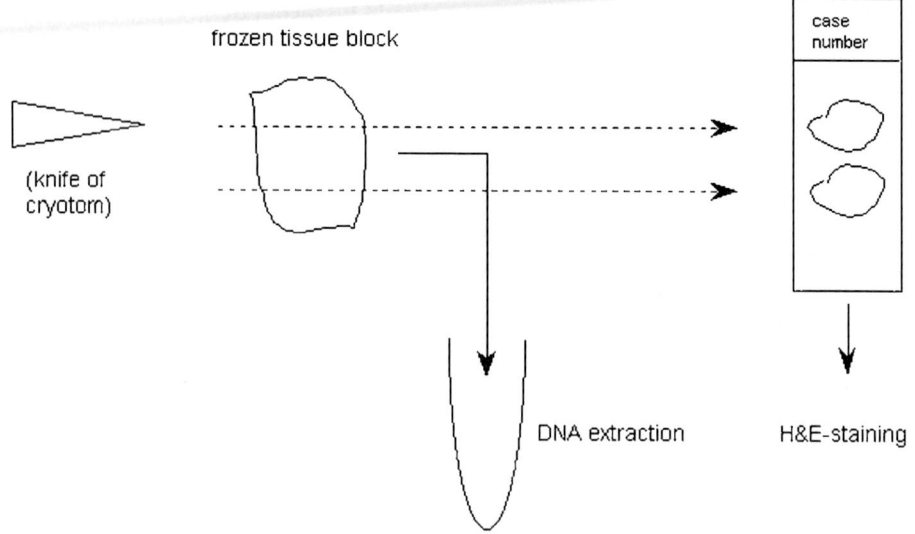

Fig. 3. DNA extraction using cryotom microdissection.

3.2. DNA Preparation

3.2.1. DNA Extraction from Frozen Tissue Using Cryotome Tissue Dissection

The main purpose of using cryostat sections for DNA extraction is the fact that it provides a control for normal cell contamination in the tumor tissue (*see* **Fig. 3**).

3.2.1.1. Tissue Dissection (*see* Note 3)

1. Freeze tissue block (0.5–1 cm^2) on the cryostat plate, e.g., by applying few drops of isotonic NaCl or water and placing the tissue immediately on the cooled tissue holder.
2. Cut the tissue block and place first section (5–8 µm) on a labeled glass slide (close to case No. label).
3. Cut about 20–30 section (20–30 µm) and transfer to labeled Eppendorf tubes containing 900 µL digestion buffer. Transfer can be facilitated by picking the cool sections up by the tip of the Pasteur pipet. Change pipet after each case to avoid cross-contamination.
4. Transfer last section (5–8 µm) onto the glass slide (distal to case No. label) and check the amount of tumor DNA in the tissue block by an H&E stain of the slide. The tumor should exceed a percentage of 70% compared to normal stroma tissue and inflammatory cells (*see* **Note 3**).

3.2.1.2. Proteinase K Digestion

1. Add 30–50 µL of Proteinase K stock solution to eppendorf tubes containing sections in digestion buffer.
2. Incubate for at least 2 h at 50°C, check the digestion for the disintegration of the tissue, and add new Proteinase K if the digestion is insufficient. Proteinase K digestion can be extended overnight or even for several days.

3.2.1.3. Phenol-Chloroform Extraction

1. Add 1000 µL of Phenol/Chloroform/Isoamylalcohol to digest, mix by inverting for about 10 min, centrifuge in a table-top centrifuge for 10–20 min until the two phases have clearly separated. Discard upper phase.
2. Repeat **step 1** at least once (twice is preferable).
3. Add 1/10 vol (about 90 µL) 3 M NaCl, mix. Add 1 vol (about 1000 µL) ice cold isopropanol. Mix by inverting. A visible white DNA precipitate is indicative of an amount and quality of DNA sufficient for nick translation and CGH.
4. Centrifuge DNA pellet for about 5 min, discard supernatant, wash once with 100% EtOH and once with 70% EtOH. Finally, air dry pellet or dry by SpeedVac (5–10 min) (*see* **Note 4**).
5. Dissolve pellet in about 100–200 µL very pure H_2O, depending on the expected amount of DNA.
6. Determine DNA concentration by photometer. The final concentration for nick translation should by higher than 150 µg/mL.

3.2.2. DNA Preparation from Paraffin-Embedded Material (see **Note 5**)

3.2.2.1. Preparation of Tissue Section and Deparaffinization

1. Use the 1st section of a paraffin block (5–8 µm) for H&E staining. Areas with tumor tissue should be marked.
2. Cut 20–40 sections (~10 µm) for DNA extraction, the number of sections being dependent on the area and density of the tumor tissue (~20 sections for 4 cm^2, ~40 for 0.5 cm^2). Place each section on a glass slide.
3. Remove the tumor tissue (with paraffin) by manual microdissection from the sections, using a clean and preferably sterile scalpel and placed in a 2-mL Eppendorf tube with 1.5 mL xylol. A fresh scalpel should be used for each new case but not necessarily for each new section.
4. Incubate tissue for 30 min at room temperature (RT), centrifuge for 5 min at 14000 rpm (16000g).
5. Remove and discard supernatant, being careful not to disturb tissue pellet.
6. Add another fresh 1.5 mL aliquot of xylol, incubate for about 30 min, centrifuge, and remove supernatant as above.
7. Remove remaining Xylol from tissue by adding 1.5 mL 100% Ethanol.
8. Incubate 10 min at RT, centrifuge 5 min, 14000 rpm (16000g), RT.
9. Repeat washing **steps 7** and **8** with ethanol.
10. Carefully remove remaining ethanol by decantation and air dry for about 30–60 min.

3.2.2.2. DIGESTION

1. Add 900 µL Digestion buffer and 20–30 µL proteinase K solution (*see* **Subheadings 2.2.1.1.** and **2.2.1.2.**) to the tissue.
2. Incubate overnight, preferably at 50°C, e.g., in a thermomixer. In the case of incomplete digestion of tissue, add more proteinase K, and extend incubation period.

3.2.2.3. DNA EXTRACTION BY PHENOL/CHLOROFORM/ISOAMYLALCOHOL

1. For DNA extraction and precipitation (*see* **Subheading 3.2.1.3.**)

3.3. Nick Translation (see Notes 6–8)

A typical pipetting scheme for nick translation of 1 and n samples, respectively. Five µg of DNA is used for each nick translation (NT) sample.

Mix (Vtotal = 50 µL):	1 probe	Mix for N probes	
NT (10X)	5 µL	(N + 1) × 5	:
β-ME	5 µL	(N + 1) × 5	: for more than 1 probe
dNTPs	5 µL	(N + 1) × 5	: pipet 19 µL to the
Bio/Dig-dUTP	2 µL	(N + 1) × 2	: DNA+H_2O
DNase (1:2000)	1 µL	(N + 1) × 1	:
Pol	1 µL	(N + 1) × 1	:
DNA + H_2O	31 µL		
Final Volume	50 µL		

1. Pipet on ice.
2. Incubate 2 h at 15°C, then place the reaction tubes on ice.
3. Store probes at –20°C while testing 5 µL of the mix by agarose electrophoresis (*see* **Fig. 4**). The optimal length of DNA fragments should be between 100–1000 bp. If necessary incubate longer with addition of fresh DNase and Pol.
4. Stop the reaction by adding 2.5 µL EDTA (0.5 *M*, pH 8.0) and 2.5 µL SDS (20%). Keep the probes at –20°C until needed for hybridization.

3.4. Hybridization

3.4.1. Precipitation of DNA and Dilution in Hybridization Solution

1. Mix the following solutions in a 1.5 mL Eppendorf tube: 10 µL tumor-DNA (Biotin-labeled) (*see* **Note 9**), 10 µL Normal-DNA (Digoxigenin-labeled), 30 µL human Cot1-DNA, 1 µL salmon sperm-DNA, 5.1 µL 3 *M* sodium acetate, 150 µL 100% ethanol (–20°C).
2. Incubate at –80°C for 30 min, centrifuge for 10 min at max. speed at 4°C. Remove the supernatant, wash the pellet with 500 µL 70% ethanol, centrifuge for 10 min at 4°C, remove the supernatant, and dry the pellet (speed vac for 5–10 min or air-dry for ~1 h).

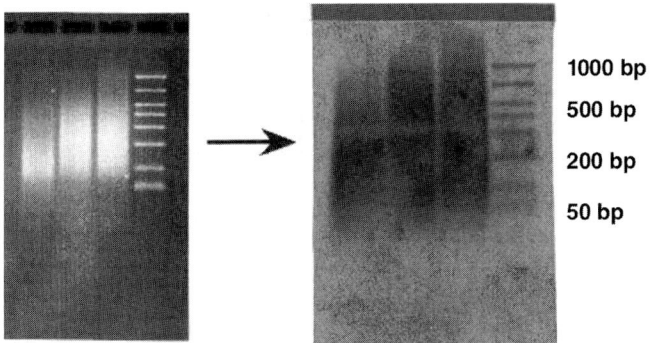

Fig. 4. Left panel, Agarose gel electrophoresis of DNA following nick translation. Right panel, To confirm the integration of biotinylated nucleotides, DNA was transferred to a nylon membrane and detection was performed by a color reaction. The optimal fragment length size following nick translation is between 100 and 1000 bp.

3. Dissolve the pellet in 5 µL formamide, incubate the probe for 10 min at 37°C, and add 10 µL master mix.

3.4.2. Denaturation of the Genomic DNA and Prehybridization

1. Denature DNA for 5 min at 77°C, centrifuge briefly.
2. Prehybridize at 37°C for at least 1 h.

3.4.3. Inspection and Denaturation of Metaphase Chromosomes

1. During the prehybridization step, inspect the slides containing the metaphase spreads under a phase-contrast microscope. Check the quality of the chromosomes and select the best region for hybridization.
2. Place 120 µL 70% FA/2xSSC on a large cover slip and place slide with chromosome spreads slowly onto the drop of denaturation solution. Turn the slide back right-side up.
3. Denature at 77°C for 90 s in a preheated oven or on a heating surface, place slide vertically to remove cover slip, then place in ice-cold 70% ethanol.
4. Dehydrate by placing slides in ascending concentrations of ethanol (70, 90, and 100% EtOH, 2 min each) and air-dry in a vertical position.

3.4.4. Addition of the Hybridization Solution to the Slide with the Metaphase Chromosomes

1. Briefly centrifuge DNA following prehybridization to remove vapor and fluid from the tube lid.
2. Place slides with the denatured and dry metaphase chromosomes on a heating plate at 37°C, label slides with case number, and add about 13 µL of hybridization solution. If air bubbles appear, remove them using the edge of a cover slip.

3. Cover the 13 µL droplet with a small (18 × 18 mm²) cover slip and seal edges with rubber cement. Alternatively, slides may be placed in a sealed, moist chamber at 37°C which negates the necessity for rubber cement. Place slides in a water tight metal box, then that box in a water bath at 37°C.
4. Hybridize for a minimum of 2 d, though three days are preferable.

3.5. DNA Detection

1. Wash slides 3 × 3 min at 37°C in FA/2X SSC. Use fresh solution for each wash.
2. Wash slides 3 × 2 min at 60°C in 0.1X SSC (fresh solution for each wash) (*see* **Note 10**).
3. Keep slides in 4X SSC/0.1% Tween 20 at 37°C (in a fresh slide holder) until the next step.
4. Cover each slide with 125 µL 3% BSA solution and a 24 × 60 mm² cover slip. Incubate 15 min in a moist chamber at 37°C for blocking.
5. Remove cover slips and place slides in 4X SSC/0.1% Tween 20 prior to the next step.
6. Centrifuge fluorochromes 3 min at 14000 rpm at 4°C (16000g) and prepare fluorochrome solution.
7. Cover each slide with 125 µL of the fluorochrome solution and a 24 × 60 mm² cover slip. Incubate 20 min in a moist chamber at 37°C.
8. Remove cover slips and wash slides 3 × 3 min at 45°C in 4X SSC/0.1% Tween 20.
9. Incubate slides 5 min in DAPI solution (room temperature).
10. Mount the slides using DABCO (about 35 µL per slide).
11. Store slides at 4°C under dark conditions until microscopy.

3.6. Image Capture and Digital Image Analysis

The main purpose of the image capture is the acquisition of three fluorescence images per metaphase, i.e., DAPI for chromosome identification, FITC/fluorescein and TRITC/rhodamine images representing the tumor and normal DNA, respectively (*see* **Note 11**). From the "tumor" and "normal" image a RATIO image is first calculated by comparing the fluorescence intensities of the FITC and TRITC images at each pixel after normalization. The normalization on the one hand guarantees that both images can be compared to each other, although the quality of the tumor and normal DNA, and thus the intensities of their fluorescence images, might be different. On the other hand, this reduces the need that exactly the same quantities are used during the preparative steps of DNA labeling, hybridization and detection. The FITC and TRITC metaphase images of a good quality hybridization are shown in **Fig. 5**. Image analysis is then applied for karyotyping, the calculation of the ratio profile and the display of CGH results.

3.6.1. Image Acquisition

Image capture is performed by fluorescence microscopy with a high-quality microscope with adequate filters and correct and sufficient illumination. A halogen lamp of 100W should be used rather than the standard 50W (*see* **Note 12**). In our laboratory we use an Axiophot epifluorescence microscope (Zeiss, Oberkochen, FRG) with the following filter sets:

- DAPI: Zeiss filter set 02, i.e., excitation G365, beamsplitter FT395, emission LP 420
- FITC: Zeiss filter set 20, i.e., excitation BP 450-490, beamsplitter FT 510, emission BP 546/12
- TRITC: Chroma filter set HQ Cy3 plus excitation filter of Zeiss filter set 15, i.e., excitation BP 546/12, beamsplitter FT 565, emission BP 570-650

Digital images are captured using a Photometrics camera with 12-bit chip enabling 4096 different gray values. The images are stored using an 8-bit format, i.e., TIFF-format (*see* **Notes 13,14**).

3.6.2. Digital Image Analysis

Several digital image analysis programs for CGH have been developed. To our knowledge, the following software packages are, at present, commercially available (*see* **Note 15**).

- KaryoMedics GbR   http://www.karyomedics.com
- Applied Imaging International Ltd.   http://www.CytoVision.com/
- MetaSystems GmbH   http://www.metasystems.de/

We and others contributed to the development of CGH software *(2,15,26–28)*. Our experience and programs, including the histogram format *(15,16,22,29)* are presented by the Karyomedics system. Digital image analysis comprises the following steps:

1. Image objects (chromosomes) of the metaphase are defined by segmentation of the inverted DAPI image.
2. The FITC and TRITC images are loaded under the DAPI segmentation mask.
3. The optical shift of the FITC and TRITC images is corrected.
4. The RATIO (FITC/TRITC) image is calculated, after normalization.
5. Touching chromosomes are separated.
6. DAPI chromosomes are karyotyped. It is very helpful if the FITC, TRITC, and RATIO chromosomes can be displayed during the karyotyping process (*see* **Note 16**).

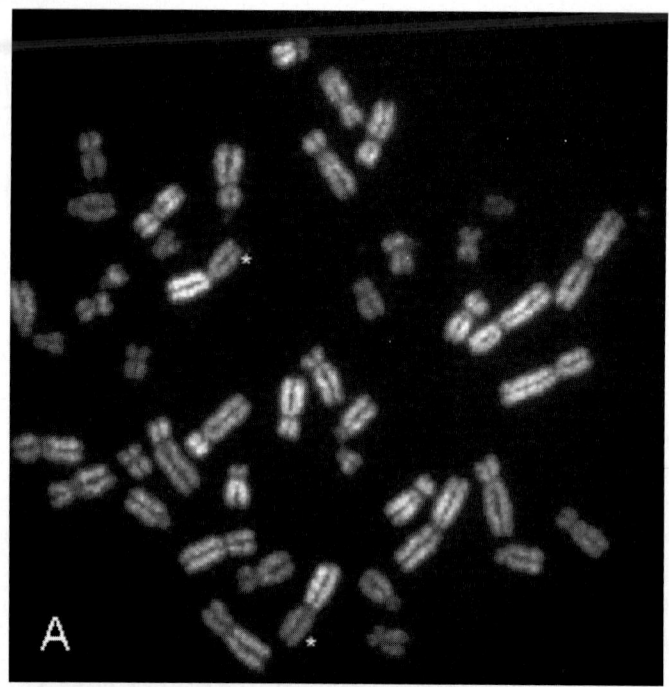

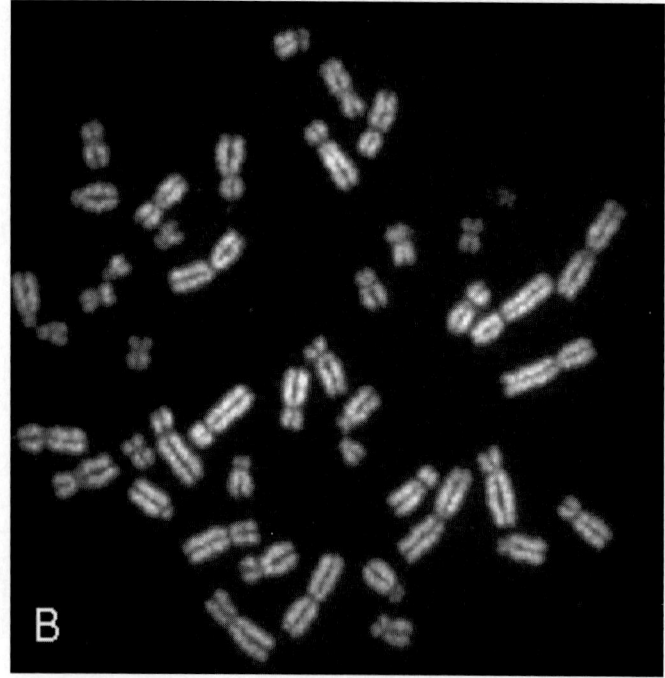

7. The mean ratio profiles are calculated by averaging the score of chromosomes of several metaphases/karyograms. These data can be complemented in some programs by the calculation of mean ratio chromosomes, i.e., the CGH sum-karyogram (*see* **Note 17**).

3.7. Determination of the Chromosomal Imbalances of a Single Case

The chromosomal imbalances are determined using the mean ratio profile and its confidence interval (*see* **Note 18**). The evaluation of a typical CGH result is exemplified in **Fig. 6**. A single mean ratio profile and 99% confidence interval of chromosome 3 of a typical squamous cell carcinoma of the lung is displayed (**Fig. 6A**). Chromosomal imbalances were determined by two different evaluation schemes, i.e., by using fixed-ratio thresholds (**Fig. 6B**) and statistical procedure (**Fig. 6C**). In the first method, only those deviations from the profile that pass exceed either the 0.75 or 1.25 ratio values are considered chromosomal loss or gain. In the second approach, all statistically significant deviations of the ratio profile from the normal value of 1.0 were scored. If the resulting ratio profile and its 99% confidence interval are on the left side of the middle vertical line this indicates DNA loss, whereas similar deviations to the right correlate to DNA gain. Obviously, the statistical approach is more sensitive, since it can detect more alterations than the quite stringent 0.75/1.25 thresholds (*see* **Note 19**). Fixed-ratio thresholds are also applied to defined pronounced alterations (*see* **Note 20**).

Fig. 5. (*see opposite page*) In addition to the DAPI image for chromosome identification (not shown), two distinct fluorescence images are captured per metaphase as gray level images. (**A**) The FITC/fluorescein image represents the tumor (SCLC in this case). (**B**) The TRITC/rhodamine image allows visualization of the normal genome. The quality of the hybridization can be judged by the strong and close-grained signal of the chromosomes in contrast to the dark background. In addition, both images are illuminated homogeneously, as is the ideal. Even at this stage, chromosomal alterations are visible in the FITC images. These are distinguished by the imbalances of the fluorescence signal between the short and the long arm of one chromosome (*, two homologs of chromosome 3 with the typical 3q isochromosome pattern of SCLC). However, to determine whether these imbalances correspond to a deletion or over-representation, digital image analysis must be performed. The first step of this process is the calculation of a RATIO image by comparing the normalized signal intensities of the FITC and TRITC images at each pixel.

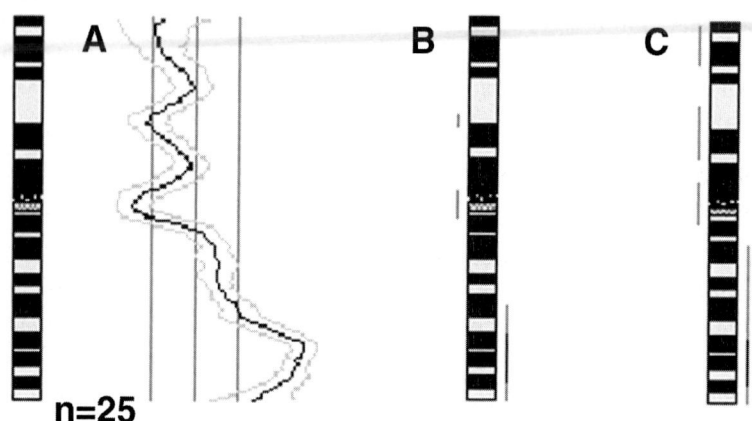

Fig. 6. Determination of chromosomal imbalances based on the mean ratio profile. (**A**) The mean ratio profile of chromosome 3 and its 99% confidence interval as determined by a Student's *t*-test. The three vertical lines along the right side of the chromosome ideogram represent various fluorescence ratios, i.e., 0.75, 1, and 1.25. The central line corresponds to the normal state (fluorescence ratio 1:1). The lines to the left and right represent theoretical values of a monosomy or trisomy in 50% of the tumor cells of an otherwise diploid tumor, respectively. The number of chromosomes that were entered into the calculation is indicated in the chromosome ideogram. (**B**) Chromosomal alterations, determined by using the fixed ratio thresholds 0.75 and 1.25, are indicated by the lines on the left and right side of the chromosome ideogram, respectively. On the short arm of 3p, only a small region is scored as a deletion. The confidence interval is not used for the evaluation. (**C**) Chromosomal alterations were determined by using a statistical procedure in which deviation from the ratio profile and its confidence interval from the normal state of 1.0 are considered. This evaluation detects larger deletions on 3p, and is thus more sensitive than the method presented in (**B**) using fixed ratio thresholds. A pronounced gain, defined by a profile exceeding the ratio value of 1.5, is visible at the telomeric 3q region (dark gray line in B and C). It correlates with a high copy amplification.

3.8. CGH Results of a Tumor Collective: Line Diagrams and Histograms

After determination of the chromosomal imbalances of a single case the results of several tumors often needs to be summarized. The classical representation of chromosomal imbalances of a tumor group is the line diagram (**Fig. 2A**). In this format, each alteration is drawn as a single line along either the right or left side of the chromosome ideograms. Although the most important changes can be identified by this method, there are several disadvantages (*see* **Note 21**). Conversely, in the histogram format, alterations are shown as frequency curves along each chromosome (**Fig. 2B**). Using this method, virtually an unlimited number of cases can be included in the histogram, and the importance of a specific chromosomal change is directly visible by observing the incidence of the respective DNA gain or loss. Importantly, the histogram approach can be used to compare the alterations of different tumor subgroups by calculating a difference histogram (**Fig. 2C**).

4. Notes

1. Cool cryostat down to –20 to –30°C about 3 h prior to dissection, and dissect only one block per time in the cryostat. Keep other tissue blocks frozen (e.g., in styrofoam box in liquid nitrogen or dry ice) during dissection. Take care to replace tissue blocks in the correct vial after dissection.
2. The acetic acid step is a washing step, particularly for removing the cytoplasm. In addition it may help in the spreading of the chromosome; the cell membranes attach to the surface of the glass slides and are disrupted by the liquid flow, and the temperature difference between the cell and the glass slides may help in the disruption of the cell membranes. If the weather conditions are favorable, the 70% acetic acid washing step may be omitted. In our experience the best metaphase spreads occur on dry and sunny days.
3. If tumor does not exceed a percentage of 70% compared to normal stroma tissue and inflammatory cells, those parts of the tissue block with the highest contamination by normal tissue might be cut off by a sterile scalpel. This step is best performed at this stage. Though it can be done after the extraction step, postponement to that stage could necessitate a second extraction, which should be preferably done from another piece of tissue.
4. If a SpeedVac is used to dry DNA, take care not to over-dry. High molecular, ultradry DNA may be difficult to dissolve.
5. For DNA from paraffin-embedded material, the conditions for nick translation, hybridization, and detection might need some modifications from the standard protocols. Generally the results from paraffin material are good if the recovery and DNA quality is sufficient, which may be checked by agarose gel electrophoresis. An ideal preparation would produce long fragments with minimal degradation. This is usually the case for biopsy material which has not been kept in formalin for very long (only hours or single days instead of several days/weeks). In the case of badly degraded DNA, chemical labeling schemes might be used instead of enzymatical reactions like nick translation *(30)*.
6. DNA labeling by nick translation is based on the incorporation of labeled nucleotides during *de novo* DNA synthesis by a DNA polymerase. The polymerase uses, as a starting point, the single-strand nicks on double stranded DNA created by Dnase. However, the action of the DNase also results in double stranded DNA cutting and long-sized fragments can be converted to smaller ones, in

a process that results in the same sort of DNA degradation induced by other means. Nick translation is most effective when long size DNA fragments are available as starting material.

7. In the standard protocol, tumor DNA is labeled with Bio-dUTP, normal DNA is labeled with Dig-dUTP. These might be interchanged in reverse labeling reactions.
8. DNA from paraffin-embedded material is often degraded. For DNA from this source, we therefore decrease the amount of Dnase in the pipetting scheme, using a 1:10000 dilution. In addition, the amount of DNA polymerase and Bio-dUTP is increased, using 2 and 3 µL, respectively. The volume of the DNA + H_2O thus needs to be reduced to 29 µL, to which a 21 µL aliquot of the reaction mix is added. In cases of severe degradation, the starting amount of DNA might be increased to 10 µg per NT instead of 5 µg.
9. For DNA from paraffin material with bad fluoresence signals, the amount of NT-labeled tumor-DNA might be increased to 20 or 30 µL.
10. For DNA probes from paraffin-embedded material, the washing temperature may need to be reduced to 50°C, since the shorter DNA fragments might be lost by very stringent washing conditions.
11. Chromosomes and nuclei should have a normal morphology (no shrinkage or eventually too pronounced a banding pattern in the DAPI image). Fluorescence should be finely dispersed and should not depend on the chromosome morphology, e.g., the DAPI banding pattern. In addition, there are specific control experiments which should be performed on a regular basis. These include test hybridization of new metaphase batches with normal versus normal genomes as DNA probes. A second hybridization with an inverse labeling scheme for the test and reference genome should be performed in cases in which the confirmation of the specific alterations is critically important. Finally, a third hybridization probe with defined chromosomal imbalances might be used an internal control within single experiments *(31)*.
12. The field of view must be homogeneously illuminated *(15)*.
13. The intensity and contrast of the images must be reasonably high. Excellent quality images have a signal to background ratio of 5:1. Good samples usually yield a ratio of 3:1. Images with a ratio of 2:1 might still be used if all other parameters are fulfilled.
14. A 8-bit camera might also be used for image aquisition *(25)*. The dynamic range of the fluorescence signal is represented by the 8-bit digital image comprising 256 gray values, i.e., the maximum value of the specific fluorescence signal on the chromosomes should be slightly below the value of 256, and should be not cutoff by overexposure.
15. Appropriate analysis software is necessary, since the comparison of fluorescence images can not be accomplished by visual inspection. These could easily lead to false results, owing to inability of the eye to discriminate differences in weak intensity signals. This is particularly true for deletions, though high

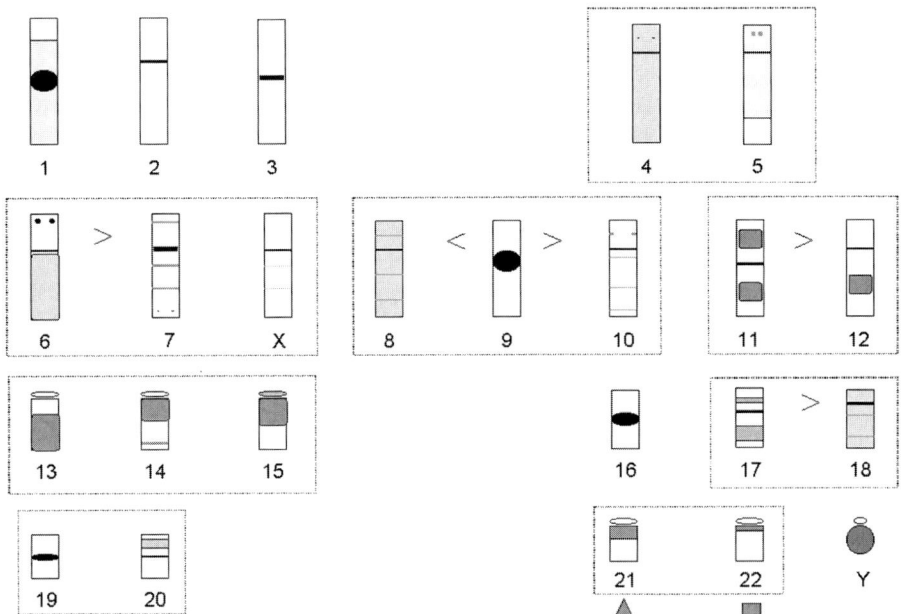

Fig. 7. Chromosome identification scheme. The dashed boxes group chromosomes of similar size that are particularly difficult to distinguish. They should thus be classified as a group. The ">" and "<" symbols indicate chromosome arms that are greater or smaller than those of the adjacent chromosome. For details for the correct identification of single chromosomes, see **Note 16**.

copy amplifications are, on occasion, easily identified visually. However, one cannot depend on digital analysis to pull data from a poor hybridization, and no hybridization should be evaluated in which the fluorescence images show a signal that is not detectable by visual inspection at the microscope.

16. Karyotyping is a particularly laborious and error-prone process. Several metaphases must be analyzed for the evaluation of one case, and DAPI banding is inferior to conventional Giemsa staining for producing good intensity and contrast in the banding pattern. To facilitate correct chromosome identification using CGH, we developed a chromosome identification scheme (**Fig. 7**). It takes advantage of the fact that the information of all 4 metaphase images (DAPI, FITC, TRITC, RATIO) can be used for correct karyotyping, compensating for some of the disadvantages of DAPI banding. For instance, the imbalance of a specific chromosome may help in its differentiation from others. Obviously, such changes need to be correctly identified. We perform the karyotyping in a two step process. Fifteen metaphases are first karyotyed by one person and this analysis is then controlled by a second investigator.

Details for correct chromosome identification

Chr. 1	Usually easy to identify, take care that the orientation (upside down?) is correct, i.e., end of the small arm brighter than the rest of the chromosome.
Chr. 2	Second largest chromosome (size similar to chr. 1), submetacentric.
Chr. 3	Third largest chromosome, metacentric, be careful with the orientation.
Chr. 4 vs Chr. 5	Both have a similar size, chr. 4 darker than 5, chr. 5 with brighter area at the end of the long chromosome arm.
Chr. 6	Dark long arm, brighter short arm with band/dots at the end appearing like "the ears of a rabbit."
Chr. 7	Three distinct bands visible, one at the short arm, two at the long arm, centromere quite pronounced.
Chr. X	Particularly difficult to distinguish from chr. 7 because it is of almost the same size. If the tumor or the normal genome originate from a male, the X-chromosome usually shows a less intense fluorescence signal in the FITC and/or TRITC image (reason: two chr. 7 in the genomes compared to one X-chromosome).
Chr. 8 vs Chr. 10	Both chromosomes particularly difficult to differentiate, use the following criteria: chr. 8 darker than chr. 10, chr. 10 shows three bands on the long arm, in particular a subcentromeric band. In contrast chr. 8 harbors only two bands on the long arm.
Chr. 9	Usually easy to identify by the intense centromeric heterochromatin (look for the suppression of the heterochromatin by the Cot1 DNA in the FITC and TRITC image). Differentiation from chr. 8 and 10 by the relatively longer p-arm of chromosome 9 compared to the p-arms of chr. 8 and chr. 10.
Chr. 11 vs Chr. 12	Chr. 11 shows two bright dark bands on the long and the short arm (looks like a butterfly). In contrast, chr. 12 show only a distinct dark band on the long arm. The short arm of chr. 11 is longer than that of chromosome 12.
Chr. 8 to 12	These are more or less of the same size, and recommended order for classification. (after Chr. 6, 7, and X): 11→12→9→8→10.
Chr. 13 vs Chr. 14,15	Might be quite difficult to differentiate from each other.
Chr. 13	Chr. 13 has a dark zone covering about 2/3 of the long arm.
Chr. 14 vs Chr. 15:	The dark zone of chr. 15 extends more telomerically than that of chr. 14.
Chr. 14	Has a distinct subtelomeric band on the long arm followed by a light zone before the chromosome ends. Chr. 15 may have a band also. However, there is usually no light zone distinguishable before the telomere.

Chr. 15	Shows a more prominent centromere (look for suppression in the FITC and TRITC image).
Chr. 16	Easy to identify as smaller than chr. 9, larger than chr. 19, with distinct centromeric heterochromatin. Check orientation.
Chr. 17 vs Chr. 18	May be difficult to differentiate.
Chr. 17	Short arm is longer than the short arm of chr. 18. Distinct bands are visible on the long arm, though mostly on the short arm.
Chr. 18	Darker than chr. 17, and there are usually two distinct bands visible on the long arm.
Chr. 19, Chr. 20	Difficult to differentiate:
Chr. 20	Has a dark band on the short arm.
Chr. 19	Centromeric heterochromatin more pronounced than chr. 20 (look for suppression in the FITC and TRITC image).
Chr. 21, Chr. 22	Together with Y-chromosome, of similar size.
Chr. 21	Has a more pronounced dark band on the long arm that takes the shape of a triangle, compared to the more distinct, rectangular-shaped staining pattern for chr. 22.
Chr. Y	Usually easy to identify, carries a lot of heterochromatin. Almost the entire chromosome shows a suppression of the fluorescence signal in the FITC and TRITC image and is therefore usually excluded from the evaluation of DNA gains or losses by CGH.
Chr. X and Y	Both should only be evaluated if the hybridizations have been performed sex neutral (male tumor DNA together with male reference DNA and female test genome with female normal DNA).

Note that heterochromatic regions, particularly of chromosomes 1, 9, and 19, must not be evaluated because of the suppression by Cot1 DNA.

17. The end result of the CGH analysis of a single case is the mean ratio profile, which should be displayed by all software. It is calculated from all chromosomes of the same class that are present in the individual karyograms. We generally evaluate 15 metaphases/karyogram, thus up to 30 chromosomes can be included in the calculation of the ratio profile (**Fig. 6**). This represents the CGH result as a one-dimensional curve *(15)*. In addition ratio sum karyograms (offered only by some CGH programs) can be calculated and chromosomal gains and losses can be displayed by pseudocolors *(17)*, nicely illustrating the chromosomal alterations of a single tumor. In general it can be stated that the more alterations that are present within a tumor the more malignant it must be. The DNA imbalances are generally determined by the ratio profile and its confidence interval.
18. The number of chromosomes used for the calculation of the confidence interval by a Student's *t*-test is provided at the ideogram (**Fig. 6A**). The confidence interval reflects the deviation of the single profiles from the mean ratio. It is

a quality parameter of the experiment, since a smooth ratio profile with close confidence intervals indicates that there was little variation of fluorescence signals from all the chromosomes that were included in the calculation of the ratio profile. This of course requires that no errors occurred during karyotyping.

19. The first approach might be viewed as biologically motivated. The ratio 1.0 represents the theoretical value for balance between tumor DNA and normal DNA. Ratio values larger than 1.0 indicate genetic gains whereas ratio values less than 1.0 indicate genetic losses. A ratio value of 0.5 would be expected if all chromosomes in a diploid tumor showed a monosomy. Similarly, a ratio value of 1.5 would indicate a trisomy in a diploid tumor. Thus, the ratios of 0.75/1.25 would theoretically be expected in a diploid tumor cell population for monosomy/trisomy of a certain chromosome in 50% of the examined cells, respectively. These thresholds reflect relatively stringent requirements, and become less sensitive when the ploidy levels increase. We therefore prefer the second approach, using common statistical methods to define conditions for gains and losses in the set of individual ratio profiles of all replicants belonging to the same chromosome type. These techniques have be shown to carry a higher sensitivity *(24,32)*. The selection of a method to detect chromosomal aberrations, however, depends on the employed CGH system and the internal standards established in individual laboratories. Thresholds such as 0.8/1.2 or 0.85/1.15 are used, and these thresholds can be used in conjunction with statistical methods, e.g., by scoring those deviations from the ratio profile and the confidence interval that exceeds 0.9/1.1 thresholds, instead of using the normal state (ratio value 1.0 corresponding to 1.0/1.0 thresholds). Because there is no consensus on how chromosomal imbalances are determined, it is critical to outline the criteria used in each study to ensure comparability of results. We present our primary data in a CGH online tumor database, in which different evaluation schemes (fixed ratio thresholds, with or without statistics, different statistical tests using either Student's *t*-test or normal distributions) can be chosen interactively (http://amba.charite.de/cgh/).

20. The ratio value of 1.5 is often used as a threshold for defining high copy amplifications. However, it is difficult to directly infer a specific amplification from the CGH profile, and occasionally these data may not correspond to amplification as defined by fluorescence *in situ* hybridization (FISH) or Southern-blot analyses. We therefore have introduced the terms of pronounced DNA gains and losses *(20)*. These define alterations based solely on ratio profiles, i.e., ratio profile values above 1.5 are considered pronounced gains whereas ratio values below 0.5 are pronounced deletions. Pronounced gains are most likely to correspond to high copy amplifications, whereas pronounced appear to represent multicopy deletions (**Fig. 6B,C**).

21. The histogram format is the method of choice for summarizing the analysis of a large collection of tumors. Unfortunately it is supported by few CGH software programs. The line formats used initially have some disadvantages. First, the space required to illustrate the alterations of a large collection of tumors is

prohibitive. Second, the frequency of a specific change cannot directly be deduced from the figure. Third, the results of statistical evaluations, e.g., the significance of a specific change by the confidence interval, is not easily visualized. Fourth, and most importantly, the line format cannot be used for statistical comparisons of tumor groups, in contrast to the utility of the differential histogram, which was first introduced by our group *(16)*. We recently expanded on this format and developed a case by case histogram format to visualize differences between tumor pairs, such as those from primary tumors and their corresponding metastases in a larger collection *(22,29)*.

References

1. Kallioniemi, O. P., Kallioniemi, A., Sudar, D., Rutovitz, D., Gray, J. W., Waldman, F., and Pinkel, D. (1992) Comparative genomic hybridization for molecular cytogenetic analysis of solid tumors. *Science* **258,** 818–821.
2. du Manoir, S., Speicher, M. R., Joos, S., Schröck, E., Popp, S., Döhner, H., et al. (1993) Detection of complete and partial chromosome gains and losses by comparative genomic in situ hybridization. *Human Genet.* **90,** 590–610.
3. Speicher, M. R., du Manoir, S., Schröck, E., Holtgreve-Grez, H., Schoell, B., Lengauer, C., et al. (1993) Molecular cytogenetic analysis of formalin-fixed, paraffin-embedded solid tumors by comparative genomic hybridization after universal DNA-amplification. *Human Mol. Genet.* **2,** 1907–1914.
4. Isola, J., DeVries, S., Chu, L., Ghazvini, S., and Waldman, F. (1994) Analysis of changes in DNA sequence copy number by comparative genomic hybridization in archival paraffin-embedded tumor samples. *Am. J. Pathol.* **145,** 1301–1308.
5. Wiltshire, R. N., Duray, P., Bittner, M. L., Visakorpi, T., Meltzer, P. S., Tuthill, R. J., et al. (1995) Direct visualization of the clonal progression of primary cuatneous melanoma: application of tissue microdissection and comparative genomic hybridization. *Cancer Res.* **55,** 3954–3957.
6. Kuukasjarvi, T., Tanner, M., Pennanen, S., Karhu, R., Visakorpi, T., and Isola, J. (1997) Optimizing DOP-PCR for universal amplification of small DNA samples in comparative genomic hybridization. *Genes Chromosomes Cancer* **18,** 94–101.
7. Kallioniemi, O. P., Kallioniemi, A., Piper, J., Isola, J., Waldman, F. M., Gray, J. W., and Pinkel, D. (1994) Optimizing comparative genomic hybridization for analysis of DNA sequence copy number changes in solid tumors. *Genes Chromosomes Cancer* **10,** 231–243.
8. Karhu, R., Kahkonen, M., Kuukasjarvi, T., Pennanen, S., Tirkkonen, M., and Kallioniemi, O. P. (1997) Quality control of CGH: impact of metaphase chromosomes and the dynamic range of hybridization. *Cytometry* **28,** 198–205.
9. Bentz, M., Plesch, A., Stilgenbauer, S., Dohner, H., and Lichter, P. (1998) Minimal sizes of deletions detected by comparative genomic hybridization. *Genes Chromosomes Cancer* **21,** 172–175.
10. Schena, M., Shalon, D., Davis, R. W., and Brown, P. O. (1995) Quantitative monitoring of gene expression patterns with a complementary DNA microarray. *Science* **270,** 467–470.

11. Kraus, J., Weber, R. G., Cremer, M., Seebacher, T., Fischer, C., Schurra, C., et al. (1997) High-resolution comparative hybridization to combed DNA fibers. *Hum. Genet.* **99**, 374–380.
12. Solinas-Toldo, S., Lampel, S., Stilgenbauer, S., Nickolenko, J., Benner, A., Dohner, H., et al. (1997) Matrix-based comparative genomic hybridization: biochips to screen for genomic imbalances. *Genes Chromosomes Cancer* **20**, 399–407.
13. Pinkel, D., Segraves, R., Sudar, D., Clark, S., Poole, I., Kowbel, D., et al. (1998) High resolution analysis of DNA copy number variation using comparative genomic hybridization to microarrays. *Nat. Genet.* **20**, 207–211.
14. Pollack, J. R., Perou, C. M., Alizadeh, A. A., Eisen, M. B., Pergamenschikov, A., Williams, C. F., et al. (1999) Genome-wide analysis of DNA copy-number changes using cDNA microarrays. *Nat. Genet.* **23**, 41–46.
15. Roth, K., Wolf, G., Dietel, M., and Petersen, I. (1997) Image analysis for Comparative Genomic Hybridization based on a karyotyping program for Windows. *Anal. Quant. Cytol. Histol.* **19**, 461–474.
16. Petersen, I., Bujard, M., Petersen, S., Wolf, G., Goeze, A., Schwendel, A., et al. (1997) Patterns of chromosomal imbalances in adenocarcinoma and squamous cell carcinoma of the lung. *Cancer Res.* **57**, 2331–2335.
17. Petersen, I., Langreck, H., Wolf, G., Schwendel, A., Psille, R., Vogt, P., et al. (1997) Small cell lung cancer is characterized by a high incidence of deletions on chromosomes 3p, 4q, 5q, 10q, 13q and 17p. *Br. J. Cancer* **75**, 79–86.
18. Schwendel, A., Langreck, H., Reichel, M. B., Schröck, E., Ried, T., Dietel, M., and Petersen, I. (1997) Comparative Genomic Hybridization reveals a common pattern of genetic changes in primary small cell lung carcinomas and their metastases. *Int. J. Cancer* **74**, 86–93.
19. Ullmann, R., Schwendel, A., Klemen, H., Wolf, G., Petersen, I., and Popper, H. H. (1998) Unbalanced chromosomal aberrations in neuroendocrine lung tumors as detected by comparative genomic hybridization. *Hum. Pathol.* **29**, 1145–1159.
20. Petersen, S., Aninat-Meyer, M., Schlüns, K., Gellert, K., Dietel, M., and Petersen, I. (2000) Chromosomal alterations in the clonal evolution to the metastatic stage of squamous cell carcinomas of the lung. *Br. J. Cancer* **82**, 65–73.
21. Petersen, I., Hidalgo, H., Petersen, S., Schlüns, K., Schewe, C., Pacyna-Gengelbach, M., et al. (2000) Chromosomal imbalances in brain metastases of solid tumors. *Brain Pathology* **10**, 395–401.
22. Goeze, A., Schlüns, K., Wolf, G., Thäsler, Z., Petersen, S., and Petersen, I. (2002) Chromosomal imbalances of primary and metastatic lung adenocarcinomas. *J. Pathol.* **196**, 8–16.
23. Ried, T., Petersen, I., Holtgreve-Grez, H., Speicher, M. R., Schröck, E., du Manoir, S., and Cremer, T. (1994) Mapping of multiple DNA gains and losses in primary small cell lung carcinomas by comparative genomic hybridization. *Cancer Res.* **54**, 1801–1806.
24. Moore, D. H., 2nd, Pallavicini, M., Cher, M. L., and Gray, J. W. (1997) A t-statistic for objective interpretation of comparative genomic hybridization (CGH) profiles. *Cytometry* **28**, 183–190.

25. Tirkkonen, M., Karhu, R., Kallioniemi, O., and Isola, J. (1996) Evaluation of camera requirements for comparative genomic hybridization. *Cytometry* **25,** 394–398.
26. du Manoir, S., Kallioniemi, O.-P., Lichter, P., Piper, J., Benedetti, P. A., Carothers, A. D., et al. (1995) Hardware and software requirements for quantitative analysis of comparative genomic hybridization. *Cytometry* **19,** 4–9.
27. Lundsteen, C., Maahr, J., Christensen, B., Bryndorf, T., Bentz, M., Lichter, P., and Gerdes, T. (1995) Image analysis in comparative genomic hybridization. *Cytometry* **19,** 42–50.
28. Piper, J., Rutowitz, D., Sudar, D., Kallioniemi, A., Kallioniemi, O. P., Waldman, F. M., et al. (1995) Computer image analysis of comparative genomic hybridization. *Cytometry* **19,** 10–26.
29. Knösel, T., Petersen, S., Schwabe, H., Schlüns, K., Stein, U., Schlag, P. M., et al. (2002) Incidence of chromosomal imbalances in advanced colorectal carcinomas and their metastases. *Virchows Arch.* **440,** 187–194.
30. Alers, J. C., Rochat, J., Krijtenburg, P. J., van Dekken, H., Raap, A. K., and Rosenberg, C. (1999) Universal linkage system: an improved method for labeling archival DNA for comparative genomic hybridization. *Genes Chromosomes Cancer* **25,** 301–305.
31. Karhu, R., Rummukainen, J., Lorch, T., and Isola, J. (1999) Four-color CGH: a new method for quality control of comparative genomic hybridization. *Genes Chromosomes Cancer* **24,** 112–118.
32. Kirchhoff, M., Gerdes, T., Rose, H., Maahr, J., Ottesen, A. M., and Lundsteen, C. (1998) Detection of chromosomal gains and losses in comparative genomic hybridization analysis based on standard reference intervals. *Cytometry* **31,** 163–173.

13

Sensitive Assays for Detection of Lung Cancer

Molecular Markers in Blood Samples

Marcia V. Fournier, Katherine J. Martin, Edgard Graner,
and Arthur B. Pardee

1. Introduction

The value of early detection is clear from current tests for specific cancers: self-examination and mammography for breast, prostate-specific antigen (PSA) for prostate, and Pap smears for cervical cancers. No such method exists for early detection of lung cancer. New methods are being designed to test for the very few tumor cells released early in cancer progression into blood (or other body fluids). Single markers for specific cancers have been reported in blood samples or other accessible sites, but these vary between patients. We propose that sets of markers provide a reproducible general pattern, more certain for cancer detection. We have found a dozen markers present in 3 mL blood samples from cancer patients that are absent in blood samples from normal individuals *(1,2)*.

Genes that are expressed early in cancer progression are the most interesting. The first question as soon as a diagnosis of cancer is made is whether the disease is localized to the primary site or has already spread to the regional lymph nodes and distant organs to form metastases. Most deaths from cancer result from metastases that are resistant to conventional therapies. The failure to reduce mortality of lung cancer patients is probably a result of the early dissemination of cancer cells to secondary sites, which are usually missed by conventional diagnostic procedures used for tumor staging. The study of solid tumor cells disseminated in blood samples from lung cancer patients provide an in vivo picture of gene expression, which potentially identifies genes associated with early stages of metastasis. Analysis of gene expression on disseminated tumor cells could enhance the sensitivity of picking up metastasis related genes. We previously demonstrated the presence of a putative molecular marker for metastatic lung cancer in two patients with localized disease *(1)*. Both cases afterwards turned out to be metastatic.

The first problem is to discover the most informative sets of mRNAs that are markers of cancer in blood. We have developed a two-step approach to identifying tumor markers in biopsy tissue or peripheral blood samples *(3)*. Our approach uses differential display (DD) as a first step to compare normal vs tumor cells and identify differentially expressed genes. As a polymerase chain reaction (PCR)-based technique, DD is highly sensitive and effective in identifying candidate cancer markers in blood samples *(4)*. PCR is highly sensitive and has been extensively applied for the detection of tumor cells dispersed in the circulation *(5–9)* or their presence in regional lymph nodes *(10)*.

The development of assays for detecting disseminated solid tumor cells in peripheral blood requires a highly sensitive high-throughput screening technique that can identify panels of informative markers. As a second step, we developed a membrane-based high-density hybridization array method, and applied cluster analysis to identify groups of genes whose expression patterns correlate with cancer. Our multifaceted blood-based expression assays have the potential to not only detect cancer cells in the blood *(2)*, but to also provide prognostic information regarding those cells. It discriminated between estrogen receptor-positive and -negative breast cancers, and therefore may predict efficacy of hormone-related therapy *(3)*.

Real-time PCR is a recent improvement of the conventional reverse transcriptase (RT)-PCR technique, which allows the detection of fluorescent PCR products while the amplification proceed. Thus, interference of limiting conditions at the end of the PCR reaction and further processing of PCR products (e.g., DNA agarose or polyacrylamide gels, staining procedures, and quantitation by densitometry) are eliminated. Real-time PCR has been successfully used to monitor minimal residual disease using blood samples from leukemia patients *(11)*.

Blood-based multifaceted expression assays have the potential to determine the site of a primary tumor, as for example in the lung. Our preliminary studies showed that some expression markers of breast cancer cells specifically recognized determinants present in the blood of only breast cancer patients, whereas others recognized determinants present in the blood of both breast cancer and lung cancer patients *(4)*.

Future applications of markers in a standard clinical diagnostic test could include: 1) earlier cancer detection and metastasis, 2) tests for screening therapeutic responses, 3) clinical outcome, 4) remission and recurrence, and 5) new target therapeutics.

2. Materials

2.1. Differential Expression in Blood Samples: Identification of Putative Markers for Early detection of Lung Cancer by DD

1. Sterile distilled H_2O (dH_2O).
2. Ficoll Paque Plus (Amersham Pharmacia Biotech, Piscataway NJ).
3. Red cell lysis solution (Ambion, Austin, TX).
4. Trizol (Gibco-BRL, Life Technologies, Rockville, MD).
5. Sensiscript RT (QIAGEN, Valencia, CA).
6. DNAse I.
7. RNA Image® kit (GenHunter Corporation, Nashville TN).
8. DD primers: three anchor primers (T11N) and eight 13–20 bp arbitrary primers.
9. Glycogen 10 mg/mL stock solution in dH_2O.
10. TE buffer: 10 mM Tris-HCl, pH 8.0, 1 mM EDTA.
11. AmpliTaq DNA polymerase (Perkin-Elmer Applied Biosystems [PE], Branchburg NJ).
12. Taq PCR buffer 10X solution: 100 mM Tris-HCl, pH 8.4, 500 mM KCl, 15 mM $MgCl_2$, and 0.01% gelatin.
13. DeoxyNTPs: stock solutions of 2.5 mM and 0.25 mM in dH_2O.
14. [α-^{33}P] dCTP, 3000 Ci/mol (DuPont NEN, Boston MA).
15. Mineral oil.
16. Isopropanol.
17. Ethanol.
18. HR-1000 6% denaturing high resolution DD gel (Genomyx, Foster City, CA).
19. 70% ethanol.
20. Genomyx LR DNA sequencer (Genomyx).
21. 3 M NaOAc.
22. Synthetic 20–30-mer oligonucleotides complementary to appropriate regions of genes of interest.
23. Access to databases.

2.2. A Two-Step Method for Identifying Tumor Markers in Biopsy or Blood Samples

2.2.1. Custom cDNA Arrays

1. Agarose.
2. PCR purification column.
3. 96-well PCR dish.
4. 10 M NaOH stock solution.
5. 0.5 M EDTA.
6. Positively charged nylon membranes (Micron Separations Inc., Westboro, MA).

7. Multiprint 96-pin replicator with 16 offset positions (V&P Scientific, Inc., San Diego, CA).
8. Phosphorimaging screen.
9. Ethanol.
10. ExpressHyb Hybridization Solution (Clontech, Palo Alto, CA).
11. Formamide.
12. 0.5 µg/µL oligo (dT)$_{12-18}$ (Gibco BRL).
13. [α-^{32}P] dCTP (DuPont NEN).
14. SuperScript II (Gibco BRL).
15. G-50 columns (Boehringer Mannheim, Indianapolis IN).
16. 1 M Tris, pH 8.0.
17. 2 N HCl.
18. Wash solution: 5 mL 20X SSC + 10 mL 20% SDS in 2 L total volume.
19. Software ImageQuant (Molecular Dynamics, Sunnyvale, CA).

2.2.2. Real Time PCR

1. DNAse I Amplification Grade, 10X DNAse I reaction buffer, 25 mM EDTA (Gibco BRL).
2. Sterile distilled H$_2$O (dH$_2$O).
3. MuLV Reverse Transcriptase (PE) or Superscript II Reverse Transcriptase (Gibco BRL).
4. RNase Inhibitor (PE).
5. 10X PCR buffer containing 15 mM MgCl$_2$ (PE).
6. Oligo dT(12-18) primers (Gibco BRL) or random hexamers (PE).
7. DeoxyNTPs: working solution of 10 mM in dH$_2$O.
8. SYBR® Green PCR Core Reagents (PE).
9. Autoclaved 1.5- and 2-mL test tubes.
10. 15-mL polypropylene cell-culture tubes.
11. 96-Well Optical Reaction Plate (PE).
12. MicroAmp Caps (8 Caps/Strip) (PE).
13. MicroAmp Cap-installing Tool (PE).
14. MicroAmp Base (PE).
15. ABI Prism 7700 Sequence Detector (PE).
16. Agarose (Gibco BRL).
17. Primers for both target and reference genes.

3. Methods

3.1. Differential Expression in Blood Samples: Identification of Putative Markers for Early Detection of Lung Cancer by Differential Display

3.1.1. Blood Collection and RNA Isolation

1. 3–10 mL of whole blood is drawn into a EDTA-collection tube. White cells are isolated within 12 h using a red cell lysis procedure (Ambion) (*see* **Note 1**) and

cell pellets can then be stored at –80°C. 5 mL of whole blood yields approx 3×10^7 white cells and 10 μg total cellular RNA (*see* **Note 2**).
2. Total cellular RNA is purified using the Trizol reagent. Add 1 mL Trizol to white cell pellet obtained from 3–10 mL of whole blood and proceed as recommended by the manufacturer. To the final pellet add 100 μL DEPC-treated dH$_2$O and dissolve RNA by incubating 10 min at 55–60°C. Extract once with phenol/chloroform (1:1) and precipitate with 2 volumes ethanol, wash, and dissolve in 10–50 μL DEPC-dH$_2$O.
3. Determine RNA concentration by measuring absorbance at 260 nm. Use agarose gel electrophoresis to determine the RNA quality and verify its concentration.

3.1.2. Comparing Blood Samples by DD

1. Limit your sample number for comparison by DD. A greater number of samples may make analysis difficult when looking for potential tumor markers in blood samples. It is suggested to compare 3–5 controls against 3–5 test samples. After choosing potential markers, extend analysis on them to as many samples as necessary.
2. Perform the RT reaction using Sensiscript RT (QIAGEN). Perform DD using the RNA Image® kit (GenHunter Corporation). All PCR should be done in duplicate. Electrophoreses duplicate PCR products in parallel on extended-format denaturing 6% polyacrylamide gels (Genomyx Corporation). Excise bands of interest from the gel (*see* **Note 3**).
3. For purification of the cDNA fragments from polyacrylamide, add 100 μL of TE buffer pH 8.0 to the fragments. Incubate 10 min at room temperature. Boil 20 min. Spin 2 min at 14000 rpm. Remove supernatant to a fresh tube, add 10 μL 3 *M* NaOAc, 3 μL 10 mg/mL glycogen, and precipitate in two volumes of ethanol.
4. Perform PCR reactions of isolated cDNA fragments. Make a 20 μL final volume PCR reaction containing 2.5 U/μL AmpliTaq DNA polymerase (PE), 100 m*M* Tris-HCl, pH 8.3, 500 m*M* KCl, 1.5 m*M* MgCl$_2$ and 250 μ*M* each of dNTP and 50 n*M* of each primer (same pair as used for DD). The PCR reaction may be programmed as follows: 95°C 1 min, 50°C 1 min, 72°C 1 min for 30 cycles; elongation at 72°C for 5 min, and refrigeration at 4°C. Purify PCR products from agarose gel (*see* **Note 4**).
5. Determine the nucleotide sequences of cDNA fragments (*see* **Note 5**). Automated sequencing is indicated when available. Sequences may now be queried against National Center for Biotechnology Information (NCBI) databases using the Basic Local Alignment Search Tool (BLAST). Confirm DD results following database verification. Design a single gene-specific 20–30-mer primer and use in combination with the appropriate DD anchor primer to PCR.
6. Perform a quantitative method for validation of potential markers. As high sensitivity is required for detection of tumor markers in blood samples, real time PCR methodology is strongly indicated (*see* **Subheading 3.2.2.**).

3.2. A Two-Step Method for Identifying Tumor Markers in Biopsy or Blood Samples

3.2.1. Custom cDNA Arrays

This procedure is composed of four parts: preparation of replicate membranes, preparation of ^{32}P-labeled first-strand cDNA, membrane hybridization, and data analysis. To produce replicate membrane arrays with tags for up to 1536 different genes, PCR products, or whole plasmids are spotted using a hand-held 96-pin spotting device. Twenty replicate membranes are conveniently made at once.

1. cDNA arrays are used to test candidate marker genes for their expression levels in the samples such as peripheral blood samples. Arrays are prepared by manually spotting PCR-amplified from cDNAs onto positively charged nylon membranes (Micron Separations Inc.) using a hand-held replicator (V-P Scientific). Thorough pin cleaning before and during arraying is critical for assay reproducibility. Dip pins in diluted detergent, apply brush, rinse well in dH$_2$O, dip in ethanol, and flame to dry.
2. Radiolabeled first-strand cDNA probes are prepared by reverse transcription of 5 µg total cellular RNA in the presence of 50 µCi [α^{32}P] dCTP. Membranes are pre-hybridized 3 h in a formamide-based buffer at 40°C. Probe is then added to buffer and membranes hybridized 18 h at 41°C. Wash membranes by first rinsing them briefly, one at a time, in 500 mL wash solution (5 mL 20X SSC + 10 mL 20% SDS in 2 L total volume). Perform three subsequent 500 mL washes, 10 min per wash, at 50°C with the remaining 1500 mL of wash solution. Agitate while washing and wash at most 3 membranes per 500 mL of wash solution. Membranes are exposed to phosphor-imaging screens and analyzed using an phosphorimager and image analysis software (Molecular Dynamics). Membranes can be stripped and reused three times. Extended membrane use is achieved with formamide-based buffer and low hybridization temperature.
3. Control experiments should be performed to test array reproducibility. Repeated analyses of a single preparation of RNA are performed on different days using different membranes. Typically 95% of data points fall within 2.5-fold limits.
4. For data analysis, integrated signal intensities background is first subtracted and if replicate spots are used these are averaged. Data from genes with high deviations within sets of spots and with consistently low signals are discarded. Each array (experiment) is then normalized to correct for differences in labeling or hybridization efficiency. If the number of genes tested is large, normalization can be performed relative to the median gene expression level. Otherwise normalization should be performed relative to several control genes (e.g., 10 or more "house-keeping" genes). Relative ratios are then calculated by comparison to a control experiment or to the mean of all experiments.

3.2.2. Real-Time PCR

Real-time PCR is a useful tool for the confirmation of array results. A complementary procedure is highly important in establishing the validity of results. We have also applied real-time PCR to determine the sensitivity limits of our arrays. Sensitivity of 1 in a million lymphatic cells is generally considered necessary for the detection of disseminated tumor cells in peripheral blood. Though maximum reported levels of tumors cells in the blood of breast cancer patients exceed 3,000 cells/mL *(12,13)*, the average levels of cells reported for different stages of breast cancer range from 0.8–6 cells/mL *(13)*.

3.2.2.1. REAL-TIME PCR CONFIRMATION OF ARRAY RESULTS

1. The primers for real-time quantitative RT-PCR must amplify a DNA fragment with 100–150 bp, and can be designed using the Primer Express 1.0 (PE Applied Biosystems) or Amplify 1.2 (University of Wisconsin, WI) software. The GC content must be in the 20–80% range; runs of identical nucleotides (mainly Gs) must be avoided. When using the Primer Express 1.0 the Tm should be between 58–60°C, and the five bases at the 3' end should contain no more than 2 Gs or Cs PE suggests (all these parameters). The reference gene must be carefully chosen for each experimental model. It may be necessary to try different genes to find a control whose expression levels are not affected in the particular experimental conditions. If possible use more than one reference (in separate experiments) to quantitate the target gene, unless a reliable control for the same experimental conditions has been already described in the literature.
2. DNAse I treatment is important to eliminate interference by the genomic DNA contamination frequently found in the RNA preparations. Take 1 µg (0.8 µg–2 µg) of each RNA sample (prepare 2 reactions for the RNA sample to be used in the standard curves), bring the volume to 8 µL with DEPC H_2O, add 1 µL DNAse I and 1 µL 10X DNAse I buffer, incubate 15 min at room temperature. Add 1 µL 25 m*M* EDTA solution and heat tubes 10 min at 65°C.
3. Prepare a 10 µL RT mix (one extra reaction to account for pipeting errors): 3 µL DEPC-dH_2O, 2 µL 10X PCR buffer (PE), 2 µL 10 m*M* dNTP solution, 1 µL MuLV RT enzyme (PE), 1 µL RNase inhibitor (PE), and 1 µL Random Hexamers (PE). Separately, prepare a no-RT mix as the negative control for the standard curve sample. Perform the RT reaction (after addition of the 10 µL DNAse I reaction to the 10 µL RT mixture and mixing by pipeting) at 42°C for 45 min, followed by 99°C for 5 min. Combine the 2 RT reactions results for the standard curves. The cDNAs can be stored at –20°C for later use. Alternatively, this step can be performed using Oligo dT (12-18) primers and the Superscript II enzyme.
4. Quantitative SYBR® Green Real-time PCR. The standard curves are generally made with 5 dilutions (e.g., 1:2.5; 1:10; 1:40; 1:160; 1:640) of cDNAs from the same sample used as the experimental control (all results will be expressed

as fold-induction relative to this particular sample). Since each experimental condition should always be performed in triplicate (including the standard curve), prepare enough standard curve cDNA dilutions for both target gene and reference gene standards. Extra volume must be prepared to account for pipeting errors. Separate PCR 1X buffers are made for the target and reference genes in 2 mL test tubes (or 15-mL polypropylene tubes, according to the number of samples): 26.25 µL dH$_2$O; 5 µL SYBR® Green 10X PCR buffer, 6 µL MgCl$_2$; 4 µL dNTP solution (from SYBR® Green kit); 0.25 µL Ampli-Taq Gold (from SYBR® Green kit); 0.5 µL Amp-Erase enzyme (from SYBR® Green kit); 3 µL forward primer (5 µ*M*); 3 µL reverse primer (5 µ*M*). Aliquot 144 µL of each PCR mix in test tubes (one tube for each standard curve point, sample, and no RT controls), add 6 µL of the RT reaction (or RT reaction dilutions), mix. Pipet 49 µL in each tube of a reaction plate (avoid air bubbles), place the optical caps with the MicroAmp Cap-installing Tool and insert the plate in the ABI 7700 Sequence Detector, according to the manufacturer's directions. Set up the Sequence Detector to read SYBR® Green fluorescence *(14)* and start the PCR reaction with the following cycling parameters: 50°C 2 min; 95°C 10 min; and 45 cycles of 95°C 15 s; 60°C 1 min. When the reaction is completed, save and export the results to a computer disc. Prepare one of each triplicate sample to be resolved in a 1.8% DNA agarose gel, to confirm the PCR specificity (since the SYBR® Green dye can bind to any double-strand DNA in the solution, unspecific PCR products can interfere in the final measurements). We also strongly recommend a regular PCR using the same cycling parameters prior to the quantitative experiments to verify the specificity of the primers.

5. Results calculation. The Threshold Cycle (Ct) is the first cycle of the PCR reaction in which fluorescence is detected (*see* **Fig. 1**). Values obtained for each cDNA dilution of standard are used to construct standard curves. For each experimental sample, the amount of target or reference gene is obtained from the standard curves, and the target values are divided by the respective reference ones *(15)*.

3.2.2.2. REAL-TIME PCR DETERMINATION OF ARRAY SENSITIVITY

1. To determine the sensitivity limits of arrays, a test gene expressed at the limit of detection is first identified. In our case, low but measurable levels of the gene maspin were detected by arrays in 33% of healthy volunteers, while others had undetectable level *(2)*. The expression level of this test gene is then measured by real-time PCR in a sample in which the arrays detected expression.
2. The absolute level of transcripts of the test gene is determined by comparison with a standard curve generated from dilutions of quantitated maspin plasmid. It is useful to establish that identical standard curves are generated when the test gene plasmid is measured alone or added to normal blood cDNA. Measurements of test gene transcripts in blood are normalized to the amount of cDNA in reverse transcription reactions, which can be quantitated by two methods: OD$_{260}$ and

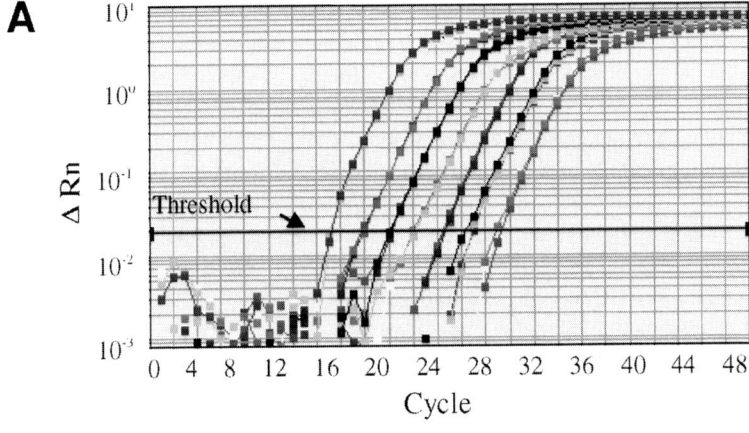

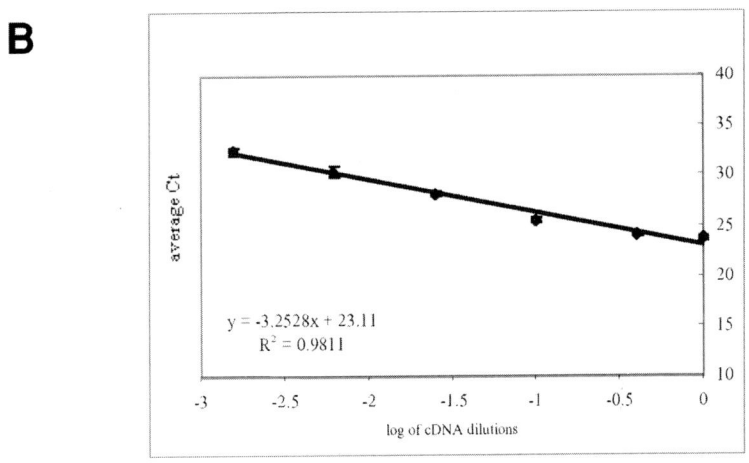

Fig. 1. Example of amplification plot using the ABI Prism 7700 sequence detector. (**A**) The graphic shows amplifications using the same set of primers and serially diluted cDNA templates, in triplicates. The number of cycles is shown in the X-axis and the normalized fluorescence intensity in the Y-axis. The Ct value (threshold cycle) is the cycle in which the fluorescent signal is first detected above the background levels (arrow). (**B**) The Ct values are used to make a standard curve and the resulting line equation for the calculation of gene expression levels of each studied sample (relative to the expression levels of the cDNA sample used in the standard curves). Each experiment has one standard curve for target gene and another for the reference gene, and the ratio between the values obtained from each curve is the normalized result.

fluorescence of the single-strand DNA dye OliGreen. Results are adjusted to account for the use of double-stranded plasmid in standard curves and for the size of the test gene cDNA, which may be longer or shorter than the average cellular transcript (2 Kb). The test gene is quantitated in cDNA preparations, and its level relative to total cDNA is assumed to be equivalent in cDNA and RNA.

3. The amount (pg) of the test gene and total cDNA in a 2 µL reverse transcription reaction is calculated, which then allows calculation of the relative level of test gene, e.g. test gene messages/total cellular messages. Since a typical mammalian cell has 3.6×10^5 RNAs in its cytoplasm, one can then calculate the copies of the test gene expressed per cell.
4. The limit of detection for our arrays was approx 1 in 2×10^8 transcripts. If one assumes that tumor cells and white blood cells express similar amounts of total RNA and that maspin is abundant in tumor cells (e.g., 1000 copies/cell), it can be calculated that the arrays can detect as few as 1 in 10^6 cells, e.g., 5 tumor cells per Ml of blood. This is two orders of magnitude less sensitive than PCR and three orders of magnitude better than oligonucleotide microarrays. The enhanced sensitivity of the arrays can be accounted for by the use of ^{32}P rather than fluorescence labeling, membranes rather than glass, and long cDNA tags rather than oligonucleotides.

4. Notes

1. White blood cells may instead be washed three times with 3–5 mL of buffer containing 10 m*M* Tris-HCl, pH 7.6, 5 m*M* MgCl$_2$, 10 m*M* NaCl. Centrifuge at 1,800*g* for 1 min between washes.
2. It is essential to treat RNA samples with DNAse I. Quantitate the RNA spectrophotometrically by making OD$_{260}$ and OD$_{280}$ readings of 1:250 dilutions (2 µL in 500 µL DEPC-ddH$_2$O). Ratio OD$_{260}$/OD$_{280}$ should be >1.6. OD$_{260}$ × 40 = µg/µL.
3. Re-expose gels after fragments are cut to confirm that the right fragment was taken.
4. Eventually, the PCR product from one reaction may not achieve the minimal concentration required for sequence analysis. In this case, it is suggested to perform multiple PCR reactions. Pool the PCR products, gel purify, and suspend in 10–30 µL sterile water. A second round of PCR reaction is not recommended, but may be performed when necessary. The PCR products of cDNA fragments should be mostly single bands.
5. Both anchor and arbitrary primers are useful for sequencing.

Acknowledgments

The authors thank Brian Kritzman and Laura Price for technical assistance, Maria G.C. Carvalho and Marcos E.M. Paschoal for helpful suggestions and fruitful discussions. The hybridization array protocol was adapted in part from methods provided by Jackson Wan. This work was supported by grant

RO-1-CA61253 from the National Institutes of Health and by the Ludwig Institute for Cancer Research. E.G. is supported by FAPESP: 99/08279-5.

References

1. Fournier, M. V., Carvalho, M. G. C., and Pardee, A. B. (1999) A strategy to identify genes associated with circulating solid tumor cell survival in peripheral blood. *Mol. Med.* **5,** 313–319.
2. Martin, K. J., Graner, E., Li, Y., Price, L. M., Kritzman, B. M., Fournier, M. V., et al. (2001) High-sensitivity array analysis of gene expression for the early detection of disseminated breast tumor cells in peripheral blood. *PNAS* **98,** 2646–2651.
3. Martin, K. J., Kritzman, B. M., Price, L. M., Koh, B., Kwan, C.-P., Zhang, X., et al. (2000) Linking gene expression patterns to therapeutic groups in breast cancer. *Cancer Res.* **60,** 2232–2238.
4. Fournier, M. V., Guimaraes, F. C., Paschoal, M. E., Ronco, L. V., Carvalho, M. G. C., and Pardee, A. B. (1999) Identification of a gene encoding a human oxysterol-binding protein-homologue: a potential general molecular marker for blood dissemination of solid tumors. *Cancer Res.* **59,** 3748–3753.
5. McKiernan, J. M., Buttyan, R., Bander, N. H., la Taille, A., Stifelman, M. D., Emanuel, E. R., et al. (1999) The detection of renal carcinoma cells in the peripheral blood with an enhanced reverse transcriptase-polymerase chain reaction assay for MN/CA9. *Cancer* **86,** 492–497.
6. Mellado, B., Gutierrez, L., Castel, T., Colomer, D., Fontanillas, M., Castro, J., and Estape, J. (1999) Prognostic significance of the detection of circulating malignant cells by reverse transcriptase-polymerase chain reaction in long-term clinically disease-free melanoma patients. *Clin. Cancer Res.* **5,** 1843–1848.
7. Hoon, D. S. B., Bostick, P., Kuo, C., Okamoto, T., Wang, H. J., Elashoff, R., and Morton, D. (2000) Molecular markers in the blood as surrogate prognostic indicators of melanoma recurrence. *Cancer Res.* **60,** 2253–2257.
8. Wong, I. H. N., Chan, A. T., and Johnson, P. J. (2000) Quantitative analysis of circulating tumor cells in peripheral blood of osteosarcoma patients using osteoblast-specific messenger RNA markers: a pilot study. *Clin. Cancer Res.* **6,** 2183–2188.
9. Sabbatini, R., Federico, M., Morselli, M., Depenni, R., Cagossi, K., Luppi, M., et al. (2000) Detection of circulating tumor cells by reverse transcriptase polymerase chain reaction of maspin in patients with breast cancer undergoing conventional-dose chemotherapy. *J. Clin. Oncol.* **18,** 1914–1920.
10. Kano, M., Shimada, Y., Kaganoi, J., Sakurai, T., Li, Z., Sato, F., et al. (2000) Detection of lymph node metastasis of oesophageal cancer by RT-nested PCR for SCC antigen gene mRNA. *Br. J. Cancer* **82,** 429–435.
11. Verhagen, O. J., Willemse, M. J., Breunis, W. B., Wijkhuijs, A. J., Jacobs, D. C., Joosten, S. A., et al. (2000). Application of germ line IGH probes in real-time quantitative PCR for the detection of minimal residual disease in acute lymphoblastic leukemia. *Leukemia* **14,** 1426–1435.

12. Kraeft, S. K., Sutherland, R., Gravelin, L., Hu, G. H., Ferland, L. H., Richardson, P., et al. (2000) Detection and analysis of cancer cells in blood and bone marrow using a rare event imaging system. *Clin. Cancer Res.* **6,** 434–442.
13. Racila, E., Euhus, D., Weiss, A. J., Rao, C., McConnell, J., Tertappen, L. W., and Uhr, J. W. (1998) Detection and characterization of carcinoma cells in the blood. *PNAS* **95,** 4589–4594.
14. User Bulletin #4, ABI Prism 7700 Sequence Detector System, PE Applied Biosystems.
15. User Bulletin #2, ABI Prism 7700 Sequence Detector System, PE Applied Biosystems.

14

Fluorescent Microsatellite Analysis in Bronchial Lavage as a Potential Diagnostic Tool for Lung Cancer

John K. Field and Triantafillos Liloglou

1. Introduction

Cancer is a multistep progressive disease of increasing genomic instability. Genomic instability is a condition where the cell looses the ability to retain the semi-conservative means of its genome replication because of vital controlling mechanisms dysfunction. Thus, replication errors as well as large chromosomal lesions occur at high rates, giving rise to genetically diverse subpopulations, some of which have an increased growth advantage. These subpopulations evolve in the tissue microenvironment through natural selection processes that will enfavor cells carrying the most "advantageous" genetic lesions. Genomic instability is a phenomenon of all cancer cells and can be detected in two forms *(1,2)*. Allelic imbalance (AI) or loss of heterozygosity (LOH) represents chromosomal instability (CIN) and involves a series of genetic phenomena like loss of chromosomal regions, duplication, DNA amplification, and aneuploidy. Solid tumor genomes exhibit gains and losses spread throughout chromosomes *(3)*. Microsatellite instability (MIN, MI, or MSI), also found in the literature as replication errors (RER) or microsatellite alterations (MA), is most often attributed to DNA repair machinery errors *(2)*.

Genomic instability is a common phenomenon in lung cancer *(5–7)* and, in some reports, has been associated with prognosis *(8–10)*. In an extensive allelotype analysis, 42/45 (93%) of NSCLC specimens were found to carry LOH or MA in at least one of the 92 markers examined *(6)*. Using fluorescent labelled primers and automated analysis on a sequencer, a panel of 12 only markers was identified to detect genomic instability in 97% of lung cancer samples while it is of note that a subset of 8 of these microsatellite markers identified LOH in 95% of lung tumors *(11)*. Genomic instability has also been detected in preneoplastic lung lesions *(13–16)*, which in some cases presented minimal atypia. Furthermore, LOH and MA have also been demonstrated in bronchial tissue specimens from chronic smokers who do not have lung cancer *(17,18)*. These findings suggest that these genetic alterations precede morphological transformation of the cells.

It is of note that genomic instability has been detected in sputum *(19,20)*, bronchial lavage *(21–23)*, and plasma/serum *(24,25)* specimens from lung cancer patients. Genomic instability may, therefore, be an important genetic marker for the identification of genetically abnormal cells by assaying biopsy material, bronchial lavage, and sputum and consequently determining an individual's risk for developing lung cancer *(26–27)*. It is of note, however, that genomic instability was detected in the bronchial lavage of individuals with nonmalignant lung disease *(28–34)*. A recent study using fluorescent microsatellite analysis has indicated that genomic instability is not an exclusive phenomenon of cancer but is also present in nonmalignant inflammatory cells *(35)*. The same study presented a number of loci demonstrating cancer-specific genomic instability (CSGI), which should be pursued towards a molecular assay for lung cancer detection.

The fact that practically all tumours display genomic instability makes it a favorable marker for the identification of neoplastic cells in clinical specimens. The PCR-based microsatellite analysis of tumor vs normal counterpart tissue has become the most widely used method to determine allelic imbalance. Microsatellites are tandem sequence repeats (2–6 nucleotides units) scattered widely in human genome. Their role is still unclear and they may have a role in homologous recombination and chromosome segregation. Microsatellite loci are highly polymorphic as the number of repeats varies between individuals. Thus many microsatellites have a high degree of heterozygosity. The relative simplicity of the method and the abundance of microsatellite sequences covering the human genome make this an attractive approach. Among the technical problems of this approach is polymerase slippage, resulting in the production of stutter bands. Moreover, Taq polymerase catalyses the addition of nontemplate A to the 3' end of the amplification products. Partial such addition results in split peaks (bands with one nucleotide difference). However, the major limitation is the accuracy of quantitation. Scoring of allelic imbalance is based on the ratio A_2/A_1 of the amplification level of the two alleles of the tumor, normalized by the same ratio of the corresponding normal DNA sample (*see* **Fig. 1**). The quantitation accuracy of image analysis and densitometry has a number of limitations, especially for gel bands of very high or low density. The use of fluorescent end-labeled markers and analysis through an automatic

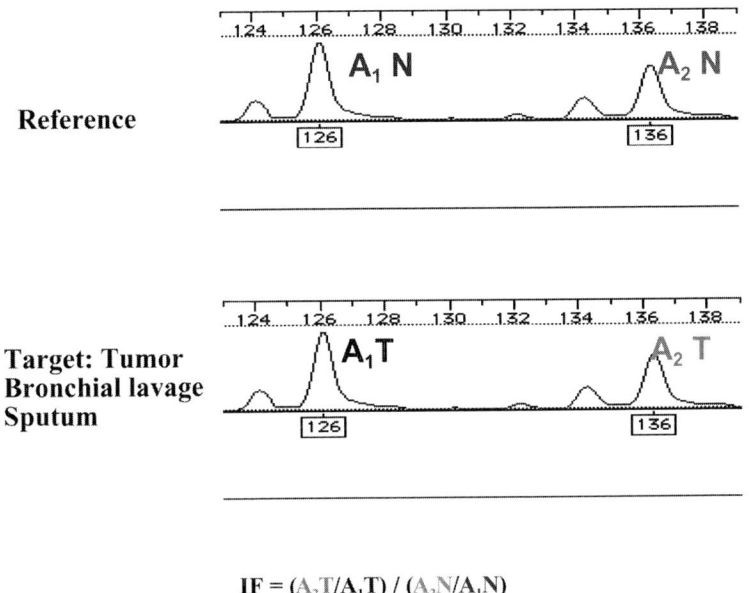

Fig. 1. Schematic representation of the calculation of imbalance factor (IF).

sequencer provides much better resolution and more accurate sizing of the amplified fragments, as it can involve internal size standards run in the same lane along with the PCR products. Moreover, accurate quantitation is mediated by sensor devices that detect the energy transferred by the fluorophore. The technological advantages of fluorescence PCR based assays provide the ability to detect DNA changes from minute amounts of starting material in multiplex reactions *(36)*. Furthermore, automated analysis on sequencers/genetic analysers not only increases throughput but also reduces operator errors during the analysis (*see* **Fig. 2**).

Unless precisely microdissected, normal contaminating cells exist in all clinical specimens. This creates a "normal" background that alters the overall allele amplification ratio of a given tumor sample. In addition, tumors are fast-replicating cell populations acquiring sequential aberrations during their multistage development. They are, therefore, composed of genetically divergent subpopulations. Consequently, allelic imbalance at a specific locus may be a feature of only a fraction of this population. It thus becomes imperative that the detection threshold of the assay has to be low in order to demonstrate such changes in a small proportion of the sample. By assessing the interassay variability, the lowest possible threshold can be set avoiding simultaneously false-positives.

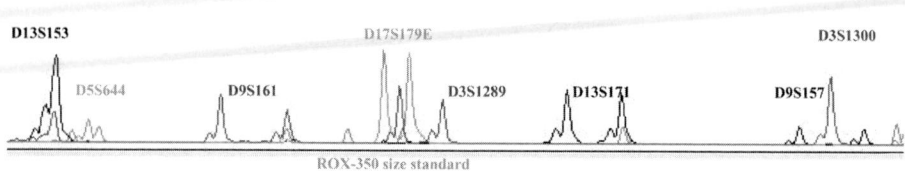

Fig. 2. Genescan™ output of a 9-plex reaction run on an ABI PRISM 377 sequencer.

The experimental platform presented in this chapter has been optimized and efficiently tested in lung tumors and bronchial lavage samples *(11,21,35) (see* **Fig. 3**). The threshold of LOH detection has been calculated by assessing the interassay variability by repeating assays of normal samples. It combines a medium to high-throughput analysis and high sensitivity and specificity of LOH detection.

2. Materials
2.1. DNA Extraction from Blood, Bronchial Lavage, and Sputum

1. Red blood cell lysis solution: 10 mM Tris-HCl, pH 8.0, 320 mM sucrose, 1% Triton, 5 mM MgCl$_2$. Adjust pH to 8.0 with 10 M NaOH. Autoclave and store at 4°C for 2–3 mo.
2. Lysis Solution: 400 mM Tris-HCl, pH 8.0, 150 mM NaCl, 60 mM EDTA. Adjust pH to 8.0 with 10 M NaOH. Autoclave, and after cooling add sodium dodecyl sulfate (SDS) to 1% (w/v). Store at room temperature for 6 mo.
3. Proteinase K Solution: Dissolve proteinase K powder in distilled H$_2$O at 10 mg/mL, aliquot in sterile tubes and store at –80°C for 12 mo.
4. 5 M Sodium perchlorate: Dissolve 61.2 g of sodium perchlorate in 100 mL distilled water. Store at room temperature for 12 mo.
5. 1 M Dithiothreitol (DTT): Dissolve 1.54 g DTT in 10 mL distilled H$_2$O, aliquot in sterile tubes and store at –20°C for 6 mo. Do not expose to freeze-thaw cycles.

2.2. PCR Amplification and Analysis

1. Primers: Primers can be purchased from Applied Biosystems (check for your local representative). The primers arrive lyophilized. Prepare a 10 mM Tris-HCl, pH 8.0, 1 mM EDTA, 50% glycerol buffer. Autoclave the buffer and keep at 4°C for 6 mo. Dissolve the lyophilized primers at 100 pmol/μL concentration in TEG buffer. Leave for 1 h at 4°C. Vortex for 1 min and spin briefly. Store at –20°C. Prepare working dilutions of 5 pmols/μL of each primer (e.g., 10 μL of each (forward and reverse) primer stock solution plus 180 μL dd H$_2$O).
2. PCR mix (reagents from Applied Biosystems, Warrington, UK): 1× Gold Buffer, 2.5 mM MgCl$_2$, 500 μM dNTPs, 0.75–2.0 μM of each primer pair (*see* **Table 1**), 0.09 U Amplitaq™ Gold per microliter of reaction. 5 ng DNA per microliter of reaction.

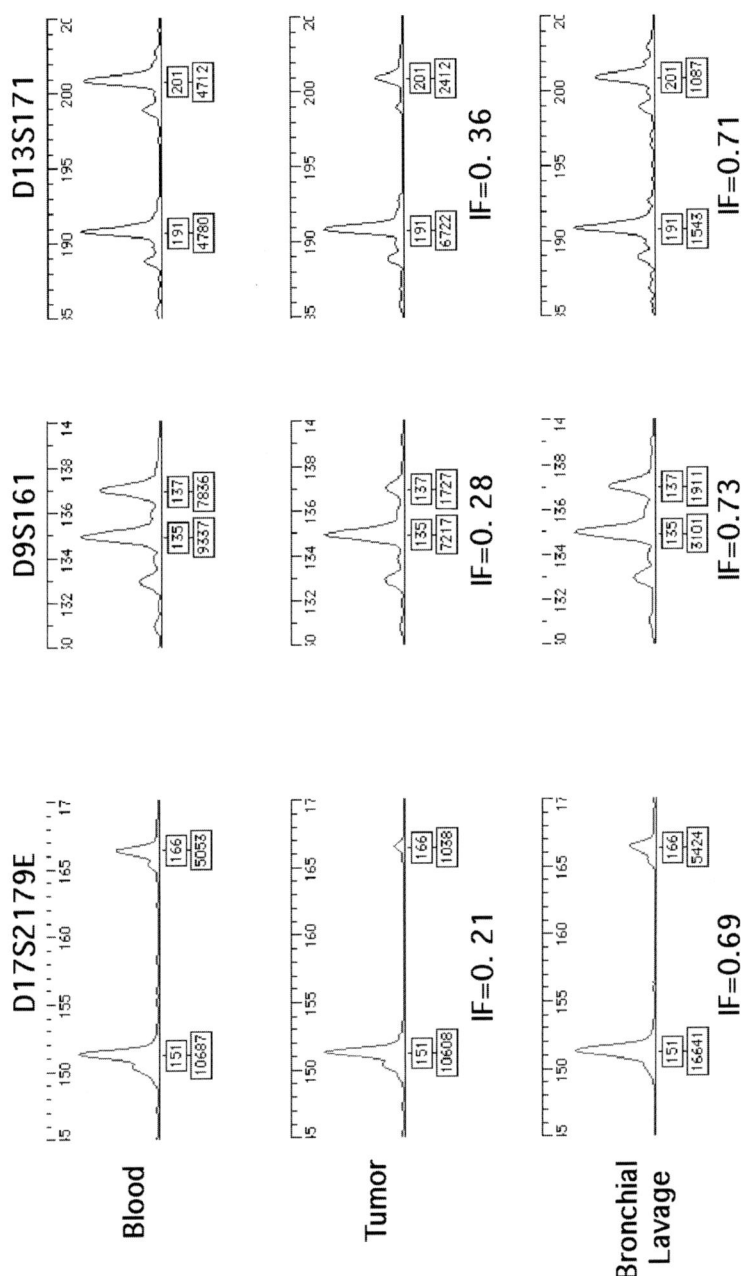

Fig. 3. Genotyper™ images demonstrating LOH in tumor and corresponding bronchial lavage sample. IF, imbalance factor.

Table 1
Microsatellite Markers Used in the Multiplex Assay

Marker	Locus	Allele size range		Dye	Concentration in PCR (nM)
D3S1300	3p14.2	235	267	FAM	75
D9S161	9p21	126	144	FAM	55
D3S1289	3p21-p23	155	182	FAM	160
D5S644	5q15	86	116	HEX	125
D17S2179E	17p13.1	131	161	HEX	90
D13S153	13q14	94	126	NED	75
D13S171	13q12.3	179	207	NED	150
D9S157	9p22-p23	229	253	NED	200

3. 10X TBE: Dissolve 108 g Tris base, 50 g boric acid in 800 mL of deionized water. Add 40 mL 0.5 M EDTA, pH 8.0. Adjust pH to 8.4. Make up to 1 L. Store at room temperature for 2–3 wk. If white precipitate is formed, discard.
4. Denaturing polyacrylamide mix (4.25%): 18 g urea, 7 mL 30% polyacrylamide stock solution (37.5:1), 0.5 g Amberlite MB-1 resin, 23 mL distilled water. Stir for 30 min, make up to 45 mL with ddH$_2$O. Using a 0.22 µm membrane filter and a vacuum pump, filter initially 5 mL 10X TBE and then the polyacrylamide mix. Leave pump on for another 15 min to degas the mix. Add 250 µL freshly prepared 10% ammonium persulfate and 35 TEMED. Mix gently and pour into the casting plates (377 gel cassette, Applied Biosystems, Warrington, UK).

3. Method
3.1. DNA Extraction
3.1.1. DNA Extraction from Blood (see **Note 1**)

1. Transfer 3 mL of blood (collected in EDTA tubes) in a 15-mL polypropylene tubes and centrifuge for 5 min, 1,000g, at 4°C. Carefully remove plasma (store at –20°C for potential further possible use).
2. Add 15 mL of red blood cell lysis buffer and mix thoroughly by hand until cell pellet is dissolved. Spin for 5 min, 1,000g, at 4°C. Carefully discard the supernatant in disinfectant.
3. Repeat **step 2** twice more until you recover a white or light pink cell pellet.
4. Resuspend the resulting pellet in 0.6 mL of lysis solution and incubate at 42°C for 12–15 h in an orbital shaker.
5. Add 0.6 mL phenol/chloroform. Mix thoroughly by hand (do not vortex).
6. Spin in a microcentrifuge at full speed for 2 min at room temperature. Transfer the aqueous (upper) phase into a clean tube.

Table 2
PCR Profile Used in the Multiplex Assay

Temperature	Minutes	Cycle no.
95°C	11:30	
94°C	00:25	
55°C	00:50	10
72°C	00:45	
93°C	00:25	
55°C	00:40	20
72°C	00:55	
72°C	20:00	

7. Precipitate DNA by adding 0.6 mL of of isopropanol. Mix well and leave samples at –20°C overnight.
8. Recover DNA by microcentrifugation for 15 min at 4°C, full speed.
9. Wash with 70% ice-cold ethanol and air (or vacuum) dry pellet.
10. Resuspend in 200–300 µL of sterile distilled H_2O or TE (see **Note 2**).

3.1.2. DNA Extraction from Bronchial Lavage and Sputum (see **Note 1**)

1. Transfer 1 mL of Bronchial lavage or sputum (collected Saccomano's fixative) in a 1.5-mL polypropylene tubes and microcentrifuge for 5 min, full speed at room temperature. Optional: to lyse mucus you can add 0.1 mL 1 M DTT and shake for 15 min prior to centrifugation.
2. Proceed as **step 4** of **Subheading 3.1.1**.
3. Resuspend in 0.05–0.1 mL of distilled water. Store at 4°C for 1–2 wk or at –20°C for 12–24 mo (see **Note 2**).

3.2. PCR Amplification and Analysis using Fluorescent-Labeled Primers

1. Set the reaction mix as described in **Subheading 2.2**. Markers are amplified in a single tube. To keep balanced amplification levels of all 6 markers in a reaction, different concentrations of each primer set are practised (see **Table 1**). The thermal profile is shown in **Table 2** (see **Notes 3–5**).
2. Mix 5 µL of the reaction to 3 µL loading solution. Denature at 95°C for 5 min (leave lid open if required to reduce volume). Chill on ice for 2 min.
3. Load onto a 4.25% polyacrylamide gel which has been prerun to warm up (all run and prerun modules available from Applied Biosystems, Warrington, UK) on a 377 ABI PRISM sequencer.

4. Run each panel of markers on a different lane. Run at 3 KV, collecting data for 3 h.
5. Analysis and interpretation of data is done using the Genescan™ and Genotyper™ software (Applied Biosystems, Warrington, UK).

3.3. Calculation of LOH, Interassay Variation, and Detection Threshold

PCR is a typical system being subject to chaotic dynamics. There is a large number of factors (reagents, pippeting errors, room temperature variation, thermocycler variation, operator errors, etc.) affecting reproducibility of PCR.

1. The detection of LOH is calculated as the imbalance factor (IF), which is the allele ratio of the target (tumor, bronchial lavage, sputum) sample normalized by the allele ratio of the normal reference sample (usually blood DNA). If we call A_1 and A_2 the allelic amplification areas of a heterozygote microsatellite PCR product (*see* **Fig. 3**) then the tumor to normal ratio will be:

$$IF = (A_2T/A_1T) / (A_2N/A_1N)$$

3. In order to calculate the threshold of LOH detection it is imperative to assess the interassay variability, in other words discriminate ratios falling into normal variability from ratios suggesting LOH in a relatively small proportion of the examined cell population *(11)*.
4. In order to achieve the above, the assay should be repeated for a number of times for a number of samples. For example, by repeating a 8-marker assay for 4 times for 24 different samples. In this case, every heterozygous marker/sample combination will be represented by 4 values (allele ratios). By calculating the ratios of those ratios we end up with 1194 *R* values. When all such R values are acquired our 99% reference range (RR) is calculated as:

$$99\% \; RR = mean \pm 3 \times [Standard\ Deviation]$$

The 99% RR is the detection threshold. Any R-value falling outside this area has a 1% false-positive probability while for replicate experiments (strongly recommended) this probability becomes 10^{-4}.

4. Notes

1. Please note that working with human tissues is a potential biohazard. An appropriate risk assessment must be prepared and operators need to follow the appropriate guidelines (vaccination, use of protective clothes, waste management etc.). The DNA extraction method in this chapter is a simple inexpensive method for extracting relatively clean DNA for PCR purposes. However, if large numbers of samples are to be examined then using a commercially available kit will be proved significantly faster. We have sufficiently tested and currently use for

routine DNA extraction from blood and sputum the DNAeasy96™ kit (Qiagen Ltd, Hilden, Germany), which can process 96 samples in 2–3 h.
2. The quality and quantity of extracted DNA should be checked by reading the $OD_{260/280}$ in a spectrophotometer or simply by running an agarose gel stained with ethidium bromide.
3. Due to the presence of a repeat sequence in the amplicon, artifacts may be a frequent problem unless the reaction is well-optimized. In general, keep Mg^{2+} and primer concentration to the lowest possible and operate at the highest possible annealing temperature. For primers purchased from Applied Biosystems annealing temperature is 55 ± 1°C for most protocols. Keep extension times relatively short, keeping in mind though that multiple fragments need to be amplified. Keep the number of cycles up to 30 and certainly below 35. The use of 5% dimethyl sulfoxide (DMSO) or 1 M Betaine or a combination of 60 mM tetramethyl-ammonium acetate (TMAC) and 5% formamide may improve specificity.
4. Please note that the DNA added in the reaction should not drop below 5 ng/µL of reaction as the possibility of false LOH and MI increases dramatically. Always repeat LOH and MI cases to reconfirm findings.
5. There are commercially available primers (Applied Biosystems, http://www.appliedbiosystems.com/molecularbiology/apply/dr/lmshd5/ and Research Genetics: http://www.resgen.com/resources/apps/mappairs/mp.php3) that have been tested and work efficiently in standard conditions. Primer stock solutions should preferably made using 50% glycerol TE (described in the materials section) instead of water. Glycerol will ensure that the solution will not freeze at –20°C. Freeze-thaw cycles gradually lead to primer and fluorescent label degradation. If not using glycerol, do not exceed 3 freeze-thaw cycles. Instead, aliquot your stocks and use one at a time. Also remember that because of the fluorescent label primer should not be exposed to light for long times.

References

1. Perucho, M. (1996) Microsatellite instability: the mutator that mutates the other mutator. *Nature Med.* **2,** 630–631.
2. Lengauer, C., Kinzler, K. W., and Vogelstein, B. (1998) Genetic instabilities in human cancers. *Nature* **396,** 643–649.
3. Mertens, F., Johansson, B., Hoglund, M., and Mitelman F. (1997) Chromosomal Imbalance maps of malignant solid tumors: a cytogenetics survey of 3185 neoplasms. *Cancer Res.* **57,** 2765–2780.
4. Tsuchiya, E., Nakamura, Y., Weng, S. Y., Nakagawa, K., Sugano, H., and Kitagawa, T. (1992) Allelotype of non-small cell lung carcinoma: comparison between loss of heterozygosity in squamous cell carcinoma and adenocarcinoma. *Cancer Res.* **52,** 2478–2481.
5. Merlo, A., Mabry, M., Gabrielson, E., Vollmer. R., Baylin, S. B., and Sidransky, D. (1994) Frequent microsatellite instability in primary small cell lung cancer. *Cancer Res.* **54,** 2098–2101.

6. Neville, E. M., Stewart, M. P., Swift, A., Risk, J. M., Liloglou, T., Ross, H., et al. (1996) Gosney, J. R., Donnelly, R. J., Field, J. K. Allelotype of non-small cell lung cancer. *Int. J. Oncol.* **9,** 533–539.
7. Field J. K., Neville E. M., Stewart M. P., Swift A., Liloglou T., Risk J. M., et al. (1996) Fractional allele loss data indicate distinct genetic populations in the development of non-small cell lung cancer. *Br. J. Cancer* **74,** 1968–1974.
8. Mitsudomi, T., Oyama, T., Nishida, K., Ogami, A., Osaki, T., Sugio, K., et al. (1996) Loss of heterozygosity at 3p in nonsmall cell lung-cancer and its prognostic implication. *Clin. Cancer. Res.* **2,** 1185–1189.
9. Pifarre, A., Rossel, R., Monzo, M., DeAnta, J. M., Moreno, I., Sanchez, J. J., et al. (1997) Prognostic value of replication errors on chromosomes 2p and 3p in non-small-cell lung cancer. *Br. J. Cancer* **75,** 184–189.
10. Zhou, X., Kemp, B. L., Khuri, F. R., Liu, D., Lee, J. J., Wu, W. G., et al. (2000) Prognostic implication of microsatellite alteration profiles in early-stage non-small cell lung cancer. *Clin. Cancer Res.* **6,** 559–565.
11. Liloglou, T., Maloney, P., Xinarianos, G., Fear, S. and Field, J. K. (2000) Sensitivity and limitations of high throughput fluorescent microsatellite analysis for the detection of allelic imbalance. Application in Lung Tumors. *Int. J. Oncol.* **16,** 5–14.
13. Hung, J., Kishimoto, Y., Sugio, K., Virmani, A. K., McIntire, D. D., Minna, J. D., and Gasdar, A. F. (1995) Allele-specific chromosome 3p deletions occur at an early-stage in the pathogenesis of lung-carcinoma. *JAMA* **273,** 558–563.
14. Kishimoto, Y., Sugio, K., Hung, J., Virmani, A. K., McIntire, D. D., Minna, J. D., and Gasdar A. F. (1995) Allele-specific loss in chromosome 9p loci in preneoplastic lesions accompanying non small cell lung-cancers. *J. Natl. Cancer Inst.* **87,** 1224–1229.
15. Kohno, H., Hiroshima, K., Toyozaki, T., Fujisawa, T. and Ohwada, H. (1999) p53 mutation and allelic loss of chromosome 3p, 9p of preneoplastic lesions in patients with nonsmall cell lung carcinoma. *Cancer* **85,** 341–347.
16. Wistuba, I. I., Behrens, C., Virmani, A. K., Mele, G., Milchgrub, S., Girard, L., et al. (2000) High resolution chromosome 3p allelotyping of human lung cancer and preneoplastic/preinvasive bronchial epithelium reveals multiple, discontinuous sites of 3p allele loss and three regions of frequent breakpoints. *Cancer Res.* **60,** 1949–1960.
17. Mao, L., Lee, J. S., Kurie, J. M., Fan, Y. H., Lippman, S. M., Lee, J. J., et al. (1997) Clonal genetic alterations in the lungs of current and former smokers. *J. Natl. Cancer. Inst.* **89,** 857–862.
18. Wistuba, I. I., Lam, S., Behrens, C., Virmani, A. K., Fong, K. M., LeRiche, J., et al. (1997) Molecular damage in the bronchial epithelium of current and former smokers. *J. Natl. Cancer Inst.* **89,** 1366–1373.
19. Mao, L., Hruban, R. H., Boyle, J. O., Tockman, M., and Sidransky, D. (1994) Detection of oncogene mutations in sputum precedes diagnosis of lung cancer. *Cancer Res.* **54,** 1634–1637.

20. Miozzo, M., Sozzi, G., Musso, K., Pilotti, S., Incarbone, M., Pastorino, U., and Pierotti, M. A. (1996) Microsatellite alterations in bronchial and sputum specimens of lung cancer patients. *Cancer Res.* **56,** 2285–2288.
21. Field, J. K., Liloglou, T., Xinarianos. G., Prime, W., Fielding, P., Walshaw, M. J., and Turnbull, L. (1999) Genetic alterations in bronchial lavage as a potential marker for individuals with a high risk of developing lung cancer. *Cancer Res.* **59,** 2690–2695.
22. Ahrendt, S. A., Chow, J. T., Xu, L. H., Yang, S. C., Eisenberger, C. F., Esteller, M., et al. (1999) Molecular detection of tumor cells in bronchoalveolar lavage fluid from patients with early stage lung cancer. *JNCI* **91,** 332–339.
23. Powell, C. A., Klares, S., O'Connor, G., and Brody, J. S. (1999) Loss of heterozygosity in epithelial cells obtained by bronchial brushing: clinical utility in lung cancer. *Clin. Cancer Res.* **5,** 2025–2034.
24. Chen, X. Q., Stroun, M., Magnenat, J. L., Nicod, L. P., Kurt, A. M., Lyautey, J., et al. (1996) Microsatellite alterations in plasma DNA of small cell lung cancer patients. *Nat. Med.* **2,** 1033–1035.
25. Sanchez-Cespedes, M., Monzo, M., Rosell, R., Pifarre, A., Calvo, R., Lopez-Cabrerizo, M. P., and Astudillo, J. (1998) Detection of chromosome 3p alterations in serum DNA of non-small-cell lung cancer patients. *Ann. Oncol.* **9,** 113–116.
26. Field, J. K. (1999) Selection and validation of new lung cancer markers for the molecular-pathological assessment of individuals with a high risk of developing lung cancer, in *Lung Tumours: Fundamental Biology and Clinical Management* (Bambilla, C. and Bambilla, E., eds.), Marcel Dekker, New York, NY, pp. 287–299.
27. Mulshine, J. L. and Henschke, C. I. (2000) Prospects for lung cancer screening. *Lancet* **355,** 592–593.
28. Siafakas, N. M., Tzortzaki, E. G., Sourvinos, G., Bouros, D., Tzanakis, N., Kafatos, A., and Spandidos D. A. (1999) Microsatellite DNA instability in COPD. *Chest* **116,** 47–51.
29. Spandidos, D. A., Ergazaki, M., Hatzistamou, J., Kiaris, H., Bouros, D., Tzortzaki, E. G., and Siafakas, N. M. (1996) Microsatellite instability in patients with chronic obstructive pulmonary disease. *Oncol. Rep.* **3,** 489–491.
30. Park, W. S., Pham, T., Wang, C. Y., Pack, S., Mueller, E., Mueller, J., et al. (1998) Loss of heterozygosity and microsatellite instability in nonneoplastic mucosa from patients with chronic ulcerative colitis. *Int. J. Mol. Med.* **2,** 221–224.
31. Suzuki, H., Harpaz, N., Tarmin, L., Yin, J., Jiang, H. Y., Bell, J. D., et al. (1994) Microsatellite instability in ulcerative colitis-associated colorectal dysplasias and cancers. *Cancer Res.* **54,** 4841–4844.
32. Brentnall T. A., Crispin D. A., Bronner M. P., Cherian S. P., Hueffed M., Rabinovitch P. S., et al. (1996) Microsatellite instability in nonneoplastic mucosa from patients with chronic ulcerative colitis. *Cancer Res.* **56,** 1237–1240.
33. Vassilakis, D. A., Sourvinos, G., Markatos, M., Psathakis, K., Spandidos, D. A., Siafakas, N. M., and Bouros, D. (1999) Microsatellite DNA instability and loss

of heterozygosity in pulmonary sarcoidosis. *Am. J. Res. Crit. Care Med.* **160,** 1729–1733.
34. Kullmann, F., Widmann, T., Kirner, A., Justen, H. P., Wessinghage, D., Dietmaier, W., et al. (2000) Microsatellite analysis in rheumatoid arthritis synovial fibroblasts. *Ann. Rheum. Dis.* **59,** 386–389.
35. Liloglou, T., Maloney, P., Xinarianos, G., Hulbert, M., Walshaw, M. J., Gosney, J. R., et al. (2001) Cancer-specific genomic instability (CSGI) in bronchial lavage: A molecular tool for lung cancer detection. *Cancer Res.* **61,** 1624–1628.
36. Kebelman-Betzing, C., Seeger, K., Dragon, S., Schmitt, G., Moricke, A., Schild, T. A., et al. (1998) Advantages of a new Taq DNA polymerase in multiplex PCR and time-release PCR. *Biotechniques* **24,** 154–158.

15

Southern Blotting of Genomic DNA from Lung and Its Tumors

Application to Analysis of Allele Loss on Chromosome 11p15.5

Diana M. Pitterle and Gerold Bepler

1. Introduction

Lung cancer is one of the most common cancers afflicting the citizens of developed countries *(1)*. While the lung is a complex tissue composed of over 40 different cell types, the most common lung cancers, large cell lung carcinoma, small cell lung carcinoma (SCLC), squamous cell carcinoma, and adenocarcinoma, are all thought to arise from bronchoepithelial cells *(2)*. Normal human cells are not easily coerced into becoming cancerous as numerous mutations are necessary to subvert the cellular processes that ensure the fidelity of DNA replication and that limit cell growth and proliferation *(3)*.

The mutations that lead to cancer can be divided into two categories, dominant and recessive *(4,5)*. A mutation is considered to be dominant if the development of cancer is promoted by the mutation of a single allele of a gene. Examples of such mutations include those that produce constitutively active forms of ras proteins or growth factor receptors. Such proteins may then act as oncogenes in the cell, forcing the cell to initiate unwarranted cycles of growth and replication. In contrast, mutations are considered to be recessive if both alleles must be affected before the development of a tumor is possible. Tumor-suppressor genes are defined as recessive. Because the product of a wild-type tumor-suppressor gene serves in a restrictive manner to limit cell proliferation and/or growth, both alleles must be mutated to abrogate that function. The classic example of a tumor-suppressor gene is the retinoblastoma gene (*Rb*). Its protein product binds the transcription factors needed to promote entry of the cell into S phase of the cell cycle. Normally, RB protein restrains the activity of those transcription factors until conditions for cell replication are met *(6,7)*. One allele of *Rb* is mutated in families with an inherited predisposition to retinoblastoma, a childhood malignancy of the eye. Sporadic mutation of the second allele is observed in all retinoblastomas.

Often the isolation of tumor suppressor genes has been aided by the identification of DNA regions that are commonly deleted in tumors from an array of patients *(8–10)*. These deletions are apparent in a Southern blot of normal and tumor DNA. Often the tumor DNA will show allele loss, also called loss of heterozygosity (LOH), when compared to DNA from normal tissue (*see* **Fig. 1**). This occurs frequently in tumors since mechanisms responsible for maintenance of DNA integrity and repair are impaired or inactivated. Other analyses have often revealed the presence of point mutations that inactivate the remaining gene in the retained allele. Thus candidate tumor-suppressor genes have been identified in regions of frequent LOH *(8–10)*.

Three factors essential to good LOH analyses are DNA quality, the identification of an informative polymorphism (for instance, a single nucleotide polymorphism [SNP] that results in an allele-specific restriction enzyme digest), and a specific nucleotide probe that hybridizes to DNA fragments from the genomic region of interest. As the first step in LOH analysis, genomic DNA must be prepared from both the tumor and a paired normal tissue. Secondly, both DNA samples are cut into fragments using an informative restriction enzyme. An enzyme is informative when one of the sites at which it cuts is affected by a polymorphism. As a result, the cut sites are not identically positioned in the maternal and paternal alleles of a gene, and fragments of different sizes are produced from the alleles. The different fragments allow one to detect maternally and paternally derived alleles. For LOH analyses, it is not necessary to assign the fragments to their maternal or paternal origin but simply to be able to distinguish the alleles. After digestion, the DNA fragments are separated by size using agarose gel electrophoresis and transferred to a membrane support. The membrane-bound DNA is hybridized with a radiolabeled DNA probe complementary to the region of interest (the classic Southern blot *[11]*). Autoradiography reveals a pattern of bands on film, allowing visualization of the loss of an allele in tumor DNA (*see* **Fig. 1**). Since tumor specimens rarely consist of a homogenous population of tumor cells but are frequently mixed with normal connective tissue and inflammatory cells, data analysis frequently requires quantitation of the band intensities in the normal and tumor samples.

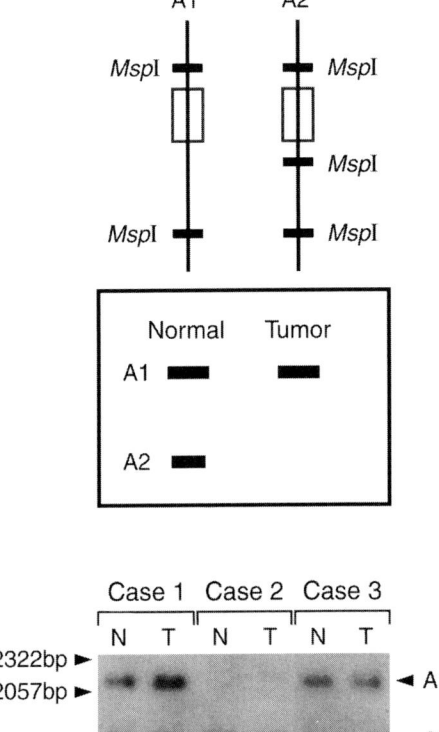

Fig. 1. Allele loss analysis. A schematic representation of the DNA in a heterozygous individual is shown on top. A1 and A2 denote the paternal and maternal alleles of a gene. Because the pattern of *MspI* sites in the two alleles is different, digesting the DNA with *MspI* will produce fragments of different sizes. The boxed area denotes the region to which the probe hybridizes. The probe will hybridize to a larger fragment from A_1 and a smaller fragment from A_2. A schematic representation of the band pattern that would be expected on a Southern blot is shown below the allele diagram. Two bands would be expected in the Southern blot of DNA from normal tissue. In contrast, tumor tissue with allele loss at A_1 will show only the band for A_2. Tumor with allele loss at A_2 will show only the band for A_1 (depicted). On the bottom, Southern blots of normal and tumor DNA samples from three patients are shown to illustrate three typical results from LOH analyses. Case 1 is heterozygous, and the tumor shows no allele loss. Case 2 is uninformative, meaning that the individual is homozygous for one of the alleles. In this example, it is A_2. Case 3 is heterozygous, and the tumor shows loss of A_2. In this example, the loss is apparent by visual inspection. Often the tumor specimens contain contaminating normal tissue, making results less striking. In such cases autoradiography bands are often quantitated using a PhosphoImager, and the results are used to assess LOH. It must also be noted that Southern blots often have additional nonspecific bands.

An outline of the procedures to be carried out follows.

1. DNA extraction.
2. Restriction enzyme digestion.
3. Agarose gel electrophoresis.
4. Transfer of DNA fragments to nylon membrane.
5. Preparation of radiolabeled probe.
6. Hybridization of probe to membrane-bound DNA fragments.
7. Data analysis.

2. Materials
2.1. DNA Extraction (see Note 1)

1. Tris-buffered saline (TBS): Add 8.0 gm NaCl, 0.2 gm KCl, 3.0 gm Tris to 800 mL distilled H_2O. Adjust pH to 7.4 using HCl. Add H_2O to 1 L. Autoclave.
2. Tris-buffered EDTA, pH 8.0 (TE, pH 8.0): 10 mM Tris-HCl, pH 8.0, 1 mM EDTA. Add 5 mL of 2 M Tris stock solution (*see* recipe in **Subheading 2.4.**) and 2 mL of 0.5 M EDTA (*see* recipe in **Subheading 2.3.**) stock solution to 800 mL distilled H_2O. Adjust pH to 8.0 at room temperature. Add H_2O to 1 L and autoclave.
3. Phenol (#100 728 Roche Molecular Biochemicals, Indianapolis, IN): Warm solid redistilled phenol in a 65°C water bath. Once liquefied, ensure phenol is colorless. (If not, discard it since this indicates impurity.) Mix 100 mL phenol with 100 mL TE, pH 8.0, and 50 mL 10% Tris (0.826 M, 10 g Tris in 100 mL distilled water, pH is 10.9 at 20°C). Mix every 15 min for 1 h. Check that the pH of the aqueous phase is >8.0. (pH paper is not accurate due to fact that phenol causes degradation of the indicators. Instead, dilute 2 mL of phenol with 8 mL of methanol and 10 mL of H_2O, and measure the pH of the total sample.) To minimize oxidation, add 0.1% 8-hydroxy-quinoline and store frozen at –20°C).

   Due to the hazardous nature of working with phenol, it is strongly recommended that a ready-to-use solution be obtained. In addition to Roche, Ambion, Inc. (Austin, TX) also has various phenol preparations. Work with phenol in a hood, wear gloves, and avoid spills. If accidental contact with skin occurs, wash with water or soap and water for 15 min.
4. Pronase (#165 921 Roche, #P6911 Sigma Chemical Co, St. Louis, MO): Dissolve 1 g in 50 mL of TE, pH 7.5 (10 mM Tris-HCl, pH 7.5, 10 mM EDTA). Final concentration is 20 mg/mL. Allow Pronase to self-digest for 1 h at 37°C. Store at –20°C in 5 mL aliquots, tightly capped. There is no need for sterile filtration.
5. RNase A (#109169 Roche, #R5500 Sigma): Dissolve 100 mg in 10 mL of 10 mM Tris-HCl, pH 7.5, 15 mM NaCl. Boil for 15 min, cool to room temperature, and aliquot in 1 mL portions. Final concentration is 10 mg/mL. Store at –20°C. There is no need for sterile filtration.
6. 10% SDS: Dissolve 100 g of electrophoresis grade SDS in 900 mL of distilled H_2O. Heat to 68°C to dissolve SDS and adjust pH to 7.2 with dilute HCl. Adjust volume to 1000 mL. Do not autoclave. Wear a mask when weighing SDS.

7. GLB: 0.1 M NaCl, 0.05 M Tris, 0.05 M EDTA, 0.5% SDS, pH 8.0. Add 20 mL of 5.0 M NaCl, 25 mL of 2 M Tris, and 100 mL of 0.5 M EDTA stock solutions to 855 mL distilled H_2O. Sterilize by filtration through 0.2 μm filter.

2.2. Restriction Enzyme Digest (see Note 2)

Restriction enzymes, 10X buffers, and 10X acetylated BSA are available from a variety of suppliers.

2.3. Agarose Gel Electrophoresis

1. EDTA: 0.5 M, pH 8.0. Dissolve 186.1 g of Na_2EDTA in 800 mL of distilled H_2O. Adjust pH to 8.0 with sodium hydroxide pellets (about 20 g). EDTA will not dissolve without NaOH. Adjust volume to 1 L and autoclave.
2. Ethidium Bromide: 10 mg/mL. For safety reasons, it is recommended that this be purchased as a 10 mg/mL solution. However, a solution can be prepared by dissolving 1 g ethidium bromide in 100 mL distilled H_2O. Allow the solution to stir for several hours to ensure dye dissolution. Store at room temperature in a dark bottle (wrap in aluminum foil).
3. TAE (50X): Dissolve 242 g Tris base in 700 mL of distilled H_2O. Add 57.1 mL glacial acetic acid and 200 mL 0.5 M EDTA, pH 8.0. Adjust volume to 1 L.
4. Sample Loading Buffer (6X): Add 0.3 mL sterile glycerol, 20 μL 50X TAE, 0.025% (w/v) bromophenol blue, 0.025% (w/v) xylene cyanol FF to 9.68 mL distilled H_2O. Use 2 μL of loading buffer per 10 μL of sample.
5. Agarose gel (20 × 25 × 1.0 cm, prepared from 500 mL 1% agarose solution): Add 10 mL 50X TAE, 5 g agarose (ultraPURE Agarose, #15510-027, Life Technologies, Gaithersburg, MD) and 300 mL of distilled H_2O to 1000 mL flask. Cover top with inverted beaker. Microwave until boiling. Using gloves, remove flask from microwave. Check for translucent spheres of undissolved agarose. Reheat as necessary until all agarose is dissolved. Add a stir bar, 50 μL Ethidium bromide, and 200 mL H_2O. Stir gently as agarose cools. Use autoclave tape to close the open sides of a gel tray. When the flask of agarose is cool enough to be handled without gloves, carefully pour agarose into gel tray taking care not to create bubbles. If any bubbles are present, remove them using a sterile pipet tip. Place comb into agarose. Allow agarose to solidify (*see* **Note 3**).
6. Agarose Gel Running Buffer: Mix 44 mL of 50X TAE with distilled H_2O to give a final volume of 2200 mL of running buffer. Mix well and pour into gel box.
7. Molecular Weight Markers: Dilute 15 μL stock 1 Kb DNA ladder (#15615-016 Life Technologies) into 500 μL 1X sample loading buffer. Load 10–20 μL on gel.

2.4. DNA Transfer to Membrane (see Note 4)

1. 5.0 M NaCl: Dissolve 292.2 g of NaCl in 800 mL of distilled H_2O and adjust volume to 1 L. Gentle heating and stirring of the solution are required. Autoclave.
2. 2.0 M Tris-HCl, pH 7.6: Dissolve 242.2 g of Tris in 800 mL distilled H_2O. Gentle heating and stirring are recommended. Adjust pH to 7.6 at room temperature by adding approx 120 mL of concentrated (37%) HCl. Adjust volume to 1 L.

Autoclave. If solution is yellow, discard it. A yellow color indicates impurities in the Tris.
3. 5.0 M NaOH: Dissolve 200 g of NaOH pellets in 800 mL distilled H_2O. Adjust volume to 1 L. Do not autoclave.
4. Solution I (Depurination): Add 20 mL concentrated (37%) HCL to 900 mL distilled H_2O and adjust volume to 1 L.
5. Solution II (Denaturation): Add 200 mL of 5 M NaCl and 100 mL of 5 M NaOH to 600 mL of distilled H_2O and adjust final volume to 1 L.
6. Solution III (Neutralization): Mix 400 mL of 2 M Tris and 600 mL of 5 M NaCl.

2.5. Radiolabeling DNA (see Notes 5–7)

1. DNA labeling kit, Prime-a-Gene Labeling System, #U1100, Promega (Madison, WI).
2. [α-^{32}P] dNTP, 50 µCi, 3000 Ci/mmol (Du Pont NEN, Boston, MA or Amersham Pharmacia Biotech Inc., Piscataway, NJ).

2.6. Hybridization (see Note 8)

1. SSC (20X): Dissolve 175.3 g NaCl and 88.2 g of sodium citrate in 800 mL distilled H_2O. Adjust pH to 7.0 with a few drops of concentrated HCl and adjust volume to 1 L with distilled H_2O. Sterilize by autoclaving.
2. Salmon Sperm DNA: Salmon testis nuclei type II-S (Sigma #S3126), Salmon testis DNA, sodium salt Type III (Sigma #D1626), or Salmon testis DNA, phenol-chloroform extracted, ethanol precipitated (Sigma #D7656).
   a. If obtained as a solution of 10 mg/mL (#D7656), boil for 15 min, then immediately place on ice for 5 min. Freeze solution in 5 mL aliquots at –20°C.
   b. If obtained as a powder (#S3126 and #D1626), dissolve in distilled H_2O to 10 mg/mL. Adjust NaCl to 0.1 M and extract once with an equal volume of phenol and once with phenol:chloroform. Recover aqueous phase and shear DNA by passing rapidly through a 17-gauge needle 10 times. Precipitate DNA by adding 2 vol of ice-cold 100% ethanol. Centrifuge and dissolve to a concentration of 10 mg/mL in distilled H_2O. For #S3126, 5 g yields 80–90 mg DNA. (Measure OD_{260}). Then boil, chill, aliquot, and store at –20°C. Just prior to use, heat an aliquot at 95°C for 5 min in a water bath or heater block, then quickly chill on ice.
3. Denhardt's Reagent (100X): For a 250 mL solution in distilled H_2O, dissolve 5 g Ficoll (Type 400 [DL]), 5 g Polyvinylpyrrolidone, and 5 g Albumin, bovine (Fraction V). Store in aliquots at –20°C. Do not autoclave. Sterile filtration is impossible.
4. 50% Dextran Sulfate: Use Sigma #D6001, average MW of approx 500,000. Add 15 mL distilled H_2O to 50 g and allow it to dissolve overnight (or longer) at 37°C. Add H_2O to final volume of 100 mL. Solution is very viscous. Do not filter or autoclave.
5. Prehybridization solution: Add 1 mL of 0.5 M EDTA, 5 mL of 10% SDS, 75 mL 20X SSC, and 50 mL 100X Denhardt's solution to 364 mL of distilled H_2O. Do

Southern Blotting of Genomic DNA

not filter or autoclave. Store at –20°C in aliquots. Just prior to use, add 5 mL of boiled and chilled salmon sperm DNA.

6. Hybridization solution: Add 1 mL of 0.5 M EDTA, 5 mL 10% SDS, 75 mL 20X SSC, 50 mL 100X Denhardt's solution, and 70 mL of 50% dextran sulfate (av. MW ~500,000) to 294 mL of distilled H_2O. Do not filter or autoclave. Store at –20°C in aliquots. Just prior to use, add 5 mL of boiled and chilled salmon sperm DNA.
7. Wash solution I: 2X SSC, 0.1% SDS, 1 mM EDTA (approx 1000 mL per wash required, up to 5 membranes per wash).
8. Wash Solution II:
   a. Low stringency: 0.5X SSC, 0.1% SDS, 1 mM EDTA (1000 mL per wash).
   b. High stringency: 0.2X SSC, 0.1% SDS, 1 mM EDTA (or lower SSC concentration as required for best results).
9. Strip Solution: 0.1X SSC, 0.5% SDS (1 L required/wash, up to 5 membranes at a time).

2.7. Data Analysis

PhosphoImager with ImageQuant software (Molecular Dynamics, Sunnyvale, CA).

3. Methods
3.1. Extracting DNA from Tissue

Briefly, this protocol has three parts. First, the majority of the protein in the tissue is digested overnight with Pronase. Second, the resulting suspension is extracted with phenol. Finally, DNA is collected from the aqueous solution on the end of a glass pipet, dissolved, phenol-extracted, collected, and dissolved in TE.

3.1.1. Tissue Protein Digestion

1. Place 0.5–1.0 g of tissue into a mortar containing a few mL of liquid nitrogen. Grind the tissue to powder keeping it frozen with liquid nitrogen. Yield of DNA from less than 0.5 g of tissue is very low (approx 200 µg DNA/g tissue).
2. Transfer ground tissue and liquid nitrogen into 50 mL polypropylene tube on dry ice. Leave for 15 min with caps loosely attached to allow liquid nitrogen to evaporate.
3. Add approx 20 mL GLB (without SDS) to 1 g tissue and keep on dry ice. Vortex to dissolve tissue in GLB (can be frozen at –20°C indefinitely).
4. Add concentrated Pronase to final concentration of 0.5 mg/mL (650 µL of 20 mg/mL stock solution to 25 mL of lysed suspension). Mix gently at room temperature on a rocker for 30 min.
5. Add 1 mL 10% SDS for final concentration of 0.5% SDS. Rock tubes overnight at 42°C.
6. Inspect solution for lumps or globs that represent undigested protein. If present, divide sample into two tubes and repeat **steps 3–6** in twice the volume. (This is *very* important).

3.1.2. Phenol Extraction

1. Add 1 volume (13 mL) of phenol to 2 volumes of DNA.
2. Rock overnight at room temperature.

3.1.3. DNA Isolation and Purification

1. Centrifuge tubes at 700g for 20 min at room temperature without brake.
2. Transfer the aqueous phase (supernatant) to a new 50-mL tube. To decrease DNA shearing, take a 10-mL sterile pipet and remove the cotton swab at the top. Use this end to take up aqueous phase. (Yield is approx 20 mL. If there is more, then two 50-mL tubes need to be used.)
3. Add 1.5 volumes (30 mL) of ice-cold 100% ethanol.
4. Gently rock suspension at room temperature until well mixed (about 15 min).
5. Allow DNA to precipitate at $-20°C$ for >2 h.
6. Spool out DNA from ethanol using a glass Pasteur pipet that has been flamed to seal the end.
7. Rinse DNA on tip of pipet with a squirt of 100% ethanol, and place pipet tip in the bottom of a 50-mL tube. Aspirate excess ethanol. Allow residual ethanol to evaporate. Do not completely dry DNA, as this will make it impossible to dissolve.
8. Add 20 mL TE to tube. Allow at least 30 min for the DNA to release from pipet and go into solution.
9. Rock tubes gently overnight (1–24 h) at 37°C to dissolve DNA.
10. Add 0.2 mL of 10 mg/mL RNase A (final concentration 0.1 mg/mL).
11. Gently rock at 42°C for 1–3 h.
12. Add 0.42 mL of 5 M NaCl and 0.22 mL of 10% SDS and invert tube gently a few times.
13. Add 0.2 mL of 20 mg/mL Pronase and rock at 42°C for >3 h.
14. Add 1 volume (11 mL) of phenol to 2 volumes of solution and rock overnight at room temperature.
15. Centrifuge at 700g for 20 min at room temperature without brake.
16. Transfer aqueous phase (supernatant) to a new 50-mL tube. Avoid taking the interphase. To decrease shearing, remove the cotton from a 10-mL sterile pipet and use it inverted to transfer the solution. (Yield is <20 mL.)
17. Extract DNA with chloroform:isoamyl alcohol (24:1). Use a 1:1 volume ratio. Mix gently on rocker at room temperature for approx 15 min, then leave upright for 5 min.
18. Centrifuge at 700g (2200 rpm) for 20 min at room temperature without brake.
19. Transfer aqueous phase to new tube (yield <20 mL).
20. Add 1.5 volumes (30 mL) of ice-cold 100% ethanol.
21. Gently rock suspension at room temperature until well-mixed (approx 15 min).
22. Allow DNA to precipitate at $-20°C$ for >4 h.
23. Collect DNA from ethanol using a pipet, as done in **step 6**.

24. Rinse DNA on tip of pipet with a squirt of 100% ethanol and place pipet in 1.5-mL tube. Aspirate excess ethanol with 200 µL pipet. Allow ethanol to evaporate at room temperature. Do not overdry DNA, as it will not go into solution afterwards.
25. Dissolve DNA in distilled water or TE to an estimated concentration of 0.5–2 mg/mL. This usually requires 0.2–1.5 mL. (Yield is about 200 µg DNA/gm of tissue.)
26. Allow DNA to dissolve at 37°C overnight. (This may require very gentle shaking in water bath or incubator.)
27. Store DNA in 1.5 mL tubes at 4°C. (It can be stored indefinitely. Remember to use wide bore or cut pipet tips to minimize shearing when working with genomic DNA.)
28. Determine concentration of DNA using OD at 260 nm. Add 5 µL DNA to 995 µL H_2O, i.e., 1:200 dilution. OD reading of 1 at 260 nm is equivalent to 50 µg DNA per mL (i.e., reading of 0.005 = 50 µg/mL in a 1:200 dilution.) Measure the OD at 280 nm and calculate the ratio of OD at 260 and 280 in order to assess the purity of the DNA. The ratio should be over 1.75. A lower ratio indicates protein contamination of the DNA.

3.2. Restriction Enzyme Digest

A sample *MspI* digest using genomic DNA at a concentration 0.5 µg/µL would be set up as follows:

1. Into a 1.5-mL microfuge tube, pipet 10 µL of DNA and 2 µL of 10X buffer.
2. Add enzyme. Use 10 U per µg genomic DNA.
3. Add x µL of water for a total volume of 20 µL.
4. Mix by pipetting up and down. Centrifuge samples briefly to collect reaction in bottom of tube.
5. Incubate tubes at 37°C (or other temperature depending on the enzyme) overnight.
6. Centrifuge briefly to collect condensation. Add 4 µL loading buffer.

3.3. Agarose Gel Electrophoresis

1. Carefully remove tape from ends of agarose gel tray. Without tearing gel, lift out comb.
2. Place tray with gel into gel apparatus. Make sure gel is submerged ~1–2 mm.
3. Load digested DNA (usually 60 µL with 5 µg) into each well. Maximum well capacity is 160 µL (10 mm gel, 3 mm tooth).
4. Load 10–20 µL (approx 500 ng) of DNA molecular weight standards.
5. Place cover on gel apparatus carefully. Plug in wires with red at bottom of apparatus. Set at approx 30–75 V (0.8–2.1 V/cm).
6. Run approx 16 h at 45 V (Bromophenol blue to ~15 cm). Bromophenol blue runs at approx 800 bp and xylene cyanol at 4000 bp if agarose concentration is 0.5–1.4%.

3.4. DNA Transfer to Membrane

1. Photograph gel under UV light with a fluorescent ruler lying next to the gel.
2. Inject India Ink into molecular markers (center of lane, just a drop).
3. Transfer gel to plastic tray.
4. Add Solution I. Float gel in plenty of fluid (~500 mL/gel). Gently shake for 10 min.
5. Rinse gel. (Pour HCl out and rinse tray twice with distilled water).
6. Pour solution II into plastic tray containing gel. Shake for 30 min.
7. Rinse gel twice with distilled water.
8. Pour solution III into plastic tray containing gel. Shake for 30 min. Repeat treatment with solution III.
9. Place Saran wrap on table.
10. Cut Whatman paper (Whatman, Clifton, NJ) to desired size.
11. Obtain a tray with distilled water, a 10-mL pipet and blotting paper.
12. Place gel on Saran wrap. Roll pipet over gel to chase out any air bubbles trapped underneath it.
13. Wet Nytran (Schleicher and Schuell, Inc., Keene, NH, 20×20 cm $\times 0.45$ μm) in distilled water. Soak Nytran in solution III for 5 min.
14. Place Nytran on gel so that the top of the Nytran paper is just under the wells. With razor blade, cut away top of gel at the wells. Trim edges to Nytran paper size. Use pipet to roll out air bubbles.
15. Pipet 10 mL solution III onto Nytran.
16. Carefully place Whatman paper over Nytran, allowing buffer to wet paper without trapping air bubbles. Use pipet to roll out air bubbles.
17. Place two pieces of dry Whatman paper on top.
18. Add a stack of blotting paper (10–15 cm, Schleicher and Schuell), glass plate, and weight (500 mL of water in a bottle) on top. Make sure there is no liquid around the paper. Allow capillary transfer from gel to membrane to proceed overnight.
19. Make marks on the Nytran where molecular markers are located, and label membrane with date of run and gel ID number.
20. Expose Nytran membrane to UV light (254 nm, Stratalinker UV crosslinker, Stratagene, La Jolla, CA) to cross-link the DNA to the nylon. Make sure that the side of the membrane carrying the DNA is exposed to UV (120,000 μJoules). Membrane is now ready for prehybridization. If not immediately used, place in plastic bag and store at room temperature.

3.5. Radiolabeling DNA Probe

1. Thaw all components on ice. (Keep Klenow fragment at –20°C until use).
2. Denature template DNA (1–25 μg/mL in H_2O or TE) for 2 min at 95–100°C, then keep on ice.
3. For a 50 μL reaction with 25 ng DNA, expect 25×10^6 cpm (specific activity is greater than 1×10^9 cpm/μg). Mix:

a. x µL nuclease-free water (to volume of 50 µL).
b. 10 µL 5X labeling buffer (1X, final concentration).
c. 2 µL mixture of unlabeled dNTPs (premix unlabeled dNTPs 1:1:1 to give 20 µM each, final concentration).
d. 25 ng denatured template DNA (500 ng/mL, final concentration).
e. 2 µL nuclease-free BSA (400 µg/mL, final concentration).
f. 5 µL [α-^{32}P]dNTP, 50 µCi, 3000 Ci/mmol (333 nM, final concentration).
g. 1 µL DNA Polymerase I (Klenow, 5 U/µL) (100 U/mL, final concentration).
4. Mix gently and incubate at room temperature for 60 min.
5. Heat to 95–100°C for 2 min. Then chill on ice.
6. Add 2 µL of 0.5 M EDTA (20 mM, final concentration).
7. Just before adding to hybridization solution, heat probe to 95–100°C for 2 min and then chill on ice. This denatures the double-stranded probe allowing the strands to hybridize to their membrane-bound counterparts.

3.6. Prehydrization, Hybridization, and Washes of Membrane

1. Place membrane in a Kapak/Scotchpak Heat Sealable bag (Kapak Corporation, Minneapolis, MN). Add approx 25 mL of 3X SSC to moisten membrane (one membrane per bag). Pour liquid out and gently compress bag by rolling a 25-mL pipet over bag. Dry outside of bag.
2. Heat prehybridization solution to 65°C in water bath, add salmon sperm DNA (SS-DNA), and place 12.5 mL/membrane (400 cm^2) in bag. Do not use less. Remove as many bubbles as possible from the bag by rolling a 25 mL pipet over bag and seal bag (18" Kapak pouch heat sealer, Kapak Corporation; can be ordered through Fischer).
3. Place sealed bag in 65°C water bath for 2 h. Place weights on corners of bag to hold it under water.
4. Cut bag and drain prehybridization solution.
5. Heat hybridization solution to 65°C in water bath, add SS-DNA and denatured probe (approx 3×10^6 cpm/mL hybridization solution).
6. Place 12.5 mL per membrane of hybridization solution with P^{32}-labeled probe in bag, and carefully seal bag with as few bubbles as possible. Put this bag inside another and seal it as well. This is done in order to make certain there is no leaking of isotope into water bath.
7. Incubate sealed bags at 65°C for at least 12 h in a gently shaking water bath.
8. Dry outside of bag.
9. Cut bag and drain P^{32} hybridization solution into radioactive waste jar. Place membrane immediately in solution I at room temperature.

3.6.1. Low Stringency Washes

1. Wash membrane once in preheated solution I at 58°C for 15 min.
2. Wash once in preheated solution II at 58°C for 15 min (make sure that membranes do not stick to each other during washes).

3.6.2. High Stringency Washes

1. Wash membrane twice in preheated solution I at 65°C for 15 min.
2. Wash membrane twice in preheated solution II at 65°C (or higher as needed) for 15 min (separate membranes between washes).
3. Blot excess liquid from membrane by placing it between paper towels briefly; wrap in saran wrap and tape to a cardboard support or old piece of film.
4. Place membrane into film cassette and expose for 2 d at –70°C (or longer as required).

   _____ Cassette
   \- Intensifier screen
   \- Film
   \- Membrane with DNA facing up
   _____ Cassette

5. Develop film.
6. Place membrane into PhosphoImager cassette and expose at room temperature overnight.
7. Detect radiolabeled bands using PhosphoImager.

3.6.3. Stripping of Membranes for Repeat Use

After membranes are exposed to X-ray film, they can be stored in a refrigerator in 3X SSC for some time. Membranes can also be stored temporarily at room temperature in 3X SSC. Do not allow them to dry out. If this happens, it is impossible to strip the probe off the membrane.

1. Wash membrane twice in preheated strip solution at 95°C for 15 min. (If doing multiple membranes at once, they will adhere to each other during the washes. Separate the membranes between washes.)
2. Wash membrane twice in 3X SSC at room temperature for 30 min.
3. Place in bag.
4. Proceed with prehybridization as described.

3.7. Data Analysis

Surgical tumor specimens contain variable amounts of normal cells. As a result, it is often difficult to assess allele loss by eye. To avoid subjectivity, it is best to quantify signal intensities using a PhosphoImager. This equipment is superior to film as film is easily saturated by strong signals.

1. Quantify the signal intensity of the bands in the normal and tumor lanes of the Southern blot using the PhosphoImager and ImageQuant software according to the manufacturer's instructions.
2. For assessment of LOH, the relative signal intensities in the tumor compared to normal are then calculated as follows:

$$[\text{Allele 1 (normal)} \times \text{Allele 2 (tumor)}] / [\text{Allele 2 (normal)} \times \text{Allele 1 (tumor)}]$$

For results greater than 1.0, the reciprocal is used. Thus, signal intensity ratios range from 0.0–1.0, i.e., total absence of one allele in the tumor specimen to equal presence of both alleles in the tumor and normal.
3. For the *D11S12* marker, samples with ratios less than 0.7 were counted as showing LOH *(12)*. In order to arrive at this cut off value, the distribution of relative signal intensity values among heterozygous normal specimens was assessed. The patient closest to the mean signal intensity value was used as reference; every other specimen was compared to it. The 10th percentile of the distribution of relative signal intensity values for *D11S12* was 0.7. The 10th percentile was selected as the cut off point for distinguishing LOH from retained heterozygosity. For any marker, the actual ratio used to distinguish LOH from retained heterozygosity will depend on the quality of the hybridization signal obtained with the marker (probe) under investigation.

4. Notes

1. The protocol described here is provided in the interest of completeness. A number of suppliers have kits available for the extraction of DNA from tissue. These kits come with most of the necessary solutions and do not require the use of phenol. Two that work well are the DNA Isolation Kit for Cells and Tissues from Roche Molecular Biochemicals and the Blood and Cell Culture Kit from Qiagen (Valencia, CA).
2. For LOH analyses of chromosome 11p15.5, one of the commonly used enzyme digests is *MspI*. There are *MspI* polymorphisms at the HRAS and D11S12 loci, detectable using American Type Culture Collection (ATCC, Rockville, MD) probes pbc-N1 and pADJ762, respectively. Other polymorphisms and probes are listed in **ref. 8**.
3. Good quality agarose is available from numerous suppliers of molecular biology reagents.
4. Numerous suppliers sell nylon membranes that are suitable for use in Southern blotting. Included among these are Nytran from Schleicher and Schuell, Hybond N+ from Amersham Pharmacia Biotech, and GeneScreen Plus from Du Pont NEN (Boston, MA). In addition to the traditional upward capillary transfer method described here, Schleicher and Schuell have developed a rapid downward transfer kit that comes with precut papers and membranes (Turboblotter). While somewhat expensive, it is convenient and works well.
5. Increasing the labeling reaction time to as much as 16 h will increase the specific activity of the probe. Avoid using more than 25 ng DNA since this will result in a lower specific activity and shorter probe length.
6. The labeling reaction can be used directly in hybridization reactions. The signal to noise ratio is slightly better, however, if the probe is purified through a Sephadex G-50 column (ProbeQuant G-50 Micro column, Amersham Pharmacia Biotech) which removes all unincorporated labeled nucleotide and all labeled oligomers shorter than 20 nucleotides. Probes generated using a mixture of random hexanucleotides and a template larger than 500 bp are usually between 250–300 bp in length *(13)*.

7. Numerous suppliers have kits that are suitable for labeling DNA probes. These include Roche Molecular Biochemicals Random Primed DNA Labeling Kit (#1 004 760).
8. Poor signal and high background are two common outcomes with Southern blotting *(14)*. In order to optimize conditions, it is recommended that low stringency conditions be used the first time any hybridization is done. It is not necessary then to repeat the hybridization steps if the background is high or the signal appears to be nonspecific. The blots can simply be unwrapped and rewashed under more stringent conditions. The binding of probe DNA to both the membrane and to its complementary membrane-bound fragment is dependent on ionic strength *(11)*. Therefore decreasing the concentration of ions makes for more stringent conditions (less nonspecific binding of probe DNA). Increasing the wash temperature also increases the stringency.

References

1. Greenlee, R. T., Hill-Harmon, M. B., Murray, T., and Thun, M. (2001) Cancer statistics, 2001. *Calif. Cancer J. Clin.* **51,** 15–36.
2. Devesa, S. S., Shaw, G. L., and Blot, W. J. (1991) Changing patterns of lung cancer incidence by histological type. *Cancer Epidemiol. Biomarker Prev.* **1,** 29–34.
3. Pitterle, D. M., Jolicoeur, E. M. C., and Bepler, G. (1998) Hot spots for molecular genetic alterations in lung cancer. *In vivo* **12,** 643–658.
4. Bishop, J. M. (1991) Molecular themes in oncogenesis. *Cell* **64,** 235–248.
5. Cantley, L. C., Auger, K. R., Carpenter, C., Duckworth, B., Graziani, A., Kapeller, R., and Soltoff, S. (1991) Oncogenes and signal transduction. *Cell* **64,** 281–302.
6. Weinberg, R. A. (1995) The retinoblastoma protein and cell cycle control. *Cell* **81,** 323–330.
7. Sager, R. (1989) Tumor suppressor genes: the puzzle and the promise. *Science* **246,** 1406–1412.
8. Bepler, G. and Garcia-Blanco, M. A. (1994) Three tumor suppressor regions on chromosome 11p identified by high-resolution deletion mapping in human non-small-cell lung cancer. *Proc. Natl. Acad. Sci. USA* **91,** 5513–5517.
9. Kok, K., Naylor, S. L., and Buys, C. H. (1997) Deletions of the short arm of chromosome 3 in solid tumors and the search for suppressor genes. *Adv. Cancer Res.* **71,** 27–92.
10. Kim, S. K., Ro, J. Y., Kemp, B. L., Lee, J. S., Kwon, T. J., Fong, K. M., et al. (1997) Identification of three distinct tumor suppressor loci on the short arm of chromosome 9 in small cell lung cancer. *Cancer Res.* **57,** 400–403.
11. Southern, E. M. (1975) Detection of specific sequences among DNA fragments separated by gel electrophoresis. *J. Mol. Biol.* **98,** 503–517.
12. Bepler, G, Fong, K. M., Johnson, B. E., O'Briant, K. C., Daly, L. A., Zimmerman, P. V., et al. (1998) Association of chromosome 11 Locus D11S12 with histology, stage, and metastases in lung cancer. *Cancer Detection and Prevention* **22,** 14–19.

13. Feinberg, A. P. and Vogelstein, B. (1983) A technique for radiolabeling DNA restriction endonuclease fragments to high specific activity. *Anal. Biochem.* **132,** 6.
14. Brown, T. (1995) Hybridization analysis of DNA blots, in *Short Protocols in Molecular Biology*, 3rd ed. (Ausubel, F. M., et al., eds.), John Wiley and Sons, Inc., New York, NY, pp. 2-36–2-40.

16

Assessment of Insulin-Like Growth Factors and Mutagen Sensitivity as Predictors of Lung Cancer Risk

Xifeng Wu, He Yu, Nimisha Makan, and Margaret R. Spitz

1. Introduction

Insulin-like growth factors (IGFs) are mitogenic peptide hormones involved in the regulation of cell proliferation, differentiation, transformation, and apoptosis. The members of the IGF family include two types of peptides (IGF-1 and IGF-2), two types of cell membrane receptors (IGF-1R and IGF-2R), and six binding proteins (IGFBP1-6) *(1,2)*. This family of growth factors has several important functions. IGF-mediated activation of the IGF-1 receptor stimulates the signal transduction pathway involving MAP kinase (i.e., mitogen-activated protein kinase) and increases the expression of cyclin D1, which accelerates cell-cycle progression from G_1 to S phase *(2,3)*. IGFs also suppress programmed cell death by increasing the synthesis of Bcl proteins and inhibiting that of Bax proteins *(4,5)*. Furthermore, IGFs counteract the actions of many antiproliferative molecules, such as retinoic acid. IGFBPs normally inhibit the mitogenic action of IGFs by blocking the binding of IGFs to their receptor *(6)*. Blocking the interaction between IGFs and their receptors can abolish IGF-stimulated proliferation of lung cancer cells *(7,8)*. IGFBP-3 is the principal IGFBP in the circulation.

Two prospective studies have shown that pre-diagnostic increased levels of IGF-1 were associated with higher risk of prostate and breast cancers; increased levels of IGFBP-3 had a protective effect *(9,10)*. We evaluated whether IGFs (specifically IGF-1, IGF-2) and their major binding protein (IGFBP-3) in plasma play a role in lung cancer, within a case-control study of 204 lung cancer patients and 218 control subjects. The results of the study demonstrated that high plasma levels of IGF-1 were associated with an increased risk of lung cancer, and this association was dose-dependent *(11,12)*. IGFBP-3 was associated with a reduced risk of lung cancer *(11)*. IGF-2, however, was not associated with lung cancer risk. If these findings can be confirmed in prospective studies, measuring levels of IGF-1 and IGFBP-3 in the blood may prove to be useful in assessing the risk of lung cancer.

On the other hand, the risk of developing lung cancer not only appears to be dependent on humoral factors (e.g., IGFs) but also on interacting host factors (e.g., intrinsic carcinogen sensitivity or mutagen sensitivity). Mutagen challenge assays using peripheral lymphocytes have been used as an indirect measure of DNA repair to estimate an individual's susceptibility to cancer *(13,14)*. Two different in vitro methods of testing the mutagen sensitivity are bleomycin (a radiomimetic agent) and benzo[α]pyrene diol epoxide (BPDE; a tobacco mutagen) *(15,16)*. Studies have shown that the mutagen-sensitive phenotype is associated with impaired DNA repair capacity and that individuals who are mutagen sensitive are at heightened risk for cancer *(15–19)*. We found that there are joint effects between proliferation potential and genetic instability and lung cancer risk. When bleomycin-sensitivity, BPDE-sensitivity and higher IGF-1 levels were collectively assessed, individuals with only one risk phenotype had 1.6- to 2.5-fold increased risk of lung cancer, those with two risk phenotypes had a 5.3- to13-fold elevated risk and individuals with all three risk phenotypes had a 17-fold increased risk of lung cancer. Therefore, the data suggest that individuals with genetic instability and higher proliferation potential are at enhanced risk for lung cancer.

IGF-1, IGF-2, and IGFBP-3 levels in plasma were measured with the use of immunoassay kits. Mutagen sensitivity was measured by quantifying bleomycin- and benzo[α]pyrene diol epoxide-induced chromatid breaks in peripheral blood lymphocyte cultures. The following protocols provide instructions on measuring mutagen sensitivity and IGF concentrations in the blood.

2. Materials
2.1. Isolation and Measurement of IGF-1, IGF-2, and IGFBP-3
1. IGF-I ELISA kit from Diagnostic Systems Laboratories (Webster, TX).
2. IGF-2 ELISA kit from Diagnostic Systems Laboratories.
3. IGFBP-3 ELISA kit from Diagnostic Systems Laboratories.
4. Automatic microplate washer.
5. Microplate shaker.
6. Microplate reader (spectrophotometer).
7. Serum or plasma (heparin, citrate, or EDTA).

2.2. Mutagen Sensitivity Assay

1. Blood medium preparation: 1X RPMI 1640 powder [GIBCO], 20% fetal bovine serum (FBS) (Gibco), 100 U/mL penicillin –100 µg/mL streptomycin, 2 mM L-glutamine (Gibco), 2 g/1000 mL sodium bicarbonate (NaHCO), 1.25% (v/v), phytohemagglutinin (PHA) (Burroughs Wellcome Co.) and 10 U/mL heparin sodium salt (Gibco) solution (reconstituted in dd H_2O).
2. Bleomycin working solution: 1.5 U/mL bleomycin (Blenoxane, Nippon Kayaku Co.) in dd H_2O. The working solution can be stored at –20°C.
3. Benzo[α]pyrene diol epoxide (BPDE) working solution: 1 mM of BPDE (Midwest Research Institute) in tetrahydrofuran (Sigma).
4. Colcemid (Demecolcine) working solution: 2 µg/mL colcemid (Gibco) in Hank's Balanced Salt Solution (HBSS) without Ca^{2+} and Mg^{2+} (*see* **Note 1**).
5. Giemsa's Stain working solution: 4% Geimsa stain (Bio/Medical Specialties) in 0.01 M PBS stock solution, pH 7.0.
6. 0.06 M KCl (hypotonic solution).
7. Sodium heparin solution: 10,000 U/mL sodium heparin (Gibco) in sterilized dd H_2O, thoroughly mixed and filtered.
8. Fixative solution: 3:1 Methanol (Fisher) and Glacial acetic acid (EM Science, Fisher), mixed. Prepare fresh every day.

3. Methods

3.1. IGF-1, IGF-2, IGFBP-3 Sample and Reagent Preparation

3.1.1. IGF-1/IGF-2

Prior to the measurement, serum or plasma samples must undergo a process of acid-ethanol extraction, which separates IGFs from their binding proteins. The sample is first mixed with Sample Buffer 1 at a ratio of 1 in 500 and incubated for 30 min at room temperature. After incubation, the solution is further mixed with an equal volume of Sample Buffer 2. Both buffers are provided in the enzyme-linked immunosorbent assay (ELISA) kits. The final dilution of the sample is 1 in 1,000. However, this dilution factor does not need to be considered in the final result because it has been taken into consideration in the assay calibration.

1. Label one 12 × 75 polypropylene culture tube per sample for the extraction.
2. Add 20 µL of sample to the bottom of the culture tube.
3. Add 990 µL of Sample Buffer 1 to each tube, vortex, and incubate at room temperature (~25°C) for 30 min.
4. Add 990 µL of Sample Buffer 2 to each tube, and mix well.

3.1.2. IGFBP-3

Prior to the measurement, serum or plasma samples need to be diluted at 1:100 in an assay buffer provided in the ELISA kit. This dilution factor needs to be considered in the final result of the measurement.

1. Label one 12 × 75 glass test tube per sample for dilution.
2. Add 10 μL of sample to the bottom of the test tube.
3. Add 1,000 μL of the zero Standard (A) to each tube, and mix well.

3.1.3. Wash Solution

1. Dilute 60 mL of the Wash Concentrate provided in the kit in 1,500 mL of deionized water in a bottle used in the automated microplate washer.

3.2. IGF-1, IGF-2, IGFBP-3 Assay Procedure

3.2.1. IGF-1

1. Mark the microplate strips to be used.
2. Add 20 μL of Standards, Quality Controls, and the acid-ethanol treated serum or plasma samples to each microplate well.
3. Add 100 μL of the Assay Buffer to each well using repeat pipetter.
4. Incubate the plate at room temperature for 2 h on a microplate shaker set at a speed of 0.84–1.21 g.
5. During the last 10 min of the 2-h incubation, prepare the Antibody-enzyme Conjugate Solution by diluting 240 μL of the Antibody-enzyme Conjugate Concentrate in 12 mL of the Assay Buffer in a test tube (*see* **Note 3**).
6. After 2-h incubation, aspirate and wash each well five times with the Wash Solution using an automated microplate washer.
7. Blot dry by inverting the plate on absorbent material.
8. Add 100 μL of the Antibody-enzyme Conjugate Solution to each well using repeat pipetter.
9. Incubate the plate at room temperature for 30 min on a microplate shaker set at a speed of 0.84–1.21 g.
10. After incubation, aspirate and wash each well five times with the Wash Solution using an automated microplate washer.
11. Blot dry by inverting the plate on absorbent material.
12. Add 100 μL of the TMB Chromogen Solution to each well using repeat pipetter.
13. Incubate the plate at room temperature for 10 min on a microplate shaker set at a speed of 0.84–1.21 g.
14. Add 100 μL of the Stopping Solution to each well using repeat pipetter, and shake the plate for 30 s on a microplate shaker.
15. Read the absorbance of the solution in the well within 30 min, using a microplate reader set at a wavelength of 450 nm (*see* **Notes 4,5**).
16. Calculate the mean absorbance for each Standard, Control and serum or plasma sample if they are tested in duplicate (*see* **Note 6**).
17. Plot the log of the mean absorbance values for each of the Standards along the y-axis vs log of the IGF-1 concentrations in ng/mL along the x-axis, using a linear curve-fit.
18. Determine the IGF-1 concentrations of the Controls and samples from the standard curve by matching their mean absorbance values with the corresponding IGF-1 concentrations.

3.2.2. IGF-2

1. Mark the microplate strips to be used.
2. Add 20 μL of standards, quality controls, and the acid-ethanol treated serum or plasma samples to each microplate well.
3. Add 200 μL of the assay buffer to each well using repeat pipetter.
4. Incubate the plate at room temperature for 2 h on a microplate shaker set at a speed of 0.84–1.21*g*.
5. During the last 10 min of the 2-h incubation, prepare the antibody-enzyme conjugate solution by diluting 210 μL of the antibody-enzyme Conjugate Concentrate in 10.5 mL of the Assay Buffer in a test tube.
6. After 2-h incubation, aspirate and wash each well five times with the Wash Solution using an automated microplate washer.
7. Blot dry by inverting the plate on absorbent material.
8. Add 100 μL of the antibody-enzyme conjugate solution to each well using repeat pipetter.
9. Incubate the plate at room temperature for 30 min on a microplate shaker set at a speed of 0.84–1.21*g*.
10. After incubation, aspirate and wash each well five times with the wash solution using an automated microplate washer.
11. Blot dry by inverting the plate on absorbent material.
12. Add 100 μL of the TMB Chromogen Solution to each well using repeat pipetter (*see* **Note 7**).
13. Incubate the plate at room temperature for 10 min on a microplate shaker set at a speed of 0.84–1.21*g*.
14. Add 100 μL of the stopping solution to each well using repeat pipetter, and shake the plate for 30 s on a microplate shaker (*see* **Note 8**).
15. Read the absorbance of the solution in the well within 30 min, using a microplate reader set at a wavelength of 450 nm.
16. Calculate the mean absorbance for each standard, control, and serum or plasma sample if they are tested in duplicate.
17. Plot the log of the mean absorbance values for each of the standards along the y-axis vs log of the IGF-2 concentrations in ng/mL along the x-axis, using a linear curve-fit.
18. Determine the IGF-2 concentrations of the Controls and samples from the standard curve by matching their mean absorbance values with the corresponding IGF-2 concentrations.

3.2.3. IGFBP-3

1. Mark the microplate strips to be used.
2. Add 25 μL of standards, quality controls, and the diluted serum or plasma samples to each microplate well.
3. Add 50 μL of the assay buffer to each well using repeat pipetter.
4. Incubate the plate at room temperature for 2 h on a microplate shaker set at a speed of 0.84–1.21*g*.

5. During the last 10 min of the 2-h incubation, prepare the antibody-enzyme conjugate solution by diluting 240 µL of the antibody-enzyme conjugate concentrate in 12 mL of the assay buffer in a test tube.
6. After 2-h incubation, aspirate and wash each well five times with the wash solution using an automated microplate washer.
7. Blot dry by inverting the plate on absorbent material.
8. Add 100 µL of the antibody-enzyme conjugate solution to each well using repeat pipetter.
9. Incubate the plate at room temperature for 60 min on a microplate shaker, which should be set at a speed of 0.84–1.21g.
10. After incubation, aspirate and wash each well five times with the wash solution using an automated microplate washer.
11. Blot dry by inverting the plate on absorbent material.
12. Add 100 µL of the TMB Chromogen Solution to each well using repeat pipetter.
13. Incubate the plate at room temperature for 10 min on a microplate shaker set at a speed of 0.84–1.21g.
14. Add 100 µL of the stopping solution to each well using repeat pipetter, and shake the plate for 30 s on a microplate shaker.
15. Read the absorbance of the solution in the well within 30 min, using a microplate reader set at a wavelength of 450 nm.
16. Calculate the mean absorbance for each standard, control, and serum or plasma sample if they are tested in duplicate.
17. Plot the log of the mean absorbance values for each of the Standards along the y-axis vs log of the IGFBP-3 concentrations in ng/ml along the x-axis, using a linear curve-fit.
18. Determine the IGFBP-3 concentrations of the Controls and samples from the standard curve by matching their mean absorbance values with the corresponding IGFBP-3 concentrations.

3.3. Mutagen Sensitivity Assay

3.3.1. Harvesting Lymphocyte Cells

1. Prepare whole blood culture by adding 1 mL whole blood to 9 mL blood medium (*see* **Note 12**) in 25 cm^2 tissue flask; incubate at 37°C for 91 h for bleomycin treatment and 72 h for BPDE treatment.
2. After incubation, add 200 µL bleomycin (final concentration is 0.03 U/mL) or 20 µL of BPDE (final concentration is 2 µM); incubate cells treated with bleomycin for 4 h and cells treated with BPDE for 23 h (*see* **Note 14**) at 37°C.
3. After incubation, add 200 µL Colcemid (final concentration is 0.04 µg/mL) (*see* **Note 15**); incubate at 37°C for 1 h.
4. After incubation, pour culture into centrifuge tube to harvest.
5. Spin for 5 min at 410g.
6. Decant supernatant.
7. Resuspend pellet in 8 mL 0.06 M KCl hypotonic solution; mix thoroughly.

IGFs and Mutagen Sensitivity

8. Leave at room temperature for 15 min.
9. Add 1.5 mL fixative solution to cells in hypotonic solution (*see* **Note 16**), mix well.
10. Spin 5 min at 410*g*.
11. Discard supernatant.
12. Resuspend pellet in fixative solution, bring volume to 10 mL; spin; discard supernatant. Repeat 3X or until pellet is white.
13. Resuspend pellet in appropriate amount of fixative solution (*see* **Note 17**) to give a slightly cloudy suspension of cells.

3.3.2. Preparing Slides

1. Rinse the slides with ddi water.
2. Drop (4–6 drops) of the suspension onto the slide; let the suspension air dry (~1 min).
3. Code the slides by Lab ID number and stain with 4% Giemsa solution for 2–3 min.
4. Score chromatid breaks (*see* **Note 18**).

4. Notes

1. Before testing, raise the temperature of samples and all reagents and components in the kit to room temperature.
2. Use of an automated microplate washer is strongly recommended. Incomplete washing may affect the assay precision.
3. The antibody-enzyme conjugate solution should be freshly made immediately before its use.
4. When reading the absorbance of the microplate wells, it is necessary to program the 0 ng/mL standard (provided in the kit) as "Blank."
5. If wavelength correction is available in the microplate reader, set the instrument to dual wavelength measurements at 450 nm with background wavelength correction set at 600 or 620 nm.
6. If any samples show the concentrations higher than the value of the highest standard, the samples need to be retested after appropriate dilution with the 0 ng/mL standard.
7. The TMB Chromogen Solution should be colorless. Development of a blue color indicates reagent contamination or instability.
8. To maintain good assay precision, avoid pipetting splash, minimize variation in the substrate incubation time, and add the stopping solution in the same order and speed as those when adding the TMB solution.
9. Avoid microbial contamination of reagents, especially the antibody-enzyme conjugate concentrate and its solution.
10. Avoid contamination of the TMB Chromogen Solution with the antibody-enzyme conjugate.
11. Avoid exposure of the reagents to excessive heat or direct sunlight during storage and incubation.

12. A premade Colcemid working solution can be used as well such as KaryoMAX Colcemid Solution (Gibco). However, the final solution should be 10 μg/mL in the blood culture.
13. Take the blood medium from freezer the day before the experiment and place it in the refrigerator. Do not let it stay in refrigerator too long.
14. The BPDE should be prepared and added to the blood culture in the dark because it is light-sensitive.
15. Do not use the Colcemid working solution for more than 20 d.
16. The solution lyses the red blood cells and the suspension will turn brown.
17. Depending on the size of the pellet, add 0.5–2 mL of the fixative solution.
18. Chromatid breaks are scored on 50 metaphases per sample. Only frank chromatid breaks or exchanges are scored. Chromatid gaps or attenuated regions are disregarded. Breaks are recorded as the average number of breaks/cell.

References

1. Macauly, V. M. (1992) Insulin-like growth factors and cancer. *Br. J. Cancer* **65,** 311–320.
2. Jones, J. I. and Clemmons, D. R. (1995) Insulin-like growth factors and their binding proteins biological actions. *Endocr. Rev.* **16,** 3–34.
3. LeRoith, D., Werner, H., Beitner-Johnson, D., and Roberts, C. T. (1995) Molecular and cellular aspects of the insulin-like growth factor I receptor. *Endocr. Rev.* **16,** 143–613.
4. Parrizas, M. and LeRoith, D. (1997) Insulin-like growth factor-I inhibition of apoptosis is associated with increased expression of the bcl-cL gene production. *Endocrinology* **138,** 1355–1358.
5. Wang, L., Ma, W., Markovich, R., Lee, W. L., and Wang, P. H. (1998) Insulin-like growth factor I modulates induction of apoptotic signaling in H9C2 cardiac muscle cells. *Endocrinology* **139,** 1354–1360.
6. Jones, J. I. and Clemmons, D. R. (1995) Insulin-like growth factors and their binding proteins biological actions. *Endocr. Rev.* **16,** 3–34.
7. Ankrapp, D. P. and Bevan, D. R. (1993) Insulin-like growth factor-I and human lung fibroblast-derived insulin-like growth factor-I stimulate the proliferation of human lung carcinoma cells in vitro. *Cancer Res.* **53,** 3399–3404.
8. Favoni, R. E., de Cupis, A., Ravera, F., Cantoni, C., Pirani, P., Ardizzoni, A., et al. (1994) Expression and function of the insulin-like growth factor I system in human non-small-cell lung cancer and normal lung cell lines. *Int. J. Cancer* **56,** 858–866.
9. Chan, J. M., Stampfer, M. J., Giovannucci, E., Gann, P. H., Ma, J., Wilkinson, P., et al. (1998) Plasma insulin-like growth factor-I and prostate cancer risk: a prospective study. *Science* **279,** 563–566.
10. Hankinson, S. E., Willett, W. C., Colditz, G. A., Hunter, D. J., Michaud, D. S., Deroo, B., et al. (1998) Circulating concentrations of insulin-like growth factor-I and risk of breast cancer. *Lancet* **351,** 1393–1396.

11. Yu, H., Spitz, M., Mistry, J., Gu, J., Hong, W. K., and Wu, X. (1999) Plasma levels of insulin-like growth factor-I and lung cancer risk: case-control analysis. *J. Natl. Cancer Inst.* **91**, 151–156.
12. Wu, X., Yu, H., Amos, C. I., Hong, W. K., and Spitz, M. R. (2000) Joint effect of insulin-like growth factors and mutagen sensitivity in lung cancer risk. *J. Natl. Cancer Inst.* **92**, 737–743.
13. Hsu, T. C., Johnston, D. A., Cherry, L. M., Ramkissoon, D., Schantz, S. P., Jessup, J. M., et al. (1989) Sensitivity to genotoxic effects of bleomycin in humans: possible relationship to environmental carcinogenesis. *Int. J. Cancer* **43**, 403–409.
14. Hsu, T. C., Spitz, M. R., and Schantz, S. P. (1991) Mutagen sensitivity: a biologic marker of cancer susceptibility. *Cancer Epidemiol. Biomarkers Prev.* **1**, 83–89.
15. Wu, X., Gu, J., Amos, C. I., Jiang, H., Hong, W. K., and Spitz, M. R. (1998) A parallel study of in vitro sensitivity to benzo[a]pyrene diol epoxide and bleomycin in lung cancer cases and controls. *Cancer* **83**, 1118–1127.
16. Wu, X., Gu, J., Hong, W. K., Lee, J. J., Amos, C. I., Jiang, H., et al. (1998) Benzo[a]pyrene diol epoxide and bleomycin sensitivity and susceptibility to cancer of upper aerodigestive tract. *J. Natl. Cancer Inst.* **90**, 1393–1399.
17. Schantz, S. P., Hsu, T. C., Ainslie, N., and Moser, R. P. (1989) Young adults with head and neck cancer express increased susceptibility to mutagen-induced chromosome damage. *JAMA* **262**, 3313–3315.
18. Spitz, M. R., Lippman, S. M., Jiang, H., Lee, J. J., Khuri, F., Hsu, T. C., et al. (1998) Mutagen sensitivity as a predictor of tumor recurrence in patients with cancer of the upper aerodigestive tract. *J. Natl. Cancer Inst.* **90**, 243–245.
19. Wei, Q., Spitz, M. R., Gu, J., Cheng, L., Xu, X., Strom, S. S., et al. (1996) DNA repair capacity correlates with mutagen sensitivity in lymphoblastoid cell lines. *Cancer Epidemiol. Biomarkers Prev.* **5**, 199–204.
20. Hsu, T. C., Johnston, D. A., Cherry, L. M., Ramkisson, D., Schantz, S. P., Jessup, J. M., et al. (1989) Sensitivity to genotoxic effects of bleomycin in humans: possible relationship to environmental carcinogenesis. *Int. J. Cancer* **43**, 403–409.
21. Lawce, H. J. and Brown, M. G. (1991) Harvesting, slide-making, and chromosome elongation techniques, in *The ATC Cytogenetics Laboratory Manual* (Barch, M. J., ed.), Raven Press Ltd., New York, NY, pp. 31–104.

II

DETECTION OF ALTERED TUMOR MARKERS IN CLINICAL SAMPLES FOR DIAGNOSIS AND PROGNOSIS

B. MOLECULAR METHODS FOR PROGNOSTIC DETECTION OF ALTERED LUNG TUMOR MARKERS IN CLINICAL SAMPLES

II. PCR-Based Analyses

17

Comparative Multiplex PCR and Allele-Specific Expression Analysis in Human Lung Cancer

Tools to Facilitate Target Identification

Jim Heighway, Daniel Betticher, Teresa Knapp, and Paul Hoban

1. Introduction

When lung cancer is detected, in the majority of cases it cannot be effectively treated and the patient will die of the disease. At presentation, most thoracic tumors are currently staged as nonresectable *(1)*. This factor, coupled with the relative resistance of the disease to chemotherapeutic agents, leads to the high mortality rate. There are therefore two clear ways in which this situation might be improved: first, enhanced diagnostic strategies might allow detection of the disease at a stage when conventional treatment is more effective, and second, improved therapeutic agents would result directly in higher cure rates.

The detection of small tumors has been revolutionized by the use of sequential, low-dose computerized tomography, coupled with comparative three-dimensional image analysis *(2,3)*. However, as effective as this approach appears to be, the analysis requires expensive equipment and involves a significant level of user interpretation. As such, on the surface, it would not appear to be a good candidate for a low-cost mass screening technology.

The transition from a normal to a malignant cell is a process that involves the loss of an appropriate response to the organisational signals that each cell receives from its local environment, its neighboring cells and the extracellular matrix (ECM) *(4)*. Simply, the abnormal cell grows in situations when it should not. Thus, the developing lesion shows an increasing level of structural disorder. This process is driven by progressively higher levels of aberrant gene expression within the preneoplastic cell, which, in turn, is generated by increasing levels of internal genetic damage. These intrinsic factors of the developing lesion hold out the promise of new ways of diagnosing lung cancer. Specifically, the increasing structural disorder of the developing neoplasm appears to lead to the exfoliation of preneoplastic cells. These shed cells may provide the basis of novel diagnostics, strategies that aim to recognize either the internal genetic disorder of the exfoliated cells *(5,6)* or their atypical gene expression *(7,8)* patterns. Gene-expression analysis can be carried out simply in sputum using standard immunocytochemical tools. With highly specific antibody targets, such strategies might be developed into robust, cheap, mass screening protocols for the detection of early disease. In the future these protocols might provide information on the state, probable fate and perhaps in the future, the in vivo location of preneoplastic areas.

Detection of disease is however only part of the story. If screening is to be justified on economic and more importantly, on humanitarian grounds, a potential must exist to effectively treat the detected disease. Thus, in addition to the identification of new diagnostic targets, our focus must also fall on the identification of new therapeutic targets. Although perhaps having high relevance in terms of diagnosis and prognosis, the identification of particular genetic damage is unlikely to generate novel therapeutic targets, nor perhaps to increase our understanding of the specific biology of the cancer cell. However, one new approach—genome-wide microarray analysis of gene expression—has greatly enhanced our ability to identify novel diagnostic and therapeutic targets and will greatly improve our understanding of the neoplastic process *(9–11)*.

A human tumor is a highly complex environment, comprising neoplastic cells, various types of normal cells and matrix components. If we are to derive gene-expression data that relates to the way that human tumors behave in the patient, then we must look at gene expression in those tumor cells in their natural state, we must look at gene expression directly in the clinical material and not in derived cell lines. We therefore will outline a simple method to isolate high-quality RNA from snap-frozen normal lung and tumor samples. However, the main focus of this article rests with downstream processes. A full expression analysis of the human genome in several hundred different lung tumor/normal pairs will generate massive lists of genes that are apparently deregulated significantly in the diseased compared to the normal tissue. These changes will occur for a number of reasons: (1) there will be a set of alterations that are characteristic of the cell type of origin of each clonal tumor; (2) there will be a set of expression changes that arise as direct consequences of local genetic damage (point mutations of control regions, chromosomal translocations, gene amplifications); (3) there will be a set of genes whose expression is altered because of damage to other genes at higher points in control pathways or altered indirectly because of the particular proliferative state of the cell. Considering the exact point of activation of critical disease-associated pathways, such genes, in specific individual lesions, may be in class (2), or (3). To identify and validate diagnostic and therapeutic targets, we will need to prioritize such gene lists for detailed analysis so that work can proceed effectively on downstream validation of a manageable number of key targets. This process will be critical, as high-throughput genomics analysis of tumors will generate massive amounts of data.

If we are focused on improving basic lung cancer diagnostics, then the prioritization criteria that we use are relatively straightforward. We are looking for genes and proteins that are overexpressed in a high frequency of preneoplastic and neoplastic lesions. Importantly, such targets should not be expressed in any of the normal cells that are likely to be present in the diagnostic compartment (sputum, peripheral blood). Such gene products might well be appropriately expressed in the target cell—that is, they may be typical of the normal cell type of origin of the lesion. Such markers might therefore present poor therapeutic targets, with high toxicity for perhaps key cells within healthy lung tissue.

To define new therapeutic targets, in addition to distinguishing genes that are differentially regulated in the tumor, the prioritization will need to be more stringent. We will need to identify genes that are specifically deregulated as a direct consequence of genetic damage, that is, gene products (pathways) that are specifically driving the neoplastic phenotype. To do this, we need to couple expression data to information on genomic damage. As the mass of data that high-throughput genomics technology can generate from single lesions is large, target validation techniques must be fast. Also, given the high value and limited nature of cDNA, DNA, and protein available from each specific analyzed case, they must use very small amounts of patient-derived material.

The methods presented herein will be illustrated using two genes already strongly implicated in human neoplasia, *EMS1 (12)* and *CCND1 (13–16)* and a third, *LPHH1 (17)*, a highly unusual gene, the products of which have potential as therapeutic targets.

The methods described:

1. Comparative multiplex polymerase chain reaction (PCR), a rapid screening protocol for the comparative detection of gene amplification and homozygous loss.
2. Allele-specific analysis of gene expression.
   a. Determination of the allelic balance of gene transcripts. Rationale: If upregulation is a direct consequence of (cis-acting) genetic damage, then this upregulation should occur for only one homolog.
   b. The observation of allele-specific imbalances in gene expression as a powerful prioritization tool for genomics data analysis.

2. Materials

2.1. Extraction of Nucleic Acids from Primary Material

2.1.1. RNA and DNA

1. Trizol (Life Technologies).
2. RNase free plastic ware (1.5, 0.5 mL microcentrifuge tubes, P200, P1000 disposable micropipet tips).
3. RNase free water.
4. Chloroform.
5. Propan-2-ol.
6. 70% (v/v) ethanol.
7. Tissue lysis buffer: 75 mM NaCl, 25 mM EDTA, proteinase K, 200 µg mL^{-1}, 1% (w/v) sodium dodecyl sulfate (SDS).
8. Tris-saturated phenol, pH 8.0 (Fisher Scientific).
9. 5 M NaCl.
10. 95% Ethanol.
11. Dialysis tubing (Medicell).
12. Sterile dH$_2$O.

2.2. Comparative Multiplex PCR

1. Taq polymerase (Roche).
2. 10X *Taq* reaction buffer (Roche).
3. Deoxynucleotide triphosphates (Roche).
4. Gene specific primers, 0.5 µg µL^{-1} (MWG Biotech).

EMSF1	5′ ACAAAGGCTGATGTCTTAACTGT
EMSR1	5′ CCATTGACAGGAGGAGGGTTC
BarxF1	5′ TGAGCCTGTCATCGCAAGCAT
BarxR1	5′ AGGATATAGGGCTTGCAACATAG
Cyclin12	5′ CTCTGTGCCACAGATGTGAAG
Cyclin30	5′ TGGCAGGCCCGGAGGCAGT
Cyclin26	5′ GTGAAGTTCATTTCCAATCCGC
Cyclin27	5′ GGGACATCACCCTCACTTAC
Gen14	5′ GAAGATCAATGATGCAGACTGATG
Gen86	5′ GTTCCCTCGGGATTATTTAGAATG
Gen50R	5′ TCGATTGTTGCACCTTTGAGTC

2.2.1. Gel Electrophoresis

1. 5X TBE buffer: 54 g Tris-OH, 27 g boric acid, 20 mL 0.5 M EDTA pH 8.0, dH$_2$O to 1 L.
2. 10X loading dye: 25% (w/v) Ficol 400, 0.01% bromophenol blue.

2.3. Allele-Specific Expression Analysis
2.3.1. First Strand cDNA Synthesis
1. Reverse Transcription System (Promega).

2.3.2. RT-PCR-RFLP Analysis
1. Restriction endocleases *Scr*FI (NEB), *Ahd* I (NEB).

2.3.3. Quantitation
1. Agilent 2100 bioanalyzer (Agilent Technologies).
2. Agilent DNA 500 LabChip® Kit (for use with the Agilent 2100 bioanalyzer).

3. Methods
3.1. Extraction of Nucleic Acids from Primary Material

Our standard protocol (*see* **Note 1**) involves the cutting of three batches of 20 × 40 μM sections (average sample ~1 cm diameter). The sections are placed into (RNase-free, where appropriate) tubes held on dry ice and are subsequently stored at –80°C to await extraction of RNA, DNA, and protein. Flanking sections to each batch are mounted and conventionally stained at the time of cutting to ensure consistency of histology throughout the sampled area.

3.1.1. RNA

We have tested a number of extraction protocols for the isolation of RNA from human lung tissue. In our hands, the most efficient procedure involved the initial extraction of total RNA using Trizol, followed, if required, by a secondary isolation of mRNA. This was particularly important for normal lung tissue where we failed completely, using a number of commercial systems, to directly extract A^+ mRNA, as a consequence of complete and irreversible coagulation of the oligo dT cellulose during the procedure. All plastic ware should be RNase-free.

1. Add 1 mL of Trizol to 20, 40 μM frozen sections (held on dry ice until addition of the Trizol) in a 1.5-mL microcentrifuge tube.
2. Homogenize for 1 min by repeatedly drawing into a 1-mL micropipet.
3. Split the lysate equally into two 1.5-mL tubes.
4. Add a further 500 µL Trizol to each tube.
5. Incubate at room temperature for 5 min.
6. Spin the tubes at 12,000*g* for 10 min at 6°C.

7. Remove supernatants using a micropipet into two new tubes and add 200 µL chloroform to each.
8. Vortex tubes for 15 s and incubate at room temperature for 3 min.
9. Spin tubes at 12,000g for 15 min at 6°C.
10. Remove upper aqueous phase to fresh tubes and add 0.5 mL propan-2-ol to each. Mix by inverting the tubes several times. Hold the tubes at room temperature for 10 min.
11. Spin tubes at 12,000g for 10 min at 6°C.
12. Remove supernatant and wash pellet by adding 1 mL of 70% ethanol and re-spinning at 7500g for 5 min.
13. Repeat **step 12**.
14. Air dry the pellet for 10 min (inside a lamina flow hood if possible).
15. Resuspend pellet in 100 µL RNase-free water. Average yield should be around 125 µg for tumor samples and 36 µg for normal lung samples.
16. Store aliquoted samples at –80°C.

3.1.2. DNA

Numerous methodologies exist for the isolation of genomic DNA from human tissue. One such method is described that generates high-quality DNA suitable for almost all applications.

1. Add 3 mL tissue lysis buffer to 20, 40 µM frozen tissue sections.
2. Homogenize for 1 min by repeatedly drawing into a 1 mL micropipet.
3. Add 3 mL Tris-saturated phenol, pH 8.0 to the tube, cap and mix sample well by repeated inversion, for 60 s.
4. Spin at 16,000g for 5 min.
5. Carefully remove upper aqueous layer using a 10 mL pipet and transfer to a disposable plastic universal.
6. Add 0.3 mL 5 M NaCl and swirl to mix.
7. Add 9 mL cold (–20°C) 95% ethanol.
8. Hook DNA precipitate out of solution with a sterile inoculating needle (or micropipet tip) and transfer to a fresh tube.
9. Re-suspend DNA in 500 µL sterile dH$_2$O.
10. To remove potential PCR inhibitors, dialyze sample for 4–16 h in clamped, boiled dialysis tubing against 4 changes of 1 L of dH$_2$O.
11. Aliquot samples, freeze bulk and store one working tube at 4°C. Average yield should lie in the region of 130–150 µg for tumor samples and 50–70 µg for normal lung samples.

3.2. Comparative Multiplex PCR

Comparative multiplex PCR (or RT-PCR) is a screening technique to examine the relative levels of particular target sequences in a DNA (or cDNA) sample. As such it provides a rapid, economical, moderate throughput methodology for the analysis of key genomic changes (gene amplification, homozygous loss) in tumors. The principle is to include four (or more) instead of two PCR primers in each reaction. One pair of primers is specific for the sequence under test, the second for a control region, which may be located on the same or any other autosome. The resultant multiplex PCR is therefore coupled to some degree, in that the amount of each respective gene product is generated through a competitive reaction between the two primer pairs. This competition is generated by a sharing of reaction components. Although it is very difficult, if not impossible, to reliably relate the level of products between two conventional PCR reactions carried out in different tubes (which do not therefore share a reaction environment), the relative band intensity in a multiplex reaction is proportional to the relative initial balance of target sequence in the DNA samples under analysis. However, care must still be taken in the interpretation of multiplex PCR data as differences in tube-specific reaction conditions may conceivably affect one primer pair more than the others. To minimize experimental variation, a number of conditions should be met.

1. Test and control primers should have broadly comparable affinities for their target sequences (similar GC content, similar length).
2. DNA quality should be comparable. For example, a multiplex comparing gene copy number in DNA extracted from paraffin-blocked material against normal peripheral blood-derived DNA would be unreliable.
3. The *Taq* polymerase used should be robust and not particularly sensitive to suboptimal reaction conditions (e.g., Roche, Qiagen products).
4. Apparent copy number changes should be confirmed using fresh primers for the test region and at least one additional control locus, located on a different chromosome to the first.

Given that the above conditions have been met, the reaction itself is a standard PCR.

1. Select primer sequences for the test and control genes (the example shown compares the *EMS1* gene with *BARX1*, both localized to 11q). Predicted PCR products should ideally be in the region of 100–300 bp, around 50 bp apart. The wider the gap between products, the more likely it is that potential differences in DNA quality (e.g., level of degredation) will skew the reaction. Primers are initially suspended to 0.5 µg µL^{-1} in dH$_2$O (*see* **Note 2**).
2. 1 µL of each sample to be analyzed (~100 ng) is placed in a PCR tube. In addition to the test samples (tumors) the reaction set should include a range (8–10) of normal (2 n) DNA samples, perhaps derived from the peripheral blood of normal donors. This will show the range of multiplex generated variability. As a PCR control, one tube is left empty of target.

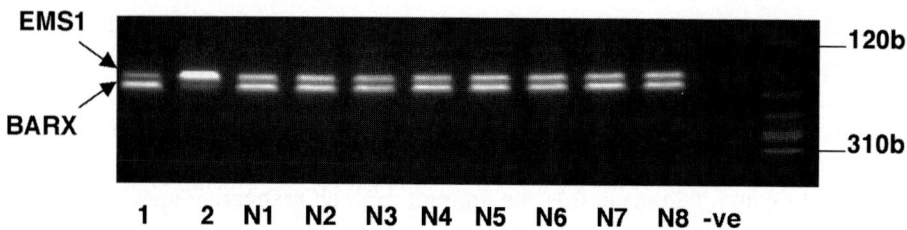

Fig. 1. A simple comparative multiplex PCR reaction (Primers EMSF1/R1, BarxF1/R1) comparing the relative genomic balance in two lung tumors of *EMS1*, located at 11q13 and a control gene, *BARX1*, mapped to 11q23. Tracks N1-8 represent products from normal genomic DNA and the relative band intensity in the tumor samples *(1,2)* is markedly shifted outside the normal range. Sample **2** has a comparative genomic hybridization (CGH) confirmed amplification of 11q13, a region which includes *EMS1* and the cyclin D1 gene, *CCND1* (which is also amplified in this lesion). Conversely, sample **1** has a presumed homozygous deletion of the same region (loss of *EMS1* signal), which, although below the resolution threshold of CGH, can be confirmed with multiple genes and multiple control loci.

3. A stock reaction mix is prepared that includes (number of tubes +1)(42 µL dH$_2$O, 5 µL 10X *Taq* reaction buffer, 0.5 µL of each primer, 0.5 µL of a 250 m*M* dNTP mix, and 1.25 U of *Taq* polymerase). The stock mix is vortexed for 10 s after *Taq* addition to mix the components.
4. Add 49 µL of the stock mix to each tube sequentially. The reaction is immediately cycled at 94°C for 2 min, 25–30 times (58°C for 1 min, 74°C for 1 min, 94°C for 1 min) with a final cycle at 58°C for 2 min and 74°C for 10 min.

3.2.1. Gel Electrophoresis

Analyze products visually under ultraviolet illumination following electrophoresis on standard 2.5% 0.5X TBE agarose gels containing ethidium bromide (0.5 µg mL^{-1}). Test sample product ratios are compared to the controls (*see* **Fig. 1**) (*see* **Note 3**).

3.3. Allele-Specific Expression Analysis

Comparative multiplex PCR is a rapid scanning methodology that can help in the association of gene expression changes with specific types of genomic damage. Allele-specific expression analysis is, in some ways, a simpler, more robust, and, above all, a more precise technique than multiplex PCR. It is an approach that facilitates the identification of genes that are linked directly to the neoplastic process by identifying those deregulated sequences, potential therapeutic targets, whose expression is altered on one chromosome copy only (or whose expression is altered disproportionately between the two

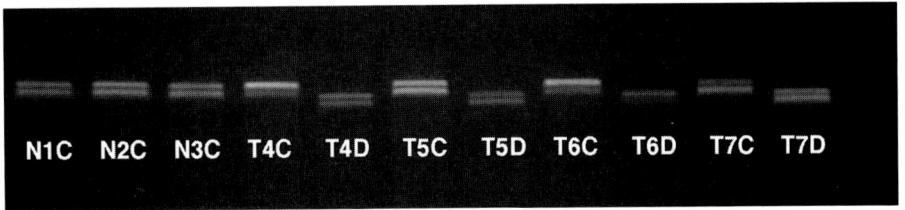

Fig. 2. (A) Allele-specific expression analysis of *LPHH1*, a novel seven-span transmembrane receptor implicated in epithelial neoplasia, in four consecutive, informative, primary lung carcinomas (track designations **T4-7**) and three normal lung tissue samples (**N1-3**). Allelic balance (relative band intensity following *AhdI* digestion) in cDNA (tracks designated **C**) was compared to the allelic balance in tumor DNA (designated **D**) for each lesion. In two samples (T4 and T7) there is an imbalance in the level of transcripts derived from the parental alleles, in relation to the allelic balance in the genomic DNA. In T6C and T6D, there is a marked allelic imbalance in both cDNA and DNA from the lesion and only in tumor T5 do both alleles appear to be present in a near-normal balance in both cDNA and tumor genomic DNA. Primers used were Gen14 and Gen86 (genomic DNA specific, tumor cDNA PCR target) and Gen14 and Gen 50R (cDNA specific). The analysis confirms this novel 7TM gene as a high-priority sequence. It encodes members of a highly drugable family of proteins, it is amplified in tumors and it shows allele-specific differences in gene expression in primary lesions. Such data clearly indicates that further target-validation studies are warranted.

parental copies of the gene). At its simplest, the technique relies on the use of transcribed sequence, single nucleotide polymorphisms (SNPs) to measure the contribution of each parental allele to the overall expression of the gene. As such, the technique is at its most powerful when, in the lesions examined, the parental alleles are represented at near-normal heterozygous levels in the tumor-derived, genomic DNA (no gene amplification). The approach therefore compliments the multiplex PCR analysis. For example, particular candidate genes within a region that has been shown to be amplified in a subset of tumors by multiplex PCR may be examined in tumors without overt amplification for evidence of allele-specific differences in expression. The observation of transcript imbalances in such genes would be a strong factor in any subsequent prioritization of such potential targets. Furthermore, this technique has an extended application in the examination of gene-expression changes in small laser microdissected groups of cells, for example in the analysis of early preneoplastic gene expression changes in histologically normal epithelia. Whilst SNPs in cDNA populations may be examined in a number of ways, the simplest method is perhaps RT-PCR-restriction fragment length polymorphism (RFLP) analysis (*see* **Fig. 2**). cDNA synthesis may be carried out simply using a number of commercially available systems. We have used the Promega, Reverse Transcription System, as we have found it to be the most robust and effective.

3.3.1. First-Strand cDNA Synthesis

1. Add 1 µL total RNA (*see* **Subheading 3.1.1.**) for each sample (approx 0.2–1 µg) to an RNase-free 0.5-mL tube.
2. Prepare a stock reaction mix from kit components, comprising [n tubes +1] (8.5 µL RNase-free dH$_2$O, 4 µL MgCl$_2$, 2 µL of 10X reaction buffer, 2 µL of dNTP mix, 1 µL oligo dT primer, 0.5 µL RNasin and 1 µL AMV reverse transcriptase). The stock is mixed by vortexing for 10 s.
3. Add 19 µL stock reaction mix to each tube and mix by once pipetting the contents up and down.
4. Incubate the synthesis reaction at 42°C for 30 min.
5. Reaction products are used immediately or stored at –20°C.

3.3.2. RT-PCR-RFLP Analysis

RT-PCR-RFLP analysis involves PCR amplification of a cDNA target and subsequent digestion of the resultant product with a restriction endonuclease, the recognition site of which spans a variable region of cDNA sequence. Generally the variant is a SNP. SNPs (*see* **Note 4**) will occur in most genes (coding, 3′ or 5′ untranslated regions) and those that do not alter a natural restriction site can generally be used to design a new variable site through the use of adjacent PCR primers with sequence mismatches near the 3′ end *(18,19)*. As most RNA contains contaminating genomic DNA, RT-PCR primers should ideally be cited in such a way that they amplify across a large region of genomic sequence (an intron), thereby making the product cDNA specific. However, they should also be tested against actual genomic DNA to exclude the possibility that they effectively amplify a processed pseudogene.

As with multiplex PCR, certain conditions should be observed. Specifically, the levels of products obtained through the RT-PCR reactions should be broadly comparable. This is to minimize and standardize the contribution to allele intensities created by heteroduplex (cut/non-cut) formation, which tends to bias digested product ratios towards the noncutting allele.

Once the primers are tested and validated, the experimental procedure is very simple.

1. RT-PCR reactions are set up as described for the multiplex PCR reaction (*see* **Subheading 3.2.**). The only difference is that only two primers are used to amplify a single locus (with 0.5 µL more dH$_2$O added to each tube) and a cDNA target (from the tumour and also from the normal tissue) is used instead of genomic DNA.

2. Similarly, a conventional PCR is set up using genomic DNA from the tumor to show the actual allelic balance at the gene level. This DNA can be derived from the tumor as described in **Subheading 3.1.2.**, amplified from contaminating genomic DNA in the RNA (or cDNA) preparation (using genomic DNA-specific primers), or it can be derived through the Trizol extraction procedure, from the material used to generate the RNA.
3. Cycle reactions as described in **Subheading 3.2.** (with longer extension times at 74°C for products longer than 1 kb).
4. Products can be checked on standard 0.5X TBE agarose gels and then an aliquot digested with the appropriate restriction enzyme, one that visualizes the RFLP. At this stage, prior to digestion, PCR products may be cleaned with systems such as the Promega PCR Prep Kit. However, this is generally not necessary for restriction enzyme digestion.
5. Digest 10 µL of the final product (~0.25–0.5 µg) in a 40 µL reaction containing 26 µL dH$_2$O, 4 µL reaction buffer and 5 U of the restriction endonuclease. Incubate the reaction for 2 h at the appropriate temperature.
6. Subsequent products may be visualized on 2.5% 0.5X TBE agarose gels (**Fig. 2**).

3.3.3. Quantitation

Although the comparison of allelic balance between tumor DNA and cDNA generally can be made clearly by visual examination of the agarose gel, a more satisfactory discrimination may be made using the Agilent 2100 bioanalyzer. This system gives both a pictorial and digital readout of the size and quantity of the DNA components of any complex solution applied to the analysis chip (*see* **Note 5**). The trace from the bioanalyzer can be compared with the same samples, which show an allele-specific expression imbalance in tumor cyclin D1 transcripts (*see* **Fig. 3**) run on a conventional agarose gel. This approach has the advantage that the machine run time is short and the data derived is directly quantitative, thereby speeding up any subsequent statistical analysis.

4. Notes

1. The most critical step in any nucleic acid or protein extraction process from solid tissue is to provide rapid access of the lysis reagents to the whole sample. This is especially important for human clinical material, which is generally limited and very valuable. The most effective way that we have identified to maximize the efficiency of such extractions is to use frozen sections. The sectioned tissue for most purposes may then be treated as equivalent to a cell suspension.
2. Initially primer pairs should be used in the multiplex reaction at equal concentrations. In many cases, such as the example shown in **Fig. 1**, this will result in equal PCR product band intensities between test and control loci when amplifying from normal DNA. However, should the relative band intensities not be matched

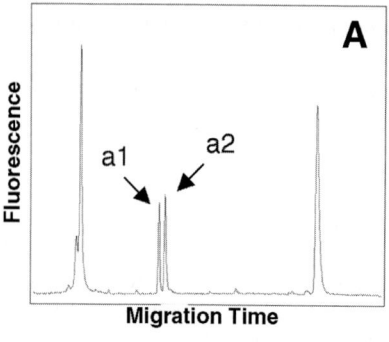
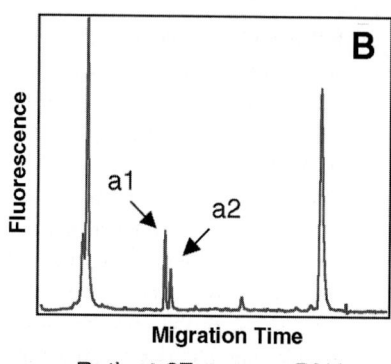
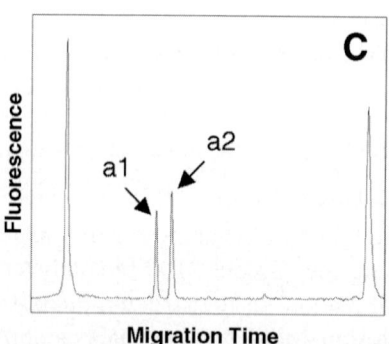
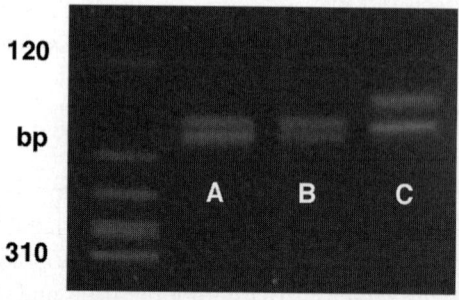

Fig. 3. An allele-specific expression analysis of the cyclin D1 gene in cDNA isolated from a patients normal lung (**A**) and tumor (**B**). The allelic balance in the tumor expression (visualized by digestion of the PCR product (cDNA specific primers cyclin12/cyclin30) with *Scr*FI can be compared to the allelic balance in genomic DNA (**C**) isolated from the lesion (primers cyclin 26/27). Although the reaction products may be visualized on a conventional gel, use of the Agilent 2100 bioanalyzer allows a rapid and accurate quantitation of that allelic balance (panels A–C). As for *LPHH1*, the allele-specific expression analysis, supported by the multiplex PCR data, has shown deregulation of *CCND1* is likely to be important in lung carcinogenesis. Had we not already known this, such priotorization tools would have been extremely valuable in moving from microarray-generated data towards the study of a very interesting and potentially clinically useful gene.

in the normal samples, then the relative primer concentration should be adjusted in the multiplex reaction. This is largely a process of trial and error and is best addressed by testing a series of dilutions of the primer pair generating the stronger band, in new multiplex reactions, on small groups of control DNA samples.
3. Alternatively, a system can be used such as the Agilent 2100 bioanalyzer, which generates an immediate quantitation of specific product levels through a chip based, computer-linked, reader.
4. SNPs within the coding sequence of genes may be identified through querying public (http://genome.ucsc.edu/goldenPath/hgTracks.html) or commercial (http://www.incyte.com) databases or identified experimentally, through sequencing cDNA, derived from high-priority candidate genes, in multiple individuals.
5. The DNA 500 LabChip® Kit was used to measure allelic balance in 1 µL aliquots of the 40 µL restriction digests, exactly as directed in the Agilent kit protocol.

References

1. Bulzebruck, H., Bopp, R., Drings, P., Bauer, E., Krysa, S., Probst, G., et al. (1992) New aspects in the staging of lung cancer. Prospective validation of the International Union Against Cancer TNM classification. *Cancer* **70,** 1102–1110.
2. Henschke, C. I., McCauley, D. I., Yankelevitz, D. F., Naidich, D. P., McGuinness, G., Miettinen, O. S., et al. (1999) Early Lung Cancer Action Project: overall design and findings from baseline screening. *Lancet* **354,** 99–105.
3. Sone, S., Takashima, S., Li, F., Yang, Z., Honda, T., Maruyama, Y., et al. (1998) Mass screening for lung cancer with mobile spiral computed tomography scanner. *Lancet* **351,** 1242–1245.
4. Huang, S. and Ingber, D. E. (1999) The structure and mechanical complexity of cell-growth control. *Nature Cell Biol.* **1,** E131–E138.
5. Tockman, M. S., Mulshine, J. L., Piantadosi, S., Erozan, Y. S., Gupta, P. K., Ruckdeschel, J. C., et al. (1997) Prospective detection of preclinical lung cancer: results from two studies of heterogeneous nuclear ribonucleoprotein A2/B1 overexpression. *Clin. Cancer Res.* **3,** 2237–2246.
6. Fielding, P., Turnbull, L., Prime, W., Walshaw, M., and Field, J. K. (1999) Heterogeneous nuclear ribonucleoprotein A2/B1 up-regulation in bronchial lavage specimens: A clinical marker of early lung cancer detection. *Clin. Cancer Res.* **5,** 4048–4052.
7. Miozzo, M., Sozzi, G., Musso, K., Pilotti, S., Incarbone, M., Pastorino, U., and Pierotti, M. A. (1996) Microsatellite alterations in bronchial and sputum specimens of lung cancer patients. *Cancer Res.* **56,** 2285–2288.
8. Palmisano, W. A., Divine, K. K., Saccomanno, G., Gilliland, F. D., Baylin, S. B., Herman, J. G., and Belinsky, S. A. (2000) Predicting lung cancer by detecting aberrant promoter methylation in sputum. *Cancer Res.* **60,** 5954–5958.
9. Alizadeh, A. A., Eisen, M. B., Davis, R. E., Ma, C., Lossos, I. S., Rosenwald, A., et al. (2000) Distinct types of diffuse large B-cell lymphoma identified by gene expression profiling. *Nature* **403,** 503–511.

10. Bittner, M., Meltzer, P., Chen, Y., Jiang, Y., Seftor, E., Hendrix, M., et al. (2000) Molecular classification of cutaneous malignant melanoma by gene expression profiling. *Nature* **406,** 536–540.
11. Celis, J. E., Kruhoffer, M., Gromova, I., Frederiksen, C., Ostergaard, M., Thykjaer, T., et al. (2000) Gene expression profiling: monitoring transcription and translation products using DNA microarrays and proteomics. *FEBS Lett.* **480,** 2–16.
12. Schuuring, E., van Damme, H., Schuuring-Scholtes, E., Verhoeven, E., Michalides, R., Geelen, E., et al. (1998) Characterization of the EMS1 gene and its product, human Cortactin. *Cell Adhes. Commun.* **61,** 85–209.
13. Sherr, C. J. (1996) Cancer cell cycles. *Science* **274,** 1672–1677.
14. Betticher, D. C., Heighway, J., Haselton, P. S., Altermatt, H. J., Ryder, W. D. J., Ceerny, T., and Thatcher, N. (1996) Prognostic significance of *CCND1* (cyclin D1) overexpression in lung cancer. *Br. J. Cancer* **73,** 294–300.
15. Betticher, D. C., Heighway, J., Thatcher, N., and Haselton, P. S. (1997) Abnormal expression of *CCND1* and *RB1* in resection margin epithelia of lung cancer patients. *Br. J. Cancer* **75,** 1761–1768.
16. Dhar, K. K., Branigan, K., Parkes, J., Howells, R. E. J., Hand, P., Musgrove, C., et al. (1999) Expression and subcellular localisation of cyclin D1 protein in ovarian epithelial tumour cells. *Br. J. Cancer* **81,** 1174–1181.
17. White, G. R. M., Varley, J. M., and Heighway, J. (2000) Genomic structure and expression profile of *LPHH1*, a 7TM gene variably expressed in breast cancer cell lines. *BBA Gene Struct. Expr.* **1491,** 75–92.
18. White, R. M. G., Stack, M. T., Santibanez-Koref, M. F., Liscia, D. S., Venesio, T., Wang, J. C., et al. (1996) Elevated levels of loss at the 17p telomere suggest the close proximity of a tumour suppressor. *Br. J. Cancer* **74,** 863–870.
19. Heighway, J. (1991) PCR primers to convert c-Ki-*ras* into a four allele RasI system. *Nucleic Acids Res.* **19,** 6966.

18

Detection of K-ras Point Mutations in Sputum from Patients with Adenocarcinoma of the Lung by Point-EXACCT

Veerle A. M. C. Somers and Frederick B. J. M. Thunnissen

1. Introduction

Lung cancer is a disease with high incidence and mortality (1). The prognosis for patients with lung cancer is most favorable when tumors are detected early in a surgically resectable stage of non-small cell lung cancer (NSCLC). Methods that can increase the percentage of cases of lung cancer detected in early stage may theoretically lead to a decrease in mortality. However, in the past, screening programs for early detection of lung cancer that used chest X-ray of the thorax with or without sputum cytology have not been successful in reducing the high numbers of lung cancer deaths (2–5). Therefore, new approaches that use genetic alterations as potential biomarkers may be beneficial for early detection.

Genetic alterations, such as point mutations in the Kirsten (K)-ras oncogene, have been suggested to have a role in the multistep process of lung carcinogenesis. Among different histologic NSCLCs, K-ras point mutations have been detected with variable frequency in adenocarcinomas, but since the use of highly sensitive detection methods, point mutations were also found in squamous cell carcinomas, although to a limited extent (6,7).

A number of reports have described the identification of point mutations in several exfoliated cells into which tumor cells are often shed from the tumor. Sidransky et al. (8,9) first showed K-ras point mutations in exfoliated cells in the stool of patients with resectable and potentially curable colon cancer and the identification of p53 mutations in urine obtained during surgery in patients with bladder cancer. Another group was successful in the detection of those mutations in pancreatic juice and peripheral blood of patients with pancreatic adenocarcinoma *(10)*.

In lung cancer, Mao et al. *(11)* first described identification of K-ras point mutations in tumor and corresponding initially stored sputum specimens from patients who later developed adenocarcinoma of the lung. Furthermore, several other groups have shown the feasibility of the detection of K-ras codon 12 mutations in individuals suspected of having lung carcinoma by point mutation analysis in sputum or bronchoalveolar lavage fluids by using sensitive techniques that enable detection of those genetic alterations when present in a low fraction of the genetically normal cells *(12–17)*. These methods include mutant-allele-specific amplification *(14)*, polymerase chain reaction (PCR)-primer-introduced restriction analysis with enrichment of mutant alleles (PCR-PIREMA) *(12)*, enriched amplification methods such as enriched PCR *(15,18)* and enriched single-strand conformational polymorphism *(17)*, allele-specific hybridization after cloning *(11,19)* and PCR-based ligase chain reaction (LCR) *(16)*. All of these methods have some flaws with respect to tediousness, requirement of relative high target frequency, or lack of internal controls. These methods use PCR to amplify (part of) the gene of interest, with further analysis of the product. In this process solution hybridization of the amplified product is an essential step for many point-mutation detection methods. DNA amplified by PCR consists of relatively short double-stranded fragments, being able to rapidly reassociate after heat denaturation, thereby reducing hybridization efficiency. Furthermore, other components of the amplification reaction including deoxynucleotides and primers could impede hybridization. Therefore, several strategies have been developed to obtain single-stranded amplified DNA as template for hybridization in contrast to traditional procedures using heat-denatured amplification products.

The "oligonucleotide ligation assay" is a point mutation detection assay developed to circumvent the need for electrophoresis and critical hybridization conditions *(20)*. Although this method is automated, the ability for analysis of samples with low target frequency detection is limited. This may be explained by possible reassociation of the heat-denatured amplification products, leading to suboptimal hybridization conditions. Thus, methods using solution hybridization can be improved if double-stranded amplification products are transferred to single-stranded DNA. As an alternative, T7 gene 6 exonuclease can be used to enhance the sensitivity of sequencing and point mutation detection procedures *(21)*. T7 gene 6 exonuclease converts double-stranded amplified DNA to single strands by digesting either 5' phosphorylated or 5' hydroxylated DNA. This enzyme can be blocked by incorporating phosphorothioate (S)-linkages into one of the PCR primers. After amplification, the strand initially primed by the phosphorothioate primer in the PCR is resistant to exonuclease digestion, whereas the other strand is selectively digested away, resulting in the single-stranded amplification products.

We have developed a highly sensitive method for the detection of known point mutations, Point-EXACCT (point mutation detection using exonuclease amplification coupled capture technique), which enables the detection of one cell that contains a mutation in more than 15,000 wild-type cells *(21–23)*. This method is rapid and uses internal controls for each base to be examined. In addition, this assay has proven succesful in the examination of sputum of patients with adenocarcinoma of the lung.

The 28 tumor samples in the current study included 18 resections, nine biopsy specimens (biopsy locations: bronchus, lung, and pleura), and one lymph node metastasis specimen. K-ras codon 12 alterations were found in 15 of 28 (54%) tumor samples. In 5 out of 11 K-ras tumor-positive patients (45%), mutations were also detected in sputum sample. Forty-four percent of the mutations identified in biopsy and tumor specimens from surgical resection were G to A transitions, whereas 56% of the mutations represent G to T transversions.

The earliest detection of a clonal population of cells that contained a mutation in at least one sputum sample that corresponded to the tumor was obtained 46 mo before histologic diagnosis of the tumor (patient no. 6; Table 2). With cytologic examination of three different sputum specimens in this patient, malignant cells were not diagnosed.

Tumor specimens obtained from patients diagnosed with an adenocarcinoma of the lung showed high-fraction mutations (data not shown), whereas point mutations identified in corresponding sputum samples showed low-fraction mutations, which referred to the presence of only a minority of tumor cells with a mutation among an excess of normal cells.

In conclusion: Point-EXACCT is a rapid, simple and efficient method for point-mutation detection with high sensitivity and specificity. This approach is useful for detection of mutations in ras genes in a heterogeneous cell population, where only a small fraction of tumor cells containing a mutation is present among much larger number of normal cells, as demonstrated in sputum of patients with adenocarcinoma of the lung.

2. Materials
2.1. Preparation of Template DNA
2.1.1. Preparation of Template DNA from Cell Lines

1. Cell line HL60 (ATCC, Manassas, VA).
2. Cell line H716 (ATCC).

Table 1
Sequences of the Probes for the Detection of Point Mutations in Codon 12 of the K-ras Gene with Point-EXACCT

Probe	Codon 12	Sequence
K-ras codon 12 biotin-labeled mutation specific capture probes[a]		
Base 1		
K12G	GGT	5'Bio-TA TAA ACT TGT GGT AGT TGG AGC TG -3'
K12A	AGT	5'Bio-TA TAA ACT TGT GGT AGT TGG AGC TA -3'
K12T	TGT	5'Bio-TA TAA ACT TGT GGT AGT TGG AGC TT -3'
K12C	CGT	5'Bio-TA TAA ACT TGT GGT AGT TGG AGC TC -3'
Base 2		
K12G	GGT	5'Bio-TA TAA ACT TGT GGT AGT TGG AGC TGG-3'
K12A	GAT	5'Bio-TA TAA ACT TGT GGT AGT TGG AGC TGA-3'
K12T	GTT	5'Bio-TA TAA ACT TGT GGT AGT TGG AGC TGT-3'
K12C	GCT	5'Bio-TA TAA ACT TGT GGT AGT TGG AGC TGC-3'
K-ras 3' digoxigenin-labeled detection probes[b]		
K-ras Dig1		5' - pG TGG CGT AGG CAA GAG TGC CTT G - Dig 3'
K-ras Dig2		5' - p TGG CGT AGG CAA GAG TGC CTT G - Dig 3'

[a]The specific mutation is underlined, the linker between biotin and first nucleotide is 8 C atoms.
[b]p = phosphorylation of 5'-end.

3. Cell line A549 (ATCC).
4. Cell line Calu-1 (ATCC).
5. FCS supplemented culture medium: Dulbecco's modified Eagle's medium (DMEM) (Gibco BRL, Life Technologies), 5% of FCS (Gibco BRL, Life Technologies).
6. Reagents for standard phenol-chloroform extraction (Sigma Chemical Co.).

2.1.2. Reconstruction Experiments

1. Trypsin: (Gibco BRL, Life Technologies).
2. 1X Phosphate-buffered saline (PBS): per L: 5.84 g NaCl, 2.64 g $NaH_2PO_4.2H_2O$, 4.72 g Na_2HPO_4, pH 7.2.
2. Burker counting chamber.

2.1.3. Preparation of Template DNA from Paraffin-Embedded Tumor and Sputum Tissue Samples

1. Paraffin-embedded archival formalin-fixed tissue blocks, sectioned and hematoxylin and eosin-stained.

2. Fixed and paraffin-embedded sputum samples corresponding to patients from whom tumor tissue is obtained.
3. Xylene.
4. Alcohol 100% (twice), Alcohol 70%, Alcohol 50%, H_2O.
5. Proteinase K: 10 mg/mL (Sigma Chemical Co.).

2.2. K-ras PCR

1. Sense primer KPR3 (5' TTT TTA TTA TAA GGC CTG CTG 3').
2. Antisense primer KI8 (5' TCA GAG AAA CCT TTA TCT GTA TCA AAG AAT GG 3').
3. PCR mix: 10 mM Tris-HCl, pH 8.3, 50 mM potassium chloride, 1.5 mM magnesium chloride, 250 pmole of each deoxyribonucleotide (dNTP), 1.1 µM of sense primer, 0.8 µM of antisense primer and 2.5 U of AmpliTaq DNA polymerase.
4. AmpliTaq DNA polymerase (Perkin-Elmer, Norwalk, CT).
5. DNA Thermal Cycler (Perkin-Elmer Cetus).
6. 2% Agarose gel (Gibco BRL ultrapure electropheresis grade) in 1X TBE buffer (per L: 10.8 g Tris, 5.5 g boric acid, 0.74 g EDTA).
7. Ethidium bromide solution: stock solution: 10 mg/mL.

2.3. Preparing Single-Stranded Target DNA

1. Dithiotreitol (DTT): stock concentration 1 M.
2. T7 gene 6 exonuclease (United States Biochemical, Cleveland, OH): Use at 3 U/µL.

2.4. Solution Hybridization

1. Hybridization buffer: 4X SSC, 20 mM HEPES, 2 mM EDTA, 0.15% Tween 20.
2. Digoxigenin-labeled detection probe (Isogen Biosciences, Maarssen, The Netherlands).
3. Biotinylated probes (Isogen Biosciences).

2.5. Coupling Products to Microtiter Plate

1. 96-well flat-bottomed microtiter plate (Falcon).
2. Bovine serum albumin (BSA), 2 µg/mL in PBS (Sigma Chemical Co.).
3. Regular washing buffer: PBS with 0.05% Tween 20.
4. Streptavidin ((Promega Corporation, Madison, USA): 1 µg/mL with 0.5% gelatin (Sigma).

2.6. Ligation Step

1. Ligase buffer: 30 mM Tris-HCl, pH 7.8, 10 mM $MgCl_2$, 10 mM DTT, 0.5 mM ATP.
2. T4 DNA ligase (Promega Corporation).
3. Formamide.

2.7. Denaturation

1. 0.07 M NaOH.
2. Denaturation washing buffer: 0.01 M NaOH/0.05% Tween 20.

2.8. Staining

1. Blocking reagent: 1% (w/v) in 100 mM Tris-HCl, pH 7.5, 150 mM NaCl (Boehringer Mannheim Biochemicals, Indianapolis, IN).
2. Peroxidase (POD)-conjugated anti-digoxigenin (DIG)-antibody Fab fragments (Boehringer Mannheim Biochemicals). Use at 1:2,000 dilution in 100 mM Tris-HCl, pH 7.5, 150 mM NaCl supplemented with 1% blocking reagent.
3. 3,3',5,5'-tetramethyl-benzidine (TMB) chromogen solution, 10 mg/mL (Sigma).
4. 2 M phosphoric acid.
5. Bio-Rad Novapath Microtiter plate Reader (Bio-Rad Laboratories).

3. Methods

3.1. Preparation of Template DNA

3.1.1. Preparation of Template DNA from Cell Lines (see **Note 1**)

DNA from 5 different cell lines with wild-type or mutated sequences for base 1 and 2 of codon 12 of the K-ras gene are used for PCR. HL60 and H716 cells have the wild type DNA for codon 12 of the K-ras gene (GGT) *(24,25)*. A549 contains a homozygous serine mutation (AGT) *(26)* and Calu-1 contains a cysteine mutation (TGT). Calu-1, harboring both a wild-type and activated allele, as demonstrated with sequencing *(27)*, is also used. SW480 is homozygous for the K-ras codon 12 valine mutation (GTT) *(27)*.

1. Cultured cells are maintained in FBS supplemented medium at 5% CO_2 and 37°C.
2. Isolate DNA using standard phenol-chloroform extraction: extract DNA 2× with phenol/chloroform/isoamylalcohol (25:24:1) and 1× with chloroform.

3.1.2. Reconstruction Experiments

Cell lines A549, Calu-1, and SW480 are used for reconstruction experiments with the wild-type cell line HL60. Cell line H716, which contains the wild-type locus for K-ras, is used for the converse reconstruction experiment with cell line A549, which has the homozygous mutation.

1. After standard culturing and a short trypsin digestion for Calu-1, A549, and SW480, centrifuge cells and resuspend the pellet in 1X PBS and determine the concentration in a Burker counting chamber.
2. Dilute all cell lines to a concentration of 2×10^6 cells in 1 mL 1X PBS.
3. The cell lines with a K-ras codon 12 mutation are diluted starting with 1 mL cells in 1X PBS (2×10^6 cells) mixed with 4 mL wild-type (for K-ras codon 12)

Point-EXACCT

cells in PBS. This 1:5 dilution is repeated several times, resulting in a stepwise reduction of the number of cells containing a mutation.
4. For the converse reconstruction of wild type cells (H716) with A549, the procedure is similar.
5. The reconstruction experiments should be repeated using new cell cultures and subsequent counting, mixing, DNA extraction, amplification, and point-mutation detection until repeating the complete procedure five times yields consistent results.

3.1.3. Preparation of Template DNA from Paraffin-Embedded Tumor and Sputum Tissue Samples (see **Note 2**)

Patients diagnosed with adenocarcinoma are chosen because this subtype of NSCLC shows a high prevalence of K-ras point mutations *(11,28)*. Only those patients from whom at least one sputum sample was collected before histologic diagnosis was available are chosen. In the current study, this resulted in a group of 22 out of 88 patients with adenocarcinoma of the lung who underwent resection with curative intent at De Wever Hospital between 1983 and 1993. Twenty-eight biopsy specimens and 54 sputum samples from these 22 patients were used. For clinical details *see* **ref. 29**.

1. Analyze biopsy or resection samples, preserved in archival paraffin-embedded tissue blocks.
2. Perform histopathologic classification after hematoxylin and eosin staining of formalin-fixed tissue sections, using the World Health Organization criteria *(30)*.
3. Retrieve stored (noninduced) sputum. This sample is stored by fixing in alcohol, embedding in paper, and transferring to paraffin overnight.
4. Cut 5-μm sections from the sputum paraffin blocks, stain with hematoxylin and eosin, and used for cytologic diagnosis.
5. Cut the remainder of the biopsy and sputum samples in the paraffin blocks for K-ras, briefly dewax in xylene and rehydrate.
6. Isolate DNA using proteinase K digestion (overnight incubation of samples at 56°C in 50–100 μL proteinase K buffer containing 50 mM Tris-HCl, pH 8.5, 1 mM EDTA and 0.5% Tween 20, depending on the size of the samples).

3.2. K-ras PCR

A 204 bp region of the K-ras gene is amplified as described previously with slight modification *(21)*.

1. Perform amplification using oligonucleotides of the first exon of K-ras flanking the codon 12 region: sense primer KPR3 and antisense primer KI8. The composition of these primers is the result of pilot studies with allele-specific amplification. Add 20 μL (1 μg) purified genomic DNA to 80 μL PCR mix.

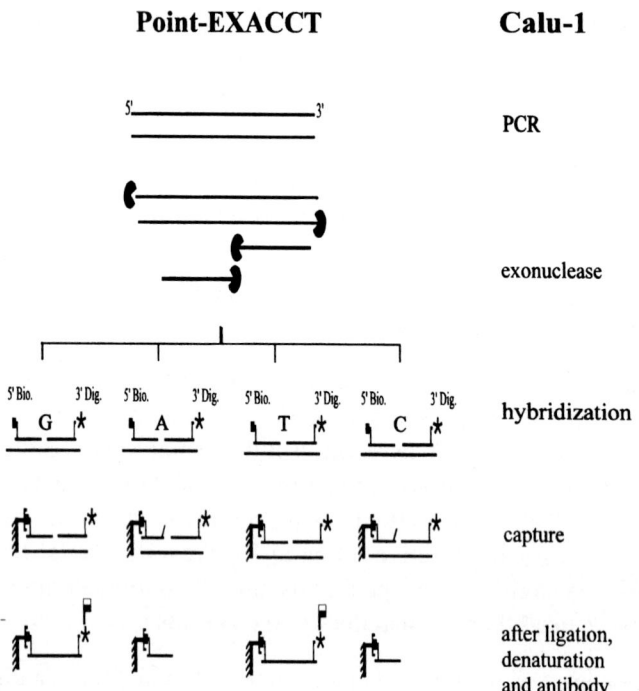

Fig. 1. After PCR a double-stranded product is present, which is digested with exonuclease to single-stranded fragments. This exonuclease only hydrolizes double-stranded DNA. The solution is then divided over different wells. These are only distinguished by a different nucleotide at the 3′ end of the biotin probe (G, A, T, or C). Biotin binds to the streptavidin coated well. All nonhybridized products are washed away. Ligase binds two adjacent probes covalently together when perfect match exists (with CALU-1 cell line G and T). After denaturation and washing digoxigenin (Dig or *) will only be present in the wells with G and T. With antibody reaction against Dig and subsequent staining reaction a color product is formed in the wells containing Dig. The amount of color is quantitatively measured.

2. Perform PCR using 40 cycles (94°C, 30 s; 55°C, 30 s; 72°C, 40 s) in an automated DNA Thermal Cycler (*see* **Note 3**).
3. After PCR take 5 µL (out of 100 µL) for electrophoresis in a 2% agarose gel. Visualize PCR product by staining with ethidium bromide (*see* **Note 4**).

3.3. Preparing Single-Stranded Target DNA

The procedure of Point-EXACCT is shown in **Fig. 1** and is slightly modified *(21)*.

1. Digest double-stranded amplification product to single strands using T7 gene 6 exonuclease. This step increases hybridization efficiency and confers increased

sensitivity and specificity of detection. Supplement PCR product with DTT to a final concentration of 1 mmol/L.
2. Digest product with 3 U/µL T7 gene 6 exonuclease for 15 min at 37°C (*see* **Notes 5–7**).
3. Heat the digested product to 75°C for 15 min to inactivate the enzyme reaction.

3.4. Solution Hybridization (see Note 8)

1. Mix the single-stranded target DNA and probes. In separate small eppendorf tubes, add 10 µL of each digested amplification product to 38 µL hybridization buffer, 1 µL (6.5 fmol) digoxigenin-labeled detection probe, and 1 µL (6.5 fmol) of one of the biotinylated probes, containing the point mutation at the 3' side (*see* **Table 1**).
2. Perform **step 1** in parallel with probes directed against base 2 of codon 12 of the K-ras gene. Thus for base 1 and 2 of codon 12 of the K-ras gene, four different biotinylated probes are separately incubated, containing either the wild-type (G) or one of the point mutations (A, T, or C).

3.5. Coupling Products to Microtiter Plate

1. Coat wells of a 96-well flat-bottomed microtiter plate with 80 µL biotinylated bovine serum albumin (BSA) for 1 h at 37°C.
2. Wash plate 3 times with 100 µL regular washing buffer and saturate with 50 µL streptavidin for 1 h at room temperature (*see* **Note 9**).
3. Transfer the complete mixture of hybridization products to an individual well in the microtiter plate (*see* **Note 10**).
4. Shake the plate carefully at room temperature at a frequency of about 60 times per minute. During this step biotinylated products are coupled to the wells (*see* **Notes 11** and **12**).
5. Wash plate three times with regular washing buffer to remove unbound products.

3.6. Ligation Step

The rational of this step is that in case of perfect complementarity, the enzyme T4 DNA ligase covalently joins the 5' biotinylated probe and the 3' detection probe. If the probes and target are mismatched at their junction, a covalent bound is not formed.

1. After the plate coupling (*see* **Subheading 3.4.**), ligate products for 15 min with 1.25 mU T4 DNA ligase in 50 µL ligase buffer supplemented with 20% formamide (*see* **Notes 13–15**).
2. Wash the wells three times with washing buffer.

3.7. Denaturation

During this step, the products that are not covalently bound to the biotinylated probe are removed.

1. Denature products by adding 50 μL 0.07 M NaOH for 2 min (*see* **Note 16**).
2. Wash wells twice with 100 μL denaturation washing buffer, and a third time with regular washing buffer.

3.8. Staining

Following denaturation, the remaining products are stained with an antibody against digoxigenin bound to an enzyme (peroxidase) which is used to induce a color reaction. All steps in the microtiter plate are performed at room temperature with shaking.

1. Add 80 μL/well POD-conjugated anti-DIG-antibody Fab fragments for 1 h.
2. Wash 3 times with washing buffer.
3. Add 100 μL/well of a TMB chromogen solution and allow color to develop for 3–5 min (*see* **Note 17**).
4. Stop color development with 50 μL/well 2 M phosphoric acid.
5. Read the plates at 450 nm in a microtiter plate reader.
6. The threshold for positivity is based on negative internal controls (A.U. relative to positive control) and should be set at the mean plus three times the standard deviation of the internal negative controls of that day's experiment. As a quality control in the daily experiments, include 4 external control samples. These include DNA of 2 different serial diluted cell lines (*see* **Subheading 2.2.**), each cell line having a ratio of mutation: wild-type of 1:3,125 and 1:15,625, respectively. If the 1:3125 sample does not work, repeat the test. Experienced hands can perform 16 samples (two microtiter plates) in a day (*see* **Notes 18–20**).

4. Notes

1. The general line of thinking for testing of the minimal target frequency starts with the amount of DNA used for PCR. Assuming a DNA content of 5–7 pg per cell, and starting with 500 ng DNA in a 50 μL reaction, in theory an equivalent of approx 50–100,000 cells is examined. Since during the procedure 8.5/50 μL reaction volume is used per well an equivalent of about 12,000–17,000 cells is present in each well. In an attempt to investigate the maximum sensitivity, up to 4 μg DNA was used in the amplification reaction. Using this high concentration of DNA, a signal above the background (10.0%) could be obtained for A549 at the 1:78,000 dilution, demonstrating the detection of 1 cell containing a mutation amidst more than 75,000 wild-type cells. For Calu-1 the relative optical density signal is close to the threshold line at this dilution (8.1% of control without ligation) (data not shown). For the converse experiment, the same results were obtained, confirming the high sensitivity of Point-EXACCT. In practice 1 cell containing a mutation amidst 75,000 and 15,000 wild-type cells was detected in cell lines A549 and Calu-1, respectively (*see* **Fig. 2A,B**). Reproducibility is high, since five times repeating of the reconstruction experiments, starting from continued cell cultures yielded the same results. The threshold line of the background values is based on internal negative controls and is set at the

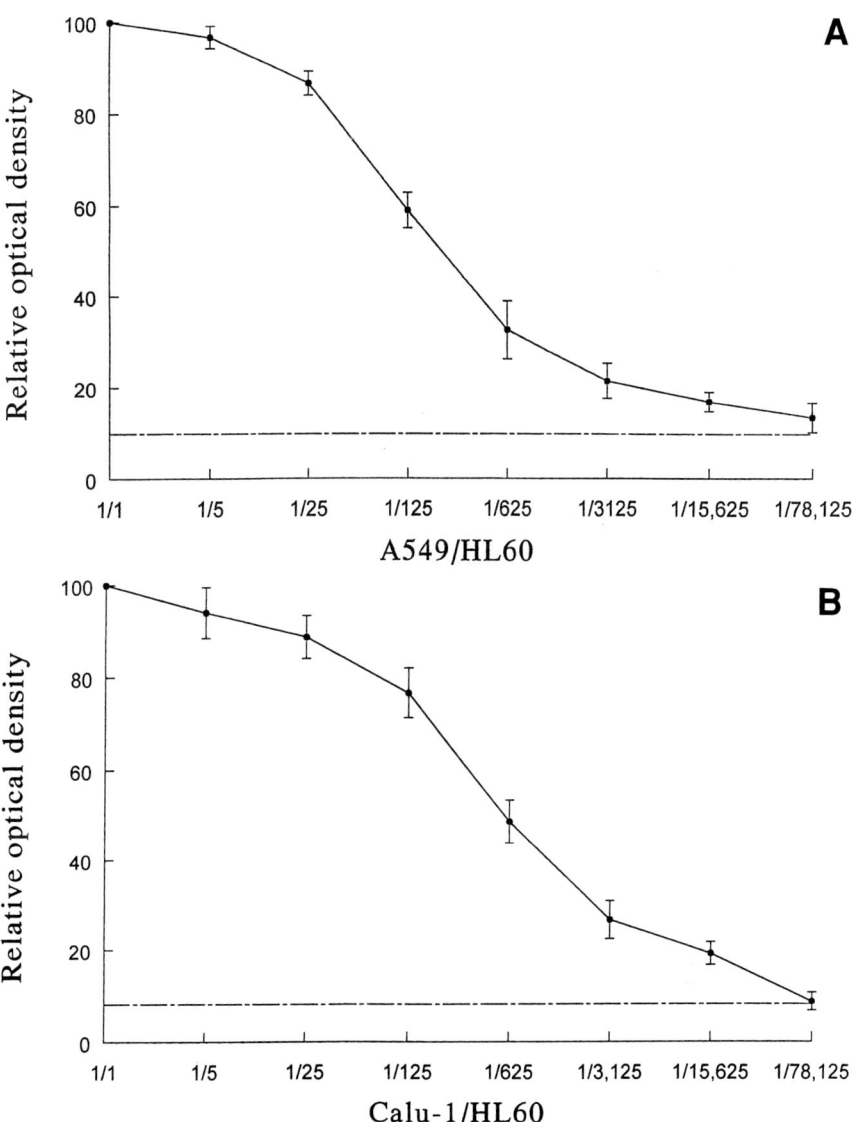

Fig. 2. Reconstruction experiments after serial dilutions of vital cells for K-ras codon 12 and wild type cell lines and testing with Point-EXACCT using 1 µg DNA in 100 µL for the amplification reaction. The results of the reconstruction experiments are shown for **(A)** cell line A549 and **(B)** cell line Calu-1. In this figure, the optical densities obtained from the different dilutions of the cell line containing a K-ras mutation in the wild type cell line HL60 were expressed relative to the (maximal) optical density of the undiluted products. The relative optical density values above the threshold values are called positive. The threshold value (represented by the broken line) is obtained by the mean plus 3x standard deviation of the relative optical density from the ligated biotinylated control probes. For A549 and Calu-1, these values are 10.5 and 8.2, respectively.

mean plus three times the standard deviation of the background values. For cell line A549 and cell line Calu-1 the threshold percentages are 10.5% and 8.0%, respectively. For these values the one sided maximal risk of false positivity is 0.5%. For higher relative optical density values this chance on false-positivity reduces rapidly. Thus, Point-EXACCT is a method able to detect samples with low target frequency.

Because the limits of the PCR reaction volume were reached using 4 µg of DNA, higher levels are not possible for the standard PCR set-up. This also gives a possible upper limit for amount of cells that can be examined in one sample. Assuming the same DNA content (pg) per cell as used in the above calculation (*see* **Note 1**) this amounts for 4 µg of DNA to an equivalent of $570–800^*10^3$ cells. This is close to, but not equal to, the detection of more then 1 in a million occasionally mentioned for other techniques. However, these usually put less template DNA in the PCR.

2. The results of the first retrospective study that examined tumor specimens and corresponding stored sputum samples obtained before clinical diagnosis for the presence of K-ras or p53 mutations were reported by Mao et al. *(11)*. The earliest detection of a clonal population of cancer cells in sputum was in a sample obtained more than 1 yr before clinical diagnosis. In our study using Point-EXACCT, the time between K-ras positivity in sputum and the first date of histologic diagnosis varied between 1 mo to almost 4 yr (46 mo), which indicates that a time period may exist for opportunity for early detection. In general, it appears that there is a period of 1–10 yr during which individuals who develop bronchogenic carcinoma exfoliate cells in their sputum, but have no clinical symptoms *(33)*. This interval, which may average 5 or more years, appears to be long enough to permit early detection and may result in improved survival. Therefore, periodic sputum analysis of groups at elevated risk of NSCLC can use this period for early detection and treatment.

Because adenocarcinomas occur in the periphery of the lung, it can be envisaged that tumor cells are less likely to be sampled in the sputum from the outer parts of the lung without induction. Nevertheless, with the use of noninduced sputum in the present study, in 45% of K-ras-positive tumors, point mutations in corresponding sputum samples were found. This was lower than that reported by Mao et al., who used sputum specimens induced by inhalation of hypertonic salt solution *(11)*.

The analysis of K-ras point mutations alone for early lung cancer diagnosis in sputum specimens may theoretically be useful for 15% of all patients with lung cancer, provided that tumor cells are present in the sample. Therefore, the investigation of more potential biomarkers for lung cancer, such as p53, is mandatory *(34,35)*. Using a combination of markers, the sensitivity of mutation detection in lung cancer diagnosis may be increased. If molecular techniques will be used to prospectively screen sputum samples in people with a high risk for lung cancer, then, e.g., K-ras mutations may be detected, whereas on radiograph and computed tomographic-scan, no tumor is visible. Because biologically K-ras is found just before the visible occurrence of lung cancer in animal studies, and in humans, it is a relatively late event in lung carcinogenesis, we believe that, when techniques with a high specificity are used, these people should be treated according to the guidelines for occult lung cancer of the American Thoracic Society and European Respiratory Society *(36)*. In addition, with the outcome of the present study, we believe that it is now worthwhile to initiate a large prospective study, in which K-ras is one of the parameters to be tested.

3. In theory, with any PCR, an error may be introduced during amplification. The chance on this error varies around 1:8,000 *(31,32)*. In order to prevent calling a sample false-positive, the whole procedure should be repeated in those cases which had a positive outcome in the first test. If the second test turns out to be positive as well, then this sample is called positive for mutation, else it is considered to be negative. In this way the chance of calling a sample false positive due to PCR error is less then approx 1^* (according to the 8,000 assumption 64^*) 10^6.

4. A strong amplified product should be obtained after PCR, reflected by a dense band on electrophoresis. This guarantees a high number of PCR products. The Point-EXACCT method is a subtle, balanced procedure requiring several steps to be adjusted to each other. Low visibility of the PCR product band in electrophoresis is suboptimal and will lead to lower optical density signals in the positive controls. The low target frequency of 1:15000 can only be reached if large optical density differences between positive and negative controls are obtained.

5. S-linkages can be directly incorporated without the use of a secondary PCR approach by incorporation of S-linkages into one of the amplification primer blocks. The desired single-stranded amplification product can then be obtained after digestion with T7 gene 6 exonuclease. This exonuclease is active in the same buffers used for PCR; thus, amplification and exonuclease digestion can be done sequentially in the same reaction tube.

6. A serial dilution of the enzyme T7 gene 6 exonuclease ranging from 0.03–4 U/µL was used for determination of exonuclease conversion to single-stranded DNA with Point-EXACCT. Even small amounts of the enzyme (0.03 U/µL) effectively converted amplification products into a detectable form, whereas higher amounts (0.3 to 3 U/µL) resulted in more sensitive detection of point mutations. The amplified products digested with 3U/µl exonuclease provided the highest optical density signal, resulting in the most sensitive detection of point mutations. Digestion with 4 U/µL exonuclease did not improve sensitivity. The hybridization efficiency after heat-denaturation alone (without exonuclease digestion) resulted in 60–70% of the relative optical density values obtained from exonuclease-treated products (*see* **Fig. 3**).

7. The oligonucleotide ligation assay was developed to circumvent the need for electrophoresis and critical hybridization conditions. As in Point-EXACCT, the oligonucleotide ligation assay is associated with a high specificity, and the time for that assay is similar to time needed for Point-EXACCT. However, in

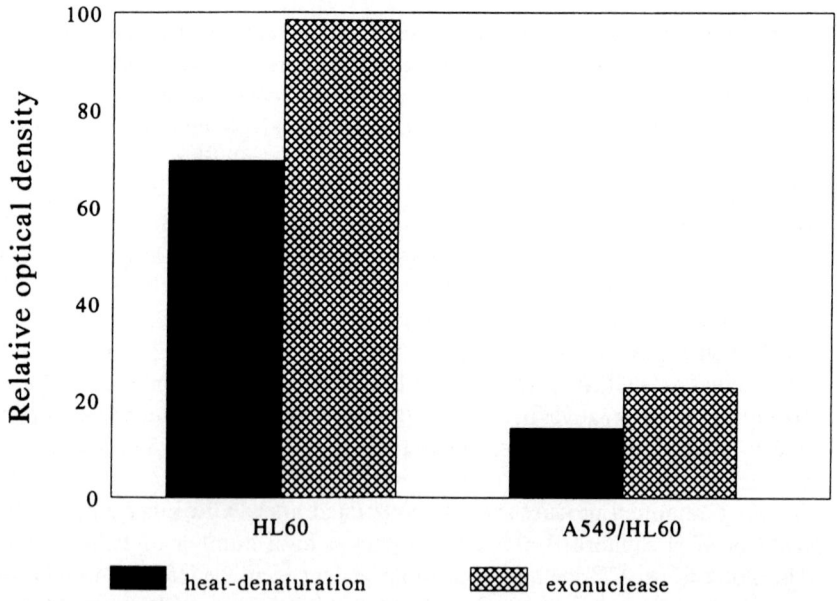

Fig. 3. Graphic display of the effect of two different approaches for denaturation: heat-induced denaturation and exonuclease digestion.

the oligonucleotide ligation assay, the sensitivity is lower, since the procedure uses heat denaturation. Using T7 gene 6 exonuclease increases hybridization efficiency. The signal difference, expressed in relative optical density, corresponds to a factor of 10–100 in cell number, demonstrating the improvement of solution hybridization after exonuclease pretreatment.
8. Compared to solid-phase hybridization, solution hybridization has a three-dimensional advantage. The Point-EXACCT procedure therefore compares favorably to Southern blotting. In fact, if the steps in **Subheadings 3.6.** and **3.7.** are omitted and one continues with the steps in **Subheading 3.7.**, the procedure will, in essence, be the same as Southern hybridization. With Point-EXACCT a high-sensitivity signal will only be obtained if the two (nested) probes are simultaneously hybridized, leading to a higher specificity. In comparison with Southern hybridization, the Point-EXACCT approach is more reliable for demonstrating that PCR products are amplified from the target gene, since two independent probes are used instead of one. Not much can go wrong during this step. Sufficient time for hybridization and the right components are the only requirements.
9. We routinely coat the wells of the microtiter plates with streptavidin just before use. We do not coat the plates beforehand, and have no experience with coated and dried plates.

10. To speed up solution handling in this step, a multichannel pipettor (8 channel) is used. Odd rows are used for sample testing, i.e., each well in this row contains a separate biotin-labeled probe, while even rows are used for additional washing of the pipet tips. In this way, possible carry over of any products from the one relevant row to the next is prevented. To the row with the "washing wells," equal volumes of washing buffer are added. Repeating the same procedure 10 times yielded a coefficient of variation (standard deviation / mean × 100) of the background values expressed in relative optical density of less than 3%.
11. More essential than originally anticipated is the shaking during this phase. In each step, the microtiter plate is slowly moved forward and back. In our hands, this mixing leads to better results then very rapid turning of the plate as is frequently used in ELISA protocols.
12. Apparently, sterical hindrance due to sequential binding of a number of molecules does not influence the outcome of this assay. The distance away from the wall of the well is created by the following molecular sequence: albumin-biotin-streptavidin- biotin-spacer-(specific)nucleotides. The spacer length is 8 C-atoms. In our hands 4 C-atoms works as well.
13. The presence of ATP is critical for ligase function. Ligase buffer should be prepared in small eppendorf tubes in advance. Preparation should be done on ice and then rapidly stored at –80°C. Just before use, a tube is thawed and then immediately stored on ice and only used once.
14. If ligase does not function properly, all mixtures may turn out to be negative. In order to check for proper functioning of the ligase, denaturation must be omitted in all probes of one additional control sample. In this sample all four probes should turn out positive, irrespective of ligase function. Optical density signals for this control should generally be between 1.1–1.3 in arbitrary units. This control will reveal the maximal optical density obtainable for the experiments on the particular day, and is dependent on the number of single stranded target DNA fragments, the two probes used for hybridization, and the staining reaction. When optical density signals for ligase function are expressed in relative values [%] compared to the same value after hybridization alone (thus without steps in **Subheadings 3.6.** and **3.7.**, maximal values set at 100%) the ligase percentage is usually higher than 85% of the maximal optical density.
15. When developing the assay, initial ligase experiments resulted in reasonable difference between positive and negative control signals. Later experiments indicated that formamide concentration needed to be slightly increased. At a still later stage, it became clear that achievement of the highest signal difference between the positive and negative controls was due to a balance of probe, ligase, and formamide concentration.
16. The NaOH stock solution should be renewed weekly, since long standing diminishes the denaturation efficiency.
17. If all steps are performed as described, clear staining should be visible within 2 min, particularly in the positive control wells. Staining may be continued up to

5–10 min, depending on the sensitivity and background of that particular day's assay. As soon as background gets slightly visible, the reaction should be ended. In the absence of distinct background staining, the reaction for color formation can be continued up to 10 min.
18. Since for each base four different biotinylated probes are tested in separate wells and a mixture of normal and possible mutated target DNA is amplified, the signal obtained from the well with the wild-type probe (guanine for base 1 and 2 of codon 12 for K-ras) is the positive internal control. One of the three remaining wells may turn out to be positive too, which reflects the specific mutation. The other two probes are internal negative controls. In theory two mutations may be present in a sample. In that case, for such a sample there will be only one negative control. As an example among 135 tumors analyzed, we found 2 patients with a double mutation.
19. The low target frequency able to be detected with PE is explained by a combination of different factors:
    a. Since, for wild-type and mutation detection different wells are used, the signals after ligation do not interfere with each other, i.e., there is no sterical competition for hybridization. We assume that one mismatch at the 3'end of the biotin probe is insufficient to prevent allele specific oligonucleotide hybridization. One limitation of the technique is more complex requirements for the amplification procedure: for 4 assay wells, 4 times the amount of PCR products are necessary.
    b. The use of exonuclease allows optimal hybridization.
    c. Before a positive signal can be obtained, several factors must converge: two different probes must hybridize to target DNA in combination with a ligation reaction between these probes. This requirement not only leads to high specificity, but is also insurance against development of PCR induced amplification errors (*see* **Note 2**).
20. The Point-EXACCT method is rapid. Starting from genomic DNA obtained from either paraffin or frozen sections, blood cells, fecal material, bronchial epithelial cells, or sputum, the detection of point mutations can be accomplished within 10 h. Therefore, analysis of point mutation detection by Point-EXACCT requires considerably less time and effort than other techniques. Another advantage of Point-EXACCT is that only small numbers of cells of DNA samples are sufficient for analysis. Therefore, this method is also useful for microdissection studies.

The high sensitivity and specificity of Point-EXACCT method therefore makes this technique highly suitable for screening purposes, where frequently less than 1 malignant cell is present in 100–1000 normal cells. Potential (pre)clinical screening applications can be performed on sputum or washings obtained during bronchoscopy from patients with lung cancer, fecal material from patients with colon cancer, or urine from patients with bladder cancer. The simplicity, reproducibility, high specificity and sensitivity of Point-EXACCT makes it a highly promising method for large-scale cancer screening. The transformation to microarray will be the next essential step.

References

1. Boring, C. C., Squires, T. S., Tong, T., et al. (1994) Cancer statistics, 1994. *CA. Cancer. J. Clin.* **44,** 7–26.
2. Frost, J. K., Ball, W. C., Levin, M. L., Tockman, M. S., Baker, R. R., Carter, D., et al. (1984) Early lung cancer detection: results of the initial (prevalence) radiologic and cytologic screening in the Johns Hopkins study. *Am. Rev. Respir. Dis.* **130,** 549–554.
3. Flehinger, B. J., Melamed, M. R., Zaman, M. B., Heelan, R. T., Perchick, W. B., and Martini, N. (1984) Early lung cancer detection: results of the initial (prevalence) radiologic and cytologic screening in the Memorial Sloan-Kettering study. *Am. Rev. Respir. Dis.* **130,** 555–560.
4. Fontana, R. S., Sanderson, D. R., Taylor, W. F., Woolner, L. B., Miller, W. E., Muhm, J. R., et al. (1984) Early lung cancer detection: results of the initial (prevalence) radiologic and cytologic screening in the Mayo Clinic study. *Am. Rev. Respir. Dis.* **130,** 561–565.
5. Frost, J. K., Ball, W. C., and Levin, W. L., (1984) Early lung cancer detection: summary and conclusions. *Am. Rev. Respir. Dis.* **130,** 565–570.
6. Vachtenheim, J., Horáková, I., Novotná, H., Opálka, P., and Roubková, H. (1995) Mutations of K-ras oncogene and absence of H-ras mutations in squamous cell carcinomas of the lung. *Clin. Cancer Res.* **1,** 359–365.
7. Nedergaard, T., Guldberg, P., Ralfklaer, E., and Zeuthen, J. (1997) A one-step DGGE scanning method for detection of mutations in the K-, N-, and H-ras oncogenes: mutations at codons 12, 13 and 61 are rare in B-cell non-Hodgkin's lymphoma. *Int. J. Cancer* **71,** 364–369.
8. Sidransky, D., Tokino, T., Hamilton, S. R., Kinzler, K. W., Levin, B., Frost, P., et al. (1992) Identification of ras oncogene mutations in the stool of patients with curable colorectal tumors. *Science* **256,** 102–105.
9. Sidransky, D., Von Eschenbach, A., Tsai, Y. C., Jones, P., Summerhayes, I., Marshall, F., et al. (1991) Identification of p53 gene mutations in bladder cancers and urine samples. *Science* **252,** 706–709.
10. Tada, M., Omata, M., Kawai, S., Saisho, H., Ohto, M., Saiki, R. K., et al. (1993) Detection of ras gene mutations in pancreatic juice and peripheral blood of patients with pancreatic adenocarcinoma. *Cancer Res.* **53,** 2472–2474.
11. Mao, L., Hruban, R. H., Boyle, J. O., Tockman, M., and Sidransky, D. (1994) Detection of oncogene mutations in sputum precedes diagnosis of lung cancer. *Cancer Res.* **54,** 1634–1637.
12. Mills, N. E., Fishman, C. L., Scholes, J., Anderson, S. E., Rom, W. N., and Jacobson, D. R., (1995) Detection of K-ras oncogene mutations in bronchoalveolar lavage fluid for lung cancer diagnosis. *J. Natl. Cancer I* **87**(14), 1056–1060.

13. Kelly, K. (1994) Evaluation of K-ras mutations in sputum samples by single-stranded conformation polymorphism. *Lung* **11(Suppl 1)**, 59.
14. Takeda, S., Ichii, S., and Nakamura, Y. (1993) Detection of K-ras mutation in sputum by mutant-allele-specific amplification (MASA). *Hum. Mutat.* **2**, 112–117.
15. Yakubovskaya, M. S., Spiegelman, V., Luo, F. C., Malaev, S., Salnev, A., Zborovskaya, I., et al. (1995) High frequency of K-ras mutations in normal appearing lung tissues and sputum of patients with lung cancer. *Int. J. Cancer* **63**, 810–814.
16. Lehman, T. A., Scott, F., Seddon, M., Kelly, K., Dempsey, E. C., Wilson, V. L., et al. (1996) Detection of K-ras oncogene mutations by polymerase chain reaction-based ligase chain reaction. *Anal. Biochem.* **239**, 153–159.
17. Marchetti, A., Buttitta, F., Carnicelli, V., Pellegrini, S., Bertacca, G., Merlo, G., et al. (1997) Enriched SSCP: A highly sensitive method for detection of unknown mutations. Application to the molecular diagnosis of lung cancer in sputum samples. *Diagn. Mol. Pathol.* **6(4)**, 185–191.
18. Ronai, Z., Yakubovskaya, M. S., Zhang, E., and Belitsky, G. A. (1996) K-ras mutation in sputum of patients with or without lung cancer. *J. Cell Biochem.* **25S**, 172–176.
19. Shiono, S., Omoe, K., and Endo A. (1996) K-ras gene mutation in sputum samples containing atypical cells and adenocarcinoma cells in the lung. *Carcinogenesis* **17**, 1683–1686.
20. Nickerson, D. A., Kaiser, R., Lappin, S., Stewart, J., and Hood, L. (1990) Automated DNA diagnostics using an ELISA-based oligonucleotide ligation assay. *Proc. Natl. Acad. Sci. USA* **87**, 8923–8927.
21. Somers, V. A. M. C., Moerkerk, P. T. M., Murtagh, J. J., Jr., and Thunnissen, F. B. J. M. (1994) A rapid, reliable method for detection of known point mutations: Point-EXACCT. *Nucleic Acids Res.* **22**, 4840–4841.
22. Somers, V. A. M. C., Leimbach, D. A., Murtagh, J. J. J., and Thunnissen, F. B. J. M. (1998) Exonuclease enhances hybridization efficiency: improved direct cycle sequencing and point mutation detection. *Biochim. Biophys. Acta* **1379(1)**, 42–52.
23. Somers, V. A. M. C., Leimbach, D. A., Theunissen, P. H. M. H., Murtagh, J. J. J., Holloway, B., Ambergen, A. W., et al. (1998) Validation of the Point-EXACCT method in non-small cell lung carcinomas. *Clin. Chem.* **44(7)**, 1404–1409.
24. Bos, J. L. (1988) The ras gene family and human carcinogenesis. *Mutat. Res.* **195**, 255–271.
25. Park, J. G., Oie, H. K., Sugarbaker, P. H., Henslee, J. G., Chen, T. R., Johnson, B. E., et al. (1987) Characteristics of cell lines established from human colorectal carcinoma. *Cancer Res.* **47**, 6710–6718.
26. Valenzuela, D. and Groffen, J. (1986) Four human carcinoma cell lines with novel mutations in positions 12 of c-K-ras oncogene. *Nucleic Acids Res.* **14(2)**, 843–851.
27. Capon, D. J., Seeburg, P. H., McGrath, J. P., Hayflick, J. S., Edma, U., Levinson, A. D., et al. (1983) Activation of Ki-ras2 gene in human colon and lung carcinomas by two different point mutations. *Nature* **304**, 507–513.

28. Mills, N. E., Fishman, C. L., Rom, W. M., Dubin, N., and Jacobson, D. R. (1995) Increased prevalence of K-ras oncogene mutations in lung adenocarcinoma. *Cancer Res.* **55,** 1444–1447.
29. Somers, V. A. M. C., Pietersen, A. M., Theunissen, P. H. M. H., and Thunnissen, F. B. J. M. (1998) Detection of K-ras point mutations in sputum from patients with adenocarcinoma of the lung by Point-EXACCT. *J. Clin. Oncol.* **16(9),** 3061–3068.
30. WHO (1982) The World Health Organization histological typing of lung tumours, second edition. *Am. J. Clin. Pathol.* **77(2),** 123–136.
31. Huang, M. M., Arnheim, N., and Goodman, M. F. (1992) Extension of base mispairs by Taq DNA polymerase: implications for single nucleotide discrimination in PCR. *Nucleic Acids Res.* **20(17),** 4567–4573.
32. Cline, J., Braman, J. C., and Hogrefe, H. H. (1996) PCR fidelity of pfu DNA polymerase and other thermostable DNA polymerases. *Nucleic Acids. Res.* **24,** 3546–3551.
33. Saccomano, G., Archer, V. E., Auerbach, O., Saunders, R. P., and Brennan, L. M. (1974) Development of carcinoma of the lung as reflected in exfoliated cells. *Cancer* **33,** 256–270.
34. Murakami, I., Fujiwara, Y., Yamaoka, N., Hiyama, K., Ishioka, S., and Yamakido, M. (1996) Detection of p53 gene mutations in cytopathology and biopsy specimens from patients with lung cancer. *Am. J. Respir. Crit. Care Med.* **154,** 1117–1123.
35. Greenblatt, M. S., Bennett, W. P., Hollstein, M., and Harris, C. C. (1994) Mutations in the p53 tumor suppressor gene: clues to cancer etiology and molecular pathogenesis. *Cancer Res.* **54,** 4855–4878.
36. Anon. (1997) Pretreatment evaluation of non-small-cell lung cancer. The American Thoracic Society and The European Respiratory Society. *Am. J. Respir. Crit. Care Med.* **156,** 320–332.

19

Detection of K-ras and p53 Mutations by "Mutant-Enriched" PCR-RFLP

Marcus Schuermann

1. Introduction

Early diagnosis of lung cancer is critical, as most cases are already inoperable at the time of diagnosis, and thus bear a grave prognosis. With increasing knowledge of the genetic aspects of lung cancer, the field has also experienced an increasing number of potential markers that might serve in the detection of changes in the lung epithelium that predispose to cancer (reviewed in **refs.** *1–3*). Of these, oncogene mutations were among the first genetic biomarkers to be studied extensively, since they fulfill many criteria that link their *de novo* appearance to the process of field cancerization *(4)*. Oncogene mutations of the *ras*- and *p53*-type are found in chronic smokers and patients with preneoplastic and neoplastic lesions but almost never in normal lung *(4)*. In animal models carcinogens present in tobacco smoke have been shown to induce typical G-T transversions that lead to missense mutations *(5)*. Moreover it has been shown that smoking leads to particular types of mutations, many of which are concentrated in particular hot spots, thus making it plausible that the same spectrum also arises in man on continuous exposure to tobacco carcinogens.

Mutations in K-*ras* are found in non-small cell lung cancer (NSCLC), predominantly in adenocarcinoma (ADC), and the rate ranges between 15% and 50%, depending on the study material and the sensitivity of the assay used *(6–12)*. The vast majority of K-*ras* mutations affect codon 12 (>90%) *(13)*. Mutations of the *p53* gene are detectable in 50–70% of lung cancers and are found in all histological types *(14–17)*. These mutations comprise both allelic loss and point mutations, the latter clustering within the "hot spot" regions of the p53 gene *(18)*. While K-*ras* mutations seem to occur late in lung cancer tumorigenesis *(19)*, somatic alterations of the *p53* gene are found in different tumor stages and may even occur in metaplastic and dysplastic states *(20–22)*.

A major problem in determining oncogene lesions in lung cancer occurs whenever tissue samples contain a large amount of genetically normal cells derived from the site of tissue sampling. This applies in particular to bronchoscopic biopsies, broncho-alveolar lavage (BAL), or brush cytology samples, which may harbor only a few cancer cells within a population of normal respiratory epithelium and alveolar macrophages. In this respect, PCR technology has generally provided a possible method for amplifying DNA from small amounts of any cell containing material, and thus, opend the gate for daily routine analysis.

Sequence alterations in oncogenes, however, comprise subtle base substitutions and therefore are difficult to detect directly. If enough clonal material is available, direct sequencing, resolution of the allele conformation by SSCP (sequence-specific conformational polymorphism) *(23)* analysis, or by denaturing gradient gel electrophoresis (DGGE) *(24)* may be advisable. With the growing knowledge on the frequency and positioning of mutations in oncogenes, a parallel development has focussed on the feasibility of direct allele detection. One method implies the selective amplification of variant alleles by using sequence-selective oligonucleotide primers, referred to as either PCR-SSO *(25,26)* or PCR-ASA (*a*llele-*s*pecific *a*mplification) *(27)*. In this method, one primer is chosen to match the sequence of an expected point mutation. Co-amplification of normal alleles is partially suppressed by adjusting primer positions and annealing temperatures in favour of the desired variant sequence serving as template. This direct PCR approach is simple, but usually requires a large set of primers, as mutations may occur in several positions of a triplicate codon. Alternatively, naturally or artificially introduced palindroms around the sequence of interest are taken as a starting point for selection of mutant alleles by restriction endonuclease digestion (referred to as PCR-RFLP: analysis by generation of artificial restriction fragment length polymorphism). A third way to amplify mutant alleles uses PCR clamping, a technique involving peptide nucleic acids which are formulated to hybridize to alleles with wild-type configuration. Due to their strong binding affinity and the inability to serve as primers for elongation by DNA-polymerase these molecules selectively suppress the amplification of normal alleles *(28,29)*. This technique can also be combined with PCR-RLFP to allow both negative and positive selection for the desired allele-status *(30,31)*.

In this chapter we describe a protocol for the enrichment of mutant alleles by a conventional, two-step, "mutant-enriched PCR" technique that has proved to be a reliable method in many laboratories *(32–34)*. Briefly, this method consists of a two-step, nested PCR in which a mismatched primer is used in the second step to introduce a restriction site into PCR products derived from normal, but not mutant, alleles. The resulting PCR product is then digested by the corresponding restriction enzyme leading to the cleavage of wild-type allelic products but not mutant ones. This second step then can then be repeated, thereby allowing the mutant alleles to serve as templates and, thus, eventually results in the accumulation of restriction endonuclease resistant PCR products. The mutation harboring fragments can be distinguished easily from remaining wild-type alleles, given their different base pair length, by the use of conventional agarose gel electrophoresis. The PCR-RFLP technique has been successfully applied to the determination of K-*ras* mutations in BAL, tumor biopsies, and microdissected tumor cell populations *(35,36)*. We have adapted these enrichment protocols and applied them to the sensitive detection of *p53* alleles mutated in several of the "hot spot" positions that are most frequently affected in lung cancer.

The protocol described below can be applied to the specific enrichment of fragments mutated in several "hot spot" regions of either the K-*ras* or the *p53* gene. The protocol ensures a comparatively high sensitivity but also provides a reliable detection of wild-type and mutant fragments in a two-step reaction of no more than 54 PCR cycles starting from genomic material.

2. Materials

2.1. DNA-Preparation

1. Qiagen Tissue Kit.
2. Scalpels.
3. Brush cytology material lysis buffer: 10 mM Tris-HCl, pH 8.3, 50 mM KCl, 2.5 mM MgCl$_2$, 0.5% Tween-20, containing proteinase K at 0.5 mg/mL.
4. 10 mM Tris-HCl, pH 8.3.

2.2. Pre-amplification

1. Genomic sample DNA dissolved in 50 µL of 10 mM Tris-HCl, pH 8.3.
2. 5X Pre-amplification buffer: 60 mM Tris-SO$_4$, pH 9.1, 18 mM (NH$_4$)$_2$SO$_4$, 1.5 mM MgCl$_2$.
3. ELONgase™ (Gibco-BRL), a mix which consists of *Taq* and *Pyrococcus species* GB-D thermostable DNA polymerases.
4. 0.2 mM of each dATP, dCTP, dGTP, dTTP dissolved in ddH$_2$O.
5. DNA-oligonucleotide primers for preamplification step.
   K-*ras* exon 1:
   K-*ras*5′: 5′ - GTATTAACCTTATGTGTGACATGTTCTAAT - 3′
   K-*ras*3′: 5′ - ACTCATGAAAATGGTCAGAGAAACCTTTATCTG - 3′

p53 exon 7:
p53p5:   5'- CCTCATCTTGGGCCTGTGTTATCTCCTAGGTTGGCT - 3' or
p53p8:   5'- CCTCCACCGCTTCTTGTCCTGCTTGCTTACCTCGCT - 3'
p53 exon 8:
p53p7:   5'- CTCTTGCTTCTCTTTTCCTATCCTGAGTAGTGGTAA - 3'
p53p2:   5'- GGTCCGTCGACTTTAGTACCTGAAGGGTGAAATAT - 3'

2.3. Purification of PCR Products

1. Quick spin columns (QIAquick PCR Purification Kit; Qiagen).

2.4. PCR-RFLP Analysis

1. *Taq* DNA polymerase (Boehringer Mannheim).
2. Reaction buffer: 2 mM Tris-HCl, 0.1 mM dithiothreitol, 10 µM EDTA, 10 µM KCl, 0.05% Nonidet P40 (v/v), 0.05% Tween 20 (v/v), 5% glycerol (v/v), pH 8.0.
3. DNA-oligonucleotide primers for PCR-RFLP step.
   K-*ras*, primer sequences (nucleotide substitutions are underlined, *see also* **Fig. 1**):
   K12MvaI: 5'-CTGCTGAAAATGACTGAATATAAACTTGTGGTAGTT GGA*C*CT- 3'
   K12as-1: 5' -CCTTTATCTGTATCAAAGAATGGTCCTGCACCAATATGC -3'
   p53, sense primer sequences:
   248CspI: 5'-ACTACATGTGTAACAGTTCCTGCATGGGCGGCA*C*GG AC - 3'
   249Bsu36I: 5'-ACATGTGTAACAGTTCCTGCATGGGCGGCATGAACC TG - 3'
   273MluI: 5'-TGGTAATCTACTGGGACGGAACAGCTTTGAG*A*CG - 3'
   p53 antisense primer sequences:
   as24x: 5'-AGAGAAAAGGAAACTGAGTGGGAGCAGTAAGATTC - 3'
   p53p8:   identical to the primer used in the pre-amplification step (*see* **Subheading 2.2.**).

2.5. Endonuclease Digestion of PCR-RFLP Products

1. *Mva*I (Boehringer Mannheim).
2. *Csp*I (Stratagene).
3. *Bsu*36I (Stratagene).
4. *Mlu*I (Boehringer Mannheim).
5. 3% ethidium bromide stained NuSieve® Agarose gel (Biozym, FMC Rockland).

3. Method
3.1. DNA-Preparation

1. For DNA analysis of sputum or bronchial lavage fluid, extract DNA from 50 µL homogenized material following the Qiagen Tissue Kit—DNA preparation protocol according to the manufactors' specifications.

"Mutant-Enriched" PCR-RFLP

A

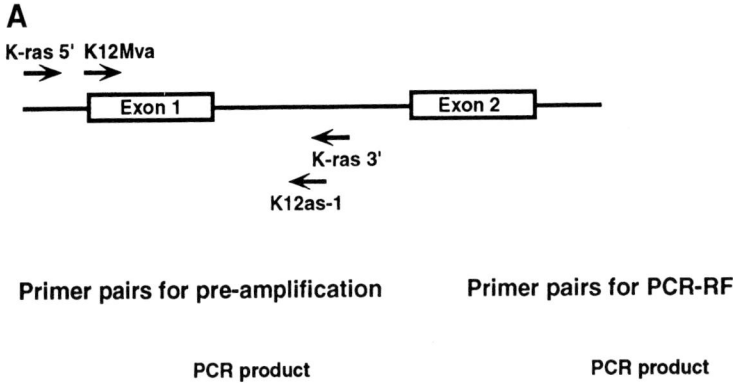

B

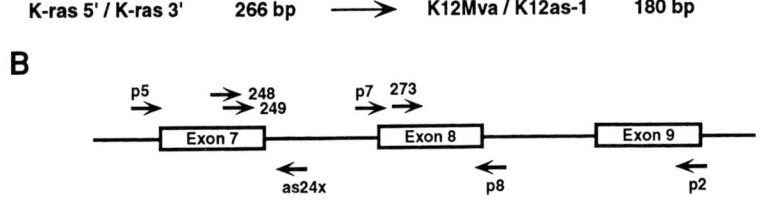

p53 codon	Primer pairs for pre-amplification		Primer pairs for PCR-RFLP	PCR product	digested PCR product
		PCR product			
248, 249	p5 / p8	647 bp	248CspI / as24x	249 bp	214 + 35 bp
			249Bsu36I / as24x	246 bp	210 + 36 bp
273	p7 / p2	350 bp	273MluI / p8	166 bp	134 + 32 bp

Fig. 1. Schematic representation of the primers used in the analysis. (**A**) location of primers used in the analysis of K-*ras*, (**B**) primers used in the detection of *p53* lesions. Shown is the exon-intron structure of the genes (not to scale), with exons given in boxes. Arrows indicate primer positions, the name of the primer is indicated above or below. The primer pairs used in the pre-amplification step (primers for 1st PCR) and in the subsequent PCR-RFLP steps (primers for PCR-RFLP) are listed below along with the size of the expected and of the digested PCR products.

2. For DNA analysis of brush cytology material, remove cell debris from the glass by scratching with a scalpel, transfer to Eppendorf tubes and incubated in 500 μL lysis buffer containing proteinase K overnight at 56°C with regular shaking. Then extract DNA using the Qiagen Tissue Kit—DNA preparation protocol.
3. DNA can be stored frozen for up to 1 yr or as a solution in 50 μL 10 m*M* Tris-HCl pH 8.3 for 4–6 wk at 4°C.

3.2. Pre-amplification

1. For the first amplification step, place 100–300 ng of either nontreated or predigested genomic DNA in a volume of 50 µL containing 5X preamplification buffer, 200 µM of each dNTP and 250 nM of each primer and 1 U ELONGase™, a mix that consists of *Taq* and *Pyrococcus species* GB-D thermostable DNA polymerases. This enzyme mix is chosen to increase the proofreading activity, leading to higher fidelity.
2. Amplification parameters are: 1 min at 94°C, 1 min at 60°C and 2 min at 72°C for a total of 30 cycles.

3.3. Purification of PCR Products

1. Purify DNA fragments resulting from the first amplification by centrifugation through quick spin columns to ensure that all residual exonuclease activity from the first amplification step are eliminated, and thus prevent any interference with the intentional introduction of site-directed mismatch positions.
2. Dilute the samples 1:100 to 1:300 and take 2 µL for PCR-RFLP analysis.

3.4. PCR-RFLP Analysis

1. Place 2 µL of an appropriate dilution from the purified products of first step PCR in a volume of 50 µL containing the following buffer components: 5X reaction buffer, 0.3 Units *Taq* DNA polymerase, 0.2 mM of each dNTP, and 350 nM of each primer.
2. Perform amplification for 24 cycles as above (*see* **Subheading 3.3.**) (*see* **Note 1**).

3.5. Endonuclease Digestion of PCR-RFLP Products

1. Digest 5 µL aliquots of the PCR-RFLP reaction with 25 U of the following respective enzymes:
   a. For K-*ras* codon 12: *Mva*I
   b. For *p53* codon 248: *Csp*I
   c. For *p53* codon 249: *Bsu*36I
   d. For *p53* codon 273: *Mlu*I
   Perform the digestion for 3 h in a total volume of 25 µL under conditions recommended by the supplier (PCR components are ignored).
2. Take 20 µL of the digestion products and electrophorese through a 3% ethidium bromide stained NuSieve® Agarose gel.

4. Notes

1. The PCR-RFLP protocol described is generally restricted to two sequential PCR steps (equaling 54 amplification cycles). For higher sensitivity subsequent enrichment steps are optional. The digestion products are then re-diluted 50-fold and 2 µL are amplified using the same conditions for PCR amplification and digestion as described (*see* **Subheading 3.4.**). In a small study of 20

bronchoscopy samples analyzed with four successive PCR steps we noted a dramatic increase in the number of mutations detected (up to 65% altered p53 alleles without knowing the significance of this finding). Although sequence analysis revealed these mutations to be natural ones, the danger arises that mutations due to *Taq*-polymerase borne errors eventually arise as soon as three to four successive PCR steps with up to 100 PCR cycles are performed. We therefore recommend to restricting to the aforementioned two-step protocol (maximum of 54 cycles), which will detect the majority of lesions. Alternatively, the pre-amplification protocol may be omitted and two subsequent PCR-RFLP steps performed allowing for an approx 10-fold enrichment. When analyzing mutations at six different p53 sites we found nearly all mutations residing in codons 248, 249, and 273, but almost no mutation in the positions 157, 175, and 245, which were analyzed in the same way. For simplicity, the experimental details for the analysis of mutations in these positions have been left out and the reader is referred to the original publication *(37)*.

References

1. Sekido, Y., Fong, K. M., and Minna, J. D. (1998) Progress in understanding the molecular pathogenesis of human lung cancer. *Biochim. Biophys. Acta* **1378,** F21–F59.
2. Salgia, R. and Skarin, A. T. (1998) Molecular abnormalities in lung cancer. *J. Clin. Oncol.* **16,** 1207–1217.
3. Sozzi, G. and Carney, D. (1998) Molecular biology of lung cancer. *Curr. Opin. Pulm. Med.* **4,** 207–212.
4. Fong, K. M., Sekido, Y., and Minna J. D. (1999) Molecular pathogenesis of lung cancer. *J. Thorac. Cardiovasc. Surg.* **118,** 1136–1152.
5. Denissenko, M. F., Pao, A., Tang, M., and Pfeifer, G. P. (1996) Preferential formation of benzo[a]pyrene adducts at lung cancer mutational hotspots in P53. *Science* **274,** 430–432.
6. Rodenhuis, S. and Slebos, R. J. (1992) Clinical significance of ras oncogene activation in human lung cancer. *Cancer Res.* **52,** 2665s–2669s.
7. Husgafvel-Pursiainen, K., Hackman, P., Ridanpaa, M., Anttila, S., Karjalainen, A., Partanen, T., et al. (1993) K-ras mutations in human adenocarcinoma of the lung: association with smoking and occupational exposure to asbestos. *Int. J. Cancer* **53,** 250–256.
8. Li, S., Rosell, R., Urban, A., Font, A., Ariza, A., Armengol, P., et al. (1994) K-ras gene point mutation: a stable tumor marker in non-small cell lung carcinoma. *Lung Cancer* **11,** 19–27.
9. Mao, L., Hruban, R. H., Boyle, J. O., Tockman, M., and Sidransky, D. (1994) Detection of oncogene mutations in sputum precedes diagnosis of lung cancer. *Cancer Res.* **54,** 1634–1637.
10. Mills, N. E., Fishman, C. L., Rom, W. N., Dubin, N., and Jacobson, D. R. (1995) Increased prevalence of K-ras oncogene mutations in lung adenocarcinoma. *Cancer Res.* **55,** 1444–1447.

11. Mitsudomi, T., Viallet, J., Mulshine, J. L., Linnoila, R. I,. Minna, J. D., and Gazdar, A. F. (1991) Mutations of ras genes distinguish a subset of non-small-cell lung cancer cell lines from small-cell lung cancer cell lines. *Oncogene* **6**, 1353–1362.
12. Rodenhuis, S., Slebos, R. J., Boot, A. J., Evers, S. G., Mooi, W. J., Wagenaar, S. S., et al. (1988) Incidence and possible clinical significance of K-ras oncogene activation in adenocarcinoma of the human lung. *Cancer Res.* **48**, 5738–5741.
13. Rodenhuis, S. and Slebos, R. J. (1990) The ras oncogenes in human lung cancer. *Am. Rev. Respir. Dis.* **142**, 27–30.
14. Chiba, I., Takahashi, T., Nau, M. M., D'Amico, D., Curiel, D. T., Mitsudomi, T., et al. (1990) Mutations in the p53 gene are frequent in primary, resected non-small cell lung cancer. Lung Cancer Study Group. *Oncogene* **5**, 1603–1610.
15. Marchetti, A., Buttitta, F., Pellegrini, S., Campani, D., Diella, F., Cecchetti, D., et al. (1993) p53 mutations and histological type of invasive breast carcinoma. *Cancer Res.* **53**, 4665–4669.
16. Takahashi, T., Suzuki, H., Hida, T., Sekido, Y., Ariyoshi, Y., and Ueda, R. (1991) The p53 gene is very frequently mutated in small-cell lung cancer with a distinct nucleotide substitution pattern. *Oncogene* **6**, 1775–1778.
17. Top, B., Mooi, W. J., Klaver, S. G., Boerrigter, L., Wisman, P., Elbers, H. R., et al. (1995) Comparative analysis of p53 gene mutations and protein accumulation in human non-small-cell lung cancer. *Int. J. Cancer* **64**, 83–91.
18. Greenblatt, M. S., Bennett, W. P., Hollstein, M., and Harris, C. C. (1994) Mutations in the p53 tumor suppressor gene: clues to cancer etiology and molecular pathogenesis. *Cancer Res.* **54**, 4855–4878.
19. Sugio, K., Kishimoto, Y., Virmani, A. K., Hung, J. Y., and Gazdar A. F. (1994) K-ras mutations are a relatively late event in the pathogenesis of lung carcinomas. *Cancer Res.* **54**, 5811–5815.
20. Fontanini, G., Vignati, S., Bigini, D., Merlo, G. R., Ribecchini, A., Angeletti, C. A., et al. (1994) Human non-small cell lung cancer: p53 protein accumulation is an early event and persists during metastatic progression. *J. Pathol.* **174**, 23–31.
21. Sozzi, G., Miozzo, M., Donghi, R., Pilotti, S., Cariani, C. T., Pastorino, U., et al. (1992) Deletions of 17p and p53 mutations in preneoplastic lesions of the lung. *Cancer Res.* **52**, 6079–6082.
22. Walker, C., Robertson, L. J., Myskow, M. W., Pendleton, N., and Dixon, G. R. (1994) p53 expression in normal and dysplastic bronchial epithelium and in lung carcinomas. *Br. J. Cancer* **70**, 297–303.
23. Orita, M., Suzuki, T., and Hayashi, K. (1989) Rapid and sensitive detection of point mutations and DNA polymorphisms using the polymerase chain reaction. *Genomics* **5**, 874–879.
24. Takahashi, N., Hiyama, K., Kodaira, M., and Satoh, C. (1990) An improved method for the detection of genetic variations in DNA with denaturing gradient gel electrophoresis. *Mutat. Res.* **234**, 61–70.
25. Anwar, K., Nakakuki, K., Shiraishi, T., Naiki, H., Yatani, R., and Inuzuka, M. (1992) Presence of ras oncogene mutations and human papillomavirus DNA in human prostate carcinomas. *Cancer Res.* **52**, 5991–5996.

26. Yashiro, T., Fulton, N., Hara, H., Yasuda, K., Montag, A., Yashiro, N., et al. (1993) Comparison of mutations of ras oncogene in human pancreatic exocrine and endocrine tumors. *Surgery* **114**, 758–763.
27. Lepage, V., Ivanova, R., Loste, M.N., Mallet, C., Douay, C., Naoumova, E., and Charron, D. (1995) Determination of DQB1 alleles using PCR amplification and allele-specific primers. *Eur. J. Immunogenet.* **22**, 413–422.
28. Ørum, H., Nielsen, P. E., Egholm, M., Berg, R. H., Buchardt, O., and Stanley, C. (1993) Single base pair mutation analysis by PNA directed PCR clamping. *Nucleic Acids Res.* **21**, 5332–5336.
29. Thiede, C., Bayerdorffer, E., Blasczyk, R., Wittig, B., and Neubauer, A. (1996) Simple and sensitive detection of mutations in the ras proto-oncogenes using PNA-mediated PCR clamping. *Nucleic Acids Res.* **24**, 983–984.
30. Behn, M., Thiede, C., Neubauer, A., Pankow, W., and Schuermann, M. (2000) Facilitated detection of oncogene mutations from exfoliated tissue material by a PNA-mediated 'enriched PCR' protocol. *J. Pathol.* **190**, 69–75.
31. Rhodes, C. H., Honsinger, C., Porter, D.M., and Sorenson, G. D. (1997) Analysis of the allele-specific PCR method for the detection of neoplastic disease. *Diagn. Mol. Pathol.* **6**, 49–57.
32. Kahn, S. M., Jiang, W., Culbertson, T. A., Weistein, I. B., Williams, G. M., Tomita, N., and Ronai, Z. (1991) Rapid and sensitive nonradioactive detection of mutant K-ras genes via 'enriched' PCR amplification. *Oncogene* **6**, 1079–1083.
33. Levi, S., Urbano-Ispizua, A., Gill, R., Thomas, D. M., Gilbertson, J., Foster, C., and Marshall, C. J. (1991) Multiple K-ras codon 12 mutations in cholangiocarcinomas demonstrated with a sensitive polymerase chain reaction technique. *Cancer Res.* **51**, 3497–3502.
34. Jacobson, D. R. and Mills, N. E. (1994) A highly sensitive assay for mutant ras genes and its application to the study of presentation and relapse genotypes in acute leukemia. *Oncogene* **9**, 553–563.
35. Sugio, K., Kishimoto, Y., Virmani, A. K., Hung, J. Y., and Gazdar, A. F. (1994) K-*ras* mutations are a relatively late event in the pathogenesis of lung carcinomas. *Cancer Res.* **54**, 5811–5815.
36. Mills, N. E., Fishman, C. L., Scholes, J., Anderson, S. E., Rom, W. N., and Jacobson, D. R. (1995) Detection of K-ras oncogene mutations in bronchoalveolar lavage fluid for lung cancer diagnosis. *J. Natl. Cancer Inst.* **87**, 1056–1060.
37. Behn, M., Qun, S., Pankow, W., Havemann, K., and Schuermann, M. (1998) Frequent detection of ras and p53 mutations in brush cytology samples from lung cancer patients by a restriction fragment length polymorphism- based "enriched PCR" technique. *Clin. Cancer Res.* **4**, 361–371.

20

Detection of Small Cell Lung Cancer by RT-PCR for Neuropeptides, Neuropeptide Receptors, or a Splice Variant of the Neuron Restrictive Silencer Factor

Judy M. Coulson, Samreen I. Ahmed, John P. Quinn, and Penella J. Woll

1. Introduction

Small cell lung cancer (SCLC) comprises a significant fraction of all lung cancers; it is most frequent in women, where it represents up to 25% *(1)*. It is characterized by neuroendocrine differentiation, with immunohistochemistry for certain neuroepithelial markers used in diagnosis to differentiate SCLC from other types of lung cancer. We have evaluated the use of reverse transcription polymerase chain reaction (RT-PCR) for selected neuroendocrine markers as a way to specifically detect SCLC cells in (1) surgical, postmortem or bronchoscopic biopsies and (2) the peripheral blood or lymph node aspirates from SCLC patients. Products are detected either directly by ethidium bromide staining of agarose gels or, with enhanced sensitivity, by transfer to a membrane followed by hybridization with a specific radiolabeled probe.

1.1. Neuropeptides and Receptors

SCLC cells express a range of neuropeptides and often also their cognate receptors (*see* Chapter 4 in volume 1.) Typically, however, a specific neuropeptide is detected in a subset of SCLC and not all express the same neuropeptide receptors. Therefore accurate phenotyping may require the use of multiple neuropeptide markers. We have evaluated differential expression of several neuropeptides and their receptors by RT-PCR in a panel of cell lines, to select those most suitable to be used as tumor markers *(2)*. From these studies, gastrin-releasing peptide (GRP) and the cholecystokinin B receptor (CCKBR) were the most commonly expressed in SCLC and most discriminating between lung tumor types. We and others have also demonstrated that arginine vasopressin (AVP) is an excellent marker, being expressed in the majority of SCLC *(3–5)*. We found AVP to be expressed in all SCLC cell lines tested and have gone on to characterize the transcriptional regulation of this neuropeptide in SCLC *(6,7)*. These studies have led to the identification of several transcription factors with differential expression between SCLC and non-small cell lung cancer (NSCLC) *(8,9)*, one of which is discussed in **Subheading 1.2.**

Nucleated cells collected from the blood of normal healthy volunteers have been evaluated for expression of the neuropeptide and receptor mRNA markers. AVP and CCKBR appeared to be the markers best suited to screen the peripheral blood of SCLC patients for circulating micrometastases, as they were not detected in the healthy control subjects *(10,11)*. In contrast, the GRP transcript was detected in one control sample. GRP is known to be expressed in neuroendocrine cells of the normal lung, but at much higher levels in fetal cells during development than in the adult *(12,13)*. In our study we did not detect GRP in primary cells derived from normal bronchial epithelium *(2)*.

1.2. A Splice Variant of NRSF

The neuron restrictive silencer factor (NRSF or REST) is a transcriptional silencer of neuronal genes *(14–16)*. We have identified a binding site for NRSF in the AVP promoter and implicated this factor in deregulation of AVP gene expression in SCLC *(7)*. We recently described a novel splice variant of the NRSF transcript in SCLC, which we named sNRSF *(8)*. The splice variant is highly expressed in SCLC and incorporates an additional 50 bp exon between exons 5 and 6 (*see* **Fig. 1**) making it easily distinguishable by RT-PCR techniques. This variant was detected at high levels in both established SCLC cell lines and primary SCLC cultures, but not in NSCLC lines or cultured normal human bronchial epithelial cells. We demonstrated high-level expression of the variant in seven out of eight SCLC cell lines tested, with expression at a lower ratio relative to the wild-type NRSF (wtNRSF) in the eighth, a variant SCLC line (*see* **Fig. 1**). sNRSF appeared to be an early marker in lymph node-derived primary SCLC cells, as it was very highly expressed over wtNRSF in a pre-treatment SCLC sample (*see* **Fig. 2**). The relative expression in the post-treatment sample from the same patient was reduced, with equivalent sNRSF and wtNRSF amplification products, similar to that seen in the majority of established SCLC cell lines.

In terms of SCLC biology, high expression of the sNRSF transcription factor splice variant may be an early step in defining the neuroendocrine phenotype of these tumors, by upregulating expression of several neuropeptide and other

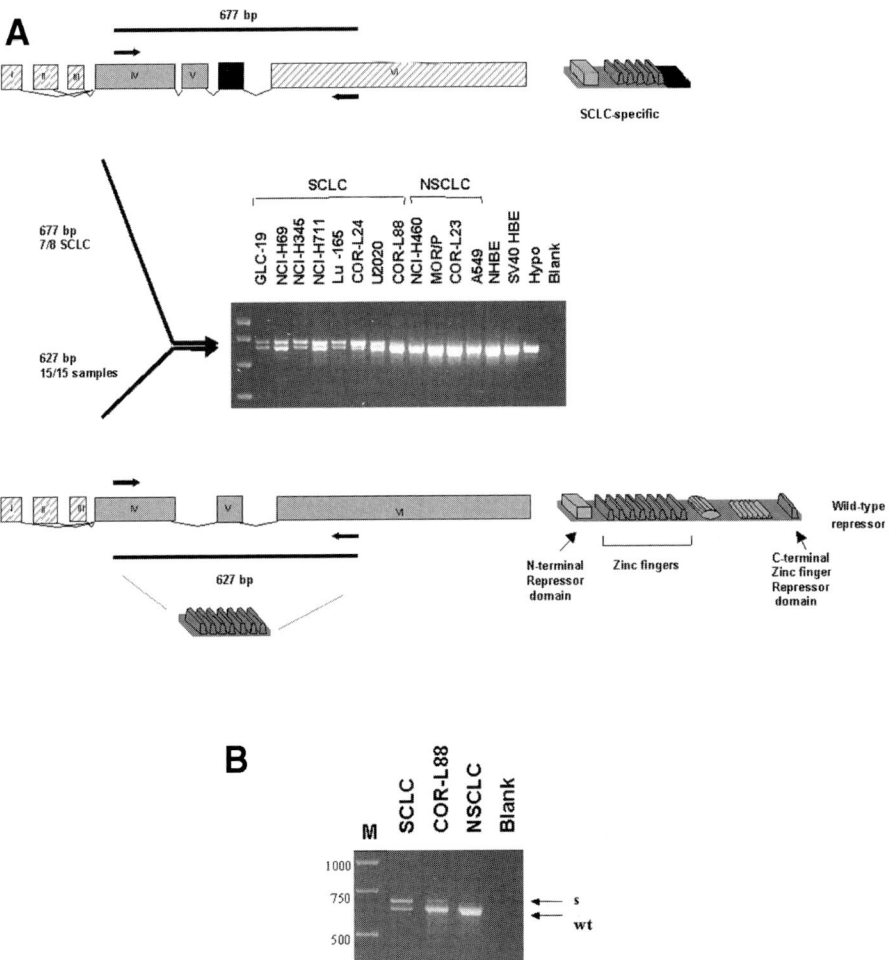

Fig. 1. Detection of the sNRSF splice variant in SCLC cell lines by RT-PCR (**A**) Amplification of NRSF mRNA generates a 627 bp product from wtNRSF transcript. Amplification of cDNA from all lung tumor and control cell lines generated this product (wt), but with an additional high-abundance product of 677 bp (s) in 7/8 SCLC. This results from an extra 50 bp exon in SCLC, encoding a truncated isoform of the NRSF protein. (**B**) The sNRSF RT-PCR product was detected using high-resolution agarose in the 8th SCLC line COR-L88 (a variant) at a lower level than in the other SCLC cell lines (Lu-165 is shown adjacent) but was not seen in NSCLC (NCI-H460). Reproduced with permission from **ref. 8**.

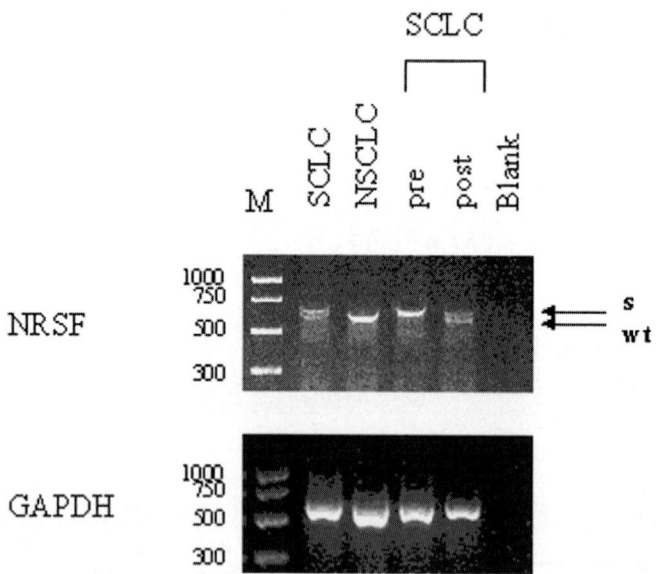

Fig. 2. sNRSF splice variant expression in SCLC tumors RT-PCR on cDNA extracted from primary SCLC cultures established from lymph node aspirates. A comparison of SCLC (011) and NSCLC (002) is shown left. A comparison of a pretreatment SCLC (pre) and relapsed post-treatment tumor (post) from the same individual is shown on the right. The wtNRSF (wt) and SCLC-specific sNRSF (s) products are indicated with arrows. Amplification of GAPDH from the same cDNAs is shown in the lower panel. Reproduced with permission from **ref. 8**.

genes. Diagnostically, it represents a specific clinical marker, which should prove useful in detection of SCLC.

2. Materials
2.1. Collection and Processing of Samples
2.1.1. Peripheral Blood
1. EDTA-coated collection tubes.

2.1.2. Biopsy Material
1. Sterile disposable scalpel and autoclaved scissors.
2. Sterile 2-mL screw-cap cryovials (Nunc).
3. Liquid nitrogen.

2.1.3. Lymph Node & Pleural Aspirates
1. Ficoll/Histopaque (Sigma).
2. HITESA selective culture medium (*17*): RPMI with L-glutamine (Gibco), supplemented with 10 n*M* hydrocortisone, 5 µg/mL insulin, 10 µg/mL transferrin,

10 n*M* estradiol, 30 n*M* selenite, and 0.25% bovine serum albumin (BSA). Prepare stock solutions of: 100 μ*M* hydrocortisone hemisuccinate in H_2O, 5 mg/mL bovine insulin in 6 m*M* HCl, 10 mg/mL transferrin in PBS, 1 m*M* estradiol in ethanol diluted to 100 μ*M* in PBS, 100 μ*M* selenite in H_2O, and 10% BSA, filter-sterilize (0.22 μm). Estradiol and selenite are toxic; suitable precautions should be taken when handling solids or concentrated solutions. Reconstitute media, filter-sterilize, and store at 4°C.

2.2. RNA Extraction

2.2.1. Peripheral Blood

1. Purescript RBC lysis kit (Gentra Systems Inc).
2. Phosphate-buffered saline (PBS): 150 m*M* sodium chloride, 20 m*M* sodium phosphate, pH 7.5.
3. 2 m*M* dithiothreitol (DTT): store as a 1 *M* stock at –20°C and thaw thoroughly before use.
4. RNase inhibitor, e.g., RNasin (Promega).
5. 0.5 *M* Tris-HCl, pH 7.5, 0.02 *M* $CaCl_2$.
6. 0.2 *M* $MgCl_2$.
7. RNase-free DNase I (Promega).
8. 500 m*M* EDTA, pH 8.0.
9. 10% sodium dodecyl sulphate (SDS).
10. Proteinase K (Sigma)..
11. Water saturated phenol/chloroform (1:1), pH 6.0.
12. Chloroform/isoamyl alcohol (24:1).
13. 3 *M* sodium acetate, pH 5.4.
14. Ice-cold ethanol (store in spark-proof freezer).
15. Glycogen (Roche Molecular Biochemicals, 20 mg/mL).
16. RNase-free water: add diethylpyrocarbonate (DEPC) to deionized distilled water (ddH_2O) to 0.01% (v/v), shake well, stand overnight, and autoclave before use. DEPC should only be handled in a fume cupboard.

2.2.2. Biopsy Material

1. RNase ZAP (Sigma).
2. DEPC-treated water (*see* **Subheading 2.2.1., item 16**).
3. Pestle and mortar: treated with RNase ZAP and rinsed with DEPC-treated water to avoid degradation of RNA by extracellular RNAses, then soaked in Trigene (Medichem International) overnight between samples.
4. Syringe, 19- and 23-gauge needles.
5. TRIzol Reagent (Life Technologies).
6. Chloroform (additive-free).
7. Isopropanol.
8. 75% ethanol (in DEPC-treated water).

2.2.3. Lymph Node and Pleural Aspirates

As described in **Subheadings 2.2.1., items 2–15**.

2.3. Quantification of RNA

1. Quartz cuvets and UV/visible spectrophotometer.
2. Molecular biology grade agarose (Biorad or other).
3. 0.5X TBE: 45 mM Tris base, 45 mM boric acid, 1 mM EDTA.
4. Ethidium bromide (10 mg/mL).
5. RNA gel loading buffer: 0.05% bromophenol blue (w/v), 10 mM sodium phosphate buffer, pH 6.8 (v/v), 50% glycerol.

2.4. Reverse Transcription and Polymerase Chain Reaction

2.4.1. Reverse Transcription

1. Reverse Transcription System (Promega).
2. 0.5-mL PCR tubes (GeneAmp or similar with bubble lids suitable for use in a PCR machine with a heated lid to prevent evaporation).

2.4.2. PCR Primers

1. Forward and reverse PCR primer pairs (MWG Biotech). Resuspend each lyophilized primer at a concentration of 100 µM in ddH$_2$O and store at –20°C. Dilute to working mixes containing 20 µM of the forward and reverse primer for each pair. The sequence of primers for housekeeping genes, and the neuroendocrine markers described here, are given in **Table 1**.

2.4.3. PCR Reactions

1. Dynazyme II DNA polymerase 2 U/µL (Flowgen).
2. Dynazyme II 10X optimized reaction buffer: 100 mM Tris-HCl, pH 8.8, 15 mM MgCl$_2$, 500 mM KCl, 1% Triton X-100 or Dynazyme II 10X Mg-free buffer: 100 mM Tris-HCl, pH 8.8, 500 mM KCl, 1% Triton X-100, and 50 mg/mL MgCl$_2$.
3. 1.25 mM dNTPs: 1 in 80 dilution of each 100 mM stock (Helena Biosciences) in ddH$_2$O, e.g., 20 µL each dATP, dCTP, dGTP, dTTP in a final volume of 1520 µL.
4. Mineral oil (Sigma).
5. 6X DNA gel loading buffer: 15% Ficoll 400, 0.03% bromophenol blue, 0.03% xylene cyanol, 0.4% orange G, 10 mM Tris-HCl, pH 7.5, 50 mM EDTA (Promega).
6. Reagents for agarose gel electrophoresis (*see* **Subheading 2.3., items 2–4**).
7. DNA ladder of appropriate range to size PCR products: PCR marker (Promega).

Table 1
RT-PCR Primer Sequences

Target mRNA	Primer sequences	Annealing temp	Product size
GAPDH	CCACCCATGGCAAATTCCATGGCA[a] TCTAGACGGCAGGTCAGGTCCACC	60°C	600 bp
β-Actin	CCTCGCCTTTGCCGATCC GGATCTTCATGAGGTAGTCAGTC	60°C	620 bp
GRP	TGCTGACCAAGATGTACCCG CTTCACGTTGAGAACCTGGA	57°C	323 bp
CCKBR	CTGAGGACTGTCACCAATGC AGAGCTCGCGAGAGATAAG	55°C	486 bp
AVP	TGCATACGGGGTCCACCTGT TAGTTCTCCTCCTGGCAGC	60°C	271 bp
wtNRSF or sNRSF	GAATCTGAAGAACAGTTTGTGCAT TTTGAAGTTGCTTCTATCTGCTGT	60°C	627 bp or 677 bp

[a]All primers are shown in the 5′ to 3′ orientation.

2.5. Blotting and Hybridization

2.5.1. Blotting

1. 3 MM filter paper (Whatmann).
2. N-Hybond filter (Amersham Pharmacia Biotech).
3. 20X SSC: 3 M NaCl, 0.3 M sodium citrate.
4. Sandwich box, gel support (e.g., Gel tray or absorbent sponge pads), paper towels, and weight.
5. 2-mL plastic pipet.
6. Saranwrap or Parafilm.
7. 5% (v/v) glacial acetic acid.
8. 0.04% (w/v) methylene blue in 0.5 M sodium citrate, pH 5.2.

2.5.2. Preparation of Radio-Labeled Probe

1. Purified nucleic acid to use as probe.
2. {α^{32}P} dCTP (>3000 Ci/mmole, 10 mCi/mL) (Amersham Pharmacia Biotech).
3. Oligonucleotide labeling kit (Amersham Pharmacia Biotech).
4. G50 NICK spin columns (Amersham Pharmacia Biotech).
5. Perspex boxes and screens for shielding.

2.5.3. Hybridization

1. Hybridization bottles and meshes (Hybaid).
2. 50X Denhardt's solution: 1% Ficoll 400, 1% polyvinylpyrrolidine, 1% BSA.
3. 10% SDS.
4. 10 mg/mL denatured salmon sperm DNA (Sigma).
5. Hybridization solution: 5X SSC, 5X Denhardt's solution, 1% SDS, 100 µg/mL denatured salmon sperm DNA.
6. Wash solutions: 5X SSC/0.1% SDS, 2X SSC/0.1% SDS, 1X SSC/0.1% SDS, 0.5X SSC/0.1% SDS, 0.2X SSC/0.1% SDS.
7. Hyperfilm βMax film (Amersham Pharmacia Biotech).
8. Saran wrap.
9. Autoradiogram cassettes.

3. Methods
3.1. Collection and Processing of Samples
3.1.1. Peripheral Blood

1. Collect 4.5 mL of blood from subjects (*see* **Note 1**) into EDTA.
2. Immediately place on ice and transfer as soon as possible to storage at –70 to –80°C until required (*see* **Notes 2** and **3**). If blood is to be processed without storage, the nucleated cells may be separated on a Ficoll-Histopaque gradient as described in **Subheading 3.1.3., step 2**.

3.1.2. Biopsy Material

1. Collect tumor tissue samples, for example from surgical lung resection or bronchoscopy (*see* **Note 1**), on ice and immediately snap freeze in cryovials using liquid nitrogen. Store at –80 to –160°C (*see* **Note 2**).
2. Where surgical specimens include tumor and surrounding normal lung: excise from the organ and cut into pieces of optimum size (~20 mg) prior to snap-freezing.

3.1.3. Lymph Node and Pleural Aspirates

1. Collect samples (*see* **Note 1**). Lymph node aspirates are not processed prior to seeding cells directly into media.
2. For pleural aspirates, immediately separate nucleated cells on a Ficoll-Histopaque gradient. Gently layer sample onto an equal volume of Ficoll-Histopaque and centrifuge at 5000*g* for 20 min. Aspirate the buffy layer containing nucleated cells from between the layers of serum / fluid and Ficoll. Wash cells in PBS.
3. Seed cells into a 25 cm^2 tissue culture flask in 5 mL HITESA serum-free culture media to select for and expand the SCLC cells recovered.
4. Maintain cells in this selective medium until there are sufficient cells to process, this is variable depending on the sample, but may require 4–6 wk. (*See* also **Note 4**.)

3.2. RNA Extraction

3.2.1. Peripheral Blood

Recover RNA from the nucleated cells of blood samples using a method modified from **ref.** *(18)* to process between 5 and 1000 cells.

1. Separate nucleated cells (comprising peripheral blood leukocytes (PBLs) together with putative circulating tumor cells) by centrifugation following red blood cell (RBC) lysis using Purescript RBC lysis buffer.
2. Wash cells twice in PBS, pelleting at 6000 rpm (2800g) in a microfuge for 5 min.
3. Resuspend in 30 µL of 2 mM DTT and lyse cells at 95°C for 60 s.
4. Immediately add 40 U of RNasin and vortex.
5. Add 5 µL of 0.5 M Tris-HCl, pH 7.5/0.02 M CaCl$_2$, 5 µL of 0.2 M MgCl$_2$ and 10 U of DNase I, digest DNA at 37°C for 15 min.
6. Add EDTA, SDS and proteinase K to final concentrations of 20 mM, 0.5% and 1 mg/mL respectively, degrade DNase and cellular proteins by incubation at 37°C for 15 min.
7. Extract RNA by adding an equal volume of phenol/chloroform, vortexing, then centrifuging at 13000 rpm (13300g) in a microfuge for 5 min. Follow this by two extractions with chloroform/isoamyl alcohol.
8. Precipitate RNA by addition of 0.1 vol 3 M sodium acetate, 3 vols cold ethanol and 1–2 µg glycogen as a carrier. Place at –70°C for 20 min, then collect RNA pellet at 13000 rpm (13300g) for 10 min.
9. Air-dry RNA and resuspend in 20 µL of DEPC-treated water and store at –80°C.

3.2.2. Biopsy Material

Prepare total RNA from human tissue using TRIzol (Life Technologies Inc), a monophasic solution of phenol and guanidine isothiocyanate, which disrupts cells and maintains the integrity of RNA.

1. Crush frozen specimens (~20 mg) in liquid nitrogen using an RNase free pestle and mortar.
2. Homogenize tissue samples with a syringe and 19-gauge needle in 1 mL of TRIzol (*see* **Note 5**).
3. Incubate samples for 5 min at room temperature to completely dissociate nucleoprotein complexes.
4. Add 0.2 mL chloroform per mL of TRIzol, shake tubes vigorously for 15 s and incubate at room temperature for 3 min.
5. Centrifuge samples at no more than 13000 rpm (13300g) for 15 min at 4°C.
6. Transfer the upper aqueous phase containing the RNA to a fresh 1.5 mL microcentrifuge tube (*see* **Note 6**) and precipitate with 0.5 mL of isopropanol

per ml of TRIzol. Incubate at room temperature for 10 min and spin at 13000 rpm (13300g) in a microfuge for 10 min at 4°C.
7. Remove supernatant and wash the RNA pellet with 1 mL 75% ethanol. Vortex samples and centrifuge at 7000 rpm (3850g) for 5 min at 4°C.
8. Air-dry RNA pellet for 15 min, then dilute in DEPC-treated water and store at –80°C.

3.2.3. Lymph Node/Pleural Aspirates

1. Total cellular RNA can be prepared from 5–1000 cultured cells. Collect these from the HITES-A culture supernatant by centrifugation at 1500 rpm (350g) for 5 min, then follow the method outlined in **Subheading 3.2.1.** starting at **step 2**.

3.3. Quantification of RNA

1. Estimate the concentration of RNA by UV/visible spectrophotometry. Dilute a small volume of the samples to between 1 in 200 and 1 in 500 in ddH$_2$O to scan absorbance from 320 to 200 nm. Use the OD$_{260}$ reading to calculate the nucleic acid concentration: an OD$_{260}$ value of 1.0 is taken to represent 40 µg/mL of RNA. The OD$_{260}$:OD$_{280}$ ratio is used to estimate the protein contamination, a ratio of 1.8–2.0 indicates an RNA preparation of high quality. Use 1 µg (or a standardized amount depending on recovery) for reverse transcription reaction.
2. Confirm the integrity of RNA samples by electrophoresis. Clean the gel tank with RNase ZAP and use DEPC-treated water to prepare the buffer. Prepare 1.5% agarose gels in 0.5% TBE containing 10 µg/mL ethidium bromide. Mix 2 µL RNA and 16 µL DEPC-treated ddH$_2$O with 2 µL RNA loading buffer. Load samples into individual wells and electrophorese at ~50 mA for 60 min. Visualize the 18S and 28S ribosomal RNA bands on a UV transilluminator; if these are not clear and there is a large amount of low molecular-weight material at the bottom of the gel, then the sample may be too degraded for accurate estimation of mRNA concentration or cDNA synthesis and subsequent PCR amplification.

3.4. Reverse Transcription and Polymerase Chain Reaction

3.4.1. Reverse Transcription

Use 1 µg RNA for reverse transcription to synthesize complementary DNA (cDNA) (*see* **Note 7**).

1. Prepare a mix of the reagents below (added in this order) sufficient for one more reaction than will be set up. For each RNA sample 10 µL of reaction mix will contain the following: 4µL 25 m*M* MgCl$_2$, 2 µL 10X reverse transcription buffer, 2 µL 10 m*M* deoxynucleotide triphosphates (dNTPs), 0.5 µL RNasin, 0.6 µL AMV reverse transcriptase, 0.5 µL oligo dT primer (*see* **Note 8**).
2. Aliquot 10 µL of reaction mix into GeneAmp 0.5-mL PCR tubes.

Detection of SCLC by RT-PCR

3. Add 1 μg RNA (to a maximum 10 μL) to each reaction and DEPC-treated water to make up a total reaction volume of 20 μL.
4. Place tubes in the PCR machine (e.g., Hybaid Touchdown) so that they are evenly distributed and ensure the lids are firmly closed. Close the lid and screw down until it touches the top of the tubes and then a further half turn, switch on the heated lid of the PCR machine.
5. Select a program to run 1 cycle of: 42°C for 1 h; 99°C for 5 min, hold at this temperature.
6. Immediately the program finishes remove tubes to ice for 5 min.
7. Dilute the denatured reactions to give a final volume of 100 μL cDNA by addition of: 4 μL 25 mM MgCl$_2$, 8 μL 10X reverse transcription buffer, 68 μL DEPC-treated water. Store at –20°C.

3.4.2. PCR Primers

1. Use RT-PCR primers designed to span an intron, such that PCR amplification products derived from any contaminating genomic DNA can be distinguished from the amplification of cDNA on the basis of size.
2. Use primers to amplify at least one house-keeping gene (*see* **Note 9**), such as β-actin *(19)* or glyceraldehyde-3-phosphate dehydrogenase (GAPDH) *(20)*, from each cDNA sample to confirm: (1) integrity of the RNA used and (2) equivalent loading on gels.
3. The sequence of primer pairs for housekeeping genes, the neuropeptides/neuropeptide receptors and wtNRSF/sNRSF are given in **Table 1** (*see* **Note 10**).

3.4.3. PCR Reactions

1. Use 3 μL (or between 1 and 10 μL) of cDNA in a PCR reaction.
2. Prepare a reaction mix for slightly more reactions than required, containing for each reaction (*see* **Note 11**): 5 μL 10X Dynazyme II optimized buffer (or Mg-free buffer supplemented with Mg^{2+}, *see* **Note 10**), 8 μL 1.25 mM dNTPs, 5 μL PCR primer mix (20 μM for each), 3 μL cDNA template, 27 μL fresh ddH$_2$O. Aliquot 48 μL into each 0.5 mL PCR tube.
3. Always include at least one blank reaction containing no template DNA as a negative control (*see* **Note 12**).
4. Pulse spin tubes in microfuge to ensure all reagents are mixed and overlay the reaction with one drop of mineral oil.
5. Allowing for 0.5 μL of enzyme per tube, dilute the Dynazyme II polymerase to 1 in 4 with ddH$_2$O. Place tubes in the PCR block and ensure the lids are firmly closed.
6. Run the following PCR program. Wait for the first hot-start at 94°C to finish, and once the tubes have returned to the annealing temperature, add 2 μL of diluted Dynazyme to the bottom of the tubes (using a clean pipet tip for each). Replace the lids and allow the rest of the program to run; 94°C 3 min, 60°C 4 min, 1 cycle hot-start, add enzyme during hold at 60°C, 94°C 1min 60°C 1min (*see* **Note 10**), 72°C 2 min, 35 cycles (*see* **Note 13**).

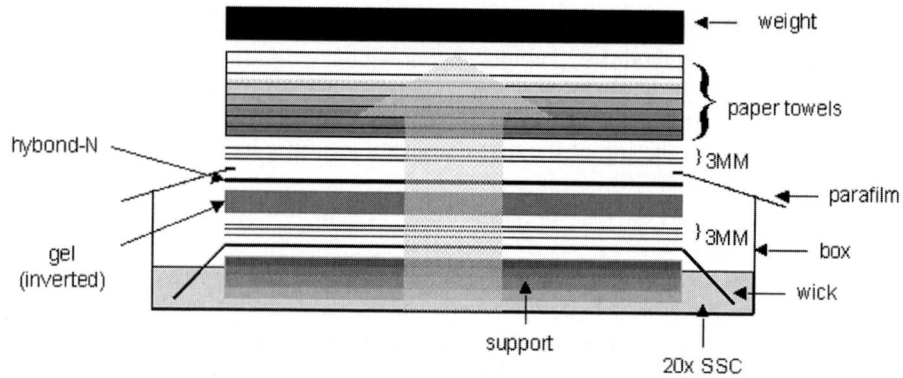

Fig. 3. Blotting apparatus.

7. Once the program is completed, remove samples from block and store at 4°C. Where the PCR machine can cool below ambient temperature a final cycle may be set to hold samples at 4°C.
8. Mix 10 µL of each reaction with 2 µL of gel loading buffer (*see* **Note 14**) and load the samples into the wells of a 2% agarose/0.5X TBE gel containing 10 µg/mL ethidium bromide. (*See* also **Note 15**).
9. Electrophorese in 0.5X TBE at ~50mA (100V) for ~45 min.
10. Visualize ethidium bromide stained DNA bands on a UV transilluminator and record the results photographically.
11. All PCR products should be purified and confirmed by direct sequencing.

3.5. Blotting and Hybridization

To confirm the identity of amplified DNA fragments and increase the sensitivity of detection (*see* **Note 16**) electrophoresed RT-PCR products may be transferred to a nylon membrane and hybridized with a specific radiolabeled probe.

3.5.1. Blotting

1. Electrophorese PCR products on 2% agarose gel as above, but omitting the ethidium bromide.
2. To set up the blotting apparatus (*see* **Fig. 3**), place the gel pouring tray upside down as a support in a large sandwich box containing ~500 mL 20X SSC (*see* also **Note 17**). Cover with two rectangular pieces of 3MM filter paper, placed at right angles to each other with either end submerged in the SSC, to form a wick all around the platform. Cover this with three more pieces of filter paper (cut to the size of the platform), soak with SSC, and remove any bubbles by rolling a plastic pipet across the surface.
3. Place the inverted gel on the filter paper and cover with a piece of Hybond-N cut to the same size as the gel. Remove any bubbles.

4. Seal around the edges of the gel with Saran wrap or Parafilm to prevent wicking of buffer around the gel. Cover with three pieces presoaked filter paper, then a stack of paper towels cut to the same size as the gel. Place an evenly distributed weight on top of the blotting apparatus and leave overnight.
5. Remove the paper towels and filter papers. Mark the position of the lanes on the membrane with pencil, then disassemble the rest of the blot, discarding everything except the membrane.
6. Fix by UV illumination for 15 s (e.g., using a Memorase model C-50, UVP Inc.) and bake at 80°C for 2 h to covalently fix the DNA. The membrane can now be stored at room temperature until required for hybridization (**Subheading 3.5.3.**).
7. To visualize the DNA ladder, cut off the marker lane and stain: soak in 5% acetic acid for 15 min, then in methylene blue for 20 min; remove excess by repeated washes in water.

3.5.2. Preparation of Radio-Labeled Probe

The probe sequence may be dsDNA, ssDNA, or RNA. For example dsDNA amplified by PCR from a positive control template such as the cloned gene, which has then been gel-purified and sequence-verified. This is generally labeled using ^{32}P (*see* **Note 18**).

1. Label DNA with (α^{32}P) dCTP by random hexanucleotide priming. Denature approximately 40 ng of DNA in a volume of 45 µL by boiling for 10 min and snap-cool on ice, before adding into a tube of freeze-dried oligonucleotide labeling mix.
2. Once dissolved, add 50 µCi (5 µL) of (α^{32}P) dCTP incubate the reaction for 1 h at 37°C.
3. Remove unincorporated radiolabeled nucleotide on a G50 spin column. Invert the column several times to resuspend the gel, allow to settle and remove top, then bottom, caps.
4. Allow the column to drain, equilibrate with 1 mL and then 2 mL H$_2$O allowing to drain each time. Then centrifuge for 4 min at 500g in swing-out rotor. Discard eluate, place column over 1.5-mL tube within a centrifuge tube. Load the sample onto the center of the column in a volume of 75–100 µL.
5. Elute the sample by centrifugation for 5 min at 500g. Discard the column containing unincorporated label.
6. Determine the specific activity of the probe by counting the d.p.m. for 1 µL using a scintillation counter; this should ideally be >10^9/µg DNA. The eluted probe may be stored at –20°C for up to 1 wk, but is best used immediately.

3.5.3. Hybridization

1. Roll the membrane up in a mesh and place in the hybridization bottle. Prehybridize in the hybridization solution (allowing 1 mL/10 cm^2 membrane) and incubate at 68°C for 2 h with rotation in a hybridization oven.

2. Add the ^{32}P-labeled DNA probe to fresh hybridization solution prewarmed to 68°C (using the same volume as in **step 1**). Use this to replace the prehybridization mix and incubate at 68°C with rotation overnight.
3. Discard the hybridization mix according to local radiation safety guidelines. Wash the membrane in two changes of 25 mL 2X SSC/0.1% SDS for 20 min at 68°C in the hybridization bottle.
4. Remove the membrane from the bottle and mesh and monitor with a Geiger counter. Use increasing stringency washes (1X, 0.5X, 0.2X SSC) at 42°C to 68°C as required, until background radioactivity on the membrane is reduced, but a differential signal can be detected with the Geiger counter at the expected position for the PCR product.
5. Expose the membrane (wrapped in Saran wrap) to autoradiography film at –80°C. Exposure times may vary from several hours to several days, a number of different exposures should be used to validate relative band intensities (*see* also **Note 19**).

4. Notes

1. Appropriate informed consent must be obtained from any subject giving blood or other tissues for biomedical research, in accordance with local institutional guidelines.
2. In whole blood that is stored immediately at –80°C, RNA should remain stable for at least 6–9 mo *(21)* and up to 12 mo *(22)*. Postmortem or surgical samples should be snap-frozen as soon as possible. Certain transcripts may have a shorter half-life and therefore be more susceptible to rapid degradation than others.
3. Blood samples may be frozen directly in collection vials. However, care should be taken on thawing samples quickly to room temperature, with vials placed inside a 25 mL universal container in case they crack.
4. The lung tumor cells recovered from lymph node or pleural aspirates are selected for and increased by expansion in HITESA. Primary cell lines may eventually be established from these cultures if required and can be stored as viable cells in media + 10% dimethyl sulfoxide (DMSO) by gradual freezing to –160°C. Total cellular RNA can then be prepared from a larger number of cultured cells, for example using a Purescript RNA kit (Gentra Systems Inc).
5. Incomplete homogenization or lysis of the samples may give rise to low yields of RNA. Additionally, samples homogenized in too small a volume of TRIzol may give rise to DNA contamination and in this case DNAse treatment of the RNA samples will be necessary (*see* **Subheading 3.2.1., steps 5–8**). Harder tissues are difficult to homogenize, so pour liquid nitrogen onto the tissue and crush again until it becomes powder-like. This process is aided by using a 23-gauge needle after the 19-gauge, and may be necessary for the majority of tissues.
6. During extraction, the aqueous phase should not be contaminated with the phenol phase to minimize protein contamination the RNA sample. Larger pieces of tissue do not necessarily yield more RNA and will require additional chloroform extractions to remove the large amounts of contaminating protein.

7. If the RNA yield from samples is not sufficient to use 1 µg for reverse transcription, use a lower but constant concentration of total RNA, and then cDNA, to ensure that RT-PCR is semi-quantitative.
8. Sensitivity and specificity of reverse transcription may be increased, if required, by using a gene specific primer rather than oligo dT to prime the first strand cDNA synthesis. This primer may be the reverse PCR primer or preferably a reverse primer 3′ to this site on the template sequence.
9. In selection of the control or housekeeping genes to normalize the RNA/cDNA used from each sample, consideration should be given to several factors: (1) transcript abundance should be similar to that of the test cDNA, and (2) the presence of pseudogenes. For example, GAPDH primers can also amplify pseudogenes, so parallel reverse transcription reactions, from which the reverse transcriptase has been omitted, should be used to determine the contribution from amplification of genomic sequence.
10. PCR conditions must be individually optimized for each primer pair, taking into consideration: the annealing temperature, extension time, number of amplification cycles, the DNA polymerase and magnesium ion concentration. In our hands the primers listed in **Table 1** all worked optimally in 1.5 mM MgCl$_2$, with the exception of the AVP primers, for which 1.0 mM MgCl$_2$ should be used. Parameters may need to be optimized for individual PCR machines.
11. If the template or primers are to be varied, make a reaction mix omitting these reagents, then use this to make submixes for each set of reactions with the same primer or template.
12. Contamination of samples producing false-positive results is a major problem in the use of PCR. Taking the following precautions can minimize this. Always include blank reactions containing no template as negative controls, these can be included between each test reaction if contamination is a recurring problem. Positive controls spiked with the template to be amplified can be used to validate the PCR reaction, but may be a source of contaminating DNA in the test reactions. The use of filter pipet tips can minimize contamination caused by aerosols, as can pulse spinning microfuge tubes before opening. Most importantly, separate areas should be designated to set up the reactions, run PCR amplification, and analyze the PCR reaction products.
13. Use constant concentrations of total RNA and cDNA at each step to ensure the RT-PCR remains semi-quantitative. For PCR to be semi-quantitative the cycle number used must not allow the amplification to reach a plateau. Both the primers and template abundance will determine the cycle number selected, for GAPDH typically use 25–28 cycles and for β-actin 28–32 cycles. Amplification with other primers may require 30–35 cycles, with higher cycle numbers used to validate negative results. For quantitative data, competitive or real-time PCR approaches may be used *(23)*.
14. Sometimes one of the dye components in the gel loading buffer co-migrates with the amplicon of interest, partially obscuring the bands on UV examination.

Where this is a problem, loading buffer can be prepared omitting or reducing the concentration of that dye to allow clear imaging.
15. Amplified products, which differ by only small numbers of bases, e.g., 50 bp for the NRSF/sNRSF splice variant products can be difficult to separate on standard agarose gels. This can be overcome by using specialist high resolution agarose, such as AMRESCO 3:1 agarose (Anachem).
16. Sensitivity of detection of tumor cells in the peripheral blood was determined by us using a serial dilution of Lu-165 SCLC cells in the blood of a healthy volunteer. Three mL of whole blood was used for each dilution of the SCLC cells to between 10^{-2}–10^{-8} SCLC per normal cell. RNA was extracted from spiked whole blood by the method described in **Subheading 2.2.1**. Sensitivity for RT-PCR alone was only 1 in 100. However, when electrophoretically separated RT-PCR products were transferred to nylon membrane and hybridized with DNA probes for AVP or GRP, the sensitivity and specificity of detection of these mRNA species was increased. This method allowed detection of one tumor cell in 10^7 peripheral blood leukocytes *(10,11)*. Sensitivity may be further increased by the use of nested PCR primers.
17. Absorbent sponges may be used as blotting apparatus to replace the invert gel tray as support and the 3MM filter paper as wick, but should be cut to the size of the gel.
18. Appropriate precautions should always be taken when handling ^{32}P radioactivity, such as the use of perspex screening. Experimental procedures, monitoring, and disposal of radioactive waste should be carried out in accordance with local guidelines. Nonradioactive probes may be substituted, for example, using the ECL Labeling and Detection System (Amersham Pharmacia Biotech).
19. Membranes may be stripped and re-probed with an alternative probe if required, for example to detect the products of different PCR amplifications on the same blot. Incubate the filter for 1 h at 65°C in 50% formamide/10 mM sodium phosphate, pH 6.8. Then wash three times in 2X SSC/0.1% SDS before rehybridization.

References

1. Ouellette, D., Desbiens, G., Emond, C., and Beauchamp, G. (1998) Lung cancer in women compared with men: Stage, treatment, and survival. *Ann. Thorac. Surg.* **66(4)**, 1140–1143.
2. Ocejo-Garcia, M., Ahmed, S. I., Coulson, J. M., and Woll, P. J. (2001) Use of RT-PCR to detect co-expression of neuropeptides and their receptors in lung cancer. *Lung Cancer* **33**, 1–9.
3. Coulson, J. M., Stanley, J., and Woll, P. J. (1999) Tumour-specific arginine vasopressin promoter activation in small-cell lung cancer. *Br. J. Cancer* **80(12)**, 1935–1944.
4. Friedmann, A. S., Malott, K. A., Memoli, V. A., Pai, S. I., Yu, X. M., and North, W. G. (1994) Products of vasopressin gene-expression in small-cell carcinoma of the lung. *Br. J. Cancer* **69(2)**, 260–263.

5. North, W. G. (2000) Gene regulation of vasopressin and vasopressin receptors in cancer. *Exp. Physiol.* **85,** 27S–40S.
6. Coulson, J. M., Fiskerstrand, C. E., Woll, P. J., and Quinn, J. P. (1999) E-box motifs within the human vasopressin gene promoter contribute to a major enhancer in small-cell lung cancer. *Biochem. J.* **344,** 961–970.
7. Coulson, J. M., Fiskerstrand, C. E., Woll, P. J., and Quinn, J. P. (1999) Arginine vasopressin promoter regulation is mediated by a neuron- restrictive silencer element in small cell lung cancer. *Cancer Res.* **59(20),** 5123–5127.
8. Coulson, J. M., Edgson, J. L., Woll, P. J., and Quinn, J. P. (2000) A splice variant of the neuron-restrictive silencer factor repressor is expressed in small cell lung cancer: a potential role in derepression of neuroendocrine genes and a useful clinical marker. *Cancer Res.* **60(7),** 1840–1844.
9. Coulson, J. M., Quinn, J. P., Edgson, J. L., Mulgrew, R. J., and Woll, P. J. (2002) Upstream stimulatory factor is an initiator of the vasopressin promoter in lung cancer cells, in press.
10. Ahmed, S. I., Coulson, J. M., and Woll, P. J. (2000) Detection of small cell lung cancer cells in the peripheral blood by RT-PCR for expressed neuropeptides. *Proc. Am. Soc. Clin. Oncol.* **19,** 1901.
11. Ahmed, S. I., Coulson, J. M., and Woll, P. J. (2001) Detection of small cell lung cancer cells in the peripheral blood by RT-PCR for expressed neuropeptides. Manuscript in preparation.
12. Wharton, J., Polak, J. M., Bloom, S. R., Ghatei, M. A., Solcia, E., Brown, M. R., and Pearse, A. G. (1978) Bombesin-like immunoreactivity in the lung. *Nature* **273(5665),** 769–770.
13. Yamaguchi, K., Abe, K., Kameya, T., Adachi, I., Taguchi, S., Otsubo, K., and Yanaihara, N. (1983) Production and molecular size heterogeneity of immunoreactive gastrin-releasing peptide in fetal and adult lungs and primary lung tumors. *Cancer Res.* **43(8),** 3932–3939.
14. Chong, J. H. A., Tapiaramirez, J., Kim, S., Toledoaral, J. J., Zheng, Y. C., Boutros, M. C., et al. (1995) REST—a mammalian silencer protein that restricts sodium-channel gene-expression to neurons. *Cell* **80(6),** 949–957.
15. Schoenherr, C. J. and Anderson, D. J. (1995) The neuron-restrictive silencer factor (NRSF): a coordinate repressor of multiple neuron-specific genes. *Science* **267(5202),** 1360–1363.
16. Chen, Z. F., Paquette, A. J., and Anderson, D. J. (1998) NRSF/REST is required in vivo for repression of multiple neuronal target genes during embryogenesis. *Nature Genet.* **20(2),** 136–142.
17. Carney, D. N., Bunn, P. A., Jr., Gazdar, A. F., Pagan, J. A., and Minna, J. D. (1981) Selective growth in serum-free hormone-supplemented medium of tumor cells obtained by biopsy from patients with small cell carcinoma of the lung. *Proc. Natl. Acad. Sci. USA* **78(5),** 3185–3189.
18. Dyanov, H. M. and Dzitoeva, S. G. (1998) Isolation of DNA-free RNA from a very small number of cells, in *The PCR Technique: RT-PCR* (Siebert, P., ed.), Eaton Publishing, Natick, MA pp. 3–9.

19. Raff, T., van der Giet, M., Endemann, D., Wiederholt, T., and Paul, M. (1998) Design and testing of beta-actin primers for RT-PCR that do not co-amplify processed pseudogenes, in *The PCR technique: RT-PCR* (Siebert, P., ed.), Eaton Publishing, Natick, MA, pp. 161–168,
20. Maier, J. A. M., Voulalas, P., Roeder, D., and Maciag, T. (1990) Extension of the life-span of human endothelial-cells by an interleukin-1-alpha antisense oligomer. *Science* **249(4976),** 1570–1574.
21. Pittman, K., Burchill, S., Smith, B., Southgate, J., Joffe, J., Gore, M., and Selby, P. (1996) Reverse transcriptase-polymerase chain reaction for expression of tyrosinase to identify malignant melanoma cells in peripheral blood. *Ann. Oncol.* **7(3),** 297–301.
22. Ahmed, S. I. (2000) The clinical significance of neuroendocrine differentiation in lung cancer. MD Thesis, University of Nottingham, UK.
23. Freeman, W. M., Walker, S. J., and Vrana, K. E. (1999) Quantitative RT-PCR: pitfalls and potential. *Biotechniques* **26(1),** 112–122, 124–125.

II

DETECTION OF ALTERED TUMOR MARKERS IN CLINICAL SAMPLES FOR DIAGNOSIS AND PROGNOSIS

B. MOLECULAR METHODS FOR PROGNOSTIC DETECTION OF ALTERED LUNG TUMOR MARKERS IN CLINICAL SAMPLES

III. Immunohistochemical Analyses

21

Utilization of Thyroid Transcription Factor-1 Immunostaining in the Diagnosis of Lung Tumors

Nelson G. Ordóñez

1. Introduction

Transcription factors play a crucial role in the determination and maintenance of differentiated cellular phenotype and their activity is considered to constitute the main switch to regulate gene expression *(1)*. Based on their localization and expression, transcription factors have been subdivided into two major groups: ubiquitous and tissue-specific. Tissue-specific transcription factors are those that present in a few cell types and are involved in the regulation of genes expressed only in those cells. Thyroid transcription factor-1 (TTF-1) is a 38-kDa homeodomain containing nuclear transcription protein of the Nkx2 gene family. It consists of a single polypeptide of 371 amino acids that is encoded by a single locus *(2)*. TTF-1 is expressed in the epithelial cells of the thyroid, lung, and diencephalon during early embryogenesis *(3–5)*. After birth, its expression is mostly maintained in the follicular and parafollicular C-cells of the thyroid and in bronchioalveolar cells. In the thyroid, TTF-1 activates the transcription of thyroglobulin, thyroperoxidase, and sodium-iodine transport protein *(6–8)*. In the lung, TTF-1 stimulates the synthesis of surfactant proteins A *(9)*, B *(10,11)*, and C *(12)*, and Clara cell secretory protein *(13)* through the binding of the corresponding gene promoter enhancers. TTF-1 gene expression is regulated in respiratory epithelial cells at the transcriptional and posttranscriptional levels *(14)* by means of cross-regulatory mechanisms involving hepatocyte nuclear factor 3, Oct-1 protein, GATA-6, and calreticulin *(15–17)*. TTF-1 plays a critical role in the normal development of embryonic epithelial cells of the thyroid and lung. Disruption of the TTF-1 locus in the mouse by homologous recombination *(4)* or TTF-1 mRNA downregulation by nitrofen *(18)* produces offspring with agenesis of the thyroid and hypoplasia of the lungs resulting in death at birth.

1.1. Expression of TTF-1 in Lung Tumors

Thyroid transcription factor-1 expression can be detected by immunohistochemistry in the columnar nonciliated cells of the fetal lung as early as 11 wk of gestation *(19)*. In normal adult lung, TTF-1 is present only in type II pneumocytes and Clara cells *(19,20)*. The cellular distribution of TTF-1 in normal lung parallels that of surfactant proteins (SPs) A, B, and C, and Clara cell secretory protein *(19,20)*. TTF-1 has been detected in all types of lung carcinomas (*see* **Table 1**) (*see* **Figs. 1** and **2**). In the large majority of studies, TTF-1 has been demonstrated in about 70–85% of pulmonary adenocarcinomas *(21,23,27–29,30,32,34)*, except for one investigation in which expression was reported in only 27% of the cases *(22)*.

Like TTF-1, SPs have also been shown to be useful immunohistochemical markers in the diagnosis of pulmonary adenocarcinomas. Using the PE-10 anti-SP-A monoclonal antibody (MAb), Noguchi et al. *(44)* reported reactivity in 61% of 21 lung adenocarcinomas in their 1989 study. In 1996, using polyclonal antibodies (PAbs) to SP-A, SP-B, and TTF-1, Bejarano et al. *(21)* obtained positivity in 54, 67, and 76% of their pulmonary adenocarcinomas, respectively. In a more recent investigation, Kaufmann and Dietel *(31)* reported reactivity for SP-A in 21 (42%) and SP-B in 19 (38%) of 50 adenocarcinomas of the lung using the PE-10 anti-SP-A monoclonal antibody and a polyclonal antibody to SP-B. They also found TTF-1 positivity in 33 (66%) of these tumors. Based on these results, and those obtained in the previously mentioned investigations, TTF-1 appears to be a more sensitive marker for adenocarcinomas of the lung than SPs.

In contrast to adenocarcinomas, squamous carcinomas of the lung rarely express TTF-1. Only 26 (8%) of 340 squamous carcinomas that have been investigated for TTF-1 expression were positive (*see* **Table 1**). The reactivity occurred more frequently using PAbs rather than MAbs. Seventeen (18.5%) of the 92 squamous carcinomas investigated using PAbs were positive *(21,22, 24,29)* compared with only 9 (3.6%) of 248 with two MAbs *(27,28,31,34)*.

The first investigation into the expression of TTF-1 in typical pulmonary carcinoid tumors was by Fabbro et al. *(22)* in 1996; all eight such cases in that study failed to show any immunoreactivity. These results differ from those obtained by Folpe et al. *(35)*, who demonstrated TTF-1 expression in 18 (35%) of 51 typical and all 9 (100%) atypical carcinoids, and in 6 (75%) of 8 large cell carcinomas investigated; and Kaufmann and Dietel *(31)*, who found reactivity in 6 (50%) of 12 typical and 2 (67%) of 3 atypical carcinoids, and in 2 (50%) of 4 large cell neuroendocrine carcinomas. Although the reasons

Table 1
Immunohistochemical Detection of TTF-1 in Lung Tumors

Adenocarcinoma	467/652	(72%)
Bejarano et al. (1996)^a *(21)*	35/46	(76%)
Fabbro et al. (1996)^a *(22)*	4/15	(27%)
Holzinger et al. (1996)^b *(23)*	5/6	(83%)
Di Loreto et al. (1997)^a *(24)*	5/8*	(63%)
Di Loreto et al. (1998)^a *(25)*	19/33	(57.5%)
Bohinski et al. (1998)^{b,d} *(26)*	12/18	(67%)
Harlamert et al. (1998)^c *(27)*	16/21*	(76%)
Khoor et al. (1999)^b *(28)*	158/208	(76%)
Puglisi et al. (1999)^a *(29)*	20/28	(71%)
Anwar (1999)^b *(30)*	18/21	(86%)
Kaufmann and Dietel (2000)^b *(31)*	67/98	(68%)
Ordóñez (2000)^b *(32)*	30/40	(75%)
Reis-Filho et al. (2000)^b *(33)*	8/13	(61.5%)
Pelosi et al. (2001)^b *(34)*	70/97	(72%)
Squamous carcinoma	26/340	(8%)
Bejarano et al. (1996)^a *(21)*	0/10	(0%)
Fabbro et al. (1996)^a *(22)*	3/13	(23%)
Di Loreto et al. (1997)^a *(24)*	1/9*	(11%)
Harlamert et al. (1998)^c *(27)*	3/8*	(38%)
Khoor et al. (1999)^b *(28)*	0/101	(0%)
Puglisi et al. (1999)^a *(29)*	13/60	(22%)
Kaufmann and Dietel (2000)^b *(31)*	0/20	(0%)
Pelosi et al. (2001)^b *(34)*	6/119	(5%)
Adenosquamous carcinoma		
Pelosi et al. (2001)^b *(34)*	2^e/2	(100%)
Large cell carcinoma	24/84	(29%)
Fabbro et al. (1996)^a *(22)*	0/1	(0%)
Khoor et al. (1999)^b *(28)*	16/61	(26%)
Kaufmann and Dietel (2000)^b *(31)*	8/20	(40%)
Pelosi et al. (2001)^b *(34)*	0/2	(0%)
Large cell neuroendocrine carcinoma	12/22	(55%)
Folpe et al. (1999)^b *(35)*	6/8	(75%)
Kaufmann and Dietel (2000)^b *(31)*	2/4	(50%)
Caccamo et al. (2000) *(36)*	4/10	(40%)
Typical carcinoid	24/71	(34%)
Fabbro et al. (1996)^a *(22)*	0/8	(0%)
Folpe et al. (1999)^b *(35)*	18/51	(35%)
Kaufmann and Dietel (2000)^b *(31)*	6/12	(50%)

(continued)

Table 1 (continued)

Atypical carcinoid	11/12	(92%)
Folpe et al. (1999)[b] *(35)*	9/9	(100%)
Kaufmann and Dietel (2000)[b] *(31)*	2/3	(67%)
Small cell carcinoma	198/218	(91%)
Fabbro et al. (1996)[a] *(22)*	10/10	(100%)
Di Loreto et al. (1997)[a] *(24)*	38/41[*]	(92.7%)
Harlamert et al. (1998)[c] *(27)*	10/12[*]	(83%)
Folpe et al. (1999)[b] *(35)*	20/21	(95%)
Byrd-Gloster et al. (2000)[b] *(37)*	35/36	(97%)
Hanly et al. (2000)[b] *(38)*	28/33	(85%)
Kaufmann and Dietel (2000)[b] *(31)*	30/37	(81%)
Ordóñez (2000)[b] *(39)*	27/28	(96%)
Alveolar adenoma		
Burke et al. (1999) *(40)*	5/5	(100%)
Peripheral papillary tumor of type II pneumocytes		
Dessy et al. (2000)[b] *(41)*	2/2	(100%)
Sclerosing hemangioma	50/53	(94%)
Chan and Chan (2000)[b] *(42)*	16/16	(100%)
Devouassoux-Shisheboran et al. (2000)[b] *(43)*	34/37	(92%)
Malignant mesothelioma	0/169	(0%)
Di Loreto et al. (1998)[a] *(25)*	0/24	(0%)
Khoor et al. (1999)[b] *(28)*	0/95	(0%)
Ordóñez (2000)[b] *(32)*	0/50	(0%)

[*]cytology specimens.
[a]Study using a polyclonal antibody.
[b]Study using 8G7G3/1 monoclonal antibody.
[c]Study using 1-2.A5.9 monoclonal antibody.
[d]Metastatic adenocarcinomas.
[e]Positive only in the adenocarcinoma component.

for the discrepancies in the results obtained in these investigations are unclear, they have been attributed to differences in the antibodies to TTF-1 and/or in the immunostaining procedure used *(35)*.

Thyroid transcription factor-1 is commonly present in small cell lung carcinomas (SCLCs). The frequency with which its expression has been reported in different series has ranged from 83.3% *(27)* up to 100% *(22)* of the cases. In a recent study by this author, TTF-1 reactivity was seen in 27 (96%) of the 28 cases investigated *(39)*. Of the 218 SCLCs that have been studied for the presence of TTF-1, 198 (91%) were positive (*see* **Table 1**).

In addition to carcinomas, TTF-1 is also expressed in some benign tumors of the lung. It has been reported in alveolar adenoma *(40)*, peripheral papillary tumor of type II pneumocytes *(41)*, and in sclerosing hemangioma *(42,43)*. The

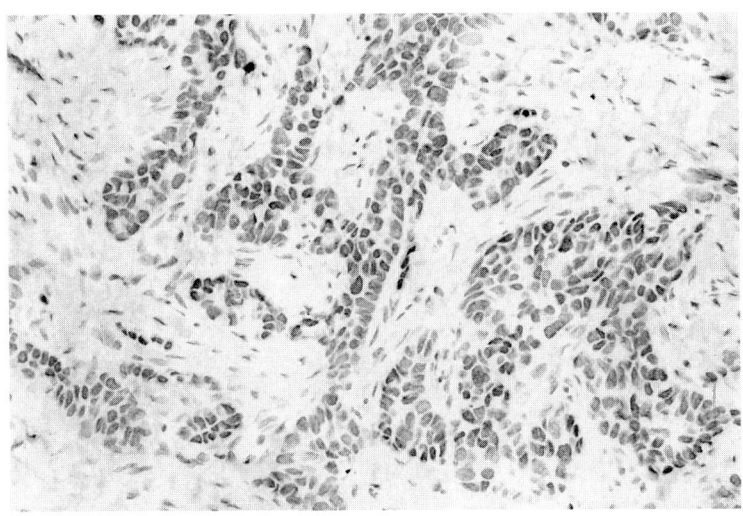

Fig. 1. Adenocarcinoma of the lung showing TTF-1 positivity in the nuclei of the cells.

finding of TTF-1 in sclerosing hemangioma is of special significance because of the continuous controversy that has existed regarding the histogenesis of this lesion since its description by Liebow and Hubbell in 1956 *(42,45–48)*. The demonstration of TTF-1 in sclerosing hemangiomas has prompted some investigators to suggest that this tumor is an epithelial neoplasm derived from primitive respiratory epithelium or incompletely differentiated type II pneumocytes or Clara cells *(42)*.

1.2. Diagnostic Applications of TTF-1 Immunostaining

Current information indicates that TTF-1 is almost exclusively expressed in adenocarcinomas of the thyroid and lung *(49)*. TTF-1 has been reported to be expressed in nearly all papillary and follicular carcinomas of the thyroid *(23,49–51)* and in about 75% of the pulmonary adenocarcinomas studied *(21,27,28,32,34)*. In contrast to adenocarcinomas of the lung and thyroid, TTF-1 has rarely been reported in adenocarcinomas originating in other organs. Only 2 of 286 adenocarcinomas of nonpulmonary or thyroid origin investigated for TTF-1 expression exhibited focal positivity (1/66 stomach, 1/8 endometrium, 0/96 breast, 0/49 colon, 0/30 ovary, 0/18 kidney, 0/19 prostate) *(21,23,27,32)*. These results indicate that TTF-1 immunostaining could be very useful in distinguishing adenocarcinomas of the lung from other adenocarcinomas, especially when it is used in conjunction with cytokeratins 7 and 20, and other tissue-associated antigens such as prostate-specific antigen, thyroglobulin, villin, and gross fibrocystic disease protein-15 (*see* **Table 2**).

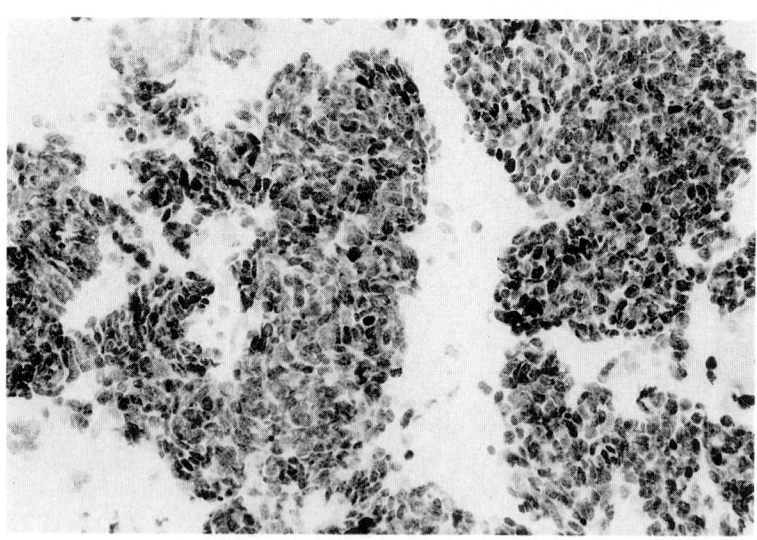

Fig. 2. SCLC reacting for TTF-1.

TTF-1 is commonly expressed in SCLCs and the frequency with which it has been demonstrated in different series has ranged from 83–100% (**Table 1**). It can also be expressed in extrapulmonary small cell carcinomas but its presence depends on the site of origin of the tumor (*see* **Table 3**). According to the literature, TTF-1 expression can occur in small cell carcinomas of the prostate *(31,52)*, urinary bladder *(31,39,52)*, gastrointestinal tract *(39)*, uterine cervix *(39,52)*, breast *(31)*, and thyroid *(31)*, but not in small cell carcinomas of the skin (Merkel cell carcinoma) *(37,39)*. Because Merkel cell carcinomas have invariably been negative for TTF-1, immunostaining for this marker can be useful in distinguishing these tumors from SCLCs, especially when it is used in conjunction with cytokeratin 20 immunostaining *(38,39)*. In contrast to TTF-1 which is commonly expressed in SCLCs but not in Merkel cell carcinomas, CK20 is often demonstrated in Merkel cell carcinomas but not in SCLCs. This distinction is important since the prognosis and treatment of both tumors may be different. SCLCs are relatively responsive to chemotherapy, whereas Merkel cell carcinomas are treated primarily by surgery, and when this tumor disseminates, chemotherapy has been shown to have little beneficial effect *(53)*.

Among the well-differentiated neuroendocrine tumors, TTF-1 expression has been reported only in carcinoid tumors of the lung and medullary carcinomas of the thyroid (*see* **Table 3**). Therefore, TTF-1 immunostaining could be useful for determining the site of origin of a metastatic well-differentiated neuroendocrine tumor *(52)*.

Table 2
Immunohistochemical Profile in Adenocarcinomas of Various Sites

Marker	Lung	Thyroid	Breast	Colon	Prostate	Pancreas	Ovary
TTF-1	+	+	–	–	–	–	–
CK7	+	+	+	–	–	+	+
CK20	–	–	–	+	(+)	+	–
PSA	–	–	(+)	–	+	–	–
GCDFP-15	–	–	+	–	(+)	–	–
Villin	(+)	–	–	+	–	+	–
Thyroglobulin	–	+	–	–	–	–	–

(+), Occasional cases positive.
TTF-1, thyroid transcription factor-1; CK, cytokeratin; PSA, prostate-specific antigen; GCDFP, gross cystic disease fluid protein.

A few studies have addressed, with contradictory results, the prognostic implications of TTF-1 immunoreactivity in non-small cell lung carcinomas (NSCLCs) *(29,34,54)*. At present, there is no conclusive evidence that TTF-1 can be used as a prognostic indicator in any subtype of lung carcinoma.

1.3. Immunostaining Technique: Streptavidin-Biotin-Peroxidase Labeling of Paraffin Sections

This is a three-step technique employing an unlabeled primary antibody, biotin-labeled secondary, and horseradish peroxidase-labeled streptavidin.

2. Materials

1. TTF-1 mouse MAb (clone 8G7G3/1, NeoMarkers, Union City, CA).
2. LSAB2 system peroxidase kit (Dako, Carpinteria, CA).
3. Antibody diluent (Dako).
4. 3% hydrogen peroxide in water.
5. Methanol, absolute.
6. Ethanol, absolute.
7. Xylene.
8. 0.02 M phosphate-buffered saline (PBS), pH 7.2.
9. 0.2% Tween 20/PBS.
10. 10 mM sodium citrate buffer, pH 6.0.
11. Liquid 3,3′-diaminobenzidine concentrated substrate pack (BioGenex, San Ramon, CA).
12. Mayer's hematoxylin.
13. Scott tap water substitute (Anapath, Lewisville, TX).
14. Permount.
15. Cover glass.

Table 3
TTF-1 Expression in Extrapulmonary Neuroendocrine Tumors

Medullary carcinoma of thyroid	30/31	(97%)
Carcinoid tumors	0/49	(0%)
Small intestine (not further specified)	0/22	(0%)
Duodenum	0/1	(0%)
Jejunum	0/1	(0%)
Ileum	0/4	(0%)
Appendix	0/1	(0%)
Colon	0/19	(0%)
Gallbladder	0/1	(0%)
Pancreatic islet cell tumors	0/15	(0%)
Paragangliomas	0/21	(0%)
Large cell neuroendocrine tumors	1/4	(25%)
Prostate	1/1	(100%)
Sinonasal	0/1	(0%)
Larynx	0/1	(0%)
Ovary	0/1	(0%)
Small cell carcinomas	20/124	(16%)
Skin (Merkel cell carcinoma)	0/61	(0%)
Gastrointestinal tract	2/12	(17%)
Sinonasal	0/8	(0%)
Urinary bladder	5/12	(42%)
Prostate	9/12	(75%)
Uterine cervix	2/11	(18%)
Salivary gland	0/3	(0%)
Breast	1/1	(100%)
Thyroid	1/3	(33%)
Pancreas	0/1	(0%)

Data accumulated from: Agoff et al. (2000) *(54)*, Bejarano and Nikiforov (1999) *(50)*, Byrd-Gloster et al. (2000) *(37)*, Di Loreto et al. (1997) *(24)*, Fabbro et al. (1996) *(22)*, Folpe et al. (1999) *(35)*, Hanly et al. (2000) *(38)*, Harlamert et al. (1998) *(27)*, Katoh et al. (2000) *(55)*, Kaufmann and Dietel (2000) *(31)*, Ordóñez (2000) *(39)*.

16. Humid chamber.
17. Black-and-Decker Handy Steamer Plus (Shelton, CT).

3. Methods

1. Prior to immunostaining, tissue specimens are fixed in 10% neutral buffered formalin, processed, and embedded in paraffin (*see* **Note 1**).
2. Four-μm thick tissue sections are cut from the paraffin blocks and mounted in poly-L-lysine coated slides or sialinized glass slides to secure the adherence of the tissue to the glass slide during the immunostaining procedure.

Utilization of TTF-1 Immunostaining

3. Tissue sections are deparaffinized in three changes of xylene for 5 min each and subsequently rehydrated in descending grades of ethanol (100% to 70%), 3 min each, to distilled water.
4. To enhance the immunostaining, place deparaffinized sections in a thermoresistant container filled with citrate buffer solution and steam for 45 min using the Black and Decker steamer, then allow to cool for 20 min.
5. Rinse in several changes of distilled water, then place in 3% hydrogen peroxide/methanol for 5 min to block endogenous peroxidase activity.
6. Rinse in several changes of distilled water, then wash in Tween 20/PBS for 5 min.
7. From this point on, do not allow sections to dry. To reduce nonspecific background, the excess PBS is quickly wiped from the periphery of the sections and normal horse serum is applied for 20 min.
8. Tilt and blot excess serum from around sections and incubate in a humid chamber with the TTF-1 antibody for 1 h at room temperature (*see* **Note 2**).
9. Wash slides in 3 changes of Tween 20/PBS for 5 min.
10. Incubate sections in biotinylated anti-mouse, biotinylated anti-rabbit immunoglobulins according to manufacturer's instructions (LSAB-2 Dako kit) for 30 min.
11. Wash slides in 3 changes of Tween 20/PBS for 5 min.
12. Incubate sections in streptavidin peroxidase conjugate according to manufacturer's instructions (LSAB-2 Dako kit) for 30 min.
13. Wash slides in 3 changes of Tween 20/PBS for 5 min.
14. Incubate sections in freshly made 3,3′-diaminobenzidine according to manufacturer's instructions for 5 min monitoring the reaction microscopically.
15. Rinse in several changes of distilled water.
16. Counterstain in Mayer's hematoxylin for 5 min.
17. Intensify nuclear staining with Scott's tap water substitute.
18. Rinse in several changes of tap water.
19. Dehydrate in ascending grades of ethanol (80, 95, 100%). Clear in xylene and mount with Permount.
20. Review slides for positive nuclear staining (*see* **Notes 3–5**).

4. Notes

1. The time of fixation depends on the size of the specimen, but ideally should be 6–24 h. It is important to keep in mind that overfixation may result in masking of the antigen and contribute to reduced immunostaining or to a false negative result.
2. The optimal dilution of the antibody should be determined using antibody diluent.
3. Only a nuclear staining signal is considered a positive reaction.
4. The specificity of the immunoreaction can be double checked by staining sections with unrelated isotypic mouse immunoglobulins at comparable solution and with normal serum alone. Multitissue blocks with known positive and negative controls stained simultaneously with primary antibody are also used.

5. Because epithelial mesotheliomas can present a wide variety of histological patterns, the distinction of these tumors from peripheral pulmonary adenocarcinomas involving the pleura or from metastatic adenocarcinomas from a distant organ can be difficult on routine histologic preparations. This differential diagnosis, however, has been greatly facilitated by the use of panels of immunohistochemical markers that are commonly expressed in mesotheliomas but not in adenocarcinomas (positive markers) or in adenocarcinomas but not in mesotheliomas (negative markers) *(55,56)*. Until recently, none of the negative markers used in the diagnosis of mesotheliomas were useful in distinguishing between adenocarcinomas of the lung and nonpulmonary adenocarcinomas. Several recent studies have demonstrated that because TTF-1 is not expressed in mesotheliomas but is almost exclusively expressed in lung adenocarcinomas this marker should be included in the battery of antibodies that are currently used in the evaluation of pleural-based epithelial malignancies *(25,28,32)*.

References

1. Mitchell, P. J. and Tjian, R. (1989) Transcriptional regulations in mammalian cells by sequence-specific DNA binding proteins. *Science* **245,** 371–378.
2. Bingle, C. D. (1997) Thyroid transcription factor-1. *Int. J. Biochem. Cell Biol.* **29,** 1471–1473.
3. Ghaffari, M., Zeng, X., Whitsett, J. A., and Yan, C. (1997) Nuclear localization domain of thyroid transcription factor-1 in respiratory epithelial cells. *Biochem. J.* **328,** 757–761.
4. Kimura, S., Hara, Y., Pineau, T., Fernandez-Salguero, P., Fox, C. H., Ward, J. M., and Gonzalez, F. J. (1996) The T/ebp null mouse: thyroid-specific enhancer-binding protein is essential for the organogenesis of the thyroid, lung, ventral forebrain, and pituitary. *Genes Dev.* **10,** 60–69.
5. Lazzaro, D., Price, M., De Felice, M., and Di Lauro, R. (1991) The transcription factor TTF-1 is expressed at the onset of thyroid and lung morphogenesis and in restricted regions of the foetal brain. *Development* **113,** 1093–1104.
6. Civitareale, D., Lonigro, R., Sinclair, A. J., and Di Lauro, R. (1989) A thyroid-specific nuclear protein essential for tissue-specific expression of the thyroglobulin promoter. *EMBO J.* **8,** 2537–2542.
7. Endo, T., Kaneshige, M., Nakazato, M., Ohmori, M., Harii, N., and Onaya, T. (1997) Thyroid transcription factor-1 activates the promoter activity of rat thyroid Na^+/I^- symporter gene. *Mol. Endocrinol.* **11,** 1747–1755.
8. Francis-Lang, H., Price, M., Polycarpou-Schwartz, M., and Di Lauro, R. (1992) Cell-type-specific expression of the rat thyroperoxidase promoter indicates common mechanism of thyroid-specific gene expression. *Mol. Cell Biol.* **12,** 576–588.
9. Bruno, M. D., Bohinski, R. J., Huelsman, K. M., Whitsett, J. A., and Korfhagen, T. R. (1995) Lung cell-specific expression of the murine surfactant protein A (SP-A) gene is mediated by interactions between the SP-A promoter and thyroid transcription factor-1. *J. Biol. Chem.* **270,** 6531–6536.

10. Bohinski, R. J., Di Lauro, R., and Whitsett, J. A. (1994) The lung-specific protein B promoter is a target for thyroid transcription factor 1 and hepatocyte nuclear factor 3, indicating common factors for organ specific gene expression along the foregut axis. *Mol. Cell. Biol.* **14**, 5671–5681.
11. Yan, C., Sever, Z., and Whitsett, J. A. (1995) Upstream enhancer activity in the human surfactant protein B gene is mediated by thryoid transcription factor 1. *J. Biol. Chem.* **270**, 24852–24857.
12. Kelly, S. E., Bachurski, C. J., Burhans, M. S., and Glasser, S. W. (1996) Transcription of the lung-specific surfactant protein C gene is mediated by thyroid transcription factor-1. *J. Biol. Chem.* **271**, 6881–6888.
13. Zhang, L., Whitsett, J. A., and Stripp, B. R. (1997) Regulation of Clara cell secretory protein gene transcription by thyroid transcription factor-1. *Biochem. Biophys. Acta.* **1350**, 359–367.
14. Lonigro, R., De Felice, M., Biffali, E., Macahia, P. E., Damante, G., Asteria, C., and Di Lauro, R. (1996) Expression of thyroid transcription factor 1 gene can be regulated at the transcriptional and posttranscriptional levels. *Cell Growth Differ.* **7**, 251–261.
15. Ikeda, K., Shaw-White, J. R., Wert, S. E., and Whitsett, J. A. (1996) Hepatocyte nuclear factor 3 activates transcription of thyroid transcription factor 1 in respiratory epithelial cells. *Mol. Cell Biol.* **16**, 3626–3636.
16. Shaw-White, J. R., Bruno, M. D., and Whitsett, J. A. (1999) GATA-6 activates transcription of thyroid transcription factor-1. *J. Biol. Chem.* **274**, 2658–2664.
17. Perrone, L., Tell, G., and Di Lauro, R. (1999) Calreticulin enhances the transcriptional activity of thyroid transcription factor-1 by binding to its homeodomain. *J. Biol. Chem.* **274**, 4640–4645.
18. Losada, A., Tovar, J. A., Xia, H. M., Diez-Pardo, J. A., and Santisteban, P. (2000) Down-regulation of thyroid transcription factor-1 gene expression in fetal lung hypoplasia is restored by glucocorticoids. *Endocrinology* **141**, 2166–2173.
19. Stahlman, M. T., Gray, M. E., and Whitsett, J. A. (1996) Expression of thyroid transcription factor-1 (TTF-1) in fetal and neonatal human lung. *J. Histochem. Cytochem.* **44**, 673–678.
20. Khoor, A., Stahlman, M. T., Gray, M. E., and Whitsett, J. A. (1994) Temporal-spatial distribution of SP-B and SP-C proteins and mRNAs in developing respiratory epithelium of human lung. *J. Histochem. Cytochem.* **42**, 1187–1199.
21. Bejarano, P. A., Baughman, R. P., Biddinger, P. W., Miller, M. A., Fenoglio-Preiser, C., Al-Kafaji, B., et al. (1996) Surfactant proteins and thyroid transcription factor-1 in pulmonary and breast carcinomas. *Mod. Pathol.* **9**, 445–452.
22. Fabbro, D., Di Loreto, C., Stamerra, O., Beltrami, C. A., Lonigro, R., and Damante, G. (1996) TTF-1 gene expression in human lung tumours. *Eur. J. Cancer* **32A**, 512–517.
23. Holzinger, A., Dingle, S., Bejarano, P. A., Miller, M. A., Weaver, T. E., Di Lauro, R., and Whitsett, J. A. (1996) Monoclonal antibody to thyroid transcription factor-1: production, characterization, and usefulness in tumor diagnosis. *Hybridoma* **15**, 49–53.

24. Di Loreto, C., Di Lauro, V., Puglisi, F., Damante, G., Fabbro, D., and Beltrami, C. A. (1997) Immunocytochemical expression of tissue specific transcription factor-1 in lung carcinoma. *J. Clin. Pathol.* **50,** 30–32.
25. Di Loreto, C., Puglisi, F., Di Lauro, V., Damante, G., and Beltrami, C. A. (1998) TTF-1 protein expression in pleural malignant mesotheliomas and adenocarcinomas of the lung. *Cancer Lett.* **124,** 73–78.
26. Bohinski, R. J., Bejarano, P. A., Balko, G., Warnick, R. E., and Whitsett, J. A. (1998) Determination of lung as the primary site of cerebral metastatic adenocarcinomas using monoclonal antibody to thyroid transcription factor-1. *J. Neuro.-Oncol.* **40,** 227–231.
27. Harlamert, H. A., Mira, J., Bejarano, P. A., Baughman, R. P., Miller, M. A., Whitsett, J. A., and Yassin, R. (1998) Thyroid transcription factor-1 and cytokeratin 7 and 20 in pulmonary and breast carcinoma. *Acta. Cytol.* **42,** 1382–1388.
28. Khoor, A., Whitsett, J. A., Stahlman, M. T., Olson, S. J., and Cagle, P. T. (1999) Utility of surfactant protein B precursor and thyroid transcription factor 1 in differentiating adenocarcinoma of the lung from malignant mesothelioma. *Hum. Pathol.* **30,** 695–700.
29. Puglisi, F., Barbone, F., Damante, G., Bruckbauer, M., Di Laura, V., Beltrami, C. A., and Di Loreto, C. (1999) Prognostic value of thyroid transcription factor-1 in primary, resected, non-small cell lung carcinoma. *Mod. Pathol.* **12,** 318–324.
30. Anwar, F. and Schmidt, R. A. (1999) Thyroid transcription factor-1 (TTF-1) distinguishes mesothelioma from pulmonary adenocarcinoma. *Lab. Invest.* **79,** 181A.
31. Kaufmann, O. and Dietel, M. (2000) Thyroid transcription factor-1 is the superior immunohistochemical marker for pulmonary adenocarcinomas and large cell carcinomas compared to surfactant proteins A and B. *Histopathology* **36,** 8–16.
32. Ordóñez, N. G. (2000) Value of thyroid transcription factor-1, E-cadherin, BG8, WT1, and CD44S immunostaining in distinguishing epithelial pleural mesothelioma from pulmonary and nonpulmonary adenocarcinoma. *Am. J. Surg. Pathol.* **24,** 598–606.
33. Reis-Filho, J. S., Carrilho, C., Valenti, C., Leitao, D., Ribeiro, C. A., Ribeiro, S. G., and Schmitt, F. C. (2000) Is TTF1 a good immunohistochemical marker to distinguish primary from metastatic lung adenocarcinomas? *Pathol. Res. Pract.* **196,** 835–840.
34. Pelosi, G., Fraggetta, F., Pasini, F., Maisonneuve, P., Sonzogni, A., Iannucci, A., et al. (2001) Immunoreactivity for thyroid transcription factor-1 in stage I non-small cell carcinomas of the lung. *Am. J. Surg. Pathol.* **25,** 363–372.
35. Folpe, A. L., Gown, A. M., Lamps, L. W., Garcia, R., Dail, D. H., Zarbo, R. J., and Schmidt, R. A. (1999) Thyroid transcription factor-1: immunohistochemical evaluation in pulmonary neuroendocrine tumors. *Mod. Pathol.* **12,** 5–8.
36. Caccamo, D. V., Linden, M. D., and Zarbo, R. J. (2000) Cytokeratin subtypes and thyroid transcription factor-1 (TTF-1) immunostaining in large cell neuroendocrine carcinoma of the lung (abstract 1222). *Mod. Pathol.* **13,** 207A.

37. Byrd-Gloster, A. L., Khoor, A., Glass, L. F., Messina, J. L., Whitsett, J. A., Livingston, S. K., and Cagle, P. T. (2000) Differential expression of thyroid transcription factor 1 in small cell lung carcinoma and Merkel cell tumor. *Hum. Pathol.* **31,** 58–62.
38. Hanly, A. J., Elgart, G. W., Jorda, M., Smith, J., and Nadji, M. (2000) Analysis of thyroid transcription factor-1 and cytokeratin 20 separates merkel cell carcinoma from small cell carcinoma of lung. *J. Cutan. Pathol.* **27,** 118–120.
39. Ordóñez, N. G. (2000) Value of thyroid transcription factor-1 immunostaining in distinguishing small cell lung carcinomas from other small cell carcinomas. *Am. J. Surg. Pathol.* **24,** 1217–1223.
40. Burke, L. M., Rush, W. I., Khoor, A., Mackay, B., Whitsett, J. A., Singh, G., et al. (1999) Alveolar adenoma: A histochemical, immunohistochemical, and ultrastructural analysis of 17 cases. *Hum. Pathol.* **30,** 158–167.
41. Dessy, E., Braidotti, P., Del Curto, B., Falleni, M., Coggi, G., Santa Cruz, G., et al. (2000) Peripheral papillary tumor of type-II pneumocytes: a rare neoplasm of undetermined malignant potential. *Virchows Archiv.* **436,** 289–295.
42. Chan, A. C. L. and Chan, J. K. C. (2000) Pulmonary sclerosing hemangioma consistently expresses thyroid transcription factor-1 (TTF-1): a new clue to its histogenesis. *Am. J. Surg. Pathol.* **24,** 1531–1536.
43. Devouassoux-Shisheboran, M., Hayashi, T., Linnoila, R. I., Koss, M. N., and Travis, W. D. (2000) A clinicopathologic study of 100 cases of pulmonary sclerosing hemangioma with immunohistochemical studies: TTF-1 is expressed in both round and surface cells, suggesting an origin from primitive respiratory epithelium. *Am. J. Surg. Pathol.* **24,** 906–916.
44. Noguchi, M., Nakajima, T., Hirohashi, S., Akiba, T., and Shimosato, Y. (1989) Immunohistochemical distinction of malignant mesothelioma from pulmonary adenocarcinoma with anti-surfactant apoprotein, anti-Lewis[a], and anti-Tn antibodies. *Hum. Pathol.* **20,** 53–57.
45. Katzenstein, A. L., Weise, D. L., Fulling, K., and Battifora, H. (1983) So-called sclerosing hemangioma of the lung. Evidence for mesothelial origin. *Am. J. Surg. Pathol.* **7,** 3–14.
46. Kay, S., Still, W. J., and Borochovitz, D. (1977) Sclerosing hemangioma of the lung: an endothelial or epithelial neoplasm? *Hum. Pathol.* **8,** 468–474.
47. Liebow, A. A. and Hubbell, D. S. (1956) Sclerosing hemangioma (histiocytoma, xanthoma) of the lung. *Cancer* **9,** 53–75.
48. Xu, H. M., Li, W. H., Hou, N., Zang, S. G., Li, H. F., Wang, S. Q., et al. (1997) Neuroendocrine differentiation in 32 cases of so-called sclerosing hemangioma of the lung: identified by immunohistochemical and ultrastructural study. *Am. J. Surg. Pathol.* **21,** 1013–1022.
49. Ordóñez, N. G. (2000) Thyroid transcription factor-1 is a marker of lung and thyroid carcinomas. *Adv. Anat. Pathol.* **2,** 123–127.
50. Bejarano, P. A. and Nikiforov, Y. E. (1999) Thyroid transcription factor-1 (TTF-1), thyroglobulin, CK7 and CK20 in neoplasms of the thyroid gland. *Lab. Invest.* **79,** 66A.

51. Fabbro, D., Di Loreto, C., Beltrami, C. A., Belfiore, A., Di Lauro, R., and Damante, G. (1994) Expression of thyroid-specific transcription factors TTF-1 and PAX-8 in human thyroid neoplasms. *Cancer Res.* **54,** 4744–4749.
52. Agoff, S. N., Lamps, L. W., Philip, A. T., Amin, M. B., Schmidt, R. A., True, L. D., and Folpe, A. L. (2000) Thyroid transcription factor-1 is expressed in extrapulmonary small cell carcinomas but not in other extrapulmonary neuroendocrine tumors. *Mod. Pathol.* **13,** 238–242.
53. Queirolo, P., Gipponi, M., Peressini, A., Dissoma, C. F., Vecchio, S., Raposio, E., et al. (1997) Merkel cell carcinoma of the skin. Treatment of primary, recurrent, and metastatic disease: review of clinical cases. *Anticancer Res.* **17,** 673–677.
54. Syed, S., Lele, S. M., Adegboyega, P. A., and Haque, A. K. (2000) TTF-1 and c-erbB-2 expression and correlation with survival in non-small cell lung cancer [abstract]. *Mod. Pathol.* **13,** 217A.
55. Ordóñez, N. G. (1999) The immunohistochemical diagnosis of epithelial mesothelioma. *Hum. Pathol.* **30,** 313–323.
56. Ordóñez, N. G. (1999) Role of immunohistochemistry in distinguishing epithelial mesothelioma from adenocarcinoma. A review and update. *Am. J. Clin. Pathol.* **112,** 75–89.

22

Molecular Biologic Substaging of Stage I NSCLC Through Immunohistochemistry Performed on Formalin-Fixed, Paraffin-Embedded Tissue

Mary-Beth Moore Joshi, Thomas A. D'Amico, and David H. Harpole, Jr.

1. Introduction

Non-small cell lung cancer (NSCLC) is the most common cause of death by malignancy in both men and women in the United States *(1)*. The current staging system for NSCLC considers the size and location of the primary tumor (T), the involvement of regional lymph nodes (N), and the presence of distant metastases (M) *(2)*. The standard treatment of patients with stage I NSCLC (T1-2N0M0) is resection of the primary tumor with no adjuvant therapy. However, even after complete resection, the 5-yr survival rate is only 55–72%, mainly due to the development of distant metastases *(2–7)*. Therefore, a significant number of patients are under-staged and may benefit from adjuvant therapy.

1.1. Factors Predicting Prognosis

In order to identify a population of stage I NSCLC patients that would benefit from adjuvant therapy, investigators have attempted to identify factors that predict poor prognosis. Conventional histopathological variables such as the analysis of performance status, subtype and size of the primary tumor, degree of tumor differentiation, mitotic rate, and the evidence of lymphatic or vascular invasion have been analyzed *(4,7–13)*. These factors indirectly measure tumor aggressiveness, but have not successfully identified a group of stage I patients who would benefit from adjuvant therapy. The use of histopathologic variables alone to construct a risk model is limited by low prevalence and the discontinuous nature of the variables *(14)*.

1.1.2. Molecular Biologic Substaging

Molecular biologic substaging refers to the use of oncogenic markers to improve risk stratification, with possible therapeutic implications *(15)*. The change from normal bronchial epithelium to carcinoma is a result of a series of genetic changes in normal cellular proteins. The major regulatory genes mutated in NSCLC include a number of protooncogenes and tumor suppressor genes. Carcinogenesis can occur either through activation or deletion of these proteins. When protooncogenes are altered, they may acquire transforming potential that can disrupt normal cell growth. Tumor-suppressor genes have been implicated in lung cancer due to elimination of both alleles, mutation, or functional abnormality of their protein products *(15)*. In lung cancer, growth-regulating proteins, cell-cycle specific proteins, apoptosis-related proteins, tumor-suppressor genes, and tumor-invasion and metastasis-related proteins have recently been studied immunohistochemically to determine prognostic importance.

1.1.3. Growth Regulating Proteins

The major growth regulating gene protein products that have been studied in NSCLC are the *ras* oncogenes, epidermal growth factor receptor (EGFr), and the c-*erb*B-2/*Her*2-neu oncogenes. Mutations that occur in the *ras* family of oncogenes results in continual cell proliferation. The three most significant members of the *ras* family are K-*ras*, H-*ras*, and N-*ras*, but most of the mutations associated with NSCLC involve K-*ras*. These mutations are commonly found in adenocarcinoma and are linked to smoking and asbestos exposure and correlate with decreased survival *(16–17)*. EGFr is a tyrosine kinase-type membrane receptor encoded by the protooncogene c-*erb*B-1. Mutations in c-*erb*B-1 ultimately result in uncontrolled tumor growth *(18)*. Elevated levels of EGFr have been found in lung cancer as compared to normal lung, in later-stage lung cancers and in those with mediastinal involvement, but no correlation between these levels and prognosis has been shown *(16,18–20)*. However, in immunohistochemical analysis of T1N0 NSCLC patients, it has been shown that overexpression of EGFr correlates with decreased survival *(14,21)*. C-*erb*B-2/*Her*2-neu encodes for a transmembrane tyrosine kinase (p185neu) that functions as a growth factor receptor. It has been shown that overexpression of c-*erb*B-2/*Her*2-neu in NSCLC is associated with decreased survival and is considered to be an independent negative prognostic factor *(13–14,22–24)*.

1.1.4. Cell-Cycle Specific Proteins

The major cell-cycle specific protein products that have been studied in NSCLC are p53, bcl-2, Retinoblastoma gene (RB), p16^{INK4a} and Ki-67. Normal bronchial epithelium is arrested in the G0 phase of the cell cycle. At the end of its life span, an apoptotic cascade is signaled, causing cell death. In NSCLC, mutations or deletions in cell-cycle specific genes can cause the tumor cells to be resistant to apoptosis therefore, making them "immortal" *(15)*. Mutations in the p53 tumor-suppressor gene are the most common mutations associated with human cancers *(25)*. In the majority of the literature, where p53 overexpression in NSCLC was studied immunohistochemically, overexpression of the p53 protein correlates with decreased survival *(3,13–14,22,26–29)*. The bcl-2 protooncogene encodes a protein localized in the mitochondrial membrane that inhibits apoptosis and thus prolonging the survival of the cell *(30,31)*.

Overexpression of bcl-2 has been shown to be associated with improved survival in NSCLC *(14,28,32–33)*. Studies suggest that this improved survival is because bcl-2 protects cells from apoptosis induced by mutations in the p53 tumor-suppressor gene *(34)*. The RB gene is a tumor-suppressor gene that encodes a nuclear protein that is involved in transcriptional events and regulates the cell cycle by keeping the cell in the G0 phase *(15)*. Loss of RB protein expression has been shown to be a common event in NSCLC and has been associated with decreased survival. It has also been shown that loss of RB protein expression in combination with overexpression of the p53 protein results in even poorer survival than loss of RB or overexpression of p53 alone *(14,35)*. P16^{INK4a} is a cyclin-dependent kinase inhibitor that plays a role in keeping the cell in the G0 phase by inhibiting cyclin-dependent kinase 4 mediated phosphorylation of the RB gene product. This results in unregulated cell growth and transformation *(36)*. In NSCLC, loss of p16INK4^a has been associated with decreased survival *(36)*. Ki-67 is a tumor proliferation marker that is used to identify rapidly proliferating cells. In NSCLC, several studies have found an association between high levels of Ki-67 expression and decreased survival *(3,13,37)*.

1.1.5. Tumor-Invasion and Metastasis-Related Proteins

The major tumor-invasion and metastasis-related proteins that have been studied immunohistochemically are the angiogenesis marker Factor VIII and the cell-cell adhesion molecule CD44. Tumor-induced neovascularization (angiogenesis) is necessary for tumor growth and metastasis. Studies in NSCLC have shown that by using an antibody against the Factor VIII protein, microvessels can be counted and angiogenesis can be assessed *(38)*. The degree of angiogenesis has been shown to correlate with negative prognostic implications *(13,14)*. CD44 is an integral membrane glycoprotein that is involved in cell-cell and cell-extracellular matrix interactions as well as metastatic spread. In NSCLC, expression of CD44 has been shown to negatively correlate with survival *(14,29)*.

1.2. Immunohistochemistry as a Tool

Immunohistochemistry is an important tool in clinical diagnostics as well as in research. It allows the location of defined tissue antigens to be visualized through the binding of antibodies to small and unique regions on the antigens (epitopes). Though easily performed on fresh, frozen tissue samples, immunohistochemistry performed on formalin-fixed paraffin-embedded tissue samples can be met with some obstacles. The major factors that effect the staining results are: 1) tissue fixation, 2) antigen unmasking, 3) sensitivity of the detection system, 4) the quality control of each assay performed, and 5) standardization. Due to the greater availability of formalin-fixed, paraffin-embedded routine and archival surgical pathology samples, as compared to fresh, frozen samples, understanding and learning to overcome these obstacles can increase research opportunities.

1.2.1. Tissue Fixation

The goal of fixation is to preserve the immunoreactivity of tissue antigens while maintaining acceptable morphology. Though substantial strides have been made to improve the methods used in localizing specific antigens, a persistent concern in immunopathology is choosing the correct fixative and duration of fixation that will provide maximal morphology preservation while minimizing the loss of antigenicity *(39)*. It is important to be aware that the effects of chemical fixation on the immunoreactivity of tissue antigens is unpredictable *(40)*. There are several factors that will influence the intensity of immunohistochemical staining: 1) intrinsic properties of the fixative, 2) pH, 3) osmolarity, 4) temperature, 5) length of treatment, and 6) type of tissue *(41)*.

The most commonly used fixative for routine surgical sample fixation is neutral buffered formalin (NBF). It maintains acceptable morphology without shrinkage artifact, but it can severely compromise the immunoreactivity of most tissue antigens through the formation of intermolecular and intramolecular crosslinks between protein end groups. These crosslinks alter the primary and tertiary structures of the antigens *(42,43)*. In order to circumvent this problem, monoclonal antibodies (MAbs) that have been developed for use in routinely fixed and processed tissues are made to recognize antigens altered by formalin fixation *(42,44–46)*.

Until recently it was believed that these alterations to the antigen structure were irreversible; however, with the development of antigen-unmasking

Molecular Biologic Substaging

techniques, previously undetectable or weakly detectable antigens may now be visualized *(43)*.

1.2.2. Antigen Unmasking

At times it may be necessary to increase assay sensitivity in order to compensate for a decrease in immunoreactivity of tissue antigens compromised by routine fixation in aldehyde-containing fixatives such as NBF. Fixation of tissue in NBF causes the formation of excess aldehyde cross-links, which leads to decreased sensitivity *(42)*. Optimally, antigen retrieval should cleave these cross-links, resulting in the reconstruction of the original three-dimensional structure of the epitope *(43)*. There are two ways to correct for decreased sensitivity due to NBF fixation: 1) proteolytic enzyme digestion and 2) heat-induced antigen retrieval applied prior to immunostaining.

1.2.2.1. PROTEOLYTIC ENZYME DIGESTION

Proteolytic enzyme digestion of tissue fixed in NBF and embedded in paraffin increases their sensitivity to that of cold acetone-fixed and processed, paraffin-embedded tissue *(41,47)*. It greatly reduces nonspecific staining while enhancing the immunoreactivity of many tissue antigens *(42,48–50)*. This allows the use of a higher antibody dilution.

There are limitations involved with the use of proteolytic enzyme digestion. Overexposure to the enzyme can cause tissue damage. This can yield false-negative results through the digestion of the target protein and can increase background through proteolytic cleavage of tissue antigens resulting in fragments common to many tissue antigens *(51,52)*. Therefore, the main disadvantage to using this technique is that titration of time and temperature is needed for different tissue types and fixation times, making it difficult to have a generally applicable, standardized procedure *(53–55)*. Therefore, each laboratory needs to develop its own protocols for proteolytic enzyme digestion.

1.2.2.2. HEAT-INDUCED ANTIGEN RETRIEVAL

Heat-induced antigen retrieval is the most commonly used form of antigen unmasking. The main principle of heat-induced antigen retrieval is the exposure of hydrated slides to boiling temperatures for approx 10 min, while immersed in fluid. The mechanism of this form of antigen retrieval is not fully understood. However, based on early studies of Fraenkel-Conrat and coworkers, the developers of this technique and others have proposed the most plausible explanation. The heat applied to the sections provides the energy needed to break the cross-links that have formed during the formalin fixation between calcium ions or other divalent metal cations and proteins *(38,56–59)*. The buffer in which the sections are heated either precipitates or chelates the released metal ions *(58)*. There are several advantages to heat-induced antigen retrieval over proteolytic enzyme digestion. It increases the reactivity of antibodies that do not benefit from proteolytic enzyme digestion, it works for a greater number antibodies, reduces background staining, works on overfixed tissues, the slides can be bulk processed, and the method is more standardized (*see* **Subheading 3.3.1.**) *(39,55)*. Several methods of heat-induced antigen retrieval have been published. Each of these protocols emphasizes different parameters such as heating duration, buffer type and pH, and source of heat *(43)*. When comparing these different methods, the use of a pressure cooker within a microwave, using a buffer in the pH range of 6.0–9.0, is the easiest and most widely used method. It attains the highest staining intensity and is the best method for a truly standardized protocol *(55)* (*see* **Subheading 3.3.2.1., 3.3.2.2.,** and **Note 14**).

1.2.3. Detection Systems

The immunodetection of antigens is a two-step process. The first step is the binding of the applied antibody to the antigen of interest and the second is the visualization of the bound antibody through an enzyme chromogenic system. The choice of a detection system will greatly impact the sensitivity of the assay. There are two main methods of immunodetection: direct and indirect. With the direct method, an enzyme is directly conjugated to the primary antibody before it is applied to the tissue. The reaction is then visualized by combining the substrate with a chromogen and applying it to the tissue. This method is not frequently used, due to the additional step of conjugating each primary antibody to be used and due to the often weak staining results produced. The indirect detection method yields improved sensitivity over the direct method. With the indirect method, the first step is the application of the primary antibody to the tissue. Following incubation in the primary antibody, a secondary antibody, or a linking antibody, is applied. The secondary antibody is raised in another animal against the immunoglobulin type of the primary antibody and is conjugated with an enzyme. The reaction is visualized by applying a mixture of substrate and chromogen that reacts with the conjugated enzyme to produce a color change. The indirect method has been further improved by separating the labeling enzyme from the link antibody providing an additional level of signal amplification.

The most commonly used indirect methods are the peroxidase-antiperoxidase (PAP), the avidin-biotin complex (ABC) and the biotin-streptavidin amplified (B-SA) system.

1.2.3.1. THE PAP SYSTEM

The PAP system uses a primary antibody, a secondary antibody, a PAP complex containing the label and the chromogen substrate. The tissue is first

Molecular Biologic Substaging

deparaffinized in xylene and rehydrated in graded alcohols. It is then treated with a hydrogen peroxide solution to block endogenous peroxidase activity. The tissue sections are then treated with a protein-blocking solution such as normal serum or a universal blocking reagent to block nonspecific binding sites. This step is followed by incubation in the primary antibody specific for the antigen to be demonstrated. The sections are then incubated in a linking or secondary antibody that binds to the primary antibody and the PAP complex. Therefore, one of the sites on the linking antibody binds to the primary antibody leaving the second binding site free for binding to the PAP complex. The PAP complex or labeling reagent is then applied to the tissue section. It has been raised in the same species as the primary antibody and is made up of the enzyme peroxidase and an antibody against peroxidase. This binds to the remaining site on the linking antibody. The reaction is then visualized by the addition of a chromogenic substrate. The sensitivity of this method is limited by the affinities of the link and label reagents for each other (generally, $10^8–10^9$ M^{-1}) *(42,60,61)*.

1.2.3.2. THE ABC SYSTEM

This system is based on the high binding affinity between avidin and biotin (10^{15} M^{-1}) and provides an increased sensitivity and a choice of labeling enzymes *(62)*. It uses a biotinylated linking antibody to covalently bind to the primary antibody and a labeling complex that contains avidin, biotin, and an enzyme label. After deparaffinization, rehydration, and blocking of nonspecific binding sites, the primary antibody is applied to the tissue followed by the application of a biotinylated secondary antibody. After incubation in the secondary antibody, the avidin-biotin complex is applied, followed by the substrate-chromogen solution for visualization.

1.2.3.3. THE B-SA SYSTEM

The B-SA detection method is based on the high binding affinity of streptavidin for biotin. This affinity is approximately 10^6 greater than that of most antibodies for their antigens, and therefore provides very specific detection and amplification of antigen-antibody binding. In fact, the use of streptavidin is preferred over avidin. Unlike avidin, streptavidin contains no carbohydrate which can nonspecifically bind to lectin-like substances found in some normal tissues *(63–66)*. Also, the isoelectric point on streptavidin is close to neutral, whereas the isoelectric point of avidin is 10. Therefore, by incorporating streptavidin conjugates, the positively charged, nonspecific binding effects that are characteristic of avidin, are greatly reduced. Unlike the ABC system, which must be prepared immediately before use, the enzyme is directly conjugated to streptavidin, this results in a highly stable reagent.

1.2.4. Quality Control

Proper quality control is the most important part of each assay performed. If each assay is not properly controlled, the results will be compromised. A known immunohistochemically positive control slide for the target protein should be run with each assay. The positive control ensures that the primary antibody and detection system reagents are working. Because many antigens are adversely affected by fixation, the positive control tissue used should be fixed and processed in the same manner as the test samples. If the control sample is optimally fixed and the test samples have been over-fixed, a false negative may result. Variable and inconsistent staining can be caused by variations in the tissue fixation and processing protocols. If the exact protocol or quality of fixation of the sample tissue is not known then a processing control should be run on each sample using the Vimentin antibody in order to determine whether or not the samples have been optimally processed. Vimentin is a molecule that is present in almost every tissue and displays partial sensitivity to formalin fixation. The longer the tissue is in fixative the more vimentin epitopes are altered or destroyed therefore, causing weak staining. A negative control slide should also be run with every sample assayed. This confirms that a positive test is the result of specific antigen-antibody binding, rather than nonspecific background. The negative control is run using the same test protocol and same tissue, excluding the primary antibody. In place of the primary antibody, apply an immunoglobulin from the same species using the same dilution as the primary antibody.

1.2.5. Standardization

Due to the great variability in specimen procurement, fixation and processing, it is not practical to attempt to standardize a single protocol across all laboratories. But, it is possible to standardize protocols within a single laboratory. The goal of standardization is to ensure assay to assay reproducibility *(67)*. In order to achieve this goal several things must be considered. First, the proper quality control standards must be rigorously followed with every assay performed. Second, reagents must be selected carefully. One way to standardize a protocol is to buy commercially available reagents that have already been through the manufacture's rigorous quality control tests rather than making each reagent from scratch. Third, calibration of the microwave for antigen retrieval protocols will standardize heat-induced antigen retrieval. As new approaches to heat-induced antigen retrieval are developed, each laboratory has adopted different procedures that yield different degrees of antigenicity restoration *(67)*. Bearing this in mind, it is extremely important to standardize this portion of the immunohistochemistry protocol so as not

Molecular Biologic Substaging

add another variable into the overall process. Fourth, when working up a new antibody, do not vary greatly from the basic immunohistochemical protocol. Once an optimal protocol and reagents have been determined, only change the antigen retrieval step (retrieval vs no retrieval, solution pH, enzyme vs heat), the primary antibody titer, and the primary antibody incubation time and temperature. The fewer portions of the protocol that are altered the easier it is to troubleshoot when staining problems arise (*see* **Subheading 3.2.** and **Note 8**). Finally, an ideal way to standardize an immunohistochemical protocol is the addition of automated immunohistochemical instrumentation. This removes the majority of human error and has been shown to outperform even the best technologist, in terms of consistency, during a sustained period of time *(67)*.

2. Materials
2.1. Slide Preparation (see Note 1)
1. Positively charged microscope slides.
2. Slide dryer.

2.2. Immunohistochemical Staining Procedure (see Note 1)
1. Warming oven.
2. Histologic grade Xylene. Storage: room temperature in a flammables cabinet. Preparation: none required.
3. 100% Ethanol: Histologic grade. Storage: room temperature in a flammables cabinet. Preparation: none required.
4. 95% Ethanol: Histologic grade. Storage: room temperature in a flammables cabinet.
5. 30% Hydrogen peroxide. Storage: 4°C. Preparation: none required.
6. Deionized water.
7. Humidified incubation chamber.
8. Endogenous peroxide blocking solution: Mix 180 mL 100% ethanol with 12 mL 30% H_2O_2.
9. Protein blocking reagent. Preparation: follow the manufacturer's recommendations. Storage: follow the manufacturer's recommendations.
10. Primary antibody. Preparation: follow the manufacturer's recommendations. Storage: follow the manufacturer's recommendations.
11. Primary antibody diluting buffer. Preparation: follow the manufacturer's recommendations. Storage: follow the manufacturer's recommendations.
12. Negative control antibody. Preparation: follow the manufacturer's recommendations. Storage: follow the manufacturer's instructions.
13. Detection system. Preparation: follow the manufacturer's recommendations. Storage: follow the manufacturer's instructions.
14. Chromogen-Diaminobenzidine (DAB) (*see* **Note 2**).

a. For a small Coplin jar: Prepare 1.25 mL DAB Stock in 50 mL 0.05 M Tris Buffer, pH 7.6. Add 500 µL 0.06% H_2O_2 just before use.
b. For a large Coplin jar: Prepare 2.5 mL DAB stock in 100 mL in 0.05 M Tris Buffer, pH 7.6. Add 1.0 mL 0.06% H_2O_2 just before use.
c. For a 25-slide Tissue-Tek® Staining Dish: Prepare 5.0 mL DAB stock in 200 mL 0.05 M Tris Buffer, pH 7.6. Add 2.0 mL 0.06% H_2O_2 just before use.

15. Hematoxylin counterstain. Preparation: filter before use. Store at room temperature.
16. Ammonia water bluing solution. Preparation: place 20 drops of ammonia in 300 mL DI water. Storage: make fresh daily.

2.2.1. Stock Solutions

1. Sodium phosphate dibasic, 0.25 M, pH 8.5: Dissolve 35.5 g sodium phosphate dibasic in 1.0 L DI H_2O, adjust the pH to 8.5. Store at room temperature for 1 mo.
2. Sodium phosphate monobasic, 0.25 M, pH 4.7: Dissolve 30.0 g of sodium phosphate monobasic in 1.0 L of DI H_2O, adjust the pH to 4.7. Store at room temperature for 1 mo.
3. Phosphate-buffered saline (PBS) pH 7.4–7.8. Preparation: 400 mL 0.25 M sodium phosphate dibasic stock solution, pH 8.5 added to 100 mL 0.25 M sodium phosphate monobasic stock solution, pH 4.7. Add to 105 g of NaCl dissolved in DI H_2O and make up final volume to 12.0 L. Adjust pH to 7.2–7.4. Store at room temperature for 1 mo.
4. 0.05 M Tris Buffer, pH 7.6: 6.06 g of Trizma HCL, 1.39 g of Trizma Base, and 3.4 g of Imidazole dissolved in 1.0 L of DI H_2O. Adjust the pH to 7.6. Store at room temperature for 1 mo.
5. Diaminobenzidine (DAB) (*see* **Note 2**) Dissolve 1 g of DAB into 50 mL of 0.05 M Tris Buffer, pH 7.6. Aliquot stock solution into 1.25-, 2.5-, or 5.0-mL vials. Store at −70°C until needed.
6. 0.6% Hydrogen Peroxide. Place 2 mL 30% hydrogen peroxide in 98 mL of DI H_2O. Store at room temperature for 1 mo.

2.3. Immunohistochemical Staining: Antigen Retrieval

1. Microwave pressure cooker.
2. Citric acid stock solution: Dissolve 21.01 g citric acid in 1.0 L DI H_2O. Store at room temperature for 1 mo.
3. Working citrate buffer, pH 6.0. Citric acid antigen retrieval solution if needed. Preparation: 9.0 mL citric acid solution mixed with 123.0 mL sodium citrate solution in 1348.0 mL DI H_2O, adjust the pH to 6.0 (+/− 0.02). Store at room temperature for 1 mo.
4. Tissue-Tek® slide staining holder.
5. Tissue-Tek® Staining Dish.

3. Methods
3.1. Slide Preparation

1. Section the formalin-fixed, paraffin-embedded sample into 4–5 micron sections (*see* **Note 3**).
2. Place each section on appropriately labeled positively charged slides. Allow to air dry overnight or place in a slide dryer for 30 min.

3.2. Immunohistochemical Staining Procedure

1. Heat slides, upright, in an oven set at 60°C for 30 min to 1 h to melt the paraffin.
2. Deparaffinize the slides in Xylene for 3 changes at 10 min each (*see* **Note 4**). Clear the slides in 100% ETOH for 3 changes at 3 min each.
3. Place the slides in an endogenous peroxidase block for 10 min (*see* **Note 5**). If using a commercially available ready-to-use peroxide block, omit this step and place slides into a Tissue-Tek container of 95% ETOH and then proceed.
4. Place container under DI water and gradually replace the 95% ETOH or the endogenous peroxide block with H_2O.
5. Antigen unmasking techniques should be performed at this point if needed. If using heat-induced antigen unmasking, allow slides to cool for about 30 min before proceeding with the protocol.
6. Wash slides with 3 changes of PBS for 5 min each. If using a commercially available ready-to-use peroxide block, apply directly to the slides after this step. Incubate in a humidified incubation chamber for 10 min. Follow this with 3 changes of PBS for 5 min each, then proceed.
7. Place slides in 200 mL of protein blocking reagent for 8–10 min (*see* **Note 6**).
8. Apply primary antibody and incubate slides for 60 min at room temperature in a humidified incubation chamber (*see* **Notes 7** and **8**).
9. Wash slides in 3 changes of PBS for 5 min each (*see* **Note 9**).
10. Apply the biotinylated secondary reagent and incubate slides for 20–30 min in a humidified incubation chamber.
11. Wash slides in 3 changes of PBS for 5 min each. At this point in the protocol the DAB must be prepared.
12. Apply conjugated streptavidin tertiary reagent and incubate slides for 20–30 min in a humidified incubation chamber (*see* **Note 10**).
13. Wash slides in 3 changes of PBS for 5 min each.
14. Apply DAB and incubate slides for 5 min in a humidity chamber.
15. Place slides in running DI water for 5 min.
16. Place slides in Hematoxylin counterstain for 5 min (*see* **Note 11**).
17. Place slides in running DI water for 5 min.
18. Dip slides 10 times in Ammonia Water Bluing Solution.
19. Run the slides through two changes of 95% ETOH, 10 dips each.
20. Run the slides through two changes of 100% ETOH, 10 dips each.

21. Run the slides through three changes of Xylene, 10 dips each.
22. Cover slip and review (*see* **Note 12**).

3.3. Immunohistochemical Staining: Microwave Antigen Retrieval

If standard immunochemical procedure yields substandard results, further steps must be taken to make tissue antigens available to binding by the primary antibody (*see* **Note 13**).

3.3.1. Standardization: Pressure Cooker Calibration Technique (see *Note 14*)

3.3.1.1. MICROWAVE CALIBRATION *(68)*

1. Add 1500 mL of citric acid antigen retrieval solution to the pressure cooker and mark the level on the inside of the pressure cooker.
2. Remove the antigen retrieval solution.
3. Add gray Tissue-Tek slide staining holder containing no less than 40 slides to the pressure cooker.
4. Add the antigen retrieval solution to the pressure cooker up to the "1500 mL" mark on the inside of the pressure cooker.
5. Dampen the gasket of the pressure cooker and put into place. Securely fasten the lid on the pressure cooker and put the weighted stopper in place.
6. Place the pressure cooker in the center of microwave. Make sure the turntable is on. Set the microwave for 30 min on High power.
7. Listen for the release of steam from the pressure cooker. At this point the solution is boiling. Allow the solution to boil for 11–13 min then turn the microwave off. Note the amount of time needed to allow slides to boil for 11–13 min. Use this time for every subsequent run.
8. Carefully remove the pressure cooker from the microwave and release the pressure. Without leaning over the pressure cooker, remove the weighted stopper. **BE CAREFUL!! STEAM WILL SHOOT UP!!** Remove the lid, turning it away from your face. Allow the slides to cool for 30 min.

3.3.1.2. TISSUE-TEK STAINING DISH CALIBRATION TECHNIQUE

1. Place a full rack of slides in a white Tissue-Tek Staining Dish.
2. Add 250 mL of working strength citric acid antigen retrieval solution to the white Tissue-Tek staining dish. Loosely secure a lid over the Tissue-Tek container with large rubber bands. This prevents boil over of the antigen retrieval solution and allows for any solution that evaporates to be collected back into the container.
3. Make sure that the turntable is on. Set the microwave on high for 2–4 min until the solution comes to a rapid boil. Then turn-off the microwave. Note the exact time for the solution to boil. Use this time setting for each subsequent assay.
4. Set the microwave at 30–50% power (240–320 watts) and heat for 7–15 min. The power setting should be adjusted so the microwave cycles on and off every

Molecular Biologic Substaging

20–30 s and the solution boils about 5–10 s each cycle. Note the power setting. Use this power setting for each subsequent assay.
5. Allow the slides to cool for 15–30 min before proceeding with the IHC protocol.

3.3.2. Methods

3.3.2.1. Pressure Cooker Technique for a Large Number of Slides

1. Follow the pressure cooker calibration technique and insert the times and power settings determined in the calibration.

3.3.2.2. Pressure Cooker Technique for a Small Number of Slides or Protocols That Call for More Than One Antigen Retrieval Solution

1. Place 2 full gray racks of slides in white Tissue-Tek staining dish (*see* **Note 15**).
2. Fill each Tissue-Tek staining dish with 250 mL of antigen retrieval solution. Loosely secure a lid(s) over the Tissue-Tek Container(s) with large rubber bands. This prevents cross contamination of antigen-retrieval solutions and prevents boil over of the antigen-retrieval solution.
3. Place the Tissue-Tek staining dish in the center of the pressure cooker. When using Tissue-Tek staining dish with two different solutions, brace either side of the containers by placing a gray rack upside down. Rapid boiling can cause them to shift, causing cross-contamination.
4. Add DI water to the pressure cooker up to the "1500 mL" mark inside the pressure cooker. Do not fill the pressure cooker with water before the full Tissue-Tek staining dishes have been placed inside.
5. Dampen the gasket and put into place. Securely fasten the lid on the pressure cooker. Put the red stopper in place.
6. Put the pressure cooker in the center of the microwave and set the microwave for time and power determined in the pressure cooker calibration.
7. Carefully remove the pressure cooker from the microwave and release the pressure.
8. Without leaning over the pressure cooker, remove the red stopper. **BE CAREFUL!! STEAM WILL SHOOT UP!!** Remove the lid, turning it away from your face.
9. Allow the slides to cool for 30 min before proceeding with the IHC protocol.

3.3.2.3. Tissue-Tek Staining Dish Technique

1. Place a full rack of slides in a white Tissue-Tek staining dish and add 250 mL of antigen-retrieval solution to the dish.
2. Loosely place a lid over the container and place in the center of the microwave.
3. Make sure that the turntable is on. Set the microwave on high using the time settings established in the calibration procedure.
4. Immediately after the timer goes off, *without opening the microwave*, reset the microwave using the power and time settings established in the calibration procedure.

5. Remove the container from the microwave and carefully remove the lid.
6. Allow to cool for 15–30 min before proceeding with the IHC protocol.

4. Notes

1. The following reagents are commercially available from several manufacturer's, and can be substituted for those described herein: PBS concentrate solution, "Ready-to-Use" peroxide block, universal protein blocking solution, pre-diluted primary antibodies, negative control antisera, pre-diluted detection kits, DAB kits, antigen-retrieval solutions of varying pHs, hematoxylin counterstain.
2. DAB is a possible carcinogen. It should be handled wearing gloves under a hood. Before using the following disposal procedure, it is important to know the laws of your state and the regulations of your institution regarding the disposal of DAB. You will need the following reagents: 0.2 M potassium permanganate (31.6 g $KMnO_4$ dissolved in 1 L of DI water) and 2.0 M sulfuric acid (112 mL concentrated acid in 1 L of DI water). Dilute the DAB solution with DI water. The final concentration of DAB should not exceed 0.9 mg/mL. For each 10 mL of DAB solution (diluted or otherwise) add 5.0 mL 0.2 M potassium permanganate and 5.0 mL 2.0 M sulfuric acid. Allow the mixture to stand to stand for at least 10 h. This will allow it to become nonmutagenic for disposal.
3. If negative or weak staining occurs in the control and specimen, check to see if the tissue been overfixed. Some epitopes are destroyed by fixatives, therefore the recommended processing for each antibody used. Conversely, if the tissue been underfixed or there a delay in fixation, this may lead to antigen loss due to autolysis.
4. Take care to completely deparaffinized sections. If the reagents are repelled by the tissue section, residual paraffin may be present, which can decrease signal.
5. Blocking endogenous peroxidase is essential when using a peroxidase detection system. Endogenous peroxidase is found in specimens that are particularly bloody (liver, kidney, spleen) and it is the most common cause of background staining.
6. To prevent background problems, the blocking reagent serum proteins must be of the same species as the linking reagent.
7. Make sure residual buffer is removed from the slide by wiping around the tissue with an absorbent material prior to applying the primary antibody. Leaving too much residual buffer can further dilute the primary antibody.
8. For best results, the titer of the primary antibody must be at a high enough concentration and the antibody should be incubated for long enough at the correct temperature. If the antibody being used is a low-affinity antibody, it is recommended that the sections be incubated overnight at 4°C. It is also important to remember that some primary antibodies are heat labile and cannot be incubated at temperatures higher than room temperature. Conversely, slides overincubated in the primary antibody (or the detection system or chromogen) can show high background or nonspecific staining. This points up the need for empiric titration of all reagents used in this method.

9. Background or nonspecific staining can result if slides are not washed thoroughly between incubations. Do not cheat on the washing steps to try to save time. Any excess reagent left behind will bind to the subsequent reagents.
10. Make sure that buffer for this step is made correctly. If using a horseradish peroxidase conjugated tertiary antibody, note that this enzyme is inhibited by sodium azide, a common preservative. Also, be sure the pH of the buffer is correct. On the other hand, blocking endogenous alkaline phosphatase is essential when using an alkaline phosphatase detection system. Endogenous alkaline phosphatase is found in all tissues except intestine and can be blocked using an acid alcohol block or by adding levamisole to the substrate. In addition, if an avidin-biotin detection system is used, an avidin-biotin block may be needed. There are many avidin-biotin blocking kits that are commercially available. Follow the manufacturer's recommended protocol for optimal results. Endogenous biotin is commonly found in liver, kidney, and breast.
11. The correct counterstain and mounting media is critical. While we use hematoxylin as a counterstain to DAB, Fast Red, and AEC require methyl green.
12. It is critical that the specimens kept properly hydrated throughout the entire staining procedure. A common occurrence is a ring of overstaining around the edge of the specimen. This is due to the tissue sections drying out during the procedure. This can be alleviated by using a humidity chamber to reduce the amount of reagent evaporation, making sure that the work surface is level, and applying an ample amount of reagent. In order to reduce the amount of reagent needed to cover a section, a pap (hydrophobic) pen can be used to circle the tissue section. This will also ensure that the tissue section will not dry out.
13. It is critical that the proper antigen-unmasking technique used. Make sure that the epitope being sought can withstand the process being used. Some antigens can be destroyed at temperatures higher than 60°C.
14. Calibrate the microwave to be used for antigen retrieval to determine the ideal time and power settings. Each microwave has different wattage therefore this is a very important step. Use a rotating platform. This improves reproducibility and ensures uniform heating of all of the slides. Use the same mass of slides and solution for each assay. This will ensure even heat exposure and little variation from assay to assay.
15. When this protocol is used, two Tissue-Tek staining dishes must be used as well as 40 slides. If only antigen retrieving a few slides, fill the container with study slides, blank slides and antigen retrieval solution. Fill the other container with only blank slides and DI water.

References

1. Landis, S. H., Murray, T., Bolden, S., and Wingo, P. A. (1999) Cancer statistics, 1999. *CA Cancer J. Clin.* **49**, 8–31.
2. Mountain, C. F. (2000) The international system for staging lung cancer. *Semin. Surg. Oncol.* **18(2)**, 106–115.

3. Harpole, D. H., Jr., Herndon, J. E., II, Young, W. G., Wolf, W. G., and Sabiston, D. C., Jr. (1995a) A prognostic model of recurrence and death in stage I non-small cell lung cancer utilizing presentation, histopathology and oncoprotein expression. *Cancer Res.* **55,** 51–56.
4. Mountain, C. F. (1997) Revisions in the international system for staging lung cancer. *Chest* **111,** 1710–1717.
5. Mountain, C. F., Lukeman, J. M., and Hammar, S. P. (1987) Lung cancer classification: the relationship of disease extent and cell type to survival in a clinical trials population. *J. Surg. Oncol.* **35,** 147–156.
6. Mountain, C. F. (1989) Value of the new TNM staging system for lung cancer. *Chest* **96,** 47S–49S.
7. Nesbitt, J. C., Putnam, J. B., Jr., Walsh, G. L., Roth, J. A., and Mountain, C. F. (1993) Survival in early-stage I non-small cell lung cancer? *J. Thorac. Cardiovasc. Surg.* **106,** 90–94.
8. Ichinose, Y., Hara, N., Ohta, M., Yano, T., Maeda, K., Asoh, H., et al. (1993) Is T factor of the TNM staging system a predominant prognostic factor in pathologic stage I non-small cell lung cancer? *J. Thorac. Cardiovasc. Surg.* **106,** 90–94.
9. Macchiarini, P., Fontanini, G., Hardin, M. J., Chuanchieh, G., Bigini, D., Vignati, S., et al. (1993) Blood vessel invasion by tumor cells predicts recurrence in completely resected T1 N0M0 non-small cell lung cancer. *J. Thorac. Cardiovasc. Surg.* **106,** 80–89.
10. Thomas, P. A. and Rubinstein, L. (1990) Cancer recurrence after resection: T1N0 non-small cell lung cancer. *Ann. Thorac. Surg.* **49,** 242–247.
11. Takise, A., Kodama, A., Shimosato, Y., Watanabe, S., and Suemasu, K. (1988) Histopathologic prognostic factors in adenocarcinoma of the lung periphery less than 2 cm in diameter. *Cancer* **61,** 2083–2088.
12. Pairolero P. C., Williams D. E., Bergstralh E. J., Piehler J. M., Bernatz P. E., and Payne W. S. (1984) Postsurgical stage I bronchogenic carcinoma: morbid implications of recurrent disease. *Ann. Thorac. Surg.* **38,** 331–336.
13. Harpole, D. H., Jr., Richards, W. G., Herndon, J. E., and Sugarbaker, D. J. (1996) Angiogenesis and molecular biologic sub-staging in patients with stage I non-small cell lung cancer. *Ann. Thorac. Surg.* **61,** 1470–1476.
14. D'Amico, T. A., Massey, M., Herndon, J. E., II, Moore, M. B., and Harpole, D. H., Jr. (1999) A Biologic risk model for Stage I lung cancer: immunohistochemical analysis of 408 patients using 10 molecular markers. *J. Thorac. Cardiovasc. Surg.* **117,** 736–743.
15. Lau, C. L., D'Amico, T. A., and Harpole, D. H. (1999) Staging and prognosis: clinical and molecular prognostic factors and modes for non-small cell lung cancer, in *Lung Cancer: Principles and Practice*, 2nd ed. (Pass, H. I., Mitchell, J. B., Johnson, D. H., Turrisi, A. T., and Minna, J. D., eds.), Lippincott, Williams and Wilkins, Philadelphia, PA, pp. 602–611.
16. Rusch, V. W. and Dmitrovsky, E. (1995) Molecular biologic features of non-small cell lung cancer: clinical implications. *Chest Surg. Clin. North Am.* **5,** 39.

17. Slebos, R. J. C., Kibbelaar, R., Dalesio, O., et al. (1990) K-ras oncogene activation as a prognostic marker in adenocarcinoma of the lung. *N. Engl. J. Med.* **323,** 561–565.
18. Pfeiffer, P., Nexo, E., and Bentzen, S. M. (1998) Enzyme-linked immunosorbent assay of epidermal growth factor receptor in lung cancer: comparisons with immunohistochemistry, clinicopathological features and prognosis. *Br. J. Cancer* **78,** 96–99.
19. Fujino, S., Enokibori, T., Tezuka, N., et al. (1996) A comparison of epidermal growth factor receptor levels and other prognostic parameters in non-small cell lung cancer. *Eur. J. Cancer* **32A,** 2070–2074.
20. Greatens, T. M., Niehans, G. A., Rubins, J. B., et al. (1998) Do molecular markers predict survival in non-small cell lung cancer? *Am. J. Respir. Crit. Care Med.* **157,** 1093–1097.
21. Pastorino, U., Andreola, S., Tagliabue, E., et al. (1997) Immunohistocehmical markers in stage I lung cancer: relevance to prognosis. *J. Clin. Oncol.* **15,** 2858.
22. Harpole, D. H., Jr., Marks, J. R., et al. (1995b) Localized adenocarcinoma of the lung: oncogene expression of erbB-2 and p53 in 150 patients. *Clin. Cancer Res.* **1,** 659–664.
23. Kern, J. A., Schwartz, D. A., Nordberg, J. E., et al. (1990) P185neu expression in human lung adenocarcinoma predicts shortened survival. *Cancer Res.* **50,** 5184–5187.
24. Tateishi, M., Ishida, T., Mitsudomi, T., et al. (1991) Prognostic value of c-erbB-2 protein expression in human lung adenocarcinoma and squamous cell carcinoma. *Eur. J. Cancer* **27,** 1372–1375.
25. Strauss, G. M. (1997) Prognostic markers in resectable non-small cell lung cancer. *Hematol. Oncol. Clin. North Am.* **11(3),** 409–434.
26. Quinlan, D. C., Davidson, A. G., Summers, C. L., et al. (1992) Accumulation of p53 protein correlates with a poor prognosis in human lung cancer. *Cancer Res.* **52,** 4828–4831.
27. Fujino, M., Dosaka-Akita, H., Harada, M., et al. (1995) Prognostic significance of p53 and ras p21 expression in non-small cell lung cancer. *Cancer* **76,** 2457–2463.
28. Kwiatkowski, D. J., Harpole, D. H., Godleski, J., et al. (1998) Molecular pathologic substaging in 244 stage I non-small cell lung cancer patients: clinical implications. *J. Clin. Oncol.* **16,** 2468–2477.
29. D'Amico, T. A., Aloia, T. A., Moore, M. B., Herndon, J. E., II, Brooks, K. R., Lau, C. L., and Harpole, D. H., Jr. (2000) Molecular biologic substaging of stage I lung cancer according to gender and histology. *Ann. Thorac. Surg.* **69,** 882–886.
30. Hockenbery, D., Nunez, G., Milliman, C., Schreiber, R. D., and Korsmeyer, S. J. (1990) Bcl-2 is an inner mitochondrial membrane protein that blocks programmed cell death. *Nature* **348,** 334–336.
31. Korsmeyer, S. J. (1992) Bcl-2 initiates a new category of oncogenes: regulators of cell death. *Blood* **80,** 879–886.
32. Pezzella, F., Turley, H., Duzu, I., et al. (1993) bcl-2 protein in non-small cell lung carcinoma. *N. Engl. J. Med.* **329,** 690–694.

33. Fontanini, G., Vignati, S., Bigini, D., et al. (1995) Bcl-2 protein: a prognostic factor inversely correlated to p53 in non-small cell lung cancer. *Br. J. Cancer* **71**, 1003–1007.
34. Wang, T., Szekely, L., Okan, I., Klein, G., and Wiman, K. G. (1993) Wild-type p53-triggered apoptosis is inhibited by bcl-2 in a v-myc-induced T-cell lymphoma line. *Oncogene* **8**, 3427–3431.
35. Xu, H. J., Quinlan, D. C., Davidson, A. G., et al. (1994) Altered retinoblastoma protein expression and prognosis in early-stage non-small cell lung carcinoma. *J. Natl. Cancer Inst.* **86**, 695–699.
36. Kratzke, R. A., Greatens, T. M., Rubins, J. B., et al. (1996) Rb and p16INK4a expression in resected non-small cell lung tumors. *Cancer Res.* **56**, 3415–3420.
37. Pence, J. C., Kerns, B. M., Dodge, R. K., et al. (1993) Prognostic significance of the proliferation index in surgically-resected non-small cell lung cancer. *Arch. Surg.* **128**, 1382–1390.
38. Macchiarini, P., Fontanini, G., Hardin, M. J., et al. (1992) Relation of neovascularisation to metastasis of non-small cell lung cancer. *Lancet* **340**, 145–146.
39. Shi, S.-R., Key, M. E., and Kalra, K. L. (1991) Antigen retrieval in formalin-fixed, paraffin-embedded tissues: an enhancement method for immunohistochemical staining based on microwave oven heating of tissue sections. *J. Histochem. Cytochem.* **39**, 741–748.
40. Brandtzaeg, P. and Rognum, T. O. (1984) Evaluation of nine different fixatives: I. Preservation of immunoglobulin isotypes, J chain and secretory components in human tissues. *Pathol. Res. Pract.* **179**, 250–266.
41. DeLellis, R. A. and Kwan, P. (1988) Technical considerations in the immunohistochemical demonstration of intermediate filaments. *Am. J. Surg. Pathol.* **12(Suppl. 1)**, 17–23.
42. Elias, J. (1990a) *Immunohistopathology: A Practical Approach to Diagnosis*. ASCP Press, Chicago, IL.
43. Werner, M., Von Wasielewski, R., and Komminoth, P. (1996) Antigen retrieval, signal amplification and intensification in immunohistochemistry. *Histochem. Cell Biol.* **105(4)**, 53–60.
44. Antel, J. P., Kuchibhotla, J., and Stefansson, K. (1985) Generation of monoclonal antibodies recognizing neuronal elements in formalin-fixed, paraffin-embedded human tissue. *J. Neuropathol. Exp. Neurol.* **44**, 533–545.
45. West, K. P., Warford, A., Fray, L., et al. (1986) The demonstration of B cell, T cell and myeloid antigens in paraffin sections. *J. Pathol.* **150(2)**, 89–101.
46. de Vente, J., Steinbusch, H. W. M., and Schipper, J. (1987) A new approach to immunocytochemistry of 3′,5′-cyclic guanosine monophosphate: preparation, specificity, and initial application of a new antiserum against formaldehyde-fixed 3′,5′-cyclic guanosine monophosphate. *Neuroscience* **22**, 361–373.
47. Kaku, T., Ekem, J. K., Lindayen, C., et al. (1983) Comparison of formalin- and acetone-fixation for immunohistochemical detection of carcinoembryonic antigen (CEA) and keratin. *Am. J. Clin. Pathol.* **80**, 806–815.

48. Reading, M. (1977) A digestion technique for the reduction of background staining in the immunoperoxidase method. *J. Clin. Pathol.* **30,** 88–90.
49. Curran, R. C. and Gregory, J. (1977) The unmasking of antigens in paraffin sections by trypsin. *Experientia* **33,** 1400–1401.
50. Jacobsen, M., Clausenper, P., and Schmidth, S. (1980) The effect of fixation and trypsinization of the immunohistochemical demonstration of intracellular immunoglobulin in paraffin embedded material. *Acta Pathol. Microbiol. Immunol. Scand. A* **88,** 368–376.
51. Elias, J. M. (1990b) Estrogen receptor localization in paraffin sections using enzyme digestion, repeated applications of primary antibody, and imidazole. *J. Histotechnol.* **13,** 29–33.
52. Heyderman, E. (1979) Immunoperoxidase technique in histopathology: applications, methods, and controls. *J. Clin. Pathol.* **32(10),** 971–978.
53. Battifora, H. (1991) Assessment of antigen damage in immunohistochemistry. *Am. J. Clin. Pathol.* **96,** 669–671.
54. Ordonez, N. G., Manning, J. T., and Brooks, T. E. (1988) Effect of trypsination on the immunostaining of formalin-fixed, paraffin-embedded tissues. *Am. J. Surg. Pathol.* **12,** 121–129.
55. Taylor, C. R, Shi, S. R., Chaiwum, B., Young, L., Imam, A., and Cote, R. J. (1994) Strategies for improving the immunohistochemical staining of various intranuclear prognostic markers in formalin-paraffin sections. *Hum. Pathol.* **25,** 163–270.
56. Fraenkel-Conrat, H. and Olcott, H. A. Reaction of formaldehyde with proteins. (1947) VI Cross-linking of amino groups with phenol, imidazole, or indole groups. *J. Biol. Chem.* **174,** 827–843.
57. Fraenkel-Conrat, H., Brandon, B. A., and Olcott, H. A. (1948) The reaction of formaldehyde with proteins. IV Participation of indole groups. Gramicidin. *J. Biol. Chem.* **168,** 99–118.
58. Morgan, J. M., Navabi, H., Schmid, K. W., and Jasani, B. (1994) Possible role of tissue-bound calcium ions in citrate-mediated high-temperature antigen retrieval. *J. Pathol.* **147,** 301–307.
59. Cattoretti, G. and Suurmeijer, A. J. H. (1995) Antigen unmasking on formalin-fixed, paraffin-embedded tissues using microwaves: a review. *Adv. Anat. Pathol.* **2,** 2–9.
60. Hsu, S. M., Raine, L., and Fanger, H. (1981a) A comparitive study of the peroxidase-anti-peroxidase method and an avidin-biotin complex method for studying polypeptide hormones with radioimmunoassay antibodies. *Am. J. Clin. Pathol.* **75(5),** 734–738.
61. Hsu, S. M., Raine, L., and Fanger, H. (1981b) Use of avidin-biotin-peroxidase complex (ABC) in immunoperoxidase techniques: a comparison between ABC and unlabled antibody (PAP) procedures. *J. Histochem. Cytochem.* **29(4),** 577–580.
62. Chaiet, L. and Wolf, F. J. (1964) The properties of streptavidin a biotin-bonding protein produced by streptomyces. *Arch. Biochem. Biophys.* **106,** 1–5.

63. Kuhlmann, W. and Krischan, R. (1981) Resin embedment of organs and postembeddment localization of antigens by immunoperoxidase methods. *Histochemistry* **72**, 377–389.
64. Naritoku, W. Y. and Taylor, C. R. (1982) A comparative study of the use of monoclonal antibodies using three different immunohistochemical methods: an evaluation of monoclonal and polyclonal antibodies against human prostatic acid phosphatase. *J. Histochem. Cytochem.* **30**, 253–260.
65. Pedraza, M. A., Mason, D., Doslu, R. A., Marsh, R. A., and Boblett, J. P. (1984) Immmunoperoxidase methods with plastic embedded materials. *Lab. Med.* **15(2)**, 113–115.
66. Woods, G. S. and Warnke, R. (1981) Suppression of endogenous avidin-binding activity in tissues and its relevance to biotin-avidin detection system. *J. Histochem. Cytochem.* **29(10)**, 1196–1204.
67. Taylor, C. R. (2000) The total test approach to standardization of immunohistochemistry. *Arch. Pathol. Lab. Med.* **124**, 945–951.
68. Tacha, D. E. and Chen, T. (1994) Modified antigen retrieval procedure calibration technique for microwave ovens. *J. Histotechnol.* **17(4)**, 365–366.

23

A Sensitive Immunofluorescence Assay for Detection of p53 Protein in the Sputum

Zumei Feng, Defa Tian, and Radoslav Goldman

1. Introduction

Lung cancer is the most common cancer with the highest mortality worldwide. An estimated 1.04 million new cases were diagnosed and 921,000 deaths occurred in 1990. In the United States, an estimated 171,600 new cases of lung cancer with 158,900 deaths were reported in 1999 (1). Most lung tumors are inoperable at the time of diagnosis, because obvious symptoms do not appear at the early stage of the disease. More than 90% of lung cancer patients die within a short time after diagnosis (2). In order to decrease the lung cancer mortality, new tools for early detection are urgently needed. Sputum samples are easy to collect and contain epithelial cells derived from the lungs and bronchial passageways. Sputum is therefore an appealing material for clinical screening of lung cancer. Sputum cytology has been used as a routine diagnostic method in the early detection of lung cancer since the last century (3), but the conventional sputum cytology used in screening of morphologic changes has not led to a decrease in lung cancer mortality (4). This is due, in part, to the lack of sensitive and specific immunohistochemical and molecular tumor markers in sputum-cells. This article describes a sensitive immunofluorescence assay for *p53* protein that might be applied to early detection and screening of individuals (5).

Mutations in the *p53* tumor-suppressor gene are among the most common gene abnormalities described in human cancers, particularly in lung cancer. The *p53* gene mutations have been detected in above 50% of the tumors and some pre-invasive lesions on paraffin-embedded lung tumor tissue sections (6). Mutant *p53* protein often loses the function of tumor suppression and can accumulate in quantities detectable by immunohistochemistry *(2)*. In sputum, both *p53* gene mutation and *p53* protein accumulation have been detected in tumor cells, as well as dysplastic cells, two or more years prior to routine histological diagnosis *(4,8–10)*. These findings suggest that the accumulation of *p53* protein in sputum cells might serve as a biomarker for early diagnosis and screening programs.

In this report we describe a sensitive immunofluorescence assay for detection of the *p53* protein in sputum-cells *(11)*. This assay is a sensitive, simple, and economical technique that can be performed in any medical laboratory. Compared with other assays, this method is typically more sensitive, does not use radioactive reagents, and can be completed in a few hours. In brief, a mouse monoclonal antibody (MAb) against p53 complexes with a biotinylated anti-mouse immunoglobulin visualized with a fluorescent FITC-streptavidin conjugate. This technique can be adopted for detection of other markers in sputum cells by using specific antibodies.

2. Materials

This assay can be performed in any medical laboratory with basic equipment, such as a light microscope, refrigerator, 37°C incubator.

2.1. Sputum Sample Preparation

1. Saccommanno's preservative solution (Shandon, Pittsburgh, PA. Cat. no. 6768001).
2. High-speed blender.
3. Trypan blue viability dye.
4. Cytospin type III with cytofunnels (Shandon, Pittsburgh, PA. Cat. nos. 5991039 or 5991040) or standard cytospin set up: clips (Shandon, Pittsburgh, PA. Cat. no. 5991053), filters (Shandon, Pittsburgh, PA. Cat. no. 190005) and chambers (Shandon, Pittsburgh, PA. Cat. no. 5991021).

2.2. Immunofluorescence Assay Procedure

1. NCI-H23, human lung adenocarcinoma cells line (ATCC, Rockville, MD) used as positive control.

2.2.1. Fixation and Re-hydration

1. Fisher Superfrost microscope slides.
2. Coplin staining jars with lantern slide racks.
3. 95% ethyl alcohol.
4. 75% ethyl alcohol.
5. Phosphate-buffered saline (PBS), pH 7.4: (DAKO, Carpinteria, CA. Cat. no. S3024).

2.2.2. Immunostaining

1. Humidity chamber with aluminum trays.
2. Blocking solution: PBS containing 2% bovine serum albumin (BSA) (Sigma, St. Louis, MO, Cat. no. B 4287).
3. Primary antibody: mouse MAb to p53 protein, clone BP53-12-1 (Biogenex, San Ramon, CA, Cat. no. MU195-UC). 10 mg/mL stock. Stock concentration diluted to an optimal concentration determined by titration (suggested dilution 1:200).
4. Antibody diluent: PBS containing 0.1% BSA.
5. Wash buffer: PBS, pH 7.4, containing 0.1% Tween 20.
6. Staining dishes.
7. Secondary antibody working solution: Biotin labeled anti-mouse immunoglobulin (DAKO, Carpinteria, CA, Cat. no. E0464) 0.6 mg/mL stock. Stock concentration diluted to an optimal concentration determined by titration (suggested dilution 1:200).
8. FITC-conjugated streptavidin (DAKO, Cat. no. F0422). 1.0 mg/mL stock. Stock concentration diluted to an optimal concentration determined by titration (suggested dilution 1:200).

2.3. Reading Slides and Scoring

1. Evans blue (Sigma, Cat. no. E0133 ST). Dilute to 1:25 dilution of 0.5% stock using ddH$_2$O.
2. Fluorescent mounting medium (DAKO, Cat. no. S3023).
3. Fluorescent microscope with excitation, 450–490 nm, and emission, 510–530 nm, filters.

3. Methods
3.1. Sputum Sample Preparation

Five days worth of first early morning sputum is collected in Saccommanno's preservative solution and stored at 4°C. Sputum samples can be stored for 6 mo (*see* **Note 1**).

1. Warm the sputum sample to room temperature, vortex, and transfer into a blender cup using a 25-mL pipet. Homogenize 10 s on high speed to break up mucus, then rest homogenate for 10 s. Carefully open blender and check the specimen for flecks or fine threads. If present, blend for another 5 s. Excessive blending should be avoided.
2. Transfer specimen to 50-mL centrifuge tubes and spin at 600g for 10 min at room temperature. Discard supernatant and re-suspend cell pellet in 5 mL Saccommanno's solution, vortex, and combine all cell pellets from one sample into one tube.
3. Add Saccommanno's solution to 40 mL for each sample and vortex to homogenize the cell pellet. Count the viable cells using trypan blue staining. Adjust the concentration of cells to 10^6 cells per mL using Saccommanno's solution as needed.

4. Assemble cytospin clip, slide, filter, and chamber (or use the disposable cytofunnel chambers). Place approx 300 µL cell suspension into each cytospin chamber and spin at 45g for 7 min to form a monolayer of cells on microscope slides. Be careful not to disrupt the cells on the slide when opening the clip. Let the slide air dry at room temperature for at least 30 min before fixation.

3.2. Immunofluorescence Assay Procedure (see Note 2)

Set up three controls in each set of experiments:

1. Positive control: Human lung carcinoma NCI-H23 cultured cells.
2. Negative control: Any cell line without *p53* mutation or sputum sample from a healthy control.
3. Blank: NCI-H23 cell line cytospin slides using the same procedure but without addition of primary antibody. Instead, incubate with PBS containing 0.1% BSA.

3.2.1. Fixation and Re-hydration

1. Place cytospin slides in a lantern staining rack and immerse in a staining dish containing 95% ethyl alcohol. Fix the slides for 10 min at room temperature.
2. Immerse the slides into 75% ethyl alcohol for 2 min, rinse in fresh 1X PBS, pH 7.4, and then wash in ddH_2O for 2 min each.

3.2.2. Immunostaining

1. Let the slides drain on a slide rack and carefully wipe around the cells using Kimwipes tissue (do not touch the cells).
2. Lay slides flat on an aluminum tray in a humidity chamber, and cover cells with 100 µL blocking solution.
3. Incubate chamber at room temperature for 20 min to block nonspecific binding. Shake off excess BSA, allow the slides to drain, and carefully wipe each slide around cells. Do not wash or allow slides to dry out.
4. Place blocked slides on the aluminum tray in a humidity chamber and cover cells with 100 µL diluted primary antibody solution.
5. Incubate at 37°C for 60 min.
6. Gently rinse the slides with wash buffer, using a wash bottle. Transfer rinsed slides into a staining dish filled with PBS for 5 min. Change PBS three times at 5 min-intervals.
7. Shake off excess PBS and carefully wipe slides around cells, without touching the cells.
8. Place slides in a humidity chamber and cover cells with 100 µL secondary antibody solution. Leave slides undisturbed at 37°C for 30 min.
9. After incubation, rinse the slides as described above.
10. Protect slides from light at all subsequent steps (*see* **Note 3**). Place slides in a humidity chamber and cover cells with 100 µL diluted FITC-conjugated streptavidin. Incubate at 37°C for 30 min in the dark.
11. Rinse the slides extensively with PBS using a wash bottle.

12. Immerse into a staining dish filled with PBS, and wash three times with changes of PBS at 5-min intervals. Wipe slides dry as before.
13. Repeat secondary antibody and FITC-conjugated streptavidin incubation (**steps 8–12**). Incubation can be reduced to 15 min and all wash steps should be performed as before. This step is crucial for improved sensitivity of staining!

3.3. Reading Slides and Scoring

1. Counterstain slides with Evans blue. Immerse slides in a staining jar for 2 min at room temperature and rinse three times with ddH_2O.
2. Stand slides on a slide rack to drain, wipe slide around cells and cover cells with a few drops of fluorescent mounting medium.
3. Place a cover slip onto the mounting media (avoid making air bubbles). If storage is required, keep the slides protected from light at 4°C. It is best to examine slides as soon as possible. The fluorescent signals begin to bleach after 2 d.
4. Examine slides using a fluorescent microscope with dark field. Use 450–490 nm excitation filter and a 510 nm emission filter (*see* **Note 4**). The yellow-green fluorescence signal is located in the nuclei of the *p53* positive tumor cells (*see* figures in **ref. *11***). Other sputum cells stain red, due to Evans blue counterstain. The intensity of the signal is graded as high (+++, brightest yellow-green color), medium (++), and low (+) (*see* **Fig. 1**).

4. Notes

1. Make sure the sputum sample is derived from the lower respiratory tract. To verify that sputum sample derives from the lung, identification of at least five alveolar macrophages in sputum by Papanicolaou staining technique is required *(3)*.
2. Keep all reagent bottles covered tightly when not in use. Prepare reagents fresh daily and discard any unused diluted reagents. Keep the FITC-conjugated streptavidin protected from light and refrigerated.
3. To prevent premature bleaching of signal, protect slides from light at all times following addition of FITC conjugated streptavidin.
4. Two investigators should independently evaluate the results.
5. If no fluorescent signal is observes at all, including the positive controls, check all reagents and dilutions. Repeat the entire procedure using fresh reagents. Protect FITC stained slides from exposure to light.
6. If the intensity of fluorescent signal is not sufficient, increase the primary antibody concentration or extend primary antibody incubation time to 120 min. Examine the slides immediately after the end of assay. Keep slides in the dark.
7. If the negative controls show fluorescent signal and strong background, dilute primary antibody or shorten the primary antibody incubation time to 30 min. Room temperature instead of 37°C can be used for incubation. In addition, do not allow the slides to dry out at any time and wash thoroughly after each incubation. Use fresh PBS for every change of wash.

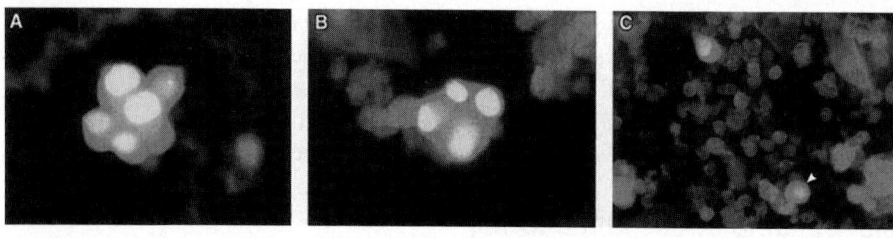

Fig. 1. **(A)** and **(B)** Immunofluorescent signals located in nuclei of tumor cells forming a cluster. The signal is graded as high (+++). **(C)** Single tumor cells show immunofluorescent signals in nuclear. The cell in the top left corner is graded medium (++), the cell marked with an arrow is graded low signal (+).

8. If cell loss is observed, pre-clean the microscope slides. It is best to use Fisher Superfrost microscope slides. Do not directly wash the cell when using a wash bottle. Just let the PBS flow over the cells.

Acknowledgments

The authors acknowledge the contributions of Drs. Judy Mumford and Mike Schmitt to the development and application of this assay and Qing Lan for providing some of the samples. Many thanks to Dr. Peter Shields for reviewing the manuscript.

References

1. Smith, R. A. and Glynn, T. J. (2000) Epidemiology of lung cancer. *Radiol. Clin. North Am.* **38,** 453–470.
2. Wiethege, T., Voss, B., and Muller, K. M. (1995) p53 accumulation and proliferating-cell nuclear antigen expression in human lung cancer. *J. Cancer Res. Clin. Oncol.* **121,** 371–377.
3. Saccomanno, G. (1986) Staining procedure for sputum smears, in *Diagnostic Pulmonary Cytology*, 2nd ed. (Saccomanno, G., ed.), American Society of Clinical Pathologists Presses, Chicago, IL, pp. 3–5.
4. Tockman, M. S., Gupta, P. K., Myers, J. D., Frost, J. K., Boylin, S. B., Gold, E. B., et al. (1988) Sensitive and specific monoclonal antibody recognition of human lung cancer antigen on preserved sputum cells: a new approach to early lung cancer detection. *J. Clin. Oncol.* **6,** 1685–1693.
5. Jacobson, D. R., Fishman, C. L., and Mills, N. E. (1995) Molecular genetic tumor markers in the early diagnosis and screening of non-small-cell lung cancer. *Ann. Oncol.* **6(Suppl. 3),** S3–S8.
6. Shields, P. G. (1999) Molecular epidemiology of lung cancer. *Ann. Oncol.* **10(Suppl. 5),** S17–S11.

7. Harris, C. C. (1996) *p53* tumor suppressor gene: from the basic research laboratory to the clinic—an abridged historical perspective. *Carcinogenesis* (Lond) **17,** 1187–1198.
8. Mao, L., Hruban, R. H., Boyle, J. O., Tockman, M., and Sidransky, D. (1994) Detection of oncogene mutations in sputum precedes diagnosis of lung cancer. *Cancer Res.* **54,** 1634–1637.
9. Anderson, M., Sladon, S., Michels, R., Davison, L., Conwell, K., Lechner, J., et al. (1996) Examinations of p53 alterations and cytokeratin expression in sputa collected from patients prior to histological diagnosis of squamous cell carcinoma. *J. Cell Biochem.* **64,** 185–190.
10. Mumford, J. L., Tian, D., Younes, M., Hu, F., Lan, Q., Ostrowski, M.L., et al. (1999) Detection of p53 protein accumulation in sputum and lung adenocarcinoma exposure to unvented coal smoke in China. *Anticancer Res.* **19,** 951–958.
11. Feng, Z., Tian, D., Lan, Q., and Mumford, J. L. (1999) A Sensitive immunofluorescence assay for detection of p53 protein accumulation in sputum. *Anticancer Res.* **19,** 3847–3852.

24

SPR1

An Early Molecular Marker for Bronchial Carcinogenesis

Derick Lau, Linlang Guo, Andrew Chan, and Reen Wu

1. Introduction

Tumorigenesis of the bronchial epithelium occurs through multiple and sequential morphological and molecular changes (1). In the respiratory tract, the earliest detectable morphological change is squamous metaplasia of the tracheobronchial epithelium upon exposure to carcinogen (2) or chronic tobacco smoke (3,4). A small proline-rich protein, SPR1, is a component of cross-linked envelopes of squamous epithelial cells of the airway (5). We have demonstrated that SPR1 is overexpressed in association with squamous differentiation in the primary culture of monkey and human bronchial epithelial cells (6,7). However, its expression is markedly diminished or lost in lung carcinoma (8,9). Furthermore, we have shown that expression and regulation of SPR1 is closely linked to multistep carcinogenesis of lung cancer in a series of human bronchial epithelial cell lines representing different stages of malignant transformation (10). Early in the transformation of bronchial epithelial cells, SPR1 expression can be upregulated by a tumor promoter, phorbol ester, and downregulated by a retinoid (6,7,10). Therefore, SPR1 appears to be a molecular marker for early transformation of the bronchial epithelium, and it may serve as an intermediate marker in chemoprevention of lung cancer.

In this chapter, we describe three methods for detecting SPR1 expression, namely:

1. Immunohistochemical staining.
2. Western blotting.
3. Reverse transcription polymerase chain reaction (RT-PCR).

2. Materials
2.1. Immunohistochemical Staining

1. 10% buffered formalin.
2. Paraffin embedding and sectioning set up.
3. Xylene.
4. 3% hydrogen peroxide.
5. Goat serum; rabbit polyclonal antiserum to the 15-amino-acid C-terminal region of human SPR1 (from Dr. Wu, University of California, Davis) *(8)*.
6. Phosphate-buffered saline (PBS).
7. Biotinylated goat anti-rabbit immunoglobulin (Bio-Rad, Hercules, CA).
8. Elite ABC kit (Vector Laboratories Inc., Burlingame, CA).
9. 0.04% 3,3-diaminobenzidine (DAB) (KPL, Gaithersburg, MD) in 100 mM Tris-HCl, pH 7.5 and 0.01% hydrogen peroxide.
10. 10 mM Tris-HCl, pH 7.5.
11. Permount.

2.2. Western Blotting
2.2.1. Cell Lines

1. Positive-control cell line, papilloma-virus-immortalized human tracheobronchial epithelial cell line (HBE1) (from Dr. Yankaskas, University of North Carolina, Chapel Hill) *(11)*.
2. Negative-control cell line, H460, derived from a human large-cell lung carcinoma (from ATCC) *(12)*.

2.2.2. Cell Culture

1. For culturing HBE1: serum-free Ham's F12 medium supplemented with insulin (5 μg/mL), transferrin (5 μg/mL), epidermal growth factor (20 ng/mL), cholera toxin (20 ng/mL), dexamethasone (0.1 μM), and bovine hypothalamus extract (30 μg/mL) as previously described *(13)*.
2. For culturing H460: Dulbecco's modified Eagle's medium (DMEM) supplemented with 10% fetal calf serum (FCS) and penicillin (100 U/mL)/streptomycin (100 mg/mL); retinoic acid (Sigma, St Louis, MO).
3. Phorbol 12-myristate 13-acetate (PMA).

2.2.3. Blotting

1. Keratin extraction buffer: 20 mM Tris-HCl, pH 7.0, 0.6 M KCl, 1% Triton X-100, and 1 mM phenylmethylsulfonyl fluoride (PMSF).
2. Bio-Rad DC Protein Microassay kit (Bio-Rad).
3. 15% sodium dodecyl sulfate (SDS)-polyacrylamide mini-gel.
4. Sample loading buffer: 100 mM Tris-HCl, 4% SDS, 10% β-mercaptoethanol, 20% glycerol, and 0.01% bromophenol blue.

5. Electrophoresis buffer: 52 mM Tris base, 52 mM glycine, and 0.1% SDS.
6. Pre-stained SDS-PAGE low-range protein standard (Bio-Rad).
7. Running buffer: 52 mM Tris base, 52 mM glycine, 1% SDS.
8. Immobilon-P nylon membrane (Millipore, Bedford, MA).
9. Dry transblotter (Bio-Rad).
10. PBS-T buffer: PBS with 0.1% Tween-20.
11. 10% nonfat dry milk in PBS-T.
12. Streptavidin-peroxidase.
13. Chemiluminescent ECL kit (Amersham Life Sicences, Arlington Heights, IL).
14. Kodak X-ray film.

2.3. Reverse Transcription/Polymerase Chain Reaction

1. Solution D: 4 M guanidinium thiocyanate, 25 mM sodium citrate, 0.5% sarcosyl, 0.1 M β-mercaptoethanol.
2. Water-saturated phenol; 2 M sodium acetate, pH 4.0.
3. Chloroform/isoamyl alcohol (49:1 v/v).
4. Isopropanol.
5. 75% ethanol.
6. RNasin, 20 µg/µL (Promega).
7. 10X PCR buffer: 500 µM KCl, 100 µM Tris-HCl, 25 mM MgCl$_2$, 0.1% gelatin, 1% Triton X-100.
8. Deoxynucleotides: dNTPs: 10 mM each of dATP, dCTP, dGTP, dTTP.
9. Random hexanucleotides: 100 pmol/µL.
10. M-MLV reverse transcriptase: 200 U/µL.
11. Taq polymerase: 5 U/µL.
12. Primers:
    SPR1 downstream primer (5'-CGTTTGCAGCATGAGTTC-3') (10 µM)
    SPR1 upstream primer (5'-TTCAGAGACTCAGAGTG-3') (10 µM)
    β-actin downstream primer (5'-GAGAAAATCTGGCACCACAC-3') (10 µM)
    β-actin upstream primer (5'-TACCCCTCGTAGATGGGCAC-3') (10 µM)
13. Mineral oil.
14. 2% agarose gel.

3. Methods
3.1. Immunohistochemical Staining

Expression of SPR1 protein can be detected in formalin-fixed and paraffin-embedded tissues.

1. Fix tissues in 10% buffered formalin overnight and then embed in paraffin (*see* **Note 1**).
2. Cut tissue block into 5-µm sections, de-paraffinized in xylene, and re-hydrate.
3. Perform immunostaining using the avidin-biotin-peroxidase complex (ABC) method *(14)*. Treat tissue sections with 3% hydrogen peroxide to block endogenous peroxidase.

4. Heat in PBS for 4 min in a microwave oven at a power setting of 6 to improve antigen retrieval *(15)*.
5. Block in 2% normal goat serum in PBS for 30 min.
6. Incubate tissue sections for 60 min at room temperature with the polyclonal antiserum to SPR1 at a dilution of 1:2,000 in PBS (*see* **Note 2**).
7. Rinse slides 3 times with PBS.
8. Incubate for 30 min with a biotinylated goat anti-rabbit secondary antibody, at a dilution of 1:500.
9. Detect SPR1 signal by using next step reagent from the Elite ABC Kit and a chromogen, 0.04% DAB, as recommended by the manufacturer.
10. Monitor the appearance of SPR1 signal under a microscope at 10X magnification.
11. Clear slides of DAB by rinsing with water, air-dry, and cover with a glass slip mounted with Permount.

3.2. Western Blotting

1. Culture HBE1, which expresses SPR1 and is used as a positive control, in Ham's F12 medium with 6 growth factors (*see* **Subheading 2.2.2.**) (*see* **Note 3**).
2. Culture H460, a cell line which does not express SPR1 and is used as a negative control, in DMEM with 10% FCS.
3. To prepare cell extract, lyse cultured cells or homogenized tissue in cold keratin extraction buffer.
4. Centrifuge lysate at 750g and determine protein concentration of the supernatant using reagents from the Bio-Rad microassay kit.
5. Mix each protein extract (15–20 µg) with an equal volume of sample loading buffer and heat at 85°C for 10 min.
6. Load samples into wells of a 15% SDS acrylamide mini-gel. A prestained protein marker standard is loaded in a separate lane.
7. Electrophorese the gel in Tris-SDS running buffer at 150 volts for approximately 1 h.
8. Transfer the proteins from the gel to an Immobilon-P membrane using a dry transblotter.
9. Rinse the membrane in PBS-T buffer and block in 10% nonfat dry milk in PBS-T for 30 min.
10. Incubate the membrane for one hr, on a shaker in a cold room, in the polyclonal antiserum to SPR1 at a concentration of 1:1,500 in 10% nonfat dry milk in PBS-T.
11. Wash 3 times in PBS-T, then incubate the membrane, at room temperature, in a secondary biotinylated goat-anti-rabbit immunoglobulin at a concentration of 1:1,000 in 10% non-fat dry milk in PBS-T.
12. Wash the membrane in PBS-T for 3 times and incubated in 0.05% streptavidin-peroxidase in PBS-T for 30 min.
13. Detect SPR1 signals using an ECL kit according to the manufacturer's instructions.

14. Expose the membrane to a Kodak X-ray film. Exposure time will vary from 30 s to 3 min.

3.3. Reverse Transcription/Polymerase Chain Reaction

1. Prepare total cell RNA according to the single-step method of Chomczynski *(16)*. In a 1.5-mL microfuge tube, lyse cell culture of approx 10^6 cells, or 50 mg of pulverized tissue, in 0.6 mL solution D with vortexing.
2. Add each of the following reagents and vortex in order: 50 µL 2 *M* sodium acetate, 0.6 mL water-saturated phenol, and 200 µL of chloroform/isoamyl alcohol.
3. Centrifuge the mixture at 14,000*g* for 10 min.
4. Remove the supernatant (~0.6 mL) and mix with 0.5 mL cold isopropanol and keep on ice for 60 min.
5. Centrifuge the tube at 1,000*g* and discard the supernatant.
6. Wash the pellet with 1 mL cold 75% ethanol, centrifuge, and dry in a hood for approx 1 h.
7. Dissolve the pellet in an appropriate volume (30–50 µL) of sterile deionized water.
8. Determine the optical density (OD) and concentration of RNA according to the formula: OD × 40 × dilution/1,000 µg/µL.
9. Reverse transcription of RNA is performed as described *(17)*. In a 500-µL tube, mix 1.0 µL RNasin, 2.0 µL 10X PCR buffer, 1.0 µL dNTPs, 1.0 µL hexanucleotides, 0.5 µL M-MLV-RT, 1.0 µg RNA and sterile deionized water to make up a total volume of 20 µL (*see* **Note 4**).
10. Place the tube in a thermal cycler programmed for 1 cycle at: 23°C, 15 min; 42°C, 30 min; 95°C, 5 min.
11. Prepare the polymerase chain reaction mixture as follows: Mix 10.0 µL 10X PCR buffer, 1.0 µL dNTPs, 1.0 µL downstream primer, 1.0 µL upstream primer, 0.3 µL Taq DNA polymerase, 2.0 µL of cDNA, water to a total volume of 100 µL. Add one drop of mineral oil as a cap to prevent evaporation.
12. Amplify the targeted gene product in a thermal cycler under the following conditions: 95°C, 30 s; 58°C, 30 s; 72°C, 30 s. For amplification of SPR1 and β-actin products, 35 and 25 cycles are required, respectively (*see* **Note 5**).
13. Analyze the PCR products on a 2% agarose gel electrophoresed for 1 h at 110 V. Stain DNA bands in the gel in ethidium bromide (*see* **Note 6**).

4. Notes

1. For fixation of tissue, avoid using formaldehyde, which contains methanol as a preservative. An alternative fixative is 4% paraformaldehyde.
2. To our knowledge, antibody to SPR1 is not commercially available. In place of the SPR1 antibody, involucrin MAb, which can be purchased from Sigma, may be used as the primary antibody at a concentration of 1:2,000. Involucrin, similar to SPR1, is believed to participate in the formation of the cornified envelope of

squamous cells *(18)*. Immunohistochemical staining of SPR1 or involucrin is observed mostly in the cytoplasm and, to a less extent, in the nuclei.

3. By Western blotting, the human SPR1 is detected as a 20-kD protein. It is recommended that the HBE1 cell line is used as a positive control. The expression of the SPR1 protein in this cell line can be upregulated by PMA treatment (100 ng/mL) for 24 h or downregulated by retinoic acid (1 μM) for 48 h. If involucrin MAb is used for primary blotting, multiple bands are observed on the membrane, indicating the existence of several isoforms.
4. The RT-PCR method is more sensitive than Northern blotting in detecting very low level of SPR1 transcripts. The primer pair for SPR1 is designed to amplify the SPR1 cDNA between nucleotides #36 and #475, yielding a PCR product of 440 base-pairs. For internal control, a primer pair is designed to amplify a 259-base-pair product of β-actin.
5. In our experience, it is best to run the amplification of SPR1 and β-actin in separate reactions.
6. Even if run separately, these PCR products from a same cDNA sample can be analyzed on the same lane of an agarose gel. For semi-quantification, relative level of SPR1 mRNA can be determined as a ratio of optical density of the SPR1 band to that of the β-actin after staining in ethidium bromide *(10)*. The cell lines HBE1 and H460 are recommended as positive and negative controls.

References

1. Gazdar, A. F. (1994) Molecular changes in preneoplastic bronchial epithelial lesions. *Proc. Am. Ass. Cancer Res.* **35,** 690.
2. Dirksen, E. R. and Crocker, T. T. (1968) Ultrastructural alterations produced by polycyclic aromatic hydrocarbons on rat tracheal epithelium in organ culture. *Cancer Res.* **28,** 906–923.
3. Davis, B. R., Whitehead, J. K., and Gill, M. E. (1975) Response of rat lung to inhaled tobacco smoke with or without prior exposure to 3,4-benzypyrene given by intratracheal instillation. *Br. J. Cancer* **31,** 469–484.
4. Kobayashi, N., Hoffman, D., and Wynder, E. L. (1974) A study of tobacco carcinogenesis. XII. Epithelial changes induced in the upper respiratory tracts of Syrian golden hamsters by cigarette smoke. *J. Natl. Cancer Inst.* **53,** 1085–1089.
5. Deng, D., Pan, R., and Wu, R. (2000) Distinct roles for amino- and carboxyl-terminal sequences of SPR1 protein in the formation of cross-linked envelopes of conducting airway epithelial cells. *J. Biol. Chem.* **275,** 5739–5747.
6. An, G., Tesfaigzi, J., Carlson, D. M., and Wu, R. (1993) Expression of a squamous cell marker, the spr1 gene, is posttranscriptionally down-regulated by retinol in airway epithelium. *J. Cell. Physiol.* **157,** 562–568.
7. An, G., Tesfaigzi, J., Chuu, Y. J., and Wu, R. (1993) Isolation and characterization of the human spr1 gene and its regulation of expression by phorbol ester and cyclic AMP. *J. Biol. Chem.* **268,** 10977–10982.
8. Hu, R., Wu, R., Deng, J., and Lau, D. H. M. (1998) A small proline-rich protein, spr1: Specific marker for squamous lung carcinoma. *Lung Cancer* **20,** 25–30.

9. DeMuth, J. P., Weaver, D. A., Crawford, E. L., Jackson, C. M., and Willey, J. C.. (1998) Loss of spr1 expression measurable by quantitative RT-PCR in human bronchogenic carcinoma cell lines. *Am. J. Respir. Cell Mol. Biol.* **19,** 25–29.
10. Lau, D., Xue, L., Hu, R., et al. (2000) Expression and regulation of a molecular marker, SPR1, in multistep bronchial carcinogenesis. *Am. J. Respir. Cell Mol. Biol.* **22,** 92–96.
11. Yankaskas, J. R., Haizlip, J. E., Conrad, M., et al. (1993) Papilloma virus immortalized tracheal epithelial cells retain a well-differentiated phenotype. *Am. J. Physiol.* **264,** c1219–c1230.
12. Phelps R. M., Johnson, B. E., Ihde, D. C., et al. (1996) NCI-Navy Medical Oncology Branch cell line data base. *J. Cell. Biochem.* **Suppl. 24,** 32–91.
13. Wu, R., Nolan, E., and Turner, C. (1985) Expression of tracheal differentiated functions in serum-free hormone-supplemented medium. *J. Cell. Physiol.* **125,** 167–181.
14. Hsu, J., Raine, L., and Fanger, H. (1981) Use of avidin-biotin-peroxidase complex (ABC) in immunoperoxidase techniques: A comparison between ABC and unlabeled antibody (PAP) procedures. *J. Histochem. Cytochem.* **29,** 577–580.
15. Shi, S. R., Key, M. E., and Kalra, K. L. (1991) Antigen retrieval in formalin-fixed, paraffin-embedded tissues: an enhancement method for immunohistochemical staining based on microwave oven heating of tissue sections. *J. Histochem. Cytochem.* **39,** 741–748.
16. Chomczynski, P. and Sacchi, N. (1987) Single-step method of RNA isolation by acid guanidinium thiocyanate-phenol-chloroform extraction. *Anal. Biochem.* **162,** 156–159.
17. Saiki, R. K., Scharf, S., Faloona, F., et al. (1985) Enzymatic amplification of β-globin genomic sequences and restriction site analysis for diagnosis of sickle cell anemia. *Science* **230,** 1350–1354.
18. Backendorf, C. and Hohl, D. (1992) A common origin for cornified envelope protein? *Nature Genet.* **2,** 91.

25

Determination of Biological Parameters on Fine-Needle Aspirates from Non-Small Cell Lung Cancer

Cecilia Bozzetti, Annamaria Guazzi, Rita Nizzoli, Nadia Naldi, Vittorio Franciosi, and Stefano Cascinu

1. Introduction

Non-small cell lung cancer (NSCLC) accounts for approximately 75% of all human lung carcinomas and is a major cause of mortality worldwide *(1)*. About 70% of the cases are diagnosed at an advanced stage, thus being suitable only for chemotherapy. Twenty-five percent of NSCLC patients are candidates for radical surgery, although even the stage I cases have a five year overall survival rate of only 40%. Unfortunately, conventional clinical and pathological factors alone cannot reliably predict the outcome of the disease, nor can they assist in the selection of patients who might require an alternative or additional therapy to surgery. Recent insights into the molecular events involved in the malignant progression of NSCLC may lead to the identification of significant predictors of prognosis and response to chemotherapy. The most widely investigated biomarkers in NSCLC *(2)* are the p53 tumor-suppressor gene, K-ras gene, bcl-2 and c-erbB-2 protein expression, Ki67 growth fraction, DNA ploidy, and S-phase fraction (SPF). Several studies have identified some of these biological markers as being independent predictors of recurrence *(3,4)*, response to chemotherapy *(5,6)*, or to radiotherapy *(7)*. Furthermore, recent reports suggest that microsatellite instability may allow for a more accurate prediction of tumor behavior and patient outcome in stage I NSCLC *(8,9)*.

Although biological parameters are usually evaluated by immunohistochemistry on paraffin sections at the time of the surgical resection of the primary tumor, the potential value of biomarkers in the choice of treatment would suggest that the presurgical knowledge of the biologic characteristics of the tumour could provide more reliable information in terms of prognosis and response to treatment. Since fine needle aspiration biopsy (FNAB) has been found to be an effective diagnostic procedure in NSCLC *(10)*, its application could be extended to the immunocytochemical detection of biological parameters at the time of diagnosis.

In our work, we evaluate Ki67 growth fraction and p53 and bcl-2 protein expression on cytologic material obtained by FNAB from surgical samples of NSCLC.

Ki67 is a proliferation-associated nuclear antigen expressed in all phases of the cell-cycle *(11)*. High levels of Ki67 antigen have been found to be associated with a poor prognosis in NSCLC *(12–14)*. p53 is a tumor-suppressor gene encoding for a protein that plays a role in the transcriptional control and repair of DNA damage *(15)*. Recent studies have reported that mutated p53 is a negative prognostic factor for survival *(16,17)*, and a negative predictive variable of response to platinum-containing chemotherapy and radiotherapy *(5,7)* but not to combined taxol-based chemotherapy and radiotherapy *(18)*.

Bcl-2 is an anti-apoptotic protein *(19)* that has been localized in the nuclear envelope, endoplasmic reticulum, and outer mitochondrial membranes of many epithelial cells. Bcl-2 is overexpressed in 20–30% of NSCLC cases and has been found to correlate with a better survival rate *(20,21)*. In addition, bcl-2 expression could be a potential predictor for response to chemotherapy, since some studies have demonstrated that taxoids and other microtubule-damaging drugs, commonly used in the treatment of human cancer, can lead to the induction of apoptotic cancer cell death through bcl-2 phosphorylation *(6,22)*.

In a recent publication *(23)*, in order to assess the reliability of Ki67, p53, and bcl-2 expression on NSCLC cytology, we compared the results determined by immunocytochemistry on fine needle aspirates (FNAs) from surgical specimens with those immunohistochemically determined on the corresponding histological sections. Concordance between FNAs and corresponding paraffin sections was 84% for Ki67, 93% for p53, and 95% for bcl-2. Good reproducibility was also found in relation to the immunocytochemical results obtained on FNAs from different areas of the same tumor, showing that tumor heterogeneity does not affect the method. The concordance between the immunocytochemical and immunohistochemical results suggests that FNAB may be a reliable procedure for the biological characterization of NSCLC. Given its limited invasiveness, FNAB could be used in vivo for the preoperative assessment of biological parameters in patients with operable or metastatic NSCLC.

Determination of Biological Parameters 407

Flow cytometric DNA ploidy and SPF are also evaluated together with the other biological parameters on FNAs obtained from surgical samples. In fact, DNA content and, to a much lesser extent, SPF have been investigated in NSCLC, but, despite some positive results, their role in predicting prognosis is still controversial *(24,25)*.

The aim of this chapter is to provide an overview of the evaluation of biological parameters and ploidy on cellular material obtained by FNAB.

2. Materials
2.1. Specimen Collection and Cytologic Staining
2.1.1. Preparation of Silanized Microscope Slides

1. 3-Aminopropyltriethoxy-silane (Code No. A3648) (Sigma).
2. 5% HCl solution in 80°C Ethanol.
3. 2% Silane solution in Acetone.
4. Microscope slides.
5. Slide rack.
6. Acetone.

2.1.2. Specimen Preparation

1. 22-gauge needle.
2. 20 mL plastic syringe.
3. Phosphate-buffered saline (Code No. BR14) (Oxoid): Prepare a 0.01 M PBS, pH 7.2 to 7.4, i.e., for each L of PBS dissolve 10 tablets of PBS in 1000 mL distilled water and store at 2–8°C. Make a fresh solution weekly.
4. Formaldehyde PBS: 3.7% Formaldehyde solution in PBS. Make a fresh daily.
5. Cytocentrifuge tubes.
6. Absolute Methanol (–10°C to –20°C).
7. Acetone (–10°C to –20°C).
8. Glass etching tool.
9. Wax pen.
10. Specimen Storage Medium: For every 1000 mL, dissolve 85.6 g Sucrose and 1.4 g Magnesium Chloride Hexahydrate in 250 mL PBS. Top up the volume to 500 mL with PBS. Add 500 mL glycerol and mix well by stirring. Store at –10°C to –20°C for up to 6 mo.
11. Propidium iodide (PI) solution.
12. May-Grunwald (Code No. 101424) (Merck, Darmstadt, Germany).
13. Giemsa (Code No. 109204) (Merck).
14. Staining solution 1: Mix 2 parts May-Grumwald with 1 part methanol.
15. Staining solution 2: Mix 1 part Giemsa with 7 parts distilled water.
16. Liquid nitrogen.
17. Entellan. Rapid mounting media for microscopy. (Code No. 107961) (Merck).

2.2. Immunocytochemistry

2.2.1. Ki67 Immunocytochemical Staining

1. Bovine Serum Albumin (BSA) Fatty Acid Free (Albumin Bovine, Code No. A 6003, Sigma).
2. 1% BSA Fatty Acid Free solution in PBS. Make a fresh solution daily.
3. Primary Antibody: Monoclonal Mouse Anti-Human Ki67 Antigen (Code No. M 0722) (Dako, Glostrup, Denmark), dilution 1:25 in 1% BSA Fatty Acid Free in PBS.
4. Bridging Secondary Antibody: Rabbit Anti-Mouse Immunoglobulins (Code No. Z 0259) (Dako) dilution 1:70 in 1% BSA Fatty Acid Free in PBS.
5. Peroxidase-Anti-Peroxidase (PAP) complex: PAP Mouse (Code No. B0650) (Dako) dilution 1:250 in 1% BSA Fatty Acid Free in PBS.
6. 3,3'-Diaminobenzidine (DAB): DAB+, Liquid (Code No. K 3468) (Dako). This substrate-chromogen system includes DAB and Buffered Substrate (containing hydrogen peroxide). For DAB preparation handle with care according to data sheet provided.

2.2.2. p53 Immunocytochemical Staining

1. Primary Antibody: Monoclonal Mouse Anti-Human p53 Protein Clone DO-7 (Code No. M 7001) (Dako) dilution 1:100 in 1% BSA Fatty Acid Free in PBS.
2. Universal DAKO Labelled Streptavidin-Biotin 2 System, Horseradish Peroxidase (LSAB 2 Kit/HRP, Rabbit/Mouse) (Code No. K0675) (Dako).

2.2.3. Bcl-2 Immunocytochemical Staining

1. Primary Antibody: Monoclonal Mouse Anti-Human bcl-2 Oncoprotein (Code No. M 0887) (Dako) dilution 1:100 in 1% BSA Fatty Acid Free in PBS.

2.2.4. Counterstaining (Common to Each Parameter Assayed)

1. Papanicolaou' s solution: 3% Harris Hematoxylin in distilled water (Harris' Hematoxylin solution, Code No. 109253, Merck).
2. 95% Ethanol.
3. 100% Ethanol.
4. 100% ethanol/xylene (1:1).
5. Xylene.
6. Entellan. Rapid mounting media for microscopy (Merck).

2.3. Flow Cytometric DNA Analysis

1. Citric Acid (Trisodium Salt Dihydrate) (Code No. C 3434) (Sigma, St. Louis, MO).
2. Propidium Iodide (Code No. P 4170) (Sigma) (Gloves should be worn while using PI, a possible mutagen).
3. Ribonuclease A Type I-A From Bovine Pancreas (Code No. R 4875) (Sigma).

4. Igepal CA-630 Nonionic Detergent (Code No. I 3021) (Sigma).
5. Propidium iodide Solution A: Dissolve 0.5 mg/mL propidium iodide in 0.1% citric acid. Store in the dark at 2–8°C for up to 1 yr.
6. Propidium iodide Solution B: Combine 50 µg/mL propidium iodide in H_2O (obtained by diluting solution A 1/10) with 1 mg/mL ribonuclease A and 0.1% Igepal CA-630. Store in the dark at 2–8°C for up to 1 wk.
7. FACScan flow cytometer (Becton-Dickinson, San Jose, CA) equipped with a doublet discrimination module.
8. Fluorescent beads for FACScan calibration.
9. Peripheral blood leukocytes from healthy donors.
10. Multicycle Cell Cycle Analysis Software (Phoenix Flow System, San Diego, CA).

3. Methods

3.1. Specimen Collection and Cytologic Staining

3.1.1. Preparation of Silanized Microscope Slides

Microscope slides are silanized in order to insure the adhesion of cytospun cells. Prepare in advance.

1. Prepare a 5% HCl solution in 80° ethanol and 2% Silane solution in Acetone.
2. Place clean microscope slides in a slide rack and immerse in 5% HCl/80° Ethanol for 10 min.
3. Repeat twice.
4. Place slides in acetone for 5 min.
5. Dip the slides rapidly 2–3 times in 2% Silane/Acetone solution.
6. Dip slides rapidly twice in Acetone.
7. Air dry and store the slides in a container at room temperature for up to 1 mo.

3.1.2. Specimen Preparation

Prior to immunocytochemical staining, cells obtained by FNAB are suspended and fixed in a Formaldehyde/PBS solution in order to preserve antigenicity, and cytospun cells are permeabilized and fixed with Absolute Methanol and Acetone.

1. Immediately after surgical excision, tumor samples are sent directly from surgical departments to the immunocytochemistry laboratory in containers without any preservative. Multidirectional FNABs are obtained from the surgical specimens in order to yield cellular material both for biomarker immunocytochemical assays and for cytofluorimetric DNA evaluation.
2. Perform FNABs with a 22-gauge needle connected to a 20-mL plastic syringe. The number of aspirates depends on the dimension of the surgical sample and on the amount of released cells.

3. Obtain a cell suspension for immunocytochemistry from the needle by rinsing it into a tube containing 2 mL 3.7% Formaldehyde/PBS.
4. Incubate the cellular suspension for 10 min at 2–8°C.
5. Dispense 0.2 mL (or any other amount according to the type of Cytocentrifuge) of cellular suspension in each cytocentrifuge tube.
6. Cytocentrifuge for 5 min at 600 rpm (45g) onto silanized glass slides.
7. Rinse the slides in PBS at room temperature for 5 min.
8. Fix in cold methanol (–10°C to –20°C) for 3–5 min.
9. Fix in cold acetone (–10°C to –20°C) for 1–3 min.
10. Rinse twice with PBS for 5 min at room temperature.
11. Drain excess buffer from slide onto a paper towel or gauze pad and then carefully wipe the area around the etched circle with an absorbent wipe or equivalent (*see* **Note 1**).
12. Mark the areas to be tested with a wax-tip pen.
13. Store cytospun slides at –10°C to –20°C in storage medium for up to 6 mo.
14. Additionally, obtain a cell suspension for DNA flow cytometry from a further FNAB by rinsing the needle into a tube containing a Propidium Iodide solution.
15. Following the aspiration procedure, obtain touch imprints from different areas of the tumor and stain the slides with May-Grunwald-Giemsa to test the representativeness of the sample: place air-dried slides in staining Solution 1 for 10 min. Rinse slides in gently running tap water for 1 min. Place slides in staining Solution 2 for 15 min. Rinse as before. Air dry slides and mount with Entellan mounting medium.
16. Freeze specimen promptly in liquid nitrogen for further assays.

3.2. Immunocytochemistry (see Note 2)

Immunocytochemical staining techniques allow for the visualization under a light microscope of cellular antigens. These techniques are based on the immunoreactivity of antibodies and the chemical properties of enzymes or enzyme complexes which react with colorless substrate-chromogen to produce a colored end-product (*see* **Fig. 1**).

3.2.1. Ki67 Immunocytochemical Staining

1. Remove slides from the storage medium and rinse in PBS for 5 min. Repeat once (*see* **Note 3**).
2. Incubate slides for 60 min with primary antibody in humidified chamber (*see* **Notes 4–6**).
3. Rinse slides in PBS bath for 3 min. Repeat twice.
4. Incubate slides for 30 min with bridging antibody in humidified chamber.
5. Rinse slides in PBS bath for 3 min. Repeat twice.
6. Incubate slides for 30 min with PAP in humidified chamber.
7. Prepare DAB solution by adding 20 µL of the DAB Chromogen per mL of Buffered Substrate. Mix well (*see* **Note 7**).

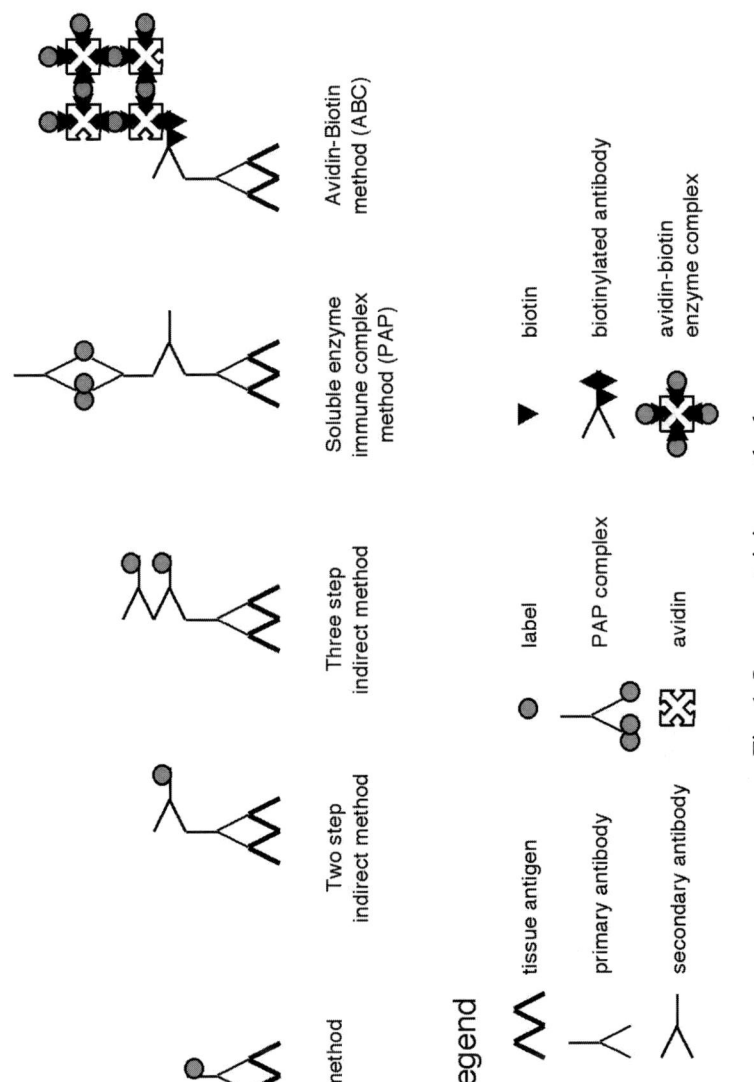

Fig. 1. Immunostaining methods.

8. Rinse slides in PBS bath for 3 min. Repeat twice using fresh PBS.
9. Incubate slides for 1 min with DAB.
10. Drain excess DAB from each slide.
11. Dip slides in distilled water. Repeat three/four times using fresh water.
12. Counterstain. Counterstaining is a step common to Ki67, p53, and bcl-2 assay. The procedure is described below in **Subheading 3.2.4.** (*see* **Note 8**).

3.2.2. p53 Immunocytochemical Staining

In contrast to the PAP system used for Ki-67 staining, the DAKO LSAB 2 System is used as revelation system both for p53 and bcl-2 assay.

1. Remove slides from the storage medium and rinse in PBS for 5 min. Repeat once.
2. Incubate slides for 60 min with primary antibody in humidified chamber.
3. Rinse slides in PBS bath for 3 min. Repeat twice using fresh PBS.
4. Incubate slides for 20 min with Biotinylated Link (*see* **Note 9**).
5. Rinse slides in PBS bath for 3 min. Repeat twice using fresh PBS.
6. Incubate slides for 20 min with Streptavidin-HPR in humidified chamber.
7. Rinse slides in PBS bath for 3 min. Repeat twice using fresh PBS.
8. Incubate slides for 1 min with DAB.
9. Drain excess DAB from each slide.
10. Dip slides in distilled water. Repeat three/four times using fresh water.
11. Counterstain (*see* **Subheading 3.2.4.**).

3.2.3. Bcl-2 Immunocytochemical Staining

1. Perform Bcl-2 Immunocytochemical staining using bcl-2 primary antibody and the same method described for p53.

3.2.4. Counterstaining (Common to Each Parameter Assayed)

1. Place the slides in a staining jar containing Papanicolaou's solution for 3 min.
2. Rinse slides in gently running tap water for 5 min.
3. Place the slides in a staining jar containing 95% ethanol for 4 min. Repeat this step once using a second jar of 100% ethanol.
4. Place the slides in a staining jar containing ethanol/xylene for 4 min.
5. Place the slides in a staining jar containing xylene for 4 min.
6. Add one drop of mounting medium to each cover slip.
7. Remove one slide at a time from xylene and immediately place cover slip over sample. Wait a minimum of 10 min before viewing under the microscope.

3.2.5. Microscopic Evaluation and Interpretation of Staining

1. Use a standard light microscope for immunoreactivity evaluation.
2. Scan the entire slide using a low-power (×250) objective, to ascertain the success of the immunocytochemical staining. The cytospun slides generally show a good

preservation of morphologic features, allowing for a reliable evaluation of the staining. An example of a cytological specimen obtained by FNAB, positively immunostained for Ki67, p53, and bcl-2, is shown in **Fig. 2**.
3. Examine the positive control slide first to verify that all reagents worked properly. The presence of a reddish-brown end-product at the site of the target antigen is indicative of positive reactivity. If the positive control fails to demonstrate a positive staining, the assay results should be considered invalid.
4. Examine the negative control to exclude the presence of nonspecific labeling to the target antigen. The absence of specific staining in the negative control confirms the lack of antibody cross-reactivity to cellular components. If specific staining occurs in the negative control, the assay is invalidated. Any diffuse staining, when present at a nonantigenic site, should be considered nonspecific. This is true for cytoplasmic staining in the case of p53 and Ki67 evaluation as well as for nuclear staining for bcl-2. Necrotic or degenerated cells often stain nonspecifically.
5. Evaluate the staining for each parameter by counting a minimum of 200 epithelial tumor cells in randomly selected fields, using a high-power (×400) objective. Results are expressed as a percentage of well-preserved stained cells out of the total number of tumor cells counted.
6. The staining should be reddish-brown and may show variable intensity. Positive staining should be evaluated within the context of any nonspecific background staining of the negative control.
7. Ki67 reactivity can be recognized as diffuse or dot-like nuclear staining. To quantify positive cells, evaluate a minimum of 500 tumor cells. Samples with 10% or less stained nuclei are classified as "negatives."
8. Expression of p53 and bcl-2 protein is indicated by nuclear and cytoplasmic staining of tumor cells respectively. Count at least 200 cancer cells for both p53 and bcl-2 assessment. The tumor samples are classified as "p53 positives" when more than 10% of the tumor cells show specific nuclear staining and as "bcl-2 positives" when more than 10% of tumor cells show cytoplasmic staining.

3.3. Flow Cytometric DNA Analysis

Flow cytometric DNA analysis is performed on fresh, unfixed cells obtained by FNAB from surgical specimens, stained with PI and analyzed on the flow cytometer (*see* **Note 10**).

1. Suspend cells obtained by a multidirectional FNAB in 1 mL propidium iodide solution B (*see* **Note 11**).
2. Stain for 30 min to 3 h in the dark at 2–8°C.
3. Filter stained cells through a 30-µm nylon mesh.
4. Analyze by means of a flow cytometer, equipped with a doublet discrimination module, that has been calibrated using fluorescence beads.

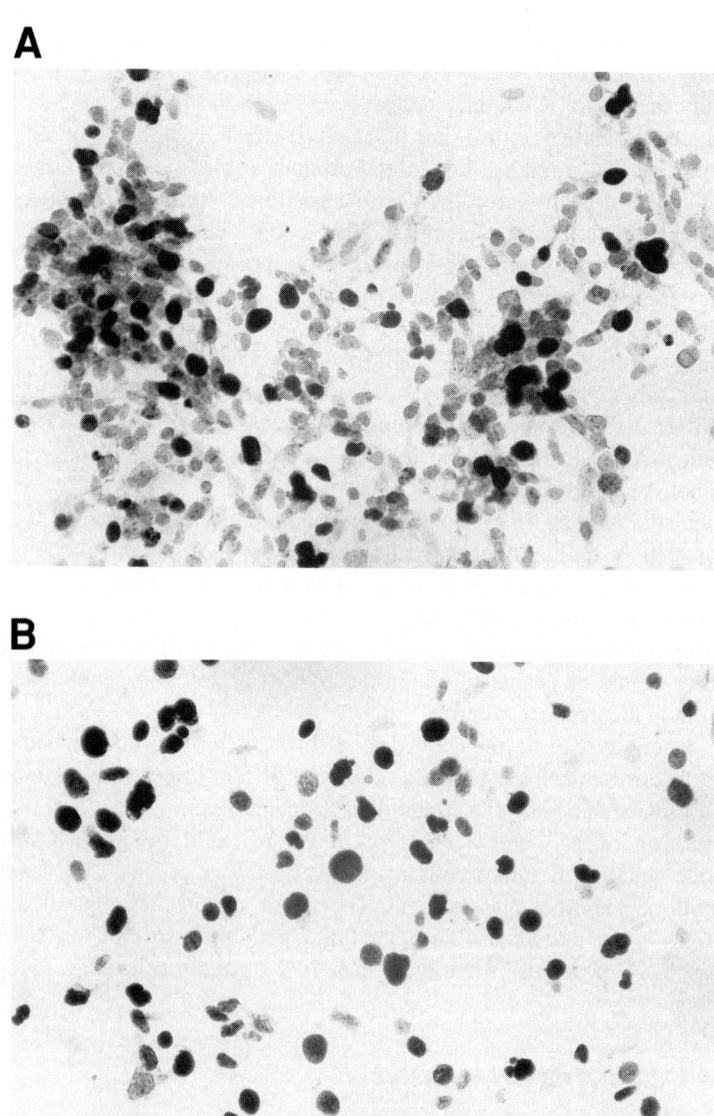

Fig. 2. Cytospin smears of lung cancer cells obtained by fine-needle aspiration biopsy positively immunostained for Ki67 (**A**), p53 (**B**), and bcl2 (**C**). Magnification, ×400.

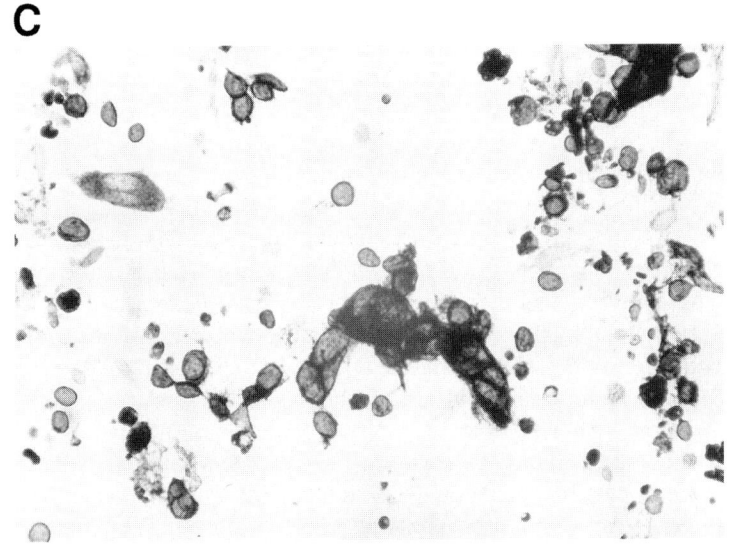

Fig. 2. *(continued)*

5. Acquire a minimum of 20,000 cells per sample at flow rate no greater than 100 events per second. If the rate exceeds this value, dilute sample with PI solution B and stain in the dark at 2–8°C for additional 10 min.
6. Cell measurements should consist of a set of samples with and without a diploid internal reference standard. Use peripheral blood leukocytes from healthy donors as a calibration standard for DNA diploidy.
7. Express the degree of DNA content aberration as a DNA index (DI). DI is the ratio between the mode (or mean) of the relative DNA content of the G0/1 cells of the sample and the mode (or mean) of the relative DNA measurements of the diploid G0/1 reference cells. Cells with a normal diploid karyotype have by definition a DI of 1 (*see* **Note 12**).

3.3.1. Interpretation of Histograms

1. Classify samples showing only one peak, even after the addition of the reference cells, as diploid.
2. Classify samples as aneuploid when two or more distinct peaks appear on the histogram, with the diploid position verified by means of the reference cells.
3. Classify samples as tetraploid when 20% or more cells are found in the $G_2 + M$ channel region of the histogram.

4. Notes

1. Do not allow the slides to dry at any time during the specimen preparation. Cytospun slides can be stored up to 6 mo without loss of immunoreactivity.

2. Initially, for immunoenzymatic staining, a direct method was utilized which conjugated enzymes directly to an antibody with known antigenic specificity (primary antibody). However, this technique lacked the sensitivity of later methods. The sensitivity of immunocytochemical stains was significantly improved by the development of an indirect technique. In two or three step methods, one or more enzyme-labelled secondary antibodies react with the antigen-bound primary antibody. Increased sensitivity was achieved with the introduction of peroxidase-anti-peroxidase (PAP) enzyme complex *(26)*. In this method the secondary antibody acts as a linking antibody between the primary antibody and the PAP complex. A further improvement was achieved by Hsu et al. *(27)* in the three-step avidin-biotin complex (ABC) method, which utilizes the strong affinity of avidin for biotin. This technique employs an enzyme labelled avidin-biotin complex which forms a complex with a biotinylated secondary antibody.
3. Use fresh PBS for every rinse.
4. Negative controls should be performed to evaluate nonspecific staining and to allow for a better interpretation of specific staining at the antigenic site. For each case, a negative control is produced by replacing the primary antibody with 1% BSA in PBS. The negative control should not exhibit specific staining. Ideally, a negative control reagent should be in the same matrix/solution as the primary antibody and contain an antibody exhibiting no specific reactivity with human tissues. Diluent alone (1% BSA in PBS) may be used as a less than ideal negative control. The incubation period for the negative control reagent should always correspond to that of the primary antibody. A known positive control cell cytospin slide should also be included with each assay run and processed as for patient's slides.
5. Perform incubation steps with primary and bridging secondary antibodies in a humidified chamber (an airtight container with a piece of damp blotting paper taped to the side of the container) at room temperature. The chamber should contain a rack on which slides can be positioned horizontally. Moisten the blotting paper with water before each use of the chamber.
6. Before applying any reagent, carefully wipe around the specimen, if necessary, to remove any remaining liquid and to keep reagents within the prescribed area. Apply enough reagents (usually a drop) to cover cytospin wells. Do not allow the slides to dry at any time during the immunocytochemical procedure.
7. The binding of the biotinylated secondary antibody is revealed by incubation with 3-3′ diaminobenzidine (DAB) substrate-chromogen, which, upon oxidation, forms a brown-colored precipitate at the site of the target antigen. It is important to prepare the chromogen substrate solution (DAB) just before use, during the PBS washes following incubation with the bridging antibody.
8. A very weak counterstaining is preferred, to avoid masking the brown nuclear DAB staining by very dark blue hematoxylin.

9. Note that we use sequential 20 min incubations in biotinylated link antibody and peroxidase-labeled streptavidin rather than the 10 min indicated by the manufacturer. Staining is completed after incubation for 1 min with DAB. The Universal DAKO Labelled Streptavidin-Biotin 2 System, Horseradish Peroxidase (DAKO LSAB 2 System, HPR) System, is based on a modified labeled avidin-biotin (LAB) technique, where a biotinylated secondary antibody forms a complex with peroxidase-conjugated streptavidin molecules *(28)*. The LAB/LSAB method has been reported to be four to eight times more sensitive than the ABC method *(29)*.

10. This procedure is based on the stoichiometric interaction of a fluorochrome (PI) with the DNA macromolecule and on the assumption that the intensity of the emitted light is directly proportional to DNA content *(30,31)*. Correct tissue sampling, staining, and cytometric operation are crucial to obtain reliable flow cytometric DNA analysis. Protocols for staining and for DNA histogram analysis have been widely reviewed in the literature *(32)*. In our experiment, flow cytometric data analysis and S-phase evaluation are performed using the Multicycle Cell Cycle Analysis Software (Phoenix Flow System, San Diego, CA), which is based on the polynomial S-phase model by Dean and Jett *(33)*. Flow cytometric DNA analysis is performed on fresh, unfixed cells obtained by FNA from surgical specimens. FNA is in fact a gentle and rapid procedure used both to obtain a representative cellular population and for the mechanical disaggregation of solid tumours *(32,34,35)*.

11. Igepal CA-630 nonionic detergent and ribonuclease A, included in PI solution B, induce cell membrane solubilization and the enzymatic digestion of RNA.

12. Most normal cells are at any one time in a quiescent or resting stage, termed G0, and are diploid. As soon as cells respond to signals to divide, they move into the G_1 stage, where RNA and proteins necessary for DNA synthesis are produced. This stage is indistinguishable from G_0 on the basis of DNA content. The replication of total cellular DNA occurs in the synthetic stage, termed S-phase. At the end of the S-phase, the cell contains twice its original DNA content (tetraploid), while in preparation for cell division further proteins are elaborated in a phase termed G_2. The stage at which the cell actually divides into two daughter cells is termed M for "mitosis." Based on DNA content, the M phase is indistinguishable from the G_2 phase. Collectively, the replication cycle is described by G_0/G_1, S, G_2/M phases. The fluorescence histograms can be analyzed for cell cycle distribution and to detect the presence of abnormal DNA stemlines (DNA aneuploidy). Cancer cells may have gross changes in their DNA content and thus would display an abnormal fluorescence intensity due to changes at the chromosome level, as shown in **Fig. 3**, where the number of cells is plotted against the intensity of the emitted light.

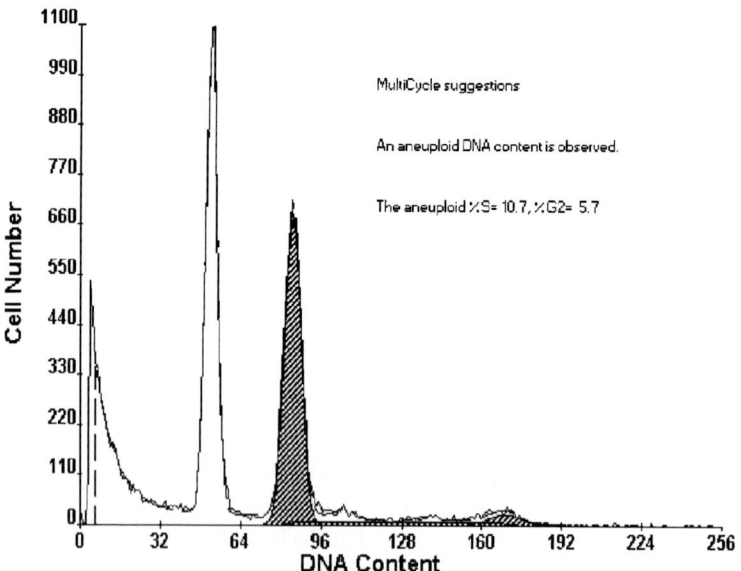

Fig. 3. Example of an aneuploid DNA histogram of lung cancer cells obtained by fine-needle aspiration biopsy.

References

1. Boring, C. C., Squires, T. S., Tong, T., and Montgomery, S. (1994) Cancer Statistics. *CA Cancer J. Clin.* **44,** 7–26.
2. Strauss, G. M., Kwiatkowski, D. J., Harpole, D. H., Lynch, T. J., Skarin, A. T., and Sugarbaker, D. J. (1995) Molecular and pathologic markers in stage I non-small-cell carcinoma of the lung. *J. Clin. Oncol.* **13,** 1265–1279.
3. Kwiatkowski, D. J., Harpole, D. H., Godleski, J., Shieh, D., Blanco, R., Richards, W., et al. (1997) Prognostic factor analysis of 250 stage I NSCLC patients: pathologic features are more important than molecular analyses. *Proc. Am. Soc. Clin. Oncol.* **16,457(Abstr.),** 1644.
4. Kim, Y. C., Park, K. O., Kern, J. A., Park, C. S., Lim, S. C., Jang, A. S., and Yang, J. B. (1998) The interactive effect of Ras, HER2, p53 and Bcl-2 expression in predicting the survival of non-small cell lung cancer patients. *Lung Cancer* **22,** 181–190.
5. Kawasaki, M., Nakanishi, Y., Kuwano, K., Takayama, K., Kiyohara, C., and Hara, N. (1998) Immunohistochemically detected p53 and P-glycoprotein predict the response to chemotherapy in lung cancer. *Eur. J. Cancer* **34,** 1352–1357.
6. Blagosklonny, M. V., Giannakakou, P., el-Deiry, W. S., Kingston, D. G., Higgs, P. I., Neckers, L., and Fojo, T. (1997) Raf-1/bcl-2 phosphorylation: a step from microtubule damage to cell death. *Cancer Res.* **57,** 130–135.

7. Matsuzoe, D., Hideshima, T., Kimura, A., Inada, K., Watanabe, K., Akita, Y., et al. (1999) p53 mutations predict non-small cell lung carcinoma response to radiotherapy. *Cancer Lett.* **135**, 189–194.
8. Pifarré, A., Rosell, R., Monzò, M., De Anta, J. M., Moreno, I., Sànchez, J. J., et al. (1997) Prognostic value of replication errors on chromosomes 2p and 3p in non-small-cell lung cancer. *Br. J. Cancer* **75**, 184–189.
9. Zhou, X., Kemp, B. L., Khuri, F. R., Liu, D., Lee, J. J., Wu, W., et al. (2000) Prognostic implication of microsatellite alteration profiles in early-stage non-small cell lung cancer. *Clin. Cancer Res.* **6**, 559–565.
10. Linsk, J. A. and Salzman, A. J. (1972) Diagnosis of intrathoracic tumors by thin needle cytologic aspiration. *Am. J. Med. Sci.* **263**, 181–195.
11. Gerdes, J., Schwab, U., Lemke, H., and Stein, H. (1983) Production of a mouse monoclonal antibody reactive with a human nuclear antigen associated with cell proliferation. *Int. J. Cancer* **31**, 13–20.
12. Scagliotti, G. V., Micela, M., Gubetta, L., Leonardo, E., Cappia, S., Borasio, P., and Pozzi, E. (1993) Prognostic significance of Ki67 labelling in resected non small cell lung cancer. *Eur. J. Cancer* **29**, 363–365.
13. Soomro, I. N., Holmes, J., and Whimster, W. F. (1998) Predicting prognosis in lung cancer: use of proliferation marker, Ki67 monoclonal antibody. *J. Pak. Med. Assoc.* **48**, 66–69.
14. Pence, J. C., Kerns, B. J., Dodge, R. K., and Iglehart, J. D. (1993) Prognostic significance of the proliferation index in surgically resected non-small-cell lung cancer. *Arch. Surg.* **128**, 1382–1390.
15. Levine, A. J., Momand, J., and Finlay, C. A. (1991) The p53 tumour suppressor gene. *Nature* **351**, 453–456.
16. Quinlan, D. C., Davidson, A. G., Summers, C. L., Warden, H. E., and Doshi, H. M. (1992) Accumulation of p53 protein correlates with a poor prognosis in human lung cancer. *Cancer Res.* **52**, 4828–4831.
17. D'Amico, T. A., Massey, M., Herndon, J. E., Moore, M. B., and Harpole, D. H. Jr. (1999) A biologic risk model for stage I lung cancer: immunohistochemical analysis of 408 patients with the use of ten molecular markers. *J. Thorac. Cardiovasc. Surg.* **117**, 736–743.
18. Safran, H., King, T., Choy, H., Gollerkeri, A., Kwakwa, H., Lopez, F., et al. (1996) p53 mutations do not predict response to paclitaxel/radiation for non-small cell lung carcinoma. Cancer 78, 1203–1210.
19. Hockenbery, D., Nunez, G., Milliman, C., Schreiber, R. D., and Korsmeyer, S. J. (1990) Bcl-2 is a inner mitochondrial membrane protein that blocks programmed cell death. *Nature* **348**, 334–336.
20. Pezzella, F., Turley, H., Kuzu, I., Tungekar, M. F., Dunnill, M. S, Pierce, C. B., et al. (1993) bcl-2 protein in non-small-cell lung carcinoma. *New Engl. J. Med.* **329**, 690–694.
21. Higashiyama, M., Doi, O., Kodama, K., Yokouchi, H., Nakamori, S., and Tateishi, R. (1997) bcl-2 oncoprotein in surgically resected non-small cell lung cancer:

possibly favorable prognostic factor in association with low incidence of distant metastasis. *J. Surg. Oncol.* **64**, 48–54.
22. Haldar, S., Basu, A., and Croce, C. M. (1997) Bcl-2 is the guardian of microtubule integrity. *Cancer Res.* **57**, 229–233.
23. Bozzetti, C., Franciosi, V., Crafa, P., Carbognani, P., Rusca, M., Nizzoli, R., et al. (2000) Biological variables in non-small cell lung cancer: comparison between immunocytochemical determination on fine needle aspirates from surgical specimens and immunohistochemical determination on tissue sections. *Lung Cancer* **29**, 33–41.
24. Volm, M., Hahn, E. W., Mattern, J., Muller, T., Vogt-Moykopf, I., and Weber E. (1988) Five-year follow-up study on independent clinical and flow cytometric prognostic factors for the survival of patients with non-small cell lung carcinoma. *Cancer Res.* **48**, 2923–2928.
25. Zimmerman, P. V., Hawson, G. A., Bint, M. H., and Parsons, P. G. (1987) Ploidy as a prognostic determinant in surgically treated lung cancer. *Lancet* **2**, 530–533.
26. Farr, A. G. and Nakane, P. K. (1981) Immunohistochemistry with enzyme labeled antibodies: a brief review. *J. Immunol. Methods* **47**, 129–144.
27. Hsu, S. M., Raine, L., and Fanger, H. (1981) Use of avidin-biotin-peroxidase complex (ABC) in immunoperoxidase techniques: a comparison between ABC and unlabeled antibody (PAP) procedures. *J. Histochem. Cytochem.* **29**, 577–580.
28. Guesdon, J. L., Ternynck, T., and Avrameas, S. (1979) The use of avidin-biotin interaction in immunoenzymatic techniques. *J. Histochem. Cytochem.* **27**, 1131–1139.
29. Giorno, R. (1984) A comparison of two immunoperoxidase staining methods based on the avidin-biotin interaction. *Diagn. Immunol.* **2**, 161–166.
30. Fried, J., Perez, A. G., and Clarkson, B. D. (1976) Flow cytofluorometric analysis of cell cycle distributions using propidium iodide. Properties of the method and mathematical analysis of the data. *J. Cell Biol.* **71**, 172–181.
31. Krishan, A. (1975) Rapid flow cytofluorometric analysis of mammalian cell cycle by propidium iodide staining. *J. Cell Biol.* **66**, 188–193.
32. Bach, B. A., Knape, W. A., Edinger, M. G., and Tubbs, R. R. (1991) Improved sensitivity and resolution in the flow cytometric DNA analysis of human solid tumor specimens. Use of in vitro fine-needle aspiration and uniform staining reagents. *Am. J. Clin. Pathol.* **96**, 615–627.
33. Dean, P. N. and Jett, J. H. (1974) Mathematical analysis of DNA distributions derived from flow microfluorometry. *J. Cell Biol.* **60**, 523–527.
34. Brown, D. C., Gatter, K. C., Dunnill, M. S., and Mason, D. Y. (1989) Immunocytochemical analysis of cytocentrifuged fine needle aspirates. A study based on lung tumors aspirated in vitro. *Anal. Quant. Cytol. Histol.* **11**, 140–145.
35. Bozzetti, C., Nizzoli, R., Camisa, R., Guazzi, A., Ceci, G., Cocconi, G., et al. (1997) Comparison between Ki-67 index and S-phase fraction on fine-needle aspiration samples from breast carcinoma. *Cancer* **81**, 287–292.

II

DETECTION OF ALTERED TUMOR MARKERS IN CLINICAL SAMPLES FOR DIAGNOSIS AND PROGNOSIS

B. MOLECULAR METHODS FOR PROGNOSTIC DETECTION OF ALTERED LUNG TUMOR MARKERS IN CLINICAL SAMPLES

IV. Imaging Analysis

26

Detection and Analysis of Lung Cancer Cells from Body Fluids Using a Rare Event Imaging System

Stine-Kathrein Kraeft, Ravi Salgia, and Lan Bo Chen

1. Introduction

Present diagnostic techniques do not allow the detection of early metastatic spread of tumor cells, although this spread largely determines the clinical course of patients with small primary cancers, especially lung cancer. To improve the diagnosis of this occult stage of early metastasis, sensitive techniques have been developed that facilitate the detection of isolated lung carcinoma cells in hematopoietic tissues (peripheral blood or bone marrow), or lymph nodes *(1–7)*. It is to be noted that the circulating lung cancer cells from the primary tumor tissue may potentially have different characteristics as compared to the original tumor.

The principal difficulty in the detection and subsequent characterization of disseminated cancer cells stems from their rarity (as little as one cell in a million *[8,9]*) and the availability of appropriate markers. Although molecular assays, such as reverse transcription polymerase chain reaction (RT-PCR), would theoretically be the most sensitive, in practice, they are often hampered by background noise from normal hematopoietic or epithelial cells. For this reason, immunocytochemical methods have become the standard in rare cancer cell detection. As an added advantage, microscopic analysis allows not only for the detection and visual confirmation of tumor cells, but also for their further characterization and the identification of eventual subpopulations, as compared to the original tumor.

Rare cell enumeration can be performed manually under the microscope, but this is a laborious task and moreover, some positive events can easily be missed, especially when present at low frequencies. Therefore, several groups, including ours have developed automated microscope systems that not only detect and quantify rare cancer cells, but also archive an image of each identified cell and allow for relocation of the object on the microscope stage *(4,10–12)*. Our Rare Event Imaging System is based on an automated fluorescence microscope and offers the possibility of multiple-marker analysis by using up to four different fluorescence filter sets.

The markers used for the detection of lung cancer cells can be classified as being general epithelial cell markers or being more specific towards particular cell populations. Antibodies to cytokeratins, which belong to the first class of markers, have been utilized for the detection of non-small cell lung cancer (NSCLC) cells in hematopoietic tissues and lymph nodes *(13–15)*. Antibodies to the neural cell adhesion molecule (NCAM/CD56), and the Le^y antigen have been used to identify tumor cells in small cell lung cancer (SCLC) patients *(16–20)*. Our laboratory has developed an immunofluorescence double-labeling technique for the phenotyping of micrometastatic SCLC cells, combining cytokeratin labeling with the detection of disialoganglioside GD2 *(4)*. GD2 is a common ganglioside found on virtually all SCLC cells, as well as on other cancer cells such as neuroblastoma, glioma, and melanoma *(21–23)* (*see* **Fig. 1**).

Below, we list some of the most common protocols employed in our laboratory for detection of circulating lung cancer cells. It is important to realize that all of the peripheral blood and bone marrow samples are obtained by Informed Consent on institutional approved IRB protocols.

2. Materials
2.1. Sample Preparation for Microscopic Analysis
1. Phosphate-buffered saline (PBS, Gibco Life Technologies, Grand Island, NY).
2. 20% human AB serum in PBS (Nabi Diagnostics, Boca Raton, FL).
3. Humidified chamber (100-mm tissue-culture dish with wet paper towel on the bottom, covered by parafilm).
4. Incubator (37°C).
5. Heparin-containing tubes for blood or bone marrow collection.
6. Isotonic ammoniumchloride: 155 mM NH$_4$Cl, 10 mM KHCO$_3$, and 0.1 mM EDTA.
7. Hematocytometer.
8. 0.4% trypan blue solution: dissolve 0.2 g trypan blue in 50 mL distilled water.
9. Adhesive slides (Paul Marienfeld GmbH & Co., KG, Bad Mergentheim, Germany).
10. Tissue-culture medium with 10% bovine calf serum (BCS).
11. Glass staining dishes.
12. Paraformaldehyde, 2% in PBS, pH 7.4.
13. Monoclonal anti-GD2 antibodies, diluted 1:500, or as necessary for the optimum result. (Mouse 1418 antibodies, Lexigen Pharmaceuticals, Lexington, MA).

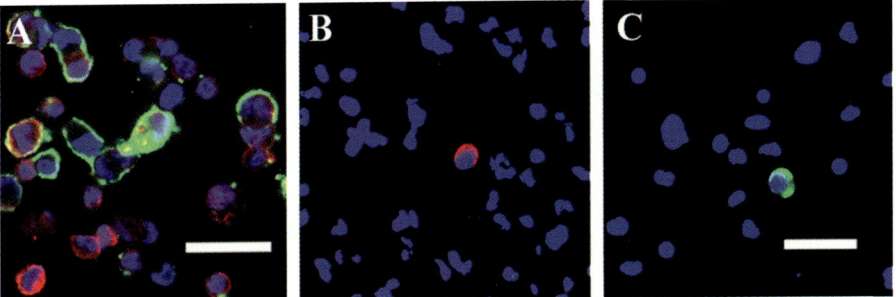

Fig. 1. Visualization of lung cancer cells using double labeling of GD2/green and cytokeratin/red. Small cell lung cancer cell line HTB-119 (**A**), peripheral blood samples of patients with NSCLC (**B**), and SCLC (**C**), respectively. All nuclei were counterstained with DAPI/blue. Bars, 20 µm.

14. FITC-conjugated anti-mouse antibodies, diluted 1:200 (Jackson Immuno Research, West Grove, PA).
15. Ice-cold methanol (–20°C).
16. Polyclonal anti-cytokeratin antibodies, diluted 1:100, or as necessary for the optimum result. (Biomedical Technologies, Stoughton, MA).
17. Rhodamine-conjugated anti-rabbit antibodies, diluted 1:200 (Jackson Immuno Research).
18. 4′,6-diamidino-2-phenylindole (DAPI), 0.5 µg/mL in PBS (Molecular Probes, Eugene, OR).
19. Gel/Mount (Biomeda, Foster City, CA).
20. Glass coverslips no.1, 24 × 60mm (Corning Inc., Big Flats, NY).
21. Nail polish.

2.2. Microscopic Detection and Total Cell Count

1. Fluorescence microscope with appropriate filters for fluorescein, rhodamine, and DAPI.

3. Method
3.1. Sample Preparation for Microscopic Analysis

All incubations are carried out in a humidified chamber at 37°C. All antibodies are diluted in PBS containing 20% human AB serum.

1. Mix 3 mL heparinized blood or bone marrow with 12 mL of isotonic ammonium chloride and incubate for 40 min at room temperature (RT).
2. Centrifuge mixture at 400*g* for 10 min at RT (*see* **Note 1**).
3. Remove supernatant and resuspend cell pellet in 10 mL PBS.
4. Centrifuge cell suspension at 400*g* for 10 min at RT.

5. Remove supernatant and resuspend cell pellet in 0.5–1 mL PBS.
6. Count living white blood cells using trypan blue.
7. Adjust white blood cell concentration to 5×10^6 living cells/ml using PBS.
8. Prepare adhesive slides by washing them in distilled water until blue layer comes off. (*see* **Note 2**).
9. Apply cell suspension to adhesive slides (100 µL per circle) and incubate for 40 min.
10. Add cell culture medium to adhesive slides (60 µL per circle) and incubate for further 20 min.
11. Place slides into a glass staining dish with 2% paraformaldehyde in PBS and fix cells at RT for 20 min.
12. Wash slides 2× in PBS, 3 min each.
13. Block slides with 20% human AB serum for 20 min.
14. Tap off excess serum. DO NOT WASH.
15. Apply mouse anti-GD2 antibody and incubate for 1 h (*see* **Note 3**).
16. Wash slides 2× in PBS, 3 min each.
17. Apply FITC conjugated anti-mouse antibodies, and incubate for 30 min (*see* **Notes 4** and **5**).
18. Wash slides 2× in PBS, 3 min each.
19. Place slides into a glass staining dish with ice-cold absolute methanol and fix cells at –20°C for 5 min.
20. Wash slides 2× in PBS, 3 min each.
21. Block slides with 20% human AB serum for 20 min.
22. Tap off excess serum. DO NOT WASH.
23. Apply rabbit anti-cytokeratin antibody and incubate for 1 h.
24. Wash slides 2× in PBS, 3 min each.
25. Apply rhodamine-conjugated anti-rabbit antibodies and incubate for 30 min.
26. Wash slides 2× in PBS, 3 min each.
27. Apply DAPI and incubate at room temperature for 10 min.
28. Wash slides in PBS, 3 min.
29. Rinse slides in distilled water.
30. Mount slides in Biomeda gel with coverslips and nail polish.
31. Store slides at 4°C in the dark (*see* **Note 6**).

3.2. Automated Microscopic Detection and Total Cell Count

Slides can be manually analyzed for fluorescently labeled cells using a fluorescence microscope equipped with the appropriate filters for fluoresccin, rhodamine, and DAPI. Alternatively, slides can be automatically scanned using a Rare Event Imaging System, developed in our laboratory and by Georgia Instruments, Inc. (Roswell, GA) *(4)*. The system employs proprietary image processing algorithms to detect rare fluorescence events and determine the total number of cells analyzed. It is comprised of an advanced computer-controlled

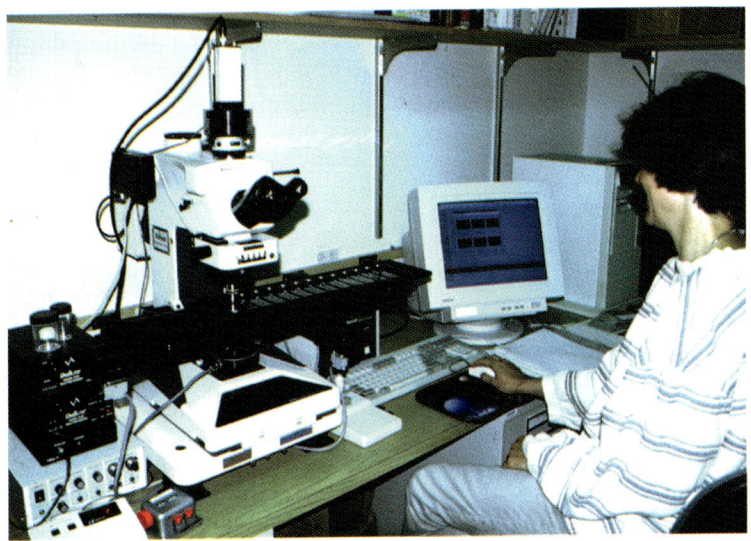

Fig. 2. Rare Event Imaging System.

microscope (Nikon Microphot-FXA, Nikon, Japan) with autofocus, motorized X, Y, and Z axis control, motorized filter selection, and electronic shuttering. Images are taken by an integrating, cooled CCD detector and processed in a Pentium imaging workstation (60 MHz) (see **Fig. 2**).

1. The slide is automatically scanned for the detection of positive events (e.g., cytokeratin-positive cells) using the rhodamine filter set. The identification of positive events is based on fluorescence intensity and area. For each positive event, the scanning coordinates are stored and the image is archived.
2. The slide is scanned for the total number of DAPI labeled nuclei per slide, representing the total cell count. The total scanned area per slide is 448 mm^2 (84% of the adhesive area) to avoid edge effects.
3. At the end of the two scans, the number of positive events and the total cell count are given, and a gallery of images containing all positive events is displayed. The user can review the images and recall any of the events for further examination using the stored coordinates attached to each image. The field of interest can then be visualized using higher microscope magnification and additional filter sets (e.g., fluorescein, or UV filter).
4. Images of different fluorescent colors can be electronically overlaid for positive confirmation of the event and for phenotypic evaluation (multiple labeling). The total scanning time (two scans) for one slide is about 1 h. The two scans can be run independently, thereby offering the option of just screening for positive events and thus shortening the scanning time to 30 min/slide.

Rare Event Imaging System is a state-of-the-art technology that helps in detecting micrometastatic disease in solid tumors such as lung cancer. With the data obtained from this unique technique, information can be gathered about prognosis and potential therapeutics for this devastating illness.

4. Notes

1. At each centrifugation step, it is important to carefully separate the supernatant from the cell pellet. We usually use a pipet tip for the manual removal of the last portion of the supernatant.
2. As an alternative to the use of adhesive slides from Marienfeld/Germany, one can attach the cells to poly-L-lysine/PBS coated slides (0.1%; Sigma), or SectionLock Slides (Polysciences, Inc., Warrington, PA) or by cytocentrifugation using a Shandon-centrifuge. In our hands, none of the alternative methods achieved a similar quality of the monolayer and high yield of cell attachment as the adhesive slides from Marienfeld did.
3. Our original protocol *(4)* was developed using the mouse monoclonal 1418 antibody directed against GD2 (antibody kindly provided by Dr. Kin-Ming Lo, Lexigen Pharmaceuticals). Recently we tested another commercially available antibody against GD2 from Matreya, Inc. (Pleasant Gap, PA). This antibody (at a 1:10 dilution) shows a similar performance on small cell cancer cell lines spiked into normal blood samples as the monoclonal 1418 antibody.
4. The protocol can be adapted for other marker combinations as antibodies to the targets are available. Each new combination should be tested in normal blood samples seeded with lung cancer cells, e.g., HTB-119 or SW-2. Those might be substituted for the secondary reagents listed in the standard protocol.
5. For automated image analysis it is crucial to work with secondary antibodies that give a bright signal while maintaining a low background. We have recently tested dyes of the Alexa-series (Molecular Probes) that give very bright and stable fluorescence signals.
6. Fluorescently labeled slides should be analyzed within 1 wk. If longer storage is desired, a mounting medium that maintains stable fluorescent signals should be used. We found that the use of the ProLong Antifade Kit (Molecular Probes) gave excellent results after 3-mo storage of the slides at 4°C.

Acknowledgment

We thank Rebecca Sutherland, Stephanie Brunelle, Gaelle Even, Anna Herlitz, and Mary Ellen Healy for their technical expertise that was so important for developing this method and testing of patients' samples.

References

1. Ohgami, A., Mitsudomi, T., Sugio, K., Tsuda, T., Oyama, T., Nishida, K., et al. (1997) Micrometastatic tumor cells in the bone marrow of patients with non-small cell lung cancer. *Ann. Thorac. Surg.* **64(2),** 363–367.

2. Pantel, K., Izbicki, J., Passlick, B., Angstwurm, M., Haussinger, K., Thetter, O., and Riethmuller, G. (1996) Frequency and prognostic significance of isolated tumour cells in bone marrow of patients with non-small-cell lung cancer without overt metastases. *Lancet* **347(9002)**, 649–653.
3. Kurusu, Y., Yamashita, J., and Ogawa, M. (1999) Detection of circulating tumor cells by reverse transcriptase-polymerase chain reaction in patients with resectable non-small-cell lung cancer. *Surgery* **126(5)**, 820–826.
4. Kraeft, S. K., Sutherland, R., Gravelin, L., Hu, G. H., Ferland, L. H., Richardson, P., et al. (2000) Detection and analysis of cancer cells in blood and bone marrow using a rare event imaging system. *Clin. Cancer Res.* **6(2)**, 434–442.
5. Cote, R. J., Beattie, E. J., Chaiwun, B., Shi, S. R., Harvey, J., Chen, S. C., et al. (1995) Detection of occult bone marrow micrometastases in patients with operable lung carcinoma. *Ann. Surg.* **222(4)**, 415–423 (discussion 423–425).
6. Chaiwun, B., Saad, A., Chatterjee, S. J., Taylor, C. R., Beattie, E. J., and Cote, R. J. (1996) Advances in the pathologic staging of lung cancer: detection of regional and systemic occult metastases. *Pathology* **4(1)**, 155–168.
7. Cote, R. J., Hawes, D., Chaiwun, B., and Beattie, E. J. (1998) Detection of occult metastases in lung carcinomas: progress and implications for lung cancer staging. *J. Surg. Oncol.* **69(4)**, 265–274.
8. Ross, A. A. (1998) Minimal residual disease in solid tumor malignancies: a review. *J. Hematother.* **7(1)**, 9–18.
9. Radbruch, A. and Recktenwald, D. (1995) Detection and isolation of rare cells. *Curr. Opin. Immunol.* **7(2)**, 270–273.
10. Bauer, K. D., de la Torre-Bueno, J., Diel, I. J., Hawes, D., Decker, W. J., Priddy, C., et al. (2000) Reliable and sensitive analysis of occult bone marrow metastases using automated cellular imaging. *Clin. Cancer Res.* **6(9)**, 3552–3559.
11. Mesker, W. E., v. d. Burg, J. M., Oud, P. S., Knepfle, C. F., Ouwerkerk-v Velzen, M. C., Schipper, N. W., and Tanke, H. J. (1994) Detection of immunocytochemically stained rare events using image analysis. *Cytometry* **17(3)**, 209–215.
12. Racila, E., Euhus, D., Weiss, A. J., Rao, C., McConnell, J., Terstappen, L. W., and Uhr, J. W. (1998) Detection and characterization of carcinoma cells in the blood. *Proc. Natl. Acad. Sci. USA* **95(8)**, 4589–4594.
13. Chen, Y. H., Gao, W., Zhou, T., Zhao, W., Zhao, H., Liu, D., and Felber, E. (1999) Detection of bone marrow micrometastasis. *Hybridoma* **18(5)**, 465–466.
14. Pantel, K., Izbicki, J. R., Angstwurm, M., Braun, S., Passlick, B., Karg, O., et al. (1993) Immunocytological detection of bone marrow micrometastasis in operable non-small cell lung cancer. *Cancer Res.* **53(5)**, 1027–1031.
15. Passlick, B., Kubuschok, B., Izbicki, J. R., Thetter, O., and Pantel, K. (1999) Isolated tumor cells in bone marrow predict reduced survival in node-negative non-small cell lung cancer. *Ann. Thorac. Surg.* **68(6)**, 2053–2058.
16. Pasini, F., Pelosi, G., Verlato, G., Guidi, G., Pavanel, F., Tummarello, D., et al. (1998) Positive immunostaining with MLuC1 of bone marrow aspirate predicts poor outcome in patients with small-cell lung cancer. *Ann. Oncol.* **9(2)**, 181–185.

17. Pelosi, G., Pasini, F., Pavanel, F., Bresaola, E., Schiavon, I., and Iannucci, A. (1999) Effects of different immunolabeling techniques on the detection of small-cell lung cancer cells in bone marrow. *J. Histochem. Cytochem.* **47(8),** 1075–1088.
18. Pasini, F., Cetto, G. L., and Pelosi, G. (1998) Does bone marrow involvement affect prognosis in small-cell lung cancer? *Ann. Oncol.* **9(3),** 247–250.
19. Pasini, F., Pelosi, G., Mostacci, R., Santo, A., Masotti, A., Spagnolli, P., et al. (1995) Detection at diagnosis of tumor cells in bone marrow aspirates of patients with small-cell lung cancer (SCLC) and clinical correlations. *Ann. Oncol.* **6(1),** 86–88.
20. Pasini, F., Pelosi, G., Ledermann, J. A., and Cetto, G. L. (1994) Detection of small-cell-lung-cancer cells in bone-marrow aspirates by monoclonal antibodies NCC-LU-243, NCC-LU-246 and MLuC1. *Int. J. Cancer Suppl.* **8(9),** 53–56.
21. Schulz, G., Cheresh, D. A., Varki, N. M., Yu, A., Staffileno, L. K., and Reisfeld, R. A. (1984) Detection of ganglioside GD2 in tumor tissues and sera of neuroblastoma patients. *Cancer Res.* **44(12 Pt 1),** 5914–5920.
22. Cheresh, D. A., Rosenberg, J., Mujoo, K., Hirschowitz, L., and Reisfeld, R. A. (1986) Biosynthesis and expression of the disialoganglioside GD2, a relevant target antigen on small cell lung carcinoma for monoclonal antibody- mediated cytolysis. *Cancer Res.* **46(10),** 5112–5118.
23. Ritter, G. and Livingston, P. O. (1991) Ganglioside antigens expressed by human cancer cells. *Semin. Cancer Biol.* **2(6),** 401–409.

27

Reconstruction of Geno-Phenotypic Evolutionary Sequences from Intracellular Patterns of Molecular Abnormalities in Human Solid Tumors

Stanley E. Shackney, Charles A. Smith, Agnese A. Pollice, and Jan F. Silverman

1. Introduction

1.1. Background and Theory

In this chapter we describe an approach to the analysis of human tumors that emphasizes the performance of multiple simultaneous measurements on each of several thousand cells in each tumor sample. This approach can be implemented using such technologies as flow cytometry, laser-scanning cytometry, or high-resolution image analysis. Here we will focus on flow cytometry studies. Before proceeding to methodological details, we would like to consider the theoretical and practical advantages of performing cell-based measurements (vis a vis gel or array based measurements, for example), and in particular, the advantages of performing multiple measurements per cell.

Individual human tumors are heterogeneous in their cellular composition. Not only do they contain variable proportions of admixed normal cells of various types, but they also contain a variety of persistent ancestral lineages of the invasive cellular components, including early metaplastic and atypical hyperplastic cellular components, and more advanced preinvasive cellular components (moderate and severe dysplasia, and carcinoma *in situ* [CIS]). This cellular heterogeneity may be compounded further by the presence of clonal variants that harbor genetic changes that may be distinctive, but are not preserved in more advanced, and presumably more aggressive clonal populations in the same tumor.

There are several types of information that can be extracted from heterogeneous cell populations using cell-based methods, that would be irretrievably lost if the cells were disrupted and their contents pooled prior to analysis.

1. Cell-based methods provide quantitative information regarding the intracellular levels of various cell constituents. For example, breast tumors containing cells with high levels of Her-2/neu amplification/overexpression, can be targeted for clinical treatment with an anti-Her-2/neu antibody. The clinical criteria that have been developed for selecting patients for anti-Her2/neu antibody administration generally rely on immunohistochemistry or fluorescence *in situ* hybridization (FISH), techniques *(1)*, both of which are cell-based tests. In the case of Her-2/neu, where it is common for a relatively large proportion (often 10–50%) of the cells in a tumor to show amplification when it is present, equivalent information could be obtained by either cell-based or non-cell-based methods *(2)*. However, if the amplifying/overexpressing cell subpopulation of interest (other than Her-2/neu alone) represents only a small proportion (e.g., <10%) of the cells present, high levels of protein expression or gene amplification in individual cells could easily be missed using non-cell-based methods, due to dilution of the sample by the contents of disrupted nonamplified, nonoverexpressing cells.

2. Cell-based methods permit the distinction between the presence of multiple abnormalities in the same cells and the presence of separate abnormalities in different cells. One might expect, for example, that genetic abnormalities that occur as a result of p53-induced genetic instability would be characterized by their obligatory co-occurrence in the same cells with the p53 abnormalities that produced them. This would not be the case for genetic abnormalities that are produced by other mechanisms.

3. Cell-based methods provide information regarding differences in the clustering patterns of multiple abnormalities in the same cells. There are distinctive clusters of geno-phenotypic abnormalities that occur within individual tumors, and are common to other tumors of similar type or subtype *(3–6)*. Tumors that arise at the same organ site can exhibit several different clustering patterns. For example, human colon cancers can be subdivided into a group of predominantly diploid, right sided tumors that often exhibit microsatellite instability *(7)* and mutations of the transforming growth factor-β (TGF-β) type II receptor and hMLH1 or hMSH2 genes, but exhibit p53 abnormalities in a minority of cases, vs a group of predominantly left sided tumors that often exhibit structural and numerical chromosomal instability, but not microsatellite instability, and frequently exhibit p53 abnormalities (reviewed in *[5]*). In human breast cancer, p53 abnormalities, aneuploidy, amplification/overexpression of Her-2/neu, and c-myc amplification have been found to cluster in infiltrating ductal carcinomas *(8)*, but these abnormalities do not occur frequently in lobular breast cancers, either alone or in combination *(8)*.

Multiparameter cell-based studies can be used to explore mixed patterns of molecular abnormalities in solid tumors. For example, what is the role of p53 abnormalities that occur in colon cancers with microsatellite instability? As a first step toward answering this question, one might wish to determine if the p53 abnormalities and the abnormalities commonly associated with microsatellite instability are found in the same cells or not.

Early cancers are thought to evolve toward advanced malignancy through the successive overgrowth of clones that accumulate genetic abnormalities and become increasingly more aggressive biologically. **Figure 1** shows how specific patterns of accumulated intracellular abnormalities might come about during the course of genetic evolution. This model suggests that patterns of accumulated geno-phenotypic abnormalities that evolve within a given tumor would be imprinted within the individual cells of the most advanced clonal populations in that tumor.

The potential importance of specific intracellular patterns of molecular abnormalities is demonstrated by studies involving four measurements per cell by flow cytometry in human breast cancers *(9–11)* that we have done over the past 10 years, and by our more recent studies in non-small cell lung cancer (NSCLC) *(24)*. In many of these tumors we found subsets of cells in which there was co-occurrence of p53 protein overexpression and/or aneuploidy, Her-2/neu protein overexpression, and ras protein overexpression in the same cells. Our early results in breast cancer indicated that patients with tumors that contained cells with three abnormalities (aneuploidy, Her-2/neu overexpression, and ras overexpression in the same cells) had a worse clinical outcome than patients whose tumors did not contain such cells *(9)*. In these studies, the fractions of triple positive cells in individual tumors were usually 10% or less, and rarely exceeded 20%. Its difficult to imagine how such small but prognostically useful cell subpopulations could be identified, much less characterized, using non-cell-based techniques.

Within specific clusters of geno-phenotypic abnormalities in human solid tumors, certain abnormalities have been found to occur in a preferred order in relation to other members of the cluster, and/or in relation to early or late clinicopathologic stages of disease *(6,11–13)*. In studies of patients with Barrett's esophagus, for example, p53 abnormalities were found to appear early in predysplastic diploid cells. During the course of evolution from the premalignant state to invasive disease, tumors with p53 abnormalities were found to progress through an intermediate stage of tetraploidy *(14)* to gross aneuploidy *(15,16)*. A similar sequence of events has been reported in human colon cancers *(17,18)* Sequential evolutionary changes have been described in NSCLC as well. Specifically, 3p abnormalities, and abnormalities involving 9p (commonly related to the p16 locus) can be found in near-normal cells adjacent to squamous cell tumors; 5q abnormalities and p53-associted abnormalities first appear in metaplastic and dysplastic cells, and Rb-associated abnormalities

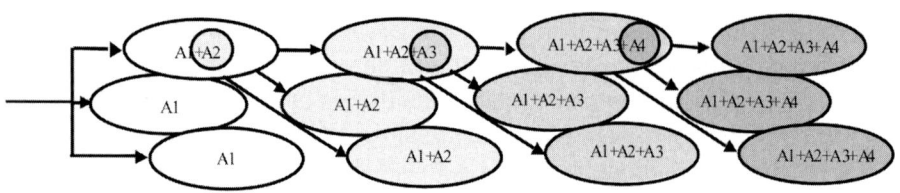

Fig. 1. A model for the accumulation of specific intracellular patterns of molecular abnormalities in subsets of tumor cells during the course of genetic evolution. An early abnormality that confers a growth or survival advantage (A1) results in the outgrowth of a clonal subpopulation in which each cell contains that abnormality. A member of this clone then develops a new abnormality (A2) that confers an additional survival advantage, resulting in the further outgrowth of a clonal subpopulation in which each cell contains abnormalities A1 and A2. Repetition of this process results in the progressive intracellular accumulation of multiple molecular abnormalities in successive clonal overgrowths. This pathway might represent one of several preferred pathways that a given tumor type could follow.

tend to occur later still *(19)*. These findings are consistent with the results of other published studies in NSCLC demonstrating the appearance in early lung cancer precursor lesions of 3p deletions *(20)*, 9p deletions *(21)* and p16^{INK4a} inactivation *(22)*, and the later appearance of p53 mutations *(23)* and ras mutations *(21)*. These results are consistent with our own published cell-based studies in NSCLC *(24)*.

Multiparameter cell-based measurements can be used to determine if the multiple geno-phenotypic abnormalities that are present in a tumor had occurred randomly, or if they might have developed in a preferred sequence. The analytical approach relies on the persistence of early and intermediate precursor cell subpopulations in the presence of more advanced cell subsets in the same tumor sample, as shown schematically in **Fig. 2**. In brief, if abnormality A1 always occurred first among those measured, then all cells containing only one measured abnormality would contain abnormality A1 *only*, and not A2, A3, or A4. If abnormality A2 always follows abnormality A1, then all cells in the tumor with two abnormalities per cell would contain abnormalities A1 and A2, and *no other* combination of two abnormalities per cell. Similarly, if abnormality A3 always follows A2, then it would always be present in cells with at least three abnormalities per cell, and would always be accompanied by abnormalities A1 and A2. If abnormality A4 always followed abnormality A3, then one would expect to see it always in the company of the other three abnormalities in the same cells. Actual multiparameter cell-based data in human solid tumors are somewhat more complex than the model shown

Evolutionary Sequences

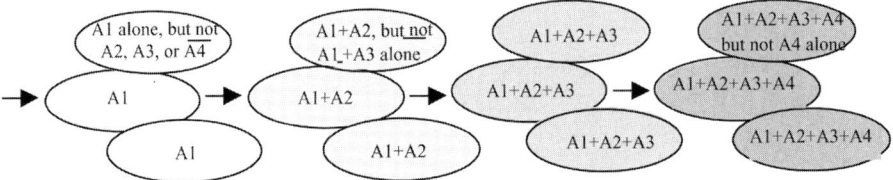

Fig. 2. A model for determining the sequence of occurrence of molecular abnormalities in a tumor from a single based on the patterns of intracellular molecular abnormalities that are found. For discussion, see text.

$\uparrow$ p53 → aneuploidy → $\uparrow$Her-2/neu → $\uparrow$ras

(often seen alone in   (usually accompanied by other
individual cells)   abnormalities in the same cells)

Fig. 3. A preferred sequence of molecular abnormalities that occurs commonly in several different types of human solid tumors.

in **Fig. 2**, often showing evidence of several potential alternative pathways in the same tumor. However, multiparameter flow cytometry studies show that certain abnormalities almost always occur early, while others occur late, and certain preferred multistep sequences of geno-phenotypic abnormalities are very common. The sequence shown in **Fig. 3** is preferred in the majority of human breast cancers *(11)*, and in substantial proportions of NSCLCs, colon cancers, and glioblastomas (*[24]* and unpublished observations).

Why would molecular abnormalities in human solid tumors occur in a preferred order? We propose the model shown in **Fig. 4** to account for these findings. p53 inhibits cell-cycle progression and induces apoptosis in response to aberrant mitogenic signaling in nontransformed cells. Thus, inactivation of this autoregulatory feedback loop by interference with wild-type p53 function might be an important step that must occur early in tumor development (*see* **ref. 5**). Abrogation of wild-type p53 function is also associated with the development of distinctive structural chromosomal changes, including gene amplification *(25)*, and Her-2/neu gene amplification in particular *(8)*. p53 dysfunction has also been linked with the development of gross numerical chromosomal abnormalities (*see* **ref. 5**). The model shown in **Fig. 4** would suggest that the development and persistence of Her-2/neu gene amplification might be more likely in the presence of pre-existing p53 dysfunction, and that this might lead to sustained and unopposed mitogenic signaling in the absence of wild-type p53. Sustained mitogenic signaling might favor the evolutionary

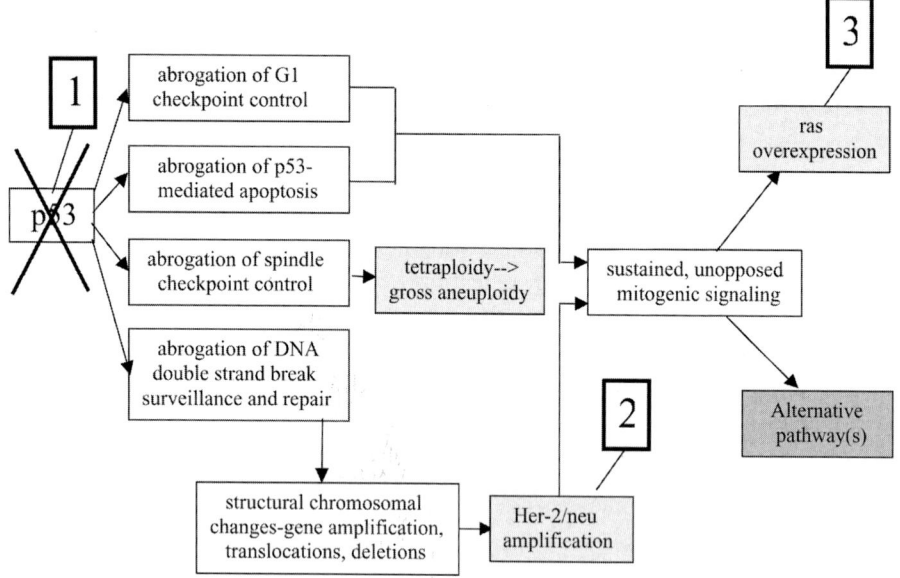

Fig. 4. A model for the sequential evolutionary consequences of the abrogation multiple p53 functions. For discussion, see text.

selection of cells with higher intracellular levels of ras, especially in breast cancers, in which activating ras mutations are rare *(26,27)*.

While the preferred sequence shown in **Fig. 4** appears to be a common one in NSCLC *(24)*, the prognostic value of intracellular ras-containing clusters of abnormalities has not been established. Just as we have found in breast cancer, there are aggressive NSCLC that contain only diploid, non-Her-2/neu overexpressing, and non-ras overexpressing cells *(24)*. Presumably, these tumors follow alternative evolutionary pathways. While there are similarities between the molecular changes that occur in NSCLC and those that occur in breast cancer, there are also substantial differences. While ras mutations are rare in breast cancer, they are common in NSCLC. It may be that both activating K-ras mutations and ras overexpression produce equivalent phenotypic effects, or that alternative non-ras mediated mitogenic signaling pathways might be especially prominent in NSCLC (*see* **Fig. 4**). There are also differences in the patterns of Her-2-neu and EGF receptor amplification and overexpression in the two tumor types. Various alternative pathways can be explored using the multiparameter cell-based approach by choosing appropriate combinations of fluorescent DNA dyes and immunofluorescent probes.

1.2. Outline of Methodology

1.2.1. Sample Preparation

All of our human tumor studies have been performed on mechanically disaggregated single cell suspensions obtained from freshly obtained primary tumor samples. The advantages of working with single-cell suspensions are that multiple quantitative cell-based measurements can be performed more accurately than in tissue sections, for example, which are subject to partial cell sectioning, to potentially uneven fixation in high concentrations of formalin, to potential difficulties with obtaining separate cell by cell measurements in highly cellular sections, and to limitations on the number of measurements that can be performed on each cell (*see* **Note 1**). In these studies the states of advancement of tumor cells are identified by the patterns of intracellular molecular abnormalities that they contain rather than by their morphologic appearance or their microarchitectural features. While correlations with cell morphology are not possible by flow cytometry, cytomorphologic information can be captured by laser-scanning cytometry and digital image-analysis technologies.

1.2.2. Event-Triggering Measurements

When one relies on a technology that performs measurements on thousands of cells, it is highly advantageous to automate the process that identifies the cells for study. Flow cytometry and laser scanning cytometry rely on "triggering" measurements, i.e., on measurements that alert the instrument that a cell is present and available for measurement. Forward (narrow angle) light scatter can be used in unfixed cells, but this signal is degraded by cell fixation. Since cells must be fixed and permeabilized in order to perform intracellular measurements, another triggering measurement must be used in fixed cells, such as cell DNA content or some ubiquitous intracellular protein (e.g., cytokeratin or tubulin). We prefer the use of cell DNA content as the triggering measurement (*see* **Note 2**).

1.2.3. Cell Fixation and Permeabilization

Cell fixation and permeabilization are required for the performance of intracellular measurements, to permit DNA-binding dyes and fluorescent antibodies against specific cell constituents to gain entrance into the cells. We have tried to avoid the use of potentially harsh antigen retrieval techniques by using mild paraformaldehyde fixation (0.5%) for crosslinking of intracellular proteins prior to subsequent permeabilization/fixation with 70% methanol *(28)* (*see* **Notes 3** and **4**).

1.2.4. Quantitative Reference Standards

If fluorescence measurements are to be quantitative, they must be anchored to known reference values. Cell DNA measurements are often referenced to the diploid G_1 peak that is almost invariably present in most clinical samples. Flow cytometers are sufficiently stable that concomitantly stained reference cells can be run in separate samples to provide quantitative benchmarks for immunofluorescence measurements. Any cell population that is stable in fixative over time (*see* **Note 5**) can serve as a relative baseline reference. One can then perform immunofluorescence measurements on this reference cell population shortly before or after measuring the test sample, and both normal and abnormal values in the test samples can expressed as ratios of this reference value. The chief advantage of this approach is its ease of application. The disadvantage of this approach is its dependence on arbitrary reference standards that are not universally accepted, which makes it difficult to perform quantitative interlaboratory comparisons.

An approach that provides absolute rather than relative quantitation is preferable. When it is possible to quantitate the average number of molecules per cell of a particular cell constituent in a reference cell population by independent means, one can then equate the mean number of molecules per cell for the reference population with the mean fluorescence value of the reference cell population obtained by flow cytometry. This absolute reference value can then be used to calculate the number of molecules per cell that corresponds to the immunofluorescence level of any cell in a concomitantly run test sample. We have stockpiled large numbers of paraformaldehyde/methanol-fixed cells from a breast tumor cell line that was established in our laboratory, of which aliquots of known numbers of cells were taken for quantitative determination of molecular concentrations of various cell constituents by enzyme-linked immunosorbent assay (ELISA) (*see* **Note 6**). In principle, measurements expressed in absolute units of molecules per cell should be independent of instrumentation, dyes, or antibody reagents, or the reference cell line used. However, the degree to which this holds in practice remains to be determined.

1.2.5. Correction for Nonspecific Antibody Binding

Immunofluorescent antibodies bind specifically to the epitopes against which they were raised, and they also bind nonspecifically to a greater or lesser extent to other intracellular components. The concentration of each immunofluorescent antibody must be titrated, and the optimum working dilution is chosen such that specific binding approaches saturation, while nonspecific binding remains minimal. Some nonspecific binding is always present, even at the optimal antibody dilutions, and with some antibodies the nonspecific antibody binding component may actually represent a substantial proportion of the total fluorescence signal. Since our purpose is to obtain quantitative cell-based measurements (as distinct from binary distinctions such as positive vs negative), we apply a correction factor for nonspecific fluorescence that is subtracted from the total fluorescence signal generated by each antibody in each cell. The method for arriving at the appropriate value for the nonspecific correction is a matter of some debate. The use of isotype-matched nonspecific antibody controls is favored by many immunologists, but these antibodies may differ in their degree of nonspecific binding from that of the specific antibody of interest, and they can be wasteful of what is often scant clinical material. Application of the specific antibody of interest to a reference cell population that has little or none of the cell constituent of interest can be used to estimate the level of nonspecific binding, but the cells in which this measurement is performed are different from the test cell population. For the present, we generally use a correction factor based on the latter approach, that is derived from measurements made on normal lymphocytes. The relationship between the two approaches is under active study.

We now describe the methods for performing quantitative four-color flow cytometric studies of cell DNA content, and cellular p53, Her-2/neu, and ras proteins on human tumor cells in greater detail.

2. Materials
2.1. Sample Preparation
1. Hank's balanced salt solution (HBSS) (Mediatech, Inc., Herndon, VA).
2. Phosphate-buffered saline (PBS) (Mediatech, Inc.).
3. 90% methanol. Stable at 4°C.
5. 0.5% paraformaldehyde in PBS. Stable at 4°C for up to 1 mo.
6. Cytospin set up.

2.2. Immunofluorescence Staining For Multiparameter Analysis
1. Paraformaldehyde/methanol-fixed JC 1939 cells.
2. Paraformaldehyde/methanol-fixed lymphocytes.
3. Methanol-fixed lymphocytes from healthy donors.
4. Bovine serum albumin (BSA) for preparation of 1.0% BSA in PBS. Stable at 4°C for 1 wk
5. Cy 5 Ab labeling kit (Amersham Pharmacia Biotech).
6. 10 µL FITC-conjugated mouse monoclonal antibody immunospecific for p53 protein /1×10^5 pelleted cells. Fluorescein-conjugated (FITC) monoclonal

antibody immunospecific for p53 protein (Clone DO-7) is purchased from Novocastra Laboratories, Ltd. (new Castle upon Tyne, UK).
7. 10 µL Cy-5-conjugated rat monoclonal antibody immunospecific for v-H-ras / 1×10^5 pelleted cells. Rat monoclonal antibody to v-H-ras (Oncogene Research Products, Boston, MA) is conjugated with Cy 5 Ab labeling kit, following manufacturer's instructions.
8. 50 µL rabbit polyclonal antibody to c-erbB-2, 1:50 dilution in PBS/1×10^5 pelleted cells. Rabbit polyclonal antibody to c-erbB-2 is purchased from Cambridge Research Biochemicals (Cambridge, UK).
9. 50 µL PE-conjugated goat anti-rabbit IgG (1:20 dilution in PBS) / 1×10^5 pelleted cells which have been stained by primary antibodies. Phycoerythrin-conjugated (PE) goat anti-rabbit IgG is purchased from Vector (Burlingame CA).

2.3. DNA Staining

1. 4',6-diamino-2-phenylindole (DAPI): Stock is 1 µg/mL in PBS. Stable stored in dark bottle at 4°C for 6 mo. Final concentration: 0.1 µg/mL.
2. Propidium Iodide (PI): 50 µg/mL in PBS + 1 mg/mL RNAse. Stable stored in dark bottle at 4°C for 6 mo.

3. Methods
3.1. Sample Preparation

1. To prepare freshly obtained tumor samples (*see* **Note 7**), trim fat and connective tissue. Finely scissor-mince in HBSS, filter through gauze, and wash in HBBS by centrifugation in 15 mL conical centrifuge tubes at 200*g* for 2 min (*see* **Note 8**).
2. Resuspend cells in PBS and wash again (*see* **Note 9**).
3. Divide cells into two aliquots: One aliquot is fixed for S-fraction analysis by adding 1 mL cold PBS (4°C) followed by 3 mL cold 90% methanol (4°C) dropwise while vortexing to prevent clumping; the second aliquot is fixed for multiparameter analysis by adding 3 mL cold 0.5% paraformaldehyde for 15 min at room temperature, washing 1X in PBS and resuspending cells in 1 mL cold PBS followed by 3 mL cold 90% methanol. Store samples at 4°C.
4. Prepare cytospin slides separately from cells from each fixed cell suspension, and evaluate cytologically for the presence of tumor cells in the sample. Estimates of the proportion of tumor cells present are recorded (*see* **Note 10**).

3.2. Immunofluorescence Staining For Multiparameter Analysis

Paraformaldhyde/methanol-fixed JC 1939 cells, are used as a reference cell line for which mean p53 protein per cell and mean Her-2/neu protein per cell has previously been determined. As assayed by ELISA in units of molecules per cell, these are used as a quantitative reference and positive staining control for p53, Her-2/neu, and ras immunofluorescence. Paraformaldhyde/methanol-fixed lymphocytes are used as baseline relative reference cells for quantitative ras immunofluorescence staining. Methanol-fixed lymphocytes from healthy donors are used as a diploid reference for S-fraction analysis.

All antibodies are titrated to determine optimum saturating dilutions (*see* **Subheading 2.2.**). Mouse IgG1-FITC, Rat IgG1-Cy5, and Rabbit IgG1-PE are used for staining isotype controls.

1. Count cells and wash 1X in PBS. 1×10^5 cells are incubated in 2 mL 1% BSA at 4°C for 1 h to block for nonspecific binding of antibodies, then pelleted by centrifugation at 200*g* for two min, leaving a small liquid residuum.
2. Add antibodies in solution directly to the cell pellet.
3. Gently vortex cells and incubate at room temperature for 1 h in the dark.
4. Wash cells in 1 mL PBS. Add PE-conjugated goat anti-rabbit IgG to the pellet and gently vortex cells. Incubate for 1 h at room temperature in the dark.
5. Wash cells 1X with PBS.

3.3. DNA Staining

1. Stain cells for multiparameter analysis with DAPI (*see* **Note 11**).
2. Cells used for S-fraction analysis are stained with PI, and incubated at 4°C for 1 h.

3.4. Flow Cytometric Analysis

Cell fluorescence measurements for multiparameter analysis are performed on a flow cytometer equipped with an argon laser emitting at the 488-nm wavelength, with a 550-nm dichroic long-pass filter, 525–nm band pass filter for FITC, and a 625-nm dichroic long-pass filter and a 575-nm band pass filter for phycoerythrin; a HeNe laser emitting at the 633-nm wavelength is used with a 675-nm band pass filter to measure Cy-5; and a water cooled argon laser with UV capabilities emitting at 325 nm with a 381-nm band pass filter for DAPI measurement.

1. Collect measurements on 10,000–20,000 cells and store in list mode.
2. Fluorescence measurements for S-fraction calculation are done separately on the methanol-fixed samples, using only the argon laser emitting at the 488-nm wavelength (*see* **Note 12**).

3.5. Data Analysis

S-fraction analysis can be performed using one of several commercially available programs, e.g., Modfit by Phoenix Flow Systems (San Diego, CA). Binary list mode files can be converted to ASCII format for entry into a spreadsheet program using one of several free software packages such as LLDATA or MFI, which can be downloaded at www.bio.umass.edu/mcbfcas. The steps that are performed to prepare the multiparameter data for analysis are as follows:

1. Perform appropriate signal crosstalk compensations on each cell if they were not already been performed in hardware.
2. For each fluorescent antibody measurement, subtract an appropriate nonspecific antibody binding correction from the total fluorescence signal of each cell (*see* **Note 13**).
3. Scale each cellular measurement either in molecules per cell when appropriate reference cells can be compared with the test cell population, or as multiples of some relative reference value, e.g., a multiple of the mean fluorescence value of concomitantly run normal lymphocytes.

3.6. Individual Cell Classifications

Each cell in the sample is classified as normal or abnormal with respect to ploidy, and with respect to its levels of p53, Her-2/neu, or ras protein expression.

3.6.1. Classification Based on Ploidy

1. The methanol-fixed samples are useful for the identification and classification of aneuploid cell populations. Distinguish between near-diploid aneuploidy, which generally involves isolated gains or losses of individual chromosomes, which is an early change that is of little prognostic value, and gross aneuploidy (hypotetraploidy, near-tetraploidy, hypertetraploidy), which can result from repeated rounds of endoreduplication and chromosome loss, and is associated with tumor progression *(29,30)*.
2. In the paraformaldehyde/methanol-fixed samples include cells that exhibit >4 N DNA content in the aneuploid cell fraction.

3.6.2. Classification Based on p53, Her-2/neu, and ras Cutoff Levels

1. Levels of p53 protein were previously determined in molecules per cell by ELISA in our reference cell line. Normal lymphocytes and other blood leukocytes generally exhibit less than 5,000 p53 molecules per cell *(11)*.
2. We use a cutoff of 10,000 molecules per cell in our studies. Others have used 9,000 p53 molecules per cell as a cutoff to distinguish normal from elevated levels.
3. Levels of Her-2/neu protein were previously determined in molecules per cell by ELISA in our reference cell line. Normal levels are generally below 100,000 molecules per cell, but we have estimated that immunohistochemical methods are generally capable of detecting levels as low as 200,000–500,000 molecules per cell *(5)*. In our hands, immunofluorescence methods are much more sensitive than immunohistochemical methods, and are capable of detecting Her-2/neu at levels as low as 20,000–50,000 molecules per cell, where sensitivity is limited largely by the magnitude of the correction for nonspecific antibody binding.
4. For purposes of comparability to immunohistochemical tests, we have adopted cutoffs ranging between 150,000 and 300,000 molecules per cell to distinguish Her-2/neu overexpressing cells from nonoverexpressing cells *(11)*.

5. Since we have not quantitated ras overexpression in molecules per cell, we use lymphocytes as a relative reference standard, and adopt a cutoff that is in the range of two- to fivefold greater than that of normal lymphocytes.

3.6.3. Classification of Cell Subpopulations

Cell subpopulations are classified by the patterns of intracellular abnormalities that they contain, and the frequencies of these cell subpopulations is calculated. From these frequencies, one can often reconstruct the order in which such abnormalities developed a particular tumor. Several examples using our analytical approach in breast cancer are described elsewhere *(11)*. Here we provide a synthetic example for illustrative purposes.

1. Suppose that the frequencies of occurrence of subpopulations exhibiting various patterns of intracellular abnormalities are as shown in **Table 1**.
2. This synthetic example includes a population of cells that do not contain any of the abnormalities measured, as is observed in most real tumor samples. Such cells could be admixed normal cells of various types, they could be tumor cells that contain abnormalities other than those measured, or they could be tumor cells in which measured abnormalities were not severe enough to meet the criteria for classification as abnormal.
3. In this example, p53 overexpression is found alone in individual cells, and one can, therefore, conclude that this is an abnormality that can occur early. Using a frequency of 1% as a cutoff, no other abnormality among those measured occurs alone in the cells of this sample. Thus, based on its relatively high frequency, the p53 overexpressing subpopulation is the most plausible precursor candidate for all cells in this sample that exhibit p53 overexpression plus additional abnormalities.
4. There are two subpopulations in this sample that exhibit two abnormalities per cell, and in both, p53 overexpression is present together with one other abnormality (either aneuploidy or Her-2/neu overexpression). Given that p53 overexpression occurred first, one can infer that either aneuploidy or Her-2/neu overexpression followed p53 in the evolutionary sequence.
5. There are two subpopulations in this sample that exhibit three abnormalities per cell, and in both, p53 overexpression is present together with two other abnormalities. Her-2/neu overexpression is generally observed in cells that overexpress p53, whether they are diploid or aneuploid. There is also a subpopulation that exhibits all four abnormalities in the same cells. Ras overexpression is observed predominantly in cell subpopulations that overexpress both p53 and Her-2/neu, but is only rarely observed in cells that express neither p53 nor Her-2/neu, or in cells that overexpress p53 alone. The most plausible evolutionary pathways in this sample, based on the observed frequencies of potential precursors in the sample, are shown in **Fig. 5**.
6. Other scenarios cannot be completely excluded, since there is the possibility that one or more precursor populations did not survive, or that they were not well-

Table 1
Frequencies of Occurence in Cell Subpopulations

Pattern of intracellular abnormalities	Frequency of cells in the sample (%)
No abnormalities	25%
↑p53 alone	10%
↑Her-2/neu alone	<1%
↑Ras alone	<1%
Aneuploidy alone	<1%
↑p53 + aneuploidy	15%
↑p53 + ↑Her-2/neu	10%
↑p53 + ↑ras	<1%
aneuploidy + ↑Her-2/neu	<1%
aneuploidy + ↑ras	<1%
↑Her-2/neu + ↑ras	<1%
↑p53 + aneuploidy + ↑Her-2/neu	15
↑p53 + ↑Her-2/neu + ↑ras	10%
aneuploidy + ↑Her-2/neu + ↑ras	<1%
↑p53 + aneuploidy + ↑Her-2/neu + ras	10%

represented in the portion of the tumor submitted for analysis. Our experience to date indicates that in almost all samples of human tumors that we have examined, all potential precursors for at least one evolutionary pathway can be found in each sample. At the same time, although converging alternative branches are often observed (as in the example given above), the number of plausible evolutionary pathways in a given tumor (based on frequencies of potential precursors), is limited. The most common convergent alternative branching patterns generally involved variability in the order of appearance of aneuploidy, as in the example given above (*see* **Note 14**).

4. Notes

1. One potential drawback to single cell analysis is the need to prepare fresh samples using methods that are not performed routinely in the pathology laboratory. As a practical matter, all studies must be done prospectively, at least at first. On the other hand, retrospective analyses performed on formalin–fixed paraffin embedded material have technical drawbacks that make it difficult to measure multiple cell constituents quantitatively, as well as statistical drawbacks that are inherent in retrospective study designs.

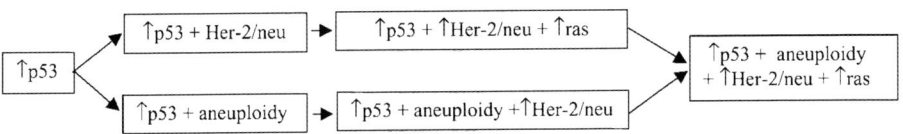

Fig. 5. Reconstruction of the genetic evolutionary sequences based on simulated data in a hypothetical tumor. For discussion, see text.

2. Most cellular constituents other than cell DNA content that might be used for event detection are present in variable amounts in cells. Thus, some cells with relevant biological information might escape detection because they fall below an arbitrary triggering threshold, while cell fragments that are of little biological interest might be included inadvertently. On the other hand, cell DNA content measurements almost always include a diploid G_1 peak, which provides a reproducible and relatively clean reference for triggering purposes. This measurement also provides additional information regarding S fraction and ploidy, which may be useful in tumor cell analysis.
3. We have found the use of methanol (or ethanol) to be superior to the use of detergents for permeabilization *(28)*, as the latter can disrupt cells, leading to unacceptable levels of cell loss.
4. The use of paraformaldehyde as a crosslinking fixative, even at low concentrations, can affect cellular DNA measurements *(28)*. The coefficient of variation of the G_1 peak is generally higher than with methanol or ethanol fixation, and S fraction calculations are not as accurate. Aneuploid cell subpopulations can still be identified and analyzed separately in multiparameter studies. We routinely fix a separate aliquot of cells from each tumor in 70% ethanol alone for definitive S fraction and ploidy analysis.
5. We have found that the levels of p53 and Her-2/neu immunofluorescence in paraformaldehyde/methanol-fixed normal lymphocytes and reference tumor cells stored at 4°C in methanol are stable for at least 3 yr when compared against fluorescent beads.
6. In order to obtain optimal estimates of antibody binding in units of molecules per cell by ELISA, both the antibody and the antigen (preferably a peptide) against which it was raised must be available, and the molar concentration of the latter must be known. Even then, the validity of the estimate of molecules per cell obtained by ELISA for immunofluorescence studies by flow cytometry hinges on the assumption that the number of antibody molecules binding to each molecule of antigen is comparable in both systems.
7. Fresh tissue/tumor samples must be transported to the laboratory in cold (4°C) HBSS on ice and processed immediately to prevent autolysis.

8. Avoid tight packing of the pellet during centrifugation, to minimize the formation of adherent cell clumps that cannot be dispersed by gentle vortexing.
9. All solutions containing phenol red (i.e., HBSS) should be washed out of the sample with PBS after scissor-mincing. Phenol red has very low-level fluorescence and may interfere with fluorescence measurements.
10. The Cytospin preparations are useful in establishing cytologically that samples analyzed by flow cytometry that contained diploid cells without any abnormalities among those measured, did in fact, contain diploid tumor cells.
11. Higher concentrations up to 10 µg/mL can be used depending on signal intensity. DAPI is saturating at high concentrations, but staining is nonlinear at concentrations below 0.1 µg/mL.
12. Cell DNA content measurements are commonly collected in the linear domain. We generally adjust gain settings for DNA such that the G_1 peak is positioned above channel 50 (to minimize binning effects) and well below 25% of total scale e.g., channel 100 out of 1024, in order to adequately visualize not only G_2M and tetraploid populations (whose G_2M peaks would be centered at channel 400, but hypertetraploid populations as well (which could extend to channel 800 and beyond).

    For many other cell constituents the dynamic range of the measurements is large, and it is often just as important to distinguish differences among small values as it is to distinguish differences among large values. For example, it would be useful to distinguish normally low intracellular levels of p53 protein (which are readily detected by immunofluorescent methods) from slightly elevated intracellular p53 levels as well as from intracellular levels of p53 that may be 10–20 times greater. Such differences are best appreciated when data are collected using a log amplifier, and/or displayed on a log scale. In our experience, four log cycles are ample for most biological measurements of interest, but care should be taken to center the region of interest on scale (by adjusting the amplifier gain settings and/or photomultiplier tube voltage), so that there is little if any binning of data in either the first or last channel. Of course, when gain settings are changed for the test sample, quantitative reference cell measurements should also be obtained at the new settings as well.
13. Mathematical operations on data collected in the log domain must be performed with care. Ratios (e.g., test values/normal values) are obtained by subtracting log values. To perform operations involving subtraction (e.g., correcting for nonspecific antibody binding) log data must be transformed into the linear domain prior to subtraction, and transformed back into the log domain afterward. Caveat: Mathematical manipulation of binned data can introduce artifacts, and the subtraction of a fixed average value for nonspecific fluorescence from each cell without regard for individual cell size may have adverse effects in heterogeneous cell population.
14. One commonly observed variant pathway included the Her-2/neu overexpression-> ras overexpression sequence, with or without aneuploidy, in tumors that did not exhibit p53 overexpression. In many of these tumors, mostly infiltrating

ductal cancers, FISH studies demonstrated p53 allelic loss *(8,11)*. In aneuploid tumors the p53 allelic loss was commonly present both in cells that were monosomic and disomic for chromosome 17 (and may have been diploid or near diploid), and in cells that had multiple copies of chromosome 17 (and were clearly aneuploid). These findings suggest that tumor evolution may be driven by early genotypic changes that result in a p53 dysfunctional state that is not accompanied by p53 protein overexpression. Combined cellular FISH and protein immunofluorescence measurements can be contemplated as more feasible by image analysis techniques than by flow cytometry.

References

1. Pauletti, G., Dandekar, S., Rong, H., Ramos, L., Peng, H., Seshadri, R., and Slamon, D. (2000) Assessment of methods for tissue-based detection of the HER-2/*neu* alteration in human breast cancer: a direct comparison of fluorescence in situ hybridization and immunohistochemistry. *J. Clin. Oncol.* **18,** 3651–3664.
2. Slamon, D., Clark, G., Wong, S., Levin, W., Ullrich, A., and McGuire, W. (1987) Human breast cancer: Correlation of relapse and survival with amplification of the HER-2/neu oncogene. *Science* **235,** 177–182.
3. Tsuda, H., Fukutomi, T., and Hirohashi, S. (1995) Pattern of gene alterations in intraductal breast neoplasms associated with histological type and grade. *Clin. Cancer Res.* **1,** 261–267.
4. Courjal, F., Cuny, M., Simony-Lafontaine, J., Louason, G., Speiser, P., Zeillinger, R., et al. (1997) Mapping of DNA amplifications at 15 chromosomal localizations in 1875 breast tumors: definition of phenotypic groups. *Cancer Res.* **57,** 4360–4367.
5. Shackney, S. and Shankey, T. (1997) Common patterns of genetic evolution in human solid tumors. *Cytometry* **29,** 1–27.
6. Hovey, R., Chu, L., Balasz, M., DeVries, S., Moore, D., Sauter, G., et al. (1998) Genetic alterations in primary bladder cancers and their metastases. *Cancer Res.* **58,** 3555–3560.
7. Lengauer, C., Kinzler, K., and Vogelstein, B. (1997) Genetic instability in colorectal cancers. *Nature* **386,** 623–627.
8. Janocko, L., Brown, K., Smith, C., Gu, L., Pollice, A., Singh, S., et al. (2001) Distinctive patterns of Her-2/neu, c-myc, and cyclin D1 gene amplification by fluorescence in situ hybridization (FISH) in primary human breast cancers. *Cytometry (Communications in Clinical Cytometry)* **148,** 136–149.
9. Shackney, S., Pollice, A., Smith, C., Alston, L., Singh, S., Janocko, L., et al. (1996) The accumulation of multiple genetic abnormalities in individual tumor cells in human breast cancers: Clinical prognostic implications. *Cancer J.* **2,** 106–113.
10. Shackney, S., Pollice, A., Smith, C., Janocko, L., Sweeney, L., Brown, K., et al. (1998) Intracellular coexpression of epidermal growth factor receptor, Her-2/neu, and p21^{ras} in human breast cancers: evidence for the existence of distinctive patterns of genetic evolution that are common to tumors from different patients. *Clin. Cancer Res.* **4,** 913–928.

11. Smith, C., Pollice, A., Gu, L.-P., Brown, K., Singh, S., Janocko, L., et al. (2000) Correlations among p53, Her-2/neu, and ras overexpression and aneuploidy by multiparameter flow cytometry in human breast cancer: evidence for a common phenotypic evolutionary pattern in infiltrating ductal carcinomas. *Clin. Cancer Res.* **6**, 112–126.
12. Fearon, E. and Vogelstein, A. (1990) A genetic model for colorectal tumorigenesis. *Cell* **61**, 759–767.
13. Ried, T., Heselmeyer-Haddad, K., Blegen, H., Schrock, E., and Auer, G. (1999) Genomic changes defining the genesis, progression, and malignancy potential in solid human tumors: a phenotype/genotype correlation. *Genes Chromosomes Cancer* **25**, 195–204.
14. Galipeau, P., Cowan, D., Sanchez, C., Barrett, M., Emond, M., Levine, D., et al. (1996) 17P (p53) allelic losses, 4N (G2/tetraploid) populations and progression to aneuploidy in Barrett's esophagus. *Proc. Natl. Acad. Sci. USA*, **93**, 7081–7084.
15. Blount, P., Galipeau, P., Sanchez, C., Neshat, K., Levine, D., Yin, J., et al. (1994) 17p allelic losses in diploid cells of patients with Barrett's esophagus who develop aneuploidy. *Cancer Res.* **54**, 2292–2295.
16. Barrett, M., Sanchez, C., Prevo, L., Wong, D., Galipeau, P., Paulson, T., et al. (1999) Evolution of neoplastic cell lineages in Barrett Oesophagus. *Nature Genet.* **22**, 106–109.
17. Carder, P., Wyllie, A., Purdie, C., Morris, R., White, S., Piris, J., and Bird, C. (1993) Stabilized p53 facilitates aneuploid clonal divergence in colorectal cancer. *Oncogene* **8**, 1397–1401.
18. Carder, P., Cripps, K., Morris, R., Collins, S., White, S., Bird, C., and Wyllie, A. (1995) Mutation of the p53 gene precedes aneuploid clonal divergence in colorectal carcinoma. *Br. J. Cancer* **71**, 215–218.
19. Wistuba, I., Behrens, C., Milchgrub, S., Bryant, D., Hung, J., Minna, J., and Gazdar, A. (1999) Sequential molecular abnormalities are involved in the multistage development of squamous cell lung carcinoma. *Oncogene* **18**, 643–650.
20. Hung, J., Kishimoto, Y., Sugio, K., Virmani, A., McIntire, D., Minna, J., and Gazdar, A. (1995) Allele-specific chromosome 3p deletions occur at an early stage in the pathogenesis of lung carcinoma. *JAMA* **273**, 558–563.
21. Kishimoto, Y., Sugio, K., Hunbg, J., Virmani, A., McIntire, D., Minna, J., and Gazdar, A. (1995) allele-specific loss in chromosone 9p loci in preneoplastic lesions accompanying non-small-cell lung Cancers. *J. Nat. Cancer Insti.* **87**, 1224–1229.
22. Belinsky, S., Nikula, K., Palmisano, W., Michels, R., Saccomano, G., Gabrielson, E., et al. (1998) Aberrant methylation of p16[INK4a] is an early event in lung cancer and a potential biomarker for early diagnosis. *Proc. Natl. Acad. Sci. USA* **95**, 11891–11896.
23. Chung, G., Sundaresan, V., Hasleton, P., Rudd, R., Taylor, R., and Rabbitts, P. (1995) Sequential molecular genetic changes in lung cancer development. *Oncogene* **11**, 2591–2598.

24. Shackney, S., Smith, C., Pollice, A., Levitt, M., Magovern, J., Wiechmann, R., et al. (1999) Genetic evolutionary staging of early non-small cell lung cancer: the p53 -> HER-2/NEU -> Ras sequence. *J. Thorac. Cardiovasc. Surg.* **118,** 259–269.
25. Livingstone, L., White, A., Sprouse, J., Livanos, E., Jacks, T., and Tlsty, T. (1992) Altered cell cycle arrest and gene amplification potential accompany loss of wild-type p53. *Cell* **70,** 923–935.
26. Theillet, C., Lidereau, R., Escot, C., Hutzell, P., Brunet, M., Gest, J., et al. (1986) Loss of a c-H-ras-1 allele and aggressive human primary breast carcinomas. *Cancer Res.* **46,** 4776–4781.
27. Rochlitz, C., Scott, G., Dodson, J., Liu, E., Dollbaum, C., Smith, H., and Benz, C. (1989) Incidence of activating ras oncogene mutations associated with primary and metastatic human breast cancer. *Cancer Res.* **49,** 357–360.
28. Pollice, A., McCoy, J., J. P., Shackney, S., Smith, C., Agarwal, J., Burholt, D., et al. (1992) Sequential paraformaldehyde and methanol fixation for simultaneous flow cytometric analysis of DNA, cell surface proteins, and intracellular proteins. *Cytometry* **13,** 432–444.
29. Shackney, S., Berg G., Simon S., Cohen J., Amina S., Pommersheim W., et al. (1995) Origins and clinical implications of aneuploidy in early bladder cancer. *Cytometry* **22,** 307–316.
30. Shackney, S., Singh, S., Yakulis, R., Smith, C., Pollice, A., Petruolo, S., Waggoner, A., and Hartsock, R. (1995) Aneuploidy in breast cancer: a fluorescence in situ hybridization study. *Cytometry* **22,** 282–291.

III

NOVEL THERAPIES

A. INTRODUCTION TO CURRENT TREATMENTS

28

Surgical Treatment of Lung Cancer

Past And Present

Clifton F. Mountain and Kay E. Hermes

1. Introduction

> "The past is never past, but continues and is very active in every form and at every manifestation of the present." *(1)*

From the fortuitous beginning of surgery of the chest in the 15th century to the scientific discoveries of the 17th–19th centuries, a body of knowledge was acquired that enabled the "rise of surgery from empiric craft to scientific discipline" *(2)*. The evolution of general surgical practice and derivation of thoracic specialty practiced, was interwoven with the advancement of science in other fields. Early discoveries in anatomy and physiology, of percussion and auscultation, of Roentgen's X-rays, the introduction of bronchoscopy, the development of anesthesia, and understanding the causes of infection and antisepsis were building blocks for progress in the surgical treatment of thoracic disease. This progress, as described by Naef *(3)*, was the "end result of heroic efforts by pioneers in the field of surgery who persisted in their quest to provide a surgical solution for the difficult problems of intrathoracic disease." The time line (*see* **Fig. 1**) reflects the development of safe thoracotomy and procedures for resection of lung cancer *(4)*.

Prior to 1930, the diagnosis of lung cancer was a rare event; thus, the main emphasis in chest radiography and the proving ground for lung resection techniques were for the treatment of tuberculosis, bronchiectasis, and lung abscess. By the mid-1930s, however, a rapid increase in cases of lung cancer was observed, not only in the total number of cases, but out of proportion

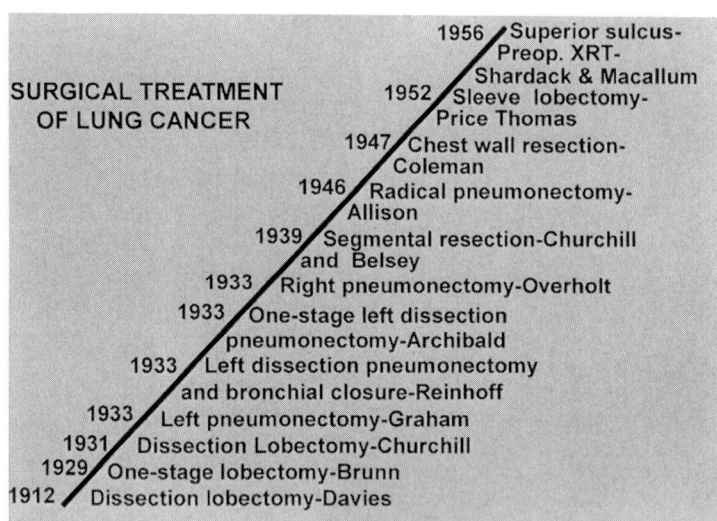

Fig. 1. Evolution of surgical treatment for lung cancer. Adapted with permission from **ref. 4**.

to cancer in general. Following the introduction of penicillin (1940) and streptomycin (1944) for infectious lung disease, surgery was no longer the primary treatment option. Now the major focus of chest surgery was cancer of the lung. Successful reports of "first" lung resections for nonmalignant conditions and the lack of other effective therapy confirmed the need for a surgical approach to the treatment of lung malignancy.

A first step toward organization of Thoracic Surgery as a new specialty was made with the creation of specific units for the care of patients with chest injuries during World War II *(5)*. Increasing knowledge of the biology of lung cancer, integrated with expectations of end results, provided a new rationale for selecting patients for surgical treatment. Discovery of the presence of occult metastatic disease, in patients thought to have had curative resections, gave impetus to studies of multimodality therapies—initially with radiotherapy, and subsequently with chemotherapy and combined programs. Today, in the 21st century, the multimodality treatment concept unites technical expertise in lung surgery with advances in biomedicine. As we examine the past and present approach to surgical treatment of lung cancer from the perspective of its history, we are reminded that "Medical history teaches us where we come from, where we stand in medicine at the present time, and in what direction we are marching" *(6)*.

2. The Proving Ground: 1880–1928

Lung cancer was diagnosed primarily at autopsy and rarely in life prior to 1900. Physicians had little with which to detect and pursue a diagnosis of chest disease. Auenbrugger described percussion in 1761 in his small book, *Inventum Novum (7)*; however, the work was only appreciated 15–20 years later. Laennec revolutionized the diagnosis of chest disease through his discovery of the stethoscope and his description of the system of auscultation *(8)*. He authored the first comprehensive discourse on the pathology of lung cancer, included in the *Dictionaire des Sciences Medicales* in 1815 under the title, *Enchephaloides (9)*.

Andral *(10)*, in 1821, suggested the examination of sputum for diagnosis of tumor cells—a technique recommended by others *(11)* from 1866 to the present. Seventy-five years after Andral's suggestion, the technique was standardized and popularized by Papanicolaou *(12)*. With the discovery of the Roentgen ray by Roentgen in 1895 *(13)*, the miracle of examining the interior of the chest cavity became a reality. Soon after publication of the discovery, the medical applications of the X-ray were recognized. Over the ensuing century the science of diagnostic and therapeutic radiology progressed rapidly to become the cornerstone of management for lung cancer. The development of bronchoscopy by Killian *(14)* in 1897, and further established by Jackson *(15)*, greatly improved clinical diagnostic capability (*see* **Fig. 2**) *(16)*.

3. Pioneers in Lung Resection

Empyema and tuberculosis were the "proving ground" for the development of techniques for lung resection from 1892–1928. The early resections employed cautery and the use of crude hilar ligation and two-stage resection. The operations were mostly unsuccessful; mortality was high due to bronchopleural fistula and infection.

The present report does not address the history of ventilatory management of intrathoracic disease. It should be noted, however, that routine operations within the thoracic cavity could only be developed after the introduction of endotracheal anesthesia in 1911.

In 1912 Morriston Davies *(17)* performed a landmark lobectomy for lung cancer using a revolutionary new dissection technique, remarkable in its similarity to present-day surgical practice. However, it received little attention and was not used again for many years—probably because lung cancer remained a rare disease for the next decade and the time-consuming nature of the procedure. In 1912, speed was an essential component of intrathoracic operation.

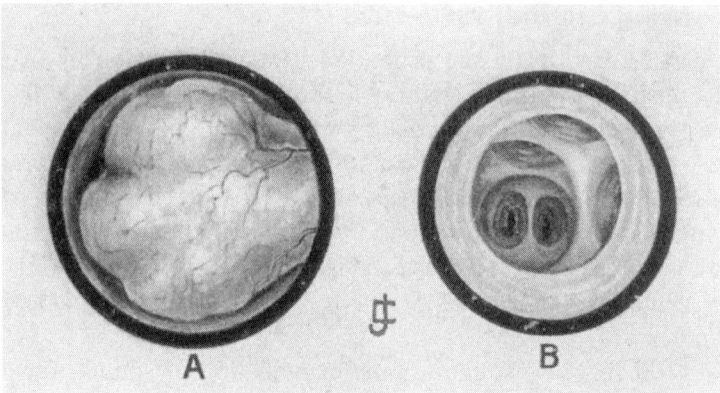

Fig. 2. Endobronchial obstructing endothelial tumor in a man, aged thirty-five yr, who complained of coughing, wheezing, and "a sensation of a ball valve shutting off his breath, sometimes on inspiratin and at other times on expiration. **(A)** Tumor presenting itself in its self-made bronchial enlargement, when the bronchoscopic tube-mouth reached the location at which **(B)** should appear. Tumor removed with forceps through a bronchoscope passed through the mouth. Patient free from symptoms and remained perfectly well a year and a half later. Adapted with permission from **ref. 16**.

4. Beginning of the Modern Era: 1929–1933
4.1. Lobectomy Techniques

In 1929, when most surgeons believed that multistage procedures were necessary for lung resection, Brunn *(18)* published a detailed, systematic description of a one-stage lobectomy and confirmed the safety of the procedure. Brunn used the "right angle Wertheim hysterectomy clamp" to clamp and suture-ligate the lobar pedicle. He emphasized the importance of obtaining early expansion of the remaining lung by immediate air-tight closure of the chest wall aided by the use of suction (*see* **Fig. 3**).

In 1932, Shenstone and Janes *(19)* emphasized the great importance of preservation of the blood supply to the divided bronchus. They reported the tourniquet technique, by a modified tonsillar snare, to occlude the hilum but not crush it. This technique provided operative speed, prevented the spilling of copious sputum, and simplified the problem of hemostasis, providing the surgeon with a feeling of security.

4.2. Successful Total Pneumonectomy

An important milestone in the evolution of lung surgery was reached in 1931 when Nissen *(20)* performed the first recorded successful total pneumonectomy

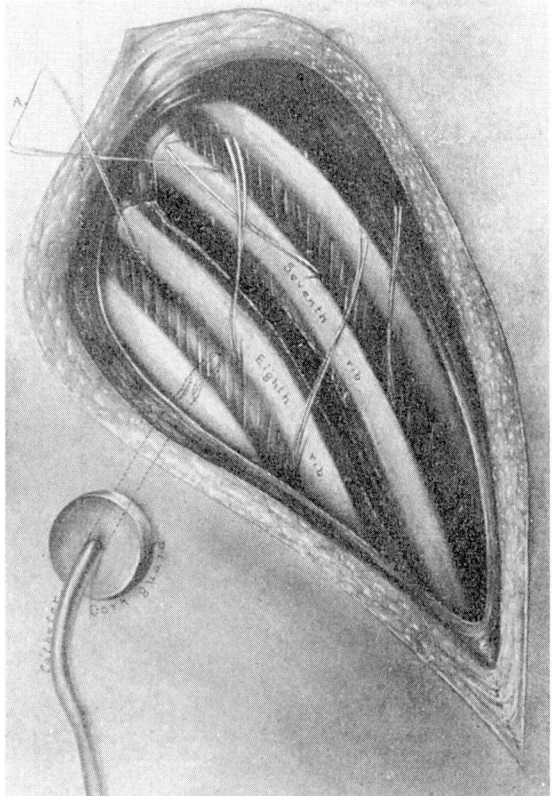

Fig. 3. Method of wound closure, the ribs being brought together. The muscles are sewn in layers and the wound is airtight. Adapted with permission from **ref. 18**.

for bronchiectasis in a 12-year-old girl. With this operation, and quite by accident, the surgeon's main concern with pneumonectomy was resolved: that sudden ligation of a major pulmonary artery would produce the clinical picture of massive pulmonary embolism. Nissen noted that, due to a temporary asystole, the first stage of the operation was discontinued before the hilum vessels were clamped. When the second stage of the operation with removal of the lung was performed, the outstanding feature of the procedure was the absence of acute symptoms following mass ligation of the hilum vessels. This was in marked contrast to the expectations in pulmonary embolism.

The second successful pneumonectomy was performed by Haight *(21)* in 1933. The patient was a 13-yr-old girl with diffuse bronchiectasis of the left lung and the procedure was the first total removal of a left lung in the United States, the second in the world. In 1949 Nissen *(20)* stated, prophetically,

Fig. 4. Evarts Graham, MD performed the first one-stage pneumonectomy for lung cancer at Barnes Hospital, April 5, 1933. Adapted with permission from **ref. 23**.

"When Haight and I ventured to remove a whole lung, it was our hope that surgical treatment of bronchogenic carcinoma might one day benefit from these experiences"—a hope that was soon realized.

5. Surgical Treatment for Lung Cancer—a Reality: 1933

In 1933, the general assumption was that only a total pneumonectomy could offer a cure for lung cancer. The idea persisted even though all patients undergoing pneumonectomy for this disease up to this point had died within 8 days of surgery. On April 5, 1933, Everts A. Graham *(22)* performed the first one-stage pneumonectomy for cancer, a carcinoma of the left upper lobe bronchus. He established surgical treatment as the standard for what had previously been regarded as a hopelessly incurable disease (*see* **Fig. 4**) *(23)*. The "proven" hilar tourniquet amputation method and endotracheal anesthesia with nitrous oxide and oxygen were used. Although the resection was initially planned as a lobectomy, Graham's account of the procedure noted,

"At the operation, however, it was possible to palpate the tumor; and I discovered that apparently it extended down into the bronchus of the lower lobe. It seemed certain, therefore, that the removal of only the upper lobe would not result in the removal of all the cancer. Moreover, a lobectomy presented some technical difficulties, because there was almost a complete absence of an interlobar fissure.

It occurred to me at once, therefore, that the only kind of an operation that might offer a chance of cure would be the total removal of the left lung. Because I was certain the patient would want me to take any chances that might effect a cure, I decided to go ahead with the total pneumonectomy."

The original operative note reveals that "7 radon seeds of 1.5 millicuries each were inserted into various parts of the stump." Pathological examination of the resected specimen showed a "1 cm × 2 cm. primary squamous cell carcinoma occluding a branch bronchus to the upper lobe of the left lung. One pulmonary hilar node contained metastatic carcinoma. Multiple mediastinal nodes did not show metastatic involvement." The patient, Dr. James L. Gilmore, was still alive in 1957, and still smoking, 24 years after this operation, at the time of Dr. Graham's death from lung cancer. Today, the anatomic classification of this case would be pStage IIB-T2 N1 M0, with an expected 40% probability of 5-yr survival.

The importance of this operative milestone was reflected in the comment of Sir Russell Brock *(24)* at Dr. Graham's memorial service (3/31/57),

"Those of us who were struggling with the surgery of bronchial carcinoma in the early 1930's remember only too well the apparent hopelessness of the problem facing the surgeon. Then, like a great light, shone out the report of Graham's success in removing the whole lung in a case of cancer ... it was the essential catalyst that set in train the development of the modern treatment of lung cancer by operation, still the most satisfactory method available...."

6. The Final Chapter: 1933–1956

Historic breakthroughs in the dissection of the hilar structures and bronchial closure, "the final chapter in modern pulmonary resection," *(25)* took place soon after Graham's publication. Although the technique of dissection lobectomy for lung cancer was reported earlier by Davies *(17)* and Churchill *(26)*, anesthetic practice at the time required that intrathoracic operations should be performed rapidly. The time-consuming dissection technique was not established or widely accepted until the anatomical dissection studies of Blades and Kent in 1940 *(27)*.

Reinhoff *(28)*, a general surgeon who was apparently unaware of previous literature, applied general surgical principles to accomplish a major breakthrough—the modern technique of dissection pneumonectomy. He reported two successful dissection pneumonectomies using a sophisticated technique

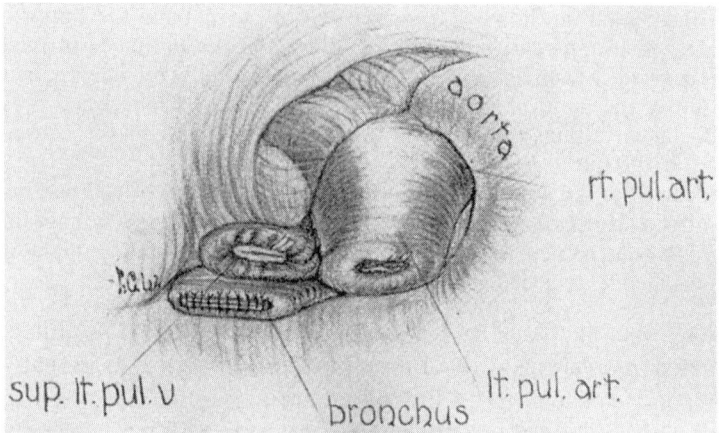

Fig. 5. Stump of the hilus of the left lung, which has been dissected, with the vessels and the bronchus ligated individually. The bronchus is closed with interrupted silk sutures. This type of closure with simple ligation of the artery and vein, is considered the method of choice. Sup. Lt. Pul. V, superior left pulmonary vein; lt. pul. art., left pulmonary artery; rt. pul. art., right pulmonary artery. Adapted with permission from **ref. 28**.

for bronchial closure, "cutting the cartilage at various points to do away with their spring-like action" and "sutured the bronchus with interrupted medium silk sutures" (*see* **Fig. 5**). These procedures, one for a child with a benign tumor and the other for a carcinoma of the left main bronchus were reported two months after Graham's landmark procedure.

Archibald *(29)*, who emphasized the differences in the technical approaches for bronchiectasis or tumor, performed a one-stage dissection pneumonectomy in July, 1933. The procedure was preceded by an "exploratory thoracotomy" to exclude the possibility of pleural or mediastinal metastasis. Overholt *(30)* accomplished the first successful right pneumonectomy for carcinoma of the lung in November, 1933.

6.1. The Concept of Limited Resection

Progress in the surgical management of tuberculosis provided new operative techniques for pulmonary resections. Churchill and Belsey *(31)* first described segmental resection in 1939. The procedure was for removal of the lingular segment of the left upper lobe, an area often involved in patients with bronchiectasis. Later work of Overholt *(32)* was responsible for the procedure's acceptance as an adequate resection for bronchiectasis—and also the proposal that the "bronchopulmonary segment may replace the lobe as the surgical unit of the lung."

The technique was later advocated for use in cancer patients with operable disease whose pulmonary reserve status precluded lobectomy. Leroux *(33)* reported that segmentectomy for treatment of bronchial carcinoma, as an alternative to no surgery, is followed by survival and recurrence rates similar to the results for patients with tumors of the same size undergoing more extensive resections. Segmentectomy as an adequate resection for lung cancer was suggested by Jensik and colleagues *(34)* in 1973. Other reports of "compromise" procedures for patients with stage I disease and limited pulmonary reserve led to continued controversy regarding indications for segmentectomy. The issue was resolved with a prospective, randomized clinical trial, conducted by the National Cancer Institute (NCI) North American Lung Cancer Study Group (LCSG) *(35)*. Significantly higher recurrence rates and death rates were observed in patients undergoing limited resection compared to those undergoing lobectomy. These results confirmed lobectomy as the surgical procedure of choice for patients with small peripheral tumors, T1 N0 M0 non-small cell lung cancer. In the treatment of synchronous or metachronous multiple primary lung tumors, however, segmentectomy may be the only surgical option.

6.2. Pneumonectomy and Lobectomy

Although pneumonectomy and lobectomy were performed regularly by 1940 for bronchial carcinoma, the operability rate remained low and the recurrence rate high. Graham *(36)* remained convinced, however that pneumonectomy was the procedure of choice for primary lung tumors,

> "Any reported successful results from radiation therapy are to be regarded with suspicion ... A bronchoscopic biopsy showing the presence of a carcinoma is to be regarded as an indication for pneumonectomy."

He reviewed his experience from 1939, the most recent 5-yr period, and reported an operative mortality of 30% (21/70) in 70 pneumonectomies for malignant neoplasms. In the most recent 25 cases, however, only 3 deaths occurred, 12%. He felt that these figures dispelled the all too common idea that total pneumonectomy for bronchiogenic carcinoma carried an enormous operative risk, and noted that it should be remembered that without operation the risk is 100%. He advised exploratory thoracotomy when a definitive diagnosis could not be ascertained preoperatively.

6.3. The Concept of Radical Pneumonectomy

Refinements in techniques for pneumonectomy were made in an effort to improve resectability and survival rates, and to adapt the method for specific types of tumor growth. The importance of the lymphatic route of metastasis in relation to the rationale of the surgical approach was described by Ochsner and DeBakey *(37)* in 1942.

A technique of individual ligation of the pulmonary vessels after wide opening of the pericardium was described by Allison *(38)* in 1946. This procedure provided a more radical approach for the treatment of central tumors, previously considered inoperable, that were located between hilar structures, or with tumor or glands adherent to pulmonary vessels or pericardium. Following this report attention was given to the block dissection of cancerous lymph nodes; variations of pneumonectomy and lymph node dissection were used and advocated. Block dissection pneumonectomy with intrapericardial exposure and ligation of the vessels in every case was recommended by Brock and Whytehead *(39)*. In response to confusion of terms regarding pneumonectomy procedures, Cahan *(40)*, suggested the following standardized definitions: 1) Simple pneumonectomy is the excision of the lung alone; 2) A radical pneumonectomy is the excision of the lung in continuity with its regional lymph nodes located in the hilar and mediastinal areas. The radical approach was in keeping with generally accepted principles of cancer surgery in other parts of the body. In 1951 Alywin *(41)* noted that the extended procedure had not provided the expected improvement in survival rates—only about one in 10 patients achieved 5-yr survival. The intrapericardial procedure provided substantial reduction in the recurrence rates, 84% following extrapericardial procedures and 42% after the intrapericardial resections. He recommended modifying the "radical pneumonectomy procedure by tying the veins inside the pericardium before manipulating the tumors and in this way inhibit tumor cell contamination of the general circulation."

It was an attractive thesis: That results might be improved by getting the last lymph node involved by lymphatic spread. The philosophy and techniques of radical pneumonectomy were widely advocated and adopted. An excellent film of the procedure, distributed by the American Cancer Society, implied that doing less than a radical pneumonectomy was not to do the best to cure patients with lung cancer *(42)*.

6.4. The Controversies

Johnson et al. *(42)* noted that reported differences in survival rates according to surgical procedure could result from patient selection; that is, good risk patients were subjected to radical pneumonectomy. A review of their own experience, and of results in the literature, led these authors to challenge the tenet that only radical pneumonectomy could provide superior results. (*see* **Table 1**) *(42,43)*. They concluded "until it can be shown that radical

Table 1
Five-Year Survival in Lung Cancer

Resection	Number	% Operative Mortality	% 5-Year Survival
Watson[a]			
Radical pneumonectomy	74	5.5	27
Simple pneumonectomy	26	7.7	23
Lobectomy	16	18.7	25
Totals	116	7.7	26
Johnson, Kirby, and Blakemore[b]			
Pneumonectomy (standard)	99	7.0	26.4
Less than pneumonectomy	17	13.0	35.0

[a]Adapted with permission from **ref. 43**.
[b]Adapted with permission from **ref. 42**.

pneumonectomy produces superior results, the thoracic surgeon is not obligated to perform the radical operation."

6.5. "Stage" of Disease, Localized and Nonlocalized

Although in the early 1960s it was generally agreed that surgical resection was the only treatment with the potential for cure of lung cancer, controversy continued regarding the optimum type of resection. Some surgeons advocated radical pneumonectomy with *en bloc* mediastinal lymph node dissection, and others recommended lesser resections, such as lobectomy with lymph node dissection as permitted by the exposure.

Shimkin et al. *(44)*, in 1962, used data contributed from the Ochsner clinic (where pneumonectomy was selected in all patients who could tolerate the procedure) and the Overholt Clinic (where lobectomy was selected as the treatment of choice) to examine the impact of the type of resection on survival (*see* **Table 2**). The three groups of patients (Overholt lobectomy, Overholt pneumonectomy and Ochsner pneumonectomy) were classified in two "stage" categories: 1) "localized", tumor confined to the lung or bronchus with no direct extension into neighboring tissues or 2) "nonlocalized" with regional lymph node metastasis or extension of the tumor beyond the lung or, in some cases, distant spread. Analysis of all patients according to the three resection groups, and according to the localized and nonlocalized categories, revealed nearly identical survival rates for all three resection groups in the localized

Table 2
Five-Year Survival Rates by Stage of Disease and Operation: Surgically Treated Lung Cancer Patients (Ochsner Clinic, 1948–1956; Overholt Clinic, 1951–1956)[a]

Stage of Disease	Lobectomy (Overholt)			Pneumonectomy (Overholt)			Pneumonectomy (Ochsner)		
	No.	%	5-Yr survival rate %	No.	%	5-Yr survival rate %	No.	%	5-Yr survival rate %
Localized	60	52	40	45	21	35	49	26	39
Not localized	56	48	12	166	79	15	142	74	7
Regional	(48)	(41)	(14)	(143)	(68)	(16)	(138)	(72)	(7)
Distant	(8)	(7)	(0)	(23)	(11)	(0)	(4)	(2)	(0)
Total, all stages	116	100	27	211	100	19	191	100	15

[a]Adapted with permission from **ref. 44**.

category (*see* **Fig. 6**) *(44)*. The authors concluded that survival after surgical resection was primarily determined by the extent of disease—survival in patients with localized cancer was similar regardless of the type of resection. Thus, it was shown that "less extensive operations do not save more patients but more extensive operations increase mortality and do not improve total survival figures." In the ensuing years the clinical experience with lobectomy greatly increased and this operation became the most frequently planned lung resection.

6.6. Introduction of Surgical Staplers

Rubio notes that the principle of mechanical suturing was practiced in ancient India by Shushruta (circa 800–600 BC) who used black ants to close traumatic intestinal tears. After the ants had bitten into the two edges of the wound, their heads were cut off and their powerful jaws were left in place as a suture *(45)*. The first stapling instrument was developed in Budapest by Hutl, a surgeon, along with an engineer, his brother, and a manufacturer of surgical instruments (1908). His interest in creating a mechanical suture was to reduce operative contamination to a minimum from an open gastric or intestinal lumen, and the first use of his device was for distal gastrectomy *(46)*. Although it was heavy and cumbersome, the instrument included all the principles of the modern stapler: 1) closure of the instrument immobilized the tissue, 2) fine wire staples were driven into the tissue in four rows, and 3) the viscus was cut, leaving two rows of staples on either side *(47–48)*. Modifications of the Hutl instrument were made by vonPetz in 1921, and others, that made it easier to

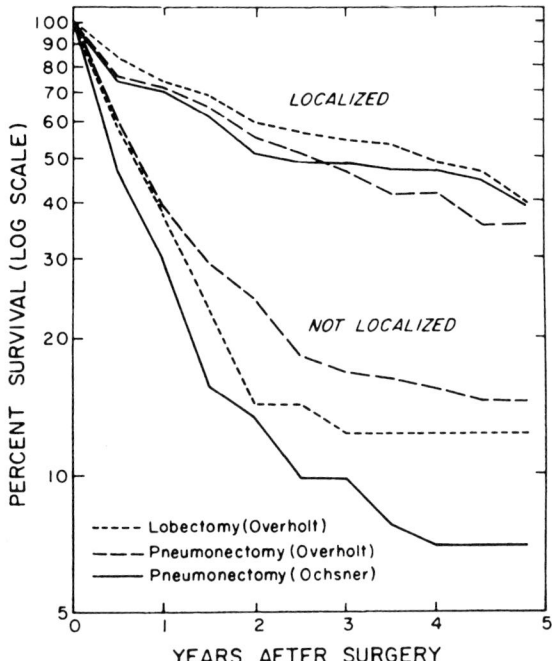

Fig. 6. Cumulative survival curves by stage of disease and operation seen in surgically treated lung cancer patients. Ochsner Clinic, 1948–1956; Overholt Clinic, 1951–1956. Adapted with permission from **ref. 44**.

handle. In the 1940s the interest and accomplishments of Russian surgeons and scientists were directed toward providing mechanical sutures for standardization of techniques for surgeons working in the vast territories of the Soviet Union. They developed linear and circular suturing and anastomosing instruments and applied linear staplers to pulmonary surgery with significant improvements in the safety and reliability of bronchial and pulmonary vascular closures. Ravitch (47) recorded his encounter with the Russian instruments and his subsequent introduction of them in the US as follows: ". . . as guests of Dr. N. M. Amosov at the Thoracic Surgical Institute, we were startled to see a series of patients who had segmental resections, lobectomies and pneumonectomies with staples instead of sutures, and saw the extraordinary simplicity and efficiency of the instruments in Dr. Amosov's hands during operations. We visited the Institute where the instruments had been developed and saw the large range of instruments they had developed for all purposes." Although the instruments were not available for sale at the institute, Ravitch managed to purchase a bronchial stapler, have it calibrated, and learned its operation and maintenance. When he returned to Baltimore, he used the instrument on a series of dogs and subsequently performed a series of pulmonary resections, mostly for tuberculosis. He persuaded American manufacturers to secure the right to reproduce or modify the Russian instruments and encouraged them to provide disposable preloaded, presterilized, and color-coded staple cartridges; the moving parts within the instrument also were a part of the disposable cartridge (United States Surgical Corporation). In 1967, when the refinements in American instruments became available, surgical stapling was revolutionized.

The first surgeons in the US to adopt stapling enthusiastically were the thoracic surgeons, who immediately saw the increase in facility and safety and the improvement in results with closures of the bronchus, parenchyma, and blood vessels—perfect suture lines, decreased anesthesia time, shortened hospitalization, and most important, quicker recovery with a decreased financial burden. The incidence of bronchopleural fistula, one of the most severe complications of lung operations, particularly after pneumonectomy, was reduced from a high of 7–18% to as low as 2–5% with the use of stapling devices *(49)*. Surgical stapling in parenchymal resections simplified the procedures and contributed to the development of new approaches; for example, the growth of video-assisted thoracic surgery. Today surgical stapling has become standard practice. **Figure 7** is a diagrammatic illustration of the use of surgical stapling in pneumonectomy *(45)*.

7. The Problem of Fruitless Thoracotomy: The Mid-20th Century

Problems of unresectability and lack of improvement in the end results of resection for lung cancer drew attention to more careful selection of patients. Greater appreciation of the anatomy and routes of lymph node metastasis provided for the introduction of mediastinoscopy—a technique for identifying the local spread of disease.

Mediastinoscopy, as introduced by Carlens *(50)* in 1959, permits direct surgical exploration and biopsy of the middle and posterior portions of the superior mediastinum (*see* **Fig. 8**). Its primary indications are: 1) for definitive diagnosis, when conventional tests have failed, and 2) to assess resectability in patients with bronchogenic carcinoma. The safety and efficacy of mediastinoscopy was confirmed by Sarin and Nohl-Oser *(51)* and others *(52)*.

Refinements in computed tomography (CT), magnetic resonance imaging (MRI), and positron emission tomography (PET) enable selective use of mediastinoscopy; however, in clinical trials populations, mediastinoscopy remains a priority examination to confirm the extent of disease. The findings from mediastinoscopy led to intense study and controversy over the concept of surgical treatment in the presence of mediastinal lymph node metastasis. The following conventional criteria of unresectability were generally accepted: 1) a diagnosis of small cell carcinoma, 2) involvement of the contralateral

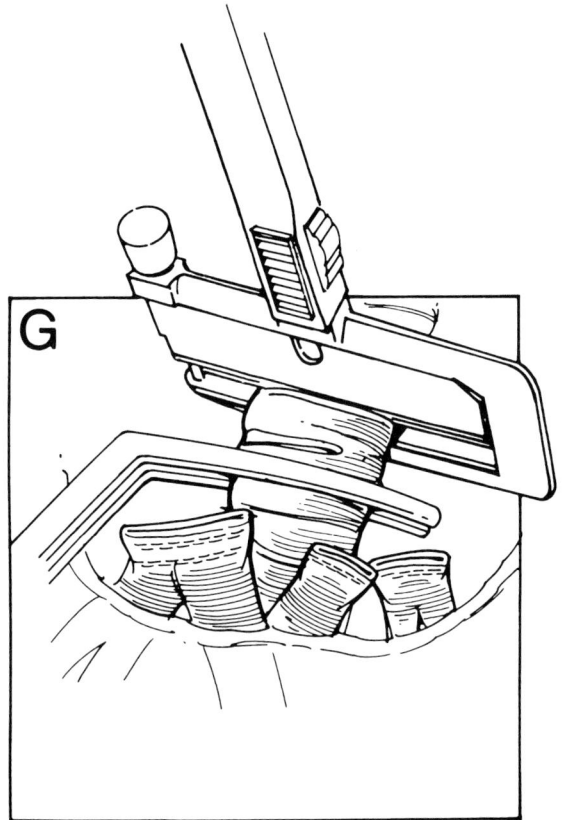

Fig. 7. A PI 55 stapler is positioned on the proximal mainstem bronchus, close to the carina, with care that the bronchus is compressed anteriorly and posteriorly to prevent cracking of the cartilage. The stapler is closed and is fired twice. A knife is then used to transect the bronchus flush with the stapler, and the specimen is removed. Adapted with permission from **ref. 45**.

paratracheal lymph nodes, 3) involvement of ipsilateral paratracheal lymph nodes in the upper half of the intrathoracic trachea, 4) direct invasion of the trachea, and 5) evidence of gross perinodal disease *(53–54)*.

8. Extended Indications for Lung Resection–Chest Wall Resections

In 1947, Coleman *(55)* challenged the concept of unresectability for lung cancer invading the bony thorax. He described the techniques of *en bloc* chest wall resection and results in seven patients—one operative death occurred and

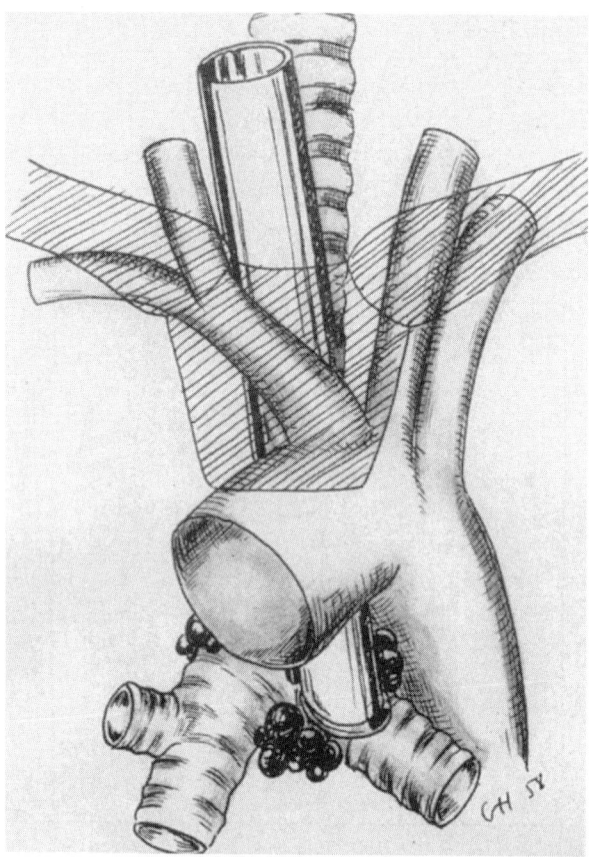

Fig. 8. The scope introduced through an incision in the suprasternal notch. Adapted with permission from **ref. 50**.

three patients were alive and well at 6 yr, 2 yr, and 5 mo. He did not use any method of chest-wall reconstruction.

Although chest-wall resection continued to be controversial, its evolution, continued use, and survival experience were reflected in later reports confirming that significant percentages of patients could achieve long-term survival free of disease (56). Negative lymph nodes were identified as the chief prognostic factor. Introduction of MRI in the early 1970s provided additional refinement in the evaluation of the extent of invasion prior to exploration. In my own experience, a 43% 5-yr survival was achieved for patients with squamous cell carcinoma who had chest-wall resections (57). Defects of the chest wall for which primary closure of the skin is possible can be very satisfactorily reconstructed using Marlex (Duval, Providence, RI) mesh or Goretex (W. L. Gore and Associates, Elkton, MD).

8.1. Superior Sulcus Tumors

In 1838, Hare *(58)* described a clinical pattern produced by carcinoma in the apex of the lung that had invaded the sympathetic ganglia and brachial plexus. In 1932, Pancoast *(59)* published the classic description of the superior sulcus tumor and syndrome that bears his name. He stated that

> "the tumors occur at a definite location at the thoracic inlet, were characterized clinically by pain around the shoulder and down the arm, Horner's syndrome and atrophy of the muscles of the hand and presented roentgenographic evidence of a small, homogenous shadow at the extreme apex, always more or less local rib destruction and often vertebral infiltration.... and that the name "superior sulcus tumor" was given to it because this term implies its approximate location and a lack of origin from the lung, pleura, ribs or mediastinum."

Pancoast acknowledged that this new designation might change with a better knowledge of the histopathology of the growth, which later proved true. The mass of a tumor in the superior pulmonary sulcus is a gross extension of the neoplasm in the lung; at least 60% is usually outside the lung, where it involves chest wall, nerve roots, the lower trunk of the brachial plexus, sympathetic chain and ganglia and the ribs or vertebrae. He concluded that the tumor is resistant to irradiation treatment and not subject to surgical removal, although it is accessible; however, most surgeons at the time continued to regard the tumors as best treated with palliative radiotherapy. Pancoast also observed that death occurred as a result of what seemed to be a comparatively trivial growth without roentgenologically detectable metastases.

Chardack and Macallum *(60)* reported the first successful result (5-yr survival) of complete *en bloc* resection of a superior sulcus tumor followed by irradiation therapy in 1956. Shaw and colleagues *(61)*, in 1961, reported on treatment of Pancoast tumors with preoperative irradiation and radical *en bloc* chest wall resection in 18 patients. An actuarial 5-yr survival of 34% for 61 patients receiving combined preoperative irradiation and extended resection was reported by Paulson *(62)* in 1975. He identified important prognostic factors as stage of nodal involvement, extent of the tumor, cell type, and pathological effects of preoperative irradiation in the resected specimens. This treatment plan, preoperative irradiation followed by surgery, was widely adopted as the model for the approach to superior sulcus tumors and it remains advocated today as conventional treatment.

8.2. The Concept of Conservative Surgery

A need for treatment of traumatic disruptions of the trachea and bronchus, due to tuberculous and traumatic strictures, was the stimulus for the early development of bronchoplastic procedures *(63)*. In 1940 Taffel *(64)* suggested sleeve resection for tumors of the bronchi; however, Price Thomas *(65)* is credited with the first sleeve resection for tumor in 1947—an adenoma occluding the right upper lobe. In 1952, Allison *(66)* accomplished the first sleeve resection for carcinoma. In the ensuing years, many other reports of experiences with bronchial sleeve resection have been published *(67–70)*. These procedures evolved from necessity and invention to solve problems not amenable to traditional procedures. The objective was to provide opportunity for cure for patients who would not withstand pneumonectomy, and for those with small tumors involving the proximal main bronchus and upper lobe bronchus.

Removal of all known disease along with conservation of lung tissue is a major issue for thoracic surgeons involved with the treatment of lung malignancy. Significant salvage of lung tissue may be provided by sleeve-type resections (sleeve lobectomy)—on the right, the middle and lower lobes are preserved, and on the left, either the upper or the lower lobe can be saved. Considerations for the procedures involve the anatomic location, histologic type, and biologic behavior of the tumor.

Sleeve lobectomy may be the only option for patients with limited pulmonary reserve who can undergo the procedure despite their inability to tolerate a pneumonectomy. Although a number of variations exist, standard sleeve resections are illustrated in **Fig. 9**. The conventional indications for tissue-preserving surgery *(71)* are: 1) elderly patients (over 70) with limited "biological reserves;" 2) limited respiratory reserve; 3) palliative surgery in bronchial carcinoma to prevent or treat complications caused by tumor; 4) localized tumor growth in the upper lobe bronchus—pneumonectomy should generally be avoided by performing sleeve resection; and 5) removal of metastasis to avoid a pneumonectomy, which might limit other treatment. An average operative mortality of 3% (from 0 to 12%) is reported *(72)*. Sleeve lobectomy as a viable alternative to more extensive procedures in order to preserve lung function has been advocated *(73)*. Detterbeck *(72)* noted that the 5-yr survivals for patients undergoing a sleeve resection are essentially the same as for those undergoing a standard resection, stage for stage, and the operative mortality and long-term survival for patients undergoing sleeve lobectomy is similar to that of standard lobectomy; the long-term survival is also similar to that of a standard lobectomy or pneumonectomy.

8.3. Video-Assisted Thoracic Surgery

Technological advances in the 1990s provided for the development of video-assisted thoracic surgery (VATS). Invention of the video monitor, instrument-tipped cameras, and percutaneous instruments brought thoracoscopy from an awkward telescope to a facile and versatile surgical approach *(74)*. Refine-

Surgical Treatment of Lung Cancer

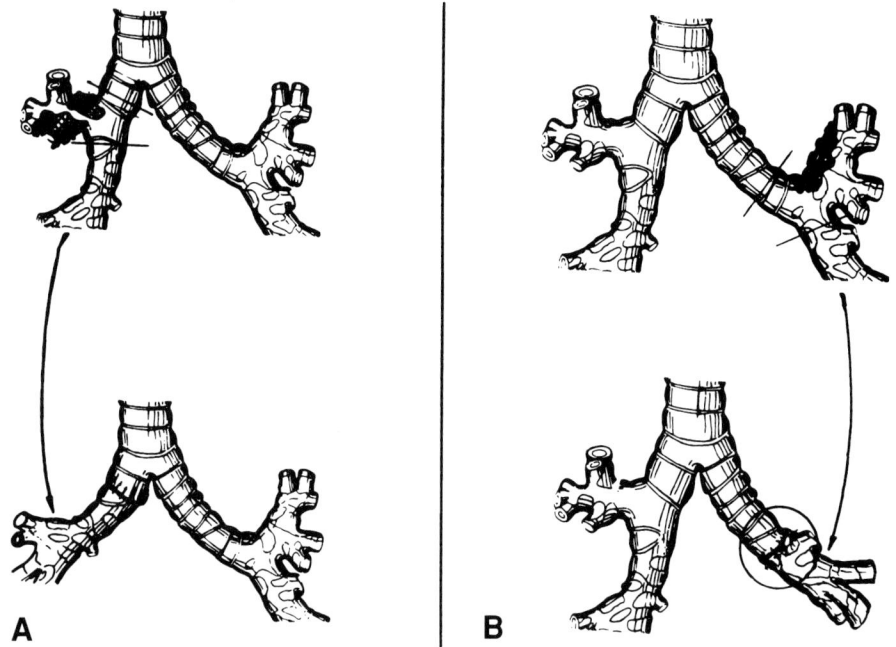

Fig. 9. (A) Standard sleeve resection of the bronchus on the right side. (B) Standard sleeve resection on the left side. Adapted with permission from **ref. 71**.

ments in ventilation and endoscopic stapling devices provided for the first thoracoscopic-assisted pulmonary resection *(75)*. The goal of VATS is to reduce postoperative pain and other post-thoracotomy related morbidity without compromising the principles of standard thoracic surgery. The technique offers the option of surgical therapy for patients who could not withstand conventional procedures. Refinements in the applications of VATS for surgical lung biopsy, for mediastinal node sampling and for lobectomy and pneumonectomy have been introduced and described by clinical studies *(76)*. Controversy exists, however, regarding the choice of this technique rather than standard procedures.

In most studies that evaluate VATS, resection of pulmonary nodules is the most frequent procedure performed *(77)*. The choice of VATS, which requires a general anesthetic, and a surgical procedure with an inpatient stay, or percutaneous needle biopsy (PNB) relates to the patient's risk for surgical intervention and the implications of the low rate of specific benign diagnoses by PNB. The introduction of high resolution computed tomography (HRCT) with nodule enhancement (done by calculating the Hounsfield Units (HU) of a nodule before and after administration of contrast) has improved the probability of predicting a malignancy and of determining that VATS excisional biopsy is indicated. Wedge biopsy of suspicious nodules is considered the primary diagnostic procedure of choice for all nodules that show nodule enhancement of 20 HU or greater. VATS allows for complete excision, eliminates the risk of false-negative results, and, reportedly, is accompanied by less pain and morbidity than open thoracotomy.

Video-assisted thoracic surgery lobectomy was first performed in 1991 *(78)*. In the ensuing years after evaluation of the procedure's feasibility, indications for its use and value compared to standard thoracotomy lobectomy have been questioned *(79)*. In a recent review, patient selection was defined as follows: 1) clinical stage I lung cancer, 2) tumor size less than 5 cm, 3) benign disease, and 4) physiologic operability. Contraindications are: 1) nodal disease, benign or malignant; 2) chest wall or mediastinal invasion (T3 or T4 tumor); 3) endo-bronchial tumor seen at bronchoscopy; 4) neoadjuvant chemotherapy, neoadjuvant radiation therapy; and 5) positive mediastinoscopy. The results of VATS lobectomy, in 1,120 patients in 7 large series of patients selected according to the listed criteria, showed a mortality rate of 0.6%; complications occurred in 10–29%; overall, 119/1239 (11.6%) were converted to thoracotomy *(78)*. Long-term, disease-free survival for VATS lobectomy for stage I lung cancer has been reported as comparable to that expected for standard thoracotomy lobectomy *(80)*. VATS lobectomy appears to be a safe procedure, with the exception of a low risk for catastrophic intraoperative bleeding or tumor implantation in the VATS incision. Thus, the procedure has been recommended as a reasonable treatment option for patients with clinical stage I lung cancer. The technical demands—VATS should be performed only by surgeons with the skills and experience to perform a complete cancer operation using the technique—and the low, but not insignificant risk of intraoperative bleeding, are factors *(80)*. Ginsberg warns against the temptation to perform less of an operation when using a lesser incision; however he confirms that VATS is a reasonable option, as long as oncologic principles are not compromised, for those technically expert with the procedure *(79)*. Other reports *(81–82)* confirm that mediastinal lymph node dissection can be accomplished during a VATS lobectomy comparable to the thoracotomy lobectomy dissection.

VATS for preoperative mediastinal exploration permits sampling of the mediastinal lymph node stations not accessible by the standard cervical mediastinoscopy *(83)*. It also allows evaluation of mediastinal adenopathy in those N2 stations inaccessible by the traditional mediastinopic techniques; that is, prevascular, retrotracheal, subaortic, paraortic, and lower mediastinal level nodes. A prospective, nonrandomized study of cervical mediastinoscopy versus VATS showed a similar diagnostic yield for both procedures; however, the complication rate and duration of hospital stay of patients undergoing VATS was greater than that for patients having cervical mediastinoscopy *(84)*. The results suggest that VATS mediastinal exploration is appropriate for those clinical situations in which the mediastinal lesions are not within the reach of the mediastinoscope, or when prior biopsy and fibrosis make repeat mediastinoscopy difficult or unsafe. The limitation of VATS mediastinal exploration is the inability to have access to and biopsy the contralateral paratracheal nodes.

9. Histologic and Anatomic Classification: 1957–2001
9.1. Histologic Classification

The importance of histopathologic classification to progress in lung cancer surgery cannot be overestimated. The availability of reproducible histopathologic classification and identification of the subtypes of lung cancer evolved from the three groups, squamous cell carcinoma, adenocarcinoma, and undifferentiated carcinomas described in 1957 by Robbins *(85)*, to the World Health Organization (WHO) Classification of Lung Tumors in 1981 *(86)*, to the present revision, in 1999, by WHO and the International Association for the Study of Lung Cancer, published in collaboration with pathologists in 14 countries *(87)*. This recent classification is the most widely accepted and used classification for lung cancer (*see* **Table 3**). From these classifications the results of surgery may be related to cell type; the prognostic significance and biological behavior of specific tumors are accurately evaluated. We arrive at the 21st century with the importance of the concept of reproducible histologic classification intact and influencing directions for the future.

9.2. Staging Classification

The relationship of the anatomic extent of disease to prognosis is the basis for the evolution of the staging concept. The invention of consistent descriptions of the primary tumor, the T factor, regional lymph nodes, the N factor and distant metastasis, the M factor, was first introduced by Denoix *(88)* in France in 1946. His purpose was to provide a common language for describing the facts in a given case of cancer. Later, the TNM categories were grouped into various combinations so that clinically meaningful stages emerged reflecting prognosis. The concept of defining, in consistent terminology, the relationship of the anatomic extent of disease to prognosis was an essential advance for lung cancer treatment. It provided a valid method for comparing the effects of differing therapeutic regimens. From 1974, when the first widely accepted recommendations for the clinical staging of lung cancer *(89)* were introduced in the United States, to the present International System for Staging the disease *(90)*, the needs of the scientific community for consistent classification

Table 3
WHO/IASLC Histologic Classification[a]

Squamous Cell Carcinoma
    Variants
        Papillary
        Clear Cell
        Small Cell
        Basaloid
Small Cell Carcinoma
    Combined small cell carcinoma
Adenocarcinoma
    Acinar
    Papillary
    Bronchioalveolar carcinoma
        Nonmucinous
        Mucinous
        Mixed mucinous and nonmucinous or indeterminate
    Adenocarcinoma with mixed subtypes
    Variants
        Well-differentiated fetal adenocarcinoma
        Mucinous ("colloid") adenocarcinoma
        Mucinous cystadenocarcinoma
        Signet ring adenocarcinoma
        Clear cell adenocarcinoma
Large Cell Carcinoma
    Variants
    Large cell neuroendocrine carcinoma
        Combined large cell neuroendocrine carcinoma
    Basaloid carcinoma
    Lymphoepithelioma-like carcinoma
    Clear cell carcinoma
    Large cell carcinoma with rhabdoid phenotype
Adenosquamous carcinoma
Carcinomas with pleomorphic, sarcomotoid, or sarcomatous elements
    Carcinomas with spindle and/or giant cells
    Carcinosarcoma
    Pulmonary blastoma
    Other
Carcinoid tumor
    Typical carcinoid
    Atypical carcinoid
Carcinomas of salivary gland type
    Mucoepidermoid carcinoma
    Adenoid cystic carcinoma
    Others
Unclassified carcinoma

WHO, World Health Organization; IASLC, International Association for the Study of Lung Cancer.
[a]Adapted with permission from **ref. 87**.

have been well served (*see* **Tables 4** and **5**). Important contributions of the staging system are, first, patients can be grouped together according to certain measurable common features of their disease, so that within each stage group treatment options and survival expectations will be relatively similar. In this manner, reliable and valid comparisons of the results of different modalities of therapy can be made. Second, survival data presented according to staging criteria are a measure of the efficacy of available therapy for lung cancer; thus the staging information serves as a valuable guide for treatment planning. Third, valid baseline criteria for entering patients into research protocols are available, and, fourth, a means is provided for communicating universal understanding of the status of lung cancer treatment. As a corollary to the present system, recommendations are made for consistent terminology in mapping the status of the regional lymph nodes *(91)*. In the 21st century biological prognostic factors that impact on the staging concept will undoubtedly be confirmed and applied to estimates of prognosis in the TNM and stage categories. Until this work becomes a reliable and proven clinical reality, anatomic staging classifications remain as a benchmark for measuring prognosis.

10. Adjuvant Therapy: 1950–Present

Efforts to improve on the results of surgical treatment for lung cancer were not forthcoming with advances in operative techniques or even in the refined selection of patients for operation. Local recurrence and distant metastasis continued to be significant unresolved problems. The use of adjuvants to surgical therapy was considered, based on the following rationale: 1) failure to achieve "cure" following apparently complete resection was due to occult metastasis not detected at the time of surgery, and 2) a combination treatment approach could enable destruction of the tumor cell population such that the disease could not re-establish itself.

The first adjuvant trials involved preoperative radiation therapy and were accomplished prior to present day concepts of the influence of staging and histologic classification on prognosis. Neither a large multicenter study, nor any other investigation, confirmed a survival benefit for preoperative adjuvant radiation *(92)*. Investigations of postoperative irradiation showed that the local recurrence rate was significantly reduced; however, this did not translate to an improvement in survival *(93)*.

The National Cancer Institute North American Lung Cancer Study Group (NCI-LGSG) investigated the use of adjuvant immunotherapy using bacillus Calmette-Guerin (BCG) postoperatively in patients with stage I non-small cell lung cancer (NSCLC) undergoing complete resection (1981) *(94)*. No benefit for the adjuvant treatment was shown; however, important criteria were

Table 4
TNM (Tumor, Metastasis, Nodes) Descriptors[a]

Primary Tumor (T)
TX   Primary tumor cannot be assessed, or tumor proven by the presence of malignant cells in sputum or bronchial washings but not visualized by imaging or bronchoscopy
T0   No evidence of primary tumor
Tis Carcinoma *in situ*
T1   Tumor 3 cm or less in greatest dimension, surrounded by lung or visceral pleura, without bronchoscopic evidence of invasion more proximal than the lobar bronchus[b] (i.e., not in the main bronchus)
T2   Tumor with any of the following features of size or extent:
     More than 3 cm in greatest dimension.
     Involves main bronchus, 2 cm or more distal to the carina.
     Invades the visceral pleura.
     Associated with atelectasis or obstructive pneumonitis that extends to the hilar region but does not involve the entire lung.
T3   Tumor of any size that directly invades any of the following: chest wall (including superior sulcus tumors), diaphragm, mediastinal pleura, parietal pericardium; or tumor in the main bronchus less than 2 cm distal to the carina, but without involvement of the carina; or associated atelectasis or obstructive pneumonitis of the entire lung.
T4   Tumor of any size that invades any of the following: mediastinum, heart, great vessels, trachea, esophagus, vertebral body, carina; or tumor with a malignant pleural or pericardial effusion[c], or with satellite tumor nodule(s) within the ipsilateral primary-tumor lobe of the lung.

Regional Lymph Nodes (N)
NX   Regional lymph nodes cannot be assessed
N0   No regional lymph node metastasis
N1   Metastasis to ipsilateral peribronchial and/or ipsilateral hilar lymph nodes, and intrapulmonary nodes involved by direct extension of the primary tumor
N2   Metastasis to ipsilateral mediastinal and/or subcarinal lymph node(s)
N3   Metastasis to contralateral mediastinal, contralateral hilar, ipsilateral or contralateral scalene, or supraclavicular lymph node(s)

M   Distant Metastasis
MX   Presence of distant metastasis cannot be assessed
M0   No distant metastasis
M1   Distant metastasis present[d]

[a]Adapted with permission from **ref. 90**.

[b]The uncommon superficial tumor of any size with its invasive component limited to the bronchial wall, which may extend proximal to the main bronchus, is also classified T1.

[c]Most pleural effusions associated with lung cancer are due to tumor. However, there are a few patients in whom multiple cytopathologic examinations of pleural fluid are negative for tumor. In these cases, the fluid is nonbloody and is not an exudate. When these elements and clinical judgment dictate that the effusion is not related to the tumor, the effusion should be excluded as a staging element and the patient should be staged T1, T2, or T3. Pericardial effusion is classified according to the same rules.

[d]Separate metastatic tumor nodule(s) in the ipsilateral nonprimary-tumor lobe(s) of the lung also are classified M1.

Table 5
Stage Grouping of the TNM Subsets

Stage	TNM subset	
Stage 0	Carcinoma *in situ*	
Stage IA	T1 N0 M0	
Stage IB	T2 N0 M0	
Stage IIA	T1 N1 M0	
Stage IIB	T2 N1 M0	
	T3 N0 M0	
Stage IIIA	T3 N1 M0	
	T1 N2 M0	
	T2 N2 M0	
	T3 N2 M0	
Stage IIIB	T4 N0 M0	T4 N1 M0
	T4 N2 M0	
	T1 N3 M0	T2 N3 M0
	T3 N3 M0	T4 N3 M0
Stage IV	ANY T ANY N M1	

Adapted with permission from **ref. 90**.
Note: Staging is not relevant for Occult Carcinoma, designated TX N0 M0.

established for objective determination of the status of regional lymph nodes in further surgical adjuvant trials for lung cancer.

Since the early 1950s, a long history of clinical trials of adjuvant chemotherapy with single agents and combinations of agents, either alone or with radiotherapy, failed to provide evidence for a survival benefit from the chemotherapy. In 1986, the NCI-LCSG reported the first positive result (prospective, randomized clinical trial) from the use of postoperative combination chemotherapy using a regimen of Cytoxan, Adriamycin, and cisplatin (CAP) compared with bacillus Calmette-Guerin (BCG) and levamisole immunotherapy in patients with stage II and IIIA NSCLC. The patients receiving chemotherapy had a significantly longer disease free interval following resection, 7 months, than the control group (who received immunotherapy) *(93)*. The effectiveness of a regimen of CAP postoperative chemotherapy and radiotherapy for patients with incompletely resected stage IIIA disease also provided significant benefit for the patients treated with the adjuvant combined regimen *(93)*.

A meta-analysis, published in 1995 *(94)*, of the role of chemotherapy in NSCLC included data from 14 trials of postoperative adjuvant chemotherapy:

five studies used alkylating agents and eight used cisplatin-based regimens. Summary of the results in trials using alkylating agents favored surgery alone, while the pattern of results for the cisplatin-based trials favored chemotherapy, a 3% benefit at 2 yr and 5% at 5 yr. Seven trials of postoperative radiotherapy vs radiotherapy plus chemotherapy suggested an absolute benefit of 2% at 5 yr for the regimens including chemotherapy. At present several ongoing international trials are evaluating the use of postoperative chemotherapy *(95)*, some of which will help define the role of "newer" drug combinations.

The advantages of a new approach to surgical adjuvant therapy for lung cancer, neoadjuvant (or preoperative or induction) chemotherapy were summarized by Johnson et al. *(96)* as follows: 1) an in vivo assessment of tumor responsiveness (which may have implications for the use of postoperative chemotherapy); 2) potential eradication of clinically occult extrathoracic micrometastases; 3) possible decrease in the incidence of hematogenous spread or local "seeding" of tumor cells caused by surgical manipulation of the primary lesion; 4) the possibility that technically unresectable lesions may become resectable; and 5) possible conservation of normal lung tissue, owing to the surgeons ability to perform a less extensive resection The results of three small clinical trials *(97–99)* showed a significant survival benefit for the chemotherapy/surgery-treated patients.

Induction therapy for locally advanced NSCLC remains a major controversy. Critical questions concerning the present status of induction therapy before surgery have been reviewed recently by Albain and Pass *(100)*. Controversy exists regarding standard of care for two groups of patients: 1) patients with early stage or nonbulky N2 disease; that is, the T2, T3N0 subset or the T1, T2, T3N1, and T1, T2, T3 minimal-N2 subsets; and 2) patients with bulky N2 disease present on CT or chest X-rays and those with T4 or N3 disease (*see* **Tables 4** and **5**). The question is

> "Can the results of review of world-wide data of second and third generation clinical trials provide standard of care recommendations for both of these groups? That is, should preoperative chemotherapy with or without radiotherapy be given routinely to patients with early-stage and non-bulky N2 disease? For patients with advanced stage or bulky disease, does subsequent resection improve survival?" *(100)*

The majority Consensus Statement of the International Association for the Study of Lung Cancer (IASLC) noted that it is premature to reach these conclusions in either disease group, based on available data, and further, completion of ongoing trials is needed to establish new practice guidelines *(101)*. Thus, surgery alone remains the recommended conventional standard for the early stage, nonbulky N2 disease group. In 1999, the results of the first large Phase III trial (The French Thoracic Cooperative Group Study), which did not reach statistical significance, were in favor of induction chemotherapy *(102)*. Therefore, continuing trials with surgery-alone control arms are justified. The IASLC consensus statement noted that feasibility and safety were shown for postinduction surgery; however, this regimen was not proven superior to chemoradiotherapy alone, which remains the conventional standard recommendation for treatment of patients in the bulky N2-advanced stage group *(101)*. Hopefully, ongoing Phase III trials (*see* **ref.** *100*, Tables 44.13 and 44.14) will provide definitive answers regarding an optimal induction approach. Albain and Pass note that many practitioners accept that the available data give sufficient support for the use of induction treatments, and they are prescribed outside the context of a clinical trials setting for different subsets of patients. This practice may jeopardize the accrual to randomized studies and delay the solution to the problem of the most effective treatment regimens for patients with non small cell lung cancer. At this time, induction therapy remains in the realm of investigational treatment, and patients should be assigned to such treatment within the context of clinical trials protocols.

Further progress in the development of surgical adjuvant treatment for lung cancer rests with bringing the newer discoveries of the biology of lung cancer to clinical trials in which surgery is an integral part of the treatment plan.

11. The Present Surgical Treatment Approach

The milestones described in the development of surgical treatment for lung cancer are culminated in today's philosophy, which remains multidisciplinary and interwoven with technical and basic scientific discoveries. The trend toward more conservative resections is tempered by adherence to

> "surgical oncologic principles *(103)*: 1) The tumor and its draining intrapulmonary lymphatic tributaries should be resected in their entirety whenever possible, 2) the tumor should be completely excised without spilling or traversing it, 3) the surgeon should resect *en bloc* any structure invaded by tumor in order to achieve negative margins, and 4) a complete mediastinal node dissection or sampling should be performed in all patients."

Resection remains the standard treatment of choice for patients with clinical stages IA, IB, IIA, and non-T3N0 stage IIB NSCLC, who are able to withstand the thoracic operation and the sacrifice of lung tissue required to remove all known disease. Some controversy exists regarding the choice of procedure and technique that will meet surgical oncologic principles; that is, conservative resection as a definitive alternative to lobectomy or pneumonectomy, or the thoracoscopic approach as discussed earlier. The T3N0M0 subset of patients classified in stage IIB includes patients with T3 tumors with no evidence of lymph node metastasis, who have, by definition, extension of disease beyond the lung—superior sulcus and other chest-wall tumors, and those with limited central extension. Tumors involving the bony thorax usually receive preoperative radiotherapy and others may be candidates for investigational radiotherapy or chemotherapy adjuvants. Patients with stage IIIa tumors, all of whom have either N1 n N2 disease, are selected for surgery on an individual basis, taking into account the potential for achieving a complete resection and the availability of potentially effective, investigational, adjuvant treatment in a clinical trials setting.

The current controversies involving neoadjuvant surgical therapy have been discussed. In summary, surgery alone is the treatment of choice for T1 N0 M0 NSCLC; surgery alone (vs neoadjuvant treatment) remains the conventional standard of care in early resectable disease (T2 or T3N0: T1, T2, or T3N1; and T1, T2, or T3 minimal-N2). The standard of care for patients with bulky N2 or T4 and N3 disease is chemoradiotherapy.

12. The Future

A new era for surgery dawns in the beginning of the 21st century. Just as the era of modern thoracic surgery was ushered in by scientific discoveries of the19th and early 20th centuries, "Information Age Surgery" will be contingent on 21st century advances in video imaging, endoscopic technology, and instrumentation. As we approach the transition to the advanced technologies for surgical practice *(104)*, it is not clear how the next generation of surgery will appear, however, a significant change will result from the use of computers and robotics. The principle behind robotics in surgery is the introduction of computer assistance to enhance dexterity and facilitate the performance of endoscopic surgery *(105)*. Creation of a completely integrated system that converts information to action is the goal of the technology. It holds promise for facilitating complex endoscopic procedures by virtue of voice control over the networked operating room. For some procedures, surgeons will no longer directly touch or see the structures on which they operate *(106)*.

> "The basic construction principle of currently used robots is that of arm manipulators with a serial architecture of joints and links. The last element in this series is the surgical instrument. Joint motion is provided by mechanical actuators that are electrically or hydraulically driven. The robotic systems currently in use are supervised by on-line systems, in which the operator is in control of the system by way of a human-machine interface, known as the master console" *(105)*.

In essence, human interface technology is the attempt to come to a natural and intuitive method of a person (an organic or biologic system) interacting with an electromechanical system (an inorganic or physics-based system), whether that be a surgical instrument, an aircraft, or a computer *(107)*.

The advent of laparoscopic surgery made surgery through a scope a reality. The first robotic device to perform laparoscopic gall bladder and gastroesophageal reflux disease surgery was reported in FDA Consumer in December 2000 *(108)*. Clinical studies showed that the robotic device was comparable in safety and effectiveness to that of standard laparoscopic surgery. FDA commissioner J.E Henny stated "This system is the first step in the development of new robotic technology that eventually could change the practice of surgery."

References

1. Castiglioni, A. (ed.) (1947) *A History of Medicine.* Knopf, New York, NY.
2. Wangensteein, O. H. and Wangensteen, S. D. (1978) *The Rise of Surgery From Empiric Concept to Scientific Discipline.* University of Minnesota Press, Minneapolis, MN, pp. 553–580.
3. Naef, A. P. (ed.) (1990) *The Story of Thoracic Surgery.* Hogrefe & Huber, Toronto, Canada, pp. 31–32.
4. Mountain, C. F. (2000) The evolution of surgical treatment for lung cancer. *Chest Surg. Clin. North Am.* **10,** 83–104.
5. Smith, R. A. (1982) Development of lung surgery in the United Kingdom. *Thorax* **37,** 161–168.
6. Gruhn, J. G. and Rosen, S. T. (1989) *Lung Cancer: The Evolution of Concepts,* vol. 1, Field and Wood, New York, NY.
7. Auenbrugger, L. (1761) Inventum Novum ex Percussione Thoracic Humani, ut Signo Abstrusos Interni Pectoris Morbos Detegendi, Vienna, 1761. (Forbes, J., tr.), Johns Hopkins University Press, Baltimore, 1936. (Cited by Gruhn, ref. 6, p. 4).
8. Laennec, R. T. H. (1816) A treatise on the diseases of the chest. (in which they are described according to their anatomical characters, and their diagnosis establishing a new principle by means of acoustick instruments). (Forbes, J. (tr.) T and G Underwood, London, 1921. (Cited in Gruhn Ref. 6, p. 29).
9. Laennec, R. T. H. (1815) "Encephaloides," Dictionaire des sciences medicales, vol. 12, Panckoucke, Paris: p. 165. (Cited in Rosenblatt, M.B. (1964) Bull. Hist. Med. **38,** 395, ref. List p. 340).
10. Andral, G. (1821) Recherches sur l'expectoration dans les differentes maladies de poitrine, vol. 87 Didat, Paris. (Cited by Brewer, L.A. [1982] Historical notes on lung cancer before and after Graham's successful pneumonectomy in 1933. *Am. J. Surg.* **143,** 650).
11. Flint, A. (1866) A practical treatise on the physical exploration of the chest and the diagnosis of diseases affecting the respiratory organs, 2nd ed., HC Lea, Philadelphia, p. 489. (Cited by Oschner, A. (1978) The development of pulmonary surgery, with special emphasis on carcinoma and bronchiectasis. *Am. J. Surg.* **135,** 734).
12. Papanicolaou, G. N. and Trent, H. F. (1946) Diagnostic value of exfoliated cells from cancerous tissue. *JAMA 131* **13,** 372–378.

13. Roentgen, W. C. (1895) Uber eine neue Art von Strahlen, Erste Mitteilung. Sitzgsber Physikal Gesellschaft, Wurtzberg, p. 132.
14. Killian, G. (1898) Ueber Directe Bronchoskopie. *Munch. Med. Wochenschr.* **45**, 844.
15. Jackson, C. (1905) Foreign bodies in the trachea, bronchus and oesophagus: the aid of oesophagoscopy, bronchoscopy and magnetism in the extraction. *Laryngoscope* **15**, 257–281.
16. Jackson, C. (1917) Endothelioima of the right bronchus removed by peroral bronchoscopy. *Am. J. Med. Sci.* **153**, 371–375.
17. Davies, H. M. (1913) Recent advances in surgery of the lung and pleura. *Br. J. Surg.* **1**, 228–258.
18. Brunn, H. (1929) Surgical principles underlying one-stage lobectomy. *Arch. Surg.* **18**, 490–515.
19. Shenstone, N. S. and Janes, R. M. (1932) Experience in pulmonary lobectomy. *Can. Med. Assoc. J.* **27**, 138–145.
20. Nissen, R. (1980) Total pneumonectomy. (Classics in Thoracic Surgery) *Ann. Thor. Surg.* **29**, 390–394.
21. Haight, C. (1934) Total removal of left lung for bronchiectasis. *Surg. Gynecol. Obstet.* **58**, 768–780.
22. Graham, E. A. and Singer J. J. (1933) Successful removal of an entire lung for carcinoma of the bronchus. *JAMA* **101**, 1371–1374.
23. Meade, R. H. (1961) *A History of Thoracic Surgery.* Charles C. Thomas, Springfield, IL, p. 589.
24. Brewer, L. A. (1984) The first pneumonectomy. *J. Thorac. Cardiovasc. Surg.* **88**, 810–826.
25. Naef, A. P. (1990) *The Story of Thoracic Surgery.* Hogrefe and Huber, Toronto, p. 40.
26. Churchill, E. D. (1932) The surgical treatment of carcinoma of the bronchus. *J. Thorac. Surg.* **2**, 254–266.
27. Hurt, R. (1996) *The History of Cardiothoracic Surgery.* Parthenon Publishing Group, New York, NY, p. 284.
28. Reinhoff, W. F. (1936) The surgical technique of total pneumonectomy. *Arch. Surg.* **32**, 218–231.
29. Archibald, E. (1934) The technique of total unilateral pneumonectomy. *Ann. Surg.* **100**, 796–811.
30. Overholt, R. H. (1934) The total removal of the right lung for carcinoma. *J. Thor. Surg.* **4**, 196–210.
31. Churchill, E. D. and Belsey, R. (1939) Segmental pneumonectomy in bronchiectasis; lingula segment of the left upper lobe. *Ann. Surg.* **109**, 481–499.
32. Overholt, A. (1947) A new technique for pulmonary segmental resection. *Surg. Gyn. Obst.* **84**, 257–268.
33. Le Roux, B. T. (1972) Management of bronchial carcinoma by segmental resection. *Thorax* **27**, 70–74.

34. Jensik, R. J. (1986) The extent of resection for localized lung cancer in segmental resection, in *Current Controversies in Thoracic Surgery* (Kittle, C. F., ed.), W.B. Saunders, Philadelphia, PA, p 175.
35. Lung Cancer Study Group (prepared by Ginsberg, R. J. and Rubenstein, L. V.) (1995) Randomized trial of lobectomy versus limited resection for T1 N0 non-small cell lung cancer. *Ann. Thor. Surg.* **60,** 615–622.
36. Graham, E. A. (1944) Indications for total pneumonectomy. *Dis. Chest.* **10,** 87–94.
37. Ochsner, A. and DeBakey, M. (1942) Significance of metastasis in primary carcinoma of the lung. *J. Thor. Surg.* **11,** 357–387.
38. Allison, P. R. (1946) Intrapericardial approach to the lung root in the treatment of bronchial carcinoma by dissection pneumonectomy. *J. Thorac. Surg.* **15,** 99–117.
39. Brock, R. and Whytehead, L. L. (1955) Radical pneumonectomy for bronchial carcinoma. *Br. J. Surg.* **43,** 8–24.
40. Cahan, W. G., Watson, W. I., and Pool, J. L. (1951) Radical pneumonectomy. *J. Thorac. Surg.* **22,** 449–473.
41. Alwyn, J. A. (1951) Avoidable vascular spread in resection for bronchial carcinoma. *Thorax* **6,** 250.
42. Johnson, J., Kirby, C. K., and Blakemore, W. S. (1958) Should we insist on "radical pneumonectomy" as a routine procedure in the treatment of carcinoma of the lung. *J. Thorac. Surg.* **36,** 309–315.
43. Watson, W. L. (1956) Carcinoma of the lung with five-year survivals: a study of 3000 cases. *Cancer* **9,** 1167–1172.
44. Shimkin, M. B., Connelly, B. S., Marcus, B. S., and Cutler, S. J. (1962) Pneumonectomy and lobectomy in bronchogenic carcinoma. *J. Thorac. Cardiovasc. Surg.* **44,** 503–519.
45. Rubio, P. A. (ed.) (1986) *Atlas of Stapling Techniques.* Aspen Publishers, Inc., Rockville, MD, p. 3.
46. Hagopian, E. J., Mann, C., Galibert, C., and Steichen F. M. (2000) The history of surgical instruments and instrumentation. *Chest Surg. Clin. North Am.* **10,** 9–43.
47. Ravitch, M. M. (1991) Historical perspective and personal viewpoint, in *Current Practice of Surgical Stapling* (Ravitch, M. M., Steichen, F. M., and Welter, R., eds.), Lea and Febiger, Philadelphia, PA, pp. 3–11.
48. Steichen, F. M. (1991) Inspiratin and rationale for the development of mechanical sutures, in *Current Practice of Surgical Stapling* (Ravitch, M. M., Steichen, F. M., and Welter, R., eds.), Lea and Febiger, Philadelphia, PA, pp. 13–22.
49. Aoki, T., Ozeki, Y., Watanabe, M., and Tanaka, S. (1998) Cartilage folding method for main bronchial stapling. *Ann. Thorac. Surg.* **65,** 1800–1801.
50. Carlens, E. (1959) Mediastinoscopy: a method for inspection and tissue biopsy in the superior mediastinum. *Dis. Chest.* **36,** 343–352.
51. Sarin, C. L. and Nohl-Oser, H. C. (1969) Mediastinoscopy: a clinical evaluation of 400 consecutive cases. *Thorax* **24,** 585–588.

52. Mountain, C. F. (1990) The role of mediasinoscopy in the management of lung cancer. Proceedings of the Sixth National Cancer Conference, Lippincott & Co., Denver, CO, pp. 829–834.
53. Mountain, C. F. (1991) The surgical treatment of lung cancer. *Crit. Rev. Oncol. Hematol.* **11,** 179–207.
54. Pearson, F. G. (1985) Management of stage III disease: Mediastinal adenopathy–the N2 lesion, in *Lung Cancer: International Trends in General Thoracic Surgery*, vol. 1, (Delarue, N. C. and Eschapasse, H., eds.), W.B. Saunders, Philadelphia, PA, pp.104–120.
55. Coleman, F. P. (1947) Primary carcinoma of the lung, with invasion of the ribs. *Ann. Surg.* **126,** 156–158.
56. Grillo, H. C. (1985) Techncal considerations in stage III disease: pleural and chest wall involvement, in *Lung Cancer: International Trends in General Thoracic Surgery*, vol. 1, (Delarue, N. C. and Eschapasse, H., eds.), W.B. Saunders, Philadelphia, PA, pp. 134–138.
57. Mountain, C. F. (1990) Expanded possibilities for surgical treatment of lung cancer. Survival in stage IIIa disease. *Chest* **97,** 1045–1051.
58. Hare, E. (1838) Tumor involving certain nerves, vol. 1, London: *Med. Gaz.* **1,** 16–18.
59. Pancoast, H. K. (1932) Superior pulmonary sulcus tumor. *JAMA* **99,** 1391–1396.
60. Chardack, W. M. and McCallum, J. D. (1956) Pancoast tumor (five year survival without recurrence or metastases following radical resection and postoperative irradiation). *J. Thorac. Surg.* **31,** 335–542.
61. Shaw, R. R., Paulson, D. L., and Kee, J. R. Treatment of the superior sulcus tumor by irradiation followed by resection. *Ann. Surg.* **154,** 29–40.
62. Paulson, D. L. (1975) Carcinomas in the superior pulmonary sulcus. *J. Thorac. Surg.* **70,** 1095–1104.
63. Kittle, F. C. (2000) History of lobectomy and segmentectomy including sleeve resection. *Chest. Surg. Clin. North Am.* **10,** 105–130.
64. Taffel, M. (1940) The Repair of tracheal and bronchial defects with free facia grafts. *Surgery* **8,** 56–71.
65. Price-Thomas, C. (1956) Conserving resection of the bronchial tree. *J. R. Coll. Surg.* (Edinburgh) **1–2,** 169–186.
66. Kittle, P. C. (2000) Allison P. R. noted in History of lobectomy and segmentectomy including sleeve resection. *Chest Surg. Clin. N. Am.* **10,** 126.
67. Grillo, H. C., Bendexen, H. H., and Gephart, T. (1963) Resection of the carina and lower trachea. *Ann. Surg.* **158,** 889–893.
68. Deslauriers, J., Gaulin, P., and Beaulier, M., et al. (1986) Long-term clinical and functional results of sleeve lobectomy for primary lung cancer. *J. Thorac. Cardiovasc. Surg.* **92,** 871–89.
69. Jensik, R. J., Faber, L. P., Milloy, F. J., and Amot, J. J. (1972) Sleeve lobectomy for carcinoma, a ten-year experience. *J. Thorac. Cardiovasc. Surg.* **64,** 400–412.

70. Redina, E. A., Venuta, F., and De Giacomo, T. (1999) Sleeve resection and prosthetic reconstrction of the pulmonary artery for lung cancer. *Ann. Thorac. Surg.* **68,** 995–1002.
71. Toomes, H. and Vogt-Mycopf, I. (1985) Conservative resection for lung cancer, in *Lung Cancer: International Trends in General Thoracic Surgery* (Delarue, N. C. and Eschapasse, H., eds.), W.B. Saunders, Philadelphia, PA, pp. 88–99.
72. Detterbeck, F. C. and Egan, T. M. (2001) Surgery for stage II non-small cell lung cancer, in *Diagnosis and Treatment of Lung Cancer* (Detterbeck, F. C., Rivera, M. P., Socinski, M. A., and Rosenman, J. G., eds.), W.B. Saunders, Philadelphia, PA, pp. 191–197.
73. Jensik, R. J. (1985) Conservative resection for lung cancer. Discussion, in *Lung Cancer: International Trends in General Thoracic Surgery* (Delarue, N. C. and Eschapasse, H., eds.), W.B. Saunders, Philadelphia, PA, pp. 100–103.
74. Dewey, T. M. and Mack, M. J. (2000) Lung cancer: surgical approaches and incisions. *Chest Surg. Clin. North Am.* **10,** 803–820.
75. Smyth, W. R. and Kaiser, L. R. (1993) History of thoracoscopic surgery, in *Thoracoscopic Surgery* (Kaisser, L. R. and Daniels, T. M., eds.), Little Brown & Company, Boston, MA, pp. 1–16.
76. Lewis, R. J., Caccavale, R. J., Bocage, J. P., and Widman, W. D. (1999) Video-assisted thoracic non-rib spreading simultaneously stapled lobectomy, a more patient-friendly operation. *Chest* **116,** 119–124.
77. Hazelrigg, S. R., McGee, M. J., and Cetindag, I. B. (1998) Video-assisted thoracic surgery of the solitary pulmonary nodule. *Chest Surg. Clin. North Am.* **8,** 763–774.
78. McKenna, R. J. (1998) The current status of video-assisted thoracic surgery lobectomy. *Chest Surg. Clin. North Am.* **8,** 775–785.
79. Ginsberg, R. J. (1998) The current status of video-assisted thoracic surgery lobectomy: commentary. *Chest Surg. Clin. North Am.* **8,** 787–788.
80. Jones, D. R. and Detterbeck, F. C. (2001) Surgery for stage I non-small cell lung cancer, in *Diagnosis and Treatment of Lung Cancer* (Detterbeck, F. C., Rivera, M. P., Socinski, M. A., and Rosenman, J. G., eds.), W.B. Saunders, Philadelphia, PA, pp. 177–190.
81. Kaseda, S., Hangai, N., and Yamamoto, S., et al. (1997) Lobectomy with extended lymph node dissection by video-assisted thoracic surgery for lung cancer. *Surg. Endosc.* **11,** 703–706.
82. Iwasaki, M., Nishiumi, N., and Maitani, F., et al. (1996) Thoracoscopic surgery for lung cancer using the two small skin incisional method. Two windows method. *J. Thorac. Cardiovasc. Surg.* (Torino) **37,** 79–81.
83. Abhoda, A. and Keller, S. M. (2001) Surgical staging of the mediastinum, in *Lung Cancer Principles and Practice* (Pass, H. I., Mitchell, J. B., Turrisi, A. T., and Minna, J. D., eds.), Lippincott Williams and Wilkins, Philadelphia, PA, pp. 628–646.

84. Gosset, P., Toledo, L., Fritsch, S., et al. (1996) Mediastinoscopy vs thoracoscopy for mediastinal biopsy. Results of a prospective non-randomized study. *Chest* **110,** 1328–1331.
85. Robbins, S. L. (1957) *Textbook of Pathology.* W.B. Saunders, Philadelphia, PA.
86. Sobin, L. H. and Yesner, R. (1981) International histological classification of tumors, in *Histological Typing of Tumors*, 2nd ed., World Health Organization, Geneva, pp. 1–36.
87. Travis, W. D., Colby, T. V., Corrin, B., et al. (1999) *Histological Typing of Lung and Pleural Tumors*, 3rd ed., Springer-Verlag, Berlin, Germany, pp. 22–23.
88. Denoix, P. F. (1946) Enquete permanent dans les centres and anticanereux. *Bull. Insit. Nat. Hyg.* **1,** 70–75.
89. Mountain, C. F., Carr, D. T., and Anderson, W. A. D. (1974) A system for the clinical staging of lung cancer. *Am. J. Roentgen. Radiother. Nucl. Med.* **120,** 130–138.
90. Mountain, C. F. (1997) Revisions in the international system for staging lung cancer. *Chest* **111,** 1710–1717.
91. Mountain, C. F. and Dresler, C. M. (1997) Regional lymph node classification for lung cancer staging. *Chest* **111,** 1718–1723.
92. Warram, J. (A Collaborative Study) (1975) Preoperative irradiation of cancer of the lung: final report of a therapeutic trial. *Cancer* **36,** 914–925.
93. Holmes, E. C. (1989) Surgical adjuvant therapy of non-small cell lung cancer, in *Thoracic Oncology* (Roth, J. A., Rucksdeschel, J. C., and Weisenburger, J. C., eds.), W. B. Saunders Company, Philadelphia, p. 220.
94. Non-small Cell Lung Cancer Cooperative Group (1995) Chemotherapy in non-small cell lung cancer: a metaanalysis using updated data on individual patients from 52 randomized clinical trials. *BMJ* **311,** 899–909.
95. Pisters, K. M. W. (2000) Surgery and chemotherapy, in *Lung Cancer Principles and Practice* (Pass, H. I., Mitchell, J. B., Johnson, D. H., Turrisi, A. T., and Minna, J. D., eds.), Lippincott, Williams and Wilkins, Philadelphia PA, pp. 769–777.
96. Johnson, D. H., Turrisi, A., and Pass, H. I. (1996) Combined modality treatment for locally advanced non-small cell lung cancer, in *Lung Cancer Principles and Practice* (Pass, H. I., Mitchell, J. B., Johnson, D. H., et al., eds.), Lippincott Raven, Philadelphia, PA, p. 869.
97. Pass, H. I., Pogrebniack, H. W., Steinberg, S. M., et al. (1992) Randomized trial of neoadjuvant therapy for lung cancer. Interim analysis. *Ann. Thorac. Surg.* **53,** 992–998.
98. Rossell, R., Gomez-Codina, J., Camps, C., et al. (1999) Presectional chemotherapy in stage IIIA non-small cell lung cancer: a 7-year assessment of a randomized clinical trial. *Cancer* **26,** 7–14.
99. Roth, J. A., Fossela, F., Komaki, et al. (1998) Long-term follow-up of patients enrolled in a randomized trial comparing perioperative chemotherapy and surgery with surgery alone in resectable stage IIIA non-small cell lung cancer. *Lung Cancer* **21,** 1–6.

100. Albain, K. S. and Pass, H. I. (2000) Induction therapy for locally advanced non-small cell lung cancer, in *Lung Cancer Principles and Practice* (Pass, H. I., Mitchell, J. B., Johnson, D. H., Turrisi, A. T., and Minna, J. D., eds.), Lippincott, Williams and Wilkins, Philadelphia, PA, pp. 798–820.
101. Perry, M. C., DesLauriers, J., Albain, K. S., et al. (1997) Induction treatment for resectable non-small cell lung cancer: a consensus report. *Lung Cancer* **17(Suppl. 1),** S15–S18.
102. DePierre, A., Milleron, B., Moro, D., et al. (1999) Phase III trial of neoadjuvant chemotherapy in resectable stage I (except T1N0), II, IIIa non-small cell lung cancer: the French experience. *Proc. Am. Soc. Clin. Oncol.* **18,** 46, 5a.
103. Ginsberg, R. J. and Port, J. L. (2000) Surgical therapy of stage I and non-T3 N0 stage II non-small cell lung cancer, in *Lung Cancer Principles and Practice* (Pass, H. I., Mitchell, J. B., Johnson, D. H., Turrisi, A. T., and Minna, J. D., eds.), Lippincott, Williams and Wilkins, Philadelphia PA, pp. 682–693.
104. Satova, R. M. (1998) Cybersurgery A new vision for General Surgery, in *Cybersurgery. Technologies for Surgical Practice* (Satova, R. M., ed.), Wiley-Liss, New York, NY, pp. 3–14.
105. Dewey, T. M. and Mack, M. J. (2000) Lung cancer. Surgical approaches and incisions. *Chest Surg. Clin. North Am.* **10,** 803–820.
106. Mack, M. J. Minimally invasive and robotic surgery. *JAMA* **285,** 568–572.
107. Satava, R. M. CDR Jones, S. B. (1998) Human interface technology, in *Cybersurgery. Technologies for Surgical Practice* (Satava, R. M., ed.), Wiley-Liss, New York, NY, pp. 17–32.
108. Bren, L. (2000) Robots help surgeons perform more precise surgery. *FDA Consumer* **34,** 12–13.

29

Recent Advances and Dilemmas in the Radiotherapeutic Management of Locally Advanced Non-Small Cell Lung Cancer

Benjamin Movsas

1. Introduction

During the 1990s, thoracic radiotherapy (RT) combined with chemotherapy was accepted as a "new" gold standard for patients with good performance status locally advanced/inoperable non-small cell lung cancer (NSCLC). This paradigm shift away from RT alone has raised several fundamental questions: What is the optimal sequencing of chemotherapy and radiotherapy? Should standard (once-daily) RT continue to be employed? What about altered fractionation schema? Is there a role for three-dimensional conformal RT? How can the toxicity of combined modality strategies be reduced?

2. Background

Approximately 40% of patients with NSCLC present with locally advanced and/or unresectable disease. This group typically includes those with stage IIIB disease (excluding malignant pleural effusions) and bulky stage IIIA presentations. Up until just over a decade ago, the standard management for this group of patients was conventional thoracic RT alone—60 Gy over 6 wk, administered once a day. While this dose of radiation was shown in a randomized trial by the Radiation Therapy Oncology Group (RTOG trial 73-01) to enhance short-term (3-yr) survival compared to lower doses, the 5-yr survival was disappointingly low (5%) *(1)*. The "standard of care" changed in the early 1990s with the publication of the landmark trial from the Cancer and Leukemia Group B (CALGB trial 84-33), in which patients were randomized to 2 cycles of induction chemotherapy (vinblastine and cisplatin) prior to RT vs RT alone *(2)*. In addition to a significant improvement in the median survival time from 9.6 mo to 13.7 mo, a recent update corroborated a 5-yr survival benefit of 17% vs 7% and 7-year survival benefit of 13% vs 6% favoring the chemoradiotherapy arm ($p = 0.01$) *(3)*. Subsequently, at least 10 more randomized trials have been published comparing RT alone to RT and chemotherapy, of which 5 demonstrated superiority of combined modality treatment *(4–8)*. Despite the negative trials, several meta-analyses have demonstrated a small, but statistically significant, improvement in survival for the combination regimens *(9–11)*. It should be noted that most of the above trials limited eligibility to patients with a favorable prognosis (e.g., Karnofsky score of 70 or higher and maximum weight loss of 5%).

3. What is the Optimal Sequencing of Chemotherapy and Radiotherapy?

With the exception of the CALGB 84-33 trial, the long-term results with sequential chemotherapy and RT appear disappointing (*see* **Table 1**). RTOG 88-08 (using the same regimen as the CALGB 84-33 trial) demonstrated a 5-yr survival of only 8% for the induction chemotherapy/RT arm (vs 5% for the RT alone arm) *(12)*. Similarly, in another randomized trial conducted by CALGB and ECOG (Eastern Cooperative Oncology Group), the arm utilizing the same strategy as CALGB 84-33 (induction cisplatin/vinblastine followed by RT) demonstrated a 4-yr survival rate of only 10% *(13)*. In both this trial and RTOG 88-08, more patients were treated with this regimen than in the initial CALGB 84-33 experience. In yet another trial, the French multicenter trial CEBI 138, a "sandwich" regimen of induction and post-RT chemotherapy (vindesine, lomustine, cisplatin and cyclophosphamide) was utilized. In this trial, the 5-yr survival rate was only 6% for the sequential RT/chemotherapy arm (vs 3% for RT alone) *(14)*. These studies suggest that although the addition of sequential chemotherapy to RT enhances short-term survival by delaying distant failure, this strategy does not appear to dramatically alter the long-term results.

Recently, two randomized trials have reported, for the first time, an advantage of concurrent over sequential chemoradiation. Furuse and colleagues from Osaka, Japan evaluated 2 cycles of mitomycin, vindesine and cisplatin (MVP)—either concurrently with or prior to thoracic RT—in unresectable stage III NSCLC. In the sequential arm, after completion of MVP, RT was administered to a total dose of 56 Gy. In the concurrent arm, a split-course of RT was administered (2 Gy/fraction × 14 d followed a 10-d rest, then an additional 2 Gy × 14 fractions). Despite the split-course RT in the concurrent arm, the overall response rate was superior for the concurrent arm (84 vs 66%) with a commensurate improvement in median survival (16.5 vs 13.3 mo,

Table 1
Sequential CT + RT (Cisplatin/Vinblastine)

Study/Reference	N	5-yr OS
CALGB 84-33 *(3)*	78	17%
RTOG 88-08/ECOG 4588 *(12)*	151	8%
CALGB/ECOG *(13)*	130	10% (4-yr)

OS, overall survival; CT, chemotherapy; RT, radiation; N, # of patients.

respectively). More importantly, the 5-yr results continued to demonstrate a significant survival benefit for the concurrent arm of 15.8 vs 8.9% for the sequential arm *(15)*. One caveat with this trial is that the total number of chemotherapy cycles administered was based on the response to the initial two cycles. Indeed, only 25% of patients on the sequential arm received 3–4 cycles of chemotherapy compared to 59% of patients on the concurrent arm ($p = 0.0001$).

This important sequencing issue (i.e., concurrent vs sequential therapy) was also addressed in a larger ($n > 600$) randomized trial, RTOG 94-10. In this trial, the "gold standard" treatment (arm I) of induction chemotherapy (with cisplatin and vinblastine) followed by standard RT (ala CALGB 84-33) was compared to the same chemotherapy and RT delivered concurrently starting on day 1 (arm 2). This study also included a third arm of hyperfractionated (twice-daily) RT and concomitant cisplatin and oral etoposide. The preliminary results have recently been reported with median survival times of 14.6 mo (arm 1), 17.0 mo (arm 2), and 16.0 mo (arm 3), respectively. The median survival in the concurrent chemoradiation arm (using daily RT) was significantly better than in the sequential arm ($p = 0.03$) *(16)*.

While we await the long-term results of RTOG 94-10, it appears, based on this study, as well as the Japanese study, that concurrent chemoradiation has become "a new gold standard" in patients with good performance status locally advanced/inoperable NSCLC. This latest paradigm shift, of course, raises further questions, many of which are addressed in a recent review *(17)*. For example, what is the optimal chemotherapy to integrate with RT? Is there a role for additional doses of chemotherapy either before or after concurrent chemoradiation? If so, should a different chemotherapy regimen be utilized in this setting? Recently, the South West Oncology Group (SWOG 9504) reported promising results (median survival 27 mo) of a Phase II trial in which stage IIIB patients were treated with cisplatin/etoposide and concurrent daily RT followed by consolidation docetaxel *(18)*. As the current review is focusing primarily on RT issues, the next issue to be addressed is the question of altered fractionation schema.

4. Altered Fractionation Schema

There are two types of altered fractionation strategies, conceptually distinct, that can be utilized in the hope of enhancing local control and survival. Hyperfractionation involves smaller than conventional RT doses (e.g., 1–1.2 Gy) administered multiple times daily (typically 2–3 times) to achieve a higher cumulative dose over the course of therapy. Radiobiologically, hyperfractionation yields differential sparing of late-reacting normal tissue compared with acute reacting malignant tissues. By contrast, with accelerated (hyperfractionated) RT, a more conventional radiation fraction size is utilized (e.g., 1.5–2 Gy), but as multiple fractions are administered daily, the overall treatment time is significantly shorter. Unlike hyperfractionated radiotherapy (HFRT), the aim of accelerated RT is to reduce the tumor cell repopulation in rapidly proliferating neoplasms by shortening the overall treatment time. Indeed, tumor cell kinetic studies of human NSCLC cell lines have demonstrated short potential doubling times *(19,20)*. In order to tease out the impact of altering the RT fractionation schema, studies involving altered RT fractionation alone will be reviewed first, followed by studies that have attempted to integrate these strategies with chemotherapy.

4.1. Hyperfractionation

RTOG trial 83-11 was a multi-institutional, prospective, dose-seeking randomized Phase II study of hyperfractionation in patients with NSCLC. Patients were randomized to receive total doses of 62 Gy, 64.8 Gy, 69.6 Gy, 74.4 Gy, or 79.2 Gy in fractions of 1.2 Gy administered twice daily. Among the 248 patients with favorable prognostic factors (i.e., Karnofsky performance status of 70 or higher and weight loss of less than 5%), there was a survival benefit for the tumor dose level of 69.6 Gy compared with lower doses *(21)*.

The 69.6 Gy arm of RTOG 83-11 was tested in a subsequent 3-arm Phase III trial in patients with locally advanced NSCLC (RTOG 8808/Eastern Cooperative Oncology Group [ECOG] 4588). This study involved 490 patients who were randomized to standard RT (2 Gy once daily to 60 Gy for 6 wk) vs hyperfractionated RT (1.2 Gy BID to 69.6 Gy) vs induction vinblastine and cisplatin followed by standard RT. In the preliminary report *(7)*, the median survival for the chemoradiotherapy arm (13.8 mo) was found to be statistically significantly superior to the HFRT arm (12.3 mo) or the standard RT arm (11.4 mo). Based on these preliminary results, the chemoradiotherapy arm was widely viewed as the new standard against which future strategies should be compared.

With longer follow-up, however, the curves for the chemoradiotherapy and hyperfractionation arms began to overlap, with 3-yr survivals of 13% vs 14% and 5-yr survivals of 8% vs 6%, respectively *(12)*.

What about HFRT in combination with chemotherapy? While the preliminary results of RTOG 94-10 did not favor the altered fractionation/chemotherapy arm, the long-term results of other studies support this strategy. Jeremic et al. randomized patients with stage III NSCLC to HFRT alone to 69.6 Gy vs HFRT and low-dose daily carboplatin plus VP-16. The combined HFRT/chemotherapy arm demonstrated significantly superior median survival of 22 mo (vs 14 mo) and an improved 4-yr survival of 23% (vs 9%), $p = 0.021$ *(5)*. In another randomized trial with almost 200 patients, Jeremic et al. compared the impact of HFRT (to 69.6 Gy) and concurrent low-dose daily carboplatin/etoposide (C/E) with or without weekend C/E. While they found no benefit to the addition of weekend C/E, both arms had promising median survivals of 20 vs 22 mo and excellent 5-yr survival rates of 20% and 23% *(22)*.

4.2. Accelerated RT

The Medical Research Council in London ran a randomized trial of 563 patients comparing accelerated RT with standard RT (to 66 Gy) in a strategy termed CHART (Continuous Hyperfractionated Accelerated RT) *(23)*. CHART, designed to counteract tumor cell repopulation, involved 1.5 Gy administered 3 times per day for 12 consecutive days, to a total dose of 54 Gy. Patients treated with CHART had a significant improvement in 2-yr survival (29%) vs those treated with standard RT (20%), $p = 0.04$. While the rate of severe dysphagia was higher during the first 3 mo in the CHART arm (19% vs 3%), this mostly occurred after completion of RT. A more "user-friendly" modification of this schema has recently been developed in which patients (and physicians) are given the weekend off, nicknamed "CHART-WEL" (CHART weekend less), or "HART" (i.e., no longer continuous). ECOG started a study randomizing patients following induction chemotherapy to either HART versus standard (once-daily) RT. However, this study recently closed due to limited patient accrual. The ability to integrate chemotherapy with such intensive accelerated RT regimens requires further investigation.

5. Three-Dimensional Conformal RT

Three-dimensional (3D) conformal RT is external beam RT in which the radiation "dose-cloud" or prescribed dose volume is made to conform closely to the target volume (*see* **Fig. 1**), thereby facilitating dose escalation. High-resolution computed tomography scans are used to acquire precise anatomical

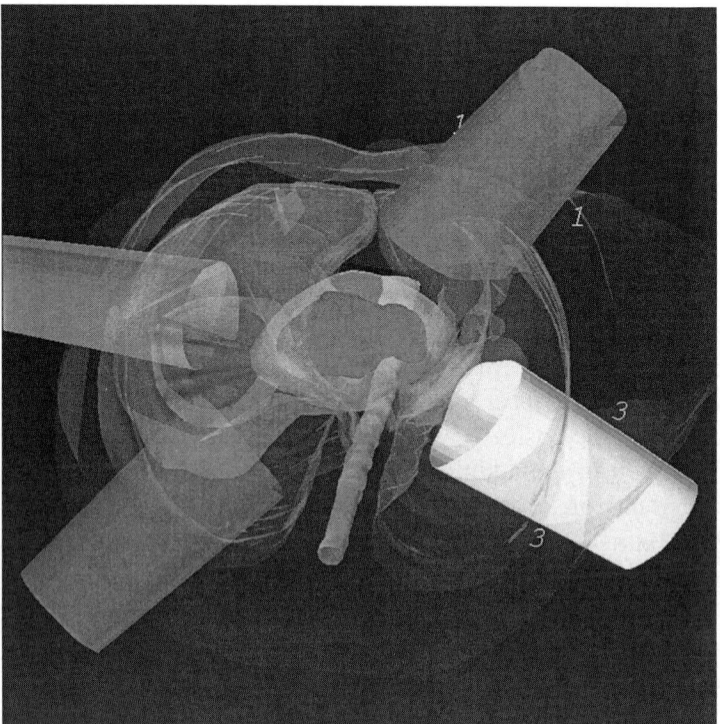

Fig. 1. Multiple radiation beams are utilized to create a three-dimensional conformal radiation "dose cloud" around the gross tumor volume (GTV) within the lung. This technique can be utilized to treat the target while minimizing radiation to the normal surrounding tissues (such as the lung and spinal cord).

data from which a computerized 3D image of the patient's normal structures and the tumor are constructed. The optimal radiation beam parameters and orientation can then be selected by comparing plans employing either multiple coplanar or noncoplanar fields. This approach facilitates dose escalation to the target volume by reducing the radiation exposure of normal tissues.

Several studies in NSCLC have shown a relationship between higher radiation doses and improved local tumor control. In RTOG 73-01, patients treated with the highest dose (60 Gy) had an intrathoracic failure rate of 33% at 3 yr compared with 42% of those treated with 50 Gy and 52% of those treated with 40 Gy in a continuous course. Patients treated with 50–60 Gy who manifested local tumor control had a 3-yr survival of 22% compared with 10% if tumor control was not achieved ($p = 0.05$) *(24)*. Similarly, Hazuka et al. demonstrated a dose response relationship for local progression-free survival and survival for the stage III subgroup (favoring patients receiving doses of 67.6 Gy or higher compared to those who received a lower dose, $p = 0.018$) *(25)*.

Radiation dose escalation can be achieved only by tailoring the treatment volume. Trials employing conformal RT typically do not attempt to treat the classic RT volumes, which encompass the regional lymph nodes. The long-term implications of this strategy are not yet known. In RTOG 73-01, approx 8% of patients relapsed at a previously uninvolved supraclavicular region at 3 yr if this region had not been treated, while 2% of patients relapsed if this region had been treated with more than 45 Gy. Improved outcome was similarly reported when the tumor-negative contralateral hilar lymph nodes were treated according to protocol (i.e., 1-cm margin) than for those treated with major variations ($p = 0.017$) *(26)*. However, more recent studies have questioned the need to treat traditional, larger thoracic RT volumes. When comparing large-volume treatment (i.e., inclusion of the uninvolved contralateral hilar and supraclavicular lymph nodes) versus small-volume treatment (exclusion of these elective nodal sites), Hazuka et al. found no difference in the local progression-free survival *(25)*. Similarly, Robinow et al. reported no failures in more than 100 patients in whom the radiographically uninvolved contralateral hilum was purposely not irradiated *(27)*. These more recent trials may reflect more accurate tumor volume definition and targeting.

Investigators at the University of Michigan have recently updated their experience with dose escalation using 3D conformal RT in a Phase I trial *(28)*. Doses were escalated based on the effective volume (V_{eff}) of both normal lungs irradiated and the risk of radiation pneumonitis (RP). Of 63 evaluable patients, grade 2 RP has occurred in 5 patients and grade 3 RP in only 2 patients. Currently, for a V_{eff} up to 18%, the level of dose escalation is up to 102.9 Gy. So far, no cases of isolated failures in clinically uninvolved nodal region (purposely not irradiated) have been found. RTOG has recently completed a study evaluating dose escalation using 3D conformal RT (to the gross tumor volume only) in patients with inoperable NSCLC (RTOG 93-11). Dose escalation was stratified by risk groups based on the percentage of total lung volume receiving more than 20 Gy. Ongoing issues for the continued development of conformal RT in lung cancer includes respiratory gating and the dosimetric implications of inhomogeneity effects, particularly with small RT fields *(29)*.

Recent studies are being developed to integrate chemotherapy with 3D conformal RT. Socinski et al. recently reported a dose escalation trial using 3D conformal RT (from 60–74 Gy) in patients receiving induction carboplatin (C) and paclitaxel (P) and concurrent weekly CP. They reported a promising median survival of 26 mo and 3-yr survival of 47% with a grade 3/4 esophagitis rate of approx 10% *(29a)*. Further escalation of the RT dose in the context of chemotherapy will need to be explored.

6. Reducing Toxicity

Although concurrent chemotherapy and thoracic RT appears to be the optimal treatment at this time, this strategy has potential disadvantages. As treatment regimens become more aggressive, the risk of normal tissue injury increases. This can potentially result in treatment breaks with dose reductions that may limit the success of therapy. Indeed, Cox et al. found that for "favorable" patients (i.e., high Karnofsky performance status, little weight loss, and less than N3 nodal disease) treated on three RTOG lung trials, interruptions in the completion of the planned RT reduced survival *(30)*. The main toxicities from thoracic RT are acute esophageal toxicity and subacute or late lung toxicity. In RTOG 94-10, significantly higher rates of acute grade ≥ 3 nonhematologic toxicity were observed in the 3rd arm, concurrent-BID RT (63%) compared to the sequential arm (31%, $p < 0.001$) or to the second arm, concurrent once-daily RT (50%, $p = 0.011$) *(16)*.

A quality-adjusted survival analysis of several RTOG chemoradiation lung studies has recently been reported *(31)*. This analysis included almost 1000 patients with locally advanced NSCLC who were treated on five Phase II or III RTOG studies employing various combinations of RT with and without sequential or concurrent chemotherapy. Quality-adjusted survival (Q-Time) was calculated by weighting the time spent with a specific toxicity and/or local distant tumor progression. Although patients receiving concurrent chemoradiation had the optimal overall survival, patients receiving induction chemotherapy followed by RT had nearly equivalent quality-adjusted survival. Reduction in esophageal, lung, and upper gastrointestinal toxicities led to the greatest improvement in the quality-adjusted survival time. Based on this analysis, a hypothesis was generated that if one could reduce the toxicity of the more intensive (concurrent) regimens, one may be able to improve not only the median survival time, but also the quality-adjusted survival time. This analysis led to the formation of a randomized trial to test the ability of a radioprotector, amifostine, to reduce the rate of esophagitis encountered in such intensive concurrent chemoradiation regimens (RTOG 98-01).

Amifostine (Ethyol, WR-2721) was originally developed by the US Army as a strategy to protect the troops from the threat of nuclear warfare. Amifostine is an organic thio-phosphate that was selected from more than 4400 compounds screened as the best radioprotective compound. This prodrug must be dephosphorylated to its active metabolite (WR-1065) by alkaline phosphatase. Once inside the cell, WR-1065, a free thiol, acts as a potent scavenger of oxygen-free radicals induced by ionizing radiation. It also provides an alternative target to DNA and RNA for the reactive molecules of alkylating or platinum agents *(32)*. Perhaps the greatest strength of amifostine lies in its potential ability to protect multiple organs, including, among others, salivary glands, bone marrow, oral mucosa, esophagus, and kidneys *(33,34)*. Several randomized trials have demonstrated the effectiveness of amifostine as a radioprotectant in specific settings *(35)*. In a study by Buntzel et al. *(36)*, 28 patients with squamous cell carcinomas of the head and neck were treated with a combination of RT (daily to 60 Gy) and concurrent carboplatin (70 mg/m^2 on d 1 through 5 and d 21 through 26). These patients were randomized to receive either amifostine (500 mg prior to each carboplatin dose) or placebo. They found a significant decrease with amifostine in not only dysphagia ($p = 0.005$), but also hematologic toxicity ($p = 0.002$) and mucositis ($p < 0.001$). Recently, Brizel et al. *(37)* confirmed the radioprotective potential of amifostine in reducing xerostomia in a large randomized trial ($n > 300$) in patients receiving RT for head and neck cancer. No difference in local-regional control was found in either study.

Preliminary data have recently become available regarding the role of amifostine in patients receiving thoracic RT (+/– chemotherapy) for locally advanced NSCLC. In one randomized trial ($n = 146$), patients receiving amifostine (340 mg/m^2) prior to daily RT had a lower rate of grade ≥ 2 esophagitis ($p < 0.001$) and post-RT pneumonitis ($p < 0.001$) *(38)*. A smaller randomized trial ($n = 68$) also suggested that amifostine (300 mg/m^2), administered prechemotherapy and daily before RT, can reduce the incidence of acute esophagitis and grade 3 pneumonitis in the context of combined modality therapy *(39)*. There were no differences in the rate of complete or partial responses in either trial. The results of these trials are being further tested in a larger trial, RTOG 98-01, in which patients with locally advanced NSCLC receive chemotherapy (paclitaxel and carboplatin) both prior to and concurrent with hyperfractionated RT. Patients are randomized to the radioprotector amifostine (during the RT treatment) to determine the ability of this agent to reduce the toxicity associated with intensive concurrent chemoradiation. As part of this trial, critical quality of life data is being collected prospectively. Another potential strategy for reducing RT treatment toxicity is the future application of 3D conformal RT. As well, recent studies suggest that gene therapy may be an emerging strategy to achieve radioprotection. Athymic mice were intratracheally injected with an adenovirus containing the transgene human manganese dismutase (MnSod) involved in scavenging free radicals.

Overexpression of MnSod in the lungs of mice prior to RT reduced the rate of organizing alveolitis. Similar gene-therapy studies support the potential for protection against RT esophagitis as well *(40)*.

7. Future Approaches

The current "state-of-the-art" in managing patients with NSCLC involves pathologic staging (e.g., histology, nodal status) and anatomic imaging (e.g., computed tomography [CT], magnetic resonance imaging [MRI]). Based upon this information, a disease stage is assigned and stage-based treatment algorithms are then recommended involving conventional therapies (e.g., surgery, RT, and/or chemotherapy). In the future, however, "molecular" strategies will improve upon these traditional modalities. In addition to pathologic staging, recent studies suggest a future role for molecular/genetic staging. Beyond anatomic imaging, there is already evidence demonstrating the usefulness of functional/molecular imaging. Rather than stage-based treatment algorithms, future promise lies in molecular-based therapies.

The goal of molecular staging is to identify predictive markers of prognosis and response to treatment. Molecular-clinical correlative studies in patients with NSCLC have identified various markers with different prognostic significance, such as P53, K-ras, Bcl2, Her2, and Ki-67, among others *(41–43)*. These studies suggest that profiles of multiple markers may be required to obtain the greatest predictive value. Recently, Cox et al. demonstrated the potential of such "molecular staging" in 168 patients with resected stage I–IIIA NSCLC *(44)*. On multivariate analysis, they identified independent poor prognostic factors, including high microvessel count ($p = 0.002$), matrix metalloproteinase (MMP) -9 ($p = 0.009$), a factor facilitating tumor invasion, nodal status ($p = 0.01$), and tumor grade ($p = 0.05$). The expression of both epidermal growth factor (EGFR) and MMP –9 was associated with a particularly poor prognosis ($p = 0.0001$). With new technological developments for screening larger numbers of genes with minute amounts of genetic material, the potential for applying this prognostic information to the clinic, even for patients with inoperable NSCLC, appears more promising. With the advent of the bioinfomatic explosion, the human genome has recently been sequenced and powerful techniques, such as DNA microarrays can perform the equivalent of 5000–10,000 individual Southern-blot experiments in only a few days *(45)*. While prospective, well-designed studies are necessary to confirm the value of potential markers, this type of "molecular fingerprinting" may pave the way, in the future, for patients with NSCLC to be candidates for molecular-based therapies targeting specific pathways.

The dramatic impact of functional imaging in the management of patients with NSCLC is only beginning to be appreciated. Prior studies have demonstrated the value of positron emission tomography (PET) scanning in staging the mediastinum as well as for systemic disease *(46)*. Recently, in a phase III study, van Tinteren et al. found that PET scanning was cost-effective by reducing in half the number of futile thoracotomies in otherwise clinically staged I–III NSCLC *(47)*. Investigators are now beginning to integrate PET scanning into radiation treatment planning for conformal therapy *(48–50)*. Functional imaging will play a critical role in the radiation oncologist's ability to accurately target the "gross tumor volume" (GTV), while sparing the normal surrounding tissues. PET scanning is also showing promise in its ability to assess response to patients treated with RT or chemoradiation. In a study of 56 patients with unresectable NSCLC, MacManus et al. found that, on multivariate analysis, the PET response (but not the CT-response) was a powerful predictor of survival ($p = 0.0004$) *(51)*. In the future, rather than using ^{18}F-fluorodeoxyglucose (FDG)-PET, functional imaging with specific molecular markers expressed by individual tumors may refine this powerful technique.

Perhaps the greatest promise lies in the advent of novel molecular-based therapies. Recently, there has been an abundance of new, relatively nontoxic, biologic therapies ranging from angiogenesis inhibitors, matrix metalloproteinase inhibitors (MMPIs), signal-transduction inhibitors, immunotherapy approaches to gene-therapy strategies *(52)*. Such biologic agents, that appear to impact on the tumor milieu (such as its vasculature), may play an important role in NSCLC. For example, Volm et al. found that the expression of vascular endothelial growth factor (VEGF), a pivotal mediator of tumor angiogenesis, was an independent prognostic factor for patients with lung cancer *(53)*. De Vore et al. recently reported the results of a randomized Phase II trial of recombinant humanized monoclonal anti-VEGF in 99 patients with advanced NSCLC *(54)*. The time to progression in the high-dose anti-VEGF plus chemotherapy arm was 7.4 mo, compared with 4.2 mo in patients receiving chemotherapy alone. While these results appear promising, 6 patients who received anti-VEGF experienced sudden life-threatening hemoptysis, which was fatal in 4. Some of these biologic agents may not be as "benign" as initially presumed, and their potential toxicities must be carefully studied. ECOG is planning a Phase III trial to compare the addition of anti-VEGF to standard chemotherapy in patients with advanced lung cancer.

Some of the emerging biologic agents appear to have radiosensitizing properties. For example, IMC-C255, a monoclonal antibody (MAb) directed

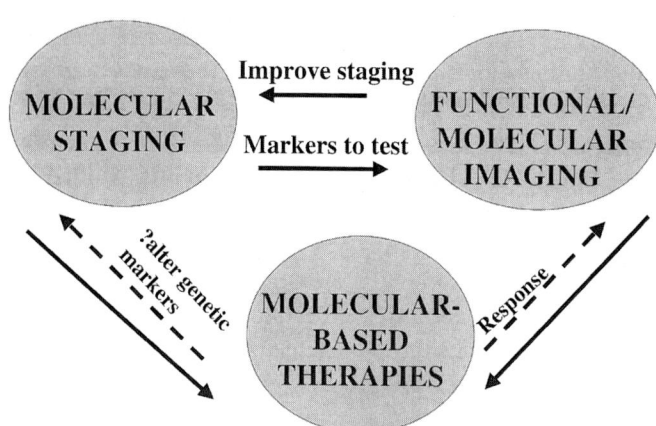

Fig. 2. Molecular-based strategies must ultimately be integrated to achieve their greatest potential.

at the EGF receptor (Cetuximab), is believed to be a potent radiosensitizer, as well as enhancing the anti-tumor activity of several chemotherapy agents, such as cisplatin, doxorubicin, and paclitaxel (55). A recent study suggests that endostatin, an anti-angiogenic factor, enhances the antitumor effects of ionizing radiation in human and murine tumors (56). Testing such agents in the setting of patients with locally advanced/inoperable NSCLC, may ultimately enhance the efficacy of more conventional therapies. In the future, the goal will be to apply "molecular fingerprinting" technology to individually tailor such biologic therapies based on unique molecular targets.

These "molecular" strategies must ultimately be integrated (see **Fig. 2**). Molecular staging will yield new biologic markers to be tested with functional molecular imaging, which will then further refine the patients' staging. Molecular staging will determine the most appropriate molecular-based therapy the response to which can be assessed using functional/molecular imaging Such novel strategies will hopefully lead to more promising results for al patients with NSCLC.

References

1. Perez, C. A., Pajak, T. F., Rubin, P., Simpson, J. R., Mohiuddin, M., Brady, L W., et al. (1987) Long-term observations of the patterns of failure in patient with unresectable non-oat cell carcinoma of the lung treated with definitiv radiotherapy. *Cancer* **59**, 1874–1881.
2. Dillman, R. O., Seagren, S. L., Propert, K. J., Guerra, J., Eaton, W. L., Perry, M. C et al. (1990) A randomized trial of induction chemotherapy plus high-dos

radiation versus radiation alone in stage III non-small cell lung cancer. *N. Engl. J. Med.* **323**, 940–945.
3. Dillman, R. O., Herndon, J., Seagren, S. L., Eaton, W., and Green, M. R. (1996) Improved survival in stage III non-small cell lung cancer: Seven-year followup of Cancer and Leukemia Group B (CALGB) 8433 trial. *J. Natl. Cancer Inst.* **88**, 1210–1215.
4. Jeremic, B., Shibamoto, Y., Acimovic, L., and Djuric, L. (1995) Randomized trial of hyperfractionated radiation therapy with or without concurrent chemotherapy for stage III non-small cell lung cancer. *J. Clin. Oncol.* **13**, 452–458.
5. Jeremic, B., Shibamoto, Y., Acimovic, L., and Milisavljevic, S. (1996) Hyperfractionated radiation therapy with or without concurrent low-dose daily carboplatin/ etoposide for stage III non-small cell lung cancer: A randomized study. *J. Clin. Oncol.* **14**, 1065–1070.
6. Le Chevalier, T., Arriagada, R., Tarayre, M., Lacombe-Terrier, M. J., LaPlanche, A., Quoix, E., et al. (1992) Significant effect of adjuvant chemotherapy on survival in locally advanced non-small cell lung carcinoma (Letter). *J. Natl. Cancer Inst.* **84**, 58.
7. Sause, W. T., Scott, C., Taylor, S., Johnson, D., Livingston, R., Komaki, R., et al. (1995) Radiation therapy oncology group (RTOG) 88-08 and eastern cooperative oncology group (ECOG) 4588: Preliminary results of a phase III trial in regionally advanced unresectable non-small cell lung cancer. *J. Natl. Cancer Inst.* **87**, 198–205.
8. Schaake-Koning, C., vanden Bogaert, W., Dalesio, O., Festen, J., Hoogenhout, J., van Houtte, P., et al. (1992) Effects of concomitant cisplatin and radiotherapy on inoperable non-small cell lung cancer. *N. Engl. J. Med.* **326**, 524–530.
9. Marino, P., Preatoni, A., and Cantoni, A. (1995) Randomized trials of radiotherapy alone vs. combined chemotherapy and radiotherapy in stages IIIa and IIIb non-small cell lung cancer: a meta-analysis. *Cancer* **76**, 593–601.
10. Non-Small Cell Lung Cancer Collaborative Group (1995) Chemotherapy in non-small cell lung cancer: a meta-analysis using updated data on individual patients from 52 randomized clinical trials. *BMJ* **311**, 899–909.
11. Pritchard, R. S. and Anthony, S. P. (1996) Chemotherapy plus radiotherapy compared with radiotherapy alone in the treatment of locally advanced, unresectable non-small cell lung cancer: A meta-analysis. *Ann. Intern. Med.* **125**, 723–729.
12. Sause, W. T., Kolesar, P., Taylor, S., Johnson, D., Livingston, R., Komaki, R., et al. (1997) Five-year results: Phase III trial of regionally advanced unresectable non-small cell lung cancer, RTOG 8808, ECOG 4588, SWOG 8892 Abstract A-1743. *Proc. Am. Soc. Clin. Oncol.* **17**, 453.
13. Clamon, G., Herndon, J., Cooper, R., Chang, A. Y., Rosenman, J., and Green, M. R. (1999) Radiosensitization with carboplatin for patients with unresectable stage III non-small cell lung cancer: a phase III trial of the Cancer and Leukemia Group B and the Eastern Cooperative Oncology Group. *J. Clin. Oncol.* **17**, 4–11.
14. Arriagada, R., Le Chevalier, T., Rekacewicz, C., Quoix, E., De Cremoux, H., Douillard, J., and Tarayre, M. (1997) Cisplatin-based chemotherapy (CT) in

patients with locally advanced non-small cell lung cancer (NSCLC): late analysis of a French randomized trial Abstract. *Proc. Ann. Meet. Am. Soc. Clin. Oncol.* **16,** 1601.
15. Furuse, K., Hosoe, S., Msuda, N., Sugiura, T., Yokota, S., Ohbayashi, K., Okazaki, M., et.al. (2000) Impact of tumor control on survival in unresectable stage III non-small cell lung cancer (NSCLC) treated with concurrent thoracic radiotherapy (TRT) and chemotherapy (CT). *Proc. ASCO* **19,** 484a.
16. Komaki, R., Seiferheld, W., Curran, W., Langer, C., Lee, J., Hauser, S., et al. (2000) Sequential vs. concurrent chemotherapy and radiation therapy for inoperable non-small cell lung cancer (NSCLC): Analysis of failures in a phase III study (RTOG 9410) Abstract 5. *Int. J. Radiat. Oncol. Biol. Phys.* **48,** 113.
17. Movsas, B. (2000) Innovative treatment strategies in locally advanced and/or unresectable non-small cell lung cancer. *Cancer Control* **7,** 25–34.
18. Gaspar, L., Gandara, D., Chansky, K., Albain K. S., Lara, P. N. Jr., Crowley, J., Livingston, R. B. (2001) Consolidation Docetaxel following concurrent chemoradiotherapy in pathologic stage IIIB non-small cell lung cancer (NSCLC) (SWOG 9504): Patterns of Failure and Updated Survival. *Proceedings ASCO* **20,** 315a. Abstract 1255.
19. Kerr, K. M. and Lamb, D. (1984) Actual growth rate and tumour cell proliferation in human pulmonary neoplasms. *Br. J. Cancer* **50,** 343–349.
20. Thames, H. D., Peters, L. J., Withers, H. R., et al. (1983) Accelerated fractionation vs hyperfractionation: rationale for several treatments per day. *Int. J. Radiat. Oncol. Biol. Phys.* **9,** 127–139.
21. Cox, J. D., Azarnia, N., Byhardt, R. W., Shin, K. H., Emami, B., and Pajak, T. F. (1990) A randomized phase I/II trial of hyperfractionated radiation therapy with total doses of 60 Gy to 79.2 Gy: possible survival benefit with >= 69.6 Gy in favorable patients with radiation therapy oncology group stage III non-small cell lung carcinoma: report of RTOG 83-11. *J. Clin. Oncol.* **8,** 1543–1553.
22. Jeremic, B., Shibamoto, Y., Radosavljevic-Asic, G., Nikolic, N., Dagovic, A., Aleksandrovic, J., et al. (2000) Hyperfractionated radiation therapy (HFXRT) and concurrent low-dose, daily carboplatin/etoposide (CE) with or without weekend CE in stage III non small cell lung cancer (NSCLC). A randomized trial. *Proc. ASCO* **19,** 504a.
23. Saunders, M. I., Dische, S., Barrett, A., et al. (1997) Continuous hyperfractionated accelerated radiotherapy (CHART) versus conventional radiotherapy in non-small cell lung cancer: A randomized multicentre trial. *Lancet* **350,** 161–165.
24. Perez, C. A., Stanley, K., Rubin, P., Kramer, S., Brady, L., Perez-Tamayo, R., et al. (1980) A prospective randomized study of various irradiation doses and fractionation schedules in the treatment of inoperable non-oat cell carcinoma of the lung. *Cancer* **45,** 2744–2753.
25. Hazuka, M. B., Turrisi, A. T., Lutz, S. T., Martel, M. K., Ten Haken, R. K., Strawderman, M., et al. (1993) Results of high-dose thoracic irradiation incorporating beam's eye view display in non-small cell lung cancer: A retrospective multivariate analysis. *Int. J. Radiat. Oncol. Biol. Phys.* **27,** 273–284.

26. Perez, C. A., Stanley, K., Grundy, G., Hanson, W., Rubin, P., Kramer, S., et al. (1982) Impact of irradiation technique and tumor extent in tumor control and survival of patients with unresectable non-oat cell carcinoma of the lung. *Cancer* **50**, 1091–1099.
27. Robinow, J. S., Shaw, E. G., Eagan, R. T., Lee, R. L., Creagan, E. T., Frytak, S., et al. (1989) Results of combination chemotherapy and thoracic radiation therapy for unresectable non-small cell carcinoma of the lung. *Int. J. Radiat. Oncol. Biol. Phys.* **17**, 1203–1210.
28. Hayman, J., Martel, M., Ten Haken, R., Normolle, D., Todd III, R., Littles, F., et al. (2001) Dose escalation in non-small cell lung cancer using three-dimensional conformal radiation therapy: update of a Phase I trial. *J. Clin. Oncol.* **19**, 127–136.
29. Mitra, R., Mah, D., Das, I., Movsas, B., and Schultheiss, T. (2000) Effect of inhomogeneity in small fields: Dosimetric implications for complex IMRT of lung 37. *Int. J. Radiat. Oncol. Biol. Phys.* **48**, 129.
29a. Socinski, M., Halle, J., Schell, M., Clark, J., Limentani, S., Mitchell, W., et al. (2000) Induction (I) and concurrent (C) carboplatin/paclitaxel (C/P) with dose-escalated thoracic conformal radiotherapy)TCRT) in stage IIIA/B non-small cell lung cancer (NSCLC): A Phase I/II trial. *Proc. ASCO* **19**, 496a.
30. Cox, J. D., Pajak, T. F., Asbell, S., Russell, A. H., Pederson, J., Byhardt, R. W., et al. (1993) Interruptions of high-dose radiation therapy decrease long-term survival of favorable patients with unresectable non-small cell carcinoma of the lung: analysis of 1244 cases from three RTOG trials. *Int. J. Radiat. Oncol. Biol. Phys.* **27**, 493–498.
31. Movsas, B., Scott, C., Sause, W., Byhardt, R., Komaki, R., Cox, J., Johnson, D., Lawton, C., Dar, A. R., Wasserman, T., Roach, M., Lee, J. S., Andras, E. (1999) The benefit of treatment intensification is age and histology-dependent in patients with locally advanced non-small cell lung cancer (NSCLC): A quality-adjusted survival analysis of Radiation Therapy Oncology Group (RTOG) chemo-radiation studies. *Int. J. Radiat. Oncol. Biol. Phys.* **45(5)**, 1143–1149.
32. Treskes, M., Holwerda, U., Nijtmans, L., Fichtinger-Schepman, A. M. J., Pinedo, H. M., and van der Vijgh, W. J. F. (1994) Modulation of cisplatin and carboplatin with WR-2721, molecular aspects. *Chem. Mod. Cancer Treat* 322–323.
33. Washburn, L. C., Rafter, J. J., and Hayes, R. L. (1976) Prediction of the effective radioprotective dose of WR-2721 in humans through an interspecies tissue distribution study. *Radiat. Res.* **55**, 100–105.
34. Yuhas, J., Spellman, J. M., and Culo, F. (1980) The role of WR-2721 in radiotherapy and/or chemotherapy. *Cancer Clin. Trials* **3**, 211–216.
35. Tannehill, S. P. and Mehta, M. P. (1996) Amifostine and radiation therapy: past, present, and future. *Semin. Oncol.* **23**, 69–77.
36. Buntzel, J., Glatzel, M., Kuttner, K., Schuth, J., Oster, J., and Frohlich, F. (1996) Selective cytoprotection with amifostine in simultaneous radiochemotherapy of head and neck cancer. *Ann. Oncol.* **7**, 81.
37. Brizel, D. M., Wasserman, T. H., Henke, M., Strnad, V., Rudat, V., Monnier, A., Eschwege, F., et al. (2000) Phase III randomized trial of amifostine as a radioprotector in head and neck cancer. *J. Clin. Oncol.* **18**, 3339-3345.

38. Antonadou, D., Coliarakis, N., Synodinou, M., Athanassiou, H., Kouveli, A., Verigos, C., Georgakopoulos, G., Panoussaki, K., Karageorgis, P., Throuvalas, N. (2001) Randomized Phase III trial of radiation treatment +/– amifostine in patients with advanced-stage lung cancer. *Int. J. Radiat. Oncol. Biol. Phys.* **51**, 915–922.
39. Antonadou, D., Synodinou, M., Bougi, M., Sagriotis, A., Paloudis, S., and Throuvalas, N. (2000) Amifostine reduces acute toxicity during radiochemotherapy in patients with localized advanced stage non small cell lung cancer. *Proc. ASCO* **19**, 501a.
40. Epperly, M. W., Bray, J. A., Krager, R. S., et al. (1999) Intratracheal injection of adenovirus containing the human MnSOD transgene protects athymic nude mice from irradiation induced organizing alveolitis. *Int. J. Radiat. Oncol. Biol. Phys.* **22**, 623–625.
41. Kim, Y.-C., Park, K.-O., Kern, J., et al. (1998) The interactive effects of Ras, HER2, p53 and Bcl2 expression in predicting the survival of non-small cell lung cancer patients. *Lung Cancer* **22**, 181–190.
42. Komaki, R., Milas, L., Ro, J., Fujii, T., Perkins, P., Allen, P., et al. (1998) Prognostic biomarker study in pathologically staged N1 non-small cell lung cancer. *Int. J. Radiat. Oncol. Biol. Phys.* **40**, 787–796.
43. Kwiatkowski, D. J., Harpole, D. H. J., Godleski, J., et.al. (1998) Molecular pathologic substaging in 244 stage I non-small cell lung cancer patients: clinical implications. *J. Clin. Oncol.* **16**, 2468–2477.
44. Cox, G., Jones, L., Andi, A., et.al. (2000) Molecular staging for operable non-small cell lung cancer Abstract 1884. *Proc. Am. Soc. Clin. Oncol.* **19**, 482a.
45. Perou, C., Sorlie, T., Elsen, M., et al. (2000) Molecular portraits of human breast tumors. *Nature* **406**, 747–752.
46. Pieterman, R., van Putten, J., Meuzelaar, J., Mooyaart, E., Vaalburg, W., Koeter, G., et al. (2000) Preoperative staging of non-small cell lung cancer with positron-emission tomography. *N. Engl. J. Med.* **343**, 254–61.
47. Van Tinteren, H., Hoekstra, O., Smit, E., et al. (2000) Randomized controlled trial (RCT) to evaluate the cost-effectiveness of positron emission tomography (PET) added to conventional diagnosis in non-small cell lung cancer (NSCLC) Abstract 1885. *Proc. Am. Soc. Clin. Oncol.* **19**, 482a.
48. Balogh, J., Caldwell, C., Ung, Y., Mah, K., Danjoux, E., Ganguli, S., and Ehrlich, L. (2000) Interobserver variation in contouring gross tumour volume in carcinoma of the lung associated with pneumonitis and atelectasis: The impact of 18 FDG-hybrid PET fusion Abstract 36. *Int. J. Radiat. Oncol. Biol. Phys.* **48**, 128.
49. Erdi, Y., Yorke, E., Erdi, A., Braban, L., Hu, Y., Macapinlac, H., et al. (2000) Radiotherapy treatment planning for patients with non-small cell lung cancer using positron emission tomography (PET) 33. *Int. J. Radiat. Oncol. Biol. Phys.* **48**, 127.
50. Schmuecking, M., Plichta, K., Lopatta, E., Przetak, C., Leonhardi, J., Gottschild, D., et al. (2000) Image fusion of F-18 FDG PET and CT: Is there a role in 3D radiation treatment planning of non-small cell lung cancer? (abstract) *Int. J. Radiat. Oncol. Biol. Phys.* **48**, 130.

51. MacManus, M., Hicks, R., Wada, M., Hogg, A., Matthews, J., Wirth, A., et al. (2000) Early F-18 FDG-PET response to radical chemoradiotherapy correlates strongly with survival in unresectable non-small cell lung cancer 1888. *Am. Soc. Clin. Oncol.* **19,** 483a.
52. Roth, J. A., Dguyen, D., Lawrence, D. D., et al. (1996) Retrovirus-mediated wild-type p53 gene transfer to tumors of patients with lung cancer. *Nat. Med.* **2,** 985–991.
53. Volm, M., Koomagi, R., and Mattern, J. (1997) Prognostic value of vascular endothelial growth factor and its receptor Flt-1 in squamous cell lung cancer. *Int. J. Cancer* **74,** 64–68.
54. DeVore, R., Fehrenbachter, L., Herbst, R., et.al. (2000) A randomized phase III trial comparing RHUMAB VEG (Recombinant humainized monoclonal antibody to vascular endothelial cell growth factor) plus carboplatin/paclitaxel (CP) to CP alone in patients with stage IIIB/IV NSCLC. *Proc. Am. Soc. Clin. Oncol.* **19,** 485a.
55. Baselga, J. (2000) New therapeutic agents targeting the epidermal growth factor receptor. *J. Clin. Oncol.* **18,** 54s–59s.
56. Hanna, N., Seetharam, S., Mauceri, H., Beckett, M., Jaskowiak, N., Salloum, R., et.al. (2000) Antitumor interaction of short-course endostatin and ionizing radiation. *Cancer J.* **6,** 287–293.

30

Photodynamic Therapy in Lung Cancer

A Review

David Ost

1. Introduction

Photodynamic therapy (PDT) involves the use of photosensitizing agents for treatment of malignant disease. These photosensitizing agents are infused intravenously and are selectively retained within tumor cells. The agents remain inactive until exposed to light of the proper wavelength. When activated by light, these compounds generate toxic oxygen radicals that result in tumor necrosis. Although several institutions worked with PDT in the 1980s, its use in the United States remained limited to research. More recently, PDT using the first FDA approved photosensitizing agent, porfimer sodium (Photofrin), has become available for routine clinical use in the United States.

PDT has the potential to complement and improve the approach to a variety of clinical problems. Potential applications include carcinoma *in situ* (CIS), early stage carcinoma in nonsurgical candidates, advanced lung cancer, and tumors metastatic to the lung. This review focuses on the mechanism of action of PDT using porfimer sodium, the technique of PDT, review of clinical studies using PDT, and an assessment of how PDT fits into a multimodality approach to clinical problems.

2. Mechanism of Action

The concept of photochemical sensitization and subsequent cell death is not a new one. Light had been used for healing by the ancient Greeks and by 1900 photochemical reactions had been used to kill paramecia. Subsequently

PDT was used to treat skin cancers in the early 1900s, leading to a variety of chemicals which were used to promote photochemical cytotoxicity *(1)*. Among the agents developed, hematoporphyrin was subsequently demonstrated to be selectively concentrated or preferentially retained within malignant cells *(2,3)*. Hematoporphyrin derivatives were subsequently shown to be retained in a large percentage of squamous and adenocarcinomas *(4–8)*. Newer hematoporphyrin derivatives were subsequently developed, with research into the possible application of these agents for use in breast cancer, bladder cancer, and other malignancies *(9–13)*. These pioneering studies highlighted the importance of several factors relating to the mechanism of PDT: membrane injury, delivery of oxygen, the role of the immune and vascular systems, the importance of the photosensitizer, and light dosimetry. Each of these factors is critical to the mechanism of action of PDT and has potential clinical implications.

2.1. Membrane Injury

The basic mechanism of cellular cytotoxicity seems to be membrane damage. The plasma membrane and mitochondrial membranes in particular are targets of PDT because of the water-lipid partition coefficient of the photosensitizing agents *(14–19)*. Porphyrin-uptake studies demonstrate initial binding with the plasma membrane with subsequent extension to the internal cellular membranes *(20)*. Damage occurs after light activation and is visible immediately. Initially it is characterized by formation of multiple areas of membrane injury or blebs. These progress to form larger balloon like areas *(16,21)*. Cellular division and normal functions cease and this is followed by cell lysis. Concurrent with injury to the plasma membrane, other internal cellular membranes are injured, including the mitochondrial membrane, nuclear membrane, Golgi apparatus, and endoplasmic reticulum *(18,22)*. With mitochondrial injury comes inhibition of oxidative phosphorylation and ATP generation, with a subsequent drop in cellular ATP *(22,23)*. Using spectroscopy with phosphate 31 to assess metabolic response to PDT, it can be shown that the fall in ATP is dramatic, becoming virtually undetectable 2–4 h after treatment *(24,25)*. Although nuclear membrane injury occurs and PDT results in DNA strand breaks, it does not appear that DNA injury plays an important role in cell death. Importantly, PDT has not been shown to be mutagenic in vitro *(16,19,26)*.

2.2. Delivery of Oxygen

PDT is dependent on the availability of oxygen at the site of injury to be effective. PDT cytotoxicity is free radical-mediated via a type II photooxidation reaction *(16,19)*. In type II reactions, light energy excites and activates the photosensitizer, energy transfer occurs from the sensitizer in its excited state to molecular oxygen, and there is generation of singlet oxygen species *(29)*. For PDT to be effective there must be oxygen available to facilitate this type II photoxidation reaction. The singlet oxygen species and resulting free radical generation leads to the membrane injury described earlier. Thus, membrane injury is predicated on the adequate availability of oxygen to generate free radicals.

In vitro, when oxygen is not present or is present at less than 2%, cells become resistant to PDT *(27,28)*. The importance of free radical-mediated injury is emphasized by the observation that free radical scavengers, such as 1,3-diphenylisobenzofuran, reduce PDT-mediated cytotoxicity *(30,31)*. Low oxygen concentrations have also been shown to lead to decreased sensitivity in vitro and tissue hypoxia models in animals support this. The clinical significance of this may be that local tumor hypoxia may account for some cases of nonresponsiveness to PDT, as has been suggested by some investigators.

2.3. Role of the Immune and Vascular Systems

Associated with this free radical injury is a complex mixture of vascular injury, coagulation, and a concurrent immune response characterized by platelet and neutrophil activation *(16,31)*. Part of the in vivo tumor destruction results from the effect of PDT on the vasculature. The neovasculature of tumors may be a target for PDT since these vessels, venous-derived, may not have sufficient strength to remain patent in the face of high extravascular pressures. Decreased flow occurs, leading to arteriolar and venular stasis, arteriolar vasoconstriction, thrombosis of venules, and increasing interstitial edema *(32,33)*.

In addition, PDT results in varying degrees of increased coagulation in the vascular bed. Studies with nuclear magnetic resonance imaging using *in situ* fluorine have demonstrated that damage to the tumor vasculature occurs prior to actual tumor necrosis *(34)*. Ben-Hur and Orenstein demonstrated increased coagulation associated with injury to the endothelium from PDT, with resultant red blood cell agglutination and thrombus formation *(35)*.

Concurrent with this free radical injury, coagulation, and vascular injury is an important localized immune response. This results in the release of vasoactive compounds including arachidonic acid derivatives, such as prostaglandin E2, I2, and thromboxane. The potential contribution of these mediators to cell death and tumor injury has been demonstrated by the observation that cyclooxygenase inhibitors reduce the effect of PDT on arterioles *(36)*. Thus, in addition to oxygen-mediated free radical membrane injury and ATP depletion, direct vascular injury, local coagulation, and immune responses play an important role in the mechanism of action of PDT.

2.4. Photosensitizers

A wide variety of photosensitizing agents have been studied and developed, including the chlorins, phthalocyanines, tetraphenylporphine sulfate and the porphines, rhodamine 123, and the porphyrin-based agents. The clinically significant aspects of photosensitizers include consideration of their biochemical properties as well as their biophysical properties. Biochemical properties that are of clinical significance include the metabolism and relative concentration of photosensitizers in tumors. Biophysical properties that are of clinical significance include their yield of singlet oxygen, the amount of tissue penetration allowed, and their photostability/lability. This discussion will examine these considerations in relation to porfimer sodium (Photofrin), since this is the only agent currently approved by the FDA for treatment of lung cancer in the United States. Assessment of these aspects, however, is equally important when examining the potential utility of other agents currently under investigation.

2.4.1. Biochemical Considerations

The concentration of the photosensitizer in the tumor tissue relative to the concentration in normal tissue is an important consideration, since this determines in part the selectivity of PDT treatment. A high tumor to normal tissue ratio is desirable, since this will minimize side effects on nearby normal tissue. Selective retention or uptake by tumors allows for a relatively high tumor:tissue concentration ratio of the photosensitizer. Most sensitizers range in concentration ratio from two to five to one. The selective retention associated with the porphyrin-based agents was initially reported by Figge in 1948 *(2)*. This was subsequently refined by Lipson and colleagues who demonstrated that derivatives of hematoporphyrin were associated with an even higher tumor:tissue concentration ratio *(4–8)*. This tumor to tissue ratio is highest at 24–48 h after intravenous injection. However, Lipson and colleagues also demonstrated that fluorescence could be picked up within 3 h of intravenous injection.

This observation of the significant difference in time between detectable tissue fluorescence and peak tumor:tissue ratio has clinical importance. When used for lung cancer treatment (i.e., PDT), the tumor:tissue ratio is critical, hence the time interval between injection and light application is typically 48–72 h. When used in investigational studies for tumor diagnosis and detection, it is more practical to have as brief a delay as possible between injection of the agent and the diagnostic procedure. In this setting, the 3-h time interval is more useful. The utility of future photoactive agents will depend in part on these considerations. Ideal agents for treatment would achieve peak tumor:tissue ratios more rapidly, allowing more rapid treatment, and would then break down, minimizing photosensitivity side effects. Similarly, shorter time to detectable fluorescence is an important practical consideration if photoactive agents are ever to be used on a routine basis for tumor diagnosis and localization. Thus, an understanding of the mechanisms resulting in high tumor:tissue concentrations may be of practical and clinical importance.

The mechanisms involved in selective retention and uptake of porphyrin-based sensitizers in tumor tissue is an ongoing area of investigation. Murine models demonstrate that sensitizers are accumulated and retained by endocytosis via the vascular endothelium *(16,31,37)*. This is in part due to the lipophilic nature of the compounds, previously described in relation to their propensity to bind to cell membranes based on their partition coefficient. The distribution pattern after intravenous infusion mirrors that of low-density lipoprotein receptors in the various organs, with the greatest amount being in the liver, followed in descending order by the adrenal glands, urinary bladder, pancreas, kidney, spleen, stomach, bone, lung, heart, muscle, and brain *(38)*. The serum half-life in humans is 20–30 h, but photosensitizing components may persist in the skin at low levels of 2–5% for up to 4–6 wk *(39)*. Possible theories for the high tumor:tissue concentration of porphyrin-based sensitizers include increased tumor uptake of low-density lipoprotein associated sensitizers, increased uptake by tumors due to their lower pH and the associated increase in water solubility of the sensitizer with low pH, poor lymphatic drainage of tumors, tumor angiogenesis factors, and changes within the stromal cells of tumors that increase uptake *(16,31)*.

2.4.2. Biophysical Considerations

As described earlier, PDT cytotoxicity is mediated via a complex set of interactions including photooxidative reactions. The photosensitizer must be able to absorb photons of appropriate wavelengths so as to become a triplet species. The light-excited sensitizer loses or accepts electrons with generation of secondary radicals to produce singlet oxygen. This is the type II photooxidation reaction described earlier. Importantly, singlet oxygen can be generated with quantum energies as low as 0.98 MeV, corresponding to a wavelength of 1220 nm. However, most available sensitizers work efficiently only with wavelengths up to 850 nm with a quantum yield of 0.2–0.6 *(40)*. The differences in quantum yield appear to be related, or at least effected by, the location of the sensitizer in the cells. In addition, for any given quantum yield, lipophilic sensitizers, such as porfimer sodium, seem to be more efficient than hydrophilic sensitizers at similar quantum yields.

The quantum energy used relates to the important clinical issue of tissue penetration. The wavelength, as determined by the quantum energy used, directly effects the maximum absorbance capacity and the depth of tissue penetration. With respect to the porphyrin family of photosensitizers, the absorption spectra demonstrate a peak at 405 nm. Use of this peak allows for fluorescence of tumors to be used as a tumor marker. However, for treatment (PDT), this wavelength of light is suboptimal, since it is nearly completely absorbed within 1 mm of the surface. Thus, for PDT, a wavelength of 630 nm is used. This allows for penetration to a depth of approx 5 mm. Obviously, the development of future photosensitizers will be greatly effected by consideration of the quantum energies used, their absorption spectra, and the resultant clinical consequences in terms of depth of penetration.

Finally, all photosensitizers can be destroyed by light and this may have an impact on PDT cytotoxicity. When exposed to light, a sensitizer may generate enough energy to destroy tumor tissue but be destroyed and lose its cytotoxic potential within normal cells, a process described as photobleaching. This effect is an important consideration, since it allows minimal damage to adjacent normal tissue while selectively destroying tumor cells.

While these considerations may seem esoteric, they have direct impact in terms of clinical efficacy and feasibility. For optimum efficacy, future photosensitizers will need to have higher quantum yields using an absorption spectra that allows a clinically useful depth of penetration while making use of photobleaching properties to minimize damage to adjacent tissues. However, to be clinically useful, there must also be careful and tightly controlled light delivery to the site of the photosensitizer to take optimum advantage of the biochemical and biophysical properties of the photosensitizer.

2.5. Light Dosimetry

PDT is dependent on accurate delivery of light to the area to be treated. It is useful to think of PDT in terms of four components: the photosensitizer's characteristics and concentration at the tumor site, the light source, the rate of energy delivery (power), and the total energy delivered. The biochemical and biophysical characteristics of photosensitizers have been described earlier. With respect to the delivery of light, any source of light with the appropriate characteristics could be used. In practice, laser light is typically used because it offers the advantage of a uniform spectrum and coherence. For lung cancer, the argon dye pump laser or the excimer laser is often used, although in theory any laser with the proper wavelength and power could be used.

Important in vivo considerations include the effect of dose rate delivery and total energy delivery. While in vitro evidence has suggested that high dose rate (power) is associated with improved cytotoxicity, in vivo data have demonstrated that lower dose rates may be more effective *(41,42)*. When treating human mesothelioma allografts in nude mice, decreasing the intensity from 200 mW/cm^2 to 50 mW/cm^2 actually improved response. The investigators hypothesized that reduction of the fluence rate or fractionation may paradoxically increase treatment effect because of an increase in singlet oxygen in regions of poor capillary flow. Whether this is the only mechanism or not, the importance of this study is the empiric observation that light dosimetry has a clinical impact on response and side effects. Thus, controlled and reproducible light dosimetry and the technology used to deliver it are an important consideration for PDT. Studies of dosimetry for future agents must study both the rate of energy delivery (power) as well as the total energy delivered.

3. Method and Technique

The method of PDT is a result of its mechanisms of action and other factors as described above. Based on the mechanism of action, we can anticipate that PDT will require at least two distinct steps: delivery of the photosensitizer and subsequent activation with laser light. In lung cancer, this is usually followed by a third step consisting of bronchoscopic removal of necrotic debris.

The first step involves infusion of the photosensitizer. This is accomplished through a simple peripheral IV with infusion of porfimer sodium (Photofrin) at a dose of 2 mg/kg. After infusion, the patient becomes photosensitive, so careful instruction must be given prior to this regarding appropriate avoidance of direct sunlight. Indoor light is no problem. Typically a handbook is given out prior to infusion with concurrent verbal instruction. The patient will be photosensitive for 1 mo.

The second step is activation of the photosensitizer by light. For porfimer sodium, the wavelength used is 630 nm. This is accomplished using an appropriate laser system delivered via the bronchoscope. The laser fibers can be tailored to fit the clinical situation, with cleaved probes for forward light projection, bulbous tips for isotropic spherical distribution, or cylindrical coatings for light perpendicular to the axis of the fiber. The fiber can be inserted into endobronchial tumors directly (interstitial delivery) or placed alongside the tumor. With a wavelength of 630 nm, the tissue effect of PDT can be anticipated to be approx 5 mm in depth.

We prefer cylindrical light diffusing fibers when working within the bronchi. The reason for this is the importance of light dosimetry. When using a forward projecting fiber or a bulbous tip spherical distribution fiber, the dosimetry is dependent upon certain assumptions which may or may not be true. For example, when using a forward-projecting fiber, the power output per square centimeter is dependent on the area illuminated. For a fixed power output, the power per square centimeter will decrease as the area illuminated increases.

While the area illuminated can be accurately measured when treating skin metastases to the chest wall, this cannot be done within the airway. If the forward projecting fiber is closer to the tumor, a smaller area would be illuminated. In this situation, a higher power per square centimeter would result. However, estimating the distance to the tumor and measurement of the area illuminated by a forward projecting fiber is difficult. This can result in highly variable power delivery per square centimeter. Therefore, although the power going to the fiber is controlled, the power per square centimeter is not. Thus, control of dosimetry may be adversely affected. Whether or not this affects clinical outcomes is unclear but is discussed below in the section on clinical studies.

When using a cylindrical light diffusing fiber, a power of 400 mW/cm is typically used, with a total energy delivery of 200–300 J/cm. Typically 200 J/cm will be used for CIS while up to 300 J/cm will be used for more advanced disease. Of the cylindrical fibers currently available for use with the bronchoscope, the 1-cm and 2.5-cm fibers are the most useful and are commercially available. Thus, the initial laser bronchoscopy will consist of simply placing a nonthermal laser either in or adjacent to the tumor for 500–750 s. Tumor necrosis will then take place over the next 24–48 h.

The third step involves removal of necrotic debris. Twenty-four to forty-eight hours after the initial laser treatment, the tumor will have become necrotic, and depending on the size of the tumor and the ability of the patient to cough, there will be obstruction of the airway. At this time repeat bronchoscopy will be needed for pulmonary hygiene. If residual tumor is present, repeat laser treatment can be done after removal of necrotic debris since the photosensitizer will be present at adequate tissue levels for a prolonged period of time. If additional laser treatment is given, repeat pulmonary hygiene bronchoscopy is mandatory 24–48 h later. Aggressive debridement of all necrotic debris is critical since this debris will absorb light and limit efficacy of any additional laser treatments in addition to leading to atelectasis and respiratory compromise. The tumor at this stage is very avascular and will appear white if PDT has been successful. It will have the consistency of very thick mucus and will be somewhat gelatinous in nature. As predicted based on the mechanism of injury (*see* **Subheading 2.3.**), there is virtually no bleeding with mechanical debridement if PDT has been successful.

4. Clinical Studies

Potential clinical applications of photosensitizers include early diagnosis and localization of lung cancer, treatment of CIS, and treatment of advanced endobronchial disease from both primary and metastatic tumors. This review will focus on PDT as a therapeutic tool. The diagnostic applications of photosensitizers and fluorescence bronchoscopy systems have been reviewed elsewhere. The diagnostic use of photosensitizers has made use of changes in the fluorescence spectrum of tumor tissue as compared to normal tissue. The hematoporphyrin family of photosensitizers have been demonstrated to successfully localize radiographically and bronchoscopically occult tumors *(43–45)*. Fluorescence bronchoscopy uses this same principle without any photosensitizer to try to localize tumors. These diagnostic uses of PDT are still in the research phase of development. PDT's use in clinical practice today can be divided into applications for CIS and for advanced endobronchial disease.

4.1. Carcinoma In Situ (CIS)

CIS represents one clinical scenario where PDT may be particularly applicable. The relatively noninvasive nature of the procedure combined with selective tumor destruction, the preservation of lung function, and the ability to repeat treatments in a group of patients at risk for second primary tumors make it intuitively appealing. Analysis of the literature for PDT in early stage lung cancer needs to take into consideration a variety of factors, including variations in case finding methods, definition of CIS used, photosensitizing agents, light dosimetry, and outcome measures (*see* **Table 1**).

Many studies with PDT for CIS include patients with early stage carcinoma who were at high risk for surgical intervention. Thus, the definition of CIS in some of these studies actually includes stage IA tumors that are more advanced than CIS and presumably have a worse prognosis. Thus, "true" CIS may actually have a better prognosis than represented in some studies. However, distinguishing "true" CIS from early stage IA disease may be difficult. Importantly, case finding, outcomes, photosensitizers, and light dosimetry vary significantly in the different studies, making generalizations difficult.

Given these limitations, several important conclusions can be drawn. First, despite the variability in study design, methods, and outcome measures, the results suggest local complete and partial remission rates of 70–100% (*see* **Table 1**) *(46–50)*. In the study by Edell and Cortese, at a follow-up of 7–49 mo, 77% of the tumors demonstrated no recurrence. In this trial, 14 early stage cancers were treated using light dosimetry of 200–400 J/cm^2 at 630 nm 2–4 d after infusion of hematoporphyrin derivative *(46)*. Similarly, in a study by Kato et al. the complete response rate was 94.2% for lesions with a longitudinal length of less than 1.0 cm but decreased to 37.5% for tumors with a longitudinal length of greater than 2 cm *(47)*. Five year survival in this study involving 95 tumors was 68.4% by Kaplan Meier analysis. Furuse et al. demonstrated

Table 1
PDT in Early Stage Lung Cancer

Reference	Tumors (n)	Clinical stage and indication	Drug	Complete response	Partial response	No response	Recurrence (%)
46	14	IA	HPD	93%	7%	0%	21%
47	95	CIS and IA	Photofrin	83%	17%	0%	6%
48	59	CIS and IA	Photofrin II	85%	10%	5%	10%
49	39	CIS (17)	Photofrin	100%	0%	0%	29%
		IA (22)			50%	45%	5%
50	23	Early stage	HPD	70%	30%	0%	48%

HPD, Hematoporphyrin derivative; CIS, Carcinoma *in situ*.

similar findings with respect to the impact of tumor size and response rates. In this series, in tumors less than 1 cm in length, the complete response rate was 97.8% as compared to 42.9% for larger tumors *(48)*.

Thus, it appears that an important predictor in terms of utility of PDT as a viable alternative for "early stage" disease is relatively small tumor size, probably less than 20 mm. Certainly, patients with CIS as detected by fluorescence bronchoscopy would be logical candidates. In the setting of early stage disease but not true CIS, PDT may still be an option in carefully selected patients with smaller, central, squamous cell carcinoma, with high surgical risk or other comorbidities *(49,50)*. In this setting, the expected response rate should be lower than indicated by studies of true CIS. The role of PDT in combination with other modalities, such as radiation, chemotherapy, or brachytherapy in this setting has not been fully defined.

Other factors that have been associated with treatment failures include tumor distal to segmental bronchial bifurcations, bronchial stump tumors, and underestimation of the true tumor size. In an analysis of 23 patients with intraluminal stage I lung cancer, Sutejda et al. demonstrated frequent treatment failures with distal segmental tumors or stump recurrences and attributed this to insufficient dosimetry and possibly inability of the light to penetrate the bronchial stump to an adequate depth *(49)*. An animal study using porcine tracheas evaluating light dosimetry supported this theory, demonstrating that there were significant variations in light dosimetry using this model *(51)*. This relates to our preference for cylindrical light diffusing fibers over forward projecting fibers. It is the high level of control and reproducible dosimetry of the cylindrical light fibers that offers a potential advantage based on these clinical studies of treatment failures.

As mentioned in the discussion on CIS, the area of tumor involvement is an important predictor of success. Unfortunately, accurately measuring the area of involvement is difficult, as is measuring the depth of tumor invasion. Difficulty in assessing depth of invasion may lead to higher recurrence rates. Hayata et al. found that PDT was less effective in obtaining the outcome of complete response when tumors had extended beyond the cartilaginous layer *(52)*. Since it has been shown previously that only 68% of radiographically occult lung cancers were truly confined to the cartilaginous layer, this finding of significant recurrence rates is consistent with previous studies *(53)*. More accurate preoperative assessment is needed, possibly with fluorescence bronchoscopy and endobronchial ultrasound. These techniques may be helpful in identifying those patients with unrecognized extension of their disease or those with residual tumor.

The side effect profile of PDT for early stage disease has been very favorable. Reactions in this low risk group (as compared to patients with advanced unresectable lung cancer) predominantly include mild sunburn type reactions. In many of these studies, patients with low Karnofsky Performance Scores (<40) and multiple other comorbidities have successfully undergone PDT. Thus, from a side effect and safety perspective, PDT certainly has advantages as a treatment for CIS.

Importantly, not all early stage carcinomas may be equally amenable to PDT and the efficacy of PDT could presumably change with the type of tumor, location, and method of screening. Much of the work with CIS and early stage tumors has been carried out on patients in whom squamous cell carcinoma was the predominant cell type. The incidence of occult lymph node metastases in cases of squamous cell carcinoma is low when these tumors are less than 2 cm in length *(54)*. This suggests that bronchoscopic treatment may be appropriate in these cases. However, occult lymph node metastases are more common with adenocarcinoma. Whether or not PDT would be less effective if CIS with adenocarcinoma features were identified is unknown given the level of evidence currently available. Thus, determining whether or not current recommendations will remain valid when applied to other populations with a higher incidence of adenocarcinoma using different case-finding methods and with different tumor locations (i.e., more peripheral), will be an important future area of investigation.

The context in which PDT is used is also important to consider. Even if PDT were effective only for early stage squamous cell carcinoma, this would represent a significant advance, but only if there was a feasible way to identify patients at an early stage. Thus, any evaluations of PDT treatment efficacy for CIS must take into account case-finding methodology, tumor type, tumor location, and treatment alternatives. If PDT is to have a significant impact for CIS, there must be a cost-effective method for screening and localizing disease. Currently there is no proven and effective method for screening, localization, and early identification of CIS.

4.2. Advanced Endobronchial Tumors

PDT has also been evaluated as a potential tool for treatment of patients with unresectable lung cancer and advanced endobronchial disease. Since there is currently no method in wide clinical use for early detection or cancer localization, this population represents a far larger group of patients than the CIS group. The standard of therapy for advanced endobronchial disease currently is Nd-YAG laser resection. Other alternatives include radiation therapy, brachytherapy, cryotherapy, and electrocautery. Most of the data with PDT comes from case series. There are no randomized controlled trials using a placebo arm and randomized trials comparing PDT with other modalities are rare. Outcome measures are also not standardized, further limiting the ability to compare the results of PDT in different trials. Reports on the efficacy of PDT in this area are therefore difficult to compare to results with other modalities.

In a large prospective series of 100 patients with advanced inoperable stage IIIA–IV bronchogenic cancer with endobronchial obstruction, PDT resulted in significant improvement in terms of endobronchial obstruction, pulmonary function testing, and palliation of symptoms *(55)*. Mean percent of endoluminal obstruction fell from 85.8% to 17.5% with an improvement in forced vital capacity (FVC) and forced expiratory volume in 1 s (FEV1) of 430 mL and 280 mL, respectively. In another series by McCaughan in patients with advanced primary lung cancer, the mean endobronchial obstruction fell from 84% prior to PDT to 18% 4 wk later *(56)*. Thus, there is evidence that PDT can perform well for palliation of advanced endobronchial obstruction.

How does PDT compare to other treatment modalities? In a study by Lam et al. of PDT combined with radiotherapy (RT) vs RT alone for patients with advanced endobronchial obstruction in 41 patients, PDT-RT was dramatically better than RT alone *(57)*. Only 10% of patients had complete reopening of the airway with RT alone, while 70% had reopening of the airway with PDT-RT.

There have been 3 randomized trials comparing PDT with Nd-YAG laser that suggest results with PDT are at least comparable in selected patients with advanced disease *(see* **Table 2**). In a prospective comparison of PDT vs Nd-YAG laser in 26 patients with stage III inoperable lung cancer, Moghissi et al. demonstrated better results with PDT than with Nd-YAG laser therapy *(58)*.

Table 2
PDT vs Nd-YAG Laser in Advanced Lung Cancer

Reference	Tumors (n)	Stage	Drug	Initial response	Prolonged response
58	26	Advanced	Photofrin	Similar for % Obstruction PDT better for FEV1, FVC	Improved with respect to % obstruction
19	211	Advanced	Photofrin	Similar	Improved with respect to % obstruction
59	31	I-IV	Photofrin	Similar	Improved in terms of time to treatment failure

The median percent obstruction was similar in the two groups prior to treatment (83.4% for Nd-YAG and 88.7% for PDT group) but after 1 mo the post-treatment percent obstruction was significantly better in the PDT group (39.1% for Nd-YAG vs 17% for PDT). There was also a significantly better change in FEV-1 and FVC in the PDT group. In another prospective randomized trial of PDT vs Nd-YAG laser in 211 patients with partially obstructive lung cancer conducted in 35 centers in Europe and the United States, response rates were similar at 1 wk but by 1 mo the PDT group demonstrated a benefit compared to the Nd-YAG group (Europe 61% vs 36%, US 42% vs 19%) *(19)*. Improvement in terms of complete response based on biopsy proven results was better in the PDT group vs the Nd-YAG groups as well (Europe 12% vs 3%, US 6% vs 5%). In a prospective randomized trial comparing PDT using the flexible bronchoscope with Nd-YAG using the rigid bronchoscope, a study by Diaze-Jimenez and colleagues demonstrated similar improvements in symptomatic relief after treatment with increased survival in the PDT group *(59)*. Conclusions from this study however are limited by the fact that by chance assignment, the PDT group contained fewer patients with advanced lung cancer.

In summary, although there is only limited data, PDT seems to have comparable efficacy to Nd-YAG and is superior to radiation therapy alone. Advantages of PDT include the technical ease of the procedure, greater margin for error, especially in smaller bronchi, less risk of bronchial perforation, decreased risk of intraoperative hemorrhage, and perhaps longer duration of response. It may be that this longer duration of response is secondary to destruction of invisible submucosal tumor that is missed with the Nd-YAG laser.

Importantly, PDT does have important limitations in terms of treatment for advanced endobronchial disease. Among these limitations are extremely high cost ($5,000 per patient for the photosensitizer alone), the need for repeat pulmonary toilet bronchoscopies, and photosensitivity of the skin. In addition, PDT is primarily effective for endobronchial tumors and is not effective for obstruction caused by extrinsic compression or submucosal spread. Notably, Nd-YAG, cryotherapy, and electrocautery are not effective interventions for these problems either. Airway stenting is probably the treatment of choice in these situations. In addition, PDT is much slower than Nd-YAG laser. For patients with acute respiratory compromise, Nd-YAG laser can achieve relief of dyspnea much more quickly. While PDT in mechanically ventilated patients has been reported, the delay in response makes it less desirable if Nd-YAG can be done with adequate safety *(60)*. Finally, because PDT causes significant tumor necrosis and mucus plugging, based on our experience we have chosen not to use it when there is significant tracheal stenosis. In these situations, the concern is that the resulting tumor necrosis and mucus plugging may result in sudden airway occlusion and respiratory failure. Thus, in the setting of tracheal obstruction, based only on anecdotal evidence, Nd-YAG is preferable.

5. Role of PDT in a Multimodality Approach

PDT is a promising approach that may complement other techniques in dealing with several clinical problems. Among the clinical problems that PDT may be effective in dealing with are those related to CIS and endobronchial tumors in patients with advanced disease. In each of these cases, the effectiveness of PDT and its appropriate use need to be considered in light of a multimodality approach to the problem.

In the case of CIS, the future impact of PDT is dependent on case-finding and screening methods. Currently, no screening method has been proven to be of benefit, although screening with spiral CT scan has had preliminary promising results. Certainly any patient with good cardiopulmonary function should be considered for surgery, but in many cases other comorbidities, such as chronic obstructive lung disease, may preclude surgery. For PDT to be effectively utilized in these cases will therefore take more than just a screening method. It will be necessary to identify those patients at high risk for occult lymph-node metastases where surgery, even if it is higher risk, may be warranted. The diagnostic approach will therefore have to take into account the treatment options available to that patient. For stage I disease, chemotherapy does not offer any additional benefit over surgery alone. Whether or not this will be true of PDT for CIS is unknown, since there are no large randomized trials available yet on whether concurrent chemotherapy or radiation is of added benefit for CIS treated with PDT. In a study by Cortese et al. in early stage squamous cell carcinomas, PDT used in a multimodality approach allowed 43% of patients to be spared an operation *(50)*.

In the case of advanced endobronchial disease, PDT has been shown to be useful in treating endobronchial tumors that are causing clinically significant dyspnea or are likely to progress and lead to further clinical complications, such as postobstructive pneumonia. Careful patient selection is critical in this setting. Since treatment with PDT in these cases is purely palliative, only lesions in the larger airways should be treated. If a patient with metastatic disease has a small lesion visible in a subsegmental airway, but it is not causing symptoms, it is not worth treating. Similarly, for patients with no viable lung distal to the obstruction, PDT is not warranted. The key question is whether or not removal of this localized obstruction will result in a meaningful change in the patient's symptoms, quality of life, or later risk of complications. When approaching these complicated patients, it will be important to combine it with other interventional tools, such as airway stents, Nd-YAG laser, and cryotherapy. Each tool will have its own place depending on the problem and in many cases multiple tools will be needed. For example, in the case of endobronchial obstruction with concurrent extrinsic compression, a combined approach using PDT with stenting may be best. Either tool alone will not work.

A second issue is integrating PDT with radiation and chemotherapy. Although it certainly appears that PDT plus RT is superior to RT alone, the optimal sequencing has yet to be refined. Similarly, PDT to date has shown remarkably few interactions with other chemotherapeutic agents. Although there are theoretical concerns with interactions with agents that may generate additional or cumulative free radical injury, whether or not this leads to a clinically significant increase in complications is unclear. The large number of patients treated with PDT to date suggests no clear trend toward increased complications with any particular chemotherapeutic agent currently in use.

There are however, theoretical advantages to earlier treatment with PDT prior to chemotherapy or radiation in patients with advanced lung cancer. By performing endobronchial interventions such as PDT up front, prior to chemotherapy, higher risk situations induced by chemotherapy side-effects can be avoided in the future. If a patient develops a postobstructive pneumonia while on chemotherapy and is pancytopenic, any laser intervention by that time will be too high-risk. By performing PDT prior to chemotherapy, this situation can be avoided. Similarly, based on anecdotal data, we initiate laser interventions, whether PDT or Nd-YAG, prior to radiation if possible, since the amount of time to restore a patent airway and reduce the risk of postobstructive pneumonia is shortened. In addition, by relieving airway obstruction, patients have an improved pulmonary reserve and are better able to deal with other complications, such as pneumonia or radiation pneumonitis.

Based on these considerations, it becomes clear that for PDT to be optimally effective, a team approach integrating PDT into a multimodality treatment program is necessary. The optimum sequencing of these modalities requires further study but will always need to be individualized to the patient.

6. Summary

PDT utilizes photosensitizing agents that are selectively retained by tumors. These agents have high resulting tumor:tissue concentrations but are inactive by themselves. When activated by light, they generate free radicals, resulting in membrane injury, vascular injury, and immune-mediated injury with relatively selective cytotoxicity to tumor cells. Clinically, PDT can be used to treat CIS as well as more advanced lung cancers with endobronchial obstruction. PDT should be viewed as one tool of many that can be used to deal with airway problems in patients with lung cancer and it will often need to be used in conjunction with other techniques, such as airway stenting, Nd-YAG, or cryotherapy. For PDT to be effective, it must be integrated into a multimodality approach, including chemotherapy and radiation therapy.

References

1. Jesionek, A and Tappeiner, V. H. (1903) Zur behandlung der hautcarcinomit mit fluoresceirenden stoffen. *Muench Med. Wochneshr.* **47,** 2042.
2. Figge, F. H. J., Weiland, G. S., and Manganiello, L. O. J. (1948) Cancer detection and therapy. Affinity of neoplastic embryonic and traumatized tissue for porphyrins and metalloporphyrins. *Proc. Soc. Exp. Biol. Med.* **68,** 640.
3. Auber, H. and Banger, G. (1942) Untersuchungen uber die Rolle der Porphyrine bei geschwulstkranken Menschen und tieren. *Z. Krebsforsch* **53,** 65.
4. Lipson, R. L. and Baldes, E. J. (1960) The photodynamic properties of a particular hematoporphyrin derivative. *Arch. Dermatol.* **82,** 508.
5. Lipson, R. L. and Baldes, E. J. (1961) The use of a derivative of hematoporphyrin in tumor detection. *J. Natl. Cancer Inst.* **26,** 1.
6. Gray, M., Lipson, R. L., Mack, J. V. S., et al. (1967) Use of hematoporphyrin derivative in detection and management of cervical cancer. *Am. J. Obstet. Gynecol.* **9,** 766.
7. Gregorie, H. G. Jr., Horger, E. O., Ward, J. L., et al. (1968) Hematoporphyrin-derivative fluorescence I malignant neoplasms. *Ann. Surg.* **167,** 820–828.
8. Lipson, R. L. and Baldes, E. J. (1961) The hematoporphyrin derivative: A new aid for endoscopic detection of malignant disease. *J. Thorac. Cardiovasc. Surg.* **42,** 623.
9. Kelly, J. F., Snell, M. E., and Berenbaum, M. C. (1975) Photodynamic destruction of human bladder carcinoma. *Br. J. Cancer* **31,** 237–244.

10. Lipson, R. L., Gray, M. J., and Baldes, E. J. (1966) Hematoporphyrin derivative for detection and management of cancer. *Proc. 9th Int. Cancer Cong. Tokyo, Japan* **393**.
11. Dougherty, T. J., Kaufman, J. E., Goldfarb, A., et al. (1978) Photoradiationtherapy for the treatment of malignant tumors. *Cancer Res.* **38**, 2628–2635.
12. Dougherty, T. J. (1974) Activated dyes as antitumor agents. *J. Natl. Cancer Inst.* **51**, 1333–1336.
13. Dougherty, T. J. (1984) Photoradiation therapy. *Urol. Suppl.* **23**, 61.
14. Christensen, T., Moan, J., Smedshammer, L., et al. (1985) Influence of hematoporphyrin derivative (HPD) and light on the attachment of cells to the substratum. *Photochem. Photobiophys.* **10**, 53.
15. Denstaman, S. C., Dillehay, L. E., and Williams, J. R. (1985) Enhanced susceptibility of HPD-sensitized phototoxicity and correlated resistance to trypsin detachment in SV40 transformed IMR-90 cell. *Photochem. Photobiophys.* **10**, 53.
16. Pass, H. I. (1993) Photodynamic therapy in oncology: mechanism and clinical use. *J. Natl. Canc Inst.* **85**, 443–456.
17. Salet, C. (1986) Hematoporphyrin and hematoporphyrin-derivative photosensitization of mitochondria. *Biochimie* **68**, 865–868.
18. Murant, R. S., Gibson, S. L., and Hilf, R. (1987) Photosensitizing effects of photofrin II on the site-selected mitochondrial enzymes adenylate kinase and monoamine oxidase. *Cancer Res.* **47**, 4323–4328.
19. Dougherty, T. J., Gomer, C. J., Henderson, B. W., et al. (1998) Photodynamic therapy. *J. Natl. Cancer Inst.* **90(12)**, 889–905.
20. Kessel, D. (1986) Sites of photosensitization by derivativesof hematoporphyrin. *Photochem. Photobiol.* **44**, 489–93.
21. Volden, G., Christensen, T., and Moan, J. (1981) Photodynamic membrane damage of hematoporphyrin derivative-treated NHIK 3025 cells in vitro. *Photochem. Photobiophys.* **3**, 105.
22. Hilf, R., Smail, D. B., and Murant, R. S. (1984) Hematoporphyrin derivative-induced photosensitivity of mitochondrial succinate dehydrogenase and selected cytosolic enzymes of R3230AC mammary adenocarcinomas of rats. *Cancer Res.* **44**, 1483–1488.
23. Hilf, R., Murant, R. S., Narayanan, U., et al. (1986) Relationship of mitochondrial function and cellular adenosine triphosphate levels to hematoporphyrin derivative-induced photosensitization in R3230AC mammary tumors. *Cancer Res.* **46**, 211–217.
24. Mattiello, J., Evelhoch, J. L., Brown, E., et al. (1990) Effect of photodynamic therapy on RIF-1 tumor metabolism and blood flow examined by 31P and 2H NMR spectroscopy. *NMR Biomed.* **3**, 64–70.
25. Dodd, N. J., Moore, J. V., Poppitt, D. G., et al. (1989) In vivo magnetic resonance imaging of the effects of photodynamic therapy. *Br. J. Cancer* **60**, 164–67.
26. Gomer, C. J., Rucker, N., Banerjee, A., et al. (1983) Comparison of mutagenicity and induction of sister chromatid exchange in Chinese hamster cells exposed to

hematoporphyrin derivative, ionizing radiation, or ultraviolet radiation. *Cancer Res.* **43,** 2622–2627.
27. Lee See, K., Forbes, I. J., and Betts, W. H. (1984) Oxygen dependency of photocytotoxicity with haematoporphyrin derivative. *Photochem. Photobiol.* **39,** 631–634.
28. Mitchell, J. B., McPherson, S., DeGraff, W., et al. (1985) Oxygen dependence of hematoporphyrin derivative-induced photoinactivation of Chinese hamster cells. *Cancer Res.* **45,** 2008–2011.
29. Foote, C. S. (1984) Mechanisms of photooxygenation, in *Porphyrin Localization and Treatment of Tumors* (Doiron, D. R. and Gomer, C. J., eds.), Alan R. Liss, New York, pp. 3.
30. Rizzoni, W. E., Matthews, K., Pass, H. I., et al. (1987) In vitro photodynamic therapy of human lung cancer. Influence of dose rate, hematoporphyrin concentration and incubation, and cellular targets. *Surg. Forum* **38,** 452–455.
31. Edell, E. S. and Cortese, D. A. (1995) Photodynamic therapy. Its use in the management of bronchogenic carcioma. *Clin. Chest Med.* **16(3),** 455–463.
32. Wieman, T. J., Mang, T. S., Fingar, V. S., et al. (1988) Effects of photodynamic therapy on blood flow in normal and tumor vessels. *Surgery* **104,** 512–517.
33. Stern, S. J., Flock, S., Small, S., et al. (1991) Chloraluminum sulphonated phthalocyanine versus dihematoporphyrin ether: early vascular events in the rat window chamber. *Laryngoscope* **101,** 1219–1225.
34. Ceckler, T. L., Gibson, S. L., Hilf, R., et al. (1990) I situ assessment of tumor vascularity using flourine NMR imaging. *Magn. Reson. Med.* **13,** 416–433.
35. Ben-Hur, E. and Orenstein, A. (1991) The endothelium and red blood cells as potential targets in PDT-induced vascular stasis. *Int. J. Radiat. Biol.* **60,** 293–301.
36. Fingar, V. H., Wieman, T. H., and Doak, K. W. (1990) Role of thromboxane and prostacycline release on photodynamic therapy-induced tumor destruction. *Cancer Res.* **50,** 2599–2603.
37. Bugelski, P. H., Porter, C. W., and Dougherty, T. J. (1981) Autoradiographic distribution of hematoporphyrin derivative in normal and tumor tissue of the mouse. *Cancer Res.* **41,** 4606–4612.
38. Barel, A., Jori, G., Perin, A., et al. (1986) Role of high-, low-, and very low-density lipoproteins in the transport and tumor-delivery of hematoporphyrin in vivo. *Cancer Lett.* **32,** 145–150.
39. Pass, H. I. (1991) Photodynamic therapy for lung cancer. *Chest Surg. Clin. North Am.* **1,** 135–151.
40. Moan. J. (1990) Properties for optimal PDT sensitizers. *J. Photochem. Photobiol.* **5,** 521–524.
41. Matthews, W., Cook, J., Mitchell, J. B., et al. (1989) In vitro photodynamic therapy of human lung cancer: inverstigation of dose-rate effects. *Cancer Res.* **49,** 1718–1721.

42. Foster, T. H., Gibson, S. L., Gao, L., et al. (1992) Analysis of photochemical oxygen consumption effects on photodynamic therapy. *Proc. Int. Soc. Optical Engin. (SPIE)* **1645,** 104–114.
43. Kato, H. and Cortese, D. A. (1985) Early detection of lung cancer by means of hematoporphyrin derivative fluorescence and laser photoradiation. *Clin. Chest Med.* **6,** 237–53.
44. Kato, H., Imaizumi, T., Aisawa, K., et al. (1990) Photodynamic diagnosis in respiratory tract malignancy using excimer laser system. *J. Photochem. Photobiol.* **6,** 189–196.
45. King, E. G., Man, G., Riche, J., et al. (1982) Fluorescence bronchoscopy in the localization of bronchogenic carcinoma. *Cancer* **49,** 777–782.
46. Edell, E. S. and Cortese, D. A. (1992) Photodynamic therapy in the management of early superficial squamous cell carcinoma as an alternative to surgical resection. *Chest* **102,** 1319–1322.
47. Kato, H., Okunaka, T., and Shimatani, H. (1996) Photodynamic therapy for early stage bronchogenic carcinoma. *J. Clin. Laser Med. Surg.* **14,** 235–238.
48. Furuse, K., Fukuoka, M., Kato, H., et al. (1993) Prospective phase II study on photodynamic therapy with photofrin II for centrally located early-stage lung cancer. The Japan lung center. *J. Clin. Oncol.* **11,** 1852–1857.
49. Sutedja, T., Lam, S., LeRiche, J. C., et al. (1994) Response and pattern of failure after photodynamic therapy for intraluminal stage I lung cancer. *J. Bronchology* **1,** 295–298.
50. Cortese, D. A., Edell, E. S., and Kinsey, J. H. (1997) Photodynamic therapy for early stage squamous cell carcinoma. *Mayo Clin. Proc.* **72,** 595–602.
51. Marijnissen, J. P. A., Baas, P., Beek, J. F., et al. (1993) Pilot study on light dosimetry for endobronchial photodynamic therapy. *Photochem. Photobiol.* **58,** 92–99.
52. Hayata, Y., Kato, H., Konaka, C., et al. (1982) Hematoporphyrin derivative and laser photoradiation in the treatment of lung cancer. *Chest* **81,** 269–277.
53. Woolner, L. B., Fontana, R. S., and Cortese, D. A. (1984) Roentgenographically occult lung cancer: Pathologic findings and frequency of multicentricity during a 10-year period. *Mayo Clin. Proc.* **59,** 453–466.
54. Nagamoto, N., Saito, Y., Ohta, S., et al. (1989) Relation of lymph node metastasis to primary tumor size and microscopic appearance of roentgengraphically occult lung cancer. *Am. J. Surg. Pathol.* **13,** 1009–1013.
55. Moghissi, K., Dixon, K., Stringer, M., et al. (1999) The place of bronchoscopic photodynamic therapy in advanced unresectable lung cancer: experience of 100 cases. *Eur. J. Cardthorac. Surg.* **15(1),** 1–6.
56. McCaughan, J. S. Jr. (1996) Photodynamic therapy of endobronchial and esophageal tumors: An overview: *J. Clin. Laser Med. Surg.* **14(5),** 223–233.
57. Lam, S., Crofton, C., and Cory, P. (1991) Combined photodynamic therapy (PDT) using Photofrin and radiotherapy (XRT) versus radiotherapy alone in patients with inoperable distribution non-small cell bronchogenic cancer. *Proc. Intl. Soc. Optical Engin. (SPIE)* 20–28.

58. Moghissi, K., Dixon, K., and Parsons, R. J. (1993) A controlled trial of Nd-YAG vs Photodynamic therapy for advanced malignant bronchial obstruction. *Lasers Med. Sci.* **8,** 269–273.
59. Diz-Jimenez, J. P., Martinez-Ballarin, J. E., Llunell, E., et al. (1999) Efficacy and safety of photodynamic therapy versus Nd-YAG laser resection in NSCLC with airway obstruction. *Eur. Respir. J.* **14,** 800–805.
60. Shah, S. and Ost, D. (2000) Photodynamic therapy: a case series demonstrating its role in patients receiving mechanical ventilation. *Chest* **118(5),** 1419–1423.

III

NOVEL THERAPIES

B. GENE THERAPIES

I. Overview: Advantages and Pitfalls of Lung-Targeted Gene Therapies

31

Gene Therapy for Lung Cancer

An Introduction

Eric B. Haura, Eduardo Sotomayor, and Scott J. Antonia

1. Introduction

In spite of the significant advances made in the conventional treatment modalities currently available for the treatment of lung cancer, this malignancy remains the most frequent cause of cancer death in North America. In recent years, therefore, much attention has been given to identify noncross-resistant modalities of treatment that could potentially be used as an adjunct in the therapy of lung cancer. Recently there has been increasing optimism that gene therapy may be utilized as an approach to develop these novel anti-cancer treatment modalities. The field of gene therapy is still in its infancy with no major successes being reported yet in the treatment of cancer patients. However the limited positive results that have been reported have demonstrated the possibilities of this approach. With regard to the treatment of cancer patients, there are four main strategies that have been devised. These include the introduction of suicide genes, the replacement of defective tumor-suppressor genes, the inactivation of oncogenes, and immunotherapy.

Vector technology for gene transfer has not yet progressed to the point where the specific targeting of tumor cells after the systemic administration of the vector is possible. The need for selective transduction of tumor cells and not normal cells is necessary in order to minimize toxicity. Furthermore, there exist good therapies for the treatment of localized disease, and most cancer patients die of systemic metastases. Therefore, the ability of a vector to target tumor cells in multiple sites when administered systemically is necessary. Currently available vectors are not capable of specifically targeting tumor cells when administered systemically, therefore most clinical trials rely on the direct injection of vectors containing the gene of interest directly into tumors. A clinical application where this approach is relevant in the treatment of lung cancer is in patients who have symptomatic locally advanced disease that is refractory to conventional treatment modalities. For instance, patients with airway compromise could benefit from the reduction in size of an endobronchial tumor by the direct intratumoral injection of a vector.

Another limitation of this technology is the efficiency of gene transfer, particularly when this approach is used for the inactivation of oncogenes, the replacement of tumor-suppressor genes, or the introduction of suicide genes. With these strategies it is necessary to deliver the gene to every cell, otherwise the remaining untransduced cells will continue to proliferate, resulting in recurrent disease in the patient. This limitation has been noted in the treatment of cystic fibrosis with adenoviral vectors where gene transfer has been reported to be less than 1% of harvested bronchial epithelial cells *(1)*. In this way gene therapy with currently available vector technology has the same limitations as chemotherapy for lung cancer where it is not possible to kill every tumor cell, and therefore patients develop recurrent disease. The efficiency of gene transfer is not as important when the use of gene therapy is designed to induce an anti-tumor immune response since the effector cells of the immune system are highly efficient and specific.

2. Introducing Suicide Genes into Tumor Cells

The genes used in the suicide gene-transfer strategy are genes that when introduced into tumor cells have the capacity to convert a nontoxic prodrug into a toxin within the tumor cell *(2–4)*. The most widely used gene in clinical trials using this approach is the herpes simplex virus thymidine kinase (HSV TK) gene. This gene is introduced into tumor cells and then the patients are given the drug ganciclovir. This drug is an acyclic nucleoside analog, which when phosphorylated by HSV TK, is incorporated into DNA resulting in the termination of DNA elongation during S-phase *(5)*. The human thymidine kinase has a low affinity for ganciclovir and therefore this drug has very little toxicity in humans *(6)*. Only the tumor cells that are forced to express HSV TK, which has a high affinity for ganciclovir, are killed. Often times the degree of tumor-size reduction is disproportionate to the expected degree of transduction efficiency, due to killing of neighboring untransduced tumor cells as a result of the so-called "bystander effect" *(7–9)*.

Due to the inability to selectively target tumor cells with the vectors that are currently available, this approach cannot yet be applied to the treatment of metastatic lung cancer. However this approach could be used with palliative intent to reduce the size of individual symptomatic tumors. Sterman and colleagues recently used this approach for the treatment of pleural mesothelioma and demonstrated minimal side effects as well as dose-related gene transfer. However, despite a strong anti-adenoviral immune response, clinical responses were uncommon *(10)*. A clinical trial using an adenoviral vector to insert the HSV TK transgene after intratumoral injection involving patients with advanced non-small cell lung carcinoma (NSCLC) will be conducted at the New York University School of Medicine *(11)*.

A novel modification of introducing suicide genes into tumor cells has been to genetically engineer replication competent adenoviruses that are tumor-selective by virtue of their ability to replicate within and lyse only cells with mutant p53 function *(12)*.

Adenoviruses express two important proteins, E1A and E1B, which allow for viral replication and lysis. The E1A protein binds to and inactivates the retinoblastoma gene product (Rb). The resulting accumulating of the E2F transcription factors allows for the induction of S-phase genes necessary for replication of adenovirus. The second protein, E1B, binds and inhibits the activity of p53, thereby preventing apoptosis resulting from deregulated E2F activity. Investigators at the ONYX corporation engineered an adenovirus with the E1B region deleted and found that this virus replicates and lyses cells with mutant p53 or that have dysregulation of the pathway controlling p53. Further, these investigators demonstrated resistance to the cytopathic effect of the virus in normal cells. Preclinical activity was demonstrated against tumor xenografts as a single agent and in combination with chemotherapy *(13)*.

A Phase I study in patients with advanced malignancies with lung involvement demonstrated safety up to 2×10^{12} viral particles given as an IV injection with the most frequent adverse event being fevers, chills, and rigors *(14)*. A study in metastatic lung cancer with the E1B deleted virus in combination with carboplatin and paclitaxel is planned. Ongoing work is also investigating the use of adenovirus with deleted E1A, which selectively replicates and kills tumors with dysregulation of the Rb pathway.

3. Replacement of Defective Tumor Suppressor Genes

The most common gene used in clinical trials utilizing the strategy of replacing defective tumor-suppressor genes is the p53 gene. This gene is frequently mutated in lung cancer *(15)* and contributes to the malignant phenotype *(16)*. Roth et al first reported the use of this strategy in the treatment of lung cancer patients *(17)*. In their original study, a retroviral vector containing the wild type p53 gene was used. Nine patients with progressive NSCLC were treated with intratumoral injections of this vector through a bronchoscope or by using CT-guided percutaneous injections. Endobronchial, chest wall, and adrenal lesions were injected. In this study they were able to demonstrate successful transfer of the transgene using polymerase chain reaction (PCR) and *in situ* hybridization, and they observed a greater degree of apoptosis in postinjection biopsies than was present in pre-injection biopsies. Of 7 patients who were evaluated for a clinical response, 3 patients had tumor regressions and 3 patients had stable disease.

A second study performed by this group was recently reported *(18,19)*. This second Phase I trial utilized an adenoviral vector, and patients received repeated injections monthly for up to a total of 6 injections. The dose was escalated in cohorts of patients from 10^6 to 10^{11} plaque-forming units (PFUs) of vector per injection. Once again the vector was injected using a bronchoscope for endobronchial lesions, or CT-guided percutaneous injections of lung, liver, axillary, or chest-wall lesions. In this study they were able to demonstrate the expression of the transgene using reverse transcription (RT)-PCR in 46% of the injected tumors. Toxicity was minimal, and of the 25 patients who were available for evaluation of a clinical response, 8% had a partial response and 64% had stable disease.

Since the studies that have been completed have demonstrated only modest response rates, there are now several protocols that have been opened which combine the p53 gene transfer approach with other treatment modalities. Preclinical studies using lung cancer cell lines demonstrated synergy between Ad-p53 and DNA damaging drugs such as cisplatin and etoposide while an additive effect was noted with the combination of Ad-p53 and taxanes *(20)*. Based on these findings, the group at M.D. Anderson sought to determine the safety and tolerability of Ad-p53 in sequence with cisplatin administration *(21)*. Twenty-four patients with NSCLC and tumor p53 mutations assayed by DNA sequencing were treated with cisplatin 80mg/m2 on d 1 and Ad-p53 is escalating dose levels on d 4 of a 28-d course. Ad-p53 was administered by fine-needle injection directly into the tumor. The most common adverse effect was transient fever associated with Ad-p53 injection occurring in nearly one-third of patients but was limited to grade 2. Treatment biopsies revealed increased tumor necrosis and apoptosis. Two patients obtained a partial response, 17 had stable disease, 4 patients had progressive disease, and 1 patient was not assessable.

Schuler et al. reported the results of another Phase I trial utilizing this strategy involving patients with advanced NSCLC *(22)*. In this study, 15 patients in each of 4 different dose levels were treated with a single intratumoral injection of a recombinant adenoviral vector containing the wild-type p53 gene using a bronchoscope or CT guided percutaneous injections. Twenty-four hours after injection, the treated lesion was biopsied and subjected to RT-PCR in order to determine if the transgene was expressed. They observed no expression in the patients treated with two of the lower doses (10^7 and 10^8 PFU), however expression was confirmed in 6 of the patients treated with the higher doses (10^9 and 10^{10} PFU). There were no clinical responses observed in the patients where there was no expression of the wild-type p53 transgene. However, of the 6 patients where the transgene was expressed, 4 of the patients had stable disease determined 28 d after injection in the treated lesion while at the same time there was progression of uninjected lesions in the same patients.

The use of adenovirus to deliver functional p53 has also been reported in the subset of patients with bronchioloalveolar carcinoma. In a Phase I trial reported in abstract form, Ad-p53 was delivered by bronchoalveolar lavage to 14 patients *(23)*. After two cycles of treatment, two patients demonstrated a pathological response on repeat biopsy, 4/9 patients had improvement in diffusion capacity, and 4/11 patients had symptomatic improvement. Only one patient had grade 4 pulmonary toxicity but other patients treated at the same viral dose did not experience this toxicity.

As is the case with using suicide genes for cancer therapy, there appears to be a significant "bystander effect," with the degree of tumor cell killing being greater than what would be expected based on transfection efficiency *(24,25)*. One study using lung cancer cell lines demonstrated that tumor cells infected with Ad-p53 inhibited the tumor growth of adjacent noninfected cells by inhibiting the expression of vascular endothelial growth factor (VEGF) *(25,26)*. Other possible mechanisms include modulation of other tumor angiogenesis pathways, activation of the Fas/Fas ligand pathway, and activation of the immune system *(1)*.

Two potential shortcomings may limit the use of adenovirus to express p53 in tumor cells; first, the expression is not regulated in tumors and second, the expression of mutant forms of p53 that can act in a dominant negative fashion are not affected. Watanabe and Sullenger describe the use of ribozymes to simultaneously restore wild-type p53 function as well as reduce the expression of mutant p53 *(27)*. Ribozymes are RNA molecules with catalytic activity and this group's use of trans-splicing ribozymes repaired mutant p53 mRNAs with high fidelity and specificity. The use of this approach in the treatment of lung cancer remains to be explored.

Similar to the strategy of introducing suicide genes into tumor cells, the p53 replacement strategy is limited to the treatment of individual tumors, as current vector technology does not allow for the specific targeting of tumor cells when administered systemically. One novel approach to this problem is to engineer adenoviral vectors, which can infect both tumor and normal cells but have promoter elements, which specifically direct expression of a transgene in tumor cells as opposed to normal cells. Parr and colleagues used this approach to engineer a recombinant adenovirus, which expresses beta-galactosidase under the control of the E2F responsive promoter *(28)*. Given that dysregulation of the G1 cyclin/Retinoblastoma/E2F pathway is a common event in tumor formation, it was surmised that transgene expression would be high in tumors while normal cells would have minimal (if any) transgene expression and no adverse effect on cell physiology.

Another group examined the ability of the human surfactant B promoter to direct transgene expression using an adenoviral vector and found expression of the transgene limited to type II pneumocytes and Clara cells *(29)*. The clinical usefulness of such an approach remains to be seen.

Other tumor-suppressor proteins are also candidates for replacement gene therapy approaches. The control of cellular proliferation and the cell cycle is highly dependent on the G_1 cyclins and their partner cyclin-dependent kinases. These complexes lead to phosphorylation of the retinoblastoma protein leading to accumulation of the E2F transcription factors which are critical for controlling genes important in regulating G1-S cell-cycle progression. The cyclins are negatively regulated by a family of proteins termed the cyclin-dependent kinase inhibitors.

The cyclin dependent kinase inhibitors p16INK4a and p27kip1 are frequently either deleted or underexpressed in lung cancer and lead to deregulation of the cell cycle *(30)*. Preclinical studies have demonstrated that replacement of these proteins using adenoviral vectors may be beneficial in the treatment of lung cancer. One group re-expressed p16INK4a in lung cancer and observed radiosensitization in a p53-dependent manner *(31)*. Another group found that overexpression of p27kip1 using adenovirus lead to growth arrest and apoptosis in lung cancer cell lines *(32)*. The clinical application of these vectors has yet to be investigated. Finally, because of the critical role for the E2F transcription factors in cellular growth control, laboratory studies have been initiated attempting to selectively block E2F function either through use of RNA ligands or antisense constructs *(33)*.

4. Inactivation of Oncogenes

In addition to the mutation of tumor-suppressor genes, the constitutive activation of oncogenes frequently occurs in lung cancers *(30)*, and contributes to the malignant phenotype *(34)*. These genes therefore present another target for the gene therapy of cancer *(35)*. There are several potential strategies that can be used to reduce the expression of activated oncogenes. One potential method is to introduce a gene that codes for a ribozyme, which is an RNA that has catalytic activity and can cleave RNA *(36)*. The ribozyme can be linked to a nucleotide sequence that is complementary to the mRNA, which codes for the oncogene in order to specifically target the oncogene mRNA *(37)*. Another method of inactivating oncogene mRNAs is to introduce a gene that codes for the oncogene antisense. When expressed in tumor cells the antisense nucleotides block translation by binding to the oncogene mRNA, and also target the mRNA for degradation by RNase H *(36)*. One could also express "dominant negative" mutant forms of the oncogene, which either bind the oncogene or bind to downstream effector molecules and prevent activation of the involved pathway. One final method of inactivating oncogenes is to introduce a gene, which codes for a portion of an antibody molecule, referred to as a single chain Fv molecule (scFv), which is specific for the oncogene product. When expressed within the tumor cell the scFv can bind to and thereby inactivate the oncogene product *(38–40)*.

While still in its infancy, the use of these approaches is becoming more common in the treatment of lung cancers. At the M.D. Anderson Cancer Center a protocol has been opened where a gene coding for the antisense of a portion of the K-ras oncogene is introduced into NSCLC cells after intratumoral injection of the vector *(41)*. Previously this group had demonstrated the suppression of lung cancer cell line growth when transduced with a retrovirus encoding the K-ras antisense cDNA. Another group has produced an adenoviral vector with a ribozyme transgene specific for the K-ras codon 12 mutant sequence and demonstrated "in vivo" antitumor effect against NSCLC xenografts with a mutant codon 12 K-ras oncogene *(42,43)*. Antisense therapy to the Raf and protein kinase C signaling pathways are also being pursued and one would anticipate studies in the future on patients with advanced lung cancer *(44,45)*.

5. Immunotherapy

Recently it has been demonstrated that cells of the immune system are capable of killing chemotherapy-resistant tumor cell lines *(46,47)*. This, together with the identification of tumor-associated antigens (TAAs) recognized by T cells *(48–52)* and the better understanding of the cellular and molecular mechanism regulating immune responses has led to a renewed interest in immunotherapy as a strategy in the management of cancer patients *(53,54)*. As a result of this renaissance of tumor immunology, several gene therapy-based strategies have been developed, a number of which are undergoing clinical evaluation as therapy for patients with lung cancer. These strategies include TAA based vaccines; tumor cell vaccines genetically engineered to secrete immunologically relevant cytokines; and co-stimulatory molecule gene-modified tumor cell vaccines.

A tumor-associated antigen that is expressed by approx 70% of NSCLCs is carcinoembrionic antigen (CEA) *(55)*. In preclinical studies, animals that received immunization with a recombinant vaccinia virus that expressed the human CEA gene (rV-CEA) showed inhibition of growth of CEA positive tumors and reject a re-challenge with the same tumor cells *(56)*. The immunogenicity and safety of this vaccine was then tested in monkeys treated with either rV-CEA or a control wild-type vaccinia virus. All immunized animals developed local skin reactions, low-grade fever, and lymphadenopathy. However, only the animals that received rV-CEA exhibited a strong DTH response to intradermal injections of purified CEA. Furthermore, rV-CEA vaccinated animals exhibited both cellular and humoral responses after challenge with CEA *(57)*. These encouraging results prompted the design of a Phase I study in patients with metastatic adenocarcinoma using escalating doses of the recombinant vaccinia-CEA vaccine. In this study, no dose-limiting toxicities were observed using doses as high as 1×10^8 PFU per vaccination. Interestingly, peripheral blood lymphocytes isolated from vaccinated patients were capable of specifically lyse CEA-expressing tumor cells "in vitro" following restimulation with CEA peptide. More recently, Cole et al. have designed a Phase I clinical trial aimed at evaluating the combination of a recombinant CEA-vaccinia virus vaccine to prime an immune response, with a postvaccination boost using CEA peptide mixed with an adjuvant. In this study, patients with NSCLC (in addition to other malignancies) that have elevated CEA serum levels or tumors which stain positively for CEA will receive r-vCEA on wk 0 and 4, followed by CEA-peptide in an adjuvant on wk 8, 12, and 16 *(58)*.

Recently, three other tumor-associated antigens have been identified in certain lung cancer cell lines, providing therefore additional targets for lung cancer vaccines. Fuc-GM1 is a ganglioside expressed by most small cell lung cancers (SCLC), but not by most normal tissues. At the Memorial Sloan Kettering Cancer Center, SCLC patients that achieved a major response with conventional treatment were vaccinated with Fuc-GM1 conjugated to keyhole limpet hemocyanin (KLH) in an adjuvant on wk 1, 2, 3, 4, 8, and 16. Among those patients entered on this study, 10 received at least five vaccinations and were therefore candidates for evaluating a response. In all of these patients, anti-FucGm1 IgM and IgG antibodies were detectable in their serum. Furthermore, these antibodies were capable of binding to tumor cells expressing Fuc-GM1 and were able to mediate complement-dependent tumor cell killing *(59)*. NY-ESO-1 is another recently identified cancer-testis antigen found in 11 of 16 SCLC lines and in 3 of 7 NSCLC cell lines. Importantly, this antigen can be presented on lung cancer cells for recognition by HLA-restricted cytotoxic T lymphocytes, pointing to NY-ESO-1 as an attractive TAA that can be exploited for immunotherapy of SCLC *(60)*. One final TAA that is being tested for the treatment of lung cancer is human epithelial mucin MUC-1. This is a large type I membrane-associated glycoprotein expressed on most epithelial cells, and is overexpressed and aberrantly glycosylated on many human epithelial tumors including NSCLC. This glycoprotein is of interest as an immunotherapeutic target because patients with epithelial malignancies overexpressing this glycoprotein have T-lymphocytes in their tumor-draining lymph nodes that recognize and lyse MUC-1 expressing tumor cells *(61)*. Based on this encouraging finding, Gitlitz et al. at the University of California Los Angeles have recently opened a Phase I/II trial of antigen-specific immunotherapy in patients with advanced NSCLC using a recombinant vaccinia-virus encoding MUC1 and the IL-2 gene *(11)*.

A critical barrier for the broader application of TAA-based vaccines is that the molecular identity of the antigens expressed by most human malignancies including lung cancer remain largely unknown. Therefore a strategy was developed by investigators at the Johns Hopkins Oncology Center that sought to utilize the whole cancer cell as a source of tumor antigens. With this approach, tumor cells are engineered to express immunomodulatory molecules such as cytokines or co-stimulatory molecules to enhance their immunogenicity *(62)*. In preclinical models, these strategies have demonstrated the ability to prime systemic immune responses capable of mediating the rejection of micrometastatic tumors at distant sites. A systematic comparison of 10 different cytokine or cell surface molecule-based tumor vaccines, showed that the cytokine granulodyte-macrophage colony stimulating factor (GM-CSF) was the most potent in augmenting a systemic anti-tumor response *(63)*. Priming with GM-CSF transduced tumor cells led to a potent, long-lived systemic anti-tumor immunity, that requires participation of both CD4 and CD8 positive T cells. Further dissection of the mechanisms mediating this strong anti-tumor response, has demonstrated that GM-CSF promotes the recruitment and activation of host antigen-presenting cells to the inoculation site. These cells process and present tumor antigens to antigen-specific T cells, which in turn mediate systemic immunity *(64)*. Multiple reports have since confirmed the bioactivity of GM-CSF transduced tumor cells in a number of different tumor model systems including lung cancer. Based on these preclinical data, a number of Phase I clinical trials involving patients with metastatic cancer have taken place. At Johns Hopkins, a Phase I trial conducted with metastatic renal cell carcinoma patients confirmed the bioactivity of GM-CSF as a molecular adjuvant *(65)*. A subsequent trial involving patients with metastatic prostate cancer or pancreatic cancer has extended these observations *(66–68)*. Currently, the Dana Farber Cancer Institute is conducting a Phase I trial of vaccination with autologous, lethally irradiated NSCLC cells engineered to secrete human GM-CSF *(11)*.

6. Summary

Incremental progress will likely continue to be made in the treatment of lung cancer using the conventional treatment modalities. However, there is now hope that the development of noncross-resistant modalities that can be accomplished using the technology of gene transfer may hasten the progress made in this disease. Although significant progress has been made, the practical application of the various strategies to the treatment of cancer patients in clinical trials has yielded only very limited results. For the strategies of introducing suicide genes, replacing defective tumor-suppressor genes, and inactivating oncogenes, considerable progress will need to be made particularly with regard to the ability of vectors to selectively and efficiently target tumor cells in order for many of these strategies to become effective. With regard to the application of gene therapy to immunotherapy, major responses are not likely at this time since these innovative therapies are being developed in patients with advanced disease. However, it is very likely that these pioneering clinical trials will provide important clinical and immunological information that will be the basis for future attempts to effectively harness the immune system against lung cancer.

References

1. Albelda, S. M., Wiewrodt, R., and Zuckerman, J. B. (2000) Gene therapy for lung disease: hype or hope? *Ann. Intern. Med.* **132,** 649–660.
2. Blaese, R. M., Ishii-Morita, H., Mullen, C., Ramsey, J., Ram, Z., Oldfield, E., and Culver, K. (1994) In situ delivery of suicide genes for cancer treatment. *Eur. J. Cancer* **8,** 1190–1193.
3. Mullen, C. A. (1994) Metabolic suicide genes in gene therapy. *Pharmacol. Ther.* **63,** 199–207.
4. Singhal, S. and Kaiser, L. R. (1998) Cancer chemotherapy using suicide genes. *Surg. Oncol. Clin. North Am.* **7,** 505–536.
5. Reid, R., Mar, E. C., Huang, E. S., and Topal, M. D. (1988) Insertion and extension of acyclic, dideoxy, and ara nucleotides by herpesviridae, human alpha and human beta polymerases. A unique inhibition mechanism for 9-(1,3-dihydroxy-2-propoxymethyl)guanine triphosphate. *J. Biol. Chem.* **263,** 3898–3904.
6. Gane, E., Saliba, F., Valdecasas, G. J., O'Grady, J., Pescovitz, M. D., Lyman, S., and Robinson, C. A. (1997) Randomised trial of efficacy and safety of oral ganciclovir in the prevention of cytomegalovirus disease in liver-transplant recipients. The Oral Ganciclovir International Transplantation Study Group [corrected] [see comments] [published erratum appears in Lancet 1998 Feb 7;351(9100):454]. *Lancet* **350,** 1729–1733.
7. Freeman, S. M., Whartenby, K. A., Freeman, J. L., Abboud, C. N., and Marrogi, A. J. (1996) In situ use of suicide genes for cancer therapy. *Semin. Oncol.* **23,** 31–45.
8. Pope, I. M., Poston, G. J., and Kinsella, A. R. (1997) The role of the bystander effect in suicide gene therapy. *Eur. J. Cancer* **33,** 1005–1016.
9. Rubsam, L. Z., Boucher, P. D., Murphy, P. J., KuKuruga, M., and Shewach, D. S. (1999) Cytotoxicity and accumulation of ganciclovir triphosphate in bystander

cells cocultured with herpes simplex virus type 1 thymidine kinase- expressing human glioblastoma cells. *Cancer Res.* **59,** 669–675.
10. Sterman, D. H., Treat, J., Litzky, L. A., Amin, K. M., Coonrod, L., Molnar-Kimber, K., et al. (1998) Adenovirus-mediated herpes simplex virus thymidine kinase/ganciclovir gene therapy in patients with localized malignancy: results of a phase I clinical trial in malignant mesothelioma. *Human Gene Ther.* **9,** 1083–1092.
11. (1999) Human gene marker/ therapy clinical protocols (complete updated listings). *Human Gene Ther.* **10,** 2037–2088.
12. Nemunaitis, J. (1999) Oncolytic viruses. *Invest. New Drugs* **17,** 375–386.
13. You, L., Yang, C. T., and Jablons, D. M. (2000) ONYX-015 works synergistically with chemotherapy in lung cancer cell lines and primary cultures freshly made from lung cancer patients. *Cancer Res.* **60,** 1009–1013.
14. Nemunaitis, J., Ganly, I., Khuri, F., Arseneau, J., Kuhn, J., McCarty, T., et al. (2000) Selective replication and oncolysis in p53 mutant tumors with ONYX-015, an E1B-55kD gene-deleted adenovirus, in patients with advanced head and neck cancer: a phase II trial. *Cancer Res.* **60,** 6359–6366.
15. Takahashi, T., Nau, M. M., Chiba, I., Birrer, M. J., Rosenberg, R. K., Vinocour, M., et al. (1989) p53: a frequent target for genetic abnormalities in lung cancer. *Science* **246,** 491–494.
16. Levine, A. J. (1997) p53, the cellular gatekeeper for growth and division. *Cell* **88,** 323–331.
17. Roth, J. A., Nguyen, D., Lawrence, D. D., Kemp, B. L., Carrasco, C. H., Ferson, D. Z., et al. (1996) Retrovirus-mediated wild-type p53 gene transfer to tumors of patients with lung cancer [see comments]. *Nat. Med.* **2,** 985–991.
18. Swisher, S. G., Roth, J. A., Nemunaitis, J., Lawrence, D. D., Kemp, B. L., Carrasco, C. H., et al. (1999) Adenovirus-mediated p53 gene transfer in advanced non-small-cell lung cancer. *J. Natl. Cancer Inst.* **91,** 763–771.
19. Roth, J. A., Swisher, S. G., Merritt, J. A., Lawrence, D. D., Kemp, B. L., Carrasco, C. H., et al. (1998) Gene therapy for non-small cell lung cancer: a preliminary report of a phase I trial of adenoviral p53 gene replacement. *Semin. Oncol.* **25,** 33–37.
20. Inoue, A., Narumi, K., Matsubara, N., Sugawara, S., Saijo, Y., Satoh, K., and Nukiwa, T. (2000) Administration of wild-type p53 adenoviral vector synergistically enhances the cytotoxicity of anti-cancer drugs in human lung cancer cells irrespective of the status of p53 gene. *Cancer Lett.* **157,** 105–112.
21. Nemunaitis, J., Swisher, S. G., Timmons, T., Connors, D., Mack, M., Doerksen, L., et al. (2000) Adenovirus-mediated p53 gene transfer in sequence with cisplatin to tumors of patients with non-small-cell lung cancer. *J. Clin. Oncol.* **18,** 609–622.
22. Schuler, M., Rochlitz, C., Horowitz, J. A., Schlegel, J., Perruchoud, A. P., Kommoss, F., et al. (1998) A phase I study of adenovirus-mediated wild-type p53 gene transfer in patients with advanced non-small cell lung cancer. *Human Gene Ther.* **9,** 2075–2082.
23. Kubba, S., Adak, S., Schiller, J., Slovis, B., Coffee, K., Worrel, J., et al. (2000) Phase I trial of adenovirus p53 in bronchioloalveolar cell lung carcinoma (BAC)

administered by bronchoalveolar lavage. *Proc. Am. Soc. Clin. Oncol.* Abstract #1904.
24. Xu, M., Kumar, D., Srinivas, S., Detolla, L. J., Yu, S. F., Stass, S. A., and Mixson, A. J. (1997) Parenteral gene therapy with p53 inhibits human breast tumors in vivo through a bystander mechanism without evidence of toxicity. *Human Gene Ther.* **8,** 177–185.
25. Bouvet, M., Ellis, L. M., Nishizaki, M., Fujiwara, T., Liu, W., Bucana, C. D., et al. (1998) Adenovirus-mediated wild-type p53 gene transfer down-regulates vascular endothelial growth factor expression and inhibits angiogenesis in human colon cancer. *Cancer Res.* **58,** 2288–2292.
26. Nishizaki, M., Fujiwara, T., Tanida, T., Hizuta, A., Nishimori, H., Tokino, T., et al. (1999) Recombinant adenovirus expressing wild-type p53 is antiangiogenic: a proposed mechanism for bystander effect. *Clin. Cancer Res.* **5,** 1015–1023.
27. Watanabe, T. and Sullenger, B. A. (2000) Induction of wild-type p53 activity in human cancer cells by ribozymes that repair mutant p53 transcripts. *Proc. Natl. Acad. Sci. USA* **97,** 8490–8494.
28. Parr, M. J., Manome, Y., Tanaka, T., Wen, P., Kufe, D. W., Kaelin, W. G., and Fine, H. A. (1997) Tumor-selective transgene expression in vivo mediated by an E2F-responsive adenoviral vector. *Nat. Med.* **3,** 1145–1149.
29. Strayer, D. S., Guttentag, S. H., and Ballard, P. L. (1998) Targeting type II and Clara cells for adenovirus-mediated gene transfer using the surfactant protein B promoter. *Am. J. Respir. Cell Mol. Biol.* **18,** 1–11.
30. Salgia, R. and Skarin, A. T. (1998) Molecular abnormalities in lung cancer. *J. Clin. Oncol.* **16,** 1207–1217.
31. Kawabe, S., Roth, J. A., Wilson, D. R., and Meyn, R. E. (2000) Adenovirus-mediated p16INK4a gene expression radiosensitizes non-small cell lung cancer cells in a p53-dependent manner. *Oncogene* **19,** 5359–5366.
32. Naruse, I., Hoshino, H., Dobashi, K., Minato, K., Saito, R., and Mori, M. (2000) Over-expression of p27kip1 induces growth arrest and apoptosis mediated by changes of pRb expression in lung cancer cell lines. *Int. J. Cancer* **88,** 377–383.
33. Ishizaki, J., Nevins, J. R., and Sullenger, B. A. (1996) Inhibition of cell proliferation by an RNA ligand that selectively blocks E2F function. *Nat. Med.* **2,** 1386–1389.
34. Haber, D. A. and Fearon, E. R. (1998) The promise of cancer genetics. *Lancet* **351(Suppl. 2),** SIII1–8.
35. Stass, S. A. and Mixson, J. (1997) Oncogenes and tumor suppressor genes: therapeutic implications. *Clin. Cancer Res.* **3,** 2687–2695.
36. Tanner, N. K. (1999) Ribozymes: the characteristics and properties of catalytic RNAs. *FEMS Microbiol. Rev.* **23,** 257–275.
37. Arndt, G. M. and Rank, G. H. (1997) Colocalization of antisense RNAs and ribozymes with their target mRNAs. *Genome* **40,** 785–797.
38. Cochet, O., Kenigsberg, M., Delumeau, I., Duchesne, M., Schweighoffer, F., Tocque, B., and Teillaud, J. L. (1998) Intracellular expression and functional properties of an anti-p21Ras scFv derived from a rat hybridoma containing specific lambda and irrelevant kappa light chains. *Mol. Immunol.* **35,** 1097–1110.

39. Cochet, O., Kenigsberg, M., Delumeau, I., Virone-Oddos, A., Multon, M. C., Fridman, W. H., et al. (1998) Intracellular expression of an antibody fragment-neutralizing p21 ras promotes tumor regression. *Cancer Res.* **58,** 1170–1176.
40. Jannot, C. B., Beerli, R. R., Mason, S., Gullick, W. J., and Hynes, N. E. (1996) Intracellular expression of a single-chain antibody directed to the EGFR leads to growth inhibition of tumor cells. *Oncogene* **13,** 275–282.
41. Roth, J. A. (1996) Modification of mutant K-ras gene expression in non-small cell lung cancer (NSCLC). *Human Gene Ther.* **7,** 875–889.
42. Zhang, Y., Nemunaitis, J., Scanlon, K. J., and Tong, A. W. (2000) Anti-tumorigenic effect of a K-ras ribozyme against human lung cancer cell line heterotransplants in nude mice. *Gene Ther.* **7,** 2041–2050.
43. Zhang, Y. A., Nemunaitis, J., and Tong, A. W. (2000) Generation of a ribozyme-adenoviral vector against K-ras mutant human lung cancer cells. *Mol. Biotechnol.* **15,** 39–49.
44. Nemunaitis, J., Holmlund, J. T., Kraynak, M., Richards, D., Bruce, J., Ognoskie, N., et al. (1999) Phase I evaluation of ISIS 3521, an antisense oligodeoxynucleotide to protein kinase C-alpha, in patients with advanced cancer. *J. Clin. Oncol.* **17,** 3586–3595.
45. Cunningham, C. C., Holmlund, J. T., Schiller, J. H., Geary, R. S., Kwoh, T. J., Dorr, A., and Nemunaitis, J. (2000) A phase I trial of c-Raf kinase antisense oligonucleotide ISIS 5132 administered as a continuous intravenous infusion in patients with advanced cancer. *Clin. Cancer Res.* **6,** 1626–1631.
46. Fuchs, E. J., Bedi, A., Jones, R. J., and Hess, A. D. (1995) Cytotoxic T cells overcome BCR-ABL-mediated resistance to apoptosis. *Cancer Res.* **55,** 463–466.
47. Shtil, A. A., Turner, J. G., Durfee, J., Dalton, W. S., and Yu, H. (1999) Cytokine-based tumor cell vaccine is equally effective against parental and isogenic multidrug-resistant myeloma cells: the role of cytotoxic T lymphocytes. *Blood* **93,** 1831–1837.
48. Boon, T. and van der Bruggen, P. (1996) Human tumor antigens recognized by T lymphocytes. *J. Exp. Med.* **183,** 725–729.
49. Boon, T. and Old, L. J. (1997) Cancer tumor antigens. *Curr. Opin. Immunol.* **9,** 681–683.
50. van der Bruggen, P., Traversari, C., Chomez, P., Lurquin, C., De Plaen, E., Van den Eynde, B., et al. (1991) A gene encoding an antigen recognized by cytolytic T lymphocytes on a human melanoma. *Science* **254,** 1643–1647.
51. Pieper, R., Christian, R. E., Gonzales, M. I., Nishimura, M. I., Gupta, G., Settlage, R. E., et al. (1999) Biochemical identification of a mutated human melanoma antigen recognized by CD4(+) T cells [see comments]. *J. Exp. Med.* **189,** 757–766.
52. Wang, R. F., Wang, X., Atwood, A. C., Topalian, S. L., and Rosenberg, S. A. (1999) Cloning genes encoding MHC class II-restricted antigens: mutated CDC27 as a tumor antigen. *Science* **284,** 1351–1354.
53. Pardoll, D. M., Golumbek, P., Levitsky, H., and Jaffee, L. (1992) Molecular engineering of the antitumor immune response. *Bone Marrow Transplant.* **9,** 182–186.

54. Leach, D. R., Krummel, M. F., and Allison, J. P. (1996) Enhancement of antitumor immunity by CTLA-4 blockade [see comments]. *Science* **271**, 1734–1736.
55. Vincent, R. G., Chu, T. M., Lane, W. W., Gutierrez, A. C., Stegemann, P. J., and Madajewicz, S. (1978) Carcinoembryonic antigen as a monitor of successful surgical resection in 130 patients with carcinoma of the lung. *J. Thorac. Cardiovasc. Surg.* **75**, 734–739.
56. Kantor, J., Irvine, K., Abrams, S., Kaufman, H., DiPietro, J., and Schlom, J. (1992) Antitumor activity and immune responses induced by a recombinant carcinoembryonic antigen-vaccinia virus vaccine [see comments]. *J. Natl. Cancer Inst.* **84**, 1084–1091.
57. Kantor, J., Irvine, K., Abrams, S., Snoy, P., Olsen, R., Greiner, J., et al. (1992) Immunogenicity and safety of a recombinant vaccinia virus vaccine expressing the carcinoembryonic antigen gene in a nonhuman primate. *Cancer Res.* **52**, 6917–6925.
58. Cole, D. J., Wilson, M. C., Baron, P. L., O'Brien, P., Reed, C., Tsang, K. Y., and Schlom, J. (1996) Phase I study of recombinant CEA vaccinia virus vaccine with post vaccination CEA peptide challenge. *Human Gene Ther.* **7**, 1381–1394.
59. Dickler, M. N., Ragupathi, G., Liu, N. X., Musselli, C., Martino, D. J., Miller, V. A., et al. (1999) Immunogenicity of a fucosyl-GM1-keyhole limpet hemocyanin conjugate vaccine in patients with small cell lung cancer. *Clin. Cancer Res.* **5**, 2773–2779.
60. Lee, L., Wang, R. F., Wang, X., Mixon, A., Johnson, B. E., Rosenberg, S. A., and Schrump, D. S. (1999) NY-ESO-1 may be a potential target for lung cancer immunotherapy. *Cancer J. Sci. Am.* **5**, 20–25.
61. Rowse, G. J., Tempero, R. M., VanLith, M. L., Hollingsworth, M. A., and Gendler, S. J. (1998) Tolerance and immunity to MUC1 in a human MUC1 transgenic murine model. *Cancer Res.* **58**, 315–321.
62. Pardoll, D. M. (1993) Genetically engineered tumor vaccines. *Ann. NY Acad. Sci.* **690**, 301–310.
63. Dranoff, G., Jaffee, E., Lazenby, A., Golumbek, P., Levitsky, H., Brose, K., et al. (1993) Vaccination with irradiated tumor cells engineered to secrete murine granulocyte-macrophage colony-stimulating factor stimulates potent, specific, and long-lasting anti-tumor immunity. *Proc. Natl. Acad. Sci. USA* **90**, 3539–3543.
64. Huang, A. Y., Golumbek, P., Ahmadzadeh, M., Jaffee, E., Pardoll, D., and Levitsky, H. (1994) Role of bone marrow-derived cells in presenting MHC class I-restricted tumor antigens. *Science* **264**, 961–965.
65. Simons, J. W., Jaffee, E. M., Weber, C. E., Levitsky, H. I., Nelson, W. G., Carducci, M. A., et al. (1997) Bioactivity of autologous irradiated renal cell carcinoma vaccines generated by ex vivo granulocyte-macrophage colony-stimulating factor gene transfer. *Cancer Res.* **57**, 1537–1546.
66. Sanda, M. G., Ayyagari, S. R., Jaffee, E. M., Epstein, J. I., Clift, S. L., Cohen, L. K., et al. (1994) Demonstration of a rational strategy for human prostate cancer gene therapy. *J. Urol.* **151**, 622–628.

67. Simons, J. W., Mikhak, B., Chang, J. F., DeMarzo, A. M., Carducci, M. A., Lim, M., et al. (1999) Induction of immunity to prostate cancer antigens: results of a clinical trial of vaccination with irradiated autologous prostate tumor cells engineered to secrete granulocyte-macrophage colony-stimulating factor using ex vivo gene transfer. *Cancer Res.* **59,** 5160–5168.
68. Jaffee, E. M., et el. (2001) Novel Allogeneic Granulocyte-Macrophase Colony-stimulating factor-secreting tumor vaccine for Pancreatic Cancer: A Phase I trial of safety and immune activation. *J. Clin. Oncol.* **19(1),** 145–156.

32

Identifying Obstacles to Viral Gene Therapy for Lung Cancer

Malignant Pleural Effusions as a Paradigm

Raj K. Batra, Raymond M. Bernal, and Sherven Sharma

1. Introduction

We have been investigating intrapleural malignancy as a "proof of concept" clinical model for the gene therapy of non-small cell lung cancer (NSCLC). Our focus has been to use malignant pleural effusions (MPE) to model an in vivo milieu, and to assess the transduction efficiency of tumor cells in this milieu by viral vectors. Consequently, our preclinical studies have used effusions derived from patients in transduction bioassays, and by targeting cell lines derived from MPE for gene-transfer studies. These studies have produced observations that have broad relevance with respect to in vivo gene therapy. Indeed, these studies assert that viral vectors confront a number of hurdles within the extracellular matrix (ECM) and the soluble component of MPE *in situ*, long before they get the opportunity to interact with the cells targeted for transduction. In essence, the extracellular milieu serves as a sink for gene-transfer vectors, thus contributing to the global inefficiency of gene transfer observed by vectors in vivo.

As an introduction, MPE from NSCLC arise in 10% of all cases of lung cancer, and mostly in patients with disseminated disease *(1,2)*. The effusions arise primarily because of ineffective lymphatic clearance of the pleural cavity owing to tumor obstructing lymphatic channels or central (mediastinal) draining lymph nodes. Characteristically, the effusions are exudative *(1)*, and because they arise due to diminished clearance from the pleural cavity, they are a stagnant reservoir, rich in soluble factors derived from plasma filtrate, metabolic-byproducts of tumor and stroma, immune cells, and the debris of dead cells. Through a series of studies that arose out of a serendipitous observation, we have identified a variety of soluble factors in MPE that inhibit gene transfer in a vector-specific manner.

1.1. Inhibition to Retroviral Gene Transfer in MPE

Chondroitin sulfate proteoglycans and glycosaminoglycans (CS-PG/GAG) were implicated in inhibiting gene transfer by a variety of retroviral vectors (RV) *(3)*. This conclusion was put forth based on the following evidence. Unfractionated MPE were observed to inhibit retroviral transduction of highly-permissive Mv1Lu cells in a dose-dependent manner. Studies to discriminate between the cellular vs soluble mediators of the inhibition were performed. Red blood cells, which are present in MPE and potentially express the amphotropic retroviral receptor (RAM-1) *(4)*, were first excluded as the source of the inhibition by competition in transduction bioassays. Next, we tested whether soluble factors within the effusions were blocking retroviral transduction. Cells and particulate debris were removed by centrifugation, and the fluid component of the effusions was sterile-filtered and used in transduction bioassays. A dose-dependent inhibition to retroviral transduction in a pattern that was virtually identical to that observed with whole effusions was evidenced. This observation suggested that the inhibition to retroviral transduction by malignant effusions was wholly mediated by soluble factors present in the fluid component of those effusions.

The inhibition of RV-transduction extended to other cell types (demonstrated using NSCLC cells and the human bronchial epithelial cell line CFT1), and other variably encapsidated RV- vectors (amphotropic, VSV-G, and GALV pseudotyped vectors). Furthermore, inactivation of Complement *(5,6)* in MPE did not reverse the observed inhibition. Because the physical properties of MPE suggested a significant contribution by PG/GAGs, we tested an alternative hypothesis that proteoglycans were blocking retroviral transduction. MPE were treated with bovine testicular hyaluronidase (BTH), a mammalian enzyme with a broad activity against PG/GAGs *(7)*. A marked reduction in the inhibitory activity of the effusions was observed following BTH treatment, implicating either hyaluronic acid or other PG/GAGs as the putative factors mediating the inhibition to retroviral transduction. On the heels of this observation, a series of experiments were next performed to specifically identify the inhibitory glycoconjugates. First, effusions were pretreated with a more specific hyaluronidase derived from *Streptomyces* (S-Hya) *(8)* to test whether hyaluronic acid was responsible for the observed inhibition. At a range of S-Hya concentrations effective in digesting hyaluronate in MPE, the block to RV-transduction persisted. Similarly, exogenous addition of hyaluronic acid to transduction medium did not inhibit RV-transduction. These data indicated that hyaluronic acid was not the GAG-component responsible for mediating the inhibition to retroviral transduction.

Next, we tested whether the inhibitory factor(s) was a member of the chondroitin sulfate family of PG/GAGs by treating effusion supernatants with specific chondroitinases *(9,10)*. Pretreatment of effusions with chondroitinases AC or ABC was as effective as BTH in abolishing the inhibition of transduction with retroviral vectors, suggesting that chondroitin sulfates are the components inhibiting retroviral transduction in pleural effusions. To verify this observation, CS-PG/GAGs were added into retroviral stock at various concentrations and were found to inhibit transduction as well. In fact, amphotropic retroviral transduction was virtually abolished when concentrations of the sulfated GAGs exceeded 50 μg/mL in the transduction-medium. We went on to document the presence of CS-PG/GAG in MPE (by immunoblots) and to quantify these elements in a panel of MPE by specific biochemical assays, and the concentrations of uronic acid and sulfated PG in the effusions fell within the range that were predicted to be inhibitory. Gel filtration (Sepharose CL-6B) analysis of effusions showed that CS-moieties started eluting with the void volume, and continued to elute until the salt peak was reached. This observation suggested that chondroitin sulfates in malignant pleural effusions are a heterogeneous group, represented by elements with molecular weights of over 1 million to smaller fragments of less than 10,000 daltons.

Lastly, we isolated PG/GAG from MPE by CsCl density gradient ultracentrifugation, and tested them in RV-transduction bioassays. The purified PG/GAG were found to be inhibitory, and BTH or chondroitinases (and not heparinase, heparitinase, keratanase, or S-Hyaluronidase) could reverse this inhibitory activity. These data enabled us to deduce that heparins or heparan sulfates, keratan sulfate, or hyaluronic acid were not the inhibitory substances in MPE for RV-transduction. Collectively, the results of biochemical assays, immunodetection, and differential enzymatic digestion analyses conclusively demonstrated that the inhibition to retroviral transduction in the pleural fluid was accounted for by the activity of chondroitin sulfates in the effusion.

1.2. Inhibition to Adenoviral Gene Transfer in MPE

The presence of significant soluble barriers to RV-transduction prompted us to consider Adenoviral (Ad)-vectors as a means of achieving gene delivery to tumor cells in this milieu. These studies demonstrate that soluble factors in effusions inhibit gene transfer by Ad-vectors as well. Heating effusions to 56°C (inactivating complement proteins) does not restore transduction efficiency, an expected finding because Ad-vectors are not encapsulated. To determine the relative molecular size of the inhibitor(s), MPE are fractionated using porous membranes that retained factors > 30 or 100 kD. The native inhibitory factors are large (>100kD) components of the effusions that fractionate with the volume retained by the filters. In contrast to RV-transduction, CS-PG/GAG or hyaluronic acid do not inhibit Ad-gene transfer, and the inhibition to Ad-transduction was not reversed by pretreatment of the effusions with BTH *(11)*.

Concomitantly, we investigated whether Ad-entry into target cells was being inhibited. The Ad-particle first interacts with the cell surface by *binding* to a high-affinity cellular receptor (e.g., CAR *(12)*. Following, Ad is *internalized* by receptor-mediated endocytosis, a process triggered by Ad-fiber-knob attachment to the cell surface and that involves interactions of the Ad-penton base with cell-surface $\alpha_v\beta_{3,5}$ integrins via RGD peptides *(13–17)*. We investigated the inhibition to Ad both in terms of binding and internalization. Binding of radiolabeled Ad to its target cell was inhibited in the presence of malignant pleural effusions. This inhibition occurred within the fluid component of the effusions rather than on the cell surface, and washing the effusions off of the target cells prior to Ad-exposure restored basal Ad-binding to the cell membrane. With respect to internalization, the $\alpha_v\beta_{3,5}$ integrins also mediate cellular adhesion to components of the ECM containing RGD moieties, including fibronectin. Because high concentrations of fibronectin (approx 200 µg/mL) are found within MPE *(18)*, we tested whether soluble fibronectin was an inhibitor that competed for Ad-interaction with cell-surface integrins. However, fibronectin, at relevant concentrations in the transduction medium does not alter Ad-transduction. Given that preformed Ad-neutralizing antibodies may be responsible for inhibiting Ad-transduction in malignant effusions, we tested whether adsorption of immunoglobulins (Ig) out of effusions (using Protein A-sepharose) results in increased gene transfer. Ig-depletion restored Ad-transduction at high Ad-concentrations, however, at Ad-MOIs of less than 1000, significant inhibition persisted, suggesting that pre-formed neutralizing antibodies constitute a significant part but not the entirety of the observed inhibition to Ad-gene transfer in malignant effusions.

In order to screen for factors within the effusions that interact with Adenoviral particles, we fractionated the constituents of the effusion by SDS-PAGE and overlaid them with a suspension of Adenovirus. These viral overlay assays suggest that adenovirus binds to factor(s) approximating 70 and 50 kD. Because the 70 kD element runs parallel with BSA, and because albumin is a known component of malignant effusions, we considered whether the 70 kD factor that bound Ad was in fact human serum albumin in the MPE. We electrophoresed a series of species-specific serum albumins by SDS-PAGE and assessed Ad-binding by viral overlays to confirm that adenovirus is ubiquitously bound to serum albumin. However, there was no difference detected in Ad-transduction efficiency in the absence or presence of albumin. Ad-binding to the 50 kD

Obstacles to Viral Gene Therapy

factor on the sodium dodecyl sulfate polyacrylamide gel electrophoresis (SDS-PAGE) resolved viral overlays was attributed to immunoglobulin-heavy chains of anti-adenoviral antibodies. This assertion stemmed from the observation that eluting the effusions through sepharose A columns leads to the removal of the 50 kD-band, although depleting the MPE of Ig does not completely restore transduction efficiency to control levels. Recognizing that Ad-binding to cells is mediated by a 47 kD receptor (CAR) on the target cell surface *(12)*, we tested whether solubilized CAR(sCAR) in MPE was a competitive inhibitor of Ad-binding to target cells. Immunoblots using anti-sera to CAR indicate that a soluble form of the receptor is present in MPE. This soluble receptor can be immunoprecipitated from MPE using the anti-CAR monoclonal RmcB *(19)*, and an affinity column generated using RmcB can be used to deplete CAR from MPE. As predicted, a significant increase in transduction efficiency at a variety of Ad-concentrations is evident when effusions are depleted both of Ig and CAR than depletion of either component alone.

2. Materials

2.1. Preparation of MPE, Determination of MPE Components, and Preparation of MPE Components for Transduction Inhibition Assays

Unless otherwise indicated, most reagents may be purchased from Sigma Biochemicals (St. Louis, MO).

2.1.1. Preparation of Malignant Pleural Effusions

1. 45-μm filters (Schleicher and Schuell, Keene, NH).

2.1.2. Enzymatic treatments of MPE

1. Bovine testicular hyaluronidase (BTH), a mammalian hyaluronidase (Sigma, St. Louis, MO), reconstituted in 50 mM sodium acetate, pH 6.0.
2. *Streptomyces* hyaluronidase (S-Hya), a bacterial hyaluronidase (Seikagaku Amano Pharmaceutical Company, Rockville, MD), reconstituted in 50 mM sodium acetate, pH 6.0.
3. Chondroitinase AC (CAC) [*Flavobacterium heparinum*] (Sigma), reconstituted in 50 mM sodium acetate/PBS, pH 7.4.
4. Chondroitinase ABC (CABC) [*Proteus vulgaris*] (Sigma), reconstituted in 50 mM sodium acetate/PBS, pH 7.4.
5. Heparinase [*Flavobacterium heparinum*] (Seikagaku Amano Pharmaceutical Company), reconstituted in 50 mM sodium acetate/PBS, pH 7.4.
6. Heparitinase [*Flavobacterium heparinum*] (Seikagaku Amano Pharmaceutical Company), reconstituted in 50 mM sodium acetate/PBS, pH 7.4.
7. Keratanase (*Pseudomonas* sp.) (Seikagaku Amano Pharmaceutical Company), reconstituted in 50 mM sodium acetate/PBS, pH 7.4.

2.1.3. Analysis of the Role of Proteoglycan (PG)/Glycosaminoglycans (GAG), and Fibronectin

1. GAGs or fibronectin are admixed with vector-containing medium at varying concentrations for transduction efficiency analyses.
2. Hyaluronic acid, reconstituted in PBS for a stock concentration of 1 mg/mL.
3. Chondroitin sulfate A, reconstituted in PBS for a stock concentration of 1 mg/mL.
4. Chondroitin sulfate B (aka: dermatan sulfate), reconstituted in PBS for a stock concentration of 1 mg/mL.
5. Chondroitin sulfate C, reconstituted in PBS for a stock concentration of 1 mg/mL.
6. Fibronectin, reconstituted in PBS for a stock concentration of 1 mg/mL.

2.1.4. Isolation of Proteoglycan (PG)/Glycosaminoglycans (GAG) in Pleural Fluid

2.1.4.1. PG/GAG Extraction and Quantitation

1. 4 M guanidium HCl CsCl density gradient.
2. Ultracentrifuge tubes.

2.1.4.2. Quantitatation of Uronic Acid Content

1. 25 mM Sodium tetraborate in sulfuric acid (H_2SO_4). Store in a dark bottle at 4°C.
2. 0.1% Carbazole in absolute ethanol. Store in a dark bottle at 4°C. Discard if solution appears yellow.
3. Glucoronic lactone acid in H_2O and store frozen in 1 mg/mL aliquots. Dilute 1/10 with water to give a 100 µg/mL working stock solution that will be used to make standards from 0 to 50 µg/mL glucuronic acid lactone.
4. Bitter-Muir reagent: 0.25 mM sodium tetraborate in sulfuric acid (H_2SO_4). Store away from light at 4°C.

2.1.4.3. DMMB Assay

1. 1,9 Dimethylene blue (DMMB): Serva or Molecular Probes.
2. DMMB working reagent: 18 mg DMMB, 1 mL formic acid, and 2 g sodium formate added to 1 L of distilled and de-ionized-H_2O. Mix formic acid and sodium formate in 1 L water. Check the pH of the solution, but do not adjust. The pH should measure approx 3.5. Dissolve dimethylmethylene blue in approx 5 mL absolute EtOH before adding to solution. Allow solution to mix for several hours. Filter through Whatman paper using a Buchner funnel. Store solution at room temperature, protected from light. Solution is good for up to 1 yr or until precipitate forms or there is a loss of dye-binding.
3. Glycosaminoglycan standard: 1 mg/mL stock solution using either water or 50 mM sodium acetate as a diluent. Standards can be purchased from Sigma. Dilute

the standard stock solution 1/10 or 100 µg/mL in the same diluent used above to make working stock.
4. Dialysis tubing, molecular-weight cutoff 3,500.

2.1.4.4. FRACTIONATION OF MPE

1. Amicon Centriplus-30 or 100 (Millipore Corporation, Bedford, MA).

2.1.5. Immunoglobulin Extraction from MPE

1. 5 mL Hi-trap columns containing Sepharose protein A beads (Amersham Pharmacia Biotech, Uppsala, Sweden).
2. Ig elution buffer: 1 M glycine, pH 2.7.
3. 1 M Tris-HCl, pH 8.0.

2.1.6. Screen for Adenoviral-Interacting Particles in MPE (Viral Overlays)

1. 10% SDS-PAGE and electrotransfer set up.
2. Nitrocellulose (Hybond, Amersham Pharmacia Biotech).
3. Transfer buffer: 39 mM glycine, 48 mM Tris base, 0.037% SDS, 20% methanol
4. TTBS: 20 mM Tris base, 137 mM NaCl, 0.1% Tween.
5. Blocking buffer: 5% (w/v) nonfat milk in TTBS.
6. Adenoviral suspension medium: 5% nonfat milk in RPMI medium.
7. Rabbit anti-adenoviral antisera: 1:1000 dilution in TTBS *(26)*.
8. HRP-linked anti-rabbit-IgG (Santa Cruz Biotechnology), 1:7000 dilution in TTBS.
9. ECL kit (Amersham Pharmacia Biotech).
10. Hyperfilm (Amersham Pharmacia Biotech).

2.1.7. Immunoprecipitation of CAR from MPE

2.1.7.1. DEPLETION OF IG FROM MPE

1. rSepharose A beads (Amersham Pharmacia Biotech).

2.1.7.2. IMMUNOPRECIPITATION

1. Gel-purified RmcB monoclonal antibody.
2. Gammabind G-Agarose (Amersham Pharmacia Biotech).
3. Washing/Binding (W/B) buffer: 10 mM sodium phosphate, pH 7.0, 150 mM sodium chloride, 10 mM EDTA.

2.1.7.3. ACRYLAMIDE GEL ELECTROPHORESIS, ELECTROTRANSFER, AND ANTI-CAR ANALYSIS

1. Laemmli sample buffer: 62.5 mM Tris-HCl pH 6.8, 2% SDS, 25% glycerol, 0.01% bromophenol blue, 0.71 M β-mercaptoethanol (BioRad).
2. 10% acrylamide Tris-HCl gel (BioRad).
3. Running buffer: 25 mM Tris, 192 mM glycine, 0.1% SDS, pH 8.3 (BioRad).

4. Transfer buffer: 25 mM tris, 192 mM glycine, 20% ethanol.
5. PBS-T: PBS, 0.1% Tween 20.
6. Blocking solution: 5% (w/v) nonfat dry milk in PBS-T (Sigma).
7. Rabbit anti-CAR antiserum (a gift from Dr. Jeffrey Bergelson) diluted 1:3000 in blocking solution.
8. Goat anti-rabbit HRP antibody, pre-adsorbed against murine and human immunoglobulins (Biosource). Use at 1:5000 dilution in blocking buffer.
9. Chemiluminescent film. (Amersham Pharmacia Biotech).

2.2. Transduction Protocols and Analysis of Gene-Transfer Efficiency

2.2.1. Transduction

All retroviral vectors used were based on the Moloney murine leukemia virus. LNPOZ is a bicistronic retrovirus vector that uses a poliovirus internal ribosome entry site sequence (IRES) to express both the neo and lacZ genes from a single mRNA. Amphotropic LNPOZ was generated from the helper-free cell line Pa-LNPOZ.1. LNPOZ, pseudotyped with the gibbon ape leukemia virus (GALV) envelope glycoproteins, was produced from PG.LNPOZ.21, a clonal producer line derived from PG13 cells.

1. Mv1Lu cell line, a lung epithelial cell line easily infectable with retroviruses (ATCC), CFT1 cell line *(21)*, a human bronchial epithelial cell line that is efficiently transduced by GALV and amphotropic retroviral vectors (Gift of Dr. James Yankaskas, UNC-Chapel Hill, NC), and NCI-H226 cell line (Acquired from Dr. Herb Oie, NCI, Bethesda, MD).
2. 12- or 6-well tissue-culture plates (Falcon or Costar).
3. Cancer cell line medium: RPMI 1640 (Gibco-BRL) supplemented with 10% fetal bovine serum (FBS; Gibco-BRL) and Penicillin (100 U/mL)/Streptomycin (100 µg/mL) (Gibco-BRL).
4. Mv1Lu cell medium: MEM (Gibco-BRL) supplemented with 10% FBS and penicillin/streptomycin (100 U/mL)/Streptomycin (100 µg/mL) (Gibco-BRL).
5. CFT1 cell medium: Defined serum-free media. (Dr. James Yankaskas, UNC-Chapel Hill, NC).
6. Adenoviral vectors (e.g., Ad5CMVLacZ): All vectors were constructed in the Vector Core at the Gene Therapy Center of the University of North Carolina School of Medicine. These Ad5 vectors were E1a/E1b and E3 deleted and expressed the *Escherichia coli lacZ* gene as the reporter under the regulation of the CMV immediate-early promoter/enhancer region. The adenoviral vector stocks ranged from 1.5×10^{12} to 3×10^{13} particles/mL as measured by OD_{260}, with titers (derived from plaque assays of 293 cells) of 1.7×10^{10} and 6.6×10^{11} pfu/mL, respectively.

2.2.2. Analysis of Transduction Efficiency

1. Phosphate-buffered saline (PBS) (Irvine Scientific).
2. 1% trypsin, 1 mM EDTA.
3. Fluorescein di-beta-D-galactopyranoside (FDG), a labeled substrate for beta-galactosidase (Molecular Probes): Use at 1 mM in FDG hypotonic shock solution.
4. FDG hypotonic shock solution: 1% dimethyl sulfoxide (DMSO) in H_2O.
5. FDG loading termination solution: Ice-cold cell medium containing 10% FBS and 1 µg/mL propidium iodide (PI).
6. FAC-Scan (Becton-Dickinson Immunocytometry Systems).
7. Fixative: 0.5% gluteraldehyde.
8. X-gal wash buffer: PBS, 2 mM $MgCl_2$.
9. X-gal staining solution: PBS containing 5 mM potassium ferrocyanide, 5 mM potassium ferricyanide, 2 mM $MgCl_2$, and 1 mg/mL X-gal.

3. Methods

3.1. Preparation of MPE, Determination of MPE Components, and Preparation of MPE Components for Transduction Inhibition Assays

3.1.1. Preparation of Malignant Pleural Effusions

1. Pleural effusions are removed using aseptic technique from patients with known pleural-based malignancy at the time of hospital admission. The use of these specimens must be approved by Institutional committee for the Protection of the Rights of Human Subjects.
2. Remove cellular debris from the effusions by centrifugation at 250g for 15 min.
3. Filter aliquots of the supernatants for storage (–70°C) or transduction bioassays.
4. Inactivate complement by heating the effusions for 1 h at 56°C.

3.1.2. Enzymatic Treatments of MPE

1. Excessive amounts of all enzymes at specified unit activities are utilized because their relative activities in biological fluids is unknown *(3)*.
2. Perform all digestions overnight at 37°C on an orbital shaker at 200 rpm.

3.1.3. Analysis of the Role of Proteoglycan (PG)/Glycosaminoglycans (GAG), and Fibronectin

Hyaluronic acid, chondroitin sulfate A, CS- B (aka: dermatan sulfate), CS-C, and fibronectin are admixed with vector-containing medium at varying concentrations for transduction efficiency analyses.

3.1.4. Isolation of Proteoglycan (PG)/Glycosaminoglycans (GAG) in Pleural Fluid

3.1.4.1. PG/GAG Extraction

1. Ultracentrifuge 10 mL of sterile filtered pleural effusion at 100,000g for 48 h) in a dissociative CsCl density gradient.
2. Snap-freeze tubes at –80°C, then cut, using a razor blade, into three equal aliquots. The bottom and middle high-density fractions (mean specific densities of 1.6 g/mL and 1.45 g/mL, respectively) contain the large majority of PG/GAG.
3. Dialyze fractions against distilled, deionized (dd) water at 4°C for 2 d with 5 exchanges.
4. Lyophilize fractions. For transduction analyses, re-solubilize the PG/GAGs in each fraction in 10 mL of dd water and sterile filter.

3.1.4.2. Quantitatation of Uronic Acid Content

1. Quantitate the uronic acid content in each fraction by the modified carbazole method *(23,24)*.
2. Set a convection oven to 100°C.
3. Place 20 µL of standard or unknown in duplicate into a 13 × 100 mm borosilicate test tube and add 10 µL of carbazole/EtOH to each tube.
4. Vortex, add 200 µL of Bitter-Muir reagent to each tube, and vortex again.
5. Cover tubes with aluminum foil and place them in the preheated 100°C oven for 15 min.
6. Remove tubes from oven and allow to cool to room temperature.
7. Remove 200 µL from each tube and place in a 96-well microtiter plate and read plate at an absorbance of 540nm. Prepare standard curve by plotting average absorbance for each standard versus its concentration in µg/mL. Plot samples against standard curve to determine sample concentration. Note: Urea does not interfere with this assay in concentrations of 0.8 M or less.

3.1.4.3. DMMB Assay

Measure the sulfated GAG concentration by the DMMB assay *(25)*. DMMB will bind to any polyanionic substance causing a metachromatic shift in the absorbance maxima, and this assay can be used to measure relative amounts of sulfated glycosaminoglycans (keratan sulfate, heparin, heparin sulfate, chondroitin sulfates A and C, and dermatan sulfate) depending on the standard used.

1. Make standard glycosaminoglycan dilution curve ranging from 0 µg/mL to 50 µg/mL and dispense 50 µL of each standard and sample in duplicate into a 96-well enzyme-linked immunosorbent assay (ELISA) plate.
2. Add 200 µL of DMMB reagent to each well and immediately read plate at an absorbance of 575 nm. Plates must be read immediately due to the formation of a precipitate in those wells containing high concentrations of GAG.

Obstacles to Viral Gene Therapy

3. Prepare a standard curve by plotting the average absorbance for each standard versus its concentration in µg/mL. Using the standard curve, determine the concentration for each unknown sample. The correlation coefficient for this assay should be no less than 0.995. Assay may not be completely linear up to 40 µg/mL. In this case, the 40 µg/mL standard should be eliminated. Only use sample concentrations that lie between the absorbance readings of 5 µg/mL and 35 µg/mL.

3.1.4.4. Fractionation of MPE

1. Fractionate sterile-filtered effusions by centrifugation at 3,000g for 4 h at 4°C through microporous membranes that enable concentration of molecules greater than 30kD or 100kD.
2. Collect both eluted and retained fractions and replace total volume to 10 mL with PBS. Repeat sterile-filtering. These fractions can be tested in transduction bioassays (50% v/v with transduction medium).

3.1.5. Immunoglobulin Extraction from MPE

1. Pass effusions through 5 mL Hi-trap columns at a flow rate of 2 mL/min.
2. Elute Ig off the column and bring the Ig-containing fractions to physiological pH with TRIS-buffer, pH 8.0. Ig-depleted specimens can be used in transduction bioassays where the control specimens are similarly diluted (50% v/v with transduction medium) with PBS.

3.1.6. Screen for Adenoviral-Interacting Particles in MPE (Viral Overlays)

1. Resolve MPE by standard 10% SDS-PAGE.
2. Transfer the separated proteins to nitrocellulose by standard electrotransfer.
3. Block the membrane in blocking buffer for 2 h at room temperature.
4. Overlay the membrane with 10^{12} adenoviral particles suspended in 5 mL of 5% nonfat milk in RPMI medium and rock gently for 2 h at room temperature.
5. Wash the membrane three times in TTBS, 10 min each, at room temperature.
6. Detect adenovirus using rabbit anti-adenoviral antisera.
7. Wash the membrane as before and expose to goat HRP-linked anti-rabbit-IgG for 1 h, RT).
8. Rinse the mebrane in TTBS, and detect the labeled proteins using the ECL detection system and exposure to hyperfilm.

3.1.7. Immunoprecipitation of CAR

3.1.7.1. Depletion of Ig from MPE

1. Pre-equilibrate rSepharose A beads in PBS. Add each effusion sample (110 µL, 0.45 µm-filtered) to 110 µL (drained gel volume) of Sepharose A beads.
2. Keep the beads in suspension atop an orbital shaker for 2.5 h at room temperature.

3. Sediment beads by brief centrifugation to harvest the antibody-depleted supernatant.

3.1.7.2. IMMUNOPRECIPITATION

1. Add 2 µg gel-purified RmcB monoclonal antibody to each sample and adjust sample volume to 200 µL with PBS.
2. Admix RmcB into the effusions on a tube inverter and incubate at 4°C for 16 h.
3. Pre-equilibrate Gammabind G-Agarose in W/B buffer and add 20 µL drained gel volume into the effusion-RmcB solution.
4. After 6 h in suspension atop an orbital shaker at room temperature, gently pellet the agarose bead complexes and sequentially wash with WS/B buffer, W/B buffer + 0.5 M NaCl, W/B buffer + 1% v/v Triton-X 100, W/B buffer twice, and water.

3.1.7.3. ACRYLAMIDE GEL ELECTROPHORESIS, ELECTROTRANSFER, AND ANTI-CAR ANALYSIS

1. After removing the water from the final wash, incubate the beads in 30 µL Laemmli sample buffer and heat to 95°C for 8 min.
2. Load the samples were loaded onto one 10% acrylamide Tris-HCl gel and electrophorese at 70 volts.
3. After electrophoresis, rock the gel for 30 min in iced transfer buffer, and subsequently transfer to nitrocellulose membrane at 110 volts for 90 min.
4. Block the blot on an orbital shaker for 1 h at room temperature.
5. Expose the blot to rabbit anti-CAR antiserum for 1 h at room temperature.
6. After washing three times in PBS-T, expose the blot was exposed to goat anti-rabbit HRP antibody for 1 h.
7. Wash the blot 6 times in PBS-T, 15 min each wash.
8. Incubate the membrane in ECL solution and exposed to chemilluminescent film.

3.2. Transduction Protocols and Analysis of Gene-Transfer Efficiency

3.2.1. Transduction

1. Culture selected target cells (*see* **Note 1**) to 70–80% confluency to vector in equivalent volumes (typically 1 or 2 mL) of test and control medium in 12- or 6-well tissue-culture plates.
2. For our studies, the vectors are always resuspended in the specified tissue culture medium for the target cells at various MOIs (lung cancer cells have a very heterogeneous transduction profile and transduction efficiencies must be determined empirically) (*see* **Note 2**).
3. Cells are typically exposed to the Ad-containing medium for 1 h, and to the RV-medium for 2 h at 37°C.

3.2.2. Analysis of Transduction Efficiency

1. Quantify transduction efficiency of transduced cells using flow cytometry to measure intracellular FDG hydrolysis.
2. Detach cells from the tissue-culture plates, sediment by centrifugation at 250g, and resuspend into 100 µL of growth medium.
3. Load cells with FDG during a brief exposure to "hypotonic shock." This step consists of incubation for 1 min at 37°C in hypotonic shock solution containing FDG.
4. Terminate FDG loading by the addition of 400 µL of ice-cold medium containing 1 µg/mL PI. Place cells on ice in the dark until FACS analysis.
5. Evaluate 12,000–16,000 viable cells for β-galactadosidase expression. Nonviable cells, defined as that population which incorporate PI, are excluded in real time as red fluorescent cells. Negative (cells that were not exposed to the vector but were loaded with FDG) and positive (cells transduced in media alone) controls are included in all experiments. For our studies, cells displaying green fluorescence (FDG) exceeding the 99th percentile of the negative control population constituted the transduced cell population *(3)*.
6. Alternatively, transduction efficiency can also be quantified histochemically by microscopy *(20)*. Fix cells for 10 min at 4°C in 0.5% gluteraldehyde.
7. Wash two times in washing buffer.
8. Stain with X-gal (briefly describe method)
9. Quantitate lacZ expression by counting 300–400 cells by hematocytometer.

4. Notes

1. NSCLC cells have a heterogeneous profile when it comes to transduction by viral vectors *(20)*. To study the effect of pleural effusions on the transduction efficiency of retroviral vectors in vitro, target cell types (Mv1Lu and CFT1) that were empirically determined to be highly permissive/transducible with amphotropic retroviral vectors were used *(3,21)*. Similarly, for gauging the inhibitory effect of effusions on Ad-gene transfer, and for isolating CAR from membranes of Ad-permissive cells, the NSCLC cell line NCI-H226 was used *(20)*. These cells bind Ad in a specific manner and exhibit an efficient transduction profile by adenoviral vectors in vitro.
2. All retroviral and adenoviral vectors were generated in specificied core facilities at the University of North Carolina at Chapel Hill or at the University of California at Los Angeles. For construction and propagation of recombinant viral vectors and the methodology used to quantify gene transfer, the reader is encouraged to gather details from the MIMM-series edited by Paul D. Robbins and entitled "Gene Therapy Protocols" *(22)*.

References

1. Sahn, S. A. (1993) Pleural effusion in lung cancer. *Clin. Chest Med.* **14,** 189–200.
2. Sahn, S. A. (1988) State of the art. The pleura. *Am. Rev. Respir. Dis.* **138,** 184–234.

3. Batra, R., Olsen, J., Hoganson, D., Caterson, B., and Boucher, R. (1997) Retroviral gene transfer is inhibited by chondroitin sulfate proteoglycans/glycosaminoglycans in malignant pleural effusions. *J. Biol. Chem.* **272,** 11736–11743.
4. Kavanaugh, M. P., Miller, D. G., Zhang, W., Law, W., Kozak, S. L., Kabat, D., and Miller, A. D. (1994) Cell-surface receptors for gibbon ape leukemia virus and amphotropic murine retrovirus are inducible sodium-dependent phosphate symporters. *Proc. Natl. Acad. Sci. USA* **91,** 7071–7075.
5. Rother, R. P., Squinto, S. P., Mason, J. M., and Rollins, S. A. (1995) Protection of retroviral vector particles in human blood through complement inhibition. *Hum. Gene Ther.* **6,** 429–435.
6. Russell, D. W., Berger, M. S., and Miller, A. D. (1995) The effects of human serum and cerebrospinal fluid on retroviral vectors and packaging cell lines. *Hum. Gene Ther.* **6,** 635–641.
7. Meyer, K., Hoffman, P., and Linker, A. (1960) In *The Enzymes. Hydrolytic Cleavage (Part A).* (Boyer, P. D., Lardy, H., and Myrbaeck, K., eds.), Academic Press, New York, p. 447–460.
8. Derby, M. A. and Pintar, J. E. (1978) The histochemical specificity of Streptomyces hyaluronidase and chondroitinase ABC. *Histochem. J.* **10,** 529–547.
9. Yamagata, T., Saito, H., Habuchi, O., and Suzuki, S. (1968) Purification and properties of bacterial chondroitinases and chondrosulfatases. *J. Biol. Chem.* **243,** 1523–1535.
10. Linhardt, R. J., Galliher, P. M., and Cooney, C. L. (1986) Polysaccharide lyases. *Appl. Biochem. Biotechnol.* **12,** 135–176.
11. Batra, R. K., Dubinett, S. M., Henkle, B. W., Sharma, S., and Gardner, B. K. (2000) Adenoviral gene transfer is inhibited by soluble factors in malignant pleural effusions. *Am. J. Respir. Cell Mol. Biol.* **22,** 613–619.
12. Bergelson, J. M., Cunningham, J. A., Droguett, G., Kurt-Jones, E. A., Krithivas, A., Hong, J. S., et al. (1997) Isolation of a common receptor for Coxsackie B viruses and adenoviruses 2 and 5. *Science* **275,** 1320–1323.
13. Wickham, T. J., Mathias, P., Cheresh, D. A., and Nemerow, G. R. (1993) Integrins alpha v beta 3 and alpha v beta 5 promote adenovirus internalization but not virus attachment. *Cell* **73,** 309–319.
14. Bai, M., Harfe, B., and Freimuth, P. (1993) Mutations that alter an Arg-Gly-Asp (RGD) sequence in the adenovirus type 2 penton base protein abolish its cell-rounding activity and delay virus reproduction in flat cells. *J. Virol.* **67,** 5198–5205.
15. Seth, P. (1994) Adenovirus-dependent release of choline from plasma membrane vesicles at an acidic pH is mediated by the penton base protein. *J. Virol.* **68,** 1204–1206.
16. Greber, U. F., Willetts, M., Webster, P., and Helenius, A. (1993) Stepwise dismantling of adenovirus 2 during entry into cells. *Cell* **75,** 477–486.
17. Goldman, M. J. and Wilson, J. M. (1995) Expression of alpha v beta 5 integrin is necessary for efficient adenovirus-mediated gene transfer in the human airway. *J. Virol.* **69,** 5951–5958.

18. Melloni, B., David, B., Lefevre, M., Clouzard, A., Trazit, M., Veziat, G., and Bonnaud, F. (1997) Fibronectin concentrations in pleural effusions (abstract). *Am. J. Respir. Crit. Care Med.* **155,** A740.
19. Hsu, K. H., Lonberg-Holm, K., Alstein, B., and Crowell, R. L. (1988) A monoclonal antibody specific for the cellular receptor for the group B coxsackieviruses. *J. Virol.* **62,** 1647–1652.
20. Batra, R., Olsen, J., Pickles, R., Hoganson, S., and Boucher, R. (1998) Transduction of non-small cell lung cancer cells by adenoviral and retroviral vectors. *Am. J. Respir. Cell Mol. Biol.* **18,** 402–410.
21. Yankaskas, J. R., Haizlip, J. E., Conrad, M., Koval, D., Lazarowski, E., Paradiso, A. M., et al. (1993) Papilloma virus immortalized tracheal epithelial cells retain a well-differentiated phenotype. *Am. J. Physiol.* **264,** C1219–C1230.
22. Robbins, P. D. (1997) *Gene Therapy Protocols.* Humana Press, Totowa, NJ.
23. Platzer, M., Ozegowski, J. H., and Neubert, R. H. (1999) Quantification of hyaluronan in pharmaceutical formulations using high performance capillary electrophoresis and the modified uronic acid carbazole reaction. *J. Pharm. Biomed. Anal.* **21,** 491–496.
24. Roden, L., Baker, J. R., Cifonelli, J. A., and Matthews, M. B. (1972) in *Methods Enzymol.* **28B,** 73.
25. Chandrasekhar, S., Esterman, M. A., and Hoffman, H. A. (1987) Microdetermination of proteoglycans and glycosaminoglycans in the presence of guanidine hydrochloride. *Anal. Biochem.* **161,** 103–108.
26. Douglas, J. T., Rogers, B. E., Rosenfeld, M. E., Michael, S. I., Feng, M., and Curiel, D. T. (1996) Targeted gene delivery by tropism-modified adenoviral vectors. *Nat. Biotechnol.* **14,** 1574–1578.

33

Topical Gene Therapy for Pulmonary Diseases with PEI-DNA Aerosol Complexes

Ajay Gautam, J. Clifford Waldrep, Frank M. Orson, Berma M. Kinsey, Bo Xu, and Charles L. Densmore

1. Introduction

Different methods for delivery of genes to the lungs such as intravenous injection, and nasal, or intratracheal instillation have been reported in animal models. However, these strategies are invasive or they may not distribute the material uniformly throughout pulmonary tissues. Currently, the technique used for administering gene delivery vectors to the lungs is mainly through injections by the use of computed tomography-guided percutaneous fine-needle injections or bronchoscopy *(1,2)*. This procedure is very invasive and targets a very restricted region in the lungs.

In contrast, aerosol delivery distributes the complexes uniformly and represents a noninvasive alternative for targeting the genes to the lungs. Aerosol gene delivery holds promise for a variety of pulmonary diseases such as cystic fibrosis, alpha-antitrypsin deficiency, pulmonary hypertension, and lung cancer. However, the levels of transgene expression have generally not been very high due, in some cases, to loss of DNA viability during nebulization. Since the first report on the potential of aerosol delivery of genes *(3)*, cationic liposomal formulations have been predominantly developed for aerosol gene therapy *(4,5)*.

Efficient gene delivery to the lungs by aerosol can be achieved by protecting the DNA during nebulization, increasing the concentration of particles in aerosol or by modifying the particle (droplet) size of aerosol. The particle size of the aerosol is an important parameter, which influences the deposition of the aerosol droplets in the various regions of the respiratory tract *(6)*. Alternate methods could involve modifying the breathing parameters such as minute volume. Breathing low levels of carbon dioxide in aerosol should theoretically increase the tidal volume and breathing frequency, and lead to greater inhalation and deposition of aerosol particles *(7)*.

Recently, we demonstrated that polyethylenimine (PEI), a cationic polymer, holds promise as a gene delivery vector by aerosol *(8)*. PEI consistently gives a higher level of transgene expression in the lungs after aerosol delivery as compared to different cationic lipids tested *(8)*. We further optimized our PEI-DNA aerosol model system, and also provided evidence for the use of 5% CO_2-in-air for nebulization to achieve enhanced gene expression in the lungs after PEI-DNA aerosol delivery *(9)*. The dose delivered by aerosol showed a therapeutic effect using the p53 tumor suppressor gene in two different animal lung metastases models *(10,11;* see Chapter 37); and also induced a robust antibody response for the purpose of genetic immunization *(8)*. Furthermore, PEI-DNA aerosol delivery is not associated with any acute inflammatory or cytokine responses *(9,12)*, in contrast to that reported for cationic liposomes and viral vectors *(12–14)*. Also, we have shown that after aerosol delivery PEI-DNA complexes transfect mainly the epithelial cells lining the airways, including the peripheral airways, with diffuse transfection in the alveolar lining cells *(15)*.

2. Materials
2.1. Preparation of PEI-DNA Complexes
1. Phosphate-buffered saline (PBS), pH 7.2 (Lifetechnologies, Grand Island, NY).
2. PEI (25 KDa, branched polymer) was purchased from Aldrich Chemical (Milwaukee, WI). A PEI stock solution is prepared at a concentration of 4.3 mg/mL (0.1 M in nitrogen) in H_2O and the pH was adjusted to 7.0–7.5 using 0.1 M HCl. The PEI stock solution can be stored at 2–4°C. Concentrated PEI is toxic and should be handled with care.
3. Plasmid DNA: All plasmids are purified commercially by Bayou Biolabs (Harahan, LA), are endotoxin free, and quantitated using UV absorbance. Agarose gel analysis revealed the plasmids to be primarily in the supercoiled form with a small amount of nicked plasmid. *(see* **Note 1**).
   a. p4119 (CAT): The bacterial chloramphenicol acetyl transferase gene *(15)* is primarily used as the reporter gene for measuring transgene expression. The CAT gene used is under the control of the human cytomegalovirus (CMV) early promoter/enhancer element.
   b. pGL3. The luciferase plasmid (Promega, Madison, WI) is also used as a control. It is modified by insertion of the CMV promoter/enhancer element and the human growth hormone polyadenylation sequence. A gift from Dr Michael Barry (Center for Cell and Gene Therapy, Baylor).

2.1.1. Aerosol Particle Size Analysis

1. Andersen/ACFM nonviable ambient particle-sizing sampler (Andersen Instruments, Atlanta, GA).
2. KaleidaGraph 2.0 software (Synergy Software, Reading, PA).
3. All-glass impinger (Ace Glass, Vineland, NJ).

2.2. Aerosol Delivery of PEI-DNA Complexes

1. Plastic cages.
2. Polyethylene tubing (Beckton Dickinson, Sparks, MD).
3. Carbon dioxide tank (Air Liquide America Corp., Houston, TX).
4. Aerotech II nebulizer (AT-II) (CIS-US, Inc., Bedford, MA).
5. A source of dry air (Aridyne 3500, Timeter, Lancaster, PA).
6. A Bird 3 M gas blender (Palm Springs, CA).
7. Fyrite CO_2 analyzer (Bacharach, Pittsburgh, PA).
8. Ventilation hood.
9. Female Balb/C mice, 7–8 wk old (Harlan Sprague Dawley, Houston, TX).

2.3. Harvesting Lungs for Reporter Gene Assays

1. Isoflurane (Vedco, St. Joseph, MO).

2.3.1. CAT Assay

1. CAT ELISA kit (Boehringer Mannheim Gmbh, Mannheim, Germany). The kit includes extraction buffer, standards and plates for enzyme-linked immunosorbent assay (ELISA).
2. Wig-L-Bug bead homogenizer (Crescent Dental Mfg., Lyons, IL).
3. 4 mm glass beads (Kimble Glass Inc., Germany).
4. 5 mL tubes (50 × 16 mm; Sarstedt, Germany) with caps for tissue homogenization.
5. 96-well plates.
6. Microtiter plate reader (Molecular Devices, Sunnyvale, CA).
7. BCA protein assay (Pierce, Rockford, IL).

2.3.2. Luciferase Assay

1. Luciferase assay kit (Promega, Madison, WI). The kit includes extraction buffer, standards and substrate for luciferase.
2. Luminometer (Microlumat LB 96 P, EG&G Berthold, Germany).

2.4. Harvesting Lungs for Histological Analysis

1. PE50 tubing (Clay Adams, Becton Dickinson).
2. Neutral buffered formalin (10%, v/v) (Richard-Allan Scientific, Kalamazoo, MI).
3. Set up for paraffin embedding and tissue sectioning.
4. Hematoxylin and eosin stain (Richard Allan Scientific).

2.4.1. Biochemical Analysis of Inflammation (see **Note 8**)

1. HTAB solution: 0.5% Hexadecyltrimethylammonium bromide in 50 mM phosphate buffer, pH 6.0. Use 5 mL HTAB/gm of tissue.
2. MPO substrate: o-diasinidine dihydrochloride (0.167 mg/mL) plus hydrogen peroxide at 0.0005%.

2.4.2. Immunohistochemistry

1. 50% (v/v) OCT in saline (optimal cutting compound, Sakura Finetek, CA).
2. Superfrost plus slides (Fisherbrand, Fisher Scientific, Pittsburg, PA).
3. 0.6% H_2O_2.
4. Blocking solution: 5% FBS in PBS.
5. Sheep polyclonal antibody to CAT, digoxigenin labeled, 20 µg/mL (Roche Diagnostics, IN).
6. Secondary antibody: anti-digoxigenin peroxidase-labeled antibody, 1:150 dilution (Roche Diagnostics, IN).
7. Aminoethylcarbazol (AEC) (Sigma, St. Louis, MO).

2.5. Evaluation of Cytokine Induction in the Lungs and Serum

1. HBSS media (Hank's balanced salt solution, Mediatech Cellgro, VA).
2. Extraction buffer: 50 mM Tris-HCl, pH 8.0, 150 mM NaCl, 1 mM EDTA, 0.5% Triton X-100, 1 µM pepstatin A, 0.25 mM PMSF and 10 µM leupeptin.
3. Specific immunoassay kits for tumor necrosis factor-α (TNF-α) and interleukin (IL)-1β (R&D Systems, Minneapolis, MN).

3. Methods
3.1. Preparation of PEI-DNA Complexes

1. Calculate quantities of PEI and DNA to add to 5 mL water required. For a 10:1 N:P ratio and 2 mg/10 mL DNA concentration, 606 µL of PEI solution (0.25 N) is added to 5 mL water and 2 mg DNA is added to another 5 mL water (*see* **Note 2**).
2. The PEI solution is then slowly vortexed and the DNA solution added drop wise to the PEI solution to make a final volume of 10 mL. Allow the mixture to stand at room temperature for about 15–20 min before nebulization. The resulting charge ratio is expressed as PEI nitrogen:DNA phosphate (N:P), which can be calculated by taking into account that DNA has 3 nmol phosphate per µg and 1 µL of 0.1 M PEI solution has 100 nmol of amine nitrogen. A 10:1 N:P ratio corresponds to a 1.29:1 PEI:DNA weight ratio (*see* **Note 3**).

3.1.1. Aerosol Particle Size Analysis

1. Determine the particle size of the prospective aerosol using a nonviable ambient particle-sizing sampler. Collect samples over a 5-min period of nebulization. (*see* **Note 4**).

2. Calculate the mass median aerodynamic diameter (MMAD) and geometric standard deviation (GSD) using KaleidaGraph 2.0 software (*see* **Notes 5** and **6**).
3. Calculate the aerosol output by drawing the PEI-DNA aerosol by vacuum through a calibrated glass tube. Hold the tip 4 mm above a 10 mL volume of water in an all-glass impinger at a flow rate of 10 L/min. Quantitate the collected fluid for DNA using a spectrophotometer.

3.1.2. Calculation of Aerosol Dosage for Mice

Based on estimates of minute volume, mice exchange 1 L/min/gm body weight of air and deposit 30% of the inhaled particles *(16)*. Thus, the estimated aerosol dosage can be calculated by the following formula:

$$D = C \times DI \times V \times T$$

where,
D = deposited dose
C = aerosol concentration of DNA
DI = deposition index (= 0.3 for mice)
V = volume of air exchanged (= 1 L/min/gm body weight for mice)
T = Time of aerosol exposure

3.2. Aerosol Delivery of PEI-DNA Complexes

Mice should be housed five to a cage and kept in a laminar-flow cabinet under specific pathogen-free conditions for 2 wk before use

1. Place mice in plastic cages and sealed with tape before aerosol delivery. This is an unrestrained, whole body aerosol exposure system (*see* **Fig. 1**).
2. Aerosolize PEI-DNA complexes using an Aerotech II nebulizer at 10 L/min flow rate using air or air containing 5% CO_2. A source of dry air is delivered to a Bird 3 *M* gas blender attached to an air compressor and a CO_2 tank. The resulting mixture of air and CO_2 is delivered to the nebulizer. The final concentration of 5% CO_2 in air was determined using a Fyrite CO_2 analyzer. The nebulization of 10 mL solution should take approx 30 min (*see* **Note 7**).

3.3. Harvesting Lungs for Reporter Gene Assays

1. Anaesthetize mice with isoflurane and sacrificed by exsanguination via the abdominal aorta. Isolate lungs and weigh.

3.3.1. CAT Assay

1. Harvest lungs and homogenize in 700 µL CAT assay extraction buffer using a Wig-L-Bug bead homogenizer.
2. Centrifuge the homogenates at 14,000*g* for 20 min at 4°C. Take 200 µL of the clarified extract for the CAT ELISA performed in a 96-well plate format according to manufacturer's instructions.

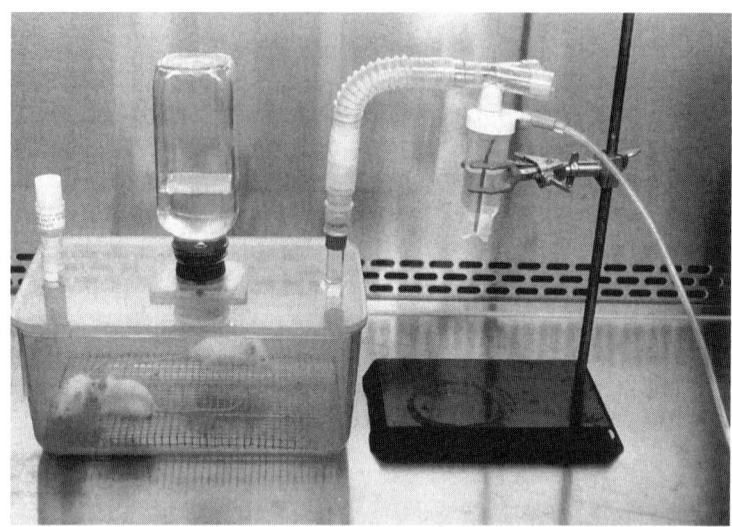

Fig. 1. Laboratory set up for aerosol exposure to the mice. The mice are placed in a plastic cage that is sealed with tape before aerosol delivery. The whole aerosol set-up is placed under a ventilated hood.

3. Read the absorbance using a microtiter plate reader.
4. Naive mice are used as controls. CAT activity is expressed as ng of CAT/mg total lung protein, using a standard curve prepared with purified CAT enzyme. The amount of CAT is normalized to the total protein content in the lungs. As shown in **Fig. 2**, the CAT expression in the lungs is persistent for about a month after a single aerosol exposure.

3.3.2. Luciferase Assay

1. Homogenize the lungs in 1 mL luciferase extraction buffer using a Wig-L-Bug bead homogenizer.
2. Centrifuge the homogenates. Take 10 µL of the extract, add to 50 µL luciferase substrate and read the luminescence for 10 s in a 96-well plate on a luminometer.
3. Naive mice are used as controls. The luciferase activity is expressed as RLU/10 sec/gm of tissue.

3.4. Harvesting Lungs for Histological Analysis

1. Anaesthetize mice with isoflurane and sacrifice by exsanguination via the abdominal aorta.
2. Isolate lungs and cannulate using a PE50 tubing. Fix by inflation with 10% neutral buffered formalin, embed in paraffin and process for histological analysis.

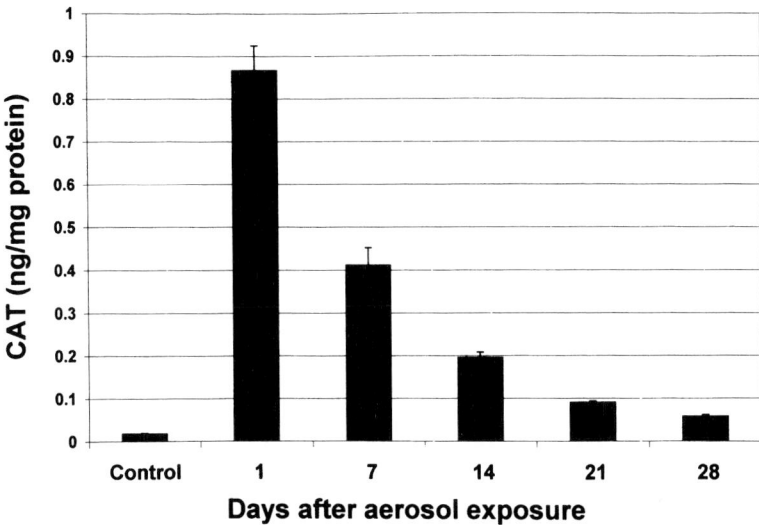

Fig. 2. Persistence of CAT expression in the lungs. Two mg of CAT plasmid was complexed with PEI at a N:P ratio of 10:1 and the complex was aerosolized to the mice using 5% CO_2-in-air. A total of 4 mg of CAT plasmid was aerosolized (20 mL aerosol exposure volume). Mice were sacrificed at different time points and the lungs were harvested and immediately frozen. The CAT ELISA was performed after the last time point (d 28). The amount of CAT was normalized to the total protein content in the lungs measured by the BCA protein assay (Pierce, Rockford, IL). Values are mean ± SD ($n = 5$ mice per time point).

3. Cut thin sections at 4 μm and observe under the microscope for any signs of inflammation or toxicity using the hematoxylin and eosin stain, or immunohistochemically stain for reporter gene expression.

3.4.1. Biochemical Analysis of Inflammation (see **Note 8**)

1. Anaesthetize mice with isoflurane and sacrifice by exsanguination via the abdominal aorta 24 h after aerosol exposure.
2. Harvest the lungs after perfusion through the heart with saline.
3. Homogenize the tissue in HTAB solution. Centrifuge homogenate at 14,000g for 10 min at room temperature.
4. Determine the myeloperoxidase (MPO) activity in the supernatant by adding MPO substrate and hydrogen peroxide. Measure absorbance at 460 nm using a microtiter plate reader. Record the absolute values after 15 min of incubation as well as the absorbance change (δA) over 2 min. Use naive mice as negative controls and mice exposed to lipopolysaccharide (LPS) aerosol or instillation as positive controls.

3.4.2. Immunohistochemistry

1. Twenty-four hours after aerosol delivery, sacrifice mice and isolate, cannulate and fix lungs by inflation with 50% (v/v) OCT.
2. Cut cryosections at 5 µm, collect on superfrost plus slides, air dry, and incubate with 0.6% H_2O_2 for 20 min to inactivate endogenous peroxidase activity.
3. Block the slides using 5% FBS for 30 min at room temperature.
4. Drain off blocking solution and incubate with anti-CAT antibody for 1 h at RT.
5. Wash slides with PBS, and incubate in secondary antibody for 1 h at RT.
6. Develop the slides using AEC for 3–5 min. Lungs from naïve mice are used as controls. As seen in **Fig. 3**, most of the intense staining for CAT is observed in the epithelial cells lining the airways, with some diffuse transfection into the alveolar lining cells.

3.5. Evaluation of Cytokine Induction in the Lungs and Serum

Mice exposed to aerosol complexes, or injected intravenously, are sacrificed at various time points. Samples from naïve mice are used as controls.

3.5.1. Serum Collection

1. Bleed mice from the abdominal aorta and collect blood. Incubate at room temperature for 2 h, then centrifuge at 14,000*g* for 10 min.
2. Collect serum and freeze at –80°C until needed.

3.5.2. Broncheoalveolar Lavage Fluid (BALF) and Lung Homogenate Collection

1. To collect BALF, expose the trachea by making a thin incision. Cannulate endotracheally and lavage with 2 ml HBSS.
2. Wash the lungs 5 times, collect BALF and freeze at –80°C.
3. Harvest the lungs and homogenize with a Wig-L-Bug bead homogenizer, using 1 mL of ice-cold extraction buffer.

Fig. 3. *(see opposite page)* Immunohistochemistry for CAT. Two mg of CAT plasmid was complexed with PEI at a N:P ratio of 10:1 and the complex was aerosolized to 3 mice using 5% CO_2-in-air. A total of 4 mg of CAT plasmid was aerosolized (20 mL aerosol exposure volume). **(A)** Twenty-four hours after the aerosol exposure, the lungs were harvested, inflated with 4% paraformaldehyde and fixed in paraformaldehyde for 1 h. The lungs were then immersed in the CAT staining solution (Roche Diagnostics, Indianapolis, IN) for 4–16 h and visualized under light for staining. The large left lobe is shown. For CAT immunohistochemistry, mice were sacrificed 24 h later, lungs harvested and processed for as described in the text. **(B)** Control Magnification 20×. **(C)** Terminal bronchioles. Magnification 10×. **(D)** Large conducting airway. Magnification 20×. **(E)** Conducting airway. Magnification 40×. **(F)** Terminal airway. Magnification 40×.

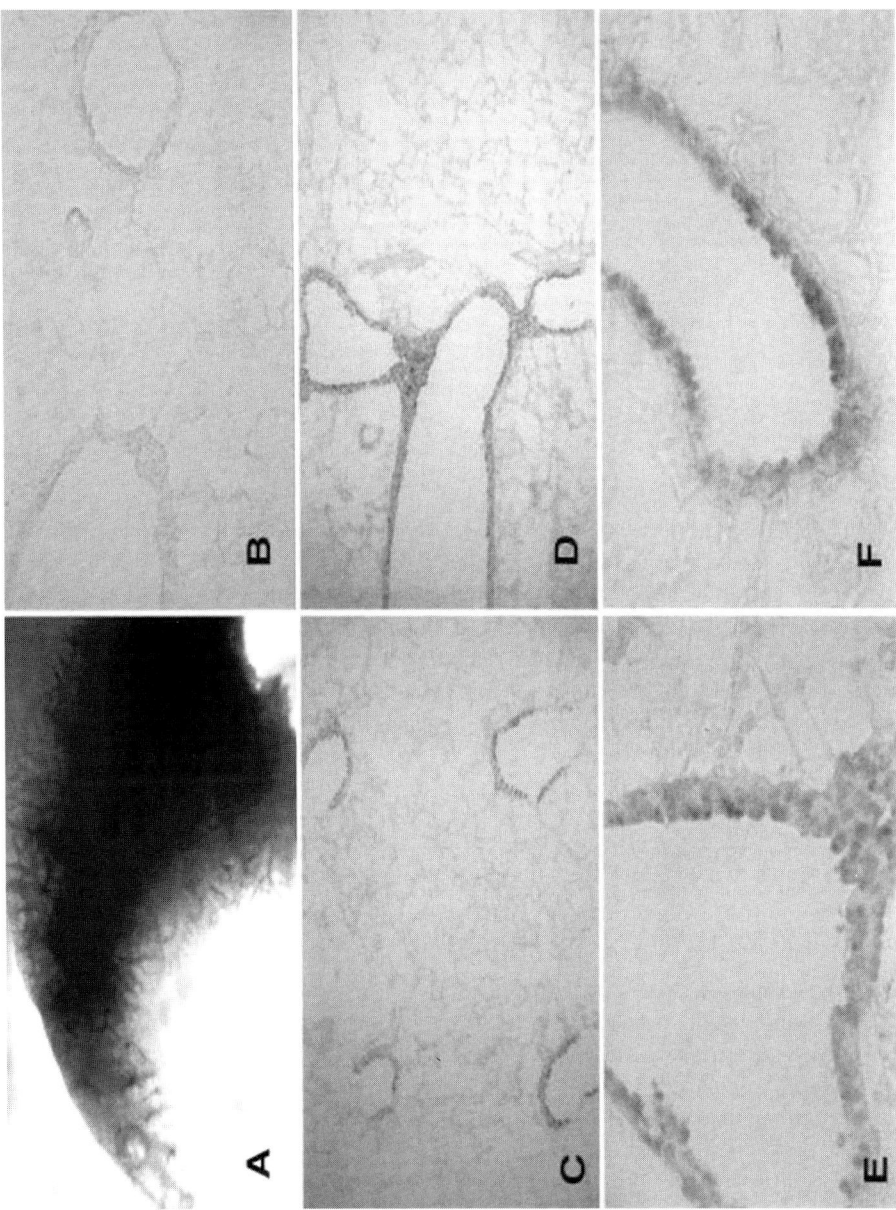

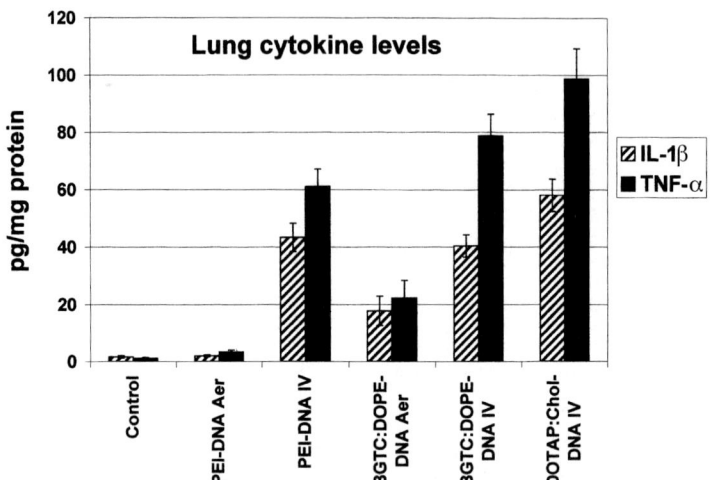

Fig. 4. Comparison of TNF-α and IL-1β induction in lungs after aerosol and intravenous delivery of cationic vector-DNA complexes. Mice were sacrificed 2 h after exposure to different complexes and lungs harvested (as described in text) for immunoassays. Samples were analyzed in duplicates on a microtiter plate reader and the cytokine levels were determined using a standard curve. The cytokine levels were normalized to the total protein content in the lungs. Values are mean ± SD ($n = 3$ to 4 mice per group). IV, intravenous; Aer, Aerosol.

4. Clarify the homogenate by centrifugation at 14,000g for 10 min at 4°C and freeze the supernatant at –80°C until needed.

3.5.3. Determination of Cytokine Levels

1. Determine TNF-α and IL-1β levels in serum, lung extract and BALF using murine specific immunoassay according to manufacturer's instructions.
2. Analyze samples in duplicate on a microtiter plate reader and determine cytokine levels using a standard curve. As shown in **Fig. 4**, the PEI-DNA aerosol delivery does not induce any significant levels of cytokines in the lungs and serum.

4. Notes

1. The structure of the plasmid is very important for transgene expression. A supercoiled plasmid generally gives better expression levels than nicked or linear plasmid. Also, frequent freezing and thawing of plasmids should be avoided since this leads to conversion of supercoiled plasmid into nicked and linear forms.
2. PEI and DNA complexes should be prepared freshly before aerosolization. The PEI-DNA solution should not be stored.

3. Transgene expression by PEI-DNA aerosol delivery is both dose and PEI:DNA (N:P) ratio dependent. The N:P ratio needs to be optimized for different plasmids. The best results should be obtained at an N:P ratio of 10:1 to 20:1.
4. Exposure to aerosol particles could be harmful for laboratory personnel. Care should be taken that the aerosol exposures to the mice are done in a ventilated hood.
5. The MMAD represents the particle size distribution of the aqueous droplets in the aerosol. The AT-II nebulizer used produces aerosols in the particle size 1–2 µm. The particles in this size range avoid oropharyngeal deposition and mainly deposit deep in the lung tissues. The size for PEI-DNA aerosol particles was calculated to be 1.6 MMAD with a GSD of 2.9.
6. The Aerotech-II (AT-II) is a high-output, efficient nebulizer demonstrated to produce aerosols in the optimal range of 1–2 µm mass median aerodynamic diameter (MMAD) for peripheral lung delivery. Using nebulizers other than AT-II will require optimization of aerosol parameters.
7. Care should be taken during aerosolization of complexes to the mice using 5% CO_2. The concentration of CO_2 should be first accurately adjusted and then the air mixture attached to the nebulizer. A high concentration of CO_2 can kill the mice quickly. The set up should be monitored periodically during nebulization for any loose connections. Mice should also be monitored for any signs of distress or becoming moribund.
8. Acute pulmonary inflammation is mediated in part by polymorphonuclear leukocytes (PMN) sequestration to the peripheral tissues. A biochemical marker for PMN is myeloperoxidase (MPO), which is a heme containing enzyme found in the azurophilic granules and is often utilized as an inflammation marker in the lungs.

Acknowledgments

The authors thank Sara Melton, Eva Golunski, and Luz Roberts for their expert technical assistance. We would also like to thank Dr. Vernon Knight, Dr. Brian Gilbert, and Dr. Nadya Koshkina for helpful suggestions and discussions. This work was supported by the Clayton Foundation for Research (Houston, TX). A.G. and C.L.D. were supported, in part, by an Advanced Technology Program Grant from the Texas Higher Education Coordinating Board (# 004949-0129-1999 to C. Densmore). Inquiries should be addressed to C. Densmore.

References

1. Swisher, S. G., Roth, J. A., Nemunaitis, J., et al. (1999) Adenovirus-mediated p53 gene transfer in advanced non-small-cell lung cancer. *J. Natl. Cancer Inst.* **91(9)**, 763–771.
2. Zuckerman, J. B., Robinson, C. B., McCoy, K. S., et al. (1999) A phase I study of adenovirus-mediated transfer of the human cystic fibrosis transmembrane conductance regulator gene to a lung segment of individuals with cystic fibrosis. *Hum. Gene Ther.* **10(18)**, 2973–2985.

3. Stribling, R., Brunette, E., Liggit, D., Gaensler, K., and Debs, R. (1992) Aerosol gene delivery *in vivo*. *Proc. Natl. Acad. Sci. USA* **89,** 11277–11281.
4. Eastman, S. J., Lukason, M. J., Tousignant, J. D., Murray, H., Lane, M. D., St. George, J. A., et al. (1997) A concentrated and stable aerosol formulation of cationic lipid:DNA complexes giving high level gene expression in mouse lung. *Hum. Gene Ther.* **8,** 765–773.
5. McDonald, R. J., Liggitt, H. D., Roche, L., Nguyen, H. T., Pearlman, R., Raabe, O. G., et al. (1998) Aerosol delivery of lipid:DNA complexes to lungs of rhesus monkeys. *Pharm. Res.* **15(5),** 671–679.
6. Gonda, I. (1992) in *Pharmaceutical Inhalation Aerosol Technology* (Hickey, A. J., ed.), Marcel Dekker, New York, pp. 61–82.
7. Koshkina, N. V., Knight, V., Gilbert, B. E., Golunski, E., Roberts, L., and Waldrep, J. C. (2001) Improved respiratory delivery of anticancer drugs, camptothecin and paclitaxel, with 5% CO_2 enriched air: Pharmacokinetic studies. *Cancer Chemother. Pharmacol.* **47,** 451–456.
8. Densmore, C. L., Orson, F. M., Xu, B., et al. (2000) Aerosol delivery of robust polyethyleneimine-DNA complexes for gene therapy and genetic immunization. *Mol. Ther.* **1(2),** 180–188.
9. Gautam, A., Densmore, C. L., Xu, B., and Waldrep, J. C. (2000) Enhanced gene expression in mouse lung after PEI-DNA aerosol delivery. *Mol. Ther.* **2(1),** 63–70.
10. Gautam, A., Densmore, C. L., and Waldrep, J. C. (2000) Inhibition of experimental lung metastasis by aerosol delivery of PEI-p53 complexes. *Mol. Ther.* **2(4),** 318–323.
11. Densmore, C. L., Kleinerman, E. S., Gautam, A., et al. (2001) Growth suppression of established human osteosarcoma lung metastases in mice by aerosol gene therapy with PEI:p53 complexes. *Cancer Gene Therapy* **8(9),** 619–627.
12. Gautam, A., Densmore, C. L., and Waldrep, J. C. Pulmonary cytokine responses associated with PEI-DNA aerosol gene therapy. *Gene Ther.* **8,** 254–257.
13. Scheule, R. K., et al. (1997) Basis of pulmonary toxicity associated with cationic lipid-mediated gene transfer to the mammalian lung. *Hum. Gene Ther.* **8(6),** 689–707.
14. Kafri, T., Morgan, D., Krahl, T., Sarvetnick, N., Sherman, L., and Verma, I. (1998) Cellular immune response to adenoviral vector infected cells does not require de novo viral gene expression: implications for gene therapy. *Proc. Natl. Acad. Sci. USA* **95(19),** 11377–11382.
15. Gautam, A., Densmore, C. L., Golunski, E., Xu, B., and Waldrep, J. C. (2001) Transgene expression in mouse airway epithelium by aerosol gene therapy with PEI-DNA complexes. *Mol. Ther.* **3(4),** 551–556.
16. Knight, V., Koshkina, N. V., Waldrep, J. C., Giovanella, B. C., and Gilbert, B. E. (1999) Anticancer effect of 9-Nitrocamptothecin liposome aerosol on human cancer xenografts in nude mice. *Cancer Chemother. Pharmacol.* **44,** 177–186.

III

NOVEL THERAPIES

B. GENE THERAPIES

II. Plasmid and Virus-Based Gene Therapies

34

Targeted Delivery of Expression Plasmids to the Lung via Macroaggregated Polyethylenimine-Albumin Conjugates

Frank M. Orson, Berma M. Kinsey, Balbir S. Bhogal, Ling Song, Charles L. Densmore, and Michael A. Barry

1. Introduction

Gene therapy for lung cancer has great potential to treat and possibly cure one of the most deadly diseases in man. Various methods are being developed to treat lung cancer by the delivery and expression of a specific gene (or genes) to the lung. These methods are intended to exploit the antineoplastic properties of the gene product, to elicit specific immune responses against cancer cells via the gene product, or to enhance immune responses elicited by the tumor itself. The advantages and disadvantages of a number of delivery methods are discussed in other chapters of this text. In this chapter, we will describe the preparation and use of macroaggregates of albumin-polyethylenimine conjugates (MAA-PEI) as a novel nonviral method of delivery of gene-expression vectors to lung tissue, and will illustrate this method as a means of immunization as well as gene delivery.

As with any new method, there are numerous issues regarding safety, ease of use, reliability, limitations on the magnitude of gene expression, and unexpected effects. Polyethylenimine itself is a DNA binding polycation that has been used by numerous investigators as a nonviral gene delivery agent *(1–3)*. It is thought to deliver DNA to endosomes, but permits escape of plasmid DNA from degradation by buffering the acidification of the endosome, leading to its disruption *(4)*. High concentrations of PEI alone tend to be toxic for cells in vitro, and rapid injection of near toxic amounts of PEI-bound plasmid in large volumes of 5% glucose is required to achieve high level expression in the lung *(2) (see* **Note 1**).

Macroaggregated albumin (MAA) is a safe clinical agent that has been used in Nuclear Medicine for many years to image pulmonary blood flow. When particles are radiolabeled, lung perfusion can be imaged because the particles collect in the pulmonary capillary bed in proportion to blood flow *(5)*. To take advantage of this property for vaccination purposes, we conjugated serum albumin to PEI and then aggregated the conjugate by heating to produce particles that avidly bind plasmid DNA *(6)*. The conceptual basis for using MAA-PEI/DNA is the targeting of pulmonary interstitial macrophages and dendritic cells *(7–9)*. The particles accumulate in peripheral lung tissue and enter the interstitium for eventual clearance by interstitial macrophages *(8,10)*. Importantly, intravenously injected MAA-PEI particles loaded with plasmid DNA selectively transfect the lung tissue of mice with no detectable gene expression in other tissues *(6)*.

In contrast to the systemic immunization elicited by i.m. or i.d. (back skin) administration of naked DNA, the expressed antigen delivered to the lung by MAA-PEI particles was found to effectively elicit high-level mucosal as well as systemic immunity *(6)*. Mucosal immune responses are more difficult to achieve in general *(11)* except with live attenuated vaccines like the polio virus vaccine, and eliciting mucosal immunity with genetic vaccines has also been difficult either with the widely used gene gun or with needle inoculation of naked DNA *(12,13)*. Since mucosal responses can be achieved with MAA-PEI bound plasmids *(6)*, focusing immune responses in the mucosal tissue of the lung could be of substantial value in the treatment and prevention of local as well as systemic pulmonary malignancy *(14,15)*.

2. Materials

2.1. Preparation of Macroaggregated Albumin-Polyethylenimine Conjugate (MAA-PEI)

1. SPDP solution: 25 mg/mL N-succinimidyl 3-(pyridyldithio)propionate (SPDP) in dimethyl sulfoxide (DMSO) (SPDP: Cat. no. 21857, Pierce, Rockford, IL, DMSO: Cat. no. 27,685-5, Aldrich, Milwaukee WI).
2. HSA solution: 10 mg/mL human serum albumin in 0.1 M NaHCO$_3$, pH 8.5 (HSA: Cat. no. A-3782, Sigma, St. Louis, MO) *(see* **Note 2**).
3. PEI solution, 50% (w/w): 72.5 mg Polyethylenimine PEI in 1 mL water. Adjust pH to 8.5 with 10% HCl. (PEI, 750,000 MW: Cat. no. P3143 Sigma) *(see* **Note 3**).
4. NAP-10 columns (Cat. no. 17-0854-02, Pharmacia, Piscataway, NJ).
5. Phosphate-buffered saline (PBS), pH 7.4 (Life Technologies, Rockville, MD).
6. Reductacryl (Cat. no. 233157, Calbiochem, San Diego, CA).
7. 0.20 µm filter for PEI solution.

8. 10% HCl.
9. 2 M NaOH.
10. 100°C oil bath.
11. 0.6 % agarose gel in TBE buffer: 0.089 M Tris, 0.089 boric acid, 0.002 EDTA, pH 8.0.

2.1.1. Plasmid Binding to MAA-PEI and HSA-PEI Suspensions

1. pCMV-Luc: luciferase (LUC) plasmid pGL3 (Promega, Madison, WI) in which the CMV enhancer/promoter region and human growth hormone polyadenylation signal from CMV1 *(16)* were inserted to replace the original SV40 promoter and polyadenylation signal, respectively.
2. pEGFP: plasmid encoding green fluorescent protein (GFP) (Clontech, Palo Alto, CA).
3. pCMV-hGH: plasmid encoding the human growth hormone gene (hGH) *(17)*.
4. pCMV-UB23: plasmid encoding a fusion product of a fragment of the human immunodeficiency virus (HIV) gp120 envelope protein and ubiquitin (a kind gift of Kathryn Sykes, University of Texas Health Science Center, Dallas, TX).

2.2. Plasmid Binding to MAA-PEI and HSA-PEI Suspensions

1. Mice: Balb/C mice, male and female (Harlan Sprague-Dawley, Houston, TX).
2. Tail-Veiner restrainer (Braintree Scientific, Braintree, MA).

2.3. Standard Assays for Temporal and Organ Specific Expression of Genes of Interest

2.3.1. Luciferase Assay

1. Luciferase Assay Kit (Promega, Madison, WI) Includes lysis buffer and substrate.
2. Conical ground glass tissue grinder (Kontes Duall 23, VWR, Houston, TX).
3. Turner TD-20e Luminometer (Turner Instruments, Sunnyvale, CA).

2.3.2. Green Fluorescent Protein Detection *(6)*

1. PE50 tubing (outer diameter 0.965 mm, Clay Adams, Sparks, MD).
2. Freezing medium: O.C.T. Compound (Miles Inc., Elkhart, IN).

2.3.3. Humoral Immune Responses And Enzyme Linked Immunosorbent Assay (ELISA) *(6)*

1. Avertin: 0.8% 2,2,2-tribromoethanol and 0.4% *tert*-amyl alcohol in saline
2. Purified hGH protein: 0.5 µg/mL (50 ng/well) in PBS buffer, pH 7.3 (CalBiochem).
3. Microtiter plates (Immunlon II, Dynex Technologies, Chantilly, VA).
4. Blocking solution: 5% nonfat milk in PBS.
5. Wash buffer: PBS-Tween (0.1%).

6. Horseradish peroxidase-conjugated goat anti mouse immunoglobulin (Bio-Rad, Hercules, CA) diluted in PBS-Tween at 1:2000.
7. Horseradish peroxidase-conjugated rabbit anti mouse immunoglobulin (Bio-Rad, Hercules, CA) diluted in PBS-Tween at 1:2000.
8. Anti-mouse IgA (Sigma) diluted in PBS-Tween at 1:2000.
9. Anti-mouse IgM (Sigma) diluted in PBS-Tween at 1:5000.
10. Anti-mouse IgG1 (Serotec, Raleigh, NC) diluted in PBS-Tween at 1:2500.
11. Anti-mouse IgG2a (Serotec) diluted in PBS-Tween at 1:2000.
12. Anti-mouse IgG2b (Serotec) diluted in PBS-Tween at 1:2000.
13. TMB substrate (Calbiochem).
14. SLT microplate reader (TELAC Inc., Research Triangle Park, NC).

2.3.4. Cytotoxicity Assay

1. Cytotox96 Assay (Promega), includes substrate mix, assay buffer, lysis buffer, stop solution.
2. P815 cell line (murine mastocytoma, ATCC).
3. p18 peptide RIQRGPGRAFVTIGK (Research Genetics, Huntsville, AL).
4. SLT microplate reader (TELAC Inc.).

3. Methods

3.1. Preparation of Macroaggregated Albumin-Polyethylenimine Conjugate

1. Add 40 µL of a freshly prepared SPDP solution with stirring to 1 mL of the HSA solution.
2. Prepare 1 ml PEI solution and add 10 µL SPDP solution with stirring.
3. After 2 h of stirring, put each solution over a NAP-10 column equilibrated in PBS.
4. Dilute the HSA effluent (1.5 mL) with PBS to 3.5 mL.
5. Adjust the PEI effluent (1.5 mL) to pH 7.0–7.4). Stir together with 20 mg of Reductacryl for 1 h.
6. Filter the PEI solution and add all at once to the HSA solution. The cloudy combined solution should be stirred overnight at room temperature.
7. Add small aliquots of 10% HCl until the cloudiness dissipates. Adjust back the pH back to 5.5–6.0 with $2\,M$ NaOH.
8. Aggregation is accomplished with vigorous magnetic stirring in a 10 mL flask at 100°C in an oil bath for 15 min (intraflask temperature should be 80–85°C under these volume and flask size conditions) (*see* **Note 4**).

3.1.1. Determination of PEI Nitrogen in Preparations

1. In order to determine the actual concentration of PEI N available in the preparation for DNA binding, dilute the particle suspension 1:100 with PBS?
2. Place 2–10 µL dilute suspension in 35 µL aliquots of PBS. Add each plasmid (1 µg in 5 µL PBS) to a separate aliquot while gently vortexing.

3. After 20 min of incubation at room temperature, run a 4 μL aliquot of each solution on a gel (0.6 % agarose in TBE buffer).
4. The lowest volume lane where no plasmid enters the gel represents a ratio of PEI N to DNA P of 2:1, where all negative charges are neutralized. Since DNA contains 0.003 μmol P (phosphate) per μg, then the molar concentration of functional PEI N (nitrogen) can be calculated according to the following equation:

$$[PEI\ N]\ (M) = (100 \times 2 \times 0.003)/\mu L \text{ of sample added to that lane}$$

Where: 100 = the initial dilution factor
2 = the 2:1 ratio for charge neutralization
0.003 = the μmol of P/μg of DNA.

If the volume added to the neutralized sample lane was 3 μL, then the concentration would be 0.2 M in PEI N. It may be more convenient to express this as 200 nmol/μL for subsequent handling, with DNA expressed as 3 nmol P/μg. Obviously the precision in this measurement depends on the dilutions selected. We have found that an N/P ratio of 15 ± 3–4 is acceptable for use, and thus slight differences in calculated concentration based on dilutions chosen are not critically important (*see* **Note 5**).

3.1.2. Plasmid Binding to MAA-PEI and HSA-PEI Suspensions

Plasmids are prepared using reagents and columns for endotoxin free DNA (Qiagen, Valencia, CA), and then dissolved in endotoxin free water at the desired concentration.

1. To mix MAA-PEI or HSA-PEI and plasmids for in vivo transfection, match the desired amount of DNA with an appropriate volume of MAA-PEI or HSA-PEI to achieve an N:P ratio of about 15:1.
2. Dilute the particles in PBS to the desired volume. For IV injection in the mouse, usually about 1 μL or less (depending on PEI N concentration) per 175 μL PBS for 1 μg of DNA is required. Add the plasmid (usually 1 μg diluted in 25 μL PBS) dropwise while gently vortexing the particle suspension.
3. Incubate the suspension at RT for a minimum of 20 min (up to a maximum of about 60 min [*see* **Note 6**]) before IV administration. The only caveats here are that high-speed vortexing seems to interfere with effective formation of transfecting particles, and excess DNA (N:P ratio ≤ 10) reduces transfection efficiency and may cause particle clumping that completely blocks transfection. Similarly, concentrations of PEI N > 5 nmol/μL or DNA > 200 ng/μL may result in clumping in the final solutions.

3.2. Injection of MAA-PEI Bound Plasmids

Balb/C mice (*see* **Note 7**) are injected intravenously with MAA-PEI particles via the tail vein, which delivers injected materials predominantly to the lung *(8,18–21)*.

1. Place in a Tail-Veiner restrainer.
2. Slowly inject particle/DNA (1 µg DNA) suspensions in PBS as a single 200 µL volume (*see* **Note 8**) into the tail vein.
3. Sacrifice mice at the desired time points by lethal anesthesia, and harvest tissues for analysis by standard techniques.

3.3. Standard Assays for Temporal and Organ Specific Expression of Genes of Interest

3.3.1. Luciferase Assay (6)

1. Place tissues of interest in 1 mL luciferase lysis buffer and homogenized using a conical ground glass tissue grinder.
2. Add a 10 µL aliquot of homogenate to 50 µL of luciferase substrate, and measure light output for 15 s in a luminometer.
3. For luciferase studies data presentation, the response can be converted to ng of luciferase/lung or other organ after calibration of the specific luminometer utilized (*see* **Note 9**).

3.3.2. Green Fluorescent Protein Detection (6)

Cells in which GFP is expressed from the pEGFP plasmid are examined by UV microscopy at appropriate time points.

1. For lung examination, surgically expose the trachea and cannulate with PE50 tubing.
2. Slowly inject a total volume of 1.0 mL of freezing medium diluted 1:1 in PBS. Freeze the inflated lungs, cut thin sections, and examined by light and UV microscopy. An example can be found in **ref. 6**.

3.3.3. Humoral Immune Responses And Enzyme-Linked Immunosorbent Assay (ELISA) (6)

1. Groups of 5 mice are exposed to MAA-PEI bound pCMV-hGH and injected via tail vein in a single 200 µL volume of PBS as described (*see* **Subheading 3.2.**). Control groups include uninjected mice, 50 µg naked plasmid DNA in 50 µL PBS injected i.m., naked plasmid DNA in 200 µL PBS injected iv, and plasmid DNA complexed with PEI alone in 200 µL PBS injected iv
2. Obtain sequential serum samples at biweekly intervals by tail bleed. Serum can be stored frozen (–20°C) until assayed.
3. At 12 wk following immunization, in addition to blood collection, bronchoalveolar lavage is performed on both untreated and immunized mice to collect pulmonary surface secretions.
   a. Sacrifice mice by Avertin anesthesia and exsanguination via cardiac puncture
   b. Surgically expose trachea and cannulate with PE50 tubing.
   c. Slowly inject 1.0 mL PBS and aspirated 3 times. Typically 80% of the bronchoalveolar lavage fluid (BALF) volume will be recovered.

4. Assay serum and BALF by ELISA for total and isotype specific antibody to human growth hormone (*see* **Note 10**). Coat purified hGH onto microtiter plates overnight at 4°C. This loading quantity was chosen as the most cost effective by comparison of various concentrations with serial dilutions of a positive control antiserum.
5. Block the wells with blocking solution for 2 h at RT.
6. Dilute sera and lung lavage fluids in PBS and add aliquots to wells. Incubate overnight at 4°C. Sera and lavage fluids from age matched, unimmunized Balb/C mice serve as negative controls in each assay (*see* **Note 11**).
7. After washing wells 5 times, bound antibodies are detected with horseradish peroxidase-conjugated goat or rabbit anti mouse immunoglobulin or anti-mouse IgA, IgM, IgG1, IgG2a, IgG2b.
8. Develop reactions using TMB substrate and read the optical density at 405 nm in a microplate reader. Use a maximal O.D. for linear reading of 1.4. The background reagent only (no serum) well should have an OD of 0.1.
9. Normalize results of the isotype specific assays for equivalent signal strength from dilution curves of bound antigen for each isotype. Group statistical comparisons are made using the one-sided Student's *t*-test in most samples, and are presented graphically showing the means and standard deviations (SD).

3.3.4. Cytotoxicity Assay

1. Expose five mice to MAA-PEI bound pCMV-UB23 by injection via tail vein in 200 µL of PBS (*see* **Subheading 3.2.**).
2. Prepare single cell suspensions from mouse spleen by mechanical disruption.
3. Load P815 target cells with p18 peptide by incubation in a 1 μM solution of the peptide at 37°C for 1 h.
4. Co-culture splenocytes (variable numbers) and targets (10,000 cells/well) at desired effector/target ratios. Control cultures at the highest ratio of splenocytes to targets are cultured with target cells without the p18 peptide.
5. Maximum release is determined by lysing target cells alone, and spontaneous release from both target cells and effector cells is measured from other wells with these cells cultured individually. After a 4-h incubation, harvest aliquots of supernatants.
6. Add substrate for measuring LDH activity released by lysed cells. Optical density measurements can be made after 30 min (*see* **Note 12**).

4. Notes

1. We have found that 10–30% of mice in the 25 gm size may die acutely (<24 h) with rapid injection of 400 µL of PEI and 100 µg of plasmid DNA at an N:P ratio of 12 or higher. Injection of 50% smaller volumes and lower amounts of PEI-DNA drastically reduce gene expression (to <10% of the maximal expression). The toxicity of conditions for high-level gene expression, and the lack of practical applications of this type of high-volume dosing in clinical practice led us to investigate other methods for nonviral gene delivery.

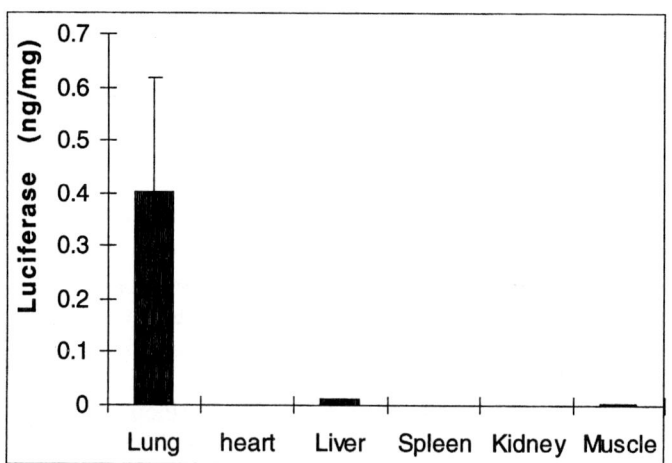

Fig. 1. Example of luciferase transfection with MAA-PEI preparations. Mice were injected with 1 µg of luciferase plasmid bound to MAA-PEI at a 15:1 N:P ratio suspended in a total volume of 200 µL. After 48 hours, the mice were sacrificed and various tissues examined for luciferase activity. The data represent the mean results from 10 experiments and 6 different MAA-PEI preparations. The results are expressed in ng of luciferase per whole organ except, in the case of skeletal muscle, per gram of tissue.

2. Human and bovine serum albumins from various sources are both suitable for aggregation under the described conditions, but most of our studies have used human serum albumin. Typical gene expression results for HSA particles are shown in **Fig. 1**. Thus for 1 µg of luciferase encoding plasmid bound to MAA-PEI injected intravenously, we usually detect at 48 h between 200 and 600 pg of enzyme.
3. Polyethylenimine is available from multiple sources in many forms. These studies emphasize the use of a branched 750 K PEI which is inexpensive and readily available (#P3143 Sigma). Other forms we have briefly evaluated include 10K and 25K PEIs (Aldrich, Milwaukee WI), a 70K and a linear 25K PEI (Polysciences, Warington PA), and a 750K PEI (Fluka, Cat. no. 03880). The 750K PEIs and the 70K PEI work in this protocol consistently, and the 25K preparation does work on occasion, but is less consistent than the 750K. Particles made using the 10K PEI and the 25K linear PEI gave unsatisfactory results. There is uncertainty in the actual characterization of these polymers. One study asserts that a putative 600-1000K polymer (Fluka, Cat. no. 03880), which may be analagous to the Sigma product actually has a MW of 155K *(22)*. Originally the Sigma product was described and sold as 50K, but this description has since been changed, based on the contract manufacturer's analysis of its product.

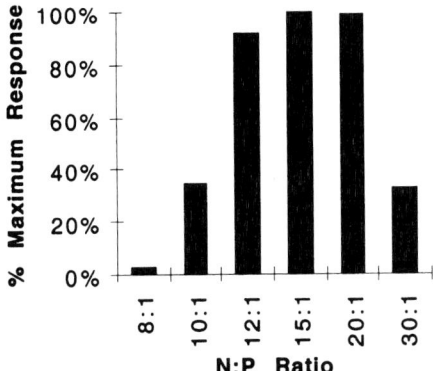

Fig. 2. N:P ratio effect on gene expression.

4. Particle sizing of such preparations has been done using a Nicomp Model 370 Submicron Particle Sizer (Particle Sizing Systems, Inc., Santa Barbara, CA). However, the lung accumulation of DNA bound MAA-PEI delivered to the lung is sufficiently high that this procedure is not essential. Even submicron particles deliver DNA mostly to the lung. As long as particles are mostly less than 100 µM by light microscopy, preparations are effective in gene delivery (see also **Note 6**).
5. We have also found that this approach for evaluating the N content of PEI solutions corresponds very well with the predicted concentration values obtained from the weight of PEI concentrate used to prepare stock solutions. The range of effective N:P ratios for particle preparations is fairly broad (10–20), but gene expression drops off precipitously at higher and lower ratios, presumably as a result of reduced transfection efficiency. In **Fig. 2**, mice were injected with particles prepared using 1 µg of pCMV-Luc at the specified N:P ratios and injected with standard techniques. After 48 h, the lungs were harvested and assayed for luciferase activity. The data is expressed as the % maximal response which was seen in this experiment at a 12:1 ratio. In most such experiments, the differences among ratios between 12:1 and 20:1 are not significant. As a result we ordinarily select 15:1 as the standard condition. We have also found that the optimal quantities of plasmid injected using these preparations is in the 1 µg/injection range, and that higher plasmid doses reduce the gene expression (**Fig. 3**; mice were injected with the specified quantities of pCMV-Luc bound to MAA-PEI with a 15:1 ratio, and luciferase activity was assayed 48 h later).
6. We have found that the minimum incubation time is 15 min for efficient delivery and expression of reporter gene plasmids. However, the maximum time of incubation is uncertain. Incubation overnight at room temperature or at 4°C results in near complete loss of gene expression in injected suspensions. In screening experiments incubation for 2 h or more also results in lower gene

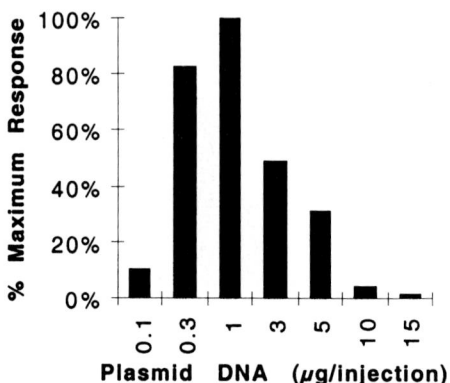

Fig. 3. Plasmid DNA dose effects.

expression than the 20–60 min time window during which gene expression appears to be consistent.

7. Mouse strain differences may exist. We have found that with this methodology, C57BL/6 mice express roughly 50% of the luciferase seen in BALB/c and SKH-Hr1 mice strains. No differences have been seen between male and female mice, and little difference has been seen between mice aged 6–52 wk.
8. We have found that injection with MAA-PEI or HSA-PEI bound plasmids can be done very slowly, thus modeling what might be used in an intravenous infusion in larger animals or humans. In contrast to the use of PEI alone with plasmid, in which very low gene expression results except with a rapid (<5 s) large bolus injection, total gene expression with the protein associated PEI is independent of injection rate between 10 s and 3 min for the 200 µL volume. For practicality, we inject in small intermittent small boluses over a period of 30–60 s as a general rule for our experiments. (**Fig. 4** shows the percent of maximum luciferase expression in the experiment 2 d after injection with different rates of injection using 1 µg of pCMV-Luc bound to MAA-PEI at a 15:1 N:P ratio.) Injections using these conditions can be done repeatedly without apparent toxicity, up to the maximum exposure conditions that we have tested (1 µg DNA injected two times per week for 12 wk). With this injection frequency, luciferase activity remains reaches a maximum at 48 h after the last injection, with a decline at 72 h to about 30% of the maximum. However, with repeated injections there is not a significant increase in luciferase activity above the single injection maximum.

We have also found that nonaggregated albumin/PEI/DNA mixtures equivalent to those described for the aggregates reliably give similar levels of gene expression in the lung using 1 µg of DNA when used with N:P ratios in the ranges described for MAA-PEI. These applications result in no apparent toxicity (injections 3 times per week for up to 3 wk), and only small amounts of expression

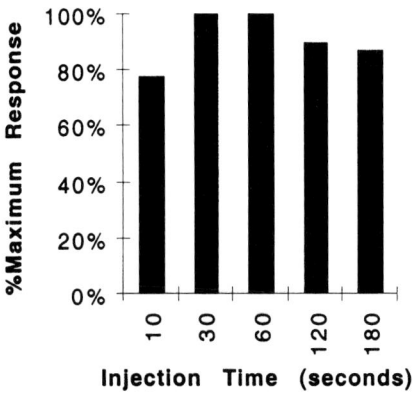

Fig. 4. Injection rate and response.

in other tissues (2–3% of the total, compared to <<1% for particles) *(6)*. HSA, BSA, mouse serum albumin, IgG, and ovalbumin (and presumably other proteins appear to function well when added to PEI without aggregation for transfection, using the same molar quantity as described for the particles. Although we have not investigated other proteins extensively, screening studies demonstrated that gene expression with these other proteins is within an order of magnitude of that achieved with HSA. However, this method has not yet been characterized in terms of immune responses.

9. For luciferase studies data presentation, we prefer to convert the response to ng of luciferase/lung or other organ (as in **Fig. 1**), since this makes it easy to compare data with other assay systems and luminometers. Using the Promega assay kit components and a TD-20e model Turner luminometer, 1 ng of purified luciferase (Sigma) produces 10,000 lumens in our laboratory.

10. As an example of the immunization results from a single i.v. particle injection, the systemic humoral immune response to MAA-PEI bound pCMV-hGH injection is compared to i.m. injection of the hGH plasmid is shown in **Fig. 5**. Mice immunized with either particle bound plasmid or intramuscular injection were significantly different from controls ($p < 0.02$ at 2 wk, and $p < 0.01$ for all subsequent time points), but did not differ from each other (panel **A**). Serial dilution of samples from the immunized groups (panel **B**) were significantly different from controls to a final dilution of 1:32,000. As shown in panel **C**, particle and naked DNA immunized mice were different from mice immunized with PEI-DNA complex alone, which had no response ($p < 0.01$ for all dilutions); the particle immunized group responses had a higher response than the naked DNA group ($p < 0.01$ at all dilutions).

11. An example of the humoral mucosal responses are elicited by MAA-PEI bound plasmid is shown in **Fig. 6**. Isotype specific anti-hGH antibody assays for IgA were performed in serial dilutions of the lavage fluid (panel **A**), and at each

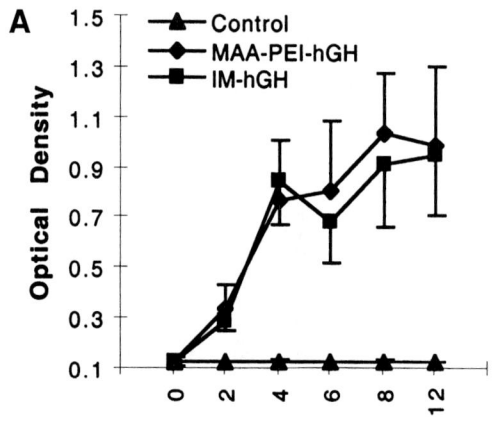
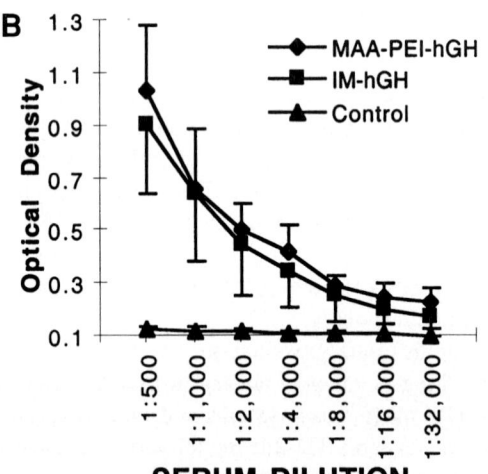
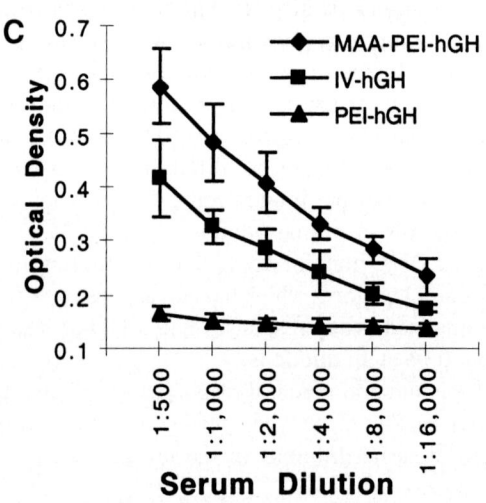

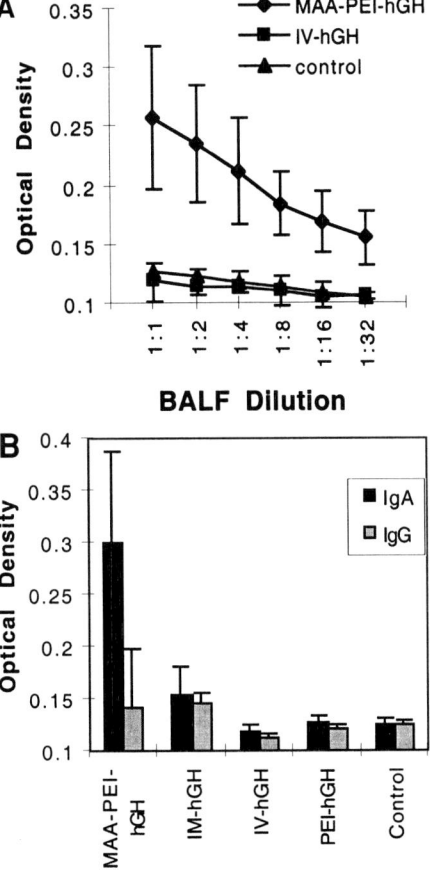

Fig. 6. Humoral mucosal immune responses to MAA-PEI bound pCMV-hGH injection. In panel **A**, control unimmunized mice and mice immunized with MAA-PEI-hGH plasmid, naked hGH plasmid, or PEI-hGH plasmid complexes (groups described in 7C) were sacrificed 8 wk after a single immunization, and bronchoalveolar lavage fluid samples were collected as described in the Methods Section. Isotype specific anti-hGH antibody assays for IgA were performed in serial dilutions of the lavage fluid. In panel **B**, the BALF from each mouse in all groups (described in **Fig. 5**) was assayed at a 1:2 dilution for both IgA and IgG antibodies to hGH, and the group means and standard deviations were plotted. Adapted with permission from **ref. 6**.

Fig. 5. *(see opposite page)* Systemic humoral immune response to MAA-PEI bound pCMV-hGH injection. In panel **A**, groups of 5 BALB/c mice were immunized IV with 5 μg pCMV-hGH plasmid bound to MAA-PEI in a final volume of 299 μL, or i.m. with 50 μg in 50 μL PBS, or were unimmunized. Serial blood samples were obtained and 1:500 dilutions were assayed by ELISA for total specific anti-hGH antibody at each time point. In panel **B**, individual serum from the each mouse in the groups described in **(A)** was tested by serial dilution, with mean and standard deviations for each group plotted. In panel **C**, serum was tested by serial dilution as in **(B)** from another set of groups (5 mice each) including MAA-PEI-hGH IV (as in **[A]**), naked hGH plasmid DNA IV (5 μg), and PEI-hGH plasmid DNA complexes (180 nmol PEI, 5 μg DNA) IV. Adapted with permission from **ref. 6**.

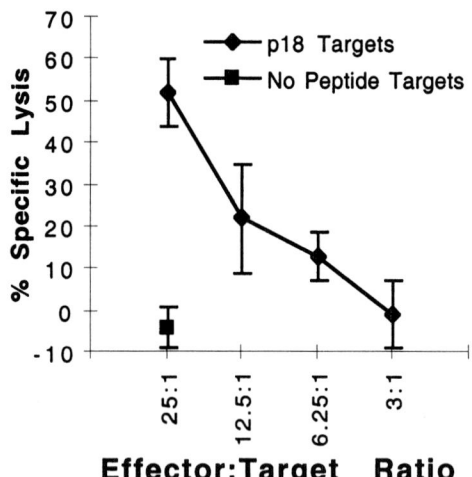

Fig. 7. P18 Specific CTL Elicited by MAA-PEI-CMV-UB23. Adapted with permission from **ref. 6**.

dilution the particle immunized group response was significantly different from the others ($p<0.01$). Panel **B** shows that IgA anti-hGH in the MAA-PEI-hGH plasmid immunized group was significantly different from all other groups and from all IgG levels as well ($p<0.01$). The i.m. group's low level IgA anti-hGH did not quite achieve statistical significance ($p=0.052$), compared with the control group baseline.

12. An example cytotoxic T lymphocyte induction by i.v. particle injection is shown in **Fig. 7**. Five Balb/C mice were injected IV with MAA-PEI-plasmid UB#23 (encoding HTLV-IIIb gp120 base pairs 6824-7429 (including p18 epitope) fused to ubiquitin). Direct CTL assays were performed with spleen cells at 8 wk, showing specific high-level splenocyte cytotoxicity to the p18 epitope in the UB#23 immunized mice, while no nonspecific activity against the target cells was present. Specific lysis was significant at all effector to target ratios shown except 3:1.

Acknowledgments

This work was supported by the Department of Veterans Affairs, the Huffington Foundation, the Center for AIDS Research, the Clayton Foundation for Research, and the Shell Center for Gene Therapy.

References

1. Boussif, O., Lezoualc'h, F., Zanta, M. A., Mergny, M. D., Scherman, D., Demeneix, B., and Behr, J.-P. (1995) A versatile vector for gene and oligonucleotide transfer

into cells in culture and *in vivo*: Polyethylenimine. *Proc. Natl. Acad. Sci. USA* **92,** 7297–7301.
2. Goula, D., Benoist, C., Mantero, S., Merlo, G., Levi, G., and Demeneix, B. A. (1998) Polyethylenimine-based intravenous delivery of transgenes to mouse lung. *Gene Ther.* **5,** 1291–1295.
3. Kichler, A., Behr, J. P., and Erbacher, P. (1999) Polyethylenimines: a family of potent polymers for nucleic acid delivery, in *Non-viral Vectors for Gene Therapy* (Huang, L., Hung, M. C., and Wagner, E., eds.), Academic Press, San Diego, CA, pp. 191–206.
4. Kichler, A., Leborgne, C., Coeytaux, E., and Danos, O. (2001) Polyethylenimine-mediated gene delivery: a mechanistic study. *J. Gene Med.* **3,** 135–144.
5. DeNardo, G. L., Goodwin, D. A., Ravasini, R., and Dietrich, P. A. (1970) The ventilatory lung scan in the diagnosis of pulmonary embolism. *N. Engl. J. Med.* **282,** 1334–1336.
6. Orson, F. M., Kinsey, B. M., Hua, P. J., Bhogal, B. S., Densmore, C. L., and Barry, M. E. (2000) Genetic immunization with lung-targeting macroaggregated polyethyleneimine-albumin conjugates elicits combined systemic and mucosal immune responses. *J. Immunol.* **164,** 6313–6321.
7. Colombetti, L. G., Moerlien, S., and Pinsky, S. (1975) Rapid and reliable preparation of macroaggregated albumin suitable for lung scintigraphy. *Intl. J. Nuclear Med. Biol.* **2,** 180–184.
8. Hapke, E. J. and Pederson, H. J. (1969) Ultrastructural changes in rat lungs induced by radioactive macroaggregated albumin. *Am. Rev. Resp. Dis.* **100,** 194–205.
9. Stauber, R. E., Mochizuki, T., Van Thiel, D. H., and Tauxe, W. N. (1992) The use of quantitative scintigraphy in the measurement of portal-systemic shunting in rats. *Ann. Nucl. Med.* **6,** 209–214.
10. DeLand, F. H. (1966) The fate of macroaggregated albumin in used in lung scanning. *J. Nucl. Med.* **7,** 883–895.
11. McGhee, J. R., Czerkinsky, C., and Mestecky, J. (1998) Mucosal vaccines: an overview, in *Mucosal Immunology*, 2nd ed. (Ogra, P. L., Mestecky, J., Lamm, M. E., Strober, W., Bienenstock, J., and McGhee, J. R., eds.), Academic Press, New York, NY, pp. 741–757.
12. Herrmann, J. E., Chen, S. C., Fynan, E. F., Santoro, J. C., Greenberg, H. B., Wang, S., and Robinson, H. L. (1996) Protection against rotavirus infections by DNA vaccination. *J. Infect. Dis.* **174,** S93–S97.
13. Pertmer, T. M., Roberts, T. R., and Haynes, J. R. (1996) Influenza virus nucleoprotein-specific immunoglobulin G subclass and cytokine responses elicited by DNA vaccination are dependent on the route of vector DNA delivery. *J. Virol.* **70,** 6119–6125.
14. Di Giorgio, A., Mingazzini, P., Sammartino, P., Canavese, A., Arnone, P., and Scarpini, M. (2000) Host defense and survival in patients with lung carcinoma. *Cancer* **89,** 2038–2045.
15. Almand, B. and Carbone, D. P. (2001) Biological considerations in lung cancer. *Cancer Treat. Res.* **105,** 1–30.

16. Andersson, S., Davis, D. N., Dählback, H., Jörnvall, H., and Russell, D. W. (1989) Cloning, structure, and expression of the mitochondrial cytochrome P-450 sterol 26 hydrolase, a bile acid biosynthetic enzyme. *J. Biol. Chem.* **264,** 8222–8229.
17. Barry, M. A., Lai, W. C., and Johnston, S. A. (1995) Protection against mycoplasma infection using expression-library immunization. *Nature,* **377,** 632–635.
18. Van Den Brenk, H. A., Burch, W. M., Kelly, H., and Orton, C. (1975) Venous diversion trapping and growth of blood-borne cancer cells en route to the lungs. *Br. J. Cancer* **31,** 46–61.
19. Heinsen, H., Mottaghy, K., and Fromel, M. (1980) Pulmonary and systemic embolism after deliberate intravenous fluorocarbon administration. *Virchows Arch. [Pathol. Anat.]* **386,** 331–341.
20. Magos, L., Clarkson, T. W., and Hudson, A. R. (1989) The effects of dose of elemental mercury and first-pass circulation time on exhalation and organ distribution of inorganic mercury in rats. *Biochim. Biophys. Acta* **991,** 85–89.
21. Watanabe, N., Shirakami, Y., Tomiyoshi, K., Oriuchi, N., Hirano, T., Higuchi, T., et al. (1997) Direct labeling of macroaggregated albumin with indium-111-chloride using acetate buffer. *J. Nucl. Med.* **38,** 1590–1592.
22. Godbey, W. T., Wu, K. K., Hirasaki, G. J., and Mikos, A. G. (1999) Improved packing of poly(ethylenimine)/DNA complexes increases transfection efficiency. *Gene Ther.* **6,** 1380–1388.

35

Preparation of Retroviral Vectors for Cell-Cycle-Targeted Gene Therapy of Lung Cancer

Pier Paolo Claudio and Antonio Giordano

1. Introduction

Lung cancer is one of the leading causes of cancer death in the world (1). The high mortality rate for lung cancer may be due to either or both the absence of standard clinical procedures for early diagnosis of tumors compared to breast, prostate, and colon cancers, and lack of effective treatments in more advanced stages of the disease (2).

Therapeutically, radiation (RT) and chemotherapy, have been the two main treatment modalities for advanced non-small cell lung cancer (NSCLC) (3). However, the natural course of the disease has changed little, even with these advances.

Various studies have indicated that several distinct chromosomal loci (3p, 9p [p16], 13q [Rb], 17p and others) are implicated in the pathogenesis of lung cancer, suggesting that sequential genetic events may occur during initiation and progression of this disease (2,4,5). Recent studies have indicated that allelic loss of heterozygosity (LOH) at several other chromosomal regions could also be involved in the pathogenesis of lung cancer. Chromosomal arms with the most frequent LOH were 1p, 2q, 3q, 4p, 4q, 5q, 6p, 6q, 7p, 8p, 8q, 9q, 10p, 10q, 11p, 11q, 14q, 15q, 16q, 17p (p53), 17q, 18q, 19p, 21q, 22q, Xp, and Xq (6–14). Because malignant transformation of pulmonary epithelial cells is the result of multistep accumulation of genetic and molecular alterations highly related to tobacco carcinogens, involving key regulatory elements of the cell cycle and mechanisms of proliferation and apoptosis, much of the recent research efforts have been focused to the understanding of the cell-cycle regulatory machinery in such a disease.

Oncogene activation (ras, myc, autocrine growth factor loops) or/and tumor-suppressor gene inactivation (p53, pRB family, cyclin-dependent-kinase-inhibitor p16) at the genetic, epigenetic, or post-translational level remove crucial constraints on cell division at the G_1 check-point and in the induction of apoptosis, accelerating cell division *(15,16)*. p53 inactivation is one of the most common alterations in lung cancer (75% of genetic alterations). Mutations of p53 have been reported with frequencies up to 50% in NSCLC and 70–80% in small cell lung cancer (SCLC) *(17,18)*. In addition, mutations or deletions of the RB gene in non-small cell carcinomas have been reported at greater than 90% of cases studied *(18)*.

The retinoblastoma (Rb) gene family includes three members: the Rb tumor suppressor Rb/p105, p107, and RB2/p130 *(19–23)*. Their protein products are termed "pocket proteins" due to a unique structural and functional domain that is composed of subdomains A and B separated by a spacer region. This domain is highly conserved among each of the proteins.

The structural homology of these proteins underlie similar functional properties. All three family members inhibit cell-cycle progression in the G_1 phase and each of their phosphorylation profiles varies in a cell-cycle dependent manner *(24)*. Interestingly, these proteins exhibit unique growth-suppressive properties that are cell type-specific, suggesting that although the pocket proteins may complement each other, their functions are not fully redundant *(25,26)*.

RB2/p130 maps to human chromosome 16q12.2, an area in which deletions are found in several human neoplasms including breast, ovarian, hepatic, and prostate cancers *(27)*. Recent immunohistochemical studies of the retinoblastoma family of proteins in 235 specimens of lung cancer show the tightest inverse association between the histological grading in the most aggressive tumor types and pRb2/p130 *(28,29)*. This prompted us to study a panel of human lung cancers for mutations in the *RB2/p130* gene. Mutations in the Rb-related gene *RB2/p130* were detected in primary lung tumors by single-strand conformation polymorphism and sequence analysis *(30)*. This led us to explore the possibility of restoration of the wild-type *RB2/p130* gene as a clinical treatment for human cancers. To establish the foundation for the potential future use of RB2/p130 in gene therapy, we have investigated an efficient means of delivering RB2/p130 gene to tumor cell lines in vitro and in vivo with retroviral vectors and have detected high levels of RB2/p130 expression and severe growth suppression in targeted tumors *(30)*.

Because lung cancer arises from a series of morphological and molecular changes that take several years to progress from normal epithelium to an invasive cancer and because those molecular changes include activation of dominant oncogenes as well as loss of recessive growth-regulatory genes or anti-oncogenes *(31)*, gene-replacement therapy is becoming one of the potential tools that will possibly support traditional therapies targeting specific molecular mediators of cancer development and progression.

The field of gene therapy has rapidly developed since the first submission of gene-therapy trials in the 1990s, which provided encouraging results *(32)*. In the last decade, much effort has been dedicated to improving protocols in human gene therapy. Many significant goals have been achieved, although there are still several issues investigators must address in order to develop new efficient therapeutic approaches to treat cancer. Cellular targeting and specificity, viral-transduction efficiency, and sustained gene expression have all been the focus of major scientific inquiries in recent years.

Various genes are currently under investigation for their use in human gene therapy such as p53, p16, p21, Bax, pRb, pRb2/p130, and others *(15,30,33–41)*. Phase I studies are now in progress, using replication-defective adenoviral-expression vector encoding wild-type p53, in patients with incurable NSCLC *(40,42)*. The viruses have been delivered by bronchoscopic intratumoral injection or by computed tomography (CT)-guided percutaneous intraluminal injection of the vector solution. Though as yet, no toxic effects have been observed using these viral vectors, some substantial pitfalls are still inherent in the technique.

There are several critical factors in the development of strategies for gene transfer. Substantial problems solving effort must go into the type and design of the vector, the delivery method, and the target organ system. In addition, a major concern in vector development is safety, and this concept must be stressed. A viral vector should be safe for the patient as well as for the operator, and at the same time, be easy to handle. To limit these concerns a variety of different techniques have been tested for production of gene therapy vectors. These range from use of crippled viral vectors, propagated in stably transfected cell lines that provide in trans the necessary tools to the viral particle to be assembled, to the use of multiple plasmids carrying the single viral parts separately. The last is, in our opinion, the safest, but can also be the most difficult method to establish in order to achieve viral production.

We have gained experience in recent years in the production and handling retroviral vectors. Retroviruses are enveloped viruses that contain a single-stranded RNA viral genome. Their high-infection efficiency, combined with their ability to stably integrate into the host genome resulting in stable exogenous gene expression, make retroviruses attractive candidates as vectors for gene delivery. They are among the most efficient vector systems for transducing genes into mammalian cells and have been used successfully to deliver therapeutic genes into humans *(43,44)*. The requirement for host cells to actively synthesize DNA to allow viral genes to stably integrate into the host genome *(44)* may be advantageous for cancer gene therapy. This would limit exogenous gene delivery to rapidly proliferating cancer cells while sparing delivery to other nonproliferating cells within the effected organ.

Retrovirus production usually requires the creation of a stable producer cell line that expresses the retroviral vector *(45)*. This can be a tedious task, which involves introducing the retroviral vector into a packaging cell line that will synthesize the necessary viral assembly proteins. Many clones must be screened in order to propagate the one clone that expresses the retroviral vector at high titer. This is important since retrovirus particles are easily inactivated during the purification and concentration process. A high multiplicity of infection (MOI) is required to achieve nearly 100% infection efficiency. The screening process alone may require more than a month's intensive labor. Additionally, during prolonged cultivation of these clones, the growth of the producer line may be retarded due to the extended high expression of retroviral mRNA which favors the propagation of poorly producing variants. In order to circumvent these problems, and minimize the time involved in creating stable producer cell lines, we utilize for our experiments a transient three-plasmids expression system for the production of high titer retroviral vectors as developed by Dr. Kingsman *(30,46)*. Using this system, retroviral-mediated delivery of wild-type *RB2/p130* to the lung tumor cell line H23 potently inhibited tumorigenesis in vitro and in vivo, as shown by the dramatic growth arrest observed in a colony assay and the suppression of anchorage-independent growth potential and tumor formation in *nude* mice. Tumors transduced with the *RB2/p130* retrovirus diminished in size after a single injection, eventually leading to a 12-fold reduction in tumor growth when compared to the control virus transduced tumors *(30)*. Materials and methods used for retroviral production and titration are outlined herein

2. Materials

2.1. Production of Retroviral Vectors by Transient Transfection (see Notes 1–5)

The system described here utilizes Moloney Leukemia Virus (MLV)-based retroviral vectors. Packaging component expression is controlled by the strong cytomegalovirus (CMV) promoter and carried on plasmids containing simian virus 40 (SV40) origins of replication, which enhances retroviral gene expression in cell lines carrying the SV40 large T antigen *(46)*. To reduce the risk of helper virus formation, the two fundamental components necessary for packaging the virus, *gag-pol* and *env*. In this system, these components are expressed from separate plasmids. The *env* and *gag-pol* plasmids encode an RNA transcript which is a substrate only for translation within 293T/17 cells. The plasmids we have used to produce retroviruses are outlined in **Table 1**.

1. 293T/17 retrovirus packaging cells (human renal carcinoma) can be purchased from the American Type Culture Collection upon authorization of Rockefeller University.
2. Dulbecco's modified Eagle medium (DMEM) supplemented with 10% heat-inactivated FBS and 2 m*M* l-glutamine for culturing the cells to be prepared for transfection. DMEM supplemented with 2% heat-inactivated FBS and 2 m*M* l-glutamine for culturing the cells following transfection.
3. Retrovirus vector plasmid DNAs (*see* **Table 1**), prepared by either C_sCl_2 gradient or by resin-exchange column system. DNA in sterile water is preferred, since excess Tris may inadvertently alter pH.
4. 10 cm tissue-culture dishes.
5. 10-mL syringes.
6. 0.45 µm low-protein-binding cellulose acetate syringe filters.
7. 0.22 µm low-protein-binding cellulose acetate 500 mL filters.
8. 4 mg/mL Polybrene (Sigma Cat. no. H9268) in phosphate-buffered saline (PBS).
9. 100 m*M* sodium butyrate (NaB) (Fluka Cat. no. 19364) in sterile ddH_2O.
10. Autoclaved sterile glycerol (Sigma Cat. no. G2025).
11. Tissue-culture grade ddH_2O, MilliQ grade, or Sigma (Cat. no. W3500).
12. Calcium phosphate precipitation reagents: 2.0 *M* $CaCl_2$ 0.22-µm filter-sterilize and freshly prepared before use and 2X HBS, pH to 7.05, 0.22 µm filter-sterilize. Correct pH is critical. Variations in the pH may result in not comparable transfection efficiencies.
13. 2X HBS 1 L: dissolve in 650 mL sterile ddH_2O: 16.4 g NaCl, 11.9g HEPES free acid, 0.21 g Na_2HPO_4. Adjust pH to 7.05 with NaOH, and filter the solution using a 0.22 µm low-protein-binding cellulose acetate 500 mL filter. Aliquot the 2X HBS solution in 15- or 50-mL polystyrene tubes and store it at –20°C for up to 2 yr.
14. 2.0 *M* $CaCl_2$ 10 mL: dissolve in sterile ddH_2O 2.94 g $CaCl_2$. After the $CaCl_2$ is dissolved, filter the solution with 0.22 µm low-protein-binding cellulose acetate syringe filters. Make the 2.0 *M* $CaCl_2$ solution fresh each time before use.

2.2. Titration of Retroviral Vectors by PRINS Technique and Evaluation of the Efficiency of Viral Transduction

1. Retroviral supernatant containing retroviruses carrying a selection marker not normally present in the target cells (for example the puromycin gene)
2. Solution of methanol and glacial acetic acid 3:1 (v:v).
3. A series of ethanol solutions in ddH_2O (70, 80, and 100%)
4. Water baths set to 95, 58, and 72°C. Alternatively, an *in situ* polymerase chain reaction (PCR) machine can be used.

Table 1
Retroviral Plasmid Sets and Their Content

Plasmid name	Content	Use
pHIT60	CMV-MLV *gag-pol*-SV40ori	*gag* encodes for capsid proteins, *pol* for reverse transcriptase and integrase. All these proteins are necessary for viral production
pHIT123	CMV-MLV ecotropic *env*-SV40ori	Ecotropic *env* encodes for ecotropic envelope proteins. These proteins are needed for viral production and for targeting viruses to rodent cellular membrane
pHIT456	CMV-MLV amphotropic *env*-SV40ori	Amphotropic *env* encodes for amphotropic envelope proteins. These proteins are needed for viral production and for targeting viruses to rodent or human cellular membrane
pHIT111	CMV-*LacZ*-SV40 promoter-NEO-SV40ori.	*LacZ* encodes for the *E. coli* β-galactosidase gene. It may be helpful in demonstrating viral transduction efficiency
MCSV-*Neo*-EB	φCMV-5′MLV-LTR-*Cassette* gene-SV40promoter-*neomycin* resistance gene-3′MLV-LTR-φ	φ is the integrase signal; the cassette is occupied by the gene subject of the study or can be empty to be used as a control; neomycin is used to select target cells and/or helper transfected cells
MCSV-*Pac*	φ-CMV-5′MLV-LTR-*Cassette* gene-SV40promoter-*puromycin* resistance gene-3′MLV-LTR- φ.	φ is the integrase signal; the cassette is occupied by the gene subject of the study or can be empty to be used as a control; puromycin is used to select target cells and/or helper transfected cells

Preparation of Retroviral Vectors

5. Oligonucleotides PUR5′ (5′-TCACCGAGCTGCAAGAAC-3′) and PUR3′ (5′-GTCCTTCGGGCACCTCGA-3′).
6. 10X PCR buffer with 15 mM MgCl$_2$ (QIAGEN, Valencia, CA).
7. A 10 mM mixture containing deoxyadenosine 5′ triphosphate (dATP), deoxycytosine 5′ triphosphate (dCTP), deoxyguanosine 5′ triphosphate (dGTP).
8. A mixture containing 6.5 mM deoxythymidine 5′ triphosphate (dTTP), and 3.5 mM digoxigenin-11-deoxyuridine 5′ triphosphate, alkali-labile, (Boheringer Mannheim, Indianapolis, IN).
9. Taq DNA polymerase (QIAGEN, Valencia, CA).
10. Sterilized glass cover slides.
11. Glass microscope slides.
12. Rubber cement glue.
13. 20X SSC 1 L: Dissolve into 800 mL of ddH$_2$O: 173.3 g of NaCl, 88.2 g NaCitrate, Adjust the pH to 7.0 with a few drops of 10 M NaOH or of citric acid. Bring the volume to 1 L with ddH$_2$O.
14. Bovine serum albumin (BSA) (Sigma Cat. no. A7030).
15. FITC-conjugated anti-digoxigenin-antibody (Sigma Cat. no. F3523).
16. Solution of propidium iodide (PI) 1 μg/mL in ddH$_2$O (F.W. 668.4) (Sigma Cat. no. P4170).
17. Confocal or fluorescence microscope.

2.3. Alternate Protocol for Retroviral Titration and Evaluation of the Efficiency of Viral Transduction by β-gal Assay for Mammalian Cells (see Note 6)

1. Fixing solution: see **Table 2**.
2. X-Gal Solution: 1 mM X-Gal, 5 mM K ferricyanide, 5 mM K ferrocyanide, 2 mM MgCl$_2$. For 10 mL of X-Gal solution mix, in 9.63 mL of ddH$_2$O, 250 μL of 40 mg/mL X-Gal in DMSO, 50 μL of 100 mM K ferricyanide (0.165 g K ferricyanide dissolved in 5 mL of PBS), 50 μL of 100 mM K ferrocyanide (0.211g K ferrocyanide dissolved in 5 mL of PBS), 20 μL of 1 M MgCl$_2$.

3. Methods

3.1. Production of Retroviral Vectors by Transient Transfection (see Notes 1–5)

*3.1.1. Basic Protocol for Calcium Phosphate-Mediated Transfection of Cultured Packaging Cells for Retroviral Production (see **Note 5**)*

Cultured packaging cells will take up precipitates of calcium, phosphates, and DNA under proper conditions. Transfected cells can then express the proteins encoded by the plasmids transferred and will assemble the viral particles. Retroviruses are recovered into the culture medium.

Table 2
Fixing Solution for Viral Titration by β-gal Assay

Volume	50 mL	25 mL	10 ml	5 mL
37% formaldehyde	2.63 mL	1.32 mL	526 µL	263 µL
25% Aqueous solution glutaraldehyde	400 µL	200 µL	80 µL	40 µL
1X PBS	46.97 mL	23.49 mL	9.39 mL	4.7 mL

1. Twenty-four hours prior to transfection, culture 293T/17 retrovirus packaging cells on 10 cm tissue-culture dishes in DMEM medium supplemented with 10% heat-inactivated FBS and 2 mM l-glutamine. Cells must be 70% confluent for transfection.
2. Change media 2–4 h before transfection. Use 9 mL D-MEM medium supplemented with 10% heat inactivated FBS and 2 mM l-glutamine. (Transfections should be performed late in the day.)
3. In a polystyrene 15-mL tube labeled (A), combine 62 µL CaCl$_2$ solution with 30 µg total DNA comprised of 10 µg each of the three plasmid DNAs carrying *gag-pol* and *env* cDNAs, and the cDNA of the gene of interest. Make volume up to 500 µL with tissue culture-grade sterile ddH$_2$O. Prepare separate tubes for each culture to be transfected (*see* **Note 7**).
4. In a second 15-mL polystyrene tube labeled (B), place 500 µL 2X HBS solution. Prepare tubes for each tissue culture dish to be transfected.
5. Add the DNA solution from tube (A) slowly and drop-wise to the 500 µL of 2X HBS in tube (B) while gently vortexing (*see* **Note 8**).
6. Let the mixture of CaCl$_2$, DNA and HBS for 20–30 min at room temperature (*see* **Note 9**).
7. Drop the 1 mL solution evenly onto the culture dish and mix gently by tilting to cover.
8. Incubate the cells with the precipitates at 37°C, 5% CO$_2$ for 12–16 h.
9. Aspirate precipitate solution and replace with 10 mL fresh DMEM medium supplemented with 2% heat-inactivated FBS and 2 mM l-Glutamine (*see* **Notes 10–12**).
10. 24–32 h after changing medium (**step 9**) collect the medium containing the retroviral particles. Filter culture medium containing viral particles through a 0.45-mm low-protein-binding cellulose acetate syringe filter to remove residual cells and debris. Alternatively, centrifuge the medium 5 min at 3000g (5000 rpm in Sorvall HB-4 rotor), 4°C, to pellet cells and debris (*see* **Note 13**).
11. Use an aliquot of the virus-containing medium to transduce recipient (target) cells for assessing viral titer and store the rest of the viruses in aliquots mixed with 10% sterile glycerol at –70°C. Retroviral supernatant can be stored in these conditions indefinitely.

3.2. Basic Protocol for Retroviral Vector Titration and Evaluation of the Efficiency of Viral Transduction by PRINS Technique

1. A suitable sample number of target cells must be prepared for the PRINS reaction. Place sterile cover slides into sterile 10 cm culture dishes. Plate target cells on sterile cover slips at a concentration of 5×10^5 cells per dish.
2. Transduce the cells with serial dilutions of the retroviral supernatant. Cells can be transduced typically with 20 µL, 50 µL, 100 µL, or 1 mL of retroviruses carrying the puromycin-resistance gene alone, or in combination with the transgene, in either the sense or the antisense orientation. As a negative control, cells can be transduced with supernatant collected from a cotransfection of 293T/17 cells with only the plasmids carrying *gag/pol* and *env* genes (empty virus).
3. Fix samples in methanol and glacial acetic acid 3:1 (v:v) for 10 min at room temperature and let them air dry for 12–24 h.
4. The next day, dehydrate samples in a series of ethanol solutions (70, 80, and 100%), each for 5 min, and air-dry again for at least 1 h.
5. At this point the samples are ready for the PRINS reaction: For the PRINS reaction, the primers PUR5′ and PUR3′ are used to amplify a stretch of 425 bp in the puromycin-resistance gene present in the plasmids MSCV*Pac*.
6. Prepare an incubation mixture containing: 100 ng of each primer, 1X PCR buffer with 1.5 m*M* MgCl$_2$, 1 m*M* deoxyadenosine 5′ triphosphate, 1 m*M* deoxycytosine 5′ triphosphate, 1 m*M* deoxyguanosine 5′ triphosphate, 0.65 m*M* deoxythymidine 5′ triphosphate, 0.35 mM digoxigenin-11- deoxyuridine 5′ triphosphate, alkali-labile, and 2.5 U Taq DNA polymerase.
7. Deposit 50 µL of the reaction mixture on individual glass microscope slides. Place a cover slip on which transduced cells have been seeded, cell-side down, onto the drop of reaction mixture. Seal top and sides with rubber cement glue.
8. Incubate each sample at 95°C (denaturation) for 5 min, 58°C (annealing) for 30 min, and 74°C (elongation) for 90 min. To do this, place slides in a humid chamber and place in a series of water baths set previously to the aforementioned temperatures.
9. Following elongation at 74°C, remove the rubber cement glue and wash the cover slips twice, 10 min each wash, in 2X SSC at room temperature to remove excess reaction mixture.
10. Incubate the slides in 2X SSC containing 2% BSA for 10 minutes at room temperature.
11. Dilute the FITC-conjugated anti-digoxigenin- antibody 1:200 into 2X SSC/2% BSA and incubate the slides for 30 min at room temperature in a dark humid chamber for digoxigenin-11-deoxyuridine 5′ triphosphate incorporation.
12. Wash the slides twice, 5 min each wash, in 2X SSC, to remove excess antibody.
13. Incubate samples for a brief period in a solution of PI (1 µg/mL) in order to stain unlabeled DNA, then rinse in ddH$_2$O.
14. Observe slides under a fluorescence microscope. Count positive versus negative cells in 10 random microscopic fields. The same process may be applied to OCT

tumor embedded (Sakura Finetek USA, Inc., Torrance, CA) frozen sections of cancer cells grown in mice.
15. Viral titer is calculated using the following formula: {[(Number plated cells) X (% FITC positive cells)/100]/(Number of µL of retroviral supernatant used to transduce)} × 10 mL.

3.3. Alternate Protocol for Retroviral Titration and Evaluation of the Efficiency of Viral Transduction by β-gal Assay for Mammalian Cells on a Tissue Culture Dish

1. Plate target cells at a concentration of 5×10^5 cells per dish (*see* **Note 14**).
2. Transduce cells with serial dilutions of the retroviral supernatant carrying the *Lac-Z* transgene. As a negative control, cells can be transduced with supernatant collected from a cotransfection of 293T/17 cells with only the plasmids carrying *gag/pol* and *env* genes (empty virus).
3. Wash cells three times in 1X PBS.
4. Fix cells for 5 min, at room temperature in 5 mL (1.5 mL) of Fixing Solution.
5. Wash cells three times in 1X PBS.
6. Stain cells in 5 mL (1.5 mL) of X-gal Solution, wrap dishes with parafilm to avoid evaporation and incubate at 37°C for 1 h to 2 d.
7. Count blue cells and determine the percentage of transfected cells.

4. Notes

1. DNA should be prepared under aseptic conditions. Proper aseptic tissue-culture protocols should be adhered to, and all culture incubations are to be performed in humidified 37°C, 5% CO_2 incubator.
2. All solutions and materials that will be employed for retroviral preparation should be sterilized before use. Solutions should be filter sterilized by 0.22-µm sterile filters. Other materials, if not already sterile, should be autoclaved.
3. Maintenance of proper pH is a critical in the preparation, handling, and storage of 2X HBS. Variations in the pH of this solution may result in poorly reproducible transfection efficiencies. 2X HBS stored long-term should be checked from time to time for pH variation.
4. The range of cells that can be transduced by a retrovirus vector is dependent on the envelope or coat protein used to make the virus. The coat protein of the virus, which binds specific cell-surface receptors, promotes vector entry into the cells. Ecotropic envelopes allow for transduction of rodent cells, while amphotropic coat proteins target mouse and human cells. Amphotropic envelopes are generally more useful because of their broad host range but at the same time require a more careful use by the operator in order to avoid personal contact with the viral particle and therefore to avoid possible injury to laboratory personnel. The 293T/17 cell line (human renal carcinoma) used herein is highly transfectable and expresses the large T antigen. It can be purchased from the American Type Culture Collection upon authorization of Rockefeller University.

Preparation of Retroviral Vectors

Table 3
Brief Protocol for Calcium-Phosphate Transfection Method

	1X	4X	6X	8X
Tube 1				
2 M CaCl$_2$	62 µL	248 µL	372 µL	496 µL
ddH$_2$O	To 0.5 mL	To 2 mL	To 3 mL	To 4 mL
DNA (do not exceed 40 µL for each tissue culture dish)	30 µg total DNA	120 µg total DNA	180 µg total DNA	240 µg total DNA
Final volume	500 µL	2 mL	3 mL	4 mL
Tube 2				
2X HBS	0.5 mL	2 mL	3 mL	4 mL

5. X-Gal solution, 100 mM K ferricyanide, and 100 mM K ferrocyanide should be prepared fresh, just before use
6. Plasmid DNAs needed in order to achieve retroviral production:
   a. Plasmid containing the *gag-pol* subunits (in our case, plasmid pHIT60).
   b. Plasmid containing either the ecotropic or the amphotropic *env* subunit (in our case pHIT123 or pHIT456, respectively).
   c. Plasmid containing the gene we would like to package and transfer to target cells (in our case either MSCVneoEB, MSCVPac, or pHIT111).
7. For scaling up transfections, the following protocol in **Table 3** can be used:
8. Following incubation at room temperature for 20–30 min, the CaCl2/DNA mixture should be faintly cloudy. If the mixture remains clear or a large precipitate develops, do not use it. A new precipitate should be prepared using a freshly prepared HBS stock. Variations in pH of HBSS is a common cause for development of wrong-sized precipitates.
9. Following removal of the calcium phosphate-DNA co-precipitate, sodium butyrate (NaB) can be added to the medium for 12–14 h at a 10 mM final concentration in order to increase viral titers by almost a log *(46)*. Fresh medium is then added, and the supernatants are harvested 12 h later. NaB increases the percentage of cells expressing exogenous DNA *(47)*, and activates several eukaryotic promoters including the CMV promoter *(48,49)*. Upon co-transfection of the three-plasmid components into 293T/17 cells via calcium phosphate treatment and exposure to NaB, helper-free viral stocks of approximately 10^7 infectious U/mL are achieved 48 h post co-transfection *(30,46)*.
10. Transduction efficiency may be improved by mixing the retroviral supernatant with 8 µg/mL polybrene, in a total volume of 3 mL for 10-cm tissue-culture dishes, and incubating the target cells with this mixture at 37°C for 2 h, followed by the addition of 7 mL of fresh DMEM medium containing 2% heat-inactivated FBS and 2 mM l-Glutamine.

11. It is important to note that complement proteins present in the FBS may inactivate retroviruses. This is the reason we recommend use of heat-inactivated FBS. Additionally, it must be stressed that the medium used to replace the DMEM used during transfection should contain only 2% heat-inactivated FBS and 2 mM l-Glutamine *(47)*.
12. Though filtering may lower viral titer due to some sticking to the membrane, filtering is preferred for viral concentration over centrifugation, since the latter may not completely remove all cells and debris.
13. This protocol is written for 10-cm dishes. For 30-mm dishes cut the solutions by approx one-third (volumes in parenthesis).

References

1. Carney, D. N. (1991) Lung cancer biology. *Eur. J. Cancer* **27(3)**, 366–369.
2. Wiest, J. S., Franklin, W. A., Drabkin, H., Gemmill, R., Sidransky, D., and Anderson, M. W. (1997) Genetic markers for early detection of lung cancer and outcome measures for response to chemoprevention. *J. Cell Biochem. Suppl.* **29(46)**, 64–73.
3. Lee, C. T. (1998) Non-surgical therapy for the patients with advanced non-small cell lung cancer. *Respirology* **3(3)**, 159–166.
4. Shimizu, E. and Sone, S. (1997) Tumor suppressor genes in human lung cancer. *J. Med. Invest.* **44(1–2)**, 15–24.
5. Todd, S., Franklin, W. A., Varella-Garcia, M., Kennedy, T., Hilliker, C. E., Hahner, L., et al. (1997). Homozygous deletions of human chromosome 3p in lung tumors. *Cancer Res.* **57(7)**, 1344–1352.
6. Bepler, G. and Garcia-Blanco, M. A. (1994) Three tumor-suppressor regions on chromosome 11p identified by high-resolution deletion mapping in human non-small-cell lung cancer. *Proc. Natl. Acad. Sci. USA* **91(12)**, 5513–5517.
7. Girard, L., Zochbauer-Muller, S., Virmani, A. K., Gazdar, A. F., and Minna, J. D. (2000) Genome-wide allelotyping of lung cancer identifies new regions of allelic loss, differences between small cell lung cancer and non-small cell lung cancer, and loci clustering. *Cancer Res.* **60(17)**, 4894–906.
8. Merlo, A., Gabrielson, E., Mabry, M., Vollmer, R., Baylin, S. B., and Sidransky, D. (1994) Homozygous deletion on chromosome 9p and loss of heterozygosity on 9q, 6p, and 6q in primary human small cell lung cancer. *Cancer Res.* **54(9)**, 2322–2326.
9. O'Briant, K. C. and Bepler, G. (1997) Delineation of the centromeric and telomeric chromosome segment 11p15.5 lung cancer suppressor regions LOH11A and LOH11B. *Genes Chromosomes Cancer* **18(2)**, 111–114.
10. Ohata, H., Emi, M., Fujiwara, Y., Higashino, K., Nakagawa, K., Futagami, R., et al. (1993) Deletion mapping the short arm of chromosome 8 in non-small cell lung carcinoma. *Genes Chromosomes Cancer* **7(2)**, 85–88.
11. Otsuka, T., Kohno, T., Mori, M., Noguchi, M., Hirohashi, S., and Yokota, J. (1996) Deletion mapping of chromosome 2 in human lung carcinoma. *Genes Chromosomes Cancer* **16(2)**, 113–119.

12. Sato, S., Nakamura, Y., and Tsuchiya, E. (1994) Difference of allelotype between squamous cell carcinoma and adenocarcinoma of the lung. *Cancer Res.* **54(21)**, 5652–5655.
13. Shiseki, M., Kohno, T., Nishikawa, R., Sameshima, Y., Mizoguchi, H., and Yokota, J. (1994) Frequent allelic losses on chromosomes 2q, 18q, and 22q in advanced non-small cell lung carcinoma. *Cancer Res.* **54(21)**, 5643–5648.
14. Virmani, A. K., Fong, K. M., Kodagoda, D., McIntire, D., Hung, J., Tonk, V., et al. (1998) Allelotyping demonstrates common and distinct patterns of chromosomal loss in human lung cancer types. *Genes Chromosomes Cancer* **21(4)**, 308–319.
15. Brambilla, E. and Brambilla, C. (1997) p53 and lung cancer. *Pathol. Biol. (Paris)* **45(10)**, 852–863.
16. Sekido, Y., Fong, K. M., and Minna, J. D. (1998) Progress in understanding the molecular pathogenesis of human lung cancer. *Biochim. Biophys. Acta* **1378(1)**, F21–59.
17. Demoly, P., Pujol, J. L., Godard, P., and Michel, F. B. (1994) [Oncogenes and anti-oncogenes in lung cancer]. *Presse Med.* **23(6)**, 291–297.
18. Salgia, R. and Skarin, A. T. (1998) Molecular abnormalities in lung cancer. *J. Clin. Oncol.* **16(3)**, 1207–1217.
19. Ewen, M. E., Xing, Y. G., Lawrence, J. B., and Livingston, D. M. (1991) Molecular cloning, chromosomal mapping, and expression of the cDNA for p107, a retinoblastoma gene product-related protein. *Cell* **66(6)**, 1155–1164.
20. Hannon, G. J., Demetrick, D., and Beach, D. (1993) Isolation of the Rb-related p130 through its interaction with CDK2 and cyclins. *Genes Dev.* **7(12A)**, 2378–2391.
21. Lee, W. H., Bookstein, R., Hong, F., Young, L. J., Shew, J. Y., and Lee, E. Y. (1987) Human retinoblastoma susceptibility gene: cloning, identification, and sequence. *Science* **235(4794)**, 1394–1399.
22. Li, Y., Graham, C., Lacy, S., Duncan, A. M., and Whyte, P. (1993) The adenovirus E1A-associated 130-kD protein is encoded by a member of the retinoblastoma gene family and physically interacts with cyclins A and E. *Genes Dev.* **7(12A)**, 2366–2377.
23. Mayol, X., Grana, X., Baldi, A., Sang, N., Hu, Q., and Giordano, A. (1993) Cloning of a new member of the retinoblastoma gene family (pRb2) which binds to the E1A transforming domain. *Oncogene* **8(9)**, 2561–2566.
24. Stiegler, P. and Giordano, A. (1999) Role of pRB2/p130 in cellular growth regulation. *Anal. Quant. Cytol. Histol.* **21(4)**, 363–366.
25. Claudio, P. P., Howard, C. M., Baldi, A., De Luca, A., Fu, Y., Condorelli, G., et al. (1994) p130/pRb2 has growth suppressive properties similar to yet distinctive from those of retinoblastoma family members pRb and p107. *Cancer Res.* **54(21)**, 5556–5560
26. Zhu, L., van den Heuvel, S., Helin, K., Fattaey, A., Ewen, M., Livingston, D., et al. (1993) Inhibition of cell proliferation by p107, a relative of the retinoblastoma protein. *Genes Dev.* **7(7A)**, 1111–1125.
27. Yeung, R. S., Bell, D. W., Testa, J. R., Mayol, X., Baldi, A., Grana, X., et al. (1993) The retinoblastoma-related gene, RB2, maps to human chromosome 16q12 and rat chromosome 19. *Oncogene* **8(12)**, 3465–3468.

28. Baldi, A., Esposito, V., De Luca, A., Fu, Y., Meoli, I., Giordano, G. G., et al. (1997) Differential expression of Rb2/p130 and p107 in normal human tissues and in primary lung cancer [In Process Citation]. *Clin, Cancer Res.* **3(10)**, 1691–1697.
29. Baldi, A., Esposito, V., De Luca, A., Howard, C. M., Mazzarella, G., Baldi, F., et al. (1996) Differential expression of the retinoblastoma gene family members pRb/p105, p107, and pRb2/p130 in lung cancer. *Clin. Cancer Res.* **2(7)**, 1239–1245.
30. Claudio, P. P., Howard, C. M., Pacilio, C., Cinti, C., Romano, G., Minimo, C., et al. (2000) Mutations in the retinoblastoma-related gene RB2/p130 in lung tumors and suppression of tumor growth in vivo by retrovirus-mediated gene transfer. *Cancer Res.* **60(2)**, 372–382.
31. Gazdar, A. F., Ihde, D. C., Saijo, N., Shimoyama, M., and Yokota, J. (1994) Report of the Seventh International Symposium of the Foundation for Promotion of Cancer Research: fundamental and clinical research in lung cancer. *Jpn. J. Clin. Oncol.* **24(4)**, 233–239.
32. Stratford-Perricaudet, L. D., Levrero, M., Chasse, J. F., Perricaudet, M., and Briand, P. (1990) Evaluation of the transfer and expression in mice of an enzyme-encoding gene using a human adenovirus vector. *Hum. Gene Ther.* **1(3)**, 241–256.
33. Bouvet, M., Bold, R. J., Lee, J., Evans, D. B., Abbruzzese, J. L., Chiao, P. J., et al. (1998) Adenovirus-mediated wild-type p53 tumor suppressor gene therapy induces apoptosis and suppresses growth of human pancreatic cancer [see comments]. *Ann. Surg. Oncol.* **5(8)**, 681–688.
34. Bouvet, M., Ellis, L. M., Nishizaki, M., Fujiwara, T., Liu, W., Bucana, C. D., et al. (1998) Adenovirus-mediated wild-type p53 gene transfer down-regulates vascular endothelial growth factor expression and inhibits angiogenesis in human colon cancer. *Cancer Res.* **58(11)**, 2288–2292.
35. Clayman, G. L. (2000) The current status of gene therapy. *Semin. Oncol.* **27(4 Suppl. 8)**, 39–43.
36. Coll, J. L., Negoescu, A., Louis, N., Sachs, L., Tenaud, C., Girardot, V., et al. (1998) Antitumor activity of bax and p53 naked gene transfer in lung cancer: in vitro and in vivo analysis. *Hum. Gene Ther.* **9(14)**, 2063–2074.
37. Gallagher, W. M. and Brown, R. (1999) p53-oriented cancer therapies: current progress. *Ann. Oncol.* **10(2)**, 139–150.
38. Joshi, U. S., Chen, Y. Q., Kalemkerian, G. P., Adil, M. R., Kraut, M., and Sarkar, F. H. (1998) Inhibition of tumor cell growth by p21WAF1 adenoviral gene transfer in lung cancer. *Cancer Gene Ther.* **5(3)**, 183–191.
39. Roth, J. A., Swisher, S. G., and Meyn, R. E. (1999) p53 tumor suppressor gene therapy for cancer. *Oncology (Huntingt.)* **13(10 Suppl. 5)**, 148–154.
40. Schuler, M., Rochlitz, C., Horowitz, J. A., Schlegel, J., Perruchoud, A. P., Kommoss, F., et al. (1998) A phase I study of adenovirus-mediated wild-type p53 gene transfer in patients with advanced non-small cell lung cancer. *Hum. Gene Ther.* **9(14)**, 2075–2082.
41. Zhang, W. W. and Roth, J. A. (1994) Anti-oncogene and tumor suppressor gene therapy—examples from a lung cancer animal model. *In Vivo* **8(5)**, 755–769.

42. Roth, J. A. (1998) Restoration of tumour suppressor gene expression for cancer. *Forum (Genova)* **8(4),** 368–376.
43. Gilboa, E. (1990) Retroviral gene transfer: applications to human therapy. *Prog. Clin. Biol. Res.* **352(4–5),** 301–311.
44. Miller, A. D. (1990) Retrovirus packaging cells. *Hum. Gene Ther.* **1(1),** 5–14.
45. Kim, Y. S., Lim, H. K., and Kim, K. J. (1998) Production of high-titer retroviral vectors and detection of replication-competent retroviruses. *Mol. Cells* **8(1),** 36–42.
46. Soneoka, Y., Cannon, P. M., Ramsdale, E. E., Griffiths, J. C., Romano, G., Kingsman, S. M., and Kingsman, A. J. (1995) A transient three-plasmid expression system for the production of high titer retroviral vectors. *Nucleic Acids Res.* **23(4),** 628–633.
47. Rosenzweig, A. and Nabel, E. G. (1996) Vectors for gene therapy, in *Current Protocols in Human Genetics* (Dracopoli, W. C., Haines, J. L., Korf, B. R., et al., eds.) John Wiley & Sons, Inc., New York, pp. Unit 12.5.1–12.5.18.
48. Radsak, K., Fuhrmann, R., Franke, R. P., Schneider, D., Kollert, A., Brucher, K. H., and Drenckhahn, D. (1989) Induction by sodium butyrate of cytomegalovirus replication in human endothelial cells. *Arch. Virol.* **107(1–2),** 151–158.
49. Tanaka, J., Sadanari, H., Sato, H., and Fukuda, S. (1991) Sodium butyrate-inducible replication of human cytomegalovirus in a human epithelial cell line. *Virology* **185(1),** 271–280.

36

Aerosol Gene Therapy for Metastatic Lung Cancer Using PEI-p53 Complexes

Ajay Gautam, J. Clifford Waldrep, Eugenie S. Kleinerman, Bo Xu, Yuen-Kai Fung, Anne T'Ang, and Charles L. Densmore

1. Introduction

Inactivation of p53wt, a tumor-suppressor gene, has been observed in a majority of small cell and non-small cell lung cancers (SCLC/NSCLC) (1,2). Many tumors that frequently metastasize to lung, such as colon, breast, liver, melanoma, and bone, also exhibit a high incidence of p53 mutations and deletions. Several studies have suggested that gene-replacement therapy with p53wt results in an antitumor response that may be of therapeutic value in these situations (3–8). Although different approaches for delivery of genes to the lungs, such as intravenous injection and nasal or intratracheal instillation, have been reported in animal models, these strategies tend to be invasive and generally do not distribute the material uniformly throughout pulmonary tissues. In contrast, aerosol delivery distributes the complexes uniformly and represents a non-invasive alternative for targeting the genes to the lungs (see Chapter 34). We have recently shown that polyethylenimine (PEI), a cationic polymer, holds promise as a gene delivery vector by aerosol (9–11), and this approach has proven to be very effective when used to deliver p53 and variant p53 genes topically to the lungs of animals bearing metastatic lung tumors (7,8).

One such mouse tumor model is B16-F10 murine melanoma which metastasizes readily to the lungs of C57BL/6 mice and grows aggressively. Most of the previous therapeutic studies with this model have involved immunomodulatory genes such as interleukin (IL)-1a, tumor necrosis factor-α (TNF-α), IL-4, IL-12, and granular macrophage colony-stimulating factor (GM-CSF) or

From: *Methods in Molecular Medicine, Vol. 75: Lung Cancer, Vol. 2:*
Diagnostic and Therapeutic Methods and Reviews
Edited by: B. Driscoll © Humana Press Inc., Totowa, NJ

suicide genes such as cytosine deaminase *(12–16)*. Also, delivery of p53 gene using cationic liposomes has been shown to have an antitumor effect, although in different models *(17–19)*. Furthermore, delivery of genes in these studies was approached through intravenous or direct intratumoral injections.

Greater than 30% of patients with osteosarcoma, the most common primary malignant bone tumor, develop lung metastases despite aggressive combination chemotherapy and surgery. Until recently, a reproducible osteosarcoma lung metastasis model has not been available. An athymic nude mouse model of human osteosarcoma lung metastases (SAOS-LM6) has now been developed by repeated cycling of SAOS-2 tumor cells in mice *(20)*, and appears to be suitable for evaluation of both chemotherapeutic *(21)* and gene therapeutic *(22)* approaches for treatment of this disease. SAOS-2, the parent line from which SAOS-LM6 cells were derived, is null for p53 and transfection with p53wt has been shown to be associated with terminal differentiation and apoptosis that inhibit progressive growth of metastasis *(23)*. In addition, induction of p53 in SAOS-2 cells stably transfected with p53 wt under the tight control of a tetracycline promoter resulted in an increased sensitivity to cisplatin cytotoxicity *(24)* and others have found that p53 transfection of SAOS-2 cells resulted in increased radiosenstivity *(25)*. SAOS-LM6 cells also lack p53 and when transfected in vitro by either p53wt or p53CD(1-366), a functional p53 deletion variant, cell proliferation is significantly inhibited.

In addition to wtp53, we have also tested a truncated p53 (p53CD1-366). Our unpublished data (Fung et al.) showed that p53CD(1-366) may have better transcriptional activation and growth suppression ability compared to wtp53 in cells at the G_0/G_1 phase of the cell cycle. This is consistent with prior observations that contact-inhibited cells lack p53 transactivation ability, which can be overcome by stimulating the cells to enter the cell cycle by overexpression of cyclin E *(26)*. This is due to the fact that wild-type p53 can assume two different conformations. Newly synthesized p53 in a cell exists in an inactive conformation and must be activated by posttranslational modification during S phase. In its inactive conformation, wtp53 cannot bind DNA and cannot transactivate. The active conformation of p53 also exists in cells by differential splicing under certain conditions, e.g., during the G2/M phase of the cell cycle or when cells are exposed to stress. P53CD(1-366) mimics this conformation and can bind specific DNA sequences and is transcriptionally active at all phases of the cell cycle. Based on in vitro cell proliferation studies with a number of tumor cell lines that have consistently shown the p53CD(1-366) variant to be more effective than the wild-type gene in arresting the growth of cells (data not shown), one might suspect the in vivo tumor-suppressive effect of p53CD(1-366) to be equal or better than that of the wild-type gene. The significant effect of p53CD(1-366) on the tumor models described here when administered by PEI aerosol is promising and may prove to be more effective than the wild-type gene at inhibiting the growth of a variety of lung tumors.

In this chapter we describe the effective treatment of two different animal lung tumor models by PEI:p53 and PEI:p53 CD(1-366) aerosol gene therapy. One model, B16-F10 murine melanoma is an aggressive tumor growing in C57-B6 mice while the other model, the human SAOS-LM-6 osteosarcoma is a much slower-growing tumor in athymic nude mice.

2. Materials
2.1. Preparation of PEI-DNA Complexes

1. Phosphate-buffered saline (PBS), pH 7.2 (GibcoBRL, Grand Island, NY).
2. PEI (25 KDa, branched polymer) was purchased from Aldrich Chemical (Milwaukee, WI). A PEI stock solution is prepared at a concentration of 4.3 mg/mL (0.1 M in nitrogen) in water, and pH adjusted to 7–7.5 using 0.1 M HCl. The PEI stock solution can be stored at 2–4°C. Concentrated PEI is toxic and should be handled with care.
3. Plasmid DNA: All plasmids are purified commercially by Bayou Biolabs (Harahan, LA), are endotoxin-free, and quantitated using UV absorbance. Agarose gel analysis revealed the plasmids to be primarily in the supercoiled form with a small amount of nicked plasmid (*see* **Note 1**).
   a. pBSH19A *(16)*. Human wild-type p53 cDNA was subcloned into pET8C and the resulting plasmid pETwtp53 was used as a template for the construction of pETp53CD (1-366) (a C-terminal deletion mutant) by polymerase chain reaction (PCR). The plasmids were then transferred to pCMV-neo-Bam, placing the p53 gene under the control of the human cytomegalovirus promoter/ enhancer element.
   b. p4119. This plasmid is used as a control. It contains the bacterial chloramphenicol acetyl transferase (CAT) gene under the control of the cytomegalovirus (CMV) promoter/enhancer element.
   c. pGL3. The luciferase plasmid (Promega, Madison, WI) is also used as a control. It is modified by insertion of the CMV promoter/enhancer element and the human growth hormone polyadenylation sequence. A gift from Dr. Michael Barry (Center for Cell and Gene Therapy, Baylor).

2.2. Aerosol Delivery of PEI-DNA Complexes

1. Plastic cages.
2. Polyethylene tubing (Beckton Dickinson, Sparks, MD).
3. Carbon dioxide tank (Air Liquide America Corp., Houston, TX).
4. Aerotech II nebulizer (AT-II) (CIS-US, Inc., Bedford, MA).
5. A source of dry air (Aridyne 3500, Timeter, Lancaster, PA).
6. A Bird 3M gas blender (Palm Springs, CA).
7. Fyrite CO_2 analyzer (Bacharach, Pittsburgh, PA).
8. Ventilation hood.

2.3. Tumor Inoculation and Dosage Regimen

2.3.1. B16-F10 Melanoma Model

1. Female C57BL/6 mice (7–8 wk old; Harlen Sprague Dawley, Houston, TX).
2. Murine B16-F10 melanoma cell line (Division of Cancer Treatment and Diagnosis Center, DCTDC, NCI, Frederick, MD).
3. Supplemented DMEM: DMEM (GibcoBRL, Grand Island, NY) with 10% fetal calf serum (FCS; HyClone, Logan, UT).

2.3.2. LM-6 Osteosarcoma Model

1. Male athymic nude mice (6–8 wk old; Charles River).
2. SAOS-LM6 is a cell line derived from parental human osteosarcoma (SAOS-2) cells by repeated cycling of tumor cells in mice *(20)*. It forms tumors in the lung following intravenous injection where microscopic pulmonary metastases are evident by 5–6 wk with macroscopic disease present by 8 wk.
3. Supplemented MEM:MEM (GibcoBRL, Grand Island, NY) with 10% FBS, glutamine, streptomycin, and antibiotics. (GibcoBRL, Grand Island, NY).
4. 0.25% trypsin/0.02% EDTA (w/v) (GibcoBRL, Grand Island, NY).
5. Supplemental EMEM (GibcoBRL, Grand Island, NY).
6. Ca^{2+} and Mg^{2+} free HBSS (Hank's balanced salt solution, Mediatech Cellgro, VA).
7. 30-G needles.
8. U-100 insulin syringes were purchased from Becton Dickinson and Co. (Franklin Lakes, NJ).

2.4. Harvesting Lungs for Tumor Analysis

1. Isoflurane (Vedco, St. Joseph, MO).
2. Fekets solution: 90% ethanol, 5% acetic acid, 5% formaldehyde.
3. Bovin's fixative: 71% Picric acid, 24% formaldehyde, 5% glacial acetic acid.
4. Dissection microscope.

2.5. Evaluation of Tumor Burden

1. Set up for tissue embedding in paraffin and sectioning.
2. Hematoxylin and eosin. (Richard Allan Scientific, Kalamazoo, MI).

2.6. P53 Assay

1. Enzyme-linked immunosorbent assay (ELISA) kit (Roche Diagnostics, Indianapolis, IN).
2. Cell lysis buffer: 20 m*M* Tris-HCl, 0.5 m*M* EDTA, 1% Nonidet P40, 0.05% sodium dodecyl sulfate (SDS), 1 m*M* PMSF, 1 mg/mL pepstatin, 2 mg/mL leupeptin.
3. Wig-L-Bug bead homogenizer (Crescent, Lyons, IL).
4. 4-mm glass beads (Kimble Glass Inc., Germany).
5. 5-mL tubes (50 × 16 mm; Sarstedt, Germany) with caps for tissue homogenization.
6. 96-well plates for ELISA.

7. Molecular Devices (Sunnyvale, CA) microtiter plate reader.
8. BCA protein assay reagents (Pierce, Rockford, IL).

2.7. P53 Immunohistochemistry

1. 50% (v/v) OCT (optimal cutting compound, Sakura Finetek, CA).
2. Saline (0.9% NaCl).
3. 0.6% H_2O_2.
4. Blocking solution: 20% normal goat serum in PBS.
5. Rabbit polyclonal antibody to p53 (1:200 dilution; Novocastra, UK).
6. Peroxidase conjugated goat anti-rabbit IgG (1:1000 dilution; Cappel, Westchester, PA).
7. Aminoethylcarbazol (AEC) (Sigma, St Louis, MO).

3. Methods

3.1. Preparation of PEI-DNA Complexes

1. Calculate quantities of PEI and DNA to add to 5 mL water required. For a 10:1 N:P ratio and 2 mg/10 mL DNA concentration, 606 µL of PEI solution (0.25 N) is added to 5 mL water and 2 mg DNA is added to another 5 mL water (*see* **Note 2**).
2. The PEI solution is then slowly vortexed and the DNA solution added drop-wise to the PEI solution to make a final volume of 10 mL. Allow the mixture to stand at room temperature for about 15–20 min before nebulization. The resulting charge ratio is expressed as PEI nitrogen:DNA phosphate (N:P), which can be calculated by taking into account that DNA has 3 nmol phosphate per µg and 1 µL of 0.1 M PEI solution has 100 nmol of amine nitrogen. A 10:1 N:P ratio corresponds to a 1.29:1 PEI:DNA weight ratio.

3.2. Aerosol Delivery of PEI-DNA Complexes

Mice should be housed five to a cage and kept in a laminar-flow cabinet under specific pathogen-free conditions for 2 wk before use

1. Place mice in plastic cages and sealed with tape before aerosol delivery. This is an unrestrained, whole-body aerosol exposure system.
2. Aerosolize PEI-DNA complexes using an Aerotech II nebulizer at 10 L/min flow rate using air or air containing 5% CO_2. A source of dry air is delivered to a Bird 3M gas blender attached to an air compressor and a CO_2 tank. The resulting mixture of air and CO_2 is delivered to the nebulizer. The final concentration of 5% CO_2 in air was determined using a Fyrite CO_2 analyzer. The nebulization of 10 mL solution should take approx 30 min (*see* **Note 3** and **4**).

3.3. Tumor Inoculation and Dosage Regimen

The dosage of treatment in both models is 2 mg plasmid/10 mL of aerosolized solution at a PEI:DNA (N:P) ratio of 10:1. This is the total amount of DNA aerosolized to the mice. The amount of DNA delivered per mouse is estimated to be about 3–5 μg in the presence of normal air, and is increased in the presence of 5% CO_2 due to the increase in tidal and minute volumes (*see* **Note 5**).

3.3.1. B16-F10 Melanoma Model

1. B16-F10 melanoma cells are maintained in supplemented DMEM. The cells are used at passages 3–12 (*see* **Note 6**).
2. C57BL/6 mice are injected through the lateral tail vein with 25,000 B16-F10 cells.
3. Treat the mice with PEI-p53 aerosol complexes twice a week starting the day after inoculation of the cancer cells. Control groups should include untreated mice, mice treated with PEI or with PEI-Luc aerosol complexes (*see* **Note 7**).
4. The experiment is terminated as soon as the control animals start dying.

3.3.2. LM-6 Osteosarcoma Model

1. SAOS-LM6 cells are maintained in supplemented MEM. Only cells younger than twelve weeks should be used for these experiments.
2. Before being inoculated, harvest the SAOS-LM6 in mid-log phase of growth by trypsinization. The cells are then suspended in supplemental EMEM and pipetted so as to produce a single-cell suspension.
3. Wash the cells and resuspend in Ca^{2+} and Mg^{2+} free HBSS at 5×10^6 cells/mL.
4. On d 0, inject single-cell suspensions of 1×10^6 cells/0.2 mL into the lateral tail vein of athymic nude mice using a 30-G needle with a 1-mL disposable syringe.
5. Treat the mice twice a week starting 6 wk after inoculation. Control groups should include untreated mice, mice treated with PEI or with PEI-Luc or with PEI complexes made with some other nonspecific plasmid.
6. Terminate the experiment at about 14 wk or as soon as the control animals start dying.

3.4. Harvesting Lungs for Tumor Analysis

1. Anaesthetize mice with isoflurane and sacrifice by exsanguination via the abdominal aorta.
2. Lungs are isolated, weighed, fixed in Fekets solution or Bovin's fixative, and examined under a dissection microscope for tumor foci.

3.5. Evaluation of Tumor Burden

1. If most of the lungs in the control groups have numerous uncountable foci, the lungs are graded based on a scale of 1–5. 1 if there are less than 10 tumor foci, 2 if there are 10–100 tumor foci, 3 if one lobe of the lung is full of tumor, 4 if both lobes are full of tumor, and 5 if the lungs are full of tumor and the tumor is growing out of the lungs and into the chest wall. The growth of tumor contiguous to the lung is considered as a part of the lung for weighing the lungs. A tumor index (an index of tumor burden) is calculated by the following formula: Tumor index = lung weight × average grade on the scale for the group (*see* **Note 8**).

2. If the tumors are discreet foci, the number counted and the size (diameter) of the foci measured. A tumor index can then be calculated by the formula: Tumor Index = Number × Size × Weight of the lungs.
3. In addition to counting the tumors visible on the surface of the resected lungs, the fixed lungs can be embedded in paraffin, sectioned, and stained with hematoxylin and eosin. The microscopic lung metastases are then quantified under a microscope.

3.6. p53 Assay

1. p53 expression is examined using the Roche ELISA kit. For measuring p53 expression in vivo, athymic nude mice (tumor-bearing or nontumor-bearing) that have been exposed to PEI:p53 or PEI-p53(CD1-366) aerosol are sacrificed at appropriate time points (24, 48, 72, and 96 h post-treatment) and the lungs harvested and weighed.
2. Homogenize the lungs in 1 mL ice-cold cell lysis buffer.
3. Clarify lysates with centrifugation at 13,000g at 4°C.
4. Take 100 µL supernatant for p53 ELISA, set up according to manufacturer's instructions, performed in a 96-well plate. The total protein content in the lungs is determined using the BCA protein assay. The absorbance (450 nm) is read in triplicate using a microtiter plate reader.
5. The amount of p53 is determined using a standard curve prepared with purified p53. The assay can detect p53 levels as low as 10 pg/mL and the linear measuring range of the assay is 50–1000 pg/mL.

3.7. p53 Immunohistochemistry

1. PEI-p53 or PEI-p53CD1 DNA complexes are prepared and delivered as indicated (*see* **Subheadings 3.1.** and **3.2.**) to either tumor-bearing or nontumor-bearing mice.
2. At appropriate time points after aerosol delivery, the mice are sacrificed and the lungs are isolated, cannulated, and fixed by inflation with 50% (v/v) OCT in saline.
3. Cut cryosections at 5 µm, collect on slides, air dry, and incubate in 0.6% H_2O_2 for 20 min to inactivate endogenous peroxidase activity.
4. Block the slides using 20% normal goat serum for 30 min at room temperature (RT).
5. The slides are then incubated with antibody to p53 for 1 h at RT.
6. After washing with PBS, the sections are incubated with a peroxidase conjugated goat anti-rabbit IgG for 1 h at RT.
7. The slides are then developed using AEC for 3–5 min. Slides where the primary antibody was excluded should be used as controls. Lungs from naïve mice should also be used as controls.

4. Notes

1. The structure of the plasmid is very important for transgene expression. A supercoiled plasmid is thought to give better expression levels than nicked or linear

plasmid. Also, frequent freezing and thawing of plasmids should be avoided since this leads to conversion of supercoiled plasmid into nicked and linear forms.
2. PEI and DNA complexes should be prepared freshly before aerosolization. The long-term stability of the PEI-DNA solution has not been adequately characterized.
3. Care should be taken during aerosolization of complexes to the mice using 5% CO_2. The concentration of CO_2 should be first accurately adjusted and then the air mixture attached to the nebulizer. A high concentration of CO_2 can kill the mice quickly. The set up should be monitored periodically during nebulization for any loose connections. Mice should also be monitored for any signs of distress or becoming moribund. Although aerosolization without CO_2 is possible, it will result in an approx threefold decrease in the level of transgene expression and this level of expression has been shown to be inadequate to result in an antitumor response in the animal tumor models described here.
4. Transgene expression by PEI-DNA aerosol delivery is both dose and PEI:DNA (N:P) ratio dependent. The N:P ratio needs to be optimized for different plasmids. The best results should be obtained at a N:P ratio of 10:1 to 20:1. The frequency of aerosol gene delivery may also need to be optimized. One treatment per week appears to be inadequate for an antitumor response in the tumor models and under the conditions described here. However, a frequency of more than twice a week may also result in decreased efficiency due to a refractory effect of aerosol PEI-mediated gene delivery that appears to persist for several days.
5. Prior to treating tumor bearing mice with p53 or other antitumor genes, it is recommended that an evaluation of gene delivery efficiency be made by exposing mice to PEI-p53 or PEI-p53(CD1-366) formulations and determining the level of expression in the lungs quantitatively by p53 ELISA and histologically by p53 immunohistochemistry as outlined in **Subheadings 3.6.** and **3.7.** The antibodies used in both of these assays recognize the p53(CD1-366) gene product and are suitable for the accurate quantitation of gene expression when this variant p53 gene is used. Also note that p53 has a relatively short half-life in the cell and expression is difficult to measure above background levels by these methods after only 4 d. Additional details pertaining to the aerosol delivery of PEI-based gene formulations are outlined in this volume in Chapter X.
6. The growth of tumors is dependent on the number of cells injected. B16-F10 is a very aggressive tumor; so 25,000 cells are inoculated into the mice. LM-6 (*see* **Subheading 3.3.2.**) is a very slow growing tumor and thus 10^6 cells are injected.
7. The antitumor efficiency of this mode of treatment appears to be enhanced when therapy is initiated at a time shortly after the injection of tumor cells and is diminished as the tumors become established and large. However, the SAOS-L LM6 osteosarcoma model (wherein therapy is initiated 6–8 wk after the injection of tumor cells, **Subheading 3.3.2.**) provides evidence that the antitumor effect is not restricted only to the earliest stages of lung tumor metastasis.
8. **Figure 1** illustrates the effect of aerosol treatment of PEI-p53 and PEI-p53(CD1-366) on B16-F10 lung metastasis. Similar results have been repeatedly

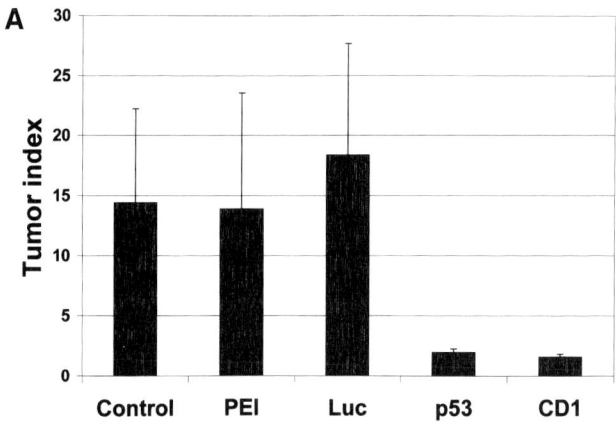
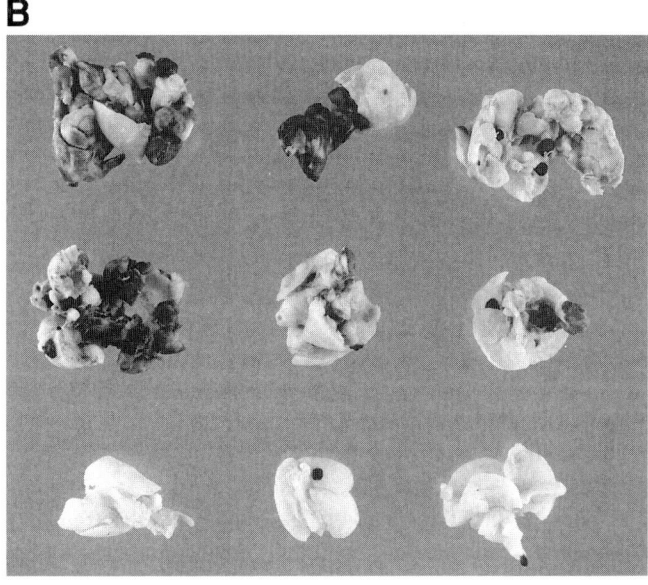

Fig. 1. Inhibition of B16-F10 lung metastasis by PEI-p53 and PEI-p53(CD1-366) aerosol delivery. Mice were injected via lateral tail vein with 25,000 B16-F10 cells on d 0. The mice were then treated with PEI alone, PEI-Luc, PEI-p53, or PEI-p53(CD1-366) aerosol complexes twice a week starting on d 1. The last treatment was on d 22-post-tumor cells injection, making it a total of 7 aerosol exposures. The mice were sacrificed on d 24, the lungs harvested, weighed, and fixed in Fekets solution. (**A**) Tumor index was calculated by the formula: Tumor index = lung weights × average grade for the group. Values are mean ± SD ($n = 10$ mice per group). (b) Representative lungs from control (top), PEI-Luc (middle) and PEI-p53 (bottom) treated mice are presented (n = 10 mice per group). Lungs from PEI treated group (not shown) were similar in shape, size and number of tumor foci to those shown for control and PEI-Luc treated groups. Lungs from PEI-p53(CD1-366) treated group (not shown) were similar in shape, size and number of tumor foci to those shown for PEI-p53 treated group. Data is representative of 2 separate experiments. Values are mean ± SD ($n = 10$ mice per group).

obtained with the SAOS-2 LM6 model with both p53 and p53(CD1-366) genes resulting in highly significant antitumor responses when compared to all control groups.

Acknowledgments

The authors would like to thank Drs. Shu-Fang Jia, and Laura L. Worth for their assistance and advice regarding the SAOS-LM6 tumor model. We would like to thank Eva Golunski, Luz Roberts, Angela Earhart, and Sara Melton for their expert technical assistance. We would also like to thank Drs. Brian Gilbert, Nadya Koshkina, Berma Kinsey, Frank Orson, Edward Hyman, and Vernon Knight for helpful suggestions and discussions. This work was supported by the Clayton Foundation for Research (Houston, TX). A.G., B.X., and C.L.D. were supported, in part, by an Advanced Technology Program Grant from the Texas Higher Education Coordinating Board (#004949-0129-1999 to C. Densmore and E. Kleinerman). Inquiries should be addressed to C. Densmore.

References

1. Brambilla, E. and Brambilla, C. (1997) p53 and lung cancer. *Pathol. Biol. (Paris)* **45**, 852–863.
2. Roth, J. A., Swisher, S. G., and Meyn, R. E. (1999) P53 tumor suppressor gene therapy for cancer. *Oncology (Huntingt.)* **13**, 148–154.
3. Lesoon-Wood, L. A., Kim, W. H., Kleinman, H. K., Weintraub, B. D., and Mixon, A. J. (1995) Systemic gene therapy with p53 reduces growth and metastasis of a malignant human breast cancer in nude mice. *Human Gene Ther.* **6**, 395–405.
4. Hsiao, M., Tse, V., Carmel, J., Tsai, Y., Felgner, P. L., Haas, M., and Silverberg, G. D. (1997) Intercavitary liposome-mediated p53 gene transfer into glioblastoma with endogenous wild-type p53 in vivo results in tumor suppression and long-term survival. *Biochem. Biophys. Res. Commun.* **233**, 359–364.
5. Xu, M., Kumar, D., Srinivas, S., Detolla, L. J., Yu, S. F., Stass, S. A., and Mixon, A. J. (1997) Parenteral gene therapy with p53 inhibits human breast tumors in vivo through a bystander mechanism without evidence of toxicity. *Human Gene Ther.* **20**, 177–185.
6. Zou, Y., Zong, G., Ling, Y.-H., and Perez-Soler, R. (2000) Development of cationic liposome formulations for intratracheal gene therapy of early lung cancer. *Cancer Gene Ther.* **7**, 683–696.
7. Gautam, A., Densmore, C. L., and C. L., Waldrep, J. C. (2000) Inhibition of experimental lung metastasis by aerosol delivery of PEI-p53 complexes. *Mol. Ther.* **2**, 318–323.
8. Densmore, C. L., Gautam, A. J., Kleinerman, E., Waldrep, J. C. Jia, S.-F., Xu, B., and Knight, V. (2000) Aerosol delivery of PEI-p53 complexes for treatment of metastatic lung cancer. *Cancer Gene Ther.* **7**, S2.

9. Densmore, C. L., Orson, F. M., Xu, B., Kinsey, B. M., Waldrep, J. C., Hua, P., et al. (2000) Aerosol delivery of robust polyethyleneimine-DNA complexes for gene therapy and genetic immunization. *Mol. Ther.* **1**, 180–188.
10. Gautam, A., Densmore, C. L., Xu, B., and Waldrep, J. C. (2000) Enhanced gene expression in mouse lung after PEI-DNA aerosol delivery. *Mol. Ther.* **2**, 63–70.
11. Gautam, A., Densmore, C. L., and Waldrep, J. C. (2001) Pulmonary cytokine responses associated with PEI-DNA aerosol delivery. *Gene Ther.* **8**, 254–257.
12. Missol, E., Sochanik, A., and Szala, S. (1995) Introduction of murine IL-4 gene into B16(F10) melanoma tumors by direct gene transfer with DNA-liposome complexes. *Cancer Lett.* **97**, 189–93.
13. Missol-Kolka, E., Sochanik, A., and Szala, S. (1998) Combined therapy of B16(F10) murine melanoma using E. coli cytosine deaminase gene and murine interleukin-4 gene. *Neoplasma* **45**, 305–311.
14. Qin, H. and Chatterjee, S. K. (1996) Cancer gene therapy using tumor cells infected with recombinant vaccinia virus expressing GM-CSF. *Hum. Gene Ther.* **7**, 1853–1860.
15. Kayaga, J., Souberbielle, B. E., Sheikh, N., Morrow, W. J., Scott-Taylor, T., Vile, R., et al. (1999) Anti-tumour activity against B16-F10 melanoma with a GM-CSF secreting allogeneic tumour cell vaccine. *Gene Ther.* **6**, 1475–1481.
16. Saito, M., Fan, D., and Lachman, L. B. (1995) Antitumor effect of liposomal IL 1 alpha and TNF alpha against the pulmonary metastases of the B16F10 murine melanoma in syngeneic mice. *Clin. Exp. Metastasis* **13**, 249–59.
17. Lesoon-Wood, L. A., Kim, W. H., Kleinman, H. K., Weintraub, B. D., and Mixson, A. J. (1995) Systemic gene therapy with p53 reduces growth and metastases of a malignant human breast cancer in nude mice. *Hum. Gene Ther.* **6**, 395–405.
18. Hsiao, M., Tse, V., Carmel, J., Tsai, Y., Felgner, P. L., Hass, M., and Silverberg, G. D. (1997) Intracavitary liposome-mediated p53 gene transfer into glioblastoma with endogenous wild-type p53 in vivo results in tumor suppression and long-term survival. *Biochem. Biophys. Res. Commun.* **233**, 359–364.
19. Xu, M., Kumar, D., Srinivas, S., Detolla, L. J., Yu, S. F., Stass, S. A., and Mixson, A. J. (1997) Parenteral gene therapy with p53 inhibits human breast tumors in vivo through a bystander mechanism without evidence of toxicity. *Hum. Gene Ther.* **8**, 177–185.
20. Jia, S-F., Worth, L. L., and Kleinerman, E. S. (1999) A nude mouse model of human osteosarcoma lung metastases for evaluation new therapeutic strategies. *Clin. Exp. Metastases* **17**, 501–506.
21. Koshkina, N. V., Kleinerman, E. S., Waldrep, J. C., Jia, S-F., Worth, L. L., Gilbert, B. E., and Knight, V. (2000) 9-Nitrocamptothecin liposome aerosol treatment of melanoma and osteosarcoma lung metastases. *Clin. Cancer Res.* **6**, 3713–3718.
22. Worth, L. L., Jia, S-F., Zhou, Z., Chen, L., and Kleinerman, E. S. (2000) Intranasal therapy with an adenoviral vector containing the murine interleukin-12 gene eradicates osteosarcoma lung metastases. *Clin. Cancer Res.* **6**, 3713–3718.

23. Radinsky, R., Fidler, I. J., Price, J. E., Esumi, N., Tsan, R., Petty, C. M., et al. (1994) Terminal differentiation and apoptosis in experimental lung metastases of human osteogenic sarcoma cells by wild type p53. *Oncogene* **9,** 1877–1883.
24. Fan, J. and Bertino, J. R. (1999) Modulation of cisplatinum cytotoxicity by p53: effect of p53-mediated apoptosis and DNA repair. *Mol. Pharmacol.* **56,** 966–972.
25. Okaichi, K., Wang, L. H., Ihara, M., and Okumura, Y. (1998) Sensitivity to ionizing radiation in Saos-2 cells transfected with mutant p53 genes depends on the mutation position. *J. Radiat. Res.* **39,** 111–118.
26. Deffie, A., Hao, M., Montes de Oca Luna, R., Hulboy, D. L., and Lozano, G. (1995) Cyclin E restores p53 activity in contact-inhibited cells. *Mol. Cell. Biol.* **15,** 3926–3933.

III

NOVEL THERAPIES

B. GENE THERAPIES

III. Antisense Oligonucleotide Gene Therapies

37

Clinical Development of Antisense Oligonucleotides as Anti-Cancer Therapeutics

Helen X. Chen

1. Introduction

Antisense oligonucleotides (AONs) are one of the several types of "antisense therapeutics" designed to modulate specific gene expression at the mRNA level. Although antisense strategies also include ribozymes (RNA enzyme) and, more recently, DNAzymes (a DNA analog of a ribozyme), AONs are by far the most advanced in clinical development and will be the focus of this review.

AONs are, by definition, short segments of single-stranded DNA ("antisense") complementary to the target mRNA sequence ("sense"). Formation of AON and mRNA hybrids results in degradation of the target mRNA molecule *(1)* and blockage of subsequent protein translation *(2)*, thereby specifically reducing the expression of the target gene. Because an antisense oligo can be designed with high specificity for virtually any sequence based on Watson and Crick's theory, AONs are perceived to be an exciting class of therapeutic agents that have the potential to capitalize rapidly on the expanding body of human genome-sequence data. The last decade has seen a rapid development of AONs for cancer and other diseases. Several AONs are being evaluated as anticancer therapeutics in different phases of clinical trials (*see* **Table 1**), including a Phase III trial for non-small cell lung cancer (NSCLC) (protein Kinase C [PKC]-α antisense). Some of these studies have shown encouraging results, supporting the rationale of further clinical development, but the data overall are preliminary and await confirmation by more extensive evaluation.

From: *Methods in Molecular Medicine, vol. 75: Lung Cancer, Vol. 2: Diagnostic and Therapeutic Methods and Reviews*
Edited by: B. Driscoll © Humana Press Inc., Totowa, NJ

Table 1
Completed and Ongoing Clinical Evaluations of Selected Antisense Oligonucleotides

Oligonucleotide	Target mRNA	Tumors	Clinical trials	Company	Reference
ISIS 3521*	PKCα	Solid tumors	• Phase II monotherapy • Phase II: + 5-FU, cisplatin • Phase III: combination with carboplatin + paclitaxel (**NSCLC**)	ISIS	(58)
ISIS 5132	Raf-1	Solid tumors	• Phase II	ISIS	(26)
ISIS 2503	H-*ras*	Solid tumors	• Phase II (**NSCLC**).	ISIS	(33)
G3139*	BCL2	Non-Hodgkin's lymphoma Leukemia Solid tumors	• Phase II monotherapy • Phase I/II: (+ paclitaxel; docetaxel; irinotecan; cytarabine) • Phase III: in combination with DTIC (melanoma)	Genta	(43) (44)
GEM®231	PKA RIα	Solid tumors	• Phase I monotherapy • Phase I/II (+chemotherapy)	Hybridon	(57)
MG98	Methyl transferase	Solid tumors	• Phase I	Methyl Gene Inc.	(59)
GTI-2040	Ribonuclease Reductase	Solid	• Phase I	Lorus Therapeutics, Inc	(60)
OL(1)p53	p53	Leukemia/MDS	• Phase I	Lynx Therapeutics	
LR-3001	c-Myb	Chronic leukemia	• Bone-marrow purge in vitro and in vivo • Phase I in myelogenous leukemia	Lynx Therapeutics	(61)

It should be recognized that there remain both technical and clinical challenges in the development and evaluation of AONs. Despite the simplicity of concept, the drug design, the mechanisms of action and, finally, the clinical translation of AON therapeutics are more complicated than initially envisioned. The following sections will review the design and general features of AONs that are relevant to clinical application as well as several examples of AON compounds in clinical trials.

2. Design of AONs

For an AON to reach therapeutic efficacy, the molecule must possess the ability to resist in vivo metabolism, must reach and enter tumor cells, and then must interact with the target mRNA sequences. These criteria demand proper selection of the AON sequence and chemical modification of the oligonucleotide backbones.

The optimal antisense sequence for a given gene is not easily predictable, since mRNAs tend to exist in secondary or tertiary structures with much of the sequence array inaccessible to AON binding. Until recently, selection of AON sequences was slow and tedious, using an empirical "gene-walking" approach for sequence generation followed by in vitro testing in cell culture. Advances in technology have led to the emergence of several efficient and exhaustive approaches. One such strategy is the RNase H mapping approach *(3)*, in which accessible mRNA sites are elucidated by combining the target mRNA with a random library of oligos in the presence of RNase H and subsequent analysis of the cleavage fragments. Another powerful method, scanning combinational oligonucleotide array, involves solid-phase immobilization of multiple AONs and hybridization of the oligo array to radiolabeled target mRNA *(4)*. A combination of the two strategies appears to be extremely efficient.

In vivo stability of AONs is conferred by chemical modifications of the native oligonucleotide, which is otherwise highy susceptible to degradation in the cellular millieu. Some of the analogs that can be used are listed in **Fig. 1**. The most widely used chemical class of AONs in clinical trials is the phosphorothioate oligonucleotides (PS-oligos), which contain a sulfa replacement at the nonbridging oxygen in the phosphate linkage. PS-oligos display sufficient nuclease resistance to reach the target cells in vivo while retaining the ability to bind and induce cleavage of mRNA. While PS-oligos, as the 1st-generation AONs, continue to hold promise for clinical development, analogs with additional modifications (e.g., at the 2′-O- residue of the sugar ring) have been developed in an attempt to reduce the polyanionic nature of PS-oligos and to further improve the pharmcodynamic of AONs. Some of these 2nd-generation AONs, or mixed backbone oligonucleotides (MBO) have recently entered clinical trails (e.g., GEM231 and MT98 in **Table 1**).

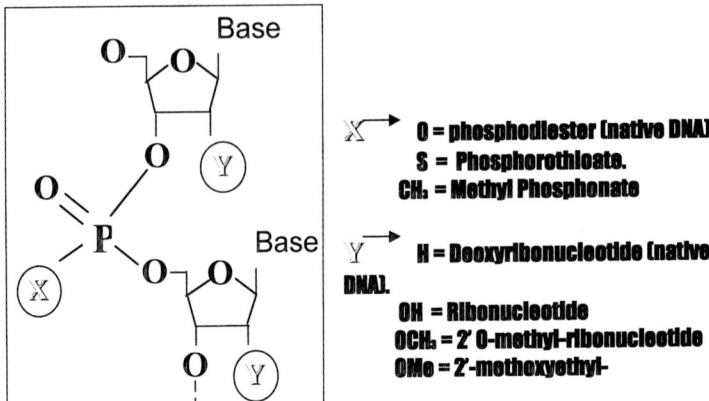

Fig. 1. Chemical Modifications of synthetic oligonucleotides. Selected chemical modifications of synthetic oligonucleotides: in native DNA, "X" represent the nonbridging oxygen (O) in the phosphodiester linkage between nucleosides and "Y" represents hydrogen (H) at the 2′ residue of the sugar ring. The first generation synthetic oligonucleotides contain modification at the X position with sulfa (Phosphorothioate) or methyl group (methyl phophonate). Additional modification are be made at the 2′- of the nucleoside ring in the mixed backbone oligos; examples of the replacement include -OH, -OCH3, -Ome, or -NH2.

Appropriately modified MBOs are superior to the 1st-generation compounds, with longer half-lives in tissue, higher affinity for mRNA, and lower toxicities *(5,6)*. Moreover, studies in mice also suggested that MBOs are suitable for oral administration with favorable bioavailability *(7)* (*see* Chapter 39, this volume).

3. Clinically Relevant Features of Pharmacodynamics and Toxicology AONs

The biophysical characteristics of synthetic oligonucleotides are the primary determinants of AON's pharmacotoxicology properties and are influenced very little by the base sequences. Understanding these features is important for design and interpretation of clinical trials of AON therapeutics.

3.1. Tissue Distribution and Cellular Uptake of AONs

Injected synthetic oligonucleotides are efficiently bound to protein in the plasma and distributed rapidly to tissues, where they undergo a slower clearance through metabolism by exonuclease. Metabolites are excreted in the urine. For PS-oligos, although the plasma $T_{1/2}$ is typically 1–2 h, $T_{1/2}$ in tissues could be 1–2 d as demonstrated in rodents. Mixed backbone oligonucleotides (MBOs) (e.g., GEM231) with appropriately positioned modifications are associated with even longer tissue $T_{1/2}$ and most of the oligos remain in the intact form for up to 7 d *(8)*. Bioavailability of AON in the tumor tissues in humans is unknown and, as suggested earlier, cannot be extrapolated from plasma concentrations and pharmacokinetics. Distribution of AONs is most predominant in highly vascular organs such as liver, kidney, and spleen, with almost none locating to the central nervous system (CNS) *(9)*.

Negatively charged molecules such as oligonucleotides are known to cross cellular membranes with low efficiency. However, while in vitro experiments using AONs commonly require lipid encapsulation for cellular delivery, systemically administered naked AONs have been demonstrated to reach tissues and lead to target gene downregulation, suggesting a more active cellular uptake in vivo. The precise mechanisms of cellular entry are unclear but may involve fluid phase pinocytosis and receptor-mediated endocytosis. The extent of AON uptake also depends on cell type and the status of cellular activation. In an in vitro experiment on patients' peripheral blood cells using fluorescent isothiocyanate (FITC)-labeled oligonucleotides *(10)*, uptake was highest in myeloid cells, intermediate in B cells and lowest in T cells. In patients with acute myelogenous leukemia (AML) or acute lymphocytic leukemia (ALL), leukemic cells were more efficient in taking up oligos than their normal counterparts from the same patient. Furthermore, activation of cultured cells lines by growth factors was associated with an increase in cellular entry of AONs.

3.2. Mechanisms of Action of AON

The primary mechanism of AON-mediated inhibition of gene expression is via degradation of the RNA strand of the RNA-DNA duplex by RNase H *(1)* and, to a lesser extent, through hybridization arrest of translation *(2,11)*. The sequence-dependence of AONs' biological effects has been validated in numerous studies, which demonstrated sequence-specific target gene downregulation and the relative inactivity of irrelevant control oligonucleotides *(1,12–14)*.

However, non-antisense activities of AON also exist. While the target gene suppression is highly specific and sequence-dependent, the antitumor activity of AONs observed in animals may be mediated by additional mechanisms besides target gene modulation. It has been shown in mice that PS-oligos containing unmethylated CpG motifs could induce polyclonal B-cell activation *(15)* and release of cytokines, including interleukin (IL)-6, IL-12, tumor necrosis factor-α (TNF-α), and interferon-γ (IFN-γ) *(15,16)*. In addition, like other polyanoids, synthetic AONs bind to proteins, especially heparin-binding proteins such as vascular endothelial growth factor (VEGF) and fibroblastic growth factors (FGF), thus modulating the biological effects of these molecules

(17). These nonspecific effects have been shown to directly or indirectly contribute to the anticancer activities in animal models *(18–20)*, and may affect the clinical outcomes of AON therapy in addition to the antisense activity.

3.3. Toxicities of AON Therapy

Most AON-related toxicities are due to the negative charge of the compound, and are dose- and infusion rate-dependent. Common side effects directly related to the plasma concentration of AONs include prolongation of activated partial thrombin time (aPTT) and activation of the complement cascade. Toxicities associated with duration of AON therapies include thrombocytopenia, fatigue, and transaminase elevation. Lethal dose of AON in primates was associated with complement activation with resultant cardiovascular collapse *(21)*.

Two schedules of AON administration have been tested in clinical trials: 2-h infusion for 2–3 times per week and continuous infusion for 14–21 d. The 2-h infusion is typically associated with concentration-related toxicities. Although transient, these side effects were dose limiting for the 1st-generation PS oligos. Consequently, most Phase II trials use the schedule of continuous infusion to avoid high peak plasma concentrations. In general, AONs are well-tolerated at doses that produce biological activities in some patients.

As mentioned, one feature of the MBOs is the reduced negative charge and hence a more favorable safety profile. As a result, much higher doses of MBOs can be safely administered by 2-h infusions. Combined with evidence of the prolonged $T_{1/2}$ of MBOs in tissues, it is conceivable that MBOs are suitable for shorter and less frequent infusions than the 1st-generation AONs.

4. Clinical Development of Therapeutic Oligonucleotide

In the last decade, an increasing number of AONs have entered clinical trials (*see* **Table 1**). Some of them are directed at targets potentially important for lung cancers, e.g., (PKC-α), bcl-2, c-raf, H-ras, and Protein Kinase A-RIα. Numerous other AONs are being explored pre-clinically, including those targeting VEGF, Her-2/Neu, E-selectin, c-myc, mdr-1, and others. The following section will only review selected AONs in clinical trials. The pathobiological importance of these targets is discussed elsewhere of this book in more detail.

4.1. C-raf-1 Antisense (ISIS 5132)

Raf-1 is a serine/threonine protein kinase encoded by *c-raf-1* and acts downstream of Ras in the MAP kinase signal-transduction pathway *(22)*. In addition, Raf-1 may be activated independently of Ras by other signaling molecules, such as Bcl-2 and PKC-α, and may promote expression of the multidrug resistance gene *mdr-1*. The establishment of Raf-1 as an important mediator of diverse signaling pathway provides a rationale for Raf-directed therapies.

ISIS 5132 (also known as CGP 69846A) is a 20-mer PS-oligo targeted at the 3′ untranslated region of *c-raf-1* mRNA. Inhibition of C-raf expression and tumor growth were observed in the A549 human lung cancer models both in vitro (IC_{50} =100 nM) and in vivo *(23)*.

Phase I dose-escalation trials of ISIS 5132 have been completed using 2 dosing schedules: 2-h infusion three times weekly for 3 out of 4 wk (dose range 0.5–6 mg/kg/dose) *(24)* and 21-d continuous infusion every 28 d (dose range 0.5–4 mg/kg/d) *(25)*. Using either schedule, ISIS 5132 was well-tolerated and the maximum tolerated dose (MTD) was not reached. However, the continuous infusion was better tolerated with respect to concentration-related side effects (aPTT prolongation and complement activation), while other toxicities associated with duration of therapy were similar for both schedules. Not surprisingly for Phase I trials, no objective responses were observed in these studies involving heterogeneous disease types and multiple dose levels, although there were cases of minor responses, prolonged stable diseases, and decline in tumor markers. Continuous infusion was chosen as the schedule for a Phase II single agent trial and for combination with 5-fluorouracil (5-FU) and leucovorin *(26)*.

Pharmacokinetics of the 21-d continuous infusion revealed that, at dose levels of 2 mg/kg/d and 4 mg/kg/d, steady-state plasma levels reached 110 nM and 420 nM, respectively. Importantly, 50–60% of the oligonucleotides in the plasma were in the intact form at all time points measured. Evaluation of the direct target effect of AON was also attempted by reverse transcription polymerase chain reaction (RT-PCR) on peripheral blood mononuclear cells (PBMN) *(27)*. With the 2-h infusion, temporal reduction of C-raf mRNA could be detected in many patients but the patterns of changes were not clearly dose- or duration- dependent.

4.2. AON against Protein Kinase C-α *(ISIS 3521)*

Protein kinase C (PKC) is a member of a family of phospholipid-dependent serine/threonine kinases, comprised of 12 distinct isozymes, which differ from one another in their primary structures and biological functions. Activation of PKC, particularly PKC-α, is associated with malignant transformation and proliferation. Several small molecules have been developed that target PKC, but they typically lack specificity for individual isozymes. ISIS 3521 (or CGP 64128A, ISIS Pharmaceuticals) is a 20-base PS-oligo directed at the 3′-untranslated region of the PKC-α mRNA. In the A549 human lung cancer cell line, ISIS 3521 inhibited PKC mRNA (IC_{50} = 200 nM) and protein expression (IC_{50} = 100 nM), while other PKC isozymes were not affected *(28)*. In vivo growth inhibition by ISIS 3521 was also demonstrated in xenograft models for A549 and other human cancer cell lines *(28)*.

Phase I trials of ISIS 3521 by both 2- and 24-h infusion have been completed *(29,30)* and the toxicity and pharmacokinetic features were similar to those of other PS-oligos. In both Phase I studies, objective tumor responses were demonstrated in patients with lymphoma and ovarian cancer. Combining ISIS 3521 with chemotherapy was explored in several clinical trials. In a Phase I/II study of ISIS 3521 (2 mg/kg/d by 14-d per cycle) in combination with standard carboplatin and paclitaxel, toxicities were similar to chemotherapy alone and no pharmacokinetic interactions were observed *(31)*. A total of 53 patients with NSCLC were enrolled to the study. Among the 48 evaluable patients (Performance Status 0–1, 83% with stage IV diseases), 42% of patients had either partial or complete response, and 41% had stable disease. Overall median time to progression was 6.6 mo, and median overall survival was 19 mo with an actuarial 1-yr survival of 75%. These results compared favorably to historical controls using the chemotherapy alone. A Phase III clinical trial is ongoing comparing standard carboplatin and paclitaxel with or without ISIS 3521 as the first-line therapy for advanced or metastatic non-small cell lung cancer (NSCLC).

4.3. H-Ras Antisense (ISIS 2503)

The RAS family consists of three highly homologous proteins (Ki-Ras, Ha-ras, and N-Ras), which play a critical role in linking growth factor-activated tyrosine kinases to downstream effectors, such as Raf and mitogen-activated protein kinase. Up to 30% of lung adenocarcinomas show mutations in the *K-ras* oncogene *(32)*, and gene mutation or overexpression are associated with decreased survival, especially in resectable cases. Conversely, inhibition of *ras* activation in preclinical models has been shown to reduce the growth of ras-overexpressing tumor cells both in vivo and in vitro.

ISIS 2503 is a 20-mer PS-oligo designed to target the initiation region of the human H-ras mRNA. Despite in vitro selectivity, ISIS 2503 also inhibits the growth of K-ras mutant and nonmutant human tumor xenografts. A Phase II evaluation of ISIS 2503 is ongoing in patients with advanced NSCLC *(33)* and the results are pending. ISIS 2503 in this case was given by continuous infusion at 6 mg/kg/d for 14 d in 21-d cycles.

4.4. BCL-2 Antisense (G3139)

BCL-2 is a prototype anti-apoptotic member of the Bcl-2 family and confer resistance to apoptosis triggered by a variety of death signals including chemotherapeutic agents. Although first described in follicular lymphoma, bcl-2 is overexpressed in most leukemia and solid tumors, including lung cancers, with a frequency of about 30% in NSCLC *(34)* and 80–100% in small cell lung cancer (SCLC) *(35)*. The clinical implication of bcl-2 overexpression is unclear, but in preclinical NSCLC and SCLC models, AONs against BCL-2 were found to reduce cellular proliferation and increase apoptosis both as single agents *(36,37)* and synergistically with etoposide, doxorubicin, and cisplatin *(38)*.

G3139 or Genesense (Genta, Inc.) is an 18-base PS-oligo complementary to the first 6 codons of the bcl-2 mRNA. It has demonstrated target gene downregulation in a dose-dependent and sequence-specific manner, and has shown activities against human cancer models both in vitro and in vivo. Furthermore, antitumor activity was significantly enhanced when G3139 is combined with a number of chemotherapeutic agents, such as docetaxel *(39)*, dacarbazine *(40)*, and anthrocycline *(41)*.

The clinical activity of G3139 was first demonstrated in non-Hodgkin's lymphoma (NHL) in a Phase I dose-escalation trial at Royal Marsden Hospital *(42,43)*. G3139 was given by 14-d subcutaneous infusion at eight dose levels (4.6–73.6 mg/m2/d). Among the 21 patients with relapsed NHL, objective responses, including one complete response, were observed at doses that were well-tolerated. Major toxicity in this trial was thrombocytopenia, especially at higher doses and in patients with extensive lymphoma involvement in the bone marrow. Importantly, Bcl-2 protein was reduced in 7 of the 16 assessable patients, as measured in tumor cells derived from lymph nodes, peripheral blood, or bone marrow. BCL-2 downregulation was also demonstrated in melanoma samples from patients after treatment with G3139 and dacarbazine *(44)*.

Given the hypothesis that anti-BCL-2 therapy may sensitize tumor cells to cytotoxic agents, combination regimens with chemotherapy have been the focus of most G3139 trials. The Phase II result of the G3139 and dacarbazine combination in advanced melanoma was encouraging *(44)*, and has led to a Phase III randomized trial comparing chemotherapy alone vs the combination regimen as the first-line treatment for metastatic melanoma. A number of Phase I trials have also established the feasibility of combining G3139 with other chemotherapeutic agents, e.g., paclitaxel *(45)*, docetaxel *(46)*, and irinotecan *(47)*. In most cases, regimens of G3139 up to 5–7 mg/kg/d for 5 d and full-dose chemotherapy could be safely given without much added toxicity. Combination of G3139 with standard chemotherapy for lung cancer and a variety of malignancies will be explored in Phase I/II trials sponsored by the National Cancer Institute (NCI).

4.5. Antisense Against Type I Protein kinase A (GEM231)

PKA is the primary mediator of cAMP activity and exists as type I (PKA-I) and type II (PKA-II). PKA-I plays an important role in G_1 to S transition *(48)* and in signal transduction of the mitogenic effects of growth factors such as epidermal growth factor (EGF) and transforming growth factor-α (TGF-α) *(49–51)*. Overexpression of PKA-I is observed in the majority of human cancers and correlates with unfavorable clinicopathological features and prognosis *(52,53)*. Selective downregulation of the regulatory subunit of PKA-I (PKA-RIα) induced growth arrest and differentiation in a large variety of human cancer cell lines *(12,14,54)*.

GEM231 is an 18-base AON targeted at PKA-RIα mRNA. Chemically, GEM231 represents a prototype MBO, which, in addition to the PS-oligo backbone, contains 2′-O-methyl modifications on the terminal 4 ribonucleosides (*see* **Fig. 1**). In preclinical studies, GEM231 exhibited sequence-specific reduction of intratumoral PKA-RIα levels and growth inhibition of a variety of human cancer xenografts, including those using lung cancer cells. Synergistic activity was also observed when GEM231 was combined with chemotherapeutic agents (e.g., taxanes, cisplatin, and topoisomerase II inhibitors) *(55)* and a monoclonal antibody against the EFG receptor (C225) *(56)*. Moreover, as a prototype MBO, GEM231 has displayed enhanced in vivo stability *(8)*, feasibility of oral administration with favorable bioavailability, and reduced concentration-related adverse effects in animal models *(5)*.

A Phase I study of GEM231 by 2-h infusion twice weekly has been completed *(57)*, which was the first clinical trial of a MBO in cancer patients. Although no antitumor activity was observed in the 13 patients with various tumor types, several features of the MBO safety profile were observed. As predicted, concentration-dependent toxicities were much lower compared to experience with the 1st-generation AONs given by the same schedule; notably, thrombocytopenia and complement activation did not appear to be a problem with GEM231. On the other hand, cumulative toxicity became prominent after 8 wk of therapy and manifested as reversible transaminase elevations. The maximal tolerated dose was determined to be 240 mg/m^2 (equivalent to 5–7 mg/kg) thrice weekly for 4-wk of administrations. Alternative schedule with rest periods is being explored to reduce the cumulative side effects.

Given the observation of minimal concentration-related toxicities of GEM231 and the preclinical evidence of prolonged tissue half-life, it is conceivable that MBOs may be the preferred chemical class of AONs for future development. Short and less frequent infusions may therefore become feasible. Based on the preclinical demonstration of synergism with chemotherapy, GEM231 is currently under Phase I/II evaluation in combination with paclitaxel.

5. Conclusions and Future Directions

As a natural extension of our increasing knowledge of the molecular basis of cancer biology, AONs have been recognized as an attractive class of anticancer therapeutics. Advances in medicinal chemistry have allowed proper sequence and chemical design of specific AONs compounds. The newer generation of AONs, using mixed backbone modifications, are expected to demonstrate even more favorable pharmacodynamic properties in patients, and have the potential for oral administration. Experience with AONs in clinical trials indicated that this therapeutic approach is well-tolerated, either alone or in combination with chemotherapy. Interesting findings have emerged from several trials, but unequivocal demonstration of the clinical activity of AONs still awaits results of randomized Phase III trials. As for the direct drug effects on targets, although some clinical studies detected target gene downregulation in tumor tissues or circulating lymphocytes, results in general have been variable and inconclusive.

Although the evaluation of the clinical efficacy of individual AONs is important, several fundamental issues of AON therapeutics should be addressed for further development of AONs in general. To date, the optimal dose and schedule of AON administration remain undetermined. Despite extensive knowledge of the plasma pharmacokinetics in humans, little is known of the intratumor bioavailability and the status of target gene expression under clinically chosen doses. Efficacy evaluation has also been challenging since most of the direct target effects of AONs are not cytotoxic and cannot be readily assessed by clinical parameters. All these factors point to the importance of incorporating molecular endpoints in clinical trials. More effort is required to improve the sensitivity and reproducibility of the bioassays for clinical samples. Ideally, such molecular analysis should be done in tumor tissues, but to circumvent the difficulty of obtaining biopsies, it is also important to explore and to validate examination of surrogate tissues, such as circulating lymphocytes and circulating malignant cells.

More AONs are expected to emerge from laboratories and await translation into clinical application. Clearly, further improvement of the AON design is warranted to optimize the bioavailability to tumor cells. Equally important are emphases on more rigorous preclinical characterization of the pharmacokinetic and pharmacodynamic features of the lead compounds. In addition, the mechanisms of action of biological activity of AONs also must be better elucidated. Certain nonantisense but biologically beneficial effects may exist in addition to the antisense mechanisms, and this phenomenon could potentially be characterized in preclinical studies using oligonucleotides with alternate sequences and modifications.

In conclusion, significant progress has been made in the development of AONs as anticancer therapeutics. However, multiple obstacles remain to be overcome before realization of the full therapeutic potential of AONs, and this can only be accomplished through the collaborative effort between basic scientists, medicinal chemists, pharmacologists as well as clinical investigators across all stages of the AON development from laboratory to clinic.

Reference

1. Monia, B. P., Sasmor, H., Johnston, J. F., Freier, S. M, Lesnik, E. A., Muller, M., et al. (1996) Sequence-specific antitumor activity of a phosphorothioate oligodeoxyribonucleotide targeted to human C-raf kinase supports an antisense mechanism of action in vivo. *Proc. Natl. Acad. Sci. USA* **93,** 15481–15484.
2. Mukhopadhyay, T. and Roth, J. A. (1996) Antisense regulation of oncogenes in human cancer. *Crit. Rev. Oncog.* **7,** 151–190.
3. Lima, W. F., Brown-Driver, V., Fox, M., Hanecak, R., and Bruice, T. W. (1997) Combinatorial screening and rational optimization for hybridization to folded hepatitis C virus RNA of oligonucleotides with biological antisense activity. *J. Biol. Chem.* **272,** 626–638.
4. Milner, N., Mir, K. U., and Southern, E. M. (1997) Selecting effective antisense reagents on combinatorial oligonucleotide arrays. *Nat. Biotechnol.* **15,** 537–541.
5. Agrawal, S. and Zhao, Q. (1998) Mixed backbone oligonucleotides: improvement in oligonucleotide-induced toxicity in vivo. *Antisense Nucleic Acid Drug. Dev.* **8,** 135–139.
6. Agrawal, S., Zhang, X., Lu, Z., Zhao, H., Tamburin, J. M., Yan, J., et al. (1995) Absorption, tissue distribution and in vivo stability in rats of a hybrid antisense oligonucleotide following oral administration. *Biochem. Pharmacol.* **50,** 571–576.
7. Wang, H., Cai, Q., Zeng, X., Yu, D., Agrawal, S., and Zhang, R. (1999) Antitumor activity and pharmacokinetics of a mixed-backbone antisense oligonucleotide targeted to the RIalpha subunit of protein kinase A after oral administration. *Proc. Natl. Acad. Sci. USA* **96,** 13989–13994.
8. Zhang, R., Lu, Z., Zhao, H., Zhang, X., Diasio, R. B., Habus, I., et al. (1995) In vivo stability, disposition and metabolism of a "hybrid" oligonucleotide phosphorothioate in rats. *Biochem. Pharmacol.* **50,** 545–556.
9. Phillips, J. A., Craig, S. J., Bayley, D., Christian, R. A., Geary, R., and Nicklin, P. L. (1997) Pharmacokinetics, metabolism, and elimination of a 20-mer phosphorothioate oligodeoxynucleotide (CGP 69846A) after intravenous and subcutaneous administration. *Biochem. Pharmacol.* **54,** 657–668.
10. Zhao, Q., Song, X., Waldschmidt, T., Fisher, E., and Krieg, A. M. (1996) Oligonucleotide uptake in human hematopoietic cells is increased in leukemia and is related to cellular activation. *Blood* **88,** 1788–1795.
11. Crooke, S. T. (1996) Progress in antisense therapeutics. *Med. Res. Rev.* **16,** 319–344.
12. Yokozaki, H., Budillon, A., Tortora, G., Meissner, S., Beaucage, S. L., Miki, K., et al. (1993) An antisense oligodeoxynucleotide that depletes RI alpha subunit of cyclic AMP-dependent protein kinase induces growth inhibition in human cancer cells. *Cancer Res.* **53,** 868–872.

13. Yazaki, T., Ahmad, S., Chahlavi, A., Zylber-Katz, E., Dean, N. M., Rabkin, S. D., et al. (1996) Treatment of glioblastoma U-87 by systemic administration of an antisense protein kinase C-alpha phosphorothioate oligodeoxynucleotide. *Mol. Pharmacol.* **50**, 236–242.
14. Nesterova, M. and Cho-Chung, Y. S. (1995) A single-injection protein kinase A-directed antisense treatment to inhibit tumour growth. *Nat. Med.* **1**, 528–533.
15. Krieg, A. M., Yi, A. K., Matson, S., Waldschmidt, T. J., Bishop, G. A., Teasdale, R., et al. (1995) CpG motifs in bacterial DNA trigger direct B-cell activation. *Nature* **374**, 546–549.
16. Klinman, D. M., Yi, A. K., Beaucage, S. L., Conover, J., and Krieg, A. M. (1996) CpG motifs present in bacteria DNA rapidly induce lymphocytes to secrete interleukin 6, interleukin 12, and interferon gamma. *Proc. Natl. Acad. Sci. USA* **93**, 2879–2883.
17. Guvakova, M. A., Yakubov, L. A., Vlodavsky, I., Tonkinson, J. L., and Stein, C. A. (1995) Phosphorothioate oligodeoxynucleotides bind to basic fibroblast growth factor, inhibit its binding to cell surface receptors, and remove it from low affinity binding sites on extracellular matrix. *J. Biol. Chem.* **270**, 2620–2627.
18. Barton, C. M. and Lemoine, N. R. (1995) Antisense oligonucleotides directed against p53 have antiproliferative effects unrelated to effects on p53 expression. *Br. J. Cancer* **71**, 429–437.
19. Wooldridge, J. E., Ballas, Z., Krieg, A. M., and Weiner, G. J. (1997) Immunostimulatory oligodeoxynucleotides containing CpG motifs enhance the efficacy of monoclonal antibody therapy of lymphoma. *Blood* **89**, 2994–2998.
20. Jansen, B., Wadl, H., Inoue, S. A., Trulzsch, B., Selzer, E., Duchene, M., et al. (1995) Phosphorothioate oligonucleotides reduce melanoma growth in a SCID-hu mouse model by a nonantisense mechanism. *Antisense Res. Dev.* **5**, 271–277.
21. Galbraith, W. M., Hobson, W. C., Giclas, P. C., Schechter, P. J., and Agrawal, S. (1994) Complement activation and hemodynamic changes following intravenous administration of phosphorothioate oligonucleotides in the monkey. *Antisense Res. Dev.* **4**, 201–206.
22. Stokoe, D., Macdonald, S. G., Cadwallader, K., Symons, M., and Hancock, J. F. (1994) Activation of Raf as a result of recruitment to the plasma membrane. *Science* **264**, 1463–1467.
23. Monia, B. P., Johnston, J. F., Geiger, T., Muller, M., and Fabbro, D. (1996) Antitumor activity of a phosphorothioate antisense oligodeoxynucleotide targeted against C-raf kinase. *Nat. Med.* **2**, 668–675.
24. Stevenson, J. P., Yao, K. S., Gallagher, M., Friedland, D., Mitchell, E. P., Cassella, A., et al. (1999) Phase I clinical/pharmacokinetic and pharmacodynamic trial of the c-raf-1 antisense oligonucleotide ISIS 5132 (CGP 69846A). *J. Clin. Oncol.* **17**, 2227–2236.
25. Cunningham, C. C., Holmlund, J. T., Schiller, J. H., Geary, R. S., Kwoh, T. J., Dorr, A., et al. (2000) A phase I trial of c-Raf kinase antisense oligonucleotide ISIS 5132 administered as a continuous intravenous infusion in patients with advanced cancer. *Clin. Cancer Res.* **6**, 1626–1631.

26. Stevenson, J. P., Gallagher, M., et al. (1999) Phase I trial of the c-Raf-1 antisense oligonucleotide ISIS 5132 administered as a 21-day continuous infusion in combination with 5-FU and Leucovorin as a daily X 5 bolus. AACR/NCI/EORTC International conference. Washington, DC, 1999, Abstract 579.
27. O'Dwyer, P. J., Stevenson, J. P., Gallagher, M., Cassella, A., Vasilevskaya, I., Monia, B. P., et al. (1999) c-raf-1 depletion and tumor responses in patients treated with the c- raf-1 antisense oligodeoxynucleotide ISIS 5132 (CGP 69846A). *Clin. Cancer Res.* **5,** 3977–3982.
28. Dean, N., McKay, R., Miraglia, L., Howard, R., Cooper, S., Giddings, J., et al. (1996) Inhibition of growth of human tumor cell lines in nude mice by an antisense of oligonucleotide inhibitor of protein kinase C-alpha expression. *Cancer Res.* **56,** 3499–3507.
29. Nemunaitis, J., Von Hoff, D. D., Holmlund, J., Dorr, A., and Eckhardt, S. G. (1998) Phase I pharmacokinetic (PK) trial of a protein kinase C-alpha antisense oligonucleotide, ISIS 3521 (CGP 64128A), administered thrice weekly. *Proc. Am. Soc. Clin. Oncol.*, Abstract 812.
30. Sikic, B. I., Yuen, A. R., Advani, R., Halsey, J., Fisher, G. A., Holmlund, J., et al. (1998) Antisense oligonucleotide targeted to protein kinase C-alpha (ISIS 3521/ CGP 64128A) by 21-day continuous intravenous infusion:results of the Phase I trail and activity in ovraina carcinomas (Meeting abstract). *Proc. Am. Soc. Clin. Oncol.*, Abstract 1654.
31. Yuen, A., Halsey, J., Fisher, G., Advani, R., Moore, M., Saleh, M., et al. (2001) A Phase I/II trial of ISIS 3521, an antisense inhibitor of protein kinase C alpha, combined with carboplatin and paclitaxel in patients with non-small cell lung cancer. *Am. Soc. Clin. Oncol.*, San Francisco, Abstract.
32. Slebos, R. J. and Rodenhuis, S. (1992) The ras gene family in human non-small-cell lung cancer. *J. Natl. Cancer Inst. Monogr.* **13,** 23–29.
33. Dang, T., Johnson, D. H., Kelly, K., Rizvi, N., Holmlund, J., and Dorr, A. (2001) Multicenter Phase II Trial of an Antisense Inhibitor of H-ras (ISIS-2503) in Advanced Non-Small Cell Lung Cancer (NSCLC). *American Society of Clinical Oncology*, San Francisco, Abstract 1325.
34. Pezzella, F., Turley, H., Kuzu, I., Tungekar, M. F., Dunnill, M. S., Pierce, C. B., et al. (1993) bcl-2 protein in non-small-cell lung carcinoma. *N. Engl. J. Med.* **329,** 690–694.
35. Yan, J. J., Chen, F. F., Tsai, Y. C., and Jin, Y. T. (1996) Immunohistochemical detection of Bcl-2 protein in small cell carcinomas. *Oncology* **53,** 6–11.
36. Koty, P. P., Zhang, H., and Levitt, M. L. (1999) Antisense bcl-2 treatment increases programmed cell death in non-small cell lung cancer cell lines. *Lung Cancer* **23,** 115–127.
37. Ziegler, A., Luedke, G. H., Fabbro, D., Altmann, K. H., Stahel, R. A., and Zangemeister-Wittke, U. (1997) Induction of apoptosis in small-cell lung cancer cells by an antisense oligodeoxynucleotide targeting the Bcl-2 coding sequence. *J. Natl. Cancer Inst.* **89,** 1027–1036.

38. Zangemeister-Wittke, U., Schenker, T., Luedke, G. H., and Stahel, R. A. (1998) Synergistic cytotoxicity of bcl-2 antisense oligodeoxynucleotides and etoposide, doxorubicin and cisplatin on small-cell lung cancer cell lines. *Br. J. Cancer* **78**, 1035–1042.
39. Yang, D., Ling, Y., Almazan, M., Guo, R., Murray, A., Brown, B., et al. (1999) Tumor regression of human breast carcinomas by combination therapy of anti-bcl-2 antisense oligoneucleotide and chemotherapeutic drugs. *Proceedings of AACR Annual Meeting*, Philadelphia, PA, Abstract #4814.
40. Jansen, B., Schlagbauer-Wadl, H., Brown, B. D., Bryan, R. N., van Elsas, A., Muller, M., et al. (1998) bcl-2 antisense therapy chemosensitizes human melanoma in SCID mice. *Nat. Med.* **4**, 232–234.
41. Kitada, S., Takayama, S., De Riel, K., Tanaka, S., and Reed, J. C. (1994) Reversal of chemoresistance of lymphoma cells by antisense-mediated reduction of bcl-2 gene expression. *Antisense Res. Dev.* **4**, 71–79.
42. Webb, A., Cunningham, D., Cotter, F., Clarke, P. A., di Stefano, F., Ross, P., et al. (1997) BCL-2 antisense therapy in patients with non-Hodgkin lymphoma. *Lancet* **349**, 1137–1141.
43. Waters, J. S., Webb, A., Cunningham, D., Clarke, P. A., Raynaud, F., di Stefano, F., et al. (2000) Phase I clinical and pharmacokinetic study of bcl-2 antisense oligonucleotide therapy in patients with non-Hodgkin's lymphoma [In Process Citation]. *J. Clin. Oncol.* **18**, 1812–1823.
44. Jansen, B., Wacheck, V., Heere-Ress, E., Schlagbauer-Wadl, H., Hollenstein, U., Lucas, T., et al. (1999) A Phase I-II Study with Dacarbazine and BCL-2 Antisense Oligonucleotide G3139 (GENTA) as a Chemosensitizer in Patients with Advanced Malignant Melanoma. *ASCO*, Atlanta, GA, Abstract #2049.
45. Morris, M. J., Tong, W., Osman, I., Maslak, P., Kelly, W. K., Terry, K., et al. (1999) A Phase I/IIA Dose-Escalating Trial of bcl-2 Antisense (G3139) Treatment by 14-Day Continuous Intravenous Infusion (CI) for Patients with Androgen-Independent Prostate Cancer or Other Advanced Solid Tumor Malignancies. *Proc. ASCO*, Atlanta, GA, Abstract #1243.
46. Chen, H. and Marshall, J. (2000).
47. Ochoa, L., Kuhn, J., Salinas, R., Hammond, L., Hao, J., Rodriguez, G., et al. (2001) G3139 downregulates the expression of BCL-2 in patients with metastatic colorectal cancer treated with irinotecan (CPT-11). *AACR*.
48. Tortora, G., Pepe, S., Bianco, C., Baldassarre, G., Budillon, A., Clair, T., et al. (1994) The RI alpha subunit of protein kinase A controls serum dependency and entry into cell cycle of human mammary epithelial cells. *Oncogene* **9**, 3233–3240.
49. Ciardiello, F., Pepe, S., Bianco, C., Baldassarre, G., Ruggiero, A., Selvam, M. P., et al. (1993) Down-regulation of RI alpha subunit of cAMP-dependent protein kinase induces growth inhibition of human mammary epithelial cells transformed by c-Ha-ras and c-erbB-2 proto-oncogenes. *Int. J. Cancer* **53**, 438–443.
50. Tortora, G., Damiano, V., Bianco, C., Baldassarre, G., Bianco, A. R., Lanfrancone, L., et al. (1997) The RIalpha subunit of protein kinase A (PKA) binds to Grb2 and allows PKA interaction with the activated EGF-receptor. *Oncogene* **14**, 923–928.

51. Rohlff, C. and Glazer, R. I. (1995) Regulation of multidrug resistance through the cAMP and EGF signalling pathways [published erratum appears in Cell Signal 1996 Feb;8(2):151]. *Cell Signal* **7**, 431–443.
52. Miller, W. R., Watson, D. M., Jack, W., Chetty, U., and Elton, R. A. (1993) Tumour cyclic AMP binding proteins: an independent prognostic factor for disease recurrence and survival in breast cancer. *Breast Cancer Res. Treat.* **26**, 89–94.
53. Simpson, B., Ramage, A., Hulme, M., Burns, D., Katsaros, D., Langdon, S., et al. (1996) Cyclic adenosine 3′,5′-monophosphage-binding proteins in human ovarian cancer: correlation with clinicopathological features. *Clinical Cancer Res.* **2**, 201–206.
54. Tortora, G., Yokozaki, H., Pepe, S., Clair, T., and Cho-Chung, Y. S. (1991) Differentiation of HL-60 leukemia by type I regulatory subunit antisense oligodeoxynucleotide of cAMP-dependent protein kinase. *Proc. Natl. Acad. Sci. USA* **88**, 2011–2015.
55. Tortora, G., Caputo, R., Damiano, V., Bianco, R., Pepe, S., Bianco, A. R., et al. (1997) Synergistic inhibition of human cancer cell growth by cytotoxic drugs and mixed backbone antisense oligonucleotide targeting protein kinase A. *Proc. Natl. Acad. Sci. USA* **94**, 12586–12591.
56. Tortora, G., Caputo, R., Pomatico, G., Pepe, S., Bianco, A. R., Agrawal, S., et al. (1999) Cooperative inhibitory effect of novel mixed backbone oligonucleotide targeting protein kinase A in combination with docetaxel and anti-epidermal growth factor-receptor antibody on human breast cancer cell growth [In Process Citation]. *Clin. Cancer Res.* **5**, 875–881.
57. Chen, H. X., Marshall, J. L., Ness, E., Martin, R. R., Dvorchik, B., Rizvi, N., et al. (2000) A safety and pharmacokinetic study of a mixed-backbone oligonucleotide (GEM231) targeting the type I protein kinase A by two-hour infusions in patients with refractory solid tumors [In Process Citation]. *Clin. Cancer Res.* **6**, 1259–1266.
58. Yuen, A., Advani, R., Fisher, G., Halsey, J., Lum, B., Geary, R., et al. (2000) A Phase I/II trial of ISIS 3521, an antisense inhibitor of protein kinase C alpha, combined with carboplatin and paclitaxel in patients with non-small cell lung cancer. *American Society of Clinical Oncology*, New Orleans, Abstract 1802.
59. Stewart, D., Donehower, R., Eisenhauer, E., Winman, N., Moore, M., Bonfils, C., et al. (2000) A Phase I and Pharmacokinetic study of MG98, a human DNA mehtyltransferase (MeTase) antisense oligonucleotide, given as a 2-hour twice weekly infusion 3 out of 4 weeks. *11th NCI-EORTC-AACR Symposium on New Drugs in Cancer Therapy*, 2000, Abstract #528.
60. Wright, J. A., Feng, N. P., Jin Hong, N., Wang, M., Lee, Y., and Young, A. (2001) GTI-2501, an outstanding antisense antitumor agent that targets the R1 component of human ribonucleotide reductase. *American Association of Cancer Research*, Abstract #4560.
61. Gewirtz, A. M., Luger, S., Sokol, D., Gowdin, B., Studmauer, E., Reecio, A., and Ratajczak, M. Z. (1996) Oligonucleotide therapeutics for human myelogenous leukemia: Interim results. *Amer. Soc. Hematol.* Abstract #1069.

38

Antisense Oligonucleotides Targeting RIα Subunit of Protein Kinase A

In Vitro and In Vivo Anti-Tumor Activity and Pharmacokinetics

Ruiwen Zhang and Hui Wang

1. Introduction
1.1. PKA as a Cancer Drug Target

cAMP-dependent protein kinase (PKA) is involved in various cellular functions such as cell proliferation, gene induction, and metabolism *(1,2)* and its regulatory subunits have been suggested as a drug target for cancer and other diseases *(3)*. PKA is composed of two catalytic (C) and two regulatory (R) subunits and has type I and type II isozymes, with different R subunits, termed RI and RII, interacting with an identical C subunit *(1)*. Thus far, four isoforms of R subunits, RIα, RIβ, RIIα, and RIIβ, have been identified. Studies have shown that the RI- and RII-regulatory subunits of PKA have opposite roles in cell growth and differentiation, with RI being growth stimulatory and RII being growth-inhibitory *(3)*. Increased expression of the RIα subunit of PKA has been shown during chemical or viral carcinogenesis and correlated with cell proliferation and neoplastic growth *(3)*. The RIα subunit of PKA is overexpressed in a variety of human tumor tissues and cell lines, including lung *(4)*, breast *(5–7)*, ovarian *(8,9)*, and colon *(10–12)*. Furthermore, RIα subunit of PKA overexpression has been shown to correlate with malignancy and worse prognosis in cancer patients *(5–9)*. More recently, studies suggest that extracellular PKA activity may serve as a diagnostic and prognostic marker for cancer *(13)*. In addition, RIα subunit has shown to be associated with multidrug resistance and decreased tumor sensitivity to chemotherapeutic agents *(14–16)*. Therefore, the RIα subunit of PKA is a potential target for human cancer therapy, with several selective type I PKA inhibitors being tested both in preclinical and clinical settings *(2,3,17–25)*. Examples of PKA RI inhibitors include 8-Cl-cAMP *(17,18)* and antisense oligonucleotides *(19–25)*.

1.2. Antisense Therapy

Since its inception over 20 years ago *(26)*, antisense technology has been extensively studied both as a research tool to investigate gene function and as a therapeutic approach to treatment of many diseases including human cancers *(27–32)*. Progress made in molecular biology, genomics, and genetics, especially in identifying, cloning, sequencing, and characterization of pathogenic genes, in the past two decades has provided a basis for genetic therapy. Major efforts in the development of genetic therapy can be summarized in two general approaches, termed gene therapy and antisense therapy. Antisense therapy is to deliver to the target cells antisense molecules that target to mRNA with which they can hybridize and specifically inhibit the expression of pathogenic genes. Although the rationale for antisense therapy seems straightforward, certain conditions must be fulfilled to achieve the goal of developing specific, rational antisense drugs *(33–35)*. Amongst many factors affecting in vivo behavior and effects of antisense oligonucleotides (oligos) are: 1) specificity and efficiency of oligos to enter the target cell that contains the target gene of interest; 2) stability of synthetic oligos in vivo after administration; 3) pharmacokinetics and metabolism of oligos in vivo after administration and the effects of metabolites on both efficacy and side effects of oligos; 4) specificity and efficiency of the interaction of antisense oligos with their mRNA targets; 5) nonsequence-specific interactions of oligos with other macromolecules in vivo ; and 6) sequence-specific non-antisense activity. Therefore, the effects of antisense oligos should be examined both in vitro and in vivo with a proper disease model and under well-controlled conditions. Examples of experiments are: 1) in vitro cell-free system to confirm that antisense oligos selectively hybridize with the specific mRNA targets, resulting in decreased expression of the targeted proteins; 2) in vitro cellular studies to determine cellular pharmacokinetic and metabolism, and in vitro biological activity; 3) animal studies to determine pharmacologic and toxicologic properties of antisense oligos including absorption, distribution, metabolism, and excretion; 4) using an animal disease model to determine in vivo biological activity such as antitumor efficacy; and 5) clinical trials in patients with targeted disease to ultimately determine therapeutic efficacy and safety of antisense oligos.

Phosphorothioate oligos (PS-oligos), representing the first generation antisense oligos, have been extensively studied both in vitro and in vivo, including clinical trials *(27,28)*. In animal pharmacokinetic studies, PS-oligos have a short distribution half-life and a longer elimination half-life in plasma, and are distributed widely into all major tissues following intravenous, intraperitoneal, or subcutaneous administration *(36–39)*. The half-life of distribution ($t_{1/2}\alpha$) is less than 1 h and the half-life of elimination ($t_{1/2}\beta$) is in the range of 40–60 h. Extensive metabolism of antisense oligos has been observed, with PS-oligos being degraded primarily from the 3′-end, but also from the 5′-end or both the 3′-and 5′-ends as well *(27,36–39)*.

In the development of next generation antisense oligos, therefore, major efforts have been devoted to stabilizing PS-oligos by various structural modifications *(40–44)*. Plasma pharmacokinetics of PS-oligo are, in general, not associated with the length or primary sequence of oligo, but associated with the modification of the backbone and specific segments at the 3′- and/or 5′-end. We and others have demonstrated that modified oligos, namely mixed-backbone oligos, have increased in vivo stability and, therefore, increased tissue uptake and decreased total elimination *(27,35,40–44)*.

Antisense approach to selectively down-regulate the expression of the RIα subunit of PKA using unmodified and phosphorothioate oligos has been shown to inhibit growth and induce differentiation of various cancer cell lines, and also has anti-tumor activity in tumor xenografts *(19–22)*. While the identified PS-oligo is selective, specific, and potent in inhibiting tumor growth, safety studies following repeated administration caused side effects in mice, thereby limiting its therapeutic utility. In contrast, a novel MBO, a modification of the selected PS-oligos by substituting four deoxynucleosides at both the 3′-end and 5′-ends by 2′-O-methylribonucleosides, provided significant improvement in safety profile compared to PS-oligos *(22–24,43,44)*. In our previous studies, these classes of MBOs have been shown to be bioavailable following oral administration *(42)*. Following extensive preclinical studies by various routes of administration, the novel MBO targeted to RIα subunit of PKA has also entered clinical Phase I study *(25)* and is presently being evaluated in Phase II trials in patients with solid tumors. In addition, one of the major applications of anti-PKA oligos is to improve therapeutic effectiveness of conventional cancer chemotherapy such as DNA damaging agents and radiation. We and others have demonstrated that the selected MBO enhanced therapeutic effectiveness of several clinically used chemotherapeutic agents including cisplatin *(23)* and Taxol *(24)*. In this chapter, we present protocols for both in vitro and in vivo testing, including an efficacy study with oral administration of the novel MBO *(23,24)*.

2. Materials

All chemicals should be analytical grade or the highest grade available. All aqueous solutions should be prepared in sterile, distilled, deionized water or phosphate-buffered saline (PBS) buffer. All cell-culture equipment, the biological hood, and culture media should be sterile or sterilized prior to use. Gloves should be worn for all operations. Toxic chemicals should be handled in a fume hood and disposed of safely. In addition, radioactive materials and waste should be handled following applicable laws and regulations.

The test MBO (**Oligo AS - GCGUGCCTCCTCACUGGC**) and its mismatch control (**Oligo ASM - GCAUGCATCCGCACAGGC**) were synthesized, purified, and analyzed using the methods previously reported *(23,40,43)* (*see* **Note 1**.) Four deoxynucleosides at both the 3'- and the 5'-ends are 2'-O-methylribonucleosides (represented by bold letters) and remaining are deoxynucleosides, and all internucleotide linkages are phosphorothioate. The underlined nucleosides are the sites of the mismatches compared to Oligo AS. The purity of the MBOs were shown to be greater than 90% by capillary gel electrophoresis and polyacrylamide gel electrophoresis (PAGE), with the remainder being n-1 and n-2 products. The integrity of internucleotide linkages was confirmed by ^{31}P nuclear magnetic resonance (NMR). ^{35}S-labeled Oligo AS used in the pharmacokinetic study was prepared as previously reported *(41,42)*, and its purity was shown to be greater than 98% by PAGE, with the remainder being n-1 and n-2 products.

2.1. Cell Culture, Lipofectin-mediated Oligonucleotide Transfection and Combination Treatment of Cultured Tumor Cells

1. Cell culture dishes (60-mm), 6-well or 24-well plates or flasks (25 mm^2 or 75 mm^2).
2. 1.5-mL microfuge tubes.
3. 15- or 50-mL centrifuge tubes.
4. Cell scrapers.
5. Hemacytometer for cell counting.
6. Penicillin/streptomycin.
7. Lipofectin (Cat. no. 18292-011 or 18292-037; Life Technologies, Gaithersburg, MD).
8. Human lung cancer cell line A549 (American Type Culture Collection, Rockville, MD).
9. Cell-culture medium: Ham's F-12K medium containing 10% fetal bovine serum (FBS) supplemented with 1% penicillin/streptomycin.
10. PBS.
11. 0.25% trypsin/0.1% EDTA.
12. 0.4% trypan blue viability stain.
13. Oligonucleotides (0.1 m*M* in PBS).
14. Chemotherapeutic agents Cisplatin, camptothecin (CPT), and 10-Hydroxycamptothecin (HCPT) *(45,46)*.

2.2. Western Blot Analysis

1. Power supply.
2. Electrophoresis apparatus (Bio-rad Protean II).
3. PAGE and protein transfer apparatus: casting stands, clamps, glass plates, spacers (0.75–1.5 mm), inner cooling core, buffer chamber, chamber lid with attached electrodes, Teflon combs with 1–20 teeth (matched to spacer width), fiber pads (2 per gel), and cooling units.
4. Vacuum pump.
5. Erlenmeyer side-arm flask.
6. Hamilton pipet or pipetter with gel loading tips (narrow diameter).
7. Blotting apparatus.
8. Membrane appropriate for application: nitrocellulose, nylon, or PVDF.
9. Whatman 3MM filter paper (precut or cut to fit gel).
10. Shallow dish for assembling gel sandwich.
11. Plastic wrap.
12. X-ray film and X-ray cassette.
13. Film developer.
14. RIPA cell lysis buffer: 150 mM NaCl, 50 mM Tris-HCl, pH 7.4, 1% sodium deoxycholate, 0.1% SDS, 1% Triton X-100, 1 mM protease inhibitor cocktail PMSF (Sigma Cat. no. p8340, 50 µL/mL; unstable in aqueous solutions; must be added fresh).
15. 4X Tris-HCl/SDS: 1.5 M Tris-HCl, pH 8.8, 0.4% SDS.
16. 4X Tris-HCl/SDS : 0.5 M Tris-HCl, pH 6.8, 0.4% SDS.
17. Acrylamide/bisacrylamide solution: 29.2% acrylamide, 0.8% bisacrylamide, (w/v).
18. 10% SDS.
19. 10% ammonium persulfate in H_2O.
20. 2X SDS sample buffer: 0.1 M/L 4X Tris-HCl/SDS, pH 6.8, 20% (w/v) glycerol, 4% (w/v) SDS, 0.5 M dithiothreitol (DTT), 0.001% (w/v) bromophenol blue.
21. 5X running buffer stock solution: 0.125 M Tris base, 0.96 M glycine, 0.5 (w/v) SDS.
22. 1X running buffer.
23. Protein molecular-weight standards for PAGE.
24. H_2O-saturated isobutyl or isoamyl alcohol.
25. 1X transfer buffer: 25 mM Tris, 192 mM glycine, 20% (v/v) methanol, pH 8.3.
26. 25 mM sodium phosphate.
27. TBS (Tris-buffered saline): 100 mM Tris-HCl, pH 7.5, 150 mM NaCl.
28. TBST (TBS-Tween 20): 100 mM Tris-HCl, pH 7.5, 150 mM NaCl, 0.1% Tween 20.
29. Blocking reagent (5% nonfat milk in TBST).
30. Primary antibodies: Anti-RIα subunit of PKA and anti-β-actin monoclonal antibodies (MAbs).

31. Secondary reagent: HRP-conjugated anti-Ig G (Cat. no. 170-6516; Bio-Rad laboratories).
32. ECL detection reagents (Amersham).

2.3. MTT Assay

1. 0.22-µm filter for reagent preparation.
2. 96-well plates.
3. 96-well microplate reader.
4. RPMI-1640 lacking phenol red.
5. Dimethyl sulfoxide (DMSO).
6. MTT (3-[4,5-dimethylthiazol-2-yl]-2,5-diphenyltetrazolium bromide; Sigma Cat. no. M5655).
7. 4X Ham's F-12K medium.

2.4. In Vivo Animal Models

1. 175-mm^2 cell-culture flasks or 100-mm dishes.
2. Microisolator cages with sterile bedding, food, and water, and cage card with animal identification information.
3. 1-cc syringes with 26-G syringe needles.
4. Sterilized animal feeding needles, 18/50 mm, curved tube style.
5. Calipers for measuring tumor diameters.
6. Balance for body weight measurement.
7. Ear tags and ear puncher for identification of animals.
8. Chlorhexiderm disinfectant.
9. Sterile gauze.
10. Culture medium without FBS or antibiotics.
11. 70% ethanol.
12. 0.9% NaCl (physiological saline, sterile).
13. Drug preparations: Oligos in physiological saline; Cisplatin in physiological saline; CPT and HCPT in cotton seed oil.
14. Solution for sample preparation for pathology analysis: add 735 mL of absolute alcohol (100%) into 315 mL of deionized H$_2$O and then add 117 mL of formalin (37% formaldehyde).
15. Matrigel® basement membrane matrix (Becton Dickinson Labware, Bedford, MA).

2.5. Pharmacokinetic Study

1. Glass metabolic cages.
2. Animal surgical instruments for blood drawing and tissue collection.
3. ^{35}S-labeled oligo.
4. Heparinized 5-cc syringe with a 23-G needle.
5. Liquid scintillation counter.
6. Tissue homogenizer.
7. Balance for tissue weight measurement.

8. Tissue solubilizer.
9. Scintillation counting vials and solvent.
10. Pharmacokinetic modeling program.
11. 0.9% NaCl (physiological saline, sterile)

3. Methods

3.1. Assay for In Vitro Effects of Oligos on Human Lung Cancer A549 Cells

3.1.1. Transfection of Cells with Oligos for Western-Blot Analysis

All procedures should be carried out under sterile conditions.

1. Trypsinize A549 cells when ~80% confluent and resuspend in Ham's F-12K medium containing 10% fetal bovine serum (FBS) with 1% penicillin/streptomycin. Plate 3–5 × 10^5 cells per 60-mm dish.
2. Incubate cells for 18 h at 37°C in 5% CO_2, then remove medium and re-feed with 2.5 mL Ham's F-12K medium containing 1% FBS without antibiotics.
3. Prepare 1 mM and 0.1 mM stock solution of oligos in sterile PBS. Keep in –20°C until use.
4. Mix 22 µL of Lipofectin with 0.25 mL of serum-free Ham's F-12K medium. Keep mixture at room temperature for 45 min prior to use (*see* **Note 2**).
5. To a separate tube, add 6 µL of 0.1 mM oligo stock solution to 0.25 mL serum-free medium (final concentration, 200 nM). For preparation of other concentrations of oligos, the 0.1 mM oligo stock solution can be diluted.
6. Combine lipofectin mixture (from **step 4**) and oligo solutions (from **step 5**) and incubate at room temperature for 10 min.
7. Add 0.5 mL of Lipofectin-oligo mixture to each 60-mm dish. Final oligo concentration is 200 nM. (For preparation of other concentrations of oligos, *see* **step 5**).
8. Incubate cells at 37°C for 20–24 h, then remove medium and wash cells twice with 1-mL ice-cold PBS for each dish.
9. Add 100–200 µL of lysis buffer (RIRA) with PMSF.
10. Scrape cells from the plate and transfer to a microfuge tube. Put the tube on ice for 10 min.
11. Centrifuge the cell suspension at 14,000g, 4°C, for 10 min.
12. Transfer the supernatant to a clean microfuge tube. The cell lysate is ready for protein assay and Western-blot analysis. If necessary, store the lysate at –80°C until analysis.

3.1.2. Transfection of Cells with Oligos in 24-Well Plates and Determination of Viable Cell Number

1. Trypsinize cells when ~80% confluent and resuspend in Ham's F-12K medium containing 10% FBS with antibiotics. Plate 3–5 × 10^4 cells per well for a 24-well plate.
2. Incubate cells at 37°C, 5% CO_2 for 18 h. Remove medium and refeed with 0.2 mL Ham's F-12K medium with 1% FBS lacking antibiotics per well.

3. Mix 2.2 µL Lipofectin with 0.05 mL serum-free medium. Incubate at room temperature for 45 min.
4. To a separate tube, add 0.6 µL of 0.1 mM oligo stock solution to 0.05 mL serum-free medium. (See above for preparation of oligo solution.)
5. Combine lipofectin mixture (from **step 3**) and oligo solutions (from **step 4**) and incubate at room temperature for 10 min.
6. Add 0.1 mL Lipofectin-oligo mixture to each well. Final oligo concentration is 200 nM and final Lipofectin concentration is 7 µg/mL. Treat control wells with Lipofectin mixture only.
7. Incubate at 37°C for 20–24 h. Remove medium and wash cells twice with 0.5 mL PBS.
8. To count viable cells, add 0.5 mL of 0.25% trypsin to each well.
9. Incubate plates at 37°C for 5–10 min or until cells detach.
10. Pipette cells up and down to dissociate.
11. Dilute 100 µL of cell suspension with 100 µL of 0.4% trypan blue viability stain. Mix well, then incubate cells 3 min at room temperature.
12. Examine the cells under a microscope and count viable cells using a hemacytometer. Count cells within 5 min of adding dye to prevent nonspecific staining.
13. Data analysis for growth inhibition study. Calculate the average cell counts for a minimum of 4 fields and multiply the average count by 2×10^4 to obtain actual cell number in each well. Calculate the mean, standard deviation (SD), and standard error (SE) for counts from replicate wells. Calculate the percentage of viable cells after drug treatment relative to the number of cells in control wells treated with Lipofectin only. Plot the cell numbers or percentage of surviving cells versus drug concentration for a dose/response curve and determine IC$_{50}$. Compare results from cells treated with test oligo with that treated with its mismatched control to determine the sequence specificity.

3.1.3. Transfection of Cells with Oligos for MTT Assay

1. Trypsinize cells and count. Adjust cell density to 2×10^4 cells/mL in Ham's F-12K medium with 10% FBS.
2. Plate 0.1 mL cell suspension per 96-well (2000 cells/well). Incubate overnight at 37°C.
3. Prepare Lipofectin and oligo solutions in Ham's F-12K medium, minus serum at 4 times final concentration.
4. Dilute Lipofectin 1:36 in Ham's F-12K medium to prepare 28 µg/mL solution (4×). Incubate 30–45 min at room temperature.
5. Prepare oligo solution in Ham's F-12K medium to obtain 4X solutions.
6. Combine equal volumes of 4X Lipofectin and 4X oligo solutions. Incubate at room temperature for 10 min.
7. Combine equal volumes of Lipofectin/oligo mixture and Ham's F-12K medium 1.6% FBS. Final serum concentration is 0.8%.
8. Add 0.1 mL of Lipofectin/oligo/medium solution to each well.

9. Incubate cells at 37°C, 5% CO_2 for desired transfection time (18–48 h). In studies with combination treatment, chemotherapeutic agents at various concentrations are added after 18–24 h transfection.
10. Incubate cells at 37°C, 5% CO_2 for additional 3–5 d.
11. Perform MTT assay when control cells reach 90% confluence.

3.1.4. MTT Assay

1. Prepare 5 mg/mL stock solution of MTT in PBS. Filter solution using 0.2-µm filter to remove undissolved particles. Store solution at 4°C in the dark for up to 1 mo.
2. Remove cell culture medium from cells in 96-well plate.
3. Replace culture medium with 0.2 mL of Ham's F-12K medium minus serum (prewarmed).
4. Multiply total number of wells by 50 µL to calculate volume of MTT solution required.
5. Dilute 5 mg/mL MTT solution in Ham's F-12K medium immediately before use to prepare a 2 mg/mL working solution.
6. Add 50 µL of 2 mg/mL MTT to each well. Two to four control wells should be incubated with Ham's F-12K medium minus MTT to determine background absorbance values.
7. Incubate plates 2–4 h at 37°C.
8. Observe plates under microscope during incubation to estimate optimum incubation time with MTT reagent. Presence of blue granules inside cells indicates viable cells.
9. Incubation time may vary with different cell lines, since the efficiency of MTT uptake by cell lines may differ. Once the optimum incubation time has been determined, use the same incubation period for all future studies.
10. Carefully remove MTT solution from wells at end of incubation.
11. Add 150 µL DMSO to each well using a multichannel pipette to solubilize dye crystals. Pipet up and down to mix samples.
12. Read plates at 540 nm using a microplate reader. Purple color will remain stable for several hours. Subtract background values obtained for control wells minus MTT reagent from sample OD values. Determine mean, SD, and SE for repeated samples.
13. Plot relative OD values against drug concentrations to illustrate the dose-dependent effect of oligos on in vitro cell growth (*see* **Note 3**).

3.2. Western-Blot Analysis to Determine the Effects of Oligos on Target Gene (RIα Subunit of PKA) Expression

3.2.1. Preparation of Cell Lysates from Cultured Cells or Tumors

1. For cultured cells, remove culture medium and wash cells twice with cold PBS. Place the cell culture dishes on ice.

2. Add 0.5 mL of cold cell lysis (RIPA) buffer per 60-mm dishes.
3. For tumor tissue, homogenize 500 mg in 2.5 mL RIPA buffer containing freshly added PMSF (50 µL/mL). Keep tissue homogenate on ice throughout procedure.
4. Centrifuge cell lysate at 14,000g for 5 min at 4°C.
5. Remove supernatant and transfer to a clean tube.
6. Quantify total protein concentration of supernatant by using a standard protein assay, e.g., modified Lowery assay (Pierce).

3.2.2. Protein Gel Electrophoresis (SDS-PAGE)

1 Fractionate proteins by 10% SDS-PAGE. Use unstained or prestained molecular-weight markers to monitor protein migration in the gel.
2. Adjust the power supply to 200 V using a constant voltage setting. Set the timer for the desired running time, generally 35–45 min.

3.2.3. Transfer Proteins to Membrane

1. Prepare and refrigerate transfer buffer and fill cooling chamber with water and freeze in advance.
2. Remove gel from apparatus and equilibrate gel in transfer buffer for 15–30 min, depending upon gel thickness. This will remove buffer salts from the gel and prevent gel from shrinking during transfer. Always wear gloves when handling membrane to prevent contamination.
3. Cut membrane to fit gel. Mark one corner of membrane with soft pencil or small cut to identify membrane orientation with respect to gel. Wet membrane by slowly sliding it into transfer buffer. Allow it to soak for 15 min.
4. Obtain precut filter paper or cut Whatman 3MM paper to fit membrane. Thoroughly wet filter paper in transfer buffer.
5. Fill the buffer tank about half full with transfer buffer and install cooling unit with ice.
6. Assemble the gel sandwich.
7. Transfer protein to the membrane at 100 V for 1 h or 30 V overnight.

3.2.4. Immunoblotting

1. Block nonspecific sites on membranes in 5% blocking solution for 1–2 h at room temperature.
2. Transfer membranes to a small container for antibody incubation. Minimize volume of antibody solution to conserve valuable reagents, but use enough solution to completely cover membrane without drying during incubation.
3. Dilute primary antibody in TBST. Optimum antibody dilution may have to be determined empirically. For commercial antibodies, consult the product data sheet for suggested antibody concentration. In this study, antibodies are diluted as follows: Anti-RIα: 1 : 1000; and Anti-β-actin: 1 : 5000.
4. Incubate membrane with primary antibody for 1–3 h at room temperature or overnight at 4°C.

5. Wash membranes three times for 15 min using a large volume of TBST (~200–400 mL per wash). For best results, do not reduce incubation time or wash volume to avoid high or uneven background.
6. Dilute HRP-conjugated secondary antibody in TBST (1:3000) or consult product data sheet for recommended dilution.
7. Incubate membrane with secondary antibody for 1 h at room temperature.
8. Wash membranes three times for 15 min using a large volume of TBST (~200–400 mL per wash).
9. Pour off the final wash solution and lightly blot excess liquid from the membrane.
10. For chemiluminescent detection, combine equal volumes of ECL solution 1 and 2.
11. Place membrane on piece of plastic wrap.
12. In the darkroom, add ECL detection reagent mix to membrane with protein side up. Incubate exactly 1 min.
13. Pour off excess detection reagent and wrap membrane with plastic wrap. Avoid wrinkles in plastic wrap, which may cause artifacts on the film.
14. Expose membrane to X-ray film for 1 min, then immediately develop film.
15. Adjust the exposure time to obtain satisfactory band intensities. Complete film exposures within a short interval, as the chemiluminescence reaction has a short half-life and the signal will begin to fade with longer delays.

3.2.5. Data Analysis: Imaging and Densitometry

1. Orient film according to position of lanes on gel and identify sample lanes. Observe band intensities and positions relative to protein standards.
2. Examine the film for artifacts that may arise during washing, detection, or film-exposure steps.
3. Note the presence of any unexpected bands that may derive from nonspecific or cross-reactive binding of primary or secondary antibody. Be aware that the possibility of nonspecific signal increases with the use of polyclonal antisera or unpurified antibody.
4. If desired, quantify protein levels using densitometry, a technique that measures optical density values for selected areas of the film. Quantitative analysis requires equivalent amounts of protein in each sample loaded on the gel.
5. It is important not to overexpose film to obtain accurate quantitative data. The film will become saturated at high band intensities and additional signal will not be resolved.

3.3. In Vivo Effects of Oligos on Human Lung Cancer A549 Xenografts in SCID Mice

3.3.1. Animal Model

1. Culture A549 cells in 175-mm^2 flasks or 100-mm dishes under aforementioned conditions.

2. Harvest tumor cells in exponential growth phase by trypsinization.
3. Combine cells and transfer to 50-mL centrifuge tubes. Centrifuge them at 500g for 5 min.
4. Remove supernatant and wash cells twice with medium lacking serum and antibiotics.
5. Remove a small aliquot of cell suspension before the last centrifugation and count the number of cells using a hemacytometer. Calculate the total cell number by multiplying cell density (cells/mL) by the volume of cell suspension.
6. Centrifuge tubes again for 5 min at 500g to collect cells.
7. Add necessary volume of culture medium lacking serum to the cells. Pipet up and down to obtain a homogeneous cell suspension (1.5×10^8 cells/mL).
8. Mix the cell suspension with Matrigel basement membrane matrix (2:1, v/v; final 1×10^8 cells/mL).
9. Transfer the cell suspension to a 1-cc syringe with a 26-G needle. Inject 0.2 mL of the cell suspension (2×10^7 cells/per animal) subcutaneously into the left inguinal area of the mouse. Use proper precaution and techniques to handle immunodeficient animals. Label each cage card with date and type of cell injected.
10. Observe animals for tumor growth and clinical signs every day. When tumors are easily visible, begin to measure tumor diameter with calipers and record the tumor measurements.
11. The following formulas are used to calculate tumor mass and volume:

$$\text{mass (mg)} = 0.5 \text{ (short diameter)(long diameter)}^2 \times 1000$$

$$\text{volume (mm}^3\text{)} = [0.25 \times \text{(short diameter + long diameter)}]^3 \times 4.18879 \times 1000$$

Note that the calculation [0.25 × (short diameter + long diameter)] corresponds to the average tumor radius (*see* **Note 4**).

3.3.2. In Vivo Chemotherapy

1. When tumors reach ~50–150 mg, initiate the treatment with oligos and chemotherapeutic agents.
2. Animals are randomly divided into treatment groups and controls (6–10 animals/group). Calculate the mean, SD, and SE in mass and volume for each group. Label the cage card with the date, experiment, and treatment group numbers, drug, dose, and frequency. In the study, the following groups are included: untreated control (saline); control oligos (3 doses); test oligos (3 doses); chemotherapeutic agents (at MTD); test oligo plus chemotherapeutic agents; and control oligo plus chemotherapeutic agents. Oligos should be given ip, iv, or po. Cisplatin is given ip. HCPT or CPT should be given orally by gavage. The injection volume should be based on body weight (5 μL/g of body weight) and the concentrations of oligos or drugs should be adjusted on the basis of the doses.
3. Tumors should be measured at least twice a week. For very fast growing tumors, it may be necessary to measure tumors every other day. Clinical observation and body weight monitoring are needed. In general, when tumor reaches 10% of the

body weight or when severe toxicity due to tumor growth or treatment toxicity develops, animals should be sacrificed.
4. Final measurements and tumor size calculations should be performed prior to sacrificing animals and removing tumors (*see* **Note 4**).
5. At the end of the experiment, confirm that final measurements and calculations have been performed for each animal. Pathology examination of tumor and host tissues may be performed.

3.3.3. Data Analysis

1. Therapeutic efficacy is determined by the ratio of %T/C (mean tumor mass of treated group divided by that of control).
2. Use ANOVA to determine the significance of differences among treated and untreated groups.
3. Body weights and survival rates should also be compared among treated and untreated groups.

3.4. Pharmacokinetic Study of Oligos in SCID Mice Bearing Human Lung Cancer A549 Xenografts

3.4.1 Animals for Pharmacokinetic Study

1. Establish xenograft model using procedures described in **Subheading 3.3.1**.
2. Animals with tumor size of ~500–1000 mg are randomly divided into treatment groups for various time points (3–5 animals/group). Calculate the mean, SD, and SE of tumor mass and body weight for each group (*see* **Note 5**). Label the cage card with the date, experiment, and treatment group numbers, and the drug, dose, and time point. In this study, the following time points are included: 5, 15, 30, and 60 min, and 2, 6, 12, 24, and 48 h.
3. Each animal should be identified by an ear tag.

3.4.2. Dosing for Pharmacokinetic Study

1. Measure body weight immediately prior to dosing.
2. Oligos should be given ip, iv (via a tail vein), or po (gavage) at one or more dose levels, e.g., 5, 10, and 20 mg/kg of body weight. The injection volume should be based on body weight (5 µL/g of body weight) and the concentrations of oligos or drugs should be adjusted on the basis of the doses.
3. If an animal is mis-dosed, it should be replaced by an animal with similar tumor size and body weight. The mis-dosed animal should be sacrificed and disposed of.
4. Each dosed animal should be placed in a glass metabolic cage and given commercial diet and water *ad libitum*.

3.4.3. Urine and Fecal Sample Collection

1. Urine samples should be collected using a metabolic cage at various intervals, e.g., 0–6, 6–12, 12–24, and 24–48 h. The urine volume should be measured and recorded.

2. Fecal samples should be collected using a metabolic cage at various intervals, e.g., 0–12, 12–24, and 24–48 h. The weight of feces should be measured and recorded.
3. At the end of each interval, the inner wall of the metabolic cage should be rinsed twice with 50 mL of sterile water. The rinse water should be collected, measured, and placed in properly labeled containers.
4. All samples and rinse water should be kept –20°C until analysis.

3.4.4. Blood and Tissue Sample Collection

1. Anesthetize mice and maintain anesthesia during blood collection.
2. Expose the abdominal aorta and insert needle to collect blood in heparinized 5-cc syringe. Withdraw blood slowly until pulsating of aortic wall ceases or until no more blood flows.
3. Remove tissues and blot lightly on filter paper before weighing and homogenizing. If necessary, tissue samples should be kept at –20°C until analysis.

3.4.5. Determination of Radioactivity in Collected Samples

1. Plasma should be obtained by centrifugation immediately after collection. 100 µL of plasma sample should be used to count radioactivity.
2. Urine samples (10–50 µL) should be used to count radioactivity.
3. Fecal samples should be homogenized in physiological saline (1:10, w/v). The homogenates should be centrifuged and supernatant (100–500 µL) used to count radioactivity.
4. Collected tissues samples should be homogenized in physiological saline (1:5, w/v). The homogenates (100–500 µL) should be mixed with tissue solubilizer to count radioactivity.
5. For any samples with color, add a decolorizing agent such as H_2O_2 prior to counting radioactivity.
6. Radioactivity counting: The above samples should be mixed with 4 mL scintillation solvent and counted in a liquid scintillation spectrometer.

3.4.6. Data Analysis

1. Plasma, urine, and tissue concentrations should be determined on the basis of quantitation of radioactivity and expressed as µg of oligo equivalents per mL or g.
2. Pharmacokinetic parameters should be estimated, including the maximal concentration (C_{max}), the time at C_{max} (T_{max}), half-life ($t_{1/2}$), the area under the curve (AUC), volume of distribution (Vd), and mean residue time (MRT) (*see* **Note 6**).

4. Notes

1. Although the design of antisense oligos is theoretically straightforward (to identify a complementary oligo on the basis of the nucleotide sequence of the

mRNA), the selection of an effective and specific antisense oligo is largely based on experience and trial-based experiments. Since certain oligo sequences such as CpG and GGGG have been shown to have sequence-dependent, nonantisense effects, these sequences should be avoided to demonstrate sequence-specific antisense effects.
2. In this protocol, Lipofectin is used to facilitate the in vitro cellular uptake of oligos and can be replaced by other lipids. Notably, the ratio of lipids/oligo will affect the results of uptake. In addition, the cytotoxicity of Lipofectin without oligo may affect the results of in vitro cell killing study of oligos. Therefore, a pilot study to determine the concentration of lipids is usually needed.
3. In general, decreased level of protein can be evidence for antisense effects. However, this is not sufficient, especially if a dose-dependent relationship is not demonstrated. Proper controls, e.g., random, sense, or mismatched sequence, should be used. Decreased and/or degraded mRNA can be further evidence for antisense effects. In addition, functional evidence is helpful such as changes in signal-transduction pathways and induction of apoptosis *(21,24)*.
4. The in vivo effects of antisense oligo can be complex. Proper control should be used and a dose-dependent relationship should be demonstrated. In vivo decreased protein or mRNA levels of target gene should be demonstrated. In addition, host conditions (strains of animals, sex, and age), tumor size, route of administration, schedule, and duration of treatment may affect the results of study.
5. Because tumor size has an impact on total uptake of oligos in tumor tissues, using animals with similar tumor sizes and randomization will be helpful to decrease the deviation.
6. In pharmacokinetic studies, concentrations of oligos and metabolites can be quantitated by various analytical methods, including high-performance liquid chromatography (HPLC) *(41,42,44)*. Qualitative analysis of oligos can be carried out by PAGE *(36–39)* to illustrate the biostability of oligos.

Acknowledgments

This study is partially supported by a grant from the National Institute of Health, National Cancer Institute to R. Zhang (R01 CA 80698).

References

1. Beebe, S. J. and Corbin, J. D. (1986) Cyclic nucleotide-dependent protein kinases, in *The Enzymes: Control by Phosphorylation*, Part A, vol. 17, (Boyer, P. D. and Kerbes, E. G., eds.), Academic Press, NY, pp. 43–111.
2. Cho-Chung, Y. S., Pepe, S., Clair, T., Budillon, A., and Nesterova, M. (1995) cAMP-dependent protein kinase: role in normal and malignant growth. *Crit. Rev. Oncol. Hematol.* **21,** 33–61.
3. Cho-Chung, Y. S. and Clair, T. (1993) The regulatory subunit of cAMP-dependent protein kinase as a target for chemotherapy of cancer and other cellular dysfunctional related diseases. *Pharmacol. Ther.* **60,** 265–288.

4. Young, M. R. I., Montpetit, M., Lozano, Y., Djordjevic, A., Devata, S., Matthews, J. P., et al. (1995) Regulation of Lewis lung carcinoma invasion and metastasis by protein kinase A. *Int. J. Cancer* **61,** 104–109.
5. Miller, W. R., Watson, D. M. A., Jack, W., Chetty, U., and Elton, R. A. (1993) Tumour cyclic AMP binding proteins: an independent prognostic factor for disease recurrence and survival in breast cancer. *Breast Cancer Res. Treat.* **26,** 89–94.
6. Gordge, P. C., Hulme, M. J., Clegg, R. A., and Miller, W. R. (1996) Elevation of protein kinase A and protein kinase C activities in malignant as compared with normal breat tissue. *Eur. J. Cancer* **32A,** 2120–2126.
7. Miller, W. R., Hulme, M. J., Bartlett, J. M., MacCallum, J., and Dixon, J. M. (1997) Changes in messenger RNA expression of protein kinase A regulatory subunit Iα in breast cancer patients treated with tamoxifen. *Clin. Cancer Res.* **3,** 2399–2404.
8. Simpson, B. J., Ramage, A. D., Hulme, M. J., Burns, D. J., Katsaros, D., Langdon, S. P., and Miller, W. R. (1996) Cyclic adenosine 3′, 5′ -monophosphate-binding proteins in human ovarian cancer: correlations with clinicopathological features. *Clin. Cancer Res.* **2,** 201–206.
9. Mcdaid, H. M., Cairns, M. T., Atkinson, R. J., McAleer, S., Harkin, D. P., Gilmore, P., and Johnston, P. G. (1999) Increased expression of the RIα subunit of the cAMP-dependent protein kinase A is associated with advanced stage ovarian cancer. *Br. J. Cancer* **79,** 933–939.
10. Bradbury, A. W., Miller, W. R., and Carter, D. C. (1991) Cyclic adenosine 3′,5′-monophosphate binding proteins in human colorectal cancer and mucosa. *Br. J. Cancer* **63,** 201–204.
11. Bradbury, A. W., Carter, D. C., Miller, W. R., Cho-Chung, Y. S., and Clair, T. (1994) Protein kinase A (PK-A) regulatory subunit expression in colorectal cancer and related mucosa. *Br. J. Cancer* **69,** 738–742.
12. Bold, R. J., Alpard, S., Ishizuka, J., Townsend, C. M. Jr., and Thompson, R. (1994) Growth-regulatory effect of gastrin on human colon cancer cell lines is determined by protein kinase A isoform content. *Regul. Pept.* **53,** 61–70.
13. Cho, Y. S., Park, Y. G., Lee, Y. N., Kim, M.-K., Bates, S., Tan, L., and Cho-Chung, Y. S. (2000) Extracellular protein kinase A as a cancer biomarker: its expression by tumor cells and reversal by a myristate-lacking Cα and RIIβ subunit overexpression. *Proc. Natl. Acad. Sci. USA* **97,** 835–840.
14. Abraham, I., Chin, K. V., Gottesman, M. M., Mayo, J. K., and Sampson, K. E. (1990) Transfection of a mutant regulatory subunit gene of cAMP-dependent protein kinase causes increased drug sensitivity and decreased expression of P-glycoprotein. *Exp. Cell Res.* **189,** 133–141.
15. Rohlff, C. and Glazer, R. T. (1995) Regulation of multidrug resistance through the cAMP and EGF signalling pathways. *Cell Signal.* **7,** 431–434.
16. Cvijic, M. E. and Chin, K.-V. (1998) Effects of RIα overexpression on cisplatin sensitivity in human ovarian carcinoma cells. *Biochem. Biophys. Res. Comm.* **249,** 723–727.

17. Glazer, R. I. and Rohlff, C. (1994) Transcriptional regulation of multidrug resistance in breast cancer. *Breast Cancer Res. Treat.* **31,** 263–271.
18. Ramage, A. D., Langdon, S. P., Ritchie, A. A., Urns, D. J., and Miller, W. R. (1995) Growth inhibition by 8-chloro-cyclic AMP of human HT29 colorectal and ZR-75-1 breast carcinoma xenografts is associated with selective modulation of protein kinase A isoenzymes. *Eur. J. Cancer* **31A,** 969–973.
19. Nesterova, M. and Cho-Chung, Y. S. (1995) A single-injection protein kinase A-directed antisense treatment to inhibit tumor growth. *Nat. Med.* **1,** 528–533.
20. Cho-Chung, Y. S., Nesterova, M., Kondrashin, A., Noguchi, K., Srivastava, R., and Pepe, S. (1997) Antisense-protein kinase A: a single-gene-based therapeutic approach. *Antisense Nucleic Acid Drug Dev.* **7,** 217–223.
21. Srivastava, R. K., Srivastava, A. R., Park, Y. G., Agrawal, S., and Cho-Chung, Y. S. (1998) Antisense depletion of RIα subunit of protein kinase A induces apoptosis and growth arrest in human breast cancer cells. *Breast Cancer Res. Treat.* **49,** 97–107.
22. Cho-Chung, Y. S., Nesterova, M., Pepe, S., Lee, G. R., Noguchi, K., Srivastava, R. K., et al. (1999) Antisense DNA-targeting protein kinase A-RIα subunit: a novel approach to cancer treatment. *Front. Biosci.* **4,** d859–868.
23. Wang, H., Cai, Q., Zeng, X., Yu, D., Agrawal, S., and Zhang, R. (1999) Anti-tumor activity and pharmacokinetics of a mixed-backbone antisense oligonucleotide targeted to RIα subunit of protein kinase A after oral administration. *Proc. Natl. Acad. Sci. USA* **96,** 13989–13994.
24 Tortora, G., Bianco, R., Damiano, V., Fontanini, G., De Placido, S., Bianco, A. R., and Ciardiello, F. (2000) oral antisense that targets protein kinase A cooperates with taxol and inhibits tumor growth, angiogenesis, and growth factor production. *Clin. Cancer Res.* **6,** 2506–2512.
25. Chen, H. X., Marchall, J. L., Ness, E., Martin, R. R., Dvorchik, B., Rizi, N., et al. (2000) A safety and pharmackinetic study of a mixed-backbone oligonucleotide (GEM231) targeting the type I protein kinase A by two-hour infusion in patients with refractory solid tumors. *Clin. Cancer Res.* **6,** 1259–1266.
26. Zamecnik, P. C. and Stephenson, M. L. (1978) Inhibition of Rous sarcoma virus replication and transformation by a specific oligonucleotide. *Proc. Natl. Acad. Sci. USA* **75,** 280–284.
27. Agrawal, S. (1996) Antisense oligonucleotides: towards clinical trial. *Trends Biotechnol.* **14,** 376–387.
28. Wickstrom, E. (1998) *Clinical Trials of Genetic Therapy with Antisense DNA and DNA Vectors.* Marcel Dekker, New York.
29. Kushner, D. M. and Silverman, R. H. (2000) Antisense cancer therapy: the state of the science. *Curr. Oncol. Rep.* **2,** 23–30.
30. Monia, B. P., Holmlund, J., and Dorr, F. A. (2000) Antisense approaches for the treatment of cancer. *Cancer Invest.* **18,** 635–650.
31. Gewirtz, A. M. (2000) Oligonucleotide therapeutics: a step forward. *J. Clin. Oncol.* **18,** 1809–1811.

32. Zhang, R. and Wang, H. (2000) Antisense oligonucleotides as anti-tumor therapeutics. *Recent Res. Dev. Cancer* **2**, 61–76.
33. Stein, C. A. and Cheng, Y. C. (1993) Antisense oligonucleotides as therapeutic agents: Is the bullet really magical? *Science* **261**, 1004–1012.
34. Diasio, R. B. and Zhang, R. (1997) Pharmacology of therapeutic oligonucleotides. *Antisense Nucleic Acid Drug Dev.* **7**, 239–243.
35. Agrawal, S. (1999) Factors affecting the specificity and mechanism of action of antisense oligonucleotide. *Antisense Nucl. Acids Drug Dev.* **9**, 371–375.
36. Zhang, R., Diasio, R. B., Lu, Z., Liu, T., Jiang, Z., Galbraith, W. M., and Agrawal, S. (1995) Pharmacokinetics and tissue disposition in rats of an oligodeoxynucleotide phosphorothioate (GEM 91) developed as a therapeutic agent for human immunodeficiency virus type-1. *Biochem. Pharmacol.* **49**, 929–939.
37. Zhang, R., Iyer, P., Yu, D., Zhang, X., Lu, Z., Zhao, H., and Agrawal, S. (1996) Pharmacokinetics and tissue disposition of a chimeric oligodeoxynucleotide phosphorothioate in rats following intravenous administration. *J. Pharm. Exp. Ther.* **278**, 971–979.
38. Agrawal, S. and Zhang, R. (1997) Pharmacokinetics of phosphorothioate oligonucleotide and its novel analogs, in *Antisense Oligodeoxynucleotides and Antisense RNA as Novel Pharmacological and Therapeutic Agents* (Weiss, B., ed.), CRC Press, Boca Raton, FL, pp. 58–78.
39. Agrawal, S. and Zhang, R. (1997) Pharmacokinetics of oligonucleotides, in *Oligonucleotides as Therapeutic Agents*. Ciba Foundation Symposium 209, Wiley, Chichester, UK, pp. 60–78.
40. Agrawal, S. and Iyer, R. P. (1995) Modified oligonucleotides as therapeutic and diagnostic agents. *Curr. Opin. Biotechnol.* **6**, 112–119.
41. Zhang, R., Lu, Z., Zhao, H., Zhang, X., Diasio, R. B., Habus, I., et al. (1995) In vivo stability, disposition, and metabolism of a "hybrid" oligonucleotide phosphorothioate in rats. *Biochem. Pharmacol.* **50**, 545–556.
42. Agrawal, S., Zhang, X., Zhao, H., Lu, Z., Yan, J., Cai, H., et al. (1995) Absorption, tissue distribution and *in vivo* stability in rats of a hybrid antisense oligonucleotide following oral administration. *Biochem. Pharm.* **50**, 571–576.
43. Agrawal, S., Jiang, Z., Zhao, Q., Shaw, D., Cai, Q., Roskey, A., et al. (1997) Mixed-backbone oligonucleotides as second generation antisense oligonucleotides: *in vitro* and *in vivo* studies. *Proc. Natl. Acad. Sci. USA* **94**, 2620–2625.
44. Agrawal, S. and Zhang, R. (1998) Pharmacokinetics and bioavailability of oligonucleotides following oral and colorectal administrations in experimental animals, in *Antisense Research and Applications* (Crooke, S., ed.), Springer-Verlag, Heidelberg, pp. 525–543.
45. Cai, Q., Lindsey, J. R., and Zhang, R. (1997) Regression of human colon cancer xenografts in SCID mice following oral administration of water-insoluble camptothecins, natural product topoisomerase I inhibitors. *Int. J. Oncol.* **10**, 953–960.
46. Zhang, R., Li, Y., Cai, Q., Liu, T., Sun, H., and Chambless, B. (1998) Preclinical pharmacology of the natural product anticancer agent 10-hydroxycamptothecin, an inhibitor of topoisomerase I. *Cancer Chemother. Pharmacol.* **41**, 257–267.

39

Use of Antisense Oligonucleotides for Therapy

Manipulation of Apoptosis Inhibitors for Destruction of Lung Cancer Cells

Siân H. Leech, Robert A. Olie, and Uwe Zangemeister-Wittke

1. Introduction

Lung cancer is the second cause of death after cardiovascular diseases and is the major cause of cancer deaths in the Western world. Lung cancer can be divided into two main categories, non-small cell lung cancer (NSCLC) and small cell lung cancer (SCLC). Due to the enormity of the problem of lung cancer throughout the world and the bad prognosis for many subtypes, mainly related to the development of therapy resistance in tumor cells, the need for novel/alternative therapies is immense. Investigations into the development of drug resistance have led to the discovery of many molecular abnormalities in lung cancer cells, including overexpression of the oncogenes ras, myc, c-erbB-2 and bcl-2, and the loss of function of the tumor-suppressor genes Rb, p53, and p16^{INK4A} *(1)*.

Apoptosis is a form of cellular suicide that is usually initiated by signaling events regulated by genes, such as tumor-repressor genes or oncogenes, which are all open to mutation or deletion. These gene alterations contribute to the development of tumors and may reduce the cells sensitivity and susceptibility to apoptosis *(2–4)*. The induction of apoptosis relies on a number of signaling pathways and essential components of the cell death machinery, such as the proteins of the Bcl-2 family, the mitochondria, cytochrome c, and the various subfamilies of death proteases named caspases *(5)*.

It is well-known that the proteins Bcl-2 and Bcl-xl are important negative regulators of apoptotic cell death *(5)* and that they regulate similar cell death pathways *(6)*, although this occurs independently of each other *(7)*. Expression levels of these two proteins within lung cancer cells varies depending on the subtype. For SCLC and several squamous carcinoma cells, it has been shown that Bcl-2 is highly expressed, but this is not the case for adenocarcinomas *(8–11)*. Bcl-xl expression is not as well-documented, but this protein has been shown to be overexpressed in both SCLC and NSCLC cells *(12)*. The expression of the anti-apoptotic protein Survivin has also been reported for lung cancers and was found to correlate with shorter overall survival *(13)*.

Because cell survival or death is determined by the competitive action of death agonists and antagonists such as Bcl-2, Bcl-xl, and Survivin, any approach that alters the balance in favour of cell death may be of therapeutic benefit and restore the sensitivity of tumor cells to chemotherapy, which is known to act via the induction of apoptosis *(14)*. Antisense oligonucleotides are useful both as biological tools and as drugs to inhibit disease-related gene expression. The improvement of these compounds over recent years, e.g., by manipulation of oligonucleotide chemistry, has resulted in compounds with increased efficacy, specificity, and stability.

The most common oligonucleotide modification is the substitution of one of the nonbridging oxygen atoms of the inter-nucleotide backbone linkage with a sulphur atom in order to produce the "first generation" of backbone-modified oligonucleotides, called phosphorothioates *(15)*. However, investigations into the stability of these phosphorothioate oligonucleotides have shown that although they retain their activity and are stable in the presence of nucleases, they also show a variety of nonantisense effects. These effects become apparent at higher oligonucleotide concentrations, and due to their unspecific toxic nature negatively influence the therapeutic index. Nonantisense effects include the inhibition of cell-matrix interaction and decreases in cell proliferation *(16)*, and these unwanted effects have been explained by the ability of phosphorothioate oligonucleotides to bind to heparin binding proteins, including basic fibroblast growth factor (bFGF) and serum albumin. Unspecific interactions with transcription factors and DNA polymerases have also been described *(17–19)*.

Despite some success in the improvement of oligonucleotide stability that was achieved by phosphorothioate modifications, several problems with regard to affinity and specificity remain. In order to overcome these problems investigations into further modifications were performed. Modifications of the ribose of the oligonucleotide have also been investigated, with various changes being made to added side groups. Some of those tested include 2′-fluororibonucleotides, 2′-propoxyribonucleotides (2′-O-propyl), and 2′-O-(2-methoxy)ethylribonucleotides (2′-MOEs) *(20,21,22)*. Although many

Antisense Oligonucleotides for Therapy

"second generation" modifications enhance the hybridization affinity to the target mRNA, 2'-O-propyl and 2'-MOE modifications have also been shown to increase resistance to nucleases *(23)*.

Upon formation of an oligonucleotide/mRNA heteroduplex, the recruitment of RNase H to this complex results in the degradation of the mRNA (preventing transcription) and the subsequent release of the oligonucleotide. Complete modification of the phosphorothioate backbone with 2'-ribose modifications, however, resulted in the inhibition of this main mechanism of action of antisense oligonucleotides, thus decreasing their efficacy *(22,24)*.

In order to rectify the loss of this important mechanism of action, "gapmer" oligonucleotides were developed with modifications only at the 5' and 3' ends of the oligonucleotide. These second-generation antisense compounds are more promising than first-generation oligonucleotides and seem to provide the required increase in stability, affinity, and efficacy compared to first-generation phosphorothioate oligonucleotides *(21,25–27)*.

Our research has focused on the use of two second-generation 2'-MOE antisense oligonucleotides: the first targeting bcl-xl and named oligonucleotide 4259 *(28,29)*, and the second targeting both bcl-xl and bcl-2 simultaneously and named oligonucleotide 4625 *(30,31)*. These compounds have been used in comparison to their corresponding scrambled sequence control oligonucleotides sc4259 and sc4625, respectively. In addition, we designed a first-generation antisense oligonucleotide targeting survivin, named 4003, which was used in comparison to a 3-base mismatch control oligonucleotide named mis4003 *(32)*.

2. Materials

2.1. Antisense Oligonucleotide Transfection of Cells

1. Oligonucleotides (*see* **Table 1**)
2. Liposomal transfection reagent, Lipofectin (Life Technologies, Glasgow, UK).
3. SW2 SCLC cells (provided by S.D. Barnal, Dana Farber Cancer Institute, Boston, MA).
4. A549 NSCLC cells (American Type Culture Collection, ATCC, Manassas, VA)
5. Antibiotics free medium RPMI-1640 (Hyclone, Cramlington, UK) containing 10% fetal calf serum (FCS; Hyclone).

2.2. Reverse Transcription Real-Time Polymerase Chain Reaction (PCK)

1. RNeasy RNA isolation kit (Qiagen AG, Basel, Switzerland).
2. Taqman Reverse Transcription Reagent kit (Applied Biosystems, Foster City, CA).
3. 96-well Optical Plate, strip caps, cap installing roller tool (Applied Biosystems).

Table 1
Target Specificity and Sequences of Antisense Oligonucleotides Used in Our Studies

Name	Target mRNA	Sequence[a]
4259	bcl-xl	AAAGTATCCCAGCCGCCGTT
sc4259	scrambled control	CATATCACGCGCGCACTATG
4625	bcl-2/bcl-xl	AAGGCATCCCAGCCTCCGTT
sc4625	scrambled control	CACGTCACGCGCGCACTATT
4003	survivin	CCCAGCCTTCCAGCTCCTTG
mis4003	3-base mismatch control	CCTAGCCTTCCAGGTCCTAG

[a]The underlined nucleotides are 2'-MOE modified.

4. Taqman Universal PCR Master Mix (Applied Biosystems).
5. Real-time PCR primers and probes (custom made by Applied Biosystems):
   bcl-xl
   - Forward: 5'-GGTCGCATTGTGGCCTTT-3'
   - Reverse: 5'- TCCTTGTCTACGCTTTCCACG-3'
   - Probe: 5'- ACAGTGCCCCGCCGAAGGAGA-3'
   bcl-2
   - Forward: 5'-CATGTGTGTGGAGAGCGTCAA-3'
   - Reverse: 5'- GCCGGTTCAGGTACTCAGTCA-3'
   - Probe: 5'- CCTGGTGGACAACATCGCCCTGT-3'
   survivin
   - Forward: 5'-ATGGGTGCCCCGACGT-3'
   - Reverse: 5'-AATGTAGAGATGCGGTGGTCCTT-3'
   - Probe: 5'-CCCCTGCCTGGCAGCCCTTTC-3'
6. Taqman Ribosomal RNA Control Reagents kit (Applied Biosystems).

2.3. Western-Blot Analysis

1. Phosphate-buffered saline (PBS).
2. Western-blot lysis buffer: 150 mM NaCl, 5 mM EDTA, 10 mM Tris-EDTA, 1% Triton X-100, Aprotinin, Leupeptin, Pepstatin A, and Chymostatin (all at 1 μg/mL), 1 mM PMSF. All protease inhibitors were from Sigma (St. Louis, MO).
2. BCA protein assay reagent (Pierce, Rockford, IL).
3. Gel buffers, Upper Tris (0.5 M Tris-HCl, 0.4% sodium dodecyl sulfate [SDS], pH 6.8), Lower Tris (1.5 M Tris-HCl, 0.4% SDS, pH 8.8).
4. Acrylamide mix: 30% acrylamide with 0.8% bisacrylamide (37.5:1) (Roth AG, Reinach, Switzerland).
5. 10% Ammonium persulphate (APS) (Life Technologies).
6. TEMED (N,N,N',N',Tetramethylethylenediamine) (Fluka Chemie AG, Buchs, Switzerland).
7. Western sample loading buffer (6X Laemmli buffer): 5 mL glycerine 87%, 2.5 mL β-mercaptoethanol, 15 mL SDS 10%, 6.25 mL Upper Tris buffer, 21 ml water.

8. Sodium dodecyl sulfate acrylamide gel electrophoresis (SDS-PAGE) running buffer: Combine 990 mL 1X Tris/glycine buffer with 10 mL 10% SDS (w/v).
9. 10X Tris Glycine buffer: 0.25 M Tris base, 1.92 M Glycine.
10. Western running buffer: 990 mL 1X Tris-glycine buffer, 10 mL 10% SDS.
11. Polyvinylidene fluoride (PVF) membrane (Immobilion P, Millipore, Bedford, CT).
12. Methanol.
13. Semi-dry blotting chamber (Schleicher & Schuell, Dassel, Germany).
14. Western blotting buffer: 850 mL 1X Tris/Gycine buffer with 150 mL of methanol.
15. Tris-buffered Saline with Tween 20 (TTBS): 20 mM Tris-HCl, pH 7.4, 500 mM NaCl, 0.1% Tween-20.
16. Blocking buffer: 5% nonfat milk powder in Tris buffered saline with 0.1% Tween 20 (TTBS).
17. Rabbit anti-human Bcl-x polyclonal antibody (PAb) (BD Transduction Laboratories, Franklin Lakes, NJ) diluted 1/1000.
18. Mouse anti-human Bcl-2 monoclonal antibody (MAb) (Dako, Glostrup, Denmark) diluted 1/1000.
19. Mouse anti-β-actin MAb (ICN Biomedicals, Inc., Aurora, OH) diluted 1/16000.
20. Goat anti-rabbit or rabbit anti-mouse immunoglobulin peroxidase conjugates (Sigma, St. Louis, MO) as secondary antibodies diluted 1/7000 or 1/10000, respectively.
21. Chemiluminescence ECL™ kit (Amersham Pharmacia Biotech UK Limited, Enhanced Little Chalfont, Buckinghamshire, UK).

2.4. Determination of Cell Viability: MTT Proliferation Assay

1. MTT reagent (3-[4,5-dimethylthiazol-2-yl]-2,5-diphenyltetrazolium bromide) (Sigma): 10 mg mL^{-1} in PBS.
2. MTT solubilisation buffer: 20% SDS in 50% dimethyl formamide, 50% water, pH 4.7 (pH adjusting solution: 80% acetate, 20% 1 M HCl).

2.5. Detection of Caspase-3-like Protease Activity

1. Cell extraction buffer (CEB): 50 mM PIPES, pH 7.0, 50 mM KCl, 5 mM EGTA, 2 mM MgCl$_2$, 2 mM PMSF, 1 mM dithiothreitol (DTT), protease inhibitors Aprotinin, Leupeptin, Pepstatin A, and Chymostatin all at 1 µg/mL.
2. Coomassie® Protein Assay Reagent (Pierce, Rockford, IL).
3. Caspase inhibitor DEVD-CHO 10 nM (Bachem, Saffron Walden, Essex, UK).
4. Caspase substrate DEVD-pNa 80 nM (Bachem).

2.6. Nuclear Morphology Analysis

1. DNA intercalating agent Hoechst 33342 (Sigma).
2. 4% paraformaldehyde (PFA).
3. Microscope glass slides.
4. Cytospin centrifuge (Shandon, Astmoor, Cheshire, UK).
5. Sample mounting agent Mowiol (Calbiochem, La Jolla, CA).

3. Methods

3.1. Antisense Oligonucleotide Transfection of Cells

To successfully transfect cells with oligonucleotides, it is necessary to use a carrier, which enables delivery of oligonucleotides into the cells. For the experiments reported here we have used the liposomal transfection reagent Lipofectin, which is a 1:1 (w/w) formulation of the two lipids N-[1-(2,3-dioleyloxy)propyl]-N,N,N-trimethylammonium chloride (DOTMA) and dioleoyl phosphotidylethanolamine (DOPE). The protocol described below is routinely used for the transfection of cells cultured in one well of a six-well plate with 600 nM oligonucleotide.

1. Plate 1.5×10^5 cells in 1 mL 10% serum containing, antibiotics-free medium 24 h prior to transfection, in order to allow adherence of anchorage-dependent cells and to limit stress-related effects following handling of cells.
2. For the transfection, and to enable liposome formation, dilute 18.6 µL Lipofectin to a concentration of 100 µg/mL^{-1} in 167.4 µL warmed (37°C) serum-free and antibiotics-free medium and incubate for 30 min at room temperature (vial A).
3. Dilute 2.4 µL of a 500 µM oligonucleotide stock in 183.6 µL serum and antibiotics-free medium to a concentration of 6.5 µM (vial B).
4. Combine and mix the contents of vials A and B by pipetting.
5. Following incubation at room temperature for 10 min, add 628 µL serum and antibiotics-free medium in order to reach an oligonucleotide concentration of 1200 nM. Addition of this mixture to the plated cells gives a final oligonucleotide concentration of 600 nM.
6. Incubate the cells at 37°C for 20 h. Then either collect the media, reclaim any floating cells by centrifugation, and add 2 mL fresh serum and antibiotics containing medium to the culture, or add 2 mL fresh serum and antibiotics containing medium to the culture in order to dilute out the Lipofectin.
7. Incubate cells as required, depending on the assay, prior to analysis or harvesting of cells. In order to validate this protocol, transfections were performed using FITC-labeled oligonucleotides to visualise oligonucleotides within the treated cells (see **Fig. 1**).

3.2. Reverse Transcription Real-Time Polymerase Chain Reaction

Real-time PCR is a quantitative method used to determine the amount of cDNA present in a sample. The PCR reaction is based on the amplification of the initial template cDNA present at the start of the reaction. Once the amplification reaction reaches its exponential phase the resulting amount of replicated template should be proportional to the number of template copies present at the start of the reaction. In order to translate this to laboratory protocols, Applied Biosystems have developed a technique of measuring

Antisense Oligonucleotides for Therapy

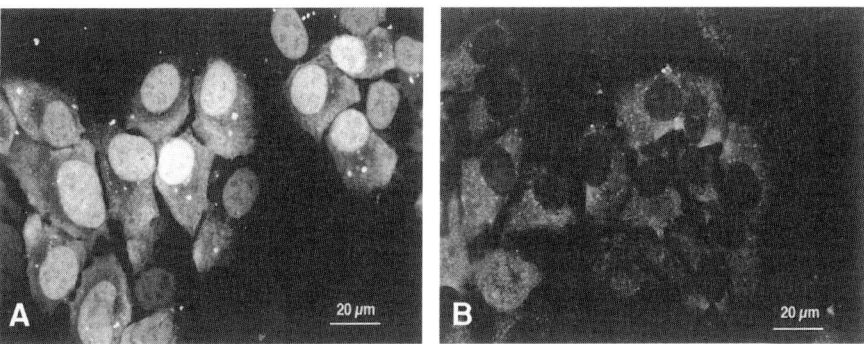

Fig. 1. A549 NSCLC cells following transfection with FITC-labeled oligonucleotide. Transfection for 4 h with (**A**), or without (**B**) the liposomal transfection reagent Lipofectin.

the number of replicated templates within the real time of the reaction, thus allowing for the determination of the exponential phase and therefore the quantification of cDNA levels within samples. The way in which the samples are monitored is based on the two primer PCR method with the addition of a third component, the probe. The probe is designed to bind to a region of the target DNA that lies between the forward and reverse primers and is labeled covalently at the 5'end with a fluorescent reporter dye and at the 3'end with a quencher dye. As the reporter and quencher dyes are in close proximity within the probe there is suppression of the fluorescence and no signal is detected. Once the PCR reaction is initiated, the probe gets displaced at the 5'end by the extending forward primer, and cleaved by the 5'-3' nuclease activity of the AmpliTaq Gold enzyme within the reaction mix. This cleavage results in the release of the reporter dye from the probe and the quencher dye, resulting in an increase in fluorescence signal. Therefore an increase in signal is representative of an increase in the amount of product (*see* **Fig. 2**).

1. Harvest cells following transfection. For suspension cells (e.g., SW2 SCLC cells), remove media from the culture plate or flask and centrifuge for 5 min at 170*g*. Trypsinize adherent cells (e.g., A549 NSCLC cells) for 3–5 min at 37°C prior to harvesting and centrifugation, as with suspension cells. Wash cells in PBS and snap-freeze in liquid nitrogen prior to immediate storage at –80°C.
2. Isolate total RNA using the RNeasy mini kit and measure the RNA concentration and purity spectrophotometrically at 260 and 280 nm. Store samples at –80°C until required.
3. Prepare complementary DNA (cDNA) by reverse transcription using 0.5 µg of isolated RNA and the Taqman Reverse Transcription Reagent kit.

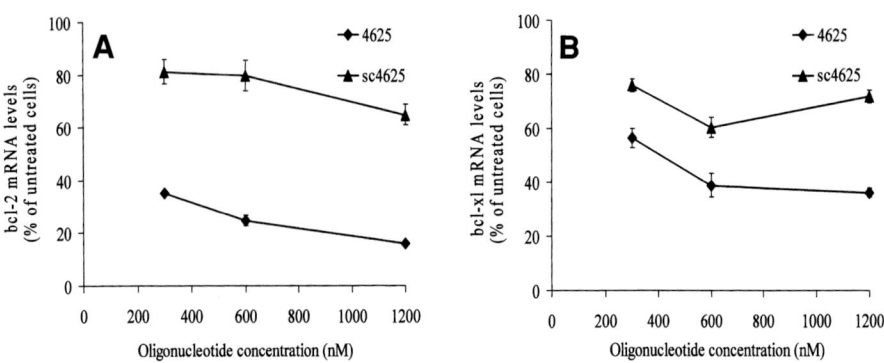

Fig. 2. Results of quantitative real-time PCR analysis of bcl-2 mRNA (**A**) and bcl-xl mRNA expression (**B**) in SW2 SCLC cells following a 6-h transfection with oligonucleotides 4625 or control oligonucleotide sc4625. Cells were treated at 1200, 600, or 300 nM oligonucleotide. Analysis was performed using the ABI Prism 7700 Sequence Detection System.

For our experiments the two targets of interest were bcl-2 and bcl-xl and the concentration of these targets was analysed alongside the concentration of ribosomal RNA (rRNA), which was used as an internal standard. The method used for analysis of our samples was the comparative C_T (threshold cycle) method (described in the ABI Prism 7700 Sequence Detection System manual, Applied Biosystems). The protocol used for real time PCR amplification was as described in the Taqman Universal Master Mix protocol (Applied Biosystems).

4. Per well of a 96-well Optical Plate, prepare a reaction volume of 25 µL, containing 12.5 µL Universal PCR Master Mix and 7.5 µL primers and probe in DEPC-treated water. Add this mixture to each well (*see* **Note 1**).
5. Prepare dilutions of sample cDNA and add 5 µL to the required wells. Seal wells with the strip caps using the cap installing roller tool. Remove air bubbles by centrifugation or by holding plates high and then lowering them quickly to the floor.
6. Place the plate into the heating block of the real time PCR machine and close and tighten the lid. Indicate target and internal standard positions on the computer software and file the run details (we routinely use 40–45 cycles). Run times are approx 2 h in length, after which files should be saved and runs analyzed. Mean C_T values can be calculated for the targets and for the rRNA standards, allowing subsequent quantification of target downregulation.

3.3. Western-Blot Analysis

Western-blot analysis is performed on transfected cell lysates in order to determine the effect of antisense oligonucleotide treatment on the expression

Antisense Oligonucleotides for Therapy

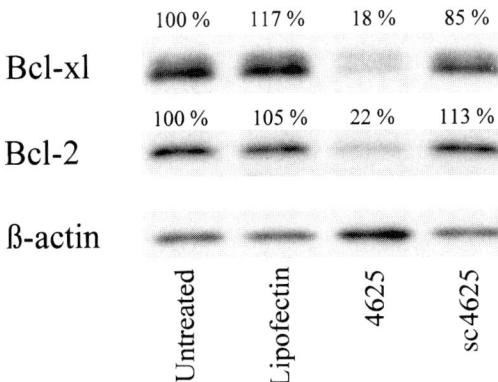

Fig. 3. Results of Western-blot analysis of Bcl-2/Bcl-xl protein expression in SW2 SCLC cells following a 20-h transfection with 600 nM bispecific oligonucleotide 4625 or scrambled control sc4625. Analysis was performed 40 h after the start of transfection. Blots were reprobed for β-actin to confirm equal protein loading. Values represent the percentage of protein expression quantified from Western blot and corrected for β-actin levels (relative to untreated cells which were set to 100%).

levels of target proteins. Quantification of Western-blot bands can demonstrate the successful and specific downregulation of the proteins (*see* **Fig. 3**).

1. Harvest cells following treatment with antisense oligonucleotides at 20, 40, or 64 h following the start of transfection. For suspension cells (e.g., SW2 SCLC cells), remove media from the culture plate or flask and centrifuge for 5 min at 170g. Trypsinize adherent cells (e.g. A549 NSCLC cells) for 3–5 min at 37°C prior to harvesting and centrifugation, as with suspension cells.
2. Wash the cells in PBS, resuspend pellet in 40–100 μL Western-blot lysis buffer, and incubate on ice for 30 min.
3. Centrifuge samples for 20 min at 17,500g, 4°C and remove supernatants prior to storage at –80°C until required for analysis.
4. Determine protein concentrations of samples using the BCA assay reagent. Prepare this reagent by mixing reagent A with reagent B (50:1) and adding to a 96-well plate (200 μL per well) together with 10 μL of sample (diluted 10X in lysis buffer).
5. Prepare a standard concentration curve for each assay using protein standards provided with the assay kit.
6. Incubate plates at 37°C for 30 minutes prior to spectrophotometrical analysis at 560 nm.
7. Prepare a 12% polyacrylamide gel for separation of the sample proteins. The separating gel consists of water (8.2 mL), acrylamide mix (10 mL), Lower Tris (6.5 mL), 10% APS (200 μL), TEMED (15 μL). The stacking gel consists

of water (6.5 mL), Acrylamide mix (1 mL), Upper Tris (2.5 mL), 10% APS (40 μL), TEMED (20 μL).
8. Assemble the gels in the electrophoresis tanks together with running buffer. Dilute 25 μg protein of each sample in lysis buffer so that the volumes to be loaded on the gel are equal for all samples. Mix each sample with 6x Laemmli buffer containing β-mercaptoethanol.
9. Heat the samples at 96°C for 5 min in order to denature the proteins and load the samples onto the gel. Run the gels.
10. In preparation for blotting, cut a PVF membrane to size and pre-wet in methanol, wash in ultra pure water for 2 min and then transfer to blotting buffer. Before running the gels, equilibrate them in blotting buffer for 15 min prior to transfer of the proteins to the PVF membrane by blotting.
11. Assemble the gel as a sandwich between membrane and blotting paper soaked in blotting buffer. Transfer by semi-dry blotting with 1 mA cm^{-2} for 1 h.
12. Block membranes in blocking buffer for 1 h at room temperature.
13. Probe with primary antibody, in 1% milk TTBS overnight at 4°C, or at room temperature for 2.5 h. Wash 3 × 5 min in TTBS before adding the secondary antibody in 1% milk TTBS. Incubate for 1 h at room temperature.
14. Wash 3 × 5 min in TTBS and visualize the immunocomplex by enhanced chemiluminescence using an ECL™ kit. Place the membrane protein side up on a clean glass plate and drip the prepared ECL™ reagent over it until it is completely covered.
15. Remove the membranes from the glass plate after 1 min, allow to drip-dry and sandwich between two sheets of plastic overhead film. Trim the film around the membrane and push excess liquid out from between the sheets of film.
16. Expose the membrane to X-ray film and develop.
17. Protein levels can be determined by density analysis of scanned images of the exposed films. Re-probing of the film for β-actin levels in the samples allows quantification of the decrease in expression of the target protein relative to a constant internal standard.

3.4. Determination of Cell Viability: MTT Proliferation Assay

To assess the dose dependence of treatment with oligonucleotides on cell viability, the colorimetric MTT (3-[4,5-dimethylthiazol-2-yl]-2,5-diphenyl-tetrazolium bromide) assay can be used. The assay is based on the principle that, in viable cells, the tetrazolium salt MTT is reduced by mitochondrial dehydrogenases, resulting in the formation of a water-insoluble formazan salt, which is dark blue in color *(33)*. This salt is solubilized in solubilization reagent and its extinction is read spectrophotometrically at 570 nm (*see* **Fig. 4**).

1. Plate 3000 cells per well in 50 μL serum containing and antibiotics free medium in a 96-well culture plate 24 h prior to treatment. Culture under normal cell-culture conditions (*see* **Note 2**).

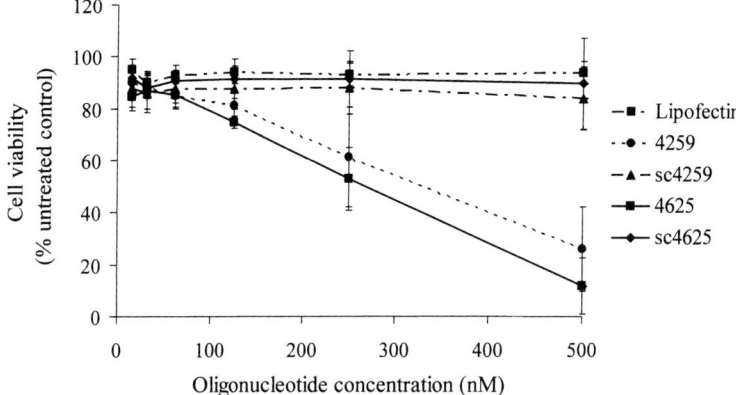

Fig. 4. Results of MTT proliferation assays of A549 NSCLC cells following treatment with oligonucleotides 4625, 4259, or control oligonucleotides sc4625 and sc4259. Analysis was performed 72 h after the start of a 20-h transfection. Treatment with oligonucleotides 4259 and 4625 decreased proliferation, while the two control oligonucleotides sc4625 and sc4259 showed very little effects.

2. Transfect the cells as described (**Subheading 3.1.**) by the addition of 50 μL transfection mixture containing oligonucleotides. An oligonucleotide concentration range can be used going from 37.5–600 nM. Include untreated wells and wells without cells, which later on serve as positive and negative controls for MTT conversion.
3. Incubate for 20 h before removing medium from adherent cells and replacing with fresh medium containing serum and antibiotics. Suspension cells may be left, or 50 μL of medium containing serum and antibiotics can be added in order to dilute the transfection reagents (*see* **Note 3**).
3. Return cells to the incubator until addition of the MTT reagent at about 72 h after the start of transfection.
4. Add 10 μL of MTT reagent to each well and incubate at 37°C for 1.5 h. Following this incubation add 100 μL solubilization reagent and incubate overnight to allow for cell lysis and salt solubilization. Measure substrate cleavage spectrophotometrically at 570 nm.
5. Calculate the reduction in viability in the treated wells relative to the viability in the untreated wells upon subtracting the background staining as measured for the wells without cells.

3.5. Detection of Caspase-3-Like Protease Activity

Caspases are proteases of the apoptosis machinery. They are engaged in the initiation and execution of cell death and activated in a coordinated fashion to form a cascade of proteolytic degradation events. Thus, activation of caspases

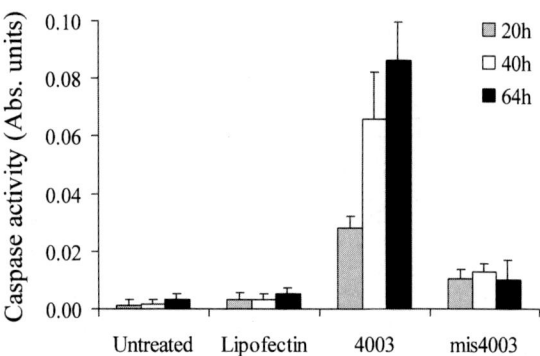

Fig. 5. Results of a caspase 3-like protease assay with A549 NSCLC cells following 20 h of treatment with oligonucleotide 4003, or control oligonucleotide mis4003. Analysis was performed 20, 40, and 64 h after the start of transfection.

in cells is an indication that death is apoptotic in nature. Caspases 3, 6, and 7 are effector caspases, those death proteases that are responsible for substrate cleavage at the end of the caspase cascade. The colorimetric substrates used for this assay detect the activity of caspases sharing homology with caspase 3, hence the name "caspase-3-like" (see **Fig. 5**).

1. Plate, transfect and incubate cells as described (see **Subheading 3.1.**). Use at least two wells of a 6-well plate per condition. As positive control for caspase activity, dying tumor cells treated with etoposide can be used (e.g., SW2 cells treated 30 µM etoposide).
2. Collect cells at 20, 40, or 64 h after start of the transfection and rinse with PBS.
3. Lyse cells in CEB and subject samples to two freeze (−80°C)/thaw cycles.
4. Centrifuge samples at 17,500g for 15 min at 4°C, remove supernatants and store at −80°C.
5. Determine protein concentration of samples by adding 2 µL of sample, diluted five times in CEB, to 100 µL of Coomassie® Protein Assay Reagent.
6. Measure samples spectrometrically immediately at 595 nm and compare with protein standards prepared in CEB.
7. For the assay itself use a total volume of 100 µL per well of a 96-well culture plate and use two wells for each sample.
8. Take a volume containing 40 µg of protein of each sample and make this up to 100 µL using CEB in each appropriate well. As a negative control add 1 µL (10 µM) of the caspase inhibitor DEVD-CHO to one of the two wells per sample.
9. Add sample extracts to the appropriate wells, mix, and incubate the plates at 37°C for 30 min.
10. Add 10 µL of 0.8 mM DEVD-pNa caspase substrate to each well to give a final concentration of 80 µM, mix and incubate at 37°C.

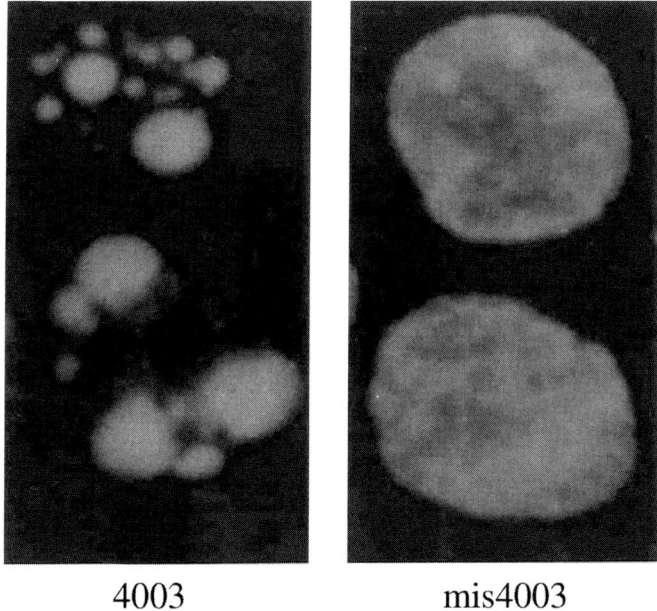

4003   mis4003

Fig. 6. Hoechst stained nuclei of A549 NSCLC cells following a 20-h treatment with oligonucleotide 4003 or control oligonucleotide 4003mis. Analysis was performed 40 h after the start of transfection. Condensed and fragmented nuclei are clearly visible following treatment with oligonucleotide 4003. Original magnification ×1000.

11. Read plate at 405 nm using a spectrophotometer for example 2, 4, and 6 h after addition of substrate.
12. Interpret the results by subtracting the negative control absorbance signal from sample absorbance value.

3.6. Nuclear Morphology Analysis

One of the hallmarks of apoptotic cell death is the appearance of condensed and fragmented nuclei due to a breakdown of the cells' nuclear lamins and degradation of DNA. This step of nuclear apoptosis can be best detected by use of a colored dye that intercalates into the DNA, such as Hoechst 33342 (*see* **Fig. 6**).

1. Plate, transfect and incubate cells as required and harvest cells as described.
2. Rinse cells with PBS and pellet by centrifugation at 170g for 5 min.
3. Fix and stain cells by resuspending in a solution of 4% PFA containing 2 µg mL^{-1} Hoechst 33342 for 15 min at room temperature.
4. Prepare slides by cytospin centrifugation at 10g for 3 min and allow cells to dry before applying cover slips with Mowiol.

4. Notes

1. For real-time PCR, the optimal concentrations of primers and probe have to be established using combinations of various concentrations of these compounds. For cDNA, we routinely use a 10–100 × dilution of the original preparation, but also here the optimal dilution has to be established depending on the level of target expression.
2. For all transfections, especially those for subsequent MTT assays, the optimal cell number to be plated has to be established. At the time of MTT addition the wells containing the untreated cells should not have grown to such a density that cells are dying because of a lack of space or nutrients.
3. The duration of transfection has to be adjusted according to the sensitivity of the cells to either Lipofectin alone and to the combination of Lipofectin and control (scrambled or mismatch) oligonucleotide. The duration of transfection and the time point of subsequent analysis has to be chosen in such a way that the window between antisense and controls is maximal.

References

1. Salgia, R. and Skarin, A. T. (1998) Molecular abnormalities in lung cancer. *J. Clin. Oncol.* **16,** 1207–1217.
2. Mashima, T., Seimiya, H., Chen, Z., Kataoka, S., and Tsuruo, T. (1998) Apoptosis resistance in tumour cells. *Cytotechnology* **27,** 293–308.
3. Reed, J. C. (1995) Bcl-2 familiy proteins: regulators of chemoresistance in cancer. *Toxicol. Lett.* **82/83,** 155–158.
4. Toft, N. J. and Arends, M. J. (1997) Apoptosis and necrosis in tumours, in *Apoptosis and Cancer* (Martin, S. J., ed.), Karger Landes Systems, Basel, Switzerland, pp. 25–45.
5. Antonsson, B. and Martinou, J. C. (2000) The Bcl-2 protein family. *Exp. Cell Res.* **256,** 50–57.
6. Chao, D. T., Linette, G. P., Boise, L. H., White, L. S., Thompson, C. B., and Korsmeyer, S. J. (1995) Bcl-xL and Bcl-2 repress a common pathway of cell death. *J. Exp. Med.* **182,** 821–828.
7. Simonian, P. L., Grillot, D. A. M., and Nunez, G. (1997) Bcl-2 and Bcl-xL can differentially block chemotherapy-induced cell death. *Blood* **90,** 1208–1216.
8. Dosaka-Akita, H., Katabami, M., Hommura, H., Fujioka, Y., Katoh, H., and Kawakami, Y. (1999) Bcl-2 expression in non-small-cell lung cancers: higher frequency of expression in squamous-cell carcinomas with earlier pT status. *Oncology* **56,** 259–264.
9. Higashiyama, M., Doi, O., Kodama, K., Yokouchi, H., and Tateishi, R. (1996) Bcl-2 oncoprotein expression is increased especially in the portion of small cell carcinoma within the combined type of small cell lung cancer. *Tumor Biol.* **17,** 341–344.
10. Pezzella, F., Turley, H., Kuzu, I., Tungekar, M. F., Dunnill, M. S., Pierce, C. B., et al. (1993) Bcl-2 protein in non-small-cell lung carcinoma. he *N. Engl. J. Med.* **329,** 690–694.

11. Reeve, J. G., Xiong, J., Morgan, J., and Bleehen, N. M. (1996) Expression of apoptosis-regulatory genes in lung tumour cell lines: relationship to p53 expression and relevance to acquired drug resistance. *Br. J. Cancer* **73**, 1193–1200.
12. Bojes, H. K., Suresh, P. K., Mills, E. M., Spitz, D. R., Sim, J. E., and Kehrer, J. P. (1998) Bcl-2 and Bcl-xL in peroxide-resistant A549 and U87MG Cells. *Toxicol. Sci.* **42**, 109–116.
13. Monzo, M., Rosell, R., Felip, E., Astudillo, J., Sanchez, J. J., Maestre, J., et al. (1999) A novel anti-apoptosis gene: re-expression of survivin messenger RNA as a prognostic marker in non-small cell lung cancers. *J. Clin. Oncol.* **17**, 2100–2104.
14. Mesner, P., Budihardjo, I., and Kaufmann, S. H. (1997) Chemotherapy-induced apoptosis. *Adv. Pharmacol.* **41**, 461–499.
15. Sharma, H. W. and Narayanan, R. (1995) The therapeutic potential of antisense oligonucleotides. *BioEssays* **17**, 1055–1063.
16. Galderisi, U., Di Bernardo, G., Melone, M. A. B., Galano, G., Cascino, A., Giordano, A., and Cipollaro, M. (1999) Antisense inhibitory effect: a comparison between 3'-partial and full phosphorothioate antisense oligonucleotides. *J. Cell. Biochem.* **74**, 31–37.
17. Crooke, S. T. (1998) Antisense therapeutics. *Biotechnol. Genet. Eng. Rev.* **15**, 121–157.
18. Crooke, S. T., Lemonidis, K. L., Neilson, L., Griffey, R., Lesnik, E. A., and Monia, B. P. (1995) Kinetic characteristics of *Escherichia coli* RNase H1: cleavage of various antisense oligonucleotide-RNA duplexes. *Biochem. J.* **312**, 599–608.
19. Guvakova, M. A., Yakubov, L. A., Vlodavsky, I., Tonkinson, J. L., and Stein, C. A. (1995) Phosphorothioate oligodeoxynucleotides bind to basic fibroblast growth factor, inhibit its binding to cell surface receptors and remove it from low affinity binding sites on extracellular matrix. *J. Biol. Chem.* **270**, 2620–2627.
20. Cooper, S. R., Taylor, J. K., Miraglia, L. J., and Dean, N. M. (1999) Pharmacology of antisense oligonucleotide inhibitors of protein expression. *Pharmacol. Therapeut.* **82**, 427–435.
21. Freier, S. M. and Altmann, K. H. (1997) The ups and downs of nucleic acid duplex stability: structure-stability studies on chemically-modified DNA:RNA duplexes. *Nucleic Acids Res.* **25**, 4429–4443.
22. Kawasaki, A. M., Casper, M. D., Freier, S. M., Lesnik, E. A., Zounes, M. C., Cummins, L. L., et al. (1993) Uniformly modified 2'-deoxy-2'-fluoro phosphorothioate oligonucleotides as nuclease resistant anti-sense compounds with high affinity and specificity for RNA targets. *J. Med. Chem.* **36**, 831–841.
23. Cooper, S. R., Taylor, J. K., Miraglia, L. J., and Dean, N. M. (1999) Pharmacology of antisense oligonucleotide inhibitors of protein expression. *Pharmacol. Therapeut.* **82**, 427–435.
24. Hanecak, R., Brown-Driver, V., Fox, M. C., Azad, R. F., Furusako, S., Nozaki, C., et al. (1996) Antisense oligonucleotide inhibition of hepatitis C virus gene expression in transformed hepatocytes. *J. Virol.* **70**, 5203–5212.
25. Crooke, S. T. (1998) Molecular mechanisms of antisense drugs: RNase H. *Antisense Nucleic Acid Drug Dev.* **8**, 133–134.

26 Monia, B. P., Lesnik, E. A., Gonzalez, C., Lima, W. F., McGee, D., Guinosso, C. J., et al. (1993) Evaluation of 2′-modified oligonucleotides containing 2′-deoxy gaps as antisense inhibitors of gene expression. *J. Biol. Chem.* **268,** 14514–14522.
27. Quartin, R. S., Brakel, C. L., and Wetmur, J. G. (1989) Number and distribution of methylphosphonate linkages in oligodeoxynucleotides affect exo- and enonuclease sensitivity and ability to form RNase H substrates. *Nucleic Acids Res.* **17,** 7253–7262.
28. Leech, S. H., Olie, R. A., Gautschi, O., Simões-Wüst, A. P., Tschopp, S., Häner, R., et al. (2000) Induction of apoptosis in lung-cancer cells following bcl-xL anti-sense treatment. *Int. J. Cancer* **86,** 570–576.
29. Simões-Wüst, A. P., Olie, R. A., Gautschi, O., Leech, S. H., Häner, R., Hall, J., et al. (2000) Bcl-xL antisense treatment induces apoptosis in breast carcinoma cells. *Int. J. Cancer* **87,** 582–590.
30. Gautschi, O., Tschopp, S., Leech, S. H., Olie, R. A., Simões-Wüst, A. P., Baumann, B., et al. (2001) Activity of a novel bcl-2/bcl-xL bispecific antisense oligonucleotide against tumors of diverse histological origins. *J. Nat. Cancer Inst.* **93,** 463–471.
31. Zangemeister-Wittke, U., Leech, S. H., Olie, R. A., Simões-Wüst, A. P., Gautschi, O., Luedke, G. H., et al. (2000) A novel bispecific antisense oligonucleotide inhibiting both bcl-2 and bcl-xL expression efficiently induces apoptosis in tumor cells. *Clin. Cancer Res.* **6,** 2547–2555.
32. Olie, R. A, Simões-Wüst, A. P., Baumann, B., Leech, S. H., Fabbro, D., Stahel, R. A., and Zangemeister-Wittke, U. (2000) A novel antisense oligonucleotide targeting survivin expression induces apoptosis and sensitizes lung cancer cells to chemotherapy. *Cancer Res.* **60,** 2805–2809.
33. Morgan, D. M. (1998) Tetrazolium (MTT) assay for cellular viability and activity. *Methods Mol. Biol.* **79,** 179–183.

40

Induction of Programmed Cell Death with an Antisense Bcl-2 Oligonucleotide

Patrick P. Koty, Wendong Lei, and Mark L. Levitt

1. Introduction

Limited advances have occurred in lung cancer prevention and treatment *(1)*, which has necessitated the investigation of other means of intervention. As cancer is a disease of altered cellular homeostasis and genetic damage, programmed cell death (PCD) has gained increasing importance as a physiologic mechanism that has relevance to cancer-treatment strategies *(2)*. Dysregulation of PCD in lung cancer renders these cells resistant to chemotherapeutic agents that induce this process *(3,4)*. Therefore, genetic manipulation of PCD regulatory genes may restore the altered homeostasis in these cells and permit these genetically damaged cells to enter PCD. A known regulator of the PCD pathway is Bcl-2, a protein capable of suppressing cell death, which is upregulated in human premalignant and malignant lung lesions *(5,6)*. Bcl-2 has been localized to the nuclear membrane, the endoplasmic reticulum, the cytosol, and the outer mitochondrial membrane *(7–11)*. The exact molecular mechanisms by which Bcl-2 inhibits PCD have not been unequivocally established. Initial results indicate Bcl-2 inactivates inducers of PCD, such as Bax, through heterodimerization and that it is the ratio of Bcl-2, or other functionally similar genes, to the expression of PCD inducers that determines whether a cell will undergo PCD *(12,13)*. When Bcl-2 heterodimerizes with Bax within the outer mitochondria membrane it forms an ion channel, which is functionally different from the ion channel formed by Bax homodimerization *(14,15)*. Overexpression of Bcl-2 prevents cytochrome c release from the outer mitochondrial membrane to the cytosol where cytochrome c activates

PCD caspases *(16,17)*. Although the mechanism(s) by which bcl-2 prevents PCD remains unclear, its manipulation may provide a means of restoring homeostasis and chemotherapeutic sensitivity in lung cancer.

One of the most promising approaches to reducing Bcl-2 levels has focused on using an antisense bcl-2 molecule. In theory, an antisense molecule should create a double stranded bond with the endogenous bcl-2 mRNA. Dependent upon where on the mRNA this bond occurs, it could result in either: 1) inhibition of translation (by causing translational arrest by hybridizing to mRNA at binding sites for ribosomes); 2) activation of RNase H (which cleaves the RNA strand of an RNA-DNA duplex); 3) inhibition of splice sites (preventing production of mature mRNA); or 4) causing destabilization of mRNA (by hybridizing to capping or polyadenylation sites) (for review *see* **refs. *18,19***). Although some studies utilize a vector containing a bcl-2 cDNA in an antisense orientation, most studies use a short synthetic antisense oligonucleotide. Selecting which area of the mRNA to target is highly accessibility dependent due to constrictive areas of secondary and tertiary RNA structures. Several strategies are currently being employed with varying success and limitation. These approaches include "gene walking" in which up to 50 oligonucleotides spanning the mRNA are tested for their antisense activity, using mRNA RNase H mapping, oligonucleotide arrays, enzymatic mapping, or computation analysis of secondary structures (for review *see* **refs. *20,21***). Modification of the antisense oligonucleotide is also necessary to increase its half-life within the cell. The most common modification incorporates phosphorothioate analogs that replace nonbridging oxygen atoms in the phosphate backbone with sulphur. These phosphorothioate backbones are negatively charged and are capable of supporting RNase H activity similar to phosphodiester oligonucleotides, but have greater resistance to nuclease degradation (for review, *see* **ref. *22***). Delivery of the modified oligonucleotide into the cell initially utilized naked oligonucleotides incorporated into the cell through the active processes of endocytosis and pinocytosis. This mechanism of cellular uptake was inefficient, therefore, antisense oligonucleotides are introduced more efficiently using vehicle-delivery molecules. Currently, the most commonly utilized delivery vehicles are cationic liposomes (for review *see* **refs. *23,24***). Electrostatic interactions allow a complex to form between the negatively charged oligonucleotides and the positively charged cationic liposomes. Once the complex enters the cytosol, the phosphorothioated oligonucleotides dissociate from the cationic lipids and are free to hinder target transcript function.

Antisense approaches are currently being explored in lung cancer cells that overexpress Bcl-2. In fact, we and others have shown that a decrease in Bcl-2 protein can increase programmed cell death in lung cancer cell lines

Induction of Programmed Cell Death

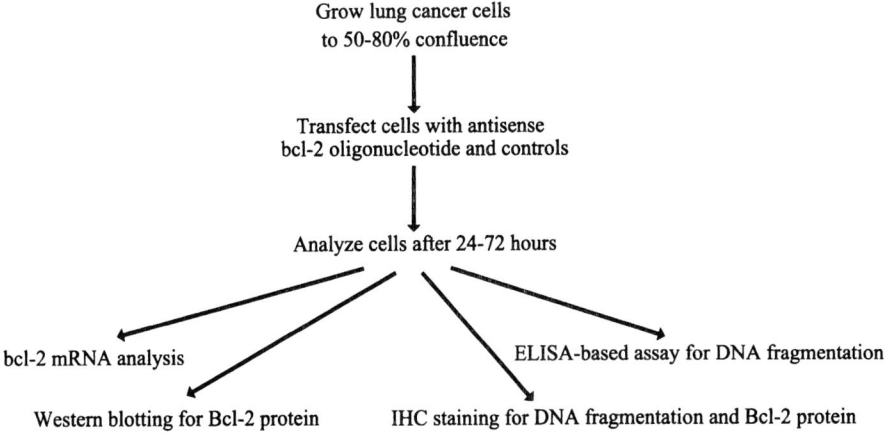

Fig 1. Scheme for treatment of lung cancer cells with a synthetic phosphorothioate antisense bcl-2 oligonucleotide.

following antisense bcl-2 treatment *(25,26)*. Although no literature currently exists regarding antisense bcl-2 clinical trials for lung cancer patients, Phase I trials were conducted in patients with non-Hodgkin's lymphoma (NHL) and malignant melanoma who were positive for Bcl-2 expression *(27,28)*. Although these initial trials primarily focused on the safety of treatment, which was well-tolerated, antisense bcl-2 therapy resulted in reduction of bcl-2 protein, a reduction in tumor burden, or decreased progression of disease in some cases.

In this chapter we present a protocol for delivering an antisense bcl-2 oligonucleotide to lung cancer cells that overexpress bcl-2 protein. This chapter includes protocols for analyzing bcl-2 mRNA and protein to determine if a decrease in bcl-2 has occurred after treatment with the antisense oligonucleotide. DNA fragmentation protocols are also provided to confirm induction of programmed cell death. An overview of this approach is summarized in **Fig. 1**.

2. Materials
2.1. Culturing Adherent Lung Cancer Cells

1. RPMI-1640 media supplemented with 0.3 µg/mL L-glutamine, 10% (v/v) heat-inactivated fetal bovine serum (FBS), 50 U/mL penicillin, 50 µg/mL streptomycin, 0.5 µg/mL fungizone (Gibco-BRL). Store at 4°C.
2. Six well plate for mRNA reverse transcription polymerase chain reaction (RT-PCR) assay, Western-blot analysis, and enzyme-linked immunosorbent assay (ELISA)-based assay for DNA fragmentation.
3. Four-chamber tissue-culture microscope slides for immunohistochemical analysis.

2.2. Antisense bcl-2 Transfection

1. RPMI-1640 media without supplements.Store at 4°C.
2. Sterile tubes (17 × 100mm) with caps.
3. LipofectAmine reagent (Gibco-BRL).Store at 4°C.
4. Synthetic phosphorothioate oligonucleotides (Store at –20°C):
   a. Nonsense control: 5'-gtatgacctagcggttgt-3'
   b. Antisense bcl-2: 5'-gttatcgtaccctgttctcc-3'
   c. Sense bcl-2: 5'-ggagaacagggtacgataac-3'

2.3. PCR-Based mRNA Analysis

2.3.1. Reverse Transcriptase-PCR (RT-PCR)

1. mRNA isolation kit (Promega, Invitrogen, Qiagen).
2. DEPC-treated H_2O.
3. poly(dT)$_{12-18}$ primer (Gibco-BRL).Store at –20°C.
4. Sterile 0.5 mL DNase- and RNase-free tubes.
5. 5X reverse transcriptase buffer: 250 mM Tris-HCl, 40 nM MgCl$_2$, 150 mM KCl, 5 mM dithiothreithol (DTT), pH 8.5.Store at –20°C.
6. Reverse transcriptase, AMV (Stratagene). Store at –20°C.
7. RNase inhibitor (Stratagene). Store at –20°C.
8. DNA thermocycler.

2.3.2. PCR Amplification of cDNA

1. 10X PCR Buffer: 100 mM Tris-HCl, pH 8.3, 500 mM KCl, 15 mM MgCl$_2$, 0.01% (w/v) gelatin. Store at –20°C.
2. dNTPs. Store at –20°C.
3. PCR primers. Store at –20°C (*see* **Note 1**).
   a. For bcl-2 (448 bp amplified product) [*forward* 5'-ccctccagatagctcatt-3', *reverse* 5'-ctagacagacaaggaaag-3']
   b. For β-actin (382 bp amplified product) [*forward* 5'-tctacaatgagctgcgtg-3', *reverse* 5'-ccttaatgtcacgcacga-3'].
4. AmpliTaq DNA polymerase (PerkinElmer). Store at –20°C.
5. Mineral oil.

2.3.3. Gel Electrophoresis

1. Agarose (molecular biology grade).
2. 10X TBE buffer (0.9 M Tris-borate, 0.02 M EDTA): Combine 108 g Tris-HCl, 55 g boric acid, 40 mL 0.5 M EDTA, pH 8.0 per liter dd H_2O.
3. Ethidium bromide stock solution (10 mg/mL). Light-sensitive, store in dark. **Caution:** Ethidium bromide is a mutagen.Handle with care, wear gloves and a mask.
4. 100bp molecular weight markers. Store at –20°C.
5. Bromophenol blue solution, 0.4% (w/v).
6. Horizontal gel electrophoresis unit.

7. UV transilluminator.
8. Polaroid photographic camera.
9. Densitometer capable of scanning Polaroid photographs.

2.4. Western-Blot Analysis for Bcl-2 Protein

2.4.1. Total Protein Isolation

1. 1X PBS buffer, pH 7.4: 137 mM NaCl, 2.7 mM KCl, 10 mM Na$_2$HPO$_4$, 2 mM KH$_2$PO$_4$.
2. Cell lysis buffer: 10 mM Tris-HCl, pH 7.4, 150 mM NaCl, 1% sodium deoxycholate, 1% Triton X-100, 0.1% SDS, 1mM EDTA. Add proteinase inhibitors prior to use: 2 µg/mL aprotinin, 2 µg/mL leupeptin, 1 µg/mL pepstatin A, 100µg/mL PMSF.Store at 4°C. **Caution:** PMSF is toxic and may be fatal if inhaled, swallowed, or absorbed through the skin. Wear gloves and a mask when using PMSF and flush eyes or skin with copious amounts of water in case of contact.
3. Sterile 1.5-mL tubes.
4. Sterile cell scraper.
5. Glycerol (Sigma).
6. 96-well plates.
7. DC Protein Assay(Bio-Rad Laboratories).
8. Microtiter plate reader (wavelength 650–750 nm).

2.4.2. Gel Electrophoresis

1. Vertical gel electrophoresis unit.
2. 30% (w/v) polyacrylamide (29.2 g/100 mL)/Bis (0.8 g/100 mL) stock solution. Store at 4°C.Light sensitive, store in dark. **Caution:** Acrylamide is a potent cumulative neurotoxin that is absorbed through the skin, wear gloves and a mask when handling.
3. Resolving gel buffer: 1.5 M Tris-HCl, pH 8.8.
4. Stacking gel buffer: 0.5 M Tris-HCl, pH 6.8.
5. 10% (w/v) ammonium persulfate (APS). Make fresh.
6. TEMED. Store at 4°C.
7. 10% sodium dodecyl sulfate (SDS) (electrophoresis grade).
8. 10% SDS-polyacrylamide gel electrophoresis (PAGE) gel. For every 10 mL of either resolving or stacking gel solution: 4 mL 30% acrylamide/Bis solution, 2.5 mL of either resolving or stacking gel buffer, 0.1 mL 10% SDS, 3.4 mL deionized H$_2$O.
9. 2X SDS gel-loading buffer: 100 mM Tris-HCl, pH 6.8, 4% (w/v) SDS, 0.2% (w/v) bromophenol blue, 20% (v/v) glycerol, 200 mM β-mercaptoethanol (**add β-mercaptoethanol just before use**).
10. Prestained molecular weight markers (Bio-Rad Laboratories). Store at –20°C.
11. 1X electrode buffer, pH 8.3: 25 mM Tris, 250 mM glycine, 0.1% SDS.

2.4.3. Western Blot Transfer

1. Electrophoretic transfer unit.
2. Nitrocellulose membrane.
3. Whatman 3MM filter paper.
4. Transfer buffer, pH 8.3: 25 mM Tris, 192 mM glycine, 20% (v/v) methanol, 0.5% SDS.

2.4.4. Probing Western Blot with Bcl-2 Antibody

1. Blocking buffer: 2% horse serum in TBS. Store at 4°C.
2. Tris-buffered saline (TBS): 10 mM Tris-HCl, pH 7.4, 150 mM NaCl. Store at 4°C.
3. Wash buffer: 0.1% Tween 20 in TBS.
4. Bcl-2 antibody (Dako) diluted 1:80 in TBS. Store at 4°C.
5. HRP-conjugated secondary antibody (Amersham Pharmacia Biotech) diluted 1:1000 in TBS. Store at 4°C.
6. ECL Western Blotting Detection Kit (Amersham Pharmacia Biotech).
7. Autoradiographic film and cassette.
8. Autoradiograph developer.
9. Densitometer capable of scanning autoradiographs.

2.5. ELISA-Based Assay for Oligonucleosomal DNA Fragmentation

1. Cell Death Detection ELISA[PLUS] Kit (Roche). Store at 4°C.
2. Microtiter plate reader (wavelength 405nm).

2.6. Immunohistochemical Dual Analysis for DNA Fragmentation and Bcl-2 Protein Expression

2.6.1. TUNEL Assay for DNA Fragmentation

1. 10% neutral buffered formalin in PBS, pH 7.4.
2. Peroxidase quenching buffer: 0.3% hydrogen peroxide in methanol.
3. TdT buffer, pH 7.2: 30 mM Tris-HCl, 140 mM sodium cacodylate. Store at 4°C.
4. TdT labeling mixture: digoxigenin-conjugated dUTP (0.5 nmoles) and TdT enzyme (5 U). Store component TdT enzyme and digoxigenin-conjugated dUTP at –20°C. Prepare TdT labeling mixture just before use.
5. Humid chamber.
6. Peroxidase conjugated anti-digoxigenin antibody (Dako) diluted 1:1000 in PBS. Store at 4°C.
7. 3-amino-9-ethyl carbazole (AEC) solution (Biomeda Corp). Store at 4°C. **Caution:** AEC is a suspected carcinogen, wear gloves.

2.6.2. Immunohistochemical Staining for Bcl-2 Protein

1. Coplin jars.
2. Blocking buffer: 10% horse serum in PBS.
3. Bcl-2 antibody (Dako) diluted 1:80 in PBS. Store stock at 4°C.

4. Biotinylated anti-mouse secondary antibody (Vector Laboratories) diluted 1:1000 in PBS. Store stock at 4°C.
5. Streptavidin-conjugated alkaline phosphatase (Vector Laboratories) diluted 1:250 in PBS. Store at 4°C.
6. Alkaline phosphatase substrate III (Vector Laboratories).
7. Mayer's hematoxylin (Sigma).
8. Aqueous-based slide mounting medium (Biomeda Corp).
9. Light microscope with photographic capabilities.

3. Methods
3.1. Tissue Culturing Adherent Lung Cancer Cells

1. Seed adherent lung cancer cell lines onto either six-well tissue-culture plates (for Western blot, mRNA analysis, ELISA-based DNA fragmentation analysis) or onto four chamber tissue-culture microscope slides (for *in situ* analysis of DNA fragmentation and Bcl-2 protein expression).
2. Grow cells to 50–80% confluence (*see* **Note 2**) in supplemented RPMI-1640 medium at 37°C in an atmosphere of 5% CO_2.

3.2. Antisense bcl-2 Transfection

1. Prepare tube A in a 17 × 100mm tube for each transfection condition (media control, nonsense control oligonucleotide, antisense bcl-2 oligonucleotide, and specificity control with equal amounts antisense bcl-2 and sense bcl-2 oligonucleotides) containing 100 µL unsupplemented RPMI-1640 at room temperature and 3–30 µM synthetic oligonucleotide. (*see* **Note 3**).
2. Prepare tube B in a 17 × 100mm tube for each transfection condition (media control, nonsense control oligonucleotide, antisense bcl-2 oligonucleotide, sense bcl-2 oligonucleotide) containing 100 µL unsupplemented RPMI-1640 at room temperature and 2–25 µL LipofectAmine (*see* **Note 4**).
3. Combine contents of tubes A and B (*see* **Note 5**) and incubate at room temperature for 5–45 min.
4. Add 0.8 mL unsupplemented RPMI-1640 media to each tube (*see* **Note 6**).
5. Wash cells once with unsupplemented RPMI-1640.
6. Overlay cells with controls and antisense oligonucleotides and incubate at 37°C in an atmosphere of 5% CO_2 for 2–24 h (*see* **Note 7**).
7. Add 1 mL supplemented RPMI-1640 with 20% fetal bovine serum (FBS) to each well.
8. Analyze at various time points after initiating transfection (*see* **Note 8**).

3.3. PCR-Based mRNA Analysis
3.3.1. Reverse Transcriptase-PCR (RT-PCR)

1. Use a commercially available kit (Promega, Invitrogen, Qiagen) to isolate mRNA from cells after treatment with antisense oligonucleotide and controls (*see* **Note 9**).

2. Add 0.5–2.5 µg total polyadenylated mRNA, 0.5 µg oligo-dT$_{12-18}$ primer and DEPC-treated H$_2$O to a final volume of 10.5 µL into a RNase-free 0.5-mL tube.
3. Denature at 70°C in a DNA thermal cycler for 2 min.
4. Place tube on ice for 2 min.
5. Add 4 µL 5X reverse transcriptase buffer, 15 U AMV reverse transcriptase, 10 mM each dNTP, 20 U RNase Inhibitor, and DEPC-treated H$_2$O for a final volume of 20 µL.
6. Incubate at 42°C in a DNA thermal cycler for 1 h.
7. Terminate reaction by heating tube in a DNA thermocycler for 2 min at 94°C.
8. Dilute RT-PCR reaction mixture to a final volume of 100 µL with sterile H$_2$O.

3.3.2. PCR Amplification of cDNA

1. To a sterile 0.2- or 0.5-mL tube, add 0.5 µL 10X PCR buffer; 2.5 mM each dATP, dTTP, dGTP, and dCTP; 20 pmole bcl-2 forward and reverse primers; 0.3 pmole β-actin forward and reverse primers; 0.5 U AmpliTaq polymerase; 1 µL diluted RT-PCR product; and sterile H$_2$O for a final volume of 5 µL. Overlay with mineral oil.
2. DNA thermal cycler conditions: 94°C, 1 min; 61°C, 1 min; 72°C, 1 min for 35 cycles.

3.3.3. Gel Electrophoresis

1. Pour a 1.5% (w/v) agarose gel in 1X TBE buffer and 10 µg/mL ethidium bromide into gel casting blocks and insert appropriate comb.
2. After gel solidifies, remove casting blocks, submerge gel in 1X TBE containing 10 µg/mL ethidium bromide, and carefully remove comb.
3. Add 1 µL bromophenol blue solution to the PCR reaction product and load sample into gel. Include a 100 bp molecular-weight markers lane.
4. Subject samples to electrophoresis at 30 mA for 1 h.
5. Visualize and photograph gel using UV transilluminator and Polaroid camera.
6. Quantify PCR products from Polaroid photograph using a densitometer.

3.4. Western-Blot Analysis for Bcl-2 Protein

3.4.1. Total Protein Isolation

1. Following treatment with antisense bcl-2 oligonucleotide and controls, wash cells once with PBS buffer (*see* **Note 10**).
2. Lyse cells in wells with 500 µL cell lysis buffer for 5 min at room temperature.
3. Scrape cells and remove cell lysate and place into sterile 1.5-mL microcentrifuge tubes.
4. Freeze lysate at –80°C for 10 min and then thaw.
5. Repeat **step 4**.
6. Centrifuge lysate for 10 min at 12,000g at 4°C.

7. Take supernatant and transfer to a sterile 1.5-mL microcentrifuge tube. Add 50 µL glycerol to 500 µL clarified lysate, mix thoroughly, and store at –80°C.
8. Determine protein concentration using a commercially available assay.

3.4.2. Gel Electrophoresis

1. Prepare 10% SDS-PAGE resolving gel solution by adding 50 µL 10% APS and 10 µL TEMED per 10 mL resolving gel solution, cast, overlay with water, and allow to polymerize for 1 h.
2. Remove water and gently dry top of resolving gel with Whatman 3MM filter paper.
3. Prepare SDS-PAGE stacking gel solution by adding 50 µL 10% APS and 10 µL TEMED per 10 mL stacking gel solution, cast on top of polymerized resolving gel, insert comb, and allow to polymerize for 45 min.
4. Remove comb and rinse wells of stacking gel with distilled water then add 1X electrode buffer to reservoirs.
5. Aliquot equal quantities of protein extracts (80 µg/lane). Mix extracts 1:1 with 2X SDS gel loading buffer. Be sure each gel includes a lane for molecular-weight markers.
6. Denature samples at 95°C for 4 min, place on ice, and load samples into wells.
7. Separate proteins by gel electrophoresis at 200V for 30–60 min.

3.4.3. Western Blot Transfer

1. Cut nitrocellulose membrane and Whatman 3MM filter paper to fit gel size (*see* **Note 11**).
2. Remove gel from electrophoresis unit and soak gel, membrane, and filter paper in transfer buffer for 1 h.
3. Prepare gel sandwich consisting of transfer cassette, fiber pad, filter paper, SDS-PAGE gel, membrane, filter paper, and fiber pad (*see* **Note 12**).
4. Assemble transfer unit (*see* **Note 13**).
5. Transfer overnight at 30V and 90mA.

3.4.4. Probing Western Blot With Bcl-2 Antibody

1. Block membrane for 20 min at room temperature with gentle agitation (*see* **Note 14**).
2. Wash membrane three times for 10 min each at room temperature.
3. Incubate membrane in diluted Bcl-2 antibody overnight at 4°C with gentle agitation.
4. Wash membrane three times for 10 min each at room temperature.
5. Incubate membrane in diluted HRP-conjugated secondary antibody for 1 h at room temperature with gentle agitation (*see* **Note 15**).
6. Visualize the antigens bound to the membranes using ECL reagents.
7. Expose blot to autoradiographic film for 1–10 min, then develop and fix film.

8. Quantify protein expression using a densitometer capable of analyzing autoradiographs.

3.5. ELISA-Based Assay for Oligonucleosomal DNA Fragmentation (see Note 16)

1. Following treatment with antisense bcl-2 and controls, wash cells once with PBS.
2. Isolate cytosolic oligonucleosomal DNA fragments using a Cell Death Detection ELISA^{PLUS} Kit according to manufacturer's instructions (*see* **Note 17**).
3. Quantify cytosolic oligonucleosomal DNA fragments photometrically on a microtiter plate reader by measuring absorbance at 405 nm.

3.6. Dual Immunohistochemical Staining for DNA Fragmentation and Bcl-2 Protein Expression

3.6.1. TUNEL Assay for DNA Fragmentation (see Note 18)

1. Following treatment with antisense bcl-2 oligonucleotide and controls, detach chambers from slides, and wash tissue slides three times with PBS for 2 min each wash. This and all subsequent steps can be performed in Coplin jars.
2. Fix cells in neutral buffered formalin for 10 min.
3. Dry slides at room temperature overnight.
4. Wash three times for 2 min each in PBS.
5. Quench endogenous peroxidase activity by incubating in quenching buffer for 30 min.
6. Wash three times for 2 min each in PBS.
7. Wash once with TdT buffer for 2 min.
8. Incubate in TdT labeling mixture containing TdT and digoxigenin-conjugated dUTP at 37°C in a high humidity chamber for 90 min.
9. Wash three times for 2 min each in PBS.
10. Incubate in diluted peroxidase conjugated anti-digoxigenin antibody for 60 min in a high humidity chamber at room temperature.
11. Visualize bound peroxidase by incubating in 3-amino-9-ethyl carbazole (AEC) solution for 5–10 min.
12. Wash in distilled water. Use these same samples for co-analysis of bcl-2 protein expression.

3.6.2. Immunohistochemical Staining for Bcl-2 Protein

1. Wash slides from previous step three times with PBS for 2 min each wash.
2. Incubate with 10% horse serum to block nonspecific proteins for 20 min in a high humidity chamber.
3. Remove horse serum, and, without washing, incubate in diluted bcl-2 antibody for 120 min at room temperature in a high humidity chamber.
4. Wash three times with PBS for 2 min each wash.

5. Incubate in diluted biotinylated secondary antibody for 1 h at room temperature in a high humidity chamber.
6. Wash three times with PBS for 2 min each wash.
7. Incubate in alkaline phosphatase-conjugated streptavidin solution for 1 h at room temperature in a high humidity chamber.
8. Wash three times with PBS for 2 min each wash.
9. Visualize bound alkaline phosphatase using an alkaline phosphatase substrate III.
10. Wash in distilled water.
11. Counterstain with Mayer's hematoxylin for 2 min.
12. Wash in tap water.
13. Incubate in PBS for 2 min.
14. Wash in distilled water.
15. Apply aqueous-based mounting medium and dry at room temperature.
16. Analyze cells for DNA fragmentation and Bcl-2 expression by counting 200 cells per field, from a minimum of three fields, using a 10× eyepiece and 20× objective.

4. Notes

1. General considerations for designing PCR primers: 1) they should be at least 20 base pairs in length to ensure specificity; 2) GC/AT content should be 50–65%; 3) avoid areas in which internal hairpinning (hybridization of primer to itself) of three or more base pairing occurs; 4) primer melting temperature should be between 60–80°C using Wallace formula: T_m (°C) = 2(NA + NT) + 4(NG + NC), N = number of A, T, G, or C bases.
2. Avoid transfecting cells that are approaching confluence. Proliferating (dividing) cells are transfected more efficiently.
3. In addition to transfecting cells with an antisense bcl-2 oligonucleotide, several controls are needed. Treatment of cells with media only provides a baseline for endogenous bcl-2 expression and DNA fragmentation. A control (nonsense oligonucleotide) for specificity and toxicity is also needed and should show less than 85% homology to any sequence within the GenBank database. A control for antisense bcl-2 specificity is to transfect equal amounts of antisense bcl-2 oligonucleotide and a sense strand bcl-2 oligonucleotide complimentary to the antisense oligonucleotide. The sense strand should compete with the bcl-2 mRNA for the antisense bcl-2 oligonucleotide and should result in a decreased effect of the antisense bcl-2 oligonucleotide alone.
4. Although 4 μL LipofectAmine works well, a range of 2–25 μL LipofectAmine should be tried to optimize transfection efficiency.
5. We prefer to use 17 × 100mm tubes (instead of smaller tubes) to provide more surface area for liposomes/DNA complexing.
6. This is a very critical step and much care should be taken to avoid disrupting the liposome/DNA complex. Add the media to the side of the tube and not directly to existing media containing complex.

7. This is also a very critical step and much care should be taken to avoid disrupting the liposome/DNA complex. Either very gently pour the liposome/DNA complex solution into the well or transfer the complex solution using a large-bore 1-mL serological pipet.
8. A time range (24–72 h) after transfection should be investigated to determine the maximum observed decrease in bcl-2 expression and the maximum observed DNA fragmentation.
9. RNA is very unstable and subject to rapid degradation by RNase. Wear sterile gloves at all times, clean bench surfaces with RNase inhibitor solutions, and only use RNase-free plasticware.
10. After treatment with antisense bcl-2 oligonucleotide, a population of cells will be floating in the media. The media containing these cells must be collected and gently centrifuged (50–100g) to pellet the cells which should be added to the lysis buffer in the appropriate well.
11. Wear gloves at all times when handling nitrocellulose membranes and use flat forceps to hold membranes.
12. Remove all bubbles between the gel and the nitrocellulose membrane by smoothing them out with a sterile glass Pasteur pipet.
13. Some transfer units utilize an ice block or a cold water circulator to cool the transfer unit. Otherwise, if necessary, transfer units can be run in a cold room. Also, place a magnetic stir bar in the buffer tank and place unit on a magnetic stirrer.
14. If it is necessary to decrease nonspecific background on the nitrocellulose membrane, use 1% (w/v) dry nonfat milk in the reaction buffer for the primary and secondary antibodies, and for the conjugated biotin and avidin-HRP solutions, if used.
15. If not using a conjugated HRP secondary antibody, it is important to dilute the concentration of avidin-conjugated HRP to 1:5000 or 1:10000.
16. Alternative assays are commercially available that identify changes associated with programmed cell death.These include but are not limited to assays for caspase activation, mitochondrial permeability changes, plasma membrane alterations, and morphological changes.
17. Increase the centrifugation speed to 3,500g for 10 minutes to pellet nucleus and fragmented plasma membrane.
18. The TUNEL protocol should always precede immunohistochemical staining.

References

1. Greenlee, R. T., Murray, T., Bolden S., and Wingo P. A. (2001) Cancer statistics, 2000. *CA Cancer J. Clin.* **50,** 7–33.
2. Thompson, H. J., Strange, R., and Schedin, P. J. (1992) Apoptosis in the genesis and prevention of cancer. *Cancer Epidemiol. Biomarkers Prev.* **1,** 597–602.
3. Shepherd, F. A. (1994) Treatment of advanced non-small cell lung cancer. *Semin. Oncol.* **21,** 7–18.

4. Reeve, J. G., Xiong, J., Morgan, J., and Bleehen, N. M. (1996) Expression of apoptosis-regulatory genes in lung tumor cell lines: relationship to p53 expression and relevance to acquired drug resistance. *Br. J. Cancer* **73**, 1193–1200.
5. Zhang, H., Yousem, S. A., Franklin, W. A., Elder, E., Landreneau, R., Ferson, P., et al. (1998) Differentiation and programmed cell death-related intermediate biomarkers for the development of non-small cell lung cancer: a pilot study. *Hum. Pathol.* **29**, 965–971.
6. Törmänen, U., Nuorva, K., Soini, Y., and Pääkkö, P. (1999) Apoptotic activity is increased in parallel with the metaplasia-dysplasia-carcinoma sequence of the bronchial epithelium. *Br. J. Cancer* **79**, 996–1002.
7. Jacobson, M. D., Burne, J. F., King, M. P., Miyashita, T., Reed, J. C., and Raff, M. C. (1993) Bcl-2 blocks apoptosis in cells lacking mitochondrial DNA. *Nature* **361**, 365–369.
8. Liu, Y. J., Mason, D. Y., Johnson, G. D., Abbot, S., Gregory, C. D., Hardie, C. D., et al. (1991) Germinal center cells express bcl-2 protein after activation by signals which prevent their entry to apoptosis. *Eur. J. Immunol.* **21**, 1905–1910.
9. Tsujimoto, Y. and Croce, C. M. (1986) Analysis of the structure, transcripts, and protein products of bcl-2, the gene involved in human follicular lymphoma. *Proc. Natl. Acad. Sci. USA* **83**, 5214–5218.
10. Bruel, A., Karsenty, E., Schmid, M., McDonnel, T. J., and Lanotte, M. (1997) Altered sensitivity to retinoid-induced apoptosis associated with changes in the subcellular distribution of Bcl-2. *Exp. Cell Res.* **233**, 281–287.
11. Hsu, Y. T., Wolter, K. G., and Youle, R. J. (1997) Cytosol-to-membrane redistribution of Bax and Bcl-X_L during apoptosis. *Proc. Natl. Acad. Sci. USA* **94**, 3668–3672.
12. Oltvai, Z. N., Milliman, C. L., and Korsmeyer, S. J. (1993) Bcl-2 heterodimerizes in vivo with a conserved homolog, Bax, that accelerates programmed cell death. *Cell* **74**, 609–619.
13. Salomons, G. S., Brady, H. J. M., Verwijs-Janssen, M., Van Den Berg, J. D., Hart, A. A. M., Van Den Berg, H., et al. (1997) The Bax(α):Bcl-2 ratio modulates the response to dexamethasone in leukaemic cells and is highly variable in childhood acute leukaemia. *Int. J. Cancer* **71**, 959–965.
14. Schendel, S. L., Xie, Z., Montal, M. O., Matsuyama, S., Montal, M., and Reed, J. C. (1997) Channel formation by antiapoptotic protein Bcl-2. *Proc. Natl. Acad. Sci. USA* **94**, 5113–5118.
15. Antonsson, B., Conti, F., Ciavatta, A. M., Montessuit, S., Lewis, S., Martinou, I., et al. (1997) Inhibition of Bax channel-forming activity by Bcl-2. *Science* **277**, 370–372.
16. Yang, J., Liu, X., Bhalla, K., Kim, C. N., Ibrado, A. M., Cai, J., et al. (1997) Prevention of apoptosis by Bcl-2: release of cytochrome c from mitochondria blocked. *Science* **275**, 1129–1132.
17. Kluck, R. M., Bossy-Wetzel, E., Green, D. R., and Newmeyer, D. D. (1997) The release of cytochrome c from mitochondria: a primary site for Bcl-2 regulation of apoptosis. *Science* **275**, 1132–1136.

18. Uhlmann, E. and Peyman, A. (1990) Antisense oligonucleotides: a new principle. *Chem. Rev.* **90,** 543–584.
19. Crooke, S. T. (1995) Molecular mechanism of action of oligonucleotides designed to interact with nucleic acids, in *Therapeutic Applications of Oligonucleotides* (Crooke, S. T., ed.), R. G. Landes, Austin, pp. 11–38.
20. Smith, L., Andersen, K. B., Hovgaard, L., and Jaroszewski, J. W. (2000) Rational selection of antisense oligonucleotide sequences. *Eur. J. Pharm. Sci.* **11,** 191–198.
21. Sohail, M. and Southern, E. M. (2000) Selecting optimal antisense reagents. *Adv. Drug Deliv. Rev.* **44,** 23–34.
22. Agrawal, S. (1999) Importance of nucleotide sequence and chemical modifications of antisense oligonucleotides. *Biochim. Biophys. Acta* **1489,** 53–68.
23. Garcia-Chaumont, C., Seksek, O., Grzybowska, J., Borowski, E., and Bolard, J. (2000) Delivery systems for antisense oligonucleotides. *Pharmacol. Ther.* **87,** 255–277.
24. Akhtar, S., Hughes, M. D., Khan, A., Bibby, M. Hussain, M., Nawaz, Q., et al. (2000) The delivery of antisense therapeutics. *Adv. Drug Deliv. Rev.* **44,** 3–21.
25. Ziegler, A., Luedke, G. H., Fabbro, D., Altmann, K. H., Stahel, R. A., and Zangmeister-Wittke, U. (1997) Induction of apoptosis in small-cell lung cancer cells by an antisense oligodeoxynucleotide targeting the Bcl-2 coding sequence. *J. Natl. Cancer Inst.* **89,** 1027–1036.
26. Koty, P. P., Zhang, H., and Levitt, M. L. (1999) Antisense bcl-2 treatment increases programmed cell death in non-small cell lung cancer cell lines. *Lung Cancer* **23,** 115–127.
27. Waters, J. S., Webb, A., Cunningham, D., Clarke, P. A., Raynaud, F., di Stefano, F., and Cotter, F. E. (2000) Phase I clinical and pharmacokinetic study of bcl-2 antisense oligonucleotide therapy in patients with non-Hodgkin's lymphoma. *J. Clin. Oncol.* **18,** 1812–1823.
28. Jansen, B., Wacheck, V., Heere-Ress, E., Schlagbauer-Wadl, H., Hoeller, C., Lucas, T., et al. (2000) Chemosensitisation of malignant melanoma by BCL2 antisense therapy. *Lancet* **356,** 1728–1733.

III

NOVEL THERAPIES

C. IMMUNE THERAPIES

41

Tumor Vaccination with Cytokine-Encapsulated Microspheres

Nejat K. Egilmez, Yong S. Jong, Edith Mathiowitz, and Richard B. Bankert

1. Introduction

The rapid advances that have taken place in tumor immunology within the past decade have fostered renewed interest in the use of immune-based therapies for the treatment of cancer (1,2). Numerous strategies, which aim to provoke and augment anti-tumor immunity in cancer patients, have been developed and are being tested in clinical trials (2–4). One strategy that has worked well in preclinical mouse models involves the use of immune stimulatory molecules, i.e., cytokines, to augment the anti-tumor activity of the immune system (5–7). The early studies focused on the systemic administration of selected cytokines to promote a nonspecific augmentation of an already existing anti-tumor immunity (5,7). While successful in murine models, particularly with interleukin (IL)-2 and -12, the systemic delivery of cytokines has not been efficacious in the clinics due to severe systemic toxicity. Systemic administration of immunotherapeutic agents ignores the paracrine nature of their activity. With the advent of molecular techniques that allowed efficient genetic modification of tumor cells in vitro, cytokine gene-modified tumor cell vaccines became the preferred alternative to systemic therapy. This approach results in the local and sustained release of cytokines by the tumor cells at the vaccine site, which induces the development of a tumor-specific, systemic anti-tumor immunity while circumventing the toxicity associated with systemic delivery (4–7). Numerous studies in syngeneic mouse tumor models demonstrated that vaccination of mice at a single subcutaneous site with cytokine gene-modified tumor cells results in the development of systemic anti-tumor immunity and rejection of later challenges with wild-type tumor cells *(8–17)*. In some cases, vaccination of tumor-bearing mice with irradiated, cytokine gene-modified tumor cells induced the eradication of established tumors with the concomitant development of systemic anti-tumor immunity *(16,18)*. As a result, clinical trials involving cytokine gene-modified tumor cell vaccines have been initiated *(3,4)*.

Despite the impressive results obtained in mice, the rapid translation of these protocols to the clinic has been viewed by some as premature *(5,6)*. In the majority of the currently approved clinical protocols, the tumor cells are removed from the patient, cultured and modified in vitro, irradiated and administered to the patient *(3)*. It is now becoming clear that, with the possible exception of melanoma, the vast majority of the human cancers are not amenable to genetic manipulation owing to the difficulty in maintaining and expanding the tumor cells in culture and to the variability in the transfectability of different tumors *(5,6)*. To circumvent these difficulties, the idea of delivering cytokines and other immunostimulatory molecules by genetically altered autologous fibroblasts admixed with tumor cells has been proposed and tested *(15–17)*. Although more feasible than the original approach, the use of autologous fibroblasts also involves labor-intensive manipulation of the cells in culture and the degree of successful genetic modification is variable from patient to patient. Alternative strategies involve *in situ* modification of tumors by in vivo gene transfer *(18–21)* and the use of gene-modified allogeneic cell lines *(22)*. These approaches also have serious drawbacks such as low-transfection efficiency or vector immunogenicity in the case of in vivo gene transfer or the possibility of a misdirected immune response in the case of allogeneic cells.

The attraction of gene therapy lies mainly in the fact that the cytokine of choice can be delivered locally to the tumor microenvironment, and in a sustained fashion, thus circumventing the severe side effects associated with systemic cytokine immunotherapy *(5–7,23)*. Studies have shown that it is the sustained presence of physiological levels of cytokine in the tumor microenvironment that is important in provoking anti-tumor immunity and not how it is delivered *(24–28)*. While gene transfer has retained its appeal as a mode of cytokine delivery, the technical difficulties associated with the current protocols has led some to ask whether simpler strategies for cytokine delivery may in fact be the more feasible choice *(5,6)*.

Local and sustained delivery of therapeutic agents can also be achieved with biodegradable controlled-release polymers *(29)*. Biodegradable polymer microspheres have been used for in vivo drug delivery *(29)*, cancer chemotherapy *(30)* and vaccination with antigenic peptides *(31)*. Very few preclinical animal studies have been done to evaluate this technology for cancer immunotherapy *(28,32,33)*. Loss of bioactivity during encapsulation of labile proteins such as cytokines has been a major obstacle in the application of controlled-release particle technology to immunotherapy. Techniques for the efficient encapsulation of a number of bioactive macromolecules, including cytokines, have recently been developed *(34)*. The use of these improved drug-delivery systems for sustained paracrine delivery of cytokines at a tumor site would significantly simplify cytokine immunotherapy and tumor vaccination protocols by eliminating the need for genetic modification of tumor cells.

A novel technology for highly efficient encapsulation of bioactive materials into polymer microspheres was recently described by us *(35)*. Phase inversion nanoencapsulation (PIN) results in the production of microspheres with a limited size distribution (1–5 microns) and labile proteins are efficiently encapsulated without significant denaturation or losses to aqueous nonsolvent baths. Moreover this process is spontaneous and does not require vigorous stirring/sonication during the formation of the microspheres (factors that have previously been shown to reduce cytokine activity) *(36)*. We have demonstrated that recombinant murine and human IL-2, IL-12 and GM-CSF-loaded polylactic acid (PLA) microspheres prepared by the PIN method release physiologically relevant quantities of bioactive cytokines for extended periods *(37–41)*. The in vivo release of recombinant human cytokines directly into the tumor microenvironment promoted the suppression of human lung tumor xenografts in severe combined immunodeficient (SCID) mice by co-engrafted human peripheral blood lymphocytes (PBL) in human/SCID mouse chimeric models *(39,40)*. In a syngeneic transplantable murine lung tumor model, a single intratumoral injection of murine IL-12-loaded microspheres promoted complete regression of established subcutaneous tumors, suppressed their metastatic spread and induced the development of potent and specific systemic anti-tumor immunity *(41)*. More recent studies demonstrated that a single intratumoral injection of IL-12-loaded microspheres into large, progressively growing subcutaneous tumors results in the suppression of primary tumor growth and the eradication of established distant micrometastases in a surgical metastasis model *(42)*.

The advantages offered by the strategy described here over current cytokine-based tumor vaccination protocols are numerous. First, owing to its simplicity, our approach does not require the establishment of specialized facilities and development of labor-intensive and expensive protocols at every hospital. Moreover, the stability of the preparations (1–4 mo at 4°C depending on the cytokine) eliminates the necessity of coordinating the effort between the laboratory and the clinic, and provides off-the-shelf convenience. In contrast to cell-based vaccines, cytokine doses can be controlled carefully with microspheres. Furthermore, this approach provides the opportunity to deliver cytokines directly to the microenvironment of established tumors in vivo. This is important because vaccination of tumors *in situ* has been shown to be superior to vaccination with irradiated or live tumor cell suspensions for the induction of an effective, systemic anti-tumor immunity *(41)*. Finally, the *in situ* tumor vaccination approach described here can be used to treat the majority of patients with solid tumors. Accessibility of tumor should not pose a major difficulty since only a single intratumoral injection is required and stereotactic injections can be employed to achieve delivery.

2. Materials

2.1. Preparation of Cytokine-Encapsulated PIN Microspheres

1. Stock solution of bovine serum albumin (BSA): 10% (w/v) in distilled water. (RIA grade BSA from Sigma Chemical Co., St. Louis, MO.)
2. Stock solution of Tween 20 (Polysorbate 20): 10% (v/v) in distilled water. (Tween 20 from Mallinckrodt, Paris, KA.)
3. Polymer solution: 50 mg polylactic acid (PLA 24K) and 50 mg polylactic acid (PLA 2K) are dissolved in 2 mL of methylene chloride. (PLA MW 24,000 and PLA MW 2,000 from Birmingham Polymers, Inc. Birmingham, AL, Methylene chloride (HPLC grade) from Fisher, Pittsburgh, PA.)
4. Cytokines. Recombinant human IL-2 (16×10^6 IU/mg), recombinant human and murine IL-12 (7.7×10^7 and 2.7×10^6 U/mg, respectively), and recombinant murine granulocyte-macrophage colony stimulating factor (GM-CSF) (7.2×10^7 U/mg) (Chiron, Inc. [Emeryville, CA], Genetics Institute, Inc. [Andover, MA] and Immunex, Inc. [Seattle, WA] respectively.)
5. Titan Cold Trap lyophilizer (–100°C) FTS Kinetics (Stone Ridge, NY).
6. Petroleum ether high-performance liquid chromatography ([HPLC]-grade) (EM Science, Gibbstown, NJ).
7. Whatman 2.7-μm filter (Clifton, NJ).

2.2. Characterization of Microspheres

2.2.1. Size Determination

1. Beckman Coulter Multisizer 3 Coulter Counter (Miami, FL).
2. ISOTON II buffer Beckman Coulter (Miami, FL).
3. Bath sonicator.
4. Hydration buffer. Hydration buffer: 1% Hydroxypropylmethylcellulose (DOW Chemical Co., Midlands, MI) and 1% Pluronic F127™ (Sigma) in phosphate-buffered saline (PBS) pH 7.2.

2.2.2. Encapsulation Efficiency

1. Acetone.

Tumor Vaccination

2.2.3. Cytokine Release Profile and Bioactivity

1. Tissue-culture medium: Dulbecco's modified Eagle medium (DMEM) supplemented with 10% fetal calf serum (FCS; Gibco/Invitrogen, Inc., Grand Island, NY).
2. 96-well culture plates
3. Micro-pipets.
4. Cytokine-specific enzyme-linked immunosorbent assay (ELISA) kits (R & D Systems, Minneapolis, MN).

2.3. Tumor Vaccination

1. Microbalance
2. Injection buffer: 1% mouse serum albumin (Sigma) in PBS.
3. Mice and tumor cells: Male or female BALB/c mice at 6–8 wk of age are obtained from Taconic Laboratories (Germantown, NY). All mice are maintained in microisolation cages (Lab Products, Federalsburg, MA) under pathogen-free conditions. Animals of both sexes are used in the studies at 8–12 wk of age. Line-1 (a BALB/c lung alveolar carcinoma cell line) was a gift from Dr. John G. Frelinger (University of Rochester, School of Medicine and Dentistry, Rochester, NY).
4. 28.5-gauge needle attached to a 0.5-mL insulin syringe

3. Methods
3.1. Preparation of Cytokine-Encapsulated PIN Microspheres

A phase inversion nanoencapsulation (PIN) technique is used for encapsulation of cytokines.

1. Vortex cytokines (0.2–1 mL), containing 10 µL of the stock BSA and 2.5 µL of the Tween 20, briefly with the polymer solution and flash-freeze in liquid nitrogen.
2. Lyophilize the frozen emulsion for 48 h. Subsequently, re-dissolve the dry matrix in methylene chloride, and discharge into petroleum ether for production of microspheres.
3. Recover microspheres by filtration through a 2.7-µm filter and lyophilized overnight for complete removal of solvent.
4. We routinely prepare formulations containing 1% BSA (w/w) with the following cytokines: human IL-2 /PEG-IL-2 (10 µg/mg PLA); human IL-12 (10 µg/mg PLA); murine IL-12 (10 µg/mg PLA), and murine GM-CSF (10 µg/mg PLA).

3.2. Characterization of Microspheres
3.2.1. Size Determination

1. Size microspheres in a Coulter Counter (Miami, FL) using ISOTON II buffer and a 50-µm aperture according to manufacturer's protocols.
2. Sonicate microspheres (~ 2 mg) in a bath sonicator in the hydration buffer for 30 s. Add the suspension to the sample reservoir. A total of 200,000 particles are counted and sized for both number and volume distributions.

3.2.2. Determination of Encapsulation Efficiency

The encapsulation efficiencies for the cytokines are extrapolated from the measurements of total protein encapsulated into the microspheres.

1. Dissolve five mg of microspheres in a 200 µL of a mixture of acetone and methylene chloride (1:1, v/v) and collect the protein precipitate by centrifugation in a microcentrifuge (12,000g for 10 min).
2. Dissolve the protein pellet in 1 mL distilled water and determine the final yield by absorbance at 280 nm (*see* **Note 1**).

3.2.3. Cytokine-Release Profile and Bioactivity

1. Suspend particles (3 mg) in 20 µL of hydration buffer as described (*see* **Subheadings 2.1.1.** and **3.1.1.**), bring up to 200 µL with tissue-culture medium and incubate in the wells of a 96-well culture plate in triplicate at 37°C.
2. Change the medium daily by carefully removing the supernatant from each well with a micro-pipet (the microspheres settle to the bottom of the well after overnight incubation) and store the supernatants frozen at −20°C.
3. Add fresh medium (200 µL) back into each well, and resuspend the microspheres by gentle pipetting. Place the plate back in the incubator.
4. Determine cytokine concentrations in the supernatants by ELISA.
5. Bioactivities of released cytokines (IL-2 and IL-12) are measured using standard cell-proliferation assays *(37,41)*. We have not tested the bioactivity of GM-CSF that is released from the microspheres (*see* **Note 2**).

3.2.4. Storage

1. Store the microspheres at −20°C under dessication as lyophilized powder (*see* **Note 3**).

3.3. Tumor Vaccination

1. Bring microspheres to room temperature, weigh on a microbalance and suspend in hydration buffer at a concentration of 200 mg/mL.
2. Vortex the suspension for 15 s and then sonicate in a water bath for an additional 15 s.
3. Add injection buffer to dilute the suspension to a final microsphere concentration of 40 mg/mL.
4. Sonicate the suspension in a bath-sonicator for 15 s (*see* **Note 4**).
5. Inject tumors (induction of tumors has been described; *41*) that are 3–4 mm in diameter with 50 µL of this suspension using a 28.5-gauge needle attached to a 0.5-mL insulin syringe. For maximum dispersal of the microspheres within the tumor, insert the needle into the center of the tumor nodule and inject about one-third of the solution. Pull the needle back slightly without exiting the tumor re-insert to one side and inject one-third more of the solution. The remaining solution is then injected to the opposite side of the tumor nodule using similar

technique. If the tumor is larger than 1 cm in diameter, the volume can be increased to 100 µL.

4. Notes

1. Losses of polymer and cytokine occur during the encapsulation process. Due to the nature of the equipment used, losses become very significant (up to 50%) when the polymer quantities are less than 100 mg. Thus it is advisable to keep the preparations to 100 mg or more. Another important yield parameter is the encapsulation efficiency. Encapsulation efficiencies have varied between 60–95% depending on the cytokine *(41)*. Whether this is related to the size and the surface properties of individual cytokines or to the presence of different excipients in the original cytokine preparations has not yet been investigated.
2. Release profiles typically display an initial burst of cytokine on d 1–3 and a rapid decline thereafter, stabilizing after d 7 *(37,40,41)*. The rate of release is dependent on the particular cytokine with low molecular-weight cytokines exhibiting a greater initial burst and a more rapid decline between d 3–7 *(37,40,41)*. Based on limited studies with two cytokines, we find that loss of bioactivity is significant and is cytokine-dependent. In the case of IL-2 the loss of specific activity is ~50% following encapsulation (37). In the case of interleukin (IL)-12, it is ~90% (unpublished observations). The increased sensitivity of IL-12 to encapsulation may be due to the fact that it is a heterodimeric molecule. In future studies, it will be important to optimize the encapsulation parameters to improve the recovery of bioactivity.
3. We demonstrated that IL-2-encapsulated microspheres remain fully active for at least 4 mo when stored at 4°C *(38)*. IL-12 microspheres on the other hand, lose their activity rapidly after 3–4 wk of storage (unpublished observations) indicating that stability characteristics can vary extensively depending on the cytokine utilized. We have recently started storing the microspheres at –20°C under dessication to determine whether the storage characteristics can be improved.
4. Proper rehydration of the microspheres is important for accurate dosing. The lyophilized material can have variable granularity and dispersal of clumps with uniform hydration is sometimes difficult to achieve. We find that increased vortexing and /or sonication times (up to 40 s) can be helpful. Another parameter that needs mention is the injection technique. We found that the single-injection technique described above is optimal since multiple injections into the same tumor nodule result in significant loss of microspheres due to leakage.

References

1. Sogn, J. A. (1998) *Tumor Immunology: The Glass is Half Full. Immunity* **9**, 757–763.
2. Davis, I. D. (2000) An overview of cancer immunotherapy. *Immunol. Cell Biol.* **78**, 179–195.

3. Minev, B. R., Chavez, F. L., and Mitchell, M. S. (1999) Cancer Vaccines: Novel approaches and new promise. *Pharmacol. Ther.* **81(2)**, 121–139.
4. Dranoff, G. (1998) Cancer gene therapy: connecting basic research with clinical inquiry. *J. Clin. Oncol.* **16**, 2548–2556.
5. Colombo, M. P. and Forni, G. (1997) Immunotherapy I: cytokine gene transfer strategies. *Cancer Met. Rev.* **16**, 421–432.
6. Gilboa, E. (1996) Immunotherapy of cancer with genetically modified tumor vaccines. *Semin. Oncol.* **23(1)**, 101–107.
7. Tuting, T., Storkus, W. J., and Lotze, M. T. (1997) Gene-based strategies for the immunotherapy of cancer. *J. Mol. Med.* **75**, 478–491.
8. Cavallo, F., Signorelli, P., Giovarelli, M., Musiani, P., Modesti, A., Brunda, M. J., et al. (1997) Anti-tumor efficacy of adenocarcinoma cells engineered to produce interleukin 12 (IL-12) or other cytokines compared with exogenous IL-12. *J. Natl. Cancer Inst.* **89(14)**, 1049–1058.
9. Dranoff, G., Jaffee, E., Lazenby, A., Golumbek, P., Levitsky, H., Brose, K., et al. (1993) Vaccination with irradiated tumor cells engineered to secrete murine granulocyte-macrophage colony stimulating factor stimulates potent, specific and long-lasting anti-tumor immunity. *Proc. Natl. Acad. Sci. USA* **90**, 3539–3543.
10. Cavallo, F., Di PIerro, F., Giovarelli, M., Gulino, A., Vacca, A., Stoppacciaro, A., et al. (1993) Protective and curative potential of vaccination with interleukin-2-gene-transfected cells from a spontaneous mouse mammary adenocarcinoma. *Cancer Res.* **53**, 5067–5070.
11. Allione, A., Consalvo, M., Nanni, P., Lollini, P. L., Cavallo, F., Giovarelli, M., et al. (1994) Immunizing and curative potential of replicating and nonreplicating murine mammary adenocarcinoma cells engineered with interleukin (IL)-2, IL-4, IL-6, IL-7, IL-10. Tumor necrosis factor-α, granulocyte-macrophage colony stimulating factor and γ-interferon gene or admixed with conventional adjuvants. *Cancer Res.* **54**, 6022–6026.
12. Gansbacher, B., Zier, K., Daniels, B. Cronin, K., Bannerji, R., and Gilboa, E. (1990) Interleukin-2 gene transfer into tumor cells abrogates tumorigenicity and induces protective immunity. *J. Exp. Med.* **172**, 1217–1224.
13. Gansbacher, B., Bannerji, R., Daniels, B. Zier, K., Cronin, K., and Gilboa, E. (1990) Retroviral vector-mediated gamma-interferon gene transfer into tumor cells generates potent and long-lasting anti-tumor immunity. *Cancer Res.* **50**, 7820–7825.
14. Golumbek, P. T., Lazenby, A. J., Levitsky, H. I., Jaffee, L. M., Karasuyama, H. Baker, M., and Pardoll, D. M. (1991) Treatment of established renal cancer by tumor cells engineered to secrete interleukin-4. *Science* **254**, 713–716.
15. Zitvogel, L., Tahara, H., Robbins, P. D., Storkus, W. J., Clarke, M. R., Nalesnik M. A., and Lotze, M. T. (1995) Cancer immunotherapy of established tumors with IL-12. Effective delivery by genetically engineered fibroblasts. *J. Immunol* **155**, 1393–1403.
16. Tahara, H., Zitvogel, L., Storkus, W. J., Zeh, H. J. 3rd, McKinney, T. G., Schreiber R. D., et al. (1995) Effective eradication of established murine tumors with IL-12 gene therapy using a polycistronic retroviral vector. *J. Immunol.* **154**, 6466–6474

17. Lotze, M. T. and Rubin, J. T. (1994) Gene therapy of cancer: a Pilot study of IL-4-gene-modified fibroblasts admixed with autologous tumor to elicit an immune response. *Human Gene Ther.* **5,** 41–55.
18. Bramson, J. L., Hitt, M., Addison, C. L., Muller, W. J., Gauldie, J., and Graham, F. L. (1996) Direct intratumoral injection of an adeno-virus expressing interleukin-12 induces regression and long-lasting immunity that is associated with highly localized expression of interleukin-12. *Hum. Gene Ther.* **7,** 1995–2002.
19. Zhang, J. F., Hu, C., Geng, Y., Selm, J., Klein, S. B., Orazi, A., and Taylor, M. W. (1996) Treatment of a human breast cancer xenograft with an adenovirus vector containing an interferon gene therapy. *Proc. Natl. Acad. Sci. USA* **93,** 4513–4518.
20. Toloza, E. M., Hunt, K., Swisher, S., McBride, W., Lau, R., Pang, S., et al. (1996) In vivo cancer gene therapy with a recombinant interleukin-2 adenovirus vector. *Cancer Gene Ther.* **3,** 11–17.
21. Hartikka, J., Sawdey, M., Cornefert-Jensen, F., Margalith, M., Barnhart, K., Nolasco, M., et al. (1996) An improved plasmid DNA expression vector for direct injection into skeletal muscle. *Hum. Gene Ther.* **7,** 1205–1217.
22. Arienti, F., Sulé-Suso, J., Belli, F., Mascheroni, L., Rivoltini, L., Melani, C., et al. (1996) Limited anti-tumor T cell response in melanoma patients vaccinated with interleukin-2 gene-transduced allogeneic melanoma cells. *Human Gene Ther.* **7,** 1955–1963.
23. Pardoll, D. M. (1995) Paracrine cytokine adjuvants in cancer immunotherapy. *Annu. Rev. Immunol.* **13,** 399–415.
24. Cavallo, F., Giovarelli, M., Gulino, A., Vacca, A., Stoppacciaro, A., Modesti, A., and Forni, G. (1992) Role of neutrophils and CD4+ T lymphocytes in the primary and memory response to non-immunogenic murine mammary adenocarcinoma made immunogenic by IL-2 gene transfection. *J. Immunol.* **149,** 3627–3635.
25. Brunda, M. J., Luistro, L., Warrier, R. R., Wright, R. B., Hubbard, B. R., Murphy, M., et al. (1993) Anti-tumor and antimetastatic activity of interleukin 12 against murine tumors. *J. Exp. Med.* **178,** 1223–1230.
26. Nastala, C. L., Edington, H. D., McKinney, T. G., Tahara, H., Nalesnik, M. A., Brunda, M. J., et al. Recombinant IL-12 administration induces tumor regression in association with IFN-g production.
27. Morikawa, K., Okada, F., Hosokawa, M., and Koybayashi, H. (1987) Enhancement of therapeutic effects of recombinant interleukin-2 on a transplantable rat fibrosarcomaby the use of a sustained release vehicle, pluronic gel. *Cancer Res.* **54,** 182–189.
28. Golumbek, P. T., Azhari, R., Jaffee, E. M., Levitsky, H. I., Lazenby, A., Leong, K., and Pardoll, D. M. (1993) Controlled release, biodegradable cytokine depots: a new approach in cancer vaccine design. *Cancer Res.* **53,** 5841–5844.
29. Langer, R. (1998) Drug delivery and targeting. *Nature* **392 (Suppl.),** 5–10.
30. Menei, P., Venier, M.-C., Gamelin, E., Saint-André, J.-P., Hayek, G., Jadaud, E., et al. (1999) Local and sustained delivery of 5-fluorouracil from biodegradable microspheres for the radiosensitization of glioblastoma. *Cancer* **86,** 325–330.

31. O'Hagan, D.T., Singh, M., and Gupta, R. K. (1998) Poly(lactide-co-glycolide) microparticles for the development of single-dose controlled-release vaccines. *Adv. Drug Del. Rev.* **32**, 225–246.
32. Liu, L.-S., Liu, S.-Q., Ng, S. Y., Froix, M., Ohno, T., and Heller, J. (1997) Controlled release of interleukin-2 for tumour immunotherapy using alginate/chitosan porous microspheres. (1997) *J. Controlled Rel.* **43**, 65–74.
33. Chen, L., Apre, R. N., and Cohen, S. (1997) Characterization of PLGA microspheres for the controlled delivery of IL-1a for tumor immunotherapy. *J. Controlled Rel.* **43**, 261–272.
34. Putney, S. D. and Burke, P. A. (1998) Improving protein therapeutics with sustained-release formulations. *Nature Biotechnol.* **16**, 153–157.
35. Mathiowitz, E., Jacob, J. S., Jong, Y. S., Carino, G. P., Chickering, D. E., Chaturvedl, P., et al. (1997) Biologically erodable microspheres as potential oral drug delivery systems. *Lett. Nature* **386**, 410–414.
36. Hora, M. S., Rana, R. K., Nunberg, J. H., Tice, T. R., Gilley, R. M., and Hudson, M. E. (1990) Controlled release of interleukin-2 from biodegradable microspheres. *Biotechnology* **8**, 755–758.
37. Egilmez, N. K., Jong, Y. S., Iwanuma, Y., Jacob, J. S., Santos, C. A., Chen, F.-A., et al. (1998) Cytokine immunotherapy of cancer with controlled release biodegradable microspheres in a human tumor xenograft/SCID mouse model. *Cancer Immunol. Immunother.* **46**, 21–24.
38. Jong, Y. S., Egilmez, N. K., Jacob, J. S., Smith, L. P., Mottl, T. S., Chen, F.-A., et al. (1999) Evaluation of cytokine delivery systems for cancer immunotherapy. *Proc. Mater. Res. Soc. Implants Tissue Engin.* **550**, 71–75.
39. Kuriakose, M. A., Chen, F.-A., Egilmez, N. K., Jong, Y. S., Mathiowitz, E., DeLacure, M. D., et al. (2000) Interleukin-12 delivered by biodegradable microspheres promotes the anti-tumor activity of human peripheral blood lymphocytes in a human head and neck tumor xenograft/SCID mouse model. *Head Neck Surg.* **22**, 57–63.
40. Egilmez, N. K., Jong, Y. S., Hess, S. D., Jacob, J. S., Mathiowitz, E., and Bankert, R. B. (2000) Cytokines delivered by biodegradable microspheres promote effective suppression of human tumors by human peripheral blood lymphocytes in the SCID/Winn model. *J. Immunother.* **23**, 190–195.
41. Egilmez, N. K., Jong, Y. S., Sabel, M. S., Jacob, J. S., Mathiowitz, E., and Bankert, R. B. (2000) *In situ* tumor vaccination with Interleukin-12 encapsulated biodegradable microspheres: induction of tumor regression and potent anti-tumor immunity. *Cancer Res.* **60**, 3832–3837.
42. Sabel, M. S., Hill, H., Jong, Y. S., Mathiowitz, E., Bankert, R. B., and Egilmez, N. K. (2001) Neoadjuvant therapy with IL-12-loaded PLA microspheres reduces local recurrence and distant metastases. *Surgery* **130(3)**, 470–478.

42

Adoptive Immunotherapy

Julian A. Kim and Suyu Shu

1. Introduction
1.1. Background

Murine models of pulmonary tumors have long provided investigators a reliable and reproducible method of testing the therapeutic efficacy of adoptive immunotherapy. Adoptive immunotherapy involves the generation of tumor-specific immune cells which, following activation and culture in vitro, can mediate regression of established tumors following transfer into a tumor-bearing host.

Adoptive immunotherapy can be tested in preclinical models in several ways: 1) transfer of human immune effector cells into immunodeficient mice bearing human tumor xenografts, 2) transfer of human immune effector cells into immunodeficient mice bearing human tumor xenografts with a reconstituted human hematopoietic system, or 3) transfer of murine immune effector cells into mice bearing syngeneic tumors. Owing to the similarities of the murine immune system to the human and the availability of immunologic reagents in mice, our laboratory has focused on the use of poorly immunogenic tumors in syngeneic mice as a preclinical model. Although these preclinical models do not predict the therapeutic efficacy of adoptive immunotherapy in humans, they do provide a method of analyzing the mechanisms involved in cell-mediated immunologic regression of established tumors and provide the rationale for translation into human clinical trials.

1.2. Adoptive Immunotherapy Using Tumor-Draining Lymph Nodes

The passive transfer of T cells but not serum from an immune mouse into a naïve host results in protective immunity. Based on this observation, the transfer of activated T cells in an attempt to treat established tumor as a form of therapy resulted in development of a process known as adoptive immunotherapy. The use of preclinical animal models to understand the mechanism of tumor rejection in vivo and the factors that influence therapeutic activity of the transferred T cells is well-established.

Adoptive immunotherapy protocols vary widely in terms of the source of immune cells, the in vitro activation of the cells, the functional characteristics of the resultant effector cells, and the therapeutic efficacy in vivo. Over 20 years ago, it was established that lymph nodes draining a progressive subcutaneous tumor were a consistent and reliable source of T cells sensitized to tumor rejection antigens in vivo *(1–5)*. These cells were termed "pre-effector" cells, because they did not exhibit stand alone effector function, but following activation and culture in vitro, could mediate regression of established tumor. Tumor-draining LN cells activated with immobilized anti-CD3 and cultured for a short term in low-dose interleukin (IL)-2 resulted in the generation of activated T cells which were immunologically specific to recognize the tumor which they originally drained but not a closely related tumor. The effector cells were > 95% TcR + and the therapeutic effects in vivo appeared to be dependent on both CD4+ and CD8+ T cells *(6–8)*. Additional studies have demonstrated that the population of cells with downregulation of expression of L-selectin (CD62L) within freshly harvested tumor-draining LNs appear to contain the fraction of pre-effector cells which contain the potential for therapeutic activity following culture activation and adoptive transfer *(9,10)*.

This chapter will provide a detailed summary of adoptive immunotherapy using lymph nodes (LN) as a source of tumor-specific T cells. This protocol is robust and is widely applicable to many different murine tumor models and animal strains *(11–15)*. This chapter will provide technical features of the generation of tumor-draining lymph nodes, the activation and culture of T cells, the adoptive transfer of activated T cells, establishment of experimental pulmonary tumors, and determination of therapeutic efficacy of the activated T cells in vivo. A notes section is provided to add technical details to the corresponding methods and to help with troubleshooting.

2. Materials

Most of the materials required will be commonly found in laboratories that perform tissue culture, lymphocyte culture, and mouse experiments. It should be emphasized that *all experiments that involve mice should be performed in a manner that is consistent with USDA/IACUC regulations on the care and handling of animals*. Most institutions require a written protocol that outlines the specific procedures that are being proposed along with justification for the use of the particular animal model and the number of animals required. Failure to comply with these regulations may be punishable by institutional and federal law.

2.1. Generation of Tumor-Draining LNs

2.1.1. Generation of Subdermal Tumors

1. Female C57BL/6 (B6) mice: Animals should be purchased at 6–8 wk of age (Jackson Laboratories, Bar Harbor, ME) and housed in a pathogen-free environment (*see* **Note 1**). Ideally mice should be maintained for 2 wk prior to tumor inoculation so that they can become acclimated to their new environment. Mice are housed in cages and given chow and fresh water ad libitum.
2. MCA 205 fibrosarcoma (Surgery Branch, National Cancer Institute): This weakly immunogenic tumor line of B6 origin was generated in 1986 by injecting 3-methylcholanthrene intramuscularly *(16)*. The tumor is maintained by serial subcutaneous transplantation in syngeneic B6 mice.
3. Hank's Balanced Salt Solution (HBSS) for suspension of cells (BioWhittaker, Walkersville, MD).
4. 1 cc tuberculin syringes for intradermal injections.

2.1.2. Tumor Cell Preparation and Inoculation for Generation of Tumor-Draining LN

1. Set up for small animal dissection.
2. Tumor digestion buffer: 40 mL HBSS supplemented with 4 mg DNAse, 100 U hyaluronidase, and 40 mg collagenase.

2.2. Culture Activation of Tumor-Draining LNs

2.2.1. Surgical Removal of Tumor-Draining LN

1. 1 cc tuberculin syringes and 20-gauge needles.
2. 100 micron nylon mesh cell strainers (Falcon Becton-Dickinson, Franklin Lakes, NJ).

2.2.2. Tumor Draining LN Cell Culture and Activation

1. Anti-CD3 monoclonal antibody from the 2C11 hybridoma (ATCC, Rockville, MD).
2. 24-Well plates (CoStar, Cambridge, MA) coated with protein A (Sigma Chemicals, St. Louis, MO).
3. Complete medium (CM): RPMI 1640 supplemented with 10% heat-inactivated fetal calf serum, 0.1 mM nonessential amino acids, 1 mM sodium pyruvate,

2 mM glutamine, 100 mg/mL streptomycin (Gibco-BRL, Grand Island, NY), and 5 × 10^{-5} M 2-mercaptoethanol. Antifungal agents such as amphotericin B at 0.5 mg/mL may also be added.
4. Sterile Pasteur pipets.
5. Gas-permeable bags (Baxter-Fenwal, Deerfield, IL).
6. Expansion medium: CM containing recombinant IL-2 at 4 1U/mL (Cetus Corp., Emeryville, CA).

2.3. Adoptive immunotherapy protocol

2.3.1. Isolation of L-Selectin^{low} Cells from TDLNs

1. Anti-L-selectin magnetic beads: magnetic microbeads directly labeled with anti-CD62L monoclonal antibody (MAb) (MiniMACS, Miltenyi Biotech, Auburn, CA).
2. Immunomagnetic column system (MiniMACS, Miltenyi Biotech, Auburn, CA).
3. MACS Elution buffer: Sterile phosphate-buffered saline (PBS) with 0.5% fetal bovine serum (FBS) (BioWhittaker, Walkersville, MD).

2.3.2. Establishment and Treatment of Experimental Pulmonary Metastases by Adoptive Transfer

1. Trypan blue.
2. 3 or 5 cc Syringes with a 27- or 30-gauge needle for intravenous injection.

2.3.3. Analysis of Results of Adoptive Transfer

1. India ink (Sigma Chemicals): Dilute in PBS to a final concentration of 15% for use in counterstaining in pulmonary metastases.

3. Methods

The following procedures are routine protocol in our laboratory and have been validated by peer-review. The procedures have been optimized for reproducibility and consistency of in vivo therapeutic results. (Note: The timeline is for experiments using d 12 tumor-draining LN cells for subsequent adoptive immunotherapy experiments. For experiments using d 9 tumor-draining LN, the sc tumor inoculation can be performed on a Monday instead of a Friday.)

3.1. Recommended Timeline for Tumor Injection and Generation of LN Cells

3.1.1. Generation of Tumor-Draining LNs

1. D 0 (Friday): Inoculation of tumor for generation of tumor-draining LN.
2. D 12 (Wednesday): Surgical removal of tumor-draining LN.

3.1.2. Culture Activation of Tumor-Draining LNs

1. D 10 (Monday): Incubate 24-well plates with Protein A.
2. D 11 (Tuesday): Wash 24-well plates and incubate with anti-CD3 MAb (2C11).
3. D 12 (Wednesday): Wash 24-well plates and overlay tumor-draining LN cells.
4. Perform FACS analysis on fresh tumor-draining LN cells.
5. D 14 (Friday): Harvest 24-well plates and resuspended LN cells in IL-2
6. FACS analysis of activated tumor-draining LN cells.

3.1.3. Adoptive Immunotherapy Protocol

1. D 14 (Friday): Inject mice with MCA 205 to establish pulmonary metastases.
2. D 17 (Monday): Harvest tumor-draining LN cells and prepare single cell suspension.
3. Calculate expansion rate; perform FACS analysis and functional tests (ELISA).
4. Transfer activated tumor-draining LN cells into tumor-bearing mice.
5. D 34–37: Harvest lungs to enumerate pulmonary metastases.

3.2. Generation of Tumor-Draining LNs

3.2.1. Generation of Subdermal Tumors

Although the adoptive immunotherapy protocol can be applied to many different tumor models, the most extensive experience is with the MCA 205 tumor model. This line grows well as a subcutaneous nodule after subdermal injection. Mice should be examined daily, and any signs of tumor overgrowth (maximum diameter>20 mm), ulceration, or disability of the mouse should prompt tumor harvest (*see* **Note 2**). Due to the fact that there are several components to the adoptive immunotherapy protocol, it is essential that attention be paid to the timing of administration of the initial sc tumor inoculation (*see* **Fig. 1**).

1. Maintain the cell line by serial subcutaneous inoculation and use in the third to seventh transplantation passage.
2. Digest the cultured tumor cells to a single-cell suspension and inoculate $1–3 \times 10^6$ MCA 205 intradermally into the flank (*see* **Note 3**). Use female B6 mice at the age of 8–10 wk, and maintain in a pathogen-free environment for tumor passaging.

3.2.2. Tumor Cell Preparation and Inoculation for Generation of Tumor-Draining LN

1. Once the subcutaneous tumors reach 10 mm in diameter, harvest the tumor under sterile conditions.
2. Prepare a single-cell suspension by digestion of minced tumors in 40 mL tumor-digestion buffer at room temperature for three hours.

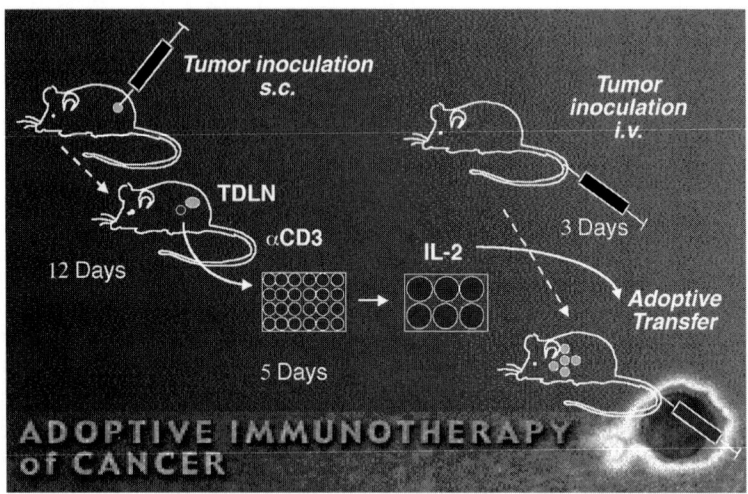

Fig. 1. Adoptive immunotherapy using tumor-draining lymph node cells. The protocol consists of sc tumor inoculation and generation of tumor-draining LNs, culture activation of tumor-draining LN cells in vitro, establishment of pulmonary metastases and adoptive transfer of activated T cells.

3. Wash the cell suspension twice and re-suspend at a concentration of 3×10^7 cells/mL in HBSS.
4. For generation of tumor-draining LN, inoculate 0.05 mL (50 µL) subcutaneously into the hind flanks, bilaterally (*see* **Fig. 2**)

3.3. Culture Activation of Tumor-Draining LNs (see Note 4)

Under sterile conditions, harvest tumor-draining LN on postinoculation d 9 or 12.

3.3.1. Surgical Removal of Tumor-Draining LN

1. Euthanize mice by CO_2 inhalation, and perform a midline incision.
2. Locate the tumor-draining lymph nodes in the iliac fossa, overlying or just lateral to the iliopsoas muscle.
3. Place LN into cold HBSS until processing. If the tumor is invading into the lymph node by direct extension, that particular lymph node should not be used.
4. Using 20-gauge needles, tease apart the LN and use the blunt end of a syringe plunger to disperse the cells out of the fibrous lymph node capsule (*see* **Fig. 3**)
5. Filter the suspension through a 100 micron nylon mesh cell strainer to obtain a single-cell suspension. This technique has been used regularly in our laboratory, and should result in an expedient retrieval of large numbers of viable cells (*see* **Note 5**).

Adoptive Immunotherapy

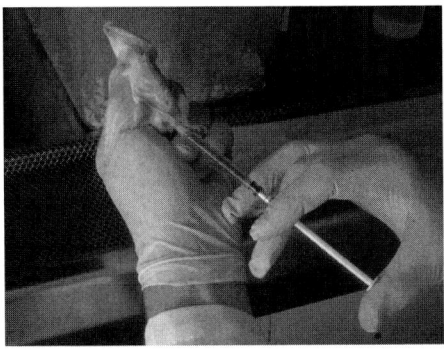
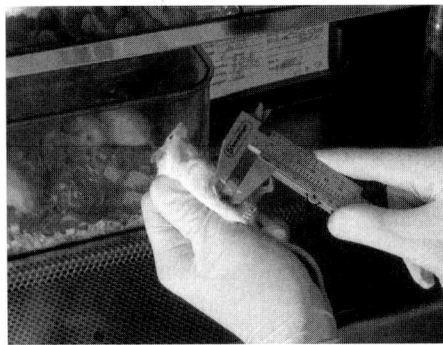

Fig. 2. Mice are inoculated with tumor-cell suspension in the flank bilaterally (left panel) for generation of tumor-draining lymph nodes. Subcutaneous tumor growth should be monitored daily and recorded prior to harvesting of lymph nodes on d 12 (right panel). Note: pictured are BALB/c mice.

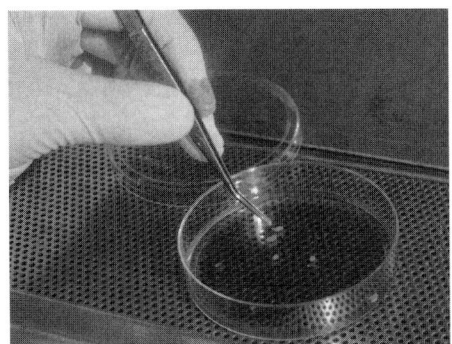

Fig. 3. Tumor-draining LNs are mechanically dissociated (left panel) and placed into a single-cell suspension. LN cells are then activated in 24-well plates (right panel) with immobilized anti-CD3 MAb for 48 h.

3.3.2. Tumor Draining LN Cell Culture and Activation

1. Plate the LN cells on immobilized anti-CD3 monoclonal antibody (MAb)-coated 24-well tissue-culture dishes in 2 mL complete media (CM) at a concentration of 4×10^6 cells per well.
2. Incubate cultures at 37°C in 5% CO_2 for 2 d (*see* **Note 6**).
3. Harvest cells from the 24-well plates using Pasteur pipets to disperse the cells into a single-cell suspension. Mechanical (scrapers, rubber policeman) or enzymatic techniques (trypsin) should not be necessary. Check plates following pipetting to ensure that all adherent and nonadherent cells are adequately retrieved. Rinse with HBSS and repeat as necessary to harvest additional cells.

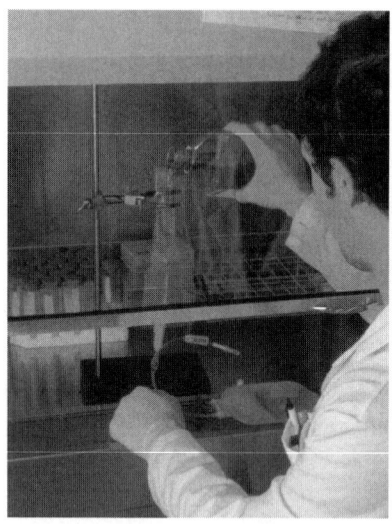 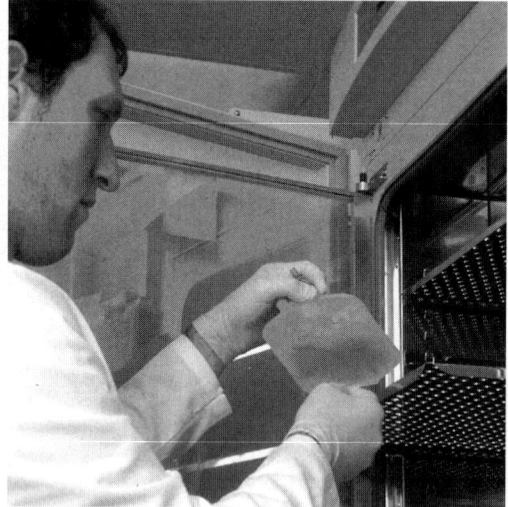

Fig. 4. Anti-CD3 activated cells are resuspended in CM containing low dose IL-2 (left panel) and transferred into gas permeable-bags. The cells are then expanded in IL-2 for 3 d prior to final harvest (right panel).

4. Wash activated tumor-draining LN cells in HBSS, count, and resuspend in expansion medium at a final cell concentration of 1×10^5 cells/mL (**Fig. 4**).
5. Transfer the cells to gas permeable bags (Baxter-Fenwal) and incubate at 37°C in 5% CO_2 for 3 d.

3.4. Adoptive Immunotherapy Protocol

A typical expansion index (EI) and phenotype of MCA 205 tumor-draining LN cells prior to and following culture activation is summarized in **Tables 1 and 2**.

3.4.1. Preparation of LNs for Adoptive Transfer

1. Following expansion in IL-2, wash cells, count, and resuspend in HBSS for adoptive transfer.
2. Preserve an aliquot of cells for routine FACS analysis to characterize the phenotype of the cells that are transferred.
   Cells may also be preserved for cytokine release assays using ELISA, cytotoxicity assays, lymphoproliferative assays, or gene-expression profiling using reverse transcriptase polymerase chain reaction (RT-PCR) for characterization of functional qualities in vitro.

Table 1
Expansion Index (E. I.) of Tumor-Draining Lymph Node Cells

Tumor used to innoculate	Cells/LN	Anti-CD3 E.I.	IL-2 E.I.	Total E.I.
MCA 205	17.4×10^6	.92	4.45	4.1

Table 2
Phenotype of Tumor-Draning Lymph Node Cells

Tumor	CD4/CD8	% T-cells	% L-Sel^{low}	% CD25$^+$
Day 0 (After harvest)				
MCA 205	25.6/23.7	59.3	43.7	14.4
Day 2 (After anti-CD3 activation)				
MCA 205	39.9/41.7	67.3	47.0	86.8
Day 5 (After IL-2 expansion)				
MCA 205	30.0/63.1	96.6	55.6	84.1

3.4.2. Isolation of L-Selectin^{low} Cells from TDLNs

Studies from our laboratory have demonstrated that the population of lymphocytes within TDLNs that express low levels of the adhesion molecule L-selectin appear to contain the majority of the therapeutic activity in vivo. Tumor-draining LN cells with both "high" and "low" expression of L-selectin may be isolated from freshly harvested TDLNs using immunomagnetic beads and a magnetic column (MACS) (*see* **Note 7**).

1. Incubate lymphocytes with anti-L-selectin magnetic beads for 1 h at room temperature.
2. Subject the column to a magnetic field. The column should be degassed with 3 exchanges (4–6 mL) of MACS elution buffer prior to loading cells.
3. Wash cells in HBSS and resuspend in MACS elution buffer at a concentration of 100×10^6 cells/mL. Slowly add cell suspension to the column.
4. Carefully add 3 exchanges of MACS buffer to elute the L-selectin^{low} expressing cells. Buffer should be added continuously such that the flow rate is constant and there are no bubbles within the column.
5. Following removal of the magnetic field, add several exchanges of MACS elution buffer to obtain the L-selectin^{high} expressing cells.

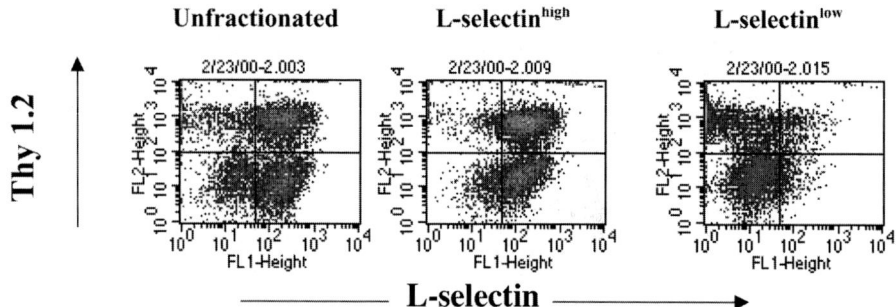

Fig. 5. FACS analysis of tumor-draining LN cells following separation of L-selectin^{high} (middle panel) and L-selectin^{low} (right panel) populations. Unfractionated LN cells are seen in the left panel.

6. Confirm the efficiency of separation by FACS analysis of both L-selectin^{low} and L-selectin^{high} populations. Using this technique the purity should be approximately 90% (see **Fig. 5**). The total yield, when both cell populations are accounted for is variable but is usually around 50–75%. This yield is acceptable as compared to other antibody-based columns or panning techniques. The yield of L-selectin^{low} cells will vary depending on the percentage of those cells in the starting unfractionated LN population (typically 25–40%).

3.4.3. Establishment and Treatment of Experimental Pulmonary Metastases by Adoptive Transfer

Adoptive transfer of activated TDLNs can be performed against 3-d established tumors of the lung, brain, and subcutaneous sites. The following applies to the treatment of established pulmonary tumors:

1. Establish pulmonary metastases by injecting 2.5×10^5 MCA 205 cells in 1cc HBSS via tail vein (see **Fig. 6**) (see **Notes 8, 9**).
2. Harvest the activated TDLN T-cells on the day of adoptive transfer (d 5 of culture).
3. Wash cells with HBSS and resuspend prior to a viability count using trypan blue and hemacytometer.
4. Dilute cells to the proper concentration dependent on the dose needed for each group of mice.
5. Administer activated TDLNs in 1 cc via tail vein injection (see **Notes 10, 11**).

3.4.4. Analysis of Results of Adoptive Transfer

3.4.4.1. ENUMERATION OF PULMONARY METASTASES

1. Mice with pulmonary are euthanized by CO_2 inhalation on d 17–20 and their lungs insufflated with India ink (see **Note 12**).

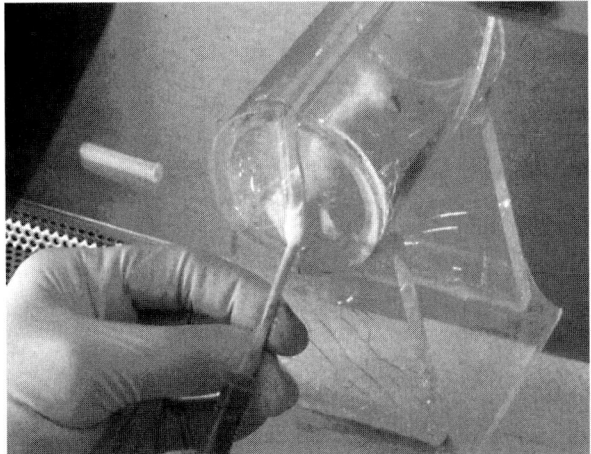

Fig. 6. iv tail vein injection of MCA 205 tumor cells to establish pulmonary metastases. Note: BALB/c mice are shown.

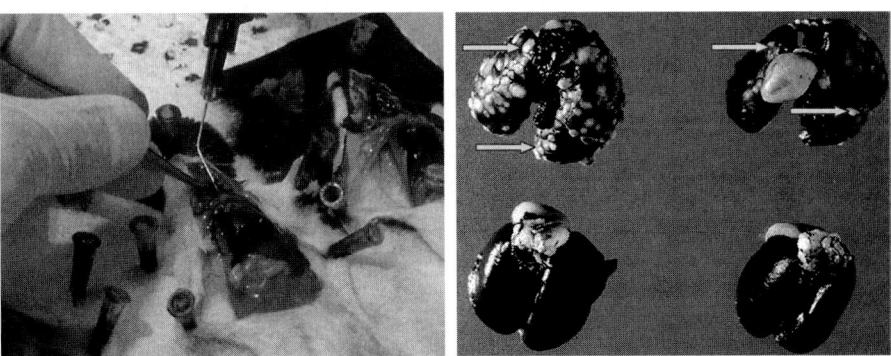

Fig. 7. Following euthanasia, intratracheal injection of India ink is performed (left panel) to assist in visualization of pulmonary metastases. Right panel shows typical findings of mice from control group (above) and mice treated with activated immune cells (below). Note multiple surface pulmonary nodules in the mice in the control group vs no tumor seen in mice treated with adoptive immunotherapy. The larger central organ is the heart and mediastinal structures.

2. Fix whole lungs by submerging and storing in Fekate's solution (*see* **Fig. 7**). Enumerate the pulmonary metastases by visual inspection of surface metastases *(17)*. Individual lungs are viewed under low-power lighted magnification and divided into five separate lobes prior to enumeration and recording of surface metastases. The individual counting the pulmonary tumors should be blinded as

to the experimental treatment group to minimize bias. Lungs that contain nodules too numerous to count should be designated as containing > 250.

3.4.4.2. STATISTICAL ANALYSIS

1. Statistical analysis using Wilcoxon rank sum may be used to determine differences in the number of pulmonary metastases in each group.
2. Five mice per group will typically result in enough power to draw statistical inferences. Use of less than five mice per group may impact on the power of the statistical analysis.

4. Notes

1. It is important to use mice that are 8–12 wk of age for generation of tumor-draining LN, as younger mice may lead to fewer numbers of cells per lymph node.
2. Care must be taken during harvesting and preparation of tumor cells, as the quality of the tumor cell preparation will directly influence the generation of tumor-draining LN cells and formation of pulmonary metastases. MCA 205 subcutaneous tumors should be harvested when there is minimal central necrosis, and viability of cells should be carefully checked following digestion. Tumor cell preparations with a high percentage (>25%) of nonviable cells typically result in poor tumor formation in vivo.
3. Inoculation of tumor cells for generation of tumor-draining LN must be performed intradermally, or the tumor may grow and invade the draining LN. Intradermal injection is best confirmed by creation of a "wheal." A 1–2 cm subcutaneous needle tract is helpful in allowing the inoculum to be retained and will contribute to consistent tumor growth and lymph node generation between experimental animals.
4. Consistent activation and growth of tumor-draining LN cells in vitro is the most important component of adoptive immunotherapy. As a general rule, poor yield from culture activation results in poor therapeutic activity in vivo following adoptive transfer.
5. Tumor-draining LN removal and processing into single-cell suspension should be performed quickly and cells kept on ice to maximize viability.
6. During activation with immobilized anti-CD3 mAb, 24-well plates should be checked for evidence of formation of large cellular aggregates that confirm T-cell activation. Cell yield following 48-h activation should be >85% of starting cell number. Expansion rate of cells incubated with IL-2 will range from three- to sixfold, with viability >90%.
7. The methods presented result in isolation of L-selectin[low] cells from a total LN population. Typical yields of L-selectin[low] cells are 8–12% of the total cells placed onto the MACS column. Isolation of L-selectin[low] T cells may be obtained by either isolating T cells from tumor-draining LN cells prior to MACS separation or by using a rat T cell column that will remove T cells labeled with rat-anti-mouse L-selectin Ig as well as antigen-presenting cells (APCs) by nonspecific means. Activation of L-selectin[low] T cells may require the addition of irradiated (500 cGy) feeder cells, usually in the form of naïve splenocytes

The recommended ratio of LN cells: feeder cells is 1:1, with the total number of cells per well not to exceed 4×10^6.

8. As mentioned previously, the viability and overall quality of the tumor cell inoculum will dramatically affect the number of pulmonary metastases formed within an experiment. The technical proficiency of the tail vein injection will affect the variability of pulmonary metastases formed in mice within each group. It is critical to inject precisely each mouse with the same volume of cell suspension, and to keep the cell suspension well mixed as the injections are being performed.
9. It is helpful to prepare the mice by applying a gentle warming using a heating lamp for 5 min prior to tail vein injection. This allows the tail vein to dilate and become more visible. The veins are paired and located on the lateral sides of the tail. The injection is performed by inserting the needle with the bevel facing upward starting at the point most distal and working towards the base of the tail. This technique allows for multiple attempts at injection for each tail vein if initial attempts are unsuccessful. Injection of the suspension should be met with little resistance, which may be a sign of extravasation of the inoculum into the extravascular space. In general, it is advised to preserve one tail vein for the injection of activated T cells on d 3 following tumor inoculation. Mice should be ear-tagged, and any difficulties with injection should be noted.

Prior to injection of activated T cells, mice should be randomized to different cages to reduce the influence of bias related to injection technique. Tumor-draining LN cells should be transferred into tumor-bearing mice shortly after harvest from culture. Again, proper mixing of cells during injection is important and will reduce intergroup variability in the results. The methods described use T-cell transfer alone. However, mice may also be treated with reagents that will modulate effector function such as MAbs or IL-2. These reagents may be administered on the same day as cell transfer, typically 2–4 h following iv administration of T cells. For 3 d established MCA 205 pulmonary tumors, T-cells doses of $5-8 \times 10^6$ L-selectin^{low} doses of $1-2 \times 10^6$ cells are usually curative.

11. Mice in the control or untreated group should be monitored daily for signs of respiratory distress, and any mice showing moderate distress should be euthanized according to institutional protocols.
12. On d 17 following tumor inoculation, a single mouse from the control group that appears to be the least compromised should be sacrificed to confirm the presence of multiple pulmonary metastases. All mice should then be euthanized for enumeration of pulmonary metastases.

References

1. Sussman, J. J., Shu, S., Sondak, V. K., and Chang, A. E. (1994) Activation of T lymphocytes for the adoptive immunotherapy of cancer. *Ann. Surg. Oncol.* **1(4),** 296–306.
2. Yoshizawa, H., Sakai, K., Chang, A. E., and Shu, S. (1991) Activation by anti-CD3 of tumor-draining lymph node cells for specific adoptive immunotherapy. *Cell Immunol.* **134(2),** 473–479.

3. Yoshizawa, H., Chang, A. E., and Shu, S. (1991) Specific adoptive immunotherapy mediated by tumor-draining lymph node cells sequentially activated with anti-CD3 and IL-2. *J. Immunol.* **147,** 729–737.
4. Chang, A. E. and Shu, S. (1996) Current status of adoptive immunotherapy of cancer. *Crit. Rev. Oncol./Hemat.* **22,** 213–228.
5. Chang, A. E., Aruga, A., Cameron, M. J., Sondak, V. K., Normolle, D. P., Fox, B. A., and Shu, S. (1997) Adoptive immunotherapy with vaccine-primed lymph node cells secondarily activated with anti-CD3 and interleukin-2. *J. Clin. Oncol.* **15,** 796–807.
6. Wahl, W. L., Strome, S. E., Nabel, G. J., Plautz, G. E., Cameron, M. J., San, H., et al. (1995) Generation of therapeutic T lymphocytes after in vivo tumor transfection with an allogeneic class I major histocompatibility complex gene. *J. Immunother.* **17,** 1–11.
7. Arca, M. J., Krauss, J. C., Aruga, A., Cameron, M. J., Shu, S., and Chang, A. E. (1996) Therapeutic efficacy of T cells derived from lymph nodes draining a poorly immunogenic tumor transduced to secrete granulocyte-macrophage colony-stimulating factor. *Cancer Gene Ther.* **3,** 39–47.
8. Krauss, J. C., Strome, S. E., Chang, A., and Shu, S. (1994) Enhancement of immune reactivity in the lymph node draining a murine melanoma engineered to elaborate IL-4. *J. Immunother.* **16,** 77–84.
9. Kagamu, H. and Shu, S. (1998) Purification of L-selectin^{low} cells promotes the generation of highly potent CD4 antitumor or effector T lymphocytes. *J. Immunol.* **160,** 3444–3452.
10. Kagamu, H., Touhalisky, J. E., Plautz, G. E., Krauss, J. C., and Shu, S. (1996) Isolation based on L-selectin expression of immune effector cells derived from tumor-draining lymph nodes. *Cancer Res.* **56,** 4338–4342.
11. Inoue, M., Plautz, G. E., and Shu, S. (1996) Treatment of intracranial tumors by systemic transfer of superantigen-activated tumor-draining lymph node T cells. *Cancer Res.* **56,** 4702–4708.
12. Krauss, J. C., Stein, J. M., and Shu, S. (1996) Tumor reactivity of immune T cells in short-term culture. *Cancer Immunol. Immunother.* **43,** 231–239.
13. Plautz, G. E., Touhalisky, J. E., and Shu, S. (1997) Treatment of murine gliomas by adoptive transfer of ex-vivo activated tumor-draining lymph node cells. *Cell Immunol.* **178,** 101–107.
14. Plautz, G. E. and Shu, S. (1997) Adoptive immunotherapy of intracranial tumors by systemic transfer of tumor-draining lymph node cells. *Intl. J. Oncol.* **11,** 389–395.
15. Peng, L., Shu, S., and Krauss, J. C. (1997) Treatment of subcutaneous tumor with adoptively transferred T cells. *Cell Immunol.* **178,** 24–32.
16. Shu, S. and Rosenberg, S. A. (1985) Adoptive immunotherapy of newly induced murine sarcomas. *Cancer Res.* **45,** 1657–1662.
17. Wexler, H. (1966) Accurate identification of experimental pulmonary metastases. *J. Natl. Cancer Inst.* **36(4),** 641–645.

43

Intratumoral Therapy with Cytokine Gene-Modified Dendritic Cells in Murine Lung Cancer Models

Sherven Sharma, Seok Chul Yang, Raj K. Batra, and Steven M. Dubinett

1. Introduction

The central importance of functional host professional antigen-presenting cell (APC) in the immune response against cancer has been well-defined *(1)*. Dendritic cells (DC) are highly specialized professional APC with potent capacity to capture, process, and present antigen to T cells *(2)*. Unfortunately, tumor cells interfere with host DC maturation and function *(3)*. To circumvent tumor-mediated inhibition of DC maturation and function in vivo, DC that have undergone cytokine-stimulated maturation ex vivo have been effectively utilized in clinical trials and murine cancer models *(4–8)* Antigen-specific cytotoxic T lymphocytes (CTL) responses have been achieved utilizing ex vivo antigen-pulsed DC *(5,9–11)*. Additionally, novel tumor antigen-delivery systems utilizing cytokine gene-transduced tumor cells and DC *(4,7)* or fusion of tumor cells with DC have resulted in induction of antitumor immunity *(12,13)*. Delivery of tumor antigens by ex vivo stimulated DC has been shown to be superior to purified peptides in avoiding CTL tolerization *(14)*. Vaccination with multiple tumor antigens may be superior to the use of a single epitope *(15,16)* and these responses can be further enhanced by the co-administration of immune-potentiating cytokines *(4,7)*. We have demonstrated previously that interleukin (IL)-7-transduced tumor cells administered intratumorally in conjunction with DC elicit potent antitumor responses in a murine model of established lung cancer *(7)*. IL-7 delivery to the tumor site may be beneficial because it enhances Ag-specific T-cell cytotoxicity *(17)* and synergizes with IL-12 in the induction of T-cell proliferation, cytotoxicity and interferon-γ (IFN-γ) release *(18)*. IL-7 also downregulates both macrophage and tumor production of transforming growth factor-β (TGF-β) and thus may serve to limit tumor-induced immune suppression *(19,20)*. However, despite these beneficial effects of IL-7, maximal antitumor benefit was demonstrated previously when bone-marrow derived DC were administered intratumorally in conjunction with IL-7 gene modified autologous tumor *(7)*.

Here we investigate an approach for achieving tumor antigen presentation utilizing the tumor as an in vivo source of antigen for DC. In contrast to in vitro immunization with purified peptide Ag, autologous tumor has the capacity to provide the activated DC administered at the tumor site access to the entire repertoire of available antigens *in situ*. This may increase the likelihood of a response and reduce the potential for tumor resistance due to phenotypic modulation. In this study we have evaluated the capacity of intratumoral DC-AdIL-7 to generate antitumor responses and systemic immunity in two murine models of lung cancer. We report that intratumoral injection of DC-AdIL-7 is effective in eradicating established tumors and generating systemic antitumor immune responses.

In two murine lung cancer models adenoviral IL-7-transduced dendritic cells (DC-AdIL-7) were administered intratumorally resulting in complete tumor regression. Intratumoral DC-AdIL-7 therapy was as effective as DC pulsed with specific tumor peptide antigens. Comparison of other intratumoral therapies including recombinant IL-7, AdIL-7 vector alone, unmodified DC, IL-7 transduced fibroblasts, or DC pulsed with tumor lysates revealed DC-AdIL-7 therapy to be superior in achieving antitumor responses and augmenting immunogenicity. Mice with complete tumor eradication as a result of either DC-AdIL-7 or AdIL-7 therapy were rechallenged with parental tumor cells 30 or more days following complete tumor eradication. All the DC-AdIL-7 treated mice completely rejected a secondary rechallenge, whereas the AdIL-7 treated mice had sustained antitumor effects in only 20–25% of the mice. DC-AdIL-7 therapy was more effective than AdIL-7 in achieving systemic antitumor responses and enhancing immunogenicity. Following complete tumor eradication, those mice treated with DC-AdIL-7 evidenced significantly greater release of splenocyte GM-CSF and IFN-γ that did controls or AdIL-7-treated mice. Following intratumoral injection, gene-modified DC trafficked from the tumor to lymph node sites and spleen. DC were detected in nodal tissues for up to 7 d following intratumoral injection. We report intratumoral DC-AdIL-7 leads to significant systemic immune responses and potent antitumor effects in murine lung cancer models.

2. Materials

2.1. Maintenance and Propagation of Cell Lines

1. Line 1 alveolar cell (L1C2, H-2^d) or Lewis lung carcinomas (3LL, H-2^b) were obtained from ATCC (Rockville, MD). The cell line was mycoplasma free and cells were utilized up to the tenth passage before thawing frozen stock cells from liquid N2.
2. Other cell lines utilized for these experiments include normal fibroblasts CRL-9392, B16 (H2^b), and EL4 (H2^b) cell lines (ATCC, Rockville, MD).
3. Cells cultured in complete medium (CM): RPMI 1640 plus glutamine (Irvine Scientific, Santa Ana, CA) supplemented with 10% fetal calf serum (FCS) (Gemini Bioproducts, Calabasas, CA), antibiotics penicillin (100 U), streptomycin (100 µg) (CM), and kept at 37°C in a humidified atmosphere containing 5% CO_2.
4. Fibroblasts were cultured in a 1:1 mixture of Dulbecco's modified Eagle's medium (DMEM) and Ham's F11, containing: 10 µg/mL bovine insulin, 15 mM HEPES buffer (Gibco); 25 µg/mL human transferrin; 20 µg/mL human high-density lipoprotein (HDL); 1 U/mL human platelet derived growth factor (PDGF); 100 ng/mL mouse epidermal growth factor (EGF) (Sigma, St. Louis, MO).

2.2. Isolation and Propagation of Bone Marrow-Derived Dendritic Cells

1. Lymphocyte and macrophage- depleting antibodies, (CD45R, anti-B cell; TIB 229, anti-Ia; TIB 150, anti-CD8 and TIB 207, anti-CD4, all obtained from ATCC).
2. Rabbit serum complement (Sigma).
3. GM-CSF and IL-4 (R & D Systems, Minneapolis, MN).
4. 12-well Costar (Cambridge, MA) plates.
5. FITC-labeled antibodies to B7-1, B7-2, CD11c and MHC II (B.D. Pharmingen, San Diego).

2.3. Transduction of Dendritic Cells and Fibroblasts with Adenoviral 5 Vector

1. The adenoviral construct (AdIL-7) is an E1-deleted, replication-deficient adenoviral type 5 vector (Ad 5) encoding a human IL-7 cDNA (24). The IL-7 cDNA was inserted into the former E1 site and driven by the CMV promoter/enhancer.
2. The control vector (Ad RR 5) is an E1-deleted, replication-deficient adenoviral type 5 vector (Ad 5) that did not contain the human IL-7 cDNA.
3. The Ad 5 Lac Z vector is an E1-deleted, replication-deficient adenoviral type 5 vector (Ad 5) that encodes the beta galactosidase gene.
4. Mouse bone marrow derived dendritic cells.
5. Complete medium RPMI 1640 plus glutamine (Irvine Scientific) supplemented with 10% fetal calf serum (FCS) (Gemini Bioproducts), antibiotics penicillin

(100 U), streptomycin (100 μg) (CM), and kept at 37°C in a humidified atmosphere containing 5% CO_2.

2.4. Measurement of IL-7 from Mouse Bone Marrow Dendritic Cells

1. mouse anti-human IL-7 mAb, 4 μg/mL (PharMingen, San Diego, CA).
2. 10% fetal bovine serum in PBS (FBS from Gemini Bioproducts).
3. Phosphate-buffered saline (PBS)/Tween: (1X PBS plus 0.05% Tween 20).
4. Recombinant IL-7 used in the assay as a standard (PharMingen).
5. Biotinylated MAb to IL-7 (PharMingen).
6. Avidin peroxidase (Sigma).

2.5. Murine Tumor Models

2.5.1. Cell Culture

1. Complete medium.
2. 75 cm^3 tissue-culture flasks.
3. L1C2 cells
4. 3LL cells.
5. Pathogen-free female BALB/c (H-2d) and C57Bl/6 (H-2b) mice (8–12 wk of age) were obtained from Simenson Laboratories (Gilroy, CA). Mice were maintained in the West LA VA Animal Research Facility.

3. Methods

3.1. Maintenance and Propagation of Cell Lines

1. Line 1 alveolar cell (L1C2, H-2d) or Lewis lung carcinomas (3LL, H-2b) were used to establish the in vivo models in BALB/c or C57Bl/6 mice, respectively. L1C2 is a weakly immunogenic, natural killer cell-resistant, highly malignant tumor with minimal metastatic potential, which arose spontaneously in a female BALB/c mouse *(21)*. 3LL is a weakly immunogenic lung carcinoma, which arose spontaneously in a C57Bl/6 mouse *(22)*.
2. Culture the L1C2, 3LL, B16 and EL4 in RPMI 1640 plus glutamine (Irvine Scientific) supplemented with 10% FCS (Gemini Bioproducts), antibiotics penicillin (100 U), streptomycin (100 μg) (CM), and kept at 37°C in a humidified atmosphere containing 5% CO_2.
3. Culture normal fibroblasts CRL-9392 in a 1:1 mixture of DMEM and Ham's F11, containing: 10 μg/mL bovine insulin, 15 mM HEPES buffer (Gibco); 25 μg/mL human transferrin; 20 μg/mL human HDL; 1 U/mL human PDGF; 100 ng/mL mouse EGF (Sigma).

3.2. Isolation and Propagation of Bone Marrow-Derived Dendritic Cells

1. Isolate DC from long bones of mice by squirting RPMI media through the bones using a insulin syringe (Becton Dickinson, Franklin Lakes, NJ).

Intratumoral Therapy

2. To deplete lymphocytes and macrophages incubate with lymphocyte and macrophage-depleting antibodies, (CD45R, anti-B cell; TIB 229, anti-Ia; TIB 150, anti-CD8 and TIB 207, anti-CD4, all obtained from ATCC, Rockville, MD), and rabbit serum complement (Sigma) for 1 h.
3. Wash the cells in CM, spin the cells down at 1500g and resuspend the cells in CM.
4. Incubate cells overnight to allow contaminating macrophages to adhere,
5. Harvest the nonadherent DC and culture in vitro for 8 d with murine GM-CSF (2 ng/mL) and IL-4 (20 ng/mL, R & D Systems) in 12-well plates as previously described *(7)*.
6. Consistent with previous studies from our laboratory as well as others *(4,7,15,23)* dendritic cells cultured using this protocol and characterized by flow cytometry have high level expression of B7-1, B7-2, CD-11c, MHC I, and MHC II. Cells cultured in this manner are >90% DC as defined by co-expression of these cell surface antigens.

3.3. Transduction of Dendritic Cells and Fibroblasts with Adenoviral 5 Vector

1. Take the adenovirus (AdIL-7) that is an E1-deleted, replication-deficient adenoviral type 5 vector (Ad 5) encoding a human IL-7 cDNA *(24)*. The IL-7 cDNA was inserted into the former E1 site and driven by the cytomegalovirus (CMV) promoter/enhancer.
2. The control vector (Ad RR 5) is an E1-deleted, replication-deficient adenoviral type 5 vector (Ad 5) that does not contain the human IL-7 cDNA.
3. To optimize the MOI for IL-7 production, in vitro propagated DC were transduced on d 7, in RPMI containing 2% FBS for 2 h with AdIL-7 (DC-AdIL-7) at a multiplicity of infection (MOI) of 10:1, 20:1, 50:1, and 100:1.
4. IL-7 protein concentrations in transduced DC or fibroblast supernatants were determined by IL-7 specific enzyme-linked immunosorbent assay (ELISA) as previously described *(25)*.
5. IL-7 protein concentrations from transduced DC supernatants or fibroblasts were assessed in vitro from d 1–17.
6. Efficiency of transduction of DC by adenoviral 5 vector was assessed utilizing an Ad 5 Lac Z vector. 5×10^6 DC were transduced with Ad 5 Lac Z vector (MOI 100:1 and 20:1) in vitro. The DC were incubated with the virus for 2 h and washed prior to x-gal staining 24 h post-transduction *(24)*.
7. X-gal staining: Wash cells with PBS. Fix cells in 0.5% glutaraldehyde for 10 min at 4°C. Wash cells twice with 2 mM MgCl$_2$/PBS. Stain in X-gal (1 mg/mL in 10 ml of staining buffer) to be made fresh each time. Incubate overnight in staining buffer at 37°C. X-gal staining solution:
    a. PBS, pH 7.4, 500 mL.
    b. 5mM Potassium ferrocyanide (1.056 g).
    c. 5 mM Potassium ferricyanide (0.823 g).
    d. 2 mM MgCl$_2$ (0.095 g).

Add stock X-gal (20 mg/mL) on day of the experiment to final concentration of 1 mg/mL.
8. For in vivo injection, an MOI of 20:1 was selected based on an efficiency of transduction of 50% and optimal IL-7 protein production post-transduction.
9. Similarly, fibroblasts were transduced with MOI of 20:1 with a similar transduction efficiency.
10. DC were washed with PBS three times prior to intratumoral injection. In control groups, mice received DC that did not undergo transduction. These cells are referred to as unmodified DC in the description of the results.

3.4. Peptide Loading of DC

1. Incubate C57Bl/6 DC with 3LL tumor-specific antigens, MUT 1 and MUT 2 *(26)*. The MUT 1 and MUT 2 peptides consist of 52-59 amino acid positions of the mutated connexin 37 protein present in the 3LL cell line.
2. Dilute peptides (Research Genetics, Inc., Huntsville, AL) to a concentration of 1 mg/mL.
3. Incubate 10 million DC in 1 mL of RPMI for 2 h with 500 µL each of 10 µM MUT 1 (H-FEQNTAQP-OH) and MUT 2 (H-FEQNTAQA-OH) at 37°C.
4. Wash the DC twice in PBS prior to injection.
5. Inject one million peptide-pulsed DC in a volume of 100 µL into the established 3LL tumor.

3.5. Treatment of Established 3LL and L1C2 Tumors In Vivo

1. To establish tumors, inoculate 10^5 3LL or L1C2 cells subcutaneously in the right flank in C57Bl/6 or BALB/c mice, respectively.
2. After 5 d, when tumors are palpable, administer 1×10^6 DC by intratumoral injection. Administer two additional injections of 1×10^6 DC on d 13 and 20 for a total of three injections as previously described *(4,7)*.
3. Treatment groups include DC-AdIL-7, AdIL-7 (10^9 pfu), recombinant IL-7 (5 ng/mL), unmodified DC, DC and recombinant IL-7 given as separate injections, empty control adenoviral vector transduced DC (DC-Ad CV), unmodified fibroblasts, AdIL-7-transduced fibroblasts, DC pulsed with MUT 1 and MUT 2 peptides (DC + MUT 1/ MUT 2) and DC pulsed with tumor lysates. Based on previous studies in our laboratory utilizing recombinant IL-7 *(27)*, an Alzet mini-osmotic pump containing 5 ng/mL of IL-7 with a delivery of 0.25 µL per hour flow rate and duration of 14 d (Alza Corporation, Palo Alto, CA) was implanted subcutaneously in the flank of the mouse. The tip of the pump was directed into the tumor site and delivered a continuous infusion of recombinant IL-7 (specific activity 5.6×10^7 U/mg/protein; Sterling Winthrop, Malvern, PA). The dose of IL-7 was selected because it corresponded to the amount of IL-7 secreted in vitro following DC transduction.
4. Assess tumor volumes three times per week following tumor implantation. Measure two bisecting diameters of each tumor were measured with calipers. The

volume is calculated using the formula (.04) (ab²), with a as the large diameter and b as the smaller diameter.

3.6. Secondary Rechallenge of Mice with Complete Tumor Regression

1. Rechallenge with a subcutaneous injection of 1×10^5 3LL cells 30 d following complete tumor regression.
2. The treatment groups (7–12 mice per group) include mice that received previous therapy with either intratumoral AdIL-7 or DC-AdIL-7.
3. Inoculate a control group of untreated C57Bl/6 mice with 3LL tumor cells.
4. Assess tumor volumes three times per week following tumor implantation.

3.7. Systemic Therapy Model

1. Establish tumors by inoculating 3LL cells sc in the right flank (10^5 cells) and 2×10^4 cells in the left flank.
2. Administer all therapies intratumorally into the right flank. All treatments are administered intratumorally one time per week for 3 wk. Treatment groups included DC-AdIL-7, DC-Ad CV (1×10^6 DC per injection) and AdIL-7 (10^9 pfu).
3. Establish 3LL tumors in mice on both flanks that will not receive treatment to serve as controls.

3.8. Measurement of Cytokines from the Spleen

1. Sacrifice mice and harvest spleens from normal, untreated tumor-bearing and cured mice (AdIL-7, DC-AdIL-7, DC and DC-mut1+mut2 treated mice)
2. Manually tease apart the spleens by crushing in 1 mL of culture medium in a 12-well plate with a 10-mL syringe. Spin the cells at $1500g$ for 7 min.
3. Deplete red blood cells (RBC) from splenocytes with distilled, de-ionized H_2O.
4. Bring the RBC depleted cell suspension to isotonicity by adding two volumes of complete media.
5. To assess the specificity of the antitumor response following secondary stimulation, isolate spleens from cured mice and culture 5×10^6 splenocytes/ml with and without irradiated 3LL tumor cells in CM. The ratio of splenocytes to irradiated tumor cells is 25:1 (25×10^6 splenocytes were cultured with 10^6 irradiated tumor cells).
6. To determine the specificity of the secondary stimulatory response, incubate splenocytes with syngeneic control tumors. Controls include irradiated B16($H2^b$) and EL4($H2^b$) tumor cells.
7. After 24 h in culture, measure IFN-γ and GM-CSF released into the supernatant by splenocytes by cytokine-specific ELISA.

3.9. Cytokine-Specific ELISA

1. Granulocyte-macrophage colony-stimulating factor (GM-CSF) and IFN-γ protein concentrations from splenocyte supernatants were determined by GM-CSF- and IFN-γ-specific ELISA as previously described.

2. Coat 96-well Costar plates overnight with 50 μL/well of 2 μg/mL murine anti-GM-CSF or anti-IFN-μ MAb (PharMingen).
3. Wash the plate three times with 200 μL/well of 1XPBS/0.05% Tween 20.
4. Block the wells of the plate with 200 μL/well 10% FBS (Gemini Bioproducts) in PBS and incubate for 30 min at room temperature.
5. Add 50 μL antigen (splenocyte supernatants or recombinant GM-CSF and IFN-γ standards obtained from PharMingen) and incubate for 1 h.
6. Wash excess Ag three times with 200 μL/well of PBS/Tween-20.
7. Add 50 μL/well of 1 μg/mL biotinylated anti-mouse GM-CSF or anti-IFN-γ (PharMingen) and incubate for 30 min at room temperature.
8. Wash excess MAb five times with 200 μL/well of PBS/Tween-20.
9. Incubate with 50 μL/well of 2.5 μg/mL avidin peroxidase for 30 min at room temperature.
10. Wash five times with 200 μL/well of PBS/Tween-20.
11. Dissolve 80 mg of OPD in 100 mL of H_2O and buffer to pH 4.35 with 0.1 M citric acid and store frozen in 10 mL aliquots. Before use, add 20 μL of 3% H_2O_2.
12. Add 100 μL of the above OPD substrate and incubate for appropriate length of time at room temperature in the dark for color development.
13. Stop the reaction with 100 μL/well of 2 M H_2SO_4.
14. Read the plate at 490 nm with a Molecular Devices spectrophotometer (Sunnyvale, CA).

N.B. The sensitivity limit of both ELISA is 15 pg/mL.

3.10. Intracellular Cytokine Analysis

1. Isolate splenocytes from naïve C57Bl/6 mice, untreated 3LL tumor-bearing mice, and mice from AdIL-7, DC-AdIL-7, DC and DC-mut1+mut2-treated groups that have completely rejected their tumors.
2. Deplete RBC from splenocytes with distilled, de-ionized H_2O.
3. Bring the RBC depleted cell suspension to isotonicity by adding two volumes of complete media.
4. To enrich T cells, antibody deplete B lymphocytes and macrophages with 200 μL TIB 229 (anti-Ia) and 200 μL TIB 146(CD45R) and 100 μL rabbit serum complement for 1 h.
5. Co-culture splenic lymphocytes (25×10^6) with irradiated 3LL tumor cells (1×10^6) for 24 h and then treat with the protein transport inhibitor kit GolgiPlug (PharMingen) according to the manufacturer's instructions.
6. Harvest cells and wash twice with 2 mL of 2% FBS/PBS.
7. Resuspend 5×10^5 cells in 200 μL of 2%FBS/PBS with 0.5 μg of fluorochrome conjugated MAb specific for cell-surface antigens CD3, CD4, CD8, and CD69 for 30 min at 4°C.
8. After two washes in 2% FBS/PBS, fix, permeabilize, and wash cells using the Cytofix/Cytoperm Kit (PharMingen) following the manufacturer's protocol.

Intratumoral Therapy

9. Resuspend the cell pellet in 100 μL of Perm/Wash solution and stain with 0.25 μg of fluorochrome conjugated anti-IFN-γ antibody for intracellular staining.
10. Incubate cells at room temperature in the dark for 30 min and wash twice with 2% FBS/PBS.
11. Resuspend in 300 μL PBS/2% paraformaldehyde solution and analyze by flow cytometry.

3.11. Trafficking Studies Utilizing Ad-GFP Transduced Dendritic Cells

1. Use Ad-GFP transduced DC (DC-Ad-GFP) to determine the trafficking of gene-modified DC to regional lymph nodes and spleen following intratumoral injection. The adenoviral construct is an E1-deleted, replication-deficient adenoviral type 5 vector (Ad 5) encoding a green fluorescent protein (Ad-GFP). The GFP cDNA is inserted into the former E1 site and driven by the CMV promoter/enhancer. Fibroblasts are used as controls in these studies.
2. Take in vitro propagated DC or fibroblasts and transduce on d 7 in RMPI medium containing 2% FBS for 2 h with Ad-GFP at an MOI of 100:1.
3. Determine, by flow cytometry, the efficiency of transduction of DC and fibroblasts with Ad-GFP. The efficiency at 100:1 MOI is 70%. Utilize Ad-GFP transduced DC or fibroblasts to adjust the gates for measuring the percent of transduced DC or fibroblasts present at the tumor site, lymph nodes, and spleen.
4. Inoculate, intratumorally into 5 d established tumors, five million Ad-GFP transduced DC (DC-Ad-GFP), Ad-GFP transduced fibroblasts or Ad-GFP vector alone (1×10^9 pfu).
5. One, three, and seven days following injection, harvest tumors and cut into small pieces and pass through a sieve (Bellco, NJ).
6. Evaluate the tumors for the presence of DC Ad-GFP, Ad-GFP-fibroblasts or other cells at the tumor site that have been transduced by Ad-GFP vector following direct intratumoral injection.
7. Harvest regional, ipsilateral and contralateral lymph nodes and spleens 1, 3, and 7 d following injection.
8. Tease apart lymph nodes and spleens to assess for the presence of Ad-GFP transduced fibroblasts, DC-Ad-GFP or other Ad-GFP positive cells.
9. Assess the percentage of positive DC-Ad-GFP, Ad-GFP fibroblasts or other cells transduced at the tumor site following intratumoral Ad-GFP vector inoculation by flow cytometric analysis after gating on the DC or fibroblast populations.

3.12. Histology

1. Isolate non-necrotic tumors d 22 and place in 10% formalin.
2. Embed in paraffin.
3. Prepare 5 μm sections were for hematoxylin/eosin staining for histopathological examination.

4. Notes

1. To isolate good-quality DC from mouse bone marrow for our experiments, we found that it was necessary to antibody deplete T, B, and macrophage subsets before incubating with GM-CSF and IL-4 for 7–8 d.
2. Transduction of DC with adenoviral vectors was performed in a small volume (200 µL) of culture medium to assure efficient transduction.

Acknowledgments

This work was supported by National Institutes of Health Grant R01 CA78654, 1P50 CA90388 (S.M D), Medical Research Funds from the Department of Veteran Affairs, the Research Enhancement Award Program in Cancer Gene Medicine and the Tobacco-Related Disease Research Program of the University of California.

References

1. Huang, A. Y. C., Golumbek, P., Ahmadzadeh, M., Jaffee, E., Pardoll, D., and Levitsky, H. (1994) Role of bone marrow-derived cells in presenting MHC class I-restricted tumor antigens. *Science* **264,** 961–965.
2. Steinman, R. M. (1991) The dendritic cell system and its role in immunogenicity. *Annu. Rev. Immunol.* **9,** 271–296.
3. Gabrilovich, D. I., Chen, H. L., Girgis, K. R., Cunningham, H. T., Meny, G. M., Nadaf, S., et al. (1996) Production of vascular endothelial growth factor by human tumors inhibits the functional maturation of dendritic cells. *Nature Med.* **2,** 1096–1103.
4. Miller, P. W., Sharma, S., Stolina, M., Chen, K., Zhu, L., Paul, R. W., and Dubinett, S. M. (1998) Dendritic cells augment granulocyte-macrophage colony-stimulating factor (GM-CSF)/herpes simplex virus thymidine kinase-mediated gene therapy of lung cancer. *Cancer Gene Ther.* **5,** 380–389.
5. Nestle, F., Alijagic, S., Gilliet, M., Sun, Y., Grabbe, S., Dummer, R., et al. (1998) Vaccination of melanoma patients with peptide- or tumor lysate-pulsed dendritic cells. *Nature Med.* **4,** 328–332.
6. Ribas, A., Butterfield, L., McBride, W., Jilani, S., Bui, L., Vollmer, C., et al. (1997) Genetic immunization for the melanoma antigen MART-1/Melan-A using recombinant adenovirus-transduced murine dendritic cells. *Cancer Res.* **57,** 2865–2869.
7. Sharma, S., Miller, P., Stolina, M., Zhu, L,. Huang, M., Paul, R., and Dubinett, S. (1997) Multi-component gene therapy vaccines for lung cancer: effective eradication of established murine tumors *in vivo* with Interleukin 7/Herpes Simplex Thymidine Kinase-transduced autologous tumor and *ex vivo*-activated dendritic cells. *Gene Ther.* **4,** 1361–1370.
8. Zitvogel, L., Mayordomo, J. I., Tjandrawan, T., DeLeo, A. B., Clarke, M. R. Lotze, M. T., and Storkus, W. J. (1996) Therapy of murine tumors with tumor

peptide-pulsed dendritic cells: dependence on T cells, B7 costimulation, and T helper cell 1-associated cytokines. *J. Exp. Med.* **183**, 87–97.
9. Boczkowski, D., Nair, S. K., Snyder, D., and Gilboa, E. (1996) Dendritic cells pulsed with RNA are potent antigen-presenting cells in vitro and in vivo. *J. Exp. Med.* **184**, 465–472.
10. Celluzzi, C. M., Mayordomo, J. I., Storkus, W. J., Lotze, M. T., and Falo Jr., L. D. (1996) Peptide-pulsed dendritic cells induce antigen-specific, CTL-mediated protective tumor immunity. *J. Exp. Med.* **183**, 283–287.
11. Mayordomo, J. I., Zorina, T., Storkus, W. J., Zitvogel, L., Celluzzi, C., Falo, L. D., et al. (1995) Bone marrow-derived dendritic cells pulsed with synthetic tumor peptides elicit protective and therapeutic antitumor immunity. *Nature Med.* **1**, 1297–1302.
12. Celluzzi, C. and Falo, L. J. (1998) Cutting edge: physical interaction between dendritic cells and tumor cells result in an immunogen that induces protective and therapeutic tumor rejection. *J. Immunol.* **160**, 3081–3085.
13. Gong, J., Chen, D., Kashiwara, M., and Kufe, D. (1997) Induction of antitumor activity by immunization with fusions of dendritic and carcinoma cells. *Nature Med.* **3**, 558–561.
14. Toes, R., van der Voort, E., Schoenberger, S., Drijfhout, J., van Bloois, L., Storm, G., et al. (1998) Enhancement of tumor outgrowth through CTL tolerization after peptide vaccination is avoided by peptide presentation on dendritic cells. *J. Immunol.* **160**, 4449–4456.
15. Fields, R. C., Shimizu, K., and Mule, J. J. (1998) Murine dendritic cells pulsed with whole tumor lysates mediate potent antitumor immune responses in vitro and in vivo. *Proc. Natl. Acad. Sci. USA* **95**, 9482–9487.
16. Toes, R. E., Hoeben, R. C., van der Voort, E. I., Ressing, M. E., van der Eb, A. J., Melief, C. J., and Offringa, R. (1997) Protective anti-tumor immunity induced by vaccination with recombinant adenoviruses encoding multiple tumor-associated cytotoxic T lymphocyte epitopes in a string-of-beads fashion. *Proc. Natl. Acad. Sci. USA* **94**, 14660–14665.
17. Kos, F. J. and Mullbacher, A. (1993) IL-2-independent activity of IL-7 in the generation of secondary antigen-specific cytotoxic T cell responses in vitro. *J. Immunol.* **150**, 387–393.
18. Mehrotra, P. T., Grant, A. J., and Siegel, J. P. (1995) Synergistic effects of IL-7 and IL-12 on human T cell activation. *J. Immunol.* **154**, 5093–5102.
19. Dubinett, S. M., Huang, M., Dhanani, S., Wang, J., and Beroiza, T., (1993) Down-regulation of macrophage transforming growth factor-β messenger RNA expression by interleukin-7. *J. Immunol.* **151**, 6670–6680.
20. Miller, A. R., McBride, W. H., Dubinett, S. M., Dougherty, G. J., Thacker, J. D., Shau, H., et al. (1993) Transduction of human melanoma cell lines with the human interleukin-7 gene using retroviral-mediated gene transfer: comparison of immunological properties with IL-2. *Blood* **82**, 3686–3694.

21. Yuhas, J., Toya, R., and Wagner, E. (1975) Specific and nonspecific stimulation of resistence to the growth and metastasis of the line 1 lung carcinoma. *Cancer Res.* **35,** 242–244.
22. Malave, I., Blanca, I., and Fuji, H. (1979) Influence of inoculation site on development of the Lewis lung carcinoma and suppressor cell activity in syngeneic mice. *J. Natl. Cancer Inst.* **62,** 83–88.
23. Inaba, K., Inaba, M., Romani, N., Aya, H., Deguchi, M., Ikehara, S., et al. (1992) Generation of large numbers of dendritic cells from mouse bone marrow cultures supplemented with granulocyte/macrophages colony-stimulated factor. *J. Exp. Med.* **176,** 1693–1703.
24. Arthur, J., Butterfield, L., Roth, M., Bui, L., Kiertscher, S., Lau, R., et al. (1997) A comparison of gene transfer methods in human dendritic cells. *Cancer Gene Ther.* **4,** 17–25.
25. Sharma, S., Wang, J., Huang, M., Paul, R., Lee, P., McBride, W., et al. (1996) Interleukin-7 gene transfer in non-small cell lung cancer decreases tumor proliferation, modifies cell surface molecule expression, and enhances antitumor reactivity. *Can. Gene Ther.* **3,** 302–313.
26. Mandelboim, O., Berke, G., Fridkin, M., Feldman, M., Eisenstein, M., and Eisenbach, L. (1994) CTL induction by a tumor-associated antigen octapeptide derived from a murine lung carcinoma. *Nature* **369,** 67–71.
27. McBride, W. H., Thacker, J. D., Comora, S., Economou, J., Kelley, D., Dubinett, S. M., and Dougherty, G. J. (1992) Genetic modification of a murine fibrosarcoma to produce IL-7 stimulates host cell infiltration and tumor immunity. *Cancer Res.* **52,** 3931–3937.

44

Cyclooxygenase 2-Dependent Regulation of Antitumor Immunity in Lung Cancer

Sherven Sharma, Min Huang, Mariam Dohadwala, Mehis Pold, Raj K. Batra, and Steven M. Dubinett

1. Introduction

Lung cancer is the leading cause of cancer death in the United States (*1*). Many tumors, including lung cancer, have the capacity to promote immune tolerance and escape host immune surveillance (*2*). Tumors utilize numerous pathways to inhibit immune responses including the elaboration of immune inhibitory cytokines. In addition to direct secretion of immunosuppressive cytokines, lung cancer cells may induce host cells to release immune inhibitors (*3–7*). We have reported previously that human lung cancer cell-derived prostaglandin E2 (PGE2) can orchestrate an imbalance in the production of interleukins (IL)-10 and -12 in lymphocytes and macrophages (*5*). IL-10 and -12 are critical regulatory elements of cell-mediated antitumor immunity. While IL-10 inhibits important aspects of cell-mediated immunity, IL-12 induces type 1 cytokine production and effective antitumor cell-mediated responses (*8–12*). IL-10 overproduction at the tumor site has been implicated in tumor-mediated immune suppression (*13,14*). In contrast, IL-12 is critical for effective antitumor immunity (*15,16*). In both tumor models and patients, the tumor-bearing state induces lymphocyte and macrophage IL-10 but inhibits macrophage IL-12 production (*5,6,17–19*). Because PGE2 appears to be pivotal in the reciprocal regulation of IL-10 and IL-12 (*20,21*), we have sought to determine the pathways responsible for its high-level production at the tumor site. We find that tumor cyclooxygenase 2 (COX-2) expression is a pivotal determinant of the expression of these cytokines in the tumor-bearing host.

From: *Methods in Molecular Medicine, vol. 75: Lung Cancer, Vol. 2:*
Diagnostic and Therapeutic Methods and Reviews
Edited by: B. Driscoll © Humana Press Inc., Totowa, NJ

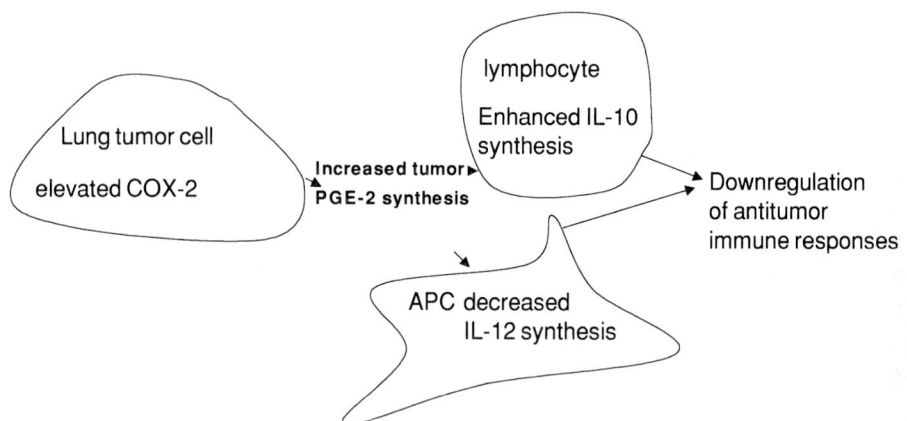

Fig. 1. Tumor COX-2 mediated dysregulation of antitumor immune responses.

The initiation of prostanoid synthesis from arachidonic acid involves the enzyme referred to as cyclooxygenase (COX) which has also been termed prostaglandin H synthase or prostaglandin-endoperoxide synthase *(22)*. Two isoenzymes have been identified: a constitutive form (COX-1) and an inducible isoenzyme (COX-2) *(23–25)*. COX-2 is upregulated in response to a variety of stimuli including growth factors and cytokines *(23)*. Because it can lead to enhanced PGE2 production and subsequent cytokine imbalance in vivo, tumor expression of COX-2 may be instrumental in the generation of tumor-induced abrogation of T cell-mediated antitumor responses *(5)*. COX-2 has been implicated in the development of colon cancer and may play a role in promoting invasion, apoptosis resistance, metastasis, and angiogenesis in established tumors *(26–29)*. In addition to lung and colon carcinomas *(5,30–32)*, COX-2 has recently been reported to be expressed in a variety of human malignancies *(33–38)*.

Here we describe experimental procedures that demonstrate that specific COX-2 inhibition serves to restore antitumor immunity. The tumor-induced imbalance in IL-10 and -12 are reversed by COX-2 inhibition in vivo. This promotes antitumor responses in an immunocompetent murine lung cancer model.

To evaluate lung tumor COX-2 modulation of antitumor immunity, we studied the antitumor effect of specific genetic or pharmacological inhibition of COX-2 in a murine Lewis lung carcinoma (3LL) model. Inhibition of COX-2 led to marked lymphocytic infiltration of the tumor and reduced tumor growth. Treatment of mice with anti-prostaglandin E2 (anti-PGE2) monoclonal antibody (mAb) replicated the growth reduction seen in tumor-bearing mice treated with COX-2 inhibitors. COX-2 inhibition was accompanied by a significant decrement in IL-10 and a concomitant restoration of IL-12 production by antigen-presenting cells (APCs). Because the COX-2 metabolite PGE2 is a potent inducer of IL-10, it was hypothesized that COX-2 inhibition led to antitumor responses by downregulating production of this potent immunosuppressive cytokine. In support of this concept, transfer of IL-10 transgenic T lymphocytes that overexpress IL-10 under control of the IL-2 promoter, reversed the COX-2 inhibitor-induced antitumor response. We conclude that abrogation of COX-2 expression promotes antitumor reactivity by restoring the balance of IL-10 and -12 in vivo.

2. Materials

2.1. Establishment of COX-2 Sense and Antisense Transfected 3LL Cells

1. Murine Lewis lung carcinoma cell line (3LL) was obtained from ATCC (Rockville, MD). The cell line was mycoplasma free and cells were utilized up to the tenth passage before thawing frozen stock 3LL cells from liquid N_2.
2. COX-2 antisense plasmid: For the antisense insert, a polymerase chain reaction (PCR) fragment was generated from the sense construct utilizing the T7 promoter as the 5' primer binding site and positions 725 - 701 of the murine cDNA as the 3' primer binding site. The PCR fragment was cloned into the pCR 3.1 TA vector.
3. COX-2 sense plasmid transfected 3LL: 4SC7-3LL. For the sense construct a 2.3 Kb *Bam*H1-*Xho*1 fragment containing the open reading frame for a polypeptide of 604 amino acids of murine COX-2 was isolated and cloned into the *Bam*H1 - *Xho*1 site of the eukaryotic expression vector, pCR 3.1 (Invitrogen, San Diego, CA).
4. Empty vector pCR 3.1 transfected 3LL: CV-3LL. In this vector, transcription of the cDNA is controlled by the CMV promoter. This vector also contains the neomycin-resistance gene that allows for selection in G418.
5. Superfect transfection reagents (Qiagen, Los Angeles, CA).
6. G418 (Gibco BRL, Gaithersburg, MD).
7. Complete medium (CM): RPMI 1640 medium (Irvine Scientific, Santa Ana, CA) supplemented with 10% fetal bovine serum (FBS) (Gemini Bioproducts, Calabasas, CA), penicillin (100 U/mL), streptomycin (0.1 mg/ml), 2 m*M* glutamine (JRH Biosciences, Lenexa, KS), and maintained at 37°C in humidified atmosphere containing 5% CO_2 in air.
8. 96-well Costar (Cambridge, MA) plates.

2.1.1. PGE2 Enzyme Immunoassay (EIA)

1. Cayman Chemical Co. (Ann Arbor, MI) EIA kit, which supplies PGE2-acetylcholinesterase conjugate.
2. Goat anti-mouse PGE2 (Biosource International, Calabasas, CA).

3. Mouse anti-PGE2 (2B5 mAb) and isotype-matched control mouse IgG1 (MOPC21) were generously provided by Searle (Skokie, IL).
4. Plate wash buffer: 1X PBS, 0.05% Tween 20 (Sigma, St. Louis, MO).
5. Dynatech MR5000 spectrophotometer (Chantilly, VA).
6. Indomethacin and aspirin were obtained from Sigma.
7. NS-398 [N-(2-Cyclohexyloxy-4-nitrophenyl) methanesulfonamide] was purchased from Cayman Chemical.
8. SC-58236 was generously provided by Searle.

2.1.2. Northern Analysis

1. Murine IL-1β (150 U/mL) from PharMingen (San Diego, CA).
2. 4 M guanidine isothiocyanate.
3. Chloroform.
4. Isopropanol.
5. 75 % ethanol.
6. 0.1% DEPC-treated double-distilled water.
7. 10% denaturing formaldehyde agarose gel.
8. Hybond nylon membrane (Amersham, Arlington Heights, IL).
9. 20X sodium chloride sodium citrate (SSC): Dissolve 175.3 g of NaCl, 27.6 g. $NaH_2PO_4 \cdot H_2O$ and 7.4 g EDTA in 800 mL of H_2O. Adjust the pH to 7.0 with a few drops of a 10 N solution of NaOH. Adjust the volume to 1 L with H_2O.
11. Ultraviolet cross-linking instrument (Stratagene, San Diego, CA).
12. ^{32}P dCTP nick translation kit (BRL, Bethesda, MD).
12. Pharmacia sephadex G50 columns.
13. Rapid hybridization solution was obtained from Amersham.
14. 2X SSC 0.1% SDS.
15. Kodak XAR-5 film.
16. Densitometry analysis performed using Ambis Image Analysis System (Bio Rad Laboratories).

2.1.3. Western Blotting

1. RIPA buffer: PBS, 1% NP40, 0.5% NP-40, 0.5% sodium deoxycholate, 0.1% sodium dodecyl sulfate (SDS), 100 ng/mL PMSF, 66 ng/mL aprotinin
2. Protein assay kit (Sigma).
3. 10% protein ready gel (Bio Rad Laboratories).
4. Nitrocellulose.
5. Anti-COX-2 MAb and rabbit anti-mouse MAb (Cayman Chemical).
6. ECL system; Amersham.
7. XAR 5 film.

2.2. 3LL Tumor Model

2.2.1. Cell Culture

1. Complete medium.
2. 25 cm^3 tissue-culture flasks.

3. 6-Well plates.
4. Aspirin (50 μg/mL).
5. Indomethacin (1 μg/mL).
6. NS-398 (5 μg/mL).
7. SC-58236 (5 μg/mL).
8. Anti-PGE2 mAb (10 μg/mL) or control Ab (10 μg/mL).
9. PBS: Dissolve 8 g of NaCl, 0.2g of KCl, 1.44 g of Na_2HPO_4, and 0.24 g of KH_2PO_4 in 800 mL of distilled H_2O. Adjust the pH to 7.4 with HCl. Add H_2O to 1 L. Sterilize the solution by autoclaving.

2.2.2. Subcutaneous Tumorigenesis Experiments

1. Mice: Pathogen-free-female C57Bl/6 mice (8–12 wk age) were obtained from Harlan (Indianapolis, IN) and maintained in the West Los Angeles VA Animal Research Facility.
2. Indomethacin (10 mg/mL in drinking water).
3. SC-58236 (3 mg/kg 3 times per week ip).
4. Anti-PGE2 MAb and control antibody (10 mg/kg, ip).

2.2.3. Measurement of Splenocyte IL-10 and IL-12 Production

1. Mouse anti-IL-10 MAb, 4 μg/mL (PharMingen, San Diego, CA).
2. 10% FBS in phosphate-buffered saline (PBS) (FBS from Gemini Bioproducts, Calabasas, CA).
3. PBS/Tween: 1X PBS plus 0.05% Tween 20).
4. Recombinant IL-10 used in the assay as a standard (PharMingen).
5. Biotinylated MAb to IL-10 (PharMingen).
6. Avidin peroxidase (Sigma).
7. IL-12 enzyme-linked immunosorbent assay (ELISA) kit from Biosource International.

2.2.4. Protocol for Lung Tumor Formation

1. 10% formalin.
2. Embedding and sectioning set up for histology.
3. Hematoxylin (H) and eosin (E) stain.
4. Microscope fitted with calibrated optical grid (a 1 cm^2 grid subdivided into 100 1.0-mm^2 squares).
5. IL-10 transgenic mice, made by standard methods at University of California Los Angeles Transgenic Mouse Core Facility (Los Angeles, CA) as previously described.
6. Dynal beads for mouse T-lymphocyte isolation (Dynal, NY).

2.2.5. Antigen Presenting Cell (APC) Isolation

1. Hybridoma cell lines were purchased from ATCC: TIB207 (anti-CD4), TIB150 (anti-CD8), and TIB146 (anti-B lymphocytes).

2. Rabbit complement (Sigma).
3. Antibodies to CD11b, CD11c and F480 (PharMingen).
4. Mouse anti-CD40 MAb (PharMingen).
5. Trypsin (Irvine Scientific).
6. PGE2 (Cayman Chemical).

3. Methods
3.1. Establishment of COX-2 Sense and Antisense Transfected 3LL Cells

To evaluate the importance of tumor COX-2 in modifying tumor growth and host immunity, a COX-2 antisense plasmid was constructed and utilized to abrogate COX-2 production in 3LL. The sense plasmid transfected 3LL (4SC7-3LL) and empty vector transfected 3LL (CV-3LL) were utilized as controls (*see* **Note 1**).

1. Prepare COX-2 sense and anti-sense oriented expression vectors as well as pCR 3.1 control vector.
2. Transfect vectors into the 3LL cell line using superfect transfection reagents.
3. Following transfection, select the 3LL clones expressing COX-2 sense, anti-sense and control vector constructs by addition of 500 µg/mL of G418 to cultures for for 10 d.
4. After selection, isolate 3LL clones expressing COX-2 sense, anti-sense and control vector constructs by limited dilutions from 96-well plates.
5. Initial screen for 3LL COX-2 sense and anti-sense is based on PGE2 production.

3.1.1. PGE2 Enzyme Immunoassay (EIA)

Using this method, it was determined that he 3LL COX-2 sense clones produced 7–9 ng/mL/10^5 cells of PGE2 whereas the COX-2 anti-sense clones produced in the range of 105–285 pg/mL/10^5 cells.

1. Determine PGE2 concentrations according to the EIA kit protocol. Pre-coat 96-well plates overnight with 4 µg/mL of goat anti-mouse PGE2.
2. Add, in succession, PGE2-acetylcholinesterase conjugate, mouse anti-PGE2 mAb, and either standard or sample to each well. At this point, to determine the relative contribution of each isoenzyme to high level PGE2 production, use the specific COX-2 inhibitors NS-398 and SC-58236, as well as indomethacin and aspirin, which block both isoenzymes.
3. Incubate plate 18-h at 25°C.
4. Wash plate using 200 µL/well wash buffer five times to remove all unbound reagents.
5. Add Ellman's reagent to each well and determine absorbance at 405 nm.

3.1.2. Northern Analysis

The clones were further characterized for COX-2 mRNA by Northern analysis, respectively. The 3LL COX-2 antisense transfected clones expressed less COX-2 mRNA than did the 3LL parental tumor cells, 3LL COX-2 sense or 3LL control vector transfected clones. Northern analysis for COX-1 gene message in the 3LL COX-2 antisense, 3LL COX-2 sense, and 3LL control vector clones showed that the COX-1 message remained unaltered.

1. To determine the time course of COX-1 and COX-2 gene expression in 3LL cells, 3LL COX-2 sense clone (4SC7-3LL) and 3LL control vector clone (CV-3LL) after IL-1β stimulation, treat 3LL cells cultures with or without murine IL-1β (150 U/mL) for 1–24 h. Cells were collected at a series of time points over the course of treatment.
2. Isolate total RNA for Northern analysis. Lyse cell pellets for each of the sample time points in 4 M guanidine isothiocyanate solution by gently pipetting.
3. Add a 0.2 vol quantity of chloroform to the cell lysate and keep on ice for 10 min.
4. Centrifuge at 12,000g for 10 min at 4°C.
5. Transfer the upper aqueous phase to a new tube.
6. Add equal volumes of isopropanol to the upper aqueous phase and keep on ice for 45 min.
7. Centrifuge at 12,000g for 10 min at 4°C.
8. Wash the RNA pellet twice with 75 % ethanol and dry at room temperature.
9. Dissolve the dried RNA pellet in 0.1 % DEPC treated double-distilled water and adjusted to a final concentration of 1 µg/mL.
10. Northern blotting analysis is done as previously described to detect COX message (5). Electrophorese 10 µg per sample for each of the conditions through a 10% denaturing formaldehyde agarose gel.
11. Transfer RNA to a Hybond nylon membrane in 20X SSC.
12. Fix the RNA to the nylon membrane by ultraviolet cross-linking.
13. Label cDNA probes for murine COX-1, COX-2 and β-actin control. Clean probes by passage through a Pharmacia sephadex G50 column.
14. Perform both prehybridization and hybridization at 68°C. Hybridize duplicate filters overnight with cDNA probes for murine COX-1, COX-2 and β-actin control.
15. Following hybridization, wash filters twice for 15 minutes at room temperature in 2X SSC 0.1% SDS and once for 10 min at 42°C and 3 times at 68°C in 2X SSC 0.1% SDS. Expose filters to Kodak XAR-5 film overnight at –80°C and perform densitometric analysis to determine alteration in COX-2 mRNA levels.

3.1.3. Western Blotting

To confirm the expression of COX-2 in sense and anti sense transfected 3 LL cells, Western Blot analysis is done.

1. Grow 3 LL cells transfected with the sense and anti sense oriented COX-2 cDNA in CM and maintain in exponential growth phase.
2. Harvest cells and lyse in RIPA buffer. Clarify lysates by centrifugation at 12000g and determine protein concentration using the Sigma protein assay kit.
3. Separate samples containing 50 μg cell lysate protein on a 10% protein ready gel.
4. Transfer proteins to nitrocellulose after electrophoresis.
5. Detect COX-2 protein levels by using an anti-COX-2 MAb and rabbit anti mouse MAb and develop binding signal using standard ECL system-detection technique according to manufacturer's instructions.
6. Expose filters to XAR 5 film and perform densitometric analysis for alterations in COX-2 protein expression.

3.2. 3LL Tumor Model

3.2.1. Cell Culture

1. Culture murine Lewis lung carcinoma cell line (3LL), the COX-2 antisense 3LL clone (1ASE7-3LL), the COX-2 sense clone (4SC7-3LL), and the control vector transfected cells (CV-3LL) as monolayers in 25 cm^3 tissue-culture flasks in CM. Cells can be utilized up to the tenth passage before thawing frozen stock 3LL cells from liquid N_2 (*see* **Note 2**).
2. For experiments utilizing 3LL, 1ASE7-3LL, 4SC7-3LL or CV-3LL cell supernatants, culture 1 × 10^5 cells/mL in 6-well plates in RPMI with or without specific COX inducers or inhibitors.
3. For experiments utilizing the COX-2 inhibitor aspirin, incubate 105 cells/mL were for 2 h in the presence of aspirin (50 μg/mL). Wash tumor cells twice in PBS and re-plate at 105 cells/mL. Following a 24-h culture period, collect tumor cell supernatants and measure PGE2 by EIA (*see* **Subheading 3.1.1.**).
4. For experiments utilizing indomethacin (1 μg/mL), 10^5 3LL cells are incubated for 24 h to condition the medium.
5. For experiments utilizing NS-398 (5 μg/mL), 10^5 3LL cells are incubated for 24 h to condition the medium.
6. For experiments utilizing SC-58236 (5 μg/mL), 10^5 3LL cells are incubated for 24 h to condition the medium.
7. For experiments utilizing anti-PGE2 mAb (10 μg/mL) or control Ab (10 μg/mL), 10^5 3LL cells are incubated for 24 h to condition the medium.

3.2.2. Subcutaneous Tumorigenesis Experiments

1. Pretreat mice with indomethacin (10 mg/mL in drinking water) or SC-58236 (3 mg/kg 3 times per week ip) 1 wk before the tumor cell inoculation and for the duration of the experiment. Pretreat with anti-PGE2 MAb or control antibody 24 h before tumor inoculation and 3 times a week for the duration of the experiment (10 mg/kg, ip).
2. Inoculate 5 × 10^5 3LL parent, 1ASE7-3LL, 4SC7-3LL, or CV-3LL cells on the right supra scapular area in C57Bl/6 mice and monitor.

3. Assess tumor volumes and survival three times per week following tumor implantation. Measure two bisecting diameters of each tumor with calipers. Calculate volume using the formula $(0.4) \times (ab2)$, with "a" as the larger diameter and "b" as the smaller diameter.
4. On d 14 after tumor inoculation, isolate non-necrotic tumors and splenic lymphocytes from tumor-bearing mice.
5. Prepare tumor homogenates by cutting tumors into small pieces and passing through a sieve (Bellco, Vineland, NJ) for determination of PGE2 using PGE2 enzyme immunoassay (EIA) (*see* **Subheading 3.1.1.**) and IL-10 concentrations by ELISA (*see* **Subheading 2.2.3.**).
5. Following RBC and macrophage depletion, assess 6-d splenic lymphocyte cultures for cytokine production. RBC was depleted using dd H_2O and bringing the cells to tonicity with PBS. Macrophages were depleted by adherence overnight to culture flasks.

3.2.3. Measurement of Splenocyte IL-10 and IL-12 Production

Previous studies indicate that the tumor-bearing state is often characterized by an increase in IL-10 production but diminished capacity to produce IL-12 *(6,18)*. One possibility is that tumor or host cell-derived PGE2 is the cause of both augmentation of IL-10 and the simultaneous decrement in IL-12 production *(5)*. To study the effect of 3LL tumor cell-derived PGE2 on splenocyte production of IL-10, splenocytes from normal C57Bl/6 mice are cultured in 3LL supernatants.

1. Harvest 3LL supernatants after a 24-h incubation with various combinations of COX-inhibitors, anti-PGE2 or control mAb (*see* **Subheading 3.2.1.**)
2. Culture splenocytes from control mice in CM for 72 h at a concentration of 2×10^6 cells /ml in tumor cell supernatants or in tumor cell supernatants from 3LL cells that were treated with COX- inhibitors (*see* **Note 3**). Following the 72-h incubation, asses IL-10 production by ELISA. The sensitivity of the IL- 10 ELISA is 15 pg/mL.
3. IL-10 protein concentrations from murine splenocytes cultured in CM or 3LL cell supernatants were determined by IL-10 specific ELISA as previously described *(4)*. Coat 96-well plates overnight at 4°C with 4 µg/mL of mouse anti-IL-10 MAb.
4. Block wells with 10% FBS in PBS for 30 min at room temperature.
5. Add antigen (recombinant IL-10 used in the assay as a standard and samples from splenic lymphocytes supernatants for 1 h at room temperature, then wash off excess with PBS/Tween.
6. Add 50 µl 1 µg/mL biotinylated MAb to IL-10 for 30 min at room temperature and wash off excess antibody with PBS/Tween.
7. Add 50 µL of 2.5 µg/mL avidin peroxidase for 30 min at room temperature.
8. Add 100 µL of OPD substrate for appropriate length of time for color development and read the color change at 490 nm with a Dynatech MR5000 spectropho-

tometer. Dissolve 80 mg of OPD in 100 mL of H_2O and buffered to pH 4.35 with 0.1 M Citric acid and stored as frozen 10-mL aliquots. Before use, 20 µL of 3% H_2O_2 is added.
9. For IL-12 measurements use the IL-12 ELISA kit from Biosource International according to the manufacturer's instructions. The sensitivity of the IL-12 ELISA is 5 pg/mL.

3.2.4. Protocol for Lung Tumor Formation

1. To compare 3LL, 4SC7-3LL, CV-3LL and 1ASE7-3LL tumor formation in the lungs of C57Bl/6 mice, inoculate 5×10^5 tumor cells iv via a lateral tail vein.
2. After 34 d, sacrifice mice, isolate lungs, and perfuse by 10% formalin for 48 h to fix.
3. Embed lungs in paraffin. Prepare 3–4 µm sections for hematoxylin (H) and eosin (E) staining and histopathological examination.
4. Evaluate tumor burden in lung sections by microscope examination using a calibrated optical grid (a 1 cm^2 grid subdivided into 100 1.0-mm^2 squares). Determine the total number of positive squares (with tumor occupying >50% of area) for 4–6 separate high-power fields from three histologic sections as previously described *(41)*.

3.2.5. Evaluation of IL-10-Mediated Immunosuppression In Vivo

1. Isolate splenic T lymphocytes from C57Bl/6 or from IL-10 transgenic mouse ice spleens. Use Dynal beads for T lymphocyte isolation according to manufacturer's protocol.
2. Transfer 5×10^7 T lymphocytes/mouse by tail vein injection to SC-58236 pretreated C57Bl/6 mice 24 h before and one week following 3LL tumor cell inoculation. (*see* **Subheading 3.2.2.**).
3. Asses tumor volumes three times per week (*see* **Subheading 3.2.2.**).

3.2.6. Antigen Presenting Cell (APC) Isolation

1. Prepare spleens from tumor bearing mice treated with COX-2 inhibitor SC-58236 and from nontreated tumor-bearing controls and control intact animals mice for splenocyte isolation. Spleens were isolated and a single-cell suspension was prepared by crushing the spleens in 1 mL of culture medium in a 12-well plate with a 10-mL syringe. Spin the cells at 1500*g* for 7 min.
2. Deplete RBCs from total splenocytes by resuspending the cells in dd H_2O and then quickly adding two volumes of the culture medium.
3. Isolate MAbs (anti-CD4, anti-CD8, anti-B lymphocytes) for antibody-mediated complement lysis of T and B cells from hybridoma cell lines by collecting supernatants from confluent hybridoma cultures. Centriguge the supernatants at 1500*g* to remove cells and debris and store antibody supernatants as frozen aliquots at –80°C.
4. Purify APC from total splenocyte suspension by antibody-mediated complement lysis of T and B cells. Incubate suspension with a cocktail of MAbs and

rabbit complement (200 μL of each Ab and 100 μL of rabbit complement) for 60 min at 37°C.
5. After antibody depletion, was APCs twice in CM.
6. Stain APCs for cell-surface markers with antibodies to CD11b, CD11c, and F480 by adding the labeled antibodies to 200,000 cells in 200 μL cell suspension in PBS for 30 min at 4°C. After staining, cells are washed twice in CM. Cells should be >95% APC following Ab depletion. Note that less than 5% of cells should stained positively for CD3 following Ab depletions.
7. To determine IL-12 levels, APCs (5×10^6 cells/ml) are stimulated with anti-CD40 (5 μg/mL) for 72-h in culture. IL-12 levels are determined by ELISA (**Subheading 3.2.3.**).
8. To purify macrophages from APC suspension, culture in 6-well plates overnight.
9. Remove nonadherent cells, then wash adherent cells twice with PBS and detach macrophages by trypsinization.
10. For IL-12 measurements from macrophages stimulate 5×10^6 cells/mL in CM with anti-CD40 (5 μg/mL in CM), 3LL supernatant, 1ASE7-3LL supernatant, 4SC7-3LL supernatant, CV-3LL supernatant, PGE2 (5 μg/mL) in CM, anti-PGE2 mAb (10 μg/mL), or isotype-matched IgG control antibody, MOPC21 (10 μg/mL in CM).

4. Notes

1. Lewis lung carcinoma (3LL) is a weakly immunogenic murine lung cancer that has been shown previously to produce prostaglandins *(42)*. Based on previous studies in a variety of host cells and tumors, we anticipated that high-level prostaglandin production by 3LL would be COX-2-dependent.
2. Since the COX-2 sense and antisense constructs are plasmid based, the 3 LL tumor cell transfectants were used to 10 passages to maintain the stability of expression from the constructs before new vials were thawed from liquid N2 storage.
3. For splenocyte IL-10 determination in the tumor supernatants, the FBS was screened for low IL-10 background levels and induction capacities.

Acknowledgments

This work was supported by National Institutes of Health Grant R01 CA78654, 1P50 CA90388 (S.M D), Medical Research Funds from the Department of Veteran Affairs, the Research Enhancement Award Program in Cancer Gene Medicine and the Tobacco-Related Disease Research Program of the University of California.

References

1. Ramanathan, R. and Belani, C. (1997) Chemotherapy for advanced non-small cell lung cancer: past, present and future. *Semin. Oncol.* **24,** 440–454.
2. Chouaib, S., Asselin-Paturel, C., Mami-Chouaib, F., Caignard, A., and Blay, J. (1997) The host-tumor immune conflict: from immunosuppression to resistance and destruction. *Immunol. Today* **18,** 493–497.

3. Alleva, D. G., Burger, C. J., and Elgert, K. D. (1994) Tumor-induced regulation of suppressor macrophage nitric oxide and TNF-alpha production: role of tumor-derived IL-10, TGF-beta and prostaglandin E2. *The Journal of Immunology* **153**, 1674–1686.
4. Huang, M., Sharma, S., Mao, J. T., and Dubinett, S. M. (1996) Non-small cell lung cancer-derived soluble mediators and prostaglandin E2 enhance peripheral blood lymphocyte IL-10 transcription and protein production. *J. Immunol.* **157**, 5512–5520.
5. Huang, M., Stolina, M., Sharma, S., Mao, J., Zhu, L., Miller, P., et al. (1998) Non-small cell lung cancer cyclooxygenase-2-dependent regulation of cytokine balance in lymphocytes and macrophages: up-regulation of interleukin 10 and down-regulation of interleukin 12 production. *Cancer Res.* **58**, 1208–1216.
6. Halak, B. K., Maguire, Jr., H. C., and Lattime, E. C. (1999) Tumor-induced interleukin-10 inhibits type 1 immune responses directed at a tumor antigen as well as a non-tumor antigen present at the tumor site. *Cancer Res.* **59**, 911–917.
7. Maeda, H. and Shiraishi, A. (1996) TGF-β contributes to the shift toward Th2-type responses through direct and IL-10 mediated pathways in tumor-bearing mice. *J. Immunol.* **156**, 73–78.
8. Beissert, S., Hosoi, J., Grabbe, S., Asahina, A., and Granstein, R. D. (1995) IL-10 inhibits tumor antigen presentation by epidermal antigen-presenting cells. *J. Immunol.* **154**, 1280–1286.
9. Nastala, C. L., Edington, H. D., McKinney, T. G., Tahara, H., Nalesnik, M. A., Brunda, M. J., et al. (1994) Recombinant IL-12 administration induces tumor regression in association with IFN-γ production. *J. Immunol.* **153**, 1697–1706.
10. Nabioullin, R., Sone, S., Nii A.,, Haku, T., and Ogura, T. (1994) Induction mechanism of human blood CD8+ T cell proliferation and cytotoxicity by natural killer cell stimulatory factor (interleukin-12). *Jpn. J. Cancer Res.* **85**, 853–861.
11. Trinchieri, G. (1993) Interleukin-12 and its role in the generation of TH1 cells. *Immunol. Today* **14**, 335–338.
12. de Waal-Malefyt, R., Haanen, J., Spits, H., Roncarolo, M. G., Te Velde, A., Fidgor, C., et al. (1991) Interleukin-10 (IL10) and viral IL-10 strongly reduce antigen-specific human T-cell proliferation by diminishing the antigen-presenting capacity of monocytes via down-regulation of class-II major histocompatibility complex expression. *J. Exp. Med.* **174**, 915–924.
13. Kim, J., Modlin, R. L., Moy, R. L., Dubinett, S. M., McHugh, T., Nickoloff, B. J., and Uyemura, K. (1995) IL-10 production in cutaneous basal and squamous cell carcinomas: A mechanism for evading the local T cell immune response. *J. Immunol.* **155**, 2240–2247.
14. Qin, Z., Noffz, G., Mohaupt, M., and Blankenstein, T. (1997) Interleukin-10 prevents dendritic cell accumulation and vaccination with granulocyte-macrophage colony-stimulating factor gene-modified tumor cells. *J. Immunol.* **159**, 770–776.
15. Bianchi, R., Grohmann, U., Belladonna, M., Silla, S., Fallarino, F., Ayroldi, E., et al. (1996) IL-12 is both required and sufficient for initiating T cell reactivity to a

class I-restricted tumor peptide (P815AB) following transfer of P815AB-pulsed dendritic cells. *J. Immunol.* **157**, 1589–1597.
16. Colombo, M., Vagliani, M., Spreafico, F., Parenza, M., Chiodoni, C., Melani, C., and Stoppacciaro, A. (1996) Amount of interleukin 12 available at the tumor site is critical for tumor regression. *Cancer Res.* **56**, 2531–2534.
17. Kobayashi, M., Kobayashi, H., Pollard, R., and Suzuki, F. (1998) A pathogenic role of Th2 cells and their cytokine products on the pulmonary metastasis of murine B16 melanoma. *J. Immunol.* **160**, 5869–5873.
18. Handel-Fernandez, M. E., Ching, X., Herbert, L. M., and Lopez, D. M. (1997) Down-regulation of IL-12, not a shift from a T helper-1 to a T helper-2 phenotype, is responsible for impaired IFN-γ production in mammary tumor-bearing mice. *J. Immunol.* **158**, 280–286.
19. Rohrer, J. W. and Coggin, Jr., J. H. (1995) CD8 T cell clones inhibit antitumor T cell function by secreting IL-10. *J. Immunol.* **155**, 5719–5727.
20. Van der Pouw Kraan, T., Boeije, L., Smeenk, R., Wijdenes, J. and Aarden, L. (1995) Prostaglandin-E2 is a potent inhibitor of human interleukin 12 production. *J. Exp. Med.* **181**, 775–779.
21. Strassmann, G., Patil-Koota V., Finkelman, F., Fong, M., and Kambayashi, T. (1994) Evidence for the involvement of interleukin 10 in the differential deactivation of murine peritoneal macrophages by prostaglandin E2. *J. Exp. Med.* **180**, 2365–2370.
22. Smith, W., Garavito, R., and DeWitt, D. (1996) Prostaglandin endoperoxide H synthases (Cyclooxygenase)-1 and -2. *J. Biol. Chem.* **271**, 33157–33160.
23. Herschman, H. (1996) Review: Prostaglandin synthase 2. *Biochim. Biophy. Acta* **1299**, 125–140.
24. Herschman, H. R., Kujubu, D. A., Fletcher, B. S., Ma, Q., Varnum, B. C., Gilbert, R. S., and Reddy, S. T. (1994) The tis genes, primary response genes induced by growth factors and tumor promoters in 3T3 cells. *Prog. Nucleic Acids Res. Mol. Biol.* **47**, 113–148.
25. Hla, T. and Neilson, K. (1992) Human Cyclooxygenase-2 cDNA. *Proc. Natl. Acad. Sci. USA* **89**, 7384–7388.
26. Tsujii, M. and Dubois, R. (1995) Alterations in cellular adhesion and apoptosis in epithelial cells overexpressing prostaglandin endoperoxide synthase-2. *Cell* **83**, 493–501.
27. Tsujii, M., Kawano, S., and DuBois, R. (1997) Cyclooxygenase-2 expression in human colon cancer cells increases metastatic potential. *Proc. Natl. Acad. Sci. USA* **94**, 3336–3340.
28. Tsujii, M., Kawano, S., Tsuji, S., Sawaoka, H., Hori, M., and DuBois, R. (1998) Cyclooxygenase regulates angiogenesis induced by colon cancer cells. *Cell* **93**, 705–716.
29. Dubois, R. N., Abramson, S. B., Crofford, L., Gupta, R. A., Simon, L. S., Van De Putte, L. B., and Lipsky, P. E. (1998) Cyclooxygenase in biology and disease. *Faseb. J.* **12**, 1063–1073.

30. Kargman, S. L., O'Neill, G. P., Vickers, P. J., Evans, J. F., Mancini, J. A., and Jothy, S. (1995) Expression of prostaglandin G/H synthase-1 and -2 protein in human colon cancer. *Cancer Res.* **55,** 2556–2559.
31. Sano, H., Kawahito, Y., Wilder, R. L., Hashiramoto, A., Mukai, S., Asai, K., et al. (1995) Expression of cyclooxygenase-1 and -2 in human colorectal cancer. *Cancer Res.* **55,** 3785–3789.
32. Wolff, H., Saukkonen, K., Anttila, S., Karjalainen, A., Vainio, H., and Ristimaki, A. (1998) Expression of cyclooxygenase-2 in human lung carcinoma. *Cancer Res.* **58.** 4997–5001.
33. Liu, X.-H. and Rose, D. P. (1996) Differential expression and regulation of cyclooxygenase-1 and -2 in two human breast cancer cell lines. *Cancer Res.* **56,** 5125–5127.
34. Ristimaki, A., Honkanen, N., Jankala, H., Sipponen, P., and Harkonen, M. (1997) Expression of cyclooxygenase-2 in human gastric carcinoma. *Cancer Res.* **57,** 1276–1280.
35. Tucker, O. N., Dannenberg, A. J., Yang, E. K., Zhang, F., Teng, L., Daly, J. M., et al. (1999) Cyclooxygenase-2 expression is up-regulated in human pancreatic cancer. *Cancer Res.* **59,** 987–990.
36. Wilson, K., Fu, S., Ramanujam, K., and Meltzer, S. (1998) Increased expression of inducible nitric oxide synthase and cyclooxygenase-2 in Barrett's esophagus and associated adenocarcinomas. *Cancer Res.* **58,** 2929–2934.
37. Chan, G., Boyle, J. O., Yang, E. K., Zhang, F., Sacks, P. G., Shah, J. P., et al. (1999) Cyclooxygenase-2 expression is up-regulated in squamous cell carcinoma of the head and neck. *Cancer Res.* **59,** 991–994.
38. Shiota, G., Okubo, M., Noumi, T., Noguchi, N., Oyama, K., Takano, Y., et al. (1999) Cyclooxygenase-2 expression in hepatocellular carcinoma. *Hepatogastroenterology* **46,** 407–412.
39. Sharma, S., Miller, P., Stolina, M., Zhu, L., Huang, M., Paul, R., and Dubinett, S. (1997) Multi-component gene therapy vaccines for lung cancer: effective eradication of established murine tumors in vivo with Interleukin 7 / Herpes Simplex Thymidine Kinase-transduced autologous tumor and ex vivo-activated dendritic cells. *Gene Ther.* **4,** 1361–1370.
40. Huang, M., Wang, J., Lee, P., Sharma, S., Mao, J. T., Meissner, H., et al. (1995) Human non-small cell lung cancer cells express a type 2 cytokine pattern. *Cancer Res.* **55,** 3847–3853.
41. Dubinett, S. M., Kurnick, J. T., and Kradin, R. L. (1989) Adoptive immunotherapy of murine pulmonary metastases with interleukin 2 and interferon-gamma. *Am. J. Respir. Cell Mol. Biol.* **1,** 361–369.
42. Chiabrando, C., Broggini, M., Castagnoli, M. N., Donelli, M. G., Noseda, A., Visintainer, M., et al. (1985) Prostaglandin and thromboxane synthesis by Lewis lung carcinoma during growth. *Cancer Res.* **45,** 3605–3608.

III

NOVEL THERAPIES

D. CHEMOPREVENTION

45

Chemoprevention of Lung Cancer

Hanspeter Witschi

1. Introduction

"Hypothesis-driven chemoprevention of lung cancer, when put to the test of randomized large-scale clinical trials, so far has been disappointing, unlike important successes with selective estrogen receptors modulators for breast cancer and non steroidal anti-inflammatory drugs and cyclooxygenase-2 inhibitors for colon cancer " *(1)*. This somewhat pessimistic statement reflects the outcome of several clinical trials that examined whether human lung cancer could effectively be prevented by dietary supplements. One such agent was beta carotene. In several epidemiological studies negative associations between estimated intakes of vitamin A or of beta carotene (provitamin A) and the risk of developing cancer at various sites seemed to provide compelling reasons for the use of beta carotene as a chemopreventive agent *(2)*. Preclinical studies with in vitro and in vivo studies looked promising. However, in large-scale clinical trials it was discovered that in active smokers, beta carotene not only failed to protect, but actually increased the lung cancer risk. The effect was so obvious that one of the clinical trials had to be halted before its scheduled end. When the experimental data base on the chemopreventive action of beta carotene was re-examined, it was found that no really convincing experiments had shown that beta carotene would interfere with tumor development in the respiratory tract. It was pointed out eventually that preclinical studies should evaluate putative chemopreventive agents not just in any system, but in animal models that would approximate the human tumor *(3)*. N-acetylcysteine (NAC) is another example of a putative chemopreventive agent whose reality did not live up to its promise. NAC does appear to protect animals against chemical carcinogenesis and against tobacco smoke toxicity *(4–6)*. While compelling mechanistic reasons exist as to why NAC should be an effective chemopreventive agent *(7)*, it was never established that NAC would inhibit tobacco smoke-induced lung tumor formation in animals, rather than just modify biomarkers of exposure. However, NAC was tested in a large-scale clinical trial. Several years later, it was found that NAC had no discernable protective effect in man *(8)*.

Although in the US the previously steep increase in lung cancer incidence has decreased in males and slowed down in females, there remains some apprehension that upwards trends in smoking among teenagers and cigar smokers might produce a new problem *(9)*. In many other countries of the world, particularly in Asia, lung cancer is a public health problem of the first order. Tobacco products eventually will kill millions of people *(10–12)*. Complete cessation of smoking would be the most effective way to prevent this burden on the public health. Unfortunately, it appears that a substantial number of smokers are unable or unwilling to quit. It is hoped that the emerging field of chemoprevention, a new basic and clinical science *(13)*, will be able to help blunt the impact of nicotine addiction in smokers. Chemoprevention might also help reduce further the risk of developing lung cancer in individuals who have quit smoking or in individuals exposed to tobacco smoke in their environment.

Attempts to find or develop drugs or naturally occurring constituents of the diet that might provide some protection to active smokers go back many years and new avenues continue to open *(14)*. Before any chemopreventive agent can be used in clinical trials, it should be rigorously evaluated in preclinical studies for possible untoward side effects or toxicity *(15)*. A considerable experimental (preclinical) database on chemoprevention of lung tumors already exists. Most preclinical studies were done on laboratory rodents treated with individual carcinogens known to occur in tobacco smoke. Frequently used were tobacco-specific nitrosamines such as 4-(methylnitrosamino)-1-(3pyridyl)-1-butanone (NNK) or aromatic polycyclic hydrocarbons such as benzo(a)pyrene. The animals were most often exposed to the carcinogen by intratracheal instillation or intraperitoneal injection *(16)*. During the last 10 years, a lung tumor assay using the A/J mouse strain has become the preferred test system. A considerable and already existing amount of basic information on molecular and biological properties of murine lung tumors makes this assay particularly attractive. It becomes more and more apparent that lung tumors in mice represent a good model for peripheral human lung adenocarcinoma *(17)*. A substantial overlap exists between man and mouse in the genetic alterations thought to be responsible for lung tumorigenesis *(18)*.

Adenocarcinoma of the lung is on the increase *(19)*. This has been attributed to the "changing cigarette." In previous decades, the most prevalent lung cancer caused by cigarette smoking was squamous cell carcinoma. The introduction of low tar, low nicotine cigarettes has influenced smoking patterns *(20,21)*. Tobacco smoke is inhaled more deeply and retained longer in the peripheral lung, where cells now become exposed predominantly to tobacco-specific nitrosamines, in addition to particle-associated polycyclic aromatic hydrocarbons. This state of affairs can be duplicated in the strain A mouse model of adenocarcinoma. It is thus not surprising that putative chemopreventive agents are screened and evaluated according to their effects on lung tumors in strain A mice. A substantial number of agents have been examined and it has been convincingly established that the carcinogenic action of tobacco-specific nitrosamines or aromatic polycyclic hydrocarbons can be counteracted by such diverse agents as lycopene *(22)*, isothiocyanates *(23)*, nonsteroidal anti-inflammatory drugs (NSAIDs) *(24)*, organoselenium compounds *(25)*, glucocorticoid hormones *(26)*, D-limonene *(27)*, perillyl alcohol *(28)*, euglobal-G1 *(29)*, black and green tea *(30)*, and many others.

There are many mechanisms by which chemopreventive agents are likely to interfere with tumor development. By inhibiting phase I enzymes, they may prevent the activation of carcinogens to proximate or ultimate carcinogens. They also may enhance detoxification by induction of phase II enzymes. Others may scavenge free radicals or enhance DNA repair mechanisms. Additional targets for chemopreventive action are signal transduction, modulation of hormonal activity, inhibitors of oncogene expression, restorers of immune response, or inducers of apoptosis. For practically all steps known to play a role in carcinogenesis, agents have been detected that may modulate such events *(7,31–33)*.

The classification scheme for chemopreventive agents, based on their mode of action, was originally developed by Wattenberg *(34,35)*. The current scheme distinguishes two classes: blocking agents and suppressing agents. Blocking agents prevent metabolic activation of carcinogens and thus bar development of DNA damage. Suppressing agents block the expansion of cells that have been initiated by carcinogens or are capable of reversing abnormal differentiation *(36)*.

For chemoprevention of lung cancer, it would be desirable to have available both blocking agents and suppressing agents. Blocking agents might be used to advantage in active smokers, where they might inhibit the activation or accelerate the detoxification of carcinogens. Effective antioxidants and free-radical scavengers might also be useful. A great variety of agents have been examined that, at least in animal models, seem able to prevent lung tumor development using such mechanisms *(7,23,37)*. Unfortunately, as will be discussed later, many agents, including the two antioxidants beta carotene and NAC, have not been found to be effective in animal models that use tobacco smoke to produce lung tumors. In addition, it might be asked to what extent blocking agents would constitute the optimal chemopreventive regimen. It has been observed that in smokers who quit, the risk to develop lung cancer actually is, for about 3–5 yr, increased rather than decreased *(38–40)*. The conventional explanation for these data is that these individuals quit smoking because they were already suffering from lung cancer or at least were bothered by chronic cough and/or pulmonary disease. In the absence of a clear indication of premalignant or malignant lung lesions needing aggressive treatment (e.g. surgery, radiation, or chemotherapy), prophylactic administration of a chemopreventive agent might be of some benefit. Unfortunately, the majority of chemopreventive agents that have been evaluated in animals models act only when present during exposure to the carcinogen. Only a few have been found to be effective when given after the last exposure to a chemical carcinogen. Examples are budesonide *(26)*, lycopene *(22)*, the Bowman-Birk protease inhibitor *(41)*, and a mixture of myoinositol-dexamethasone *(42)*.

There thus exists a large database on the effects of a variety of chemopreventive agents. In practically all studies, lung tumors were induced by specific agents thought to be important elements in tobacco smoke for induction of lung carcinogenesis. It must be pointed out, however, that on occasion some doubts have been raised as to whether studies with model compounds may substitute for exposure to the full and complex mixture of tobacco smoke *(18,43)*. As will be discussed later, these doubts were justified. In an animal model of tobacco smoke-induced lung carcinogenesis, many agents that were effective against lung tumors in strain A mice induced with model compounds eventually proved not to be as helpful as was originally anticipated.

2. Methods

During the last few years, we have consistently produced a significant increase in tumor multiplicity and incidence in the lungs of strain A mice exposed to tobacco smoke. This presented an opportunity to evaluate the effects of chemopreventive agents. The strain A mouse lung tumor assay, developed and refined by Stoner and Shimkin *(44)*, has several advantages. After exposure to a carcinogen, strain A mice usually develop lung tumors that are easily visualized on the lung surface. The average number of tumors per lung (lung tumor multiplicity) is directly proportional to the dose of carcinogen administered. After intraperitoneal or oral exposure of the animals to a carcinogen, visible tumors usually can be detected within 4–5 mo. Exposure to tobacco smoke requires a 5-mo exposure, followed by a 4 mo recovery period in air. Because lung tumor multiplicity rather than lung tumor incidence is a measure of carcinogenic effects, only about 15–30 animals per group are required to observe a statistically significant effect with a two-tailed *t*-test, whereas usually 50 animals per group are needed when differences in incidence are evaluated by Fisher's exact test. Histologically, the tumors impress as adenomas, adenomas with carcinomatous foci or fully developed adenocarcinomas. Exposure to tobacco smoke or administration of chemopreventive agents do not produce shifts in the distribution of the histological pasterns (usually 80% adenomas and 20% adenomas with adenocarcinomas or fully developed adenocarcinomas). Invasion of adjacent tissue or lymphatics by malignant cells is seen on occasion.

With this assay we evaluated the effects of selected chemopreventive agents on lung tumor development in mice exposed to tobacco smoke. In all experiments, we used a mixture of 89% sidestream and 11% mainstream tobacco smoke, generated by burning Kentucky 1R4F reference cigarettes. Although this mixture can be taken to represent environmental tobacco smoke (ETS), it contains essentially all ingredients of tobacco smoke inhaled by active smokers (differences between the two forms of smoke being quantitative rather than qualitative) *(45)*. The protocol involved exposure of mice for 5 mo to smoke, followed by a 4 mo "recovery" phase in air. In 7 independently conducted studies, we found a significant increase in lung tumor multiplicity in all and a significant increase in lung tumor incidence in 5 out of the 7 experiments *(46–51)*. The exposure protocol has thus reproducibly produced lung tumors in mice. It has also been successfully reproduced in a different laboratory *(52)*.

We used our experimental model to examine the effects of chemopreventive agents. Based on the observations made by others in preclinical studies, all agents we evaluated had been claimed to become, eventually, useful chemopreventive agents against lung cancer in man. We used the following criteria in the selection and dosage regimens: published evidence that the agents were indeed highly effective in the strain A mouse lung tumor assay against known tobacco smoke carcinogens such as NNK or benzo(a)pyrene. The same dosing regimens were then selected. The animals were given the chemopreventive agent in the diet during tobacco smoke exposure and also during the recovery period in air. In each experiment, a positive control group was included, consisting of animals treated with a model carcinogen, but fed the same chemopreventive regimen. Agents examined thus far are listed in **Table 1**. The results of our studies have been reported in detail *(49–51)*. Our observations have been summarized in **Table 2**. Essentially we found that the majority of the agents were largely useless when evaluated against tobacco smoke rather than against model compounds.

Green tea had no activity in positive control experiments. This makes the negative data with tobacco smoke difficult to interpret and opens the question as to whether different sources of tea might have different activity. It has been reported in one experiment that decaffeinated black tea extracts inhibited,

Table 1
Chemopreventive Regimens[a]

Agent	Concentration	Carcinogen against which chemical was tested originally	Reference
Green tea	1.25% extract in drinking water	NNK	*(30)*
N-acetylcysteine	2 g/kg of diet	Urethane	*(72)*
Acetylsalicylic acid	300 mg/kg of diet	NNK	*(53)*
Phenethyl isothiocyanate (PEITC)	500 mg/kg diet	NNK	*(73)*
d-Limonene	6.3 g/kg of diet	NNK	*(27)*
1,4-phenylenebis (methylene) selenoisocyanate (p-XSC)	20 mg/kg of diet (10 ppm of Se)	NNK	*(25)*
Myo-inositol and dexamethasone	10 g/kg and 0.5 mg/kg each in diet	NNK	*(42)*
Mixture of PEITC and benzyl isothiocyanate (BITC)	250 mg/kg each in diet	—	
Beta carotene	0.5 and 0.05 mg/kg	—	

[a]Reproduced with permission from **ref. 46**. Copyright 2001.

but in another two experiments stimulated, skin tumor development initiated by 7,12-dimethylbenz(a)anthracene and subsequently promoted with 12-O-tetradecanoylphorbol-13-acetate *(53)*. It is possible that different teas (green or black, decaffeinated or not) or teas from different sources may vary in their chemopreventive potencies, accounting for our failure to achieve protection against NNK-induced lung tumors in strain A mice. The question whether green tea will effectively prevent lung cancer in man must remain open to further study.

Acetylsalicylic acid is an approved human drug and has been found to be a promising inhibitor of colon carcinogenesis. In the lungs of strain A mice, N-acetylsalicylic acid inhibits NNK-induced tumor growth, because it greatly reduced the number of tumors with a diameter of more than 1 mm *(54)*. In our experiment we confirmed that in NNK-treated animals, the growth of tumors was slowed. In smoke-exposed animals no such effect was seen *(49)*.

Table 2
Efficiency of Various Chemopreventive Against Tobacco Smoke-Specific Carcinogens and Full Tobacco Smoke

Chemopreventive agent	Tumor multiplicity as percentage of controls[a] in animals exposed to:	
	Specific carcinogen[b]	Full tobacco smoke
a) Chemopreventive agents not effective against specific carcinogens or against full tobacco smoke		
Green tea	100%	100%
Acetylsalicylic acid	106%	105%
b) Chemopreventive agents moderately to highly effective against specific carcinogens, ineffective against tobacco smoke		
N-acetylcysteine (NAC)	65%[c]	123%
PEITC	14%[c]	85%
d-limonene	60%[c]	93%
p-XSC	20%[c]	86%
PEITC/BITC	—	88%
Beta carotene[d]	—	100%
c) Chemopreventive agents effective against specific carcinogens AND against tobacco smoke		
Myoinositol-dexamethasone	14%[c]	48%[c]

[a]Lung tumor multiplicities calculated as percentage of corresponding controls. Data adapted with permission from **refs. 49–51**.
[b]Urethane for N-acetylcysteine, NNK for all others.
[c]Significantly different from control values (100%).
[d]Unpublished data.

NAC, PEITC, d-limonene, and p-XSC substantially reduced lung tumor development following administration of the model carcinogens NNK or of urethane, but had no activity against the full complex mixture of tobacco smoke. Since lung tumor multiplicities following exposure to tobacco smoke are rather modest (from 1.3–2.8 tumors/lung as against 0.5–1.0 tumors per lung in controls), it could be argued that the assay does not allow detection of comparatively small reductions produced by the chemopreventive agents. The observations made with the mixture of myoinositol and dexamethasone refute this assumption. These two agents not only substantially reduced lung tumor multiplicity following treatment of the animals with NNK, but were equally effective in cutting down lung tumor multiplicity and incidence in animals exposed to tobacco smoke (*49*, and **Table 3**).

In animal models, NNK and benzo(a)pyrene are potent lung carcinogens. However, PEITC, BITC, or pXSC, although effective against these compounds, were ineffective against full tobacco smoke. It may therefore be asked how important the overall contribution of NNK and benzo(a)pyrene is to tobacco-smoke carcinogenesis. The question is not without relevance for the development of chemopreventive drugs. It is generally assumed that it is mostly the particulate (tar) phase of tobacco smoke that is responsible for its carcinogenic action. As a consequence, great efforts were made to develop cigarettes with effective filters and reduced tar content. Unfortunately, the introduction of filter cigarettes had a smaller impact on lung cancer incidence than was originally hoped for *(55)*. It is possible that the gas phase of tobacco smoke contains some as-yet-unidentified, but highly potent, carcinogens. This hypothesis is supported by animal studies. It has been repeatedly reported that the gas phase of tobacco smoke is as effective as full tobacco smoke for tumor induction *(48,56)*. The question is also of some interest inasmuch as cigarettes are currently being developed and marketed that no longer burn, but only heat, tobacco. These cigarettes are generally believed to be much less harmful than cigarettes that burn tobacco *(57)*. Unless demonstrated otherwise, they may nevertheless provide a carcinogenic risk by extracting and delivering carcinogenic ingredients in the gas phase *(58)*.

The pioneering studies by Wattenberg were the first to show that *myo*inositol and dexamethasone, alone or in combination, were highly effective against tumor development in several sites following exposure to nitrosamines or polycyclic aromatic hydrocarbons *(42,59,60,61)*. Their proven effectiveness against tobacco smoke in animal studies (*see* **Table 3**) also make them promising agents for eventual use in clinical trials. One potential drawback in their use might be untoward side effects. This is a real concern with glucocorticoids. However, newer studies show that it might be possible to greatly reduce the dose necessary for effective chemoprevention by delivering the agent directly to the target via inhalation *(62)*. No such concerns seem to exist in regard to myoinositol. It is a naturally occurring agent found in rice bran, cereals, and legumes *(63,64)* and has been administered to people at a dose of 12 g per day for 4 wk with no adverse effects in hematology, kidney, or liver function observed *(65)*.

The chemopreventive effect of the combination of *myo*inositol and dexamethasone is remarkable for yet another reason. While many chemopreventive agents fall under the category of "blocking" agents, i.e., agents that interfere with the metabolism of carcinogens, there are considerably fewer that are true "suppressing" agents, i.e., agents that interfere with tumor development even in the absence of carcinogen exposure. In earlier studies, Wattenberg and Estensen *(42,59)* had shown that in mice treated with NNK or BaP (benzo[a]pyrene), *myo*inositol or dexamethasone were highly effective in preventing lung tumor development when given in the postinitiation period, whereas they had no

Table 3
The Chemopreventive Effects of *Myoinositol-Dexamethasone*[a]

Treatment	*My*oinositol-dexamethasone diet fed during entire 9 mo	*My*oinisotol-dexamethasone diet fed only after animals were removed from tobacco smoke
	Lung tumor multiplicity	
ETS + Control diet	2.1 ± 0.3 (35)	2.4 ± 0.3 (26)
ETS + *Myoinositol*-dexamethasone	1.0 ± 0.2 (34)[b]	1.0 ± 0.2 (26)[b]
Air + Control diet	0.6 ± 0.1 (30)	1.0 ± 0.1 (54)
Air + *Myoinositol*-dexamethasone	0.6 ± 0.2 (25)	1.0 ± 0.2 (29)
	Lung tumor incidence	
ETS + Control diet	85%	89%
ETS + *Myoinositol*-dexamethasone	62%[b]	62%[b]
Air + Control diet	50%	65%
Air + *Myoinositol*-dexamethasone	44%	59%

[a]Data adapted with permission from **refs. *49,51*.**
[b]Significantly different ($p < 0.05$) from animals exposed to ETS and fed control diet.

significant effect when given during carcinogen exposure. The Bowman-Birk protease inhibitor *(41,66)*, the synthetic glucocorticoid budesonide *(23,34)*, and lycopene *(22)* are other agents found to be effective during the post-initiation period. Because epidemiological investigations suggest that after quitting smoking there is a temporarily increased risk in lung cancer development, suppressing agents would be of considerable value.

3. Outlook

There are at present numerous agents available that in animal models have been found to prevent effectively the development of lung tumors. As has been discussed above, however, effectiveness in model compounds does not necessarily translate into effectiveness against full tobacco smoke. Tobacco smoke-induced lung carcinogenesis depends on many factors, and it is therefore doubtful whether agents with one particular mode of action, i.e., inhibition of activation or enhancement of detoxification of specific carcinogens or carcinogen classes, would be sufficient for overall protection. Unless we can pinpoint one specific mechanisms for tobacco-smoke carcinogenesis, it will probably be prudent to evaluate putative chemopreventive agents in an animal model of full tobacco-smoke carcinogenesis, rather than with model compounds, to produce lung tumors. Unfortunately, this is a much less expedient procedure than a straightforward assay with a specific chemical. In such assays, results can usually be obtained within 5–6 months, as opposed to the 10–12 mo necessary to conduct a proper tobacco-smoke inhalation bioassay. The number of animals required to obtain statistically significant results in an inhalation assay is also about two to three times greater than in the assay recommended with model compounds thought to mimic cigarette smoke *(67)*.

Chemopreventive agents, as opposed to chemotherapeutic agents, will likely need to be administered to healthy people over prolonged periods of time. They should therefore be devoid of any untoward side effects or toxicity. Most agents, particularly those occurring as natural constituents in plants, fruits, and vegetables, easily fulfill these criteria. Unfortunately, two of the more promising chemopreventive agents may have serious side effects: the steroid hormones and the retinoids. Attempts have been made to reduce toxicity of these agents by direct delivery to the putative target sites. Both budesonide and cis-retinoic acid have recently been administered via inhalation from nebulizers to strain A mice treated with tobacco smoke-specific carcinogens such as NNK or benzo(a)pyrene. In both instances, significant reductions in tumor development were observed. For budesonide to be effective, the aerosol was administered 3 times a week. delivering in any given treatment a dose of 10 or 25 µg/kg. A second synthetic glucocorticosteroid, beclomethasone, was equally protective *(62)*. The doses the mice were exposed to approximated doses that are delivered to humans via nebulizers in the chronic treatment of asthma. Exposure to isotretinoin (13-cis retinoic acid) was at dose levels of 0.24, 1.6, and 24.9 mg/kg per week, with an estimated retention in the lung of 16%. Development of urethane and benzo(a)pyrene induced lung tumors was significantly reduced *(68)*. It was pointed out that the effective doses were more than 100 times lower than were oral doses given to mice without any noticeable effect on lung tumor development. Administration of these compounds by inhalation thus appears to be a viable way to prevent or circumvent potential undesirable side effects. The concept of field cancerization *(69–71)* supports the hypothesis that administration of chemopreventive agents over the entire surface of the respiratory tract might constitute an effective mode to reduce the risk of lung cancer development in populations at risk.

Acknowledgments

I wish to thank Imelda Espiritu, Dale Uyeminami, and Robert R. Maronpot for their help in performing the experiments described herein. This publication was made possible by grants number ES07908, ES07499, and ES05707 from the National Institute of Environmental Health Sciences (NIEHS). Its contents are solely the responsibility of the authors and do not necessarily represent the official views of the NIEHS, NIH.

References

1. Omenn, G. S. (2000) Chemoprevention of lung cancer is proving difficult and frustrating, requiring new approaches. *J. Natl. Cancer Inst.* **92,** 959–960.
2. Omenn, G. S. (1998) Chemoprevention of lung cancer: the rise and demise of beta-carotene. *Annu. Rev. Public Health* **19,** 73–99.
3. De Luca, L. M. and Ross, S. A. (1996) Beta-carotene increases lung cancer incidence in cigarette smokers. *Nutr. Rev.* **54,** 178–180.
4. De Flora, S., Astengo, M., Serra, D., and Bennicelli, C. (1986) Inhibition of urethan-induced lung tumors in mice by dietary N-acetylcysteine. *Cancer Lett.* **32,** 235–241.
5. Izzotti, A., Balansky, R., Scatolini, L., Rovida, A., and De Flora, S. (1995) Inhibition by N-acetylcysteine of carcinogen-DNA adducts in the tracheal epithelium of rats exposed to cigarette smoke. *Carcinogenesis* **16,** 669–672.
6. Izzotti, A., Balansky, R. M., Coscia, N., Scatolini, L., D'Agostini, F., and De Flora, S. (1992) Chemoprevention of smoke-related DNA adduct formation in rat lung and heart. *Carcinogenesis* **13,** 2187–2190.
7. De Flora, S., Cesarone, C. F., Balansky, R. M., Albini, A., D'Agostini, F., Bennicelli, C., et al. (1995) Chemopreventive properties and mechanisms of N-Acetylcysteine. The experimental background. *J. Cell Biochem. Suppl.* **22,** 33–41.

8. van Zandwijk, N., Dalesio, O., Pastorino, U., De Vries, N., and van Tinteren, H. (2000) EUROSCAN, a randomized trial of vitamin A and N-acetylcysteine in patients with head and neck cancer or lung cancer. For the European Organization for Research and Treatment of Cancer Head and Neck and Lung Cancer Cooperative Groups. *J. Natl. Cancer Inst.* **92,** 977–986.
9. Wingo, P. A., Ries, L. A., Giovino, G. A., Miller, D. S., Rosenberg, H. M., Shopland, D. R., et al. (1999) Annual report to the nation on the status of cancer, 1973–1996, with a special section on lung cancer and tobacco smoking. *J. Natl. Cancer Inst.* **91,** 675–690.
10. Pisani, P., Parkin, D. M., Bray, F., and Ferlay, J. (1999) Estimates of the worldwide mortality from 25 cancers in 1990. *Int. J. Cancer* **83,** 18–29.
11. Liu, B. Q., Peto, R., Chen, Z. M., Boreham, J., Wu, Y. P., Li, J. Y., et al. (1998) Emerging tobacco hazards in China: 1. Retrospective proportional mortality study of one million deaths. *BMJ* **317,** 1411–1422.
12. Yang, G., Fan, L., Tan, J., Qi, G., Zhang, Y., Samet, J. M., et al. (1999) Smoking in China: findings of the 1996 National Prevalence Survey. *JAMA* **282,** 1247–1253.
13. Hong, W. and Sporn, M. B. (1999) Prevention of cancer in the next millenium: Report of the chemoprevention working group to the American Association for Cancer Research. *Cancer Res.* **59,** 4743–4758.
14. Goodman, G. E. (2000) Prevention of lung cancer. *Crit. Rev. Oncol. Hematol.* **33,** 187–197.
15. Johnson, K. A., Beitz, J., Justice, R., Schmidt, W., Andrews, P., and DeLap, R. (1997) Protocol design considerations that relate to demonstrating the safety and effectiveness of chemopreventive agents. *J. Cell Biochem. Suppl.* **27,** 1–6.
16. Moon, R. C., Rao, K. V. N., Detrisac, C. J., and Kelloff, G. J. (1992) Hamster lung cancer model of carcinogenesis and chemoprevention, in *The Biology and Prevention of Aerodigestive Tract Cancers* (Newell, G. R. and Hong, W. K., eds.), Plenum Press, New York, pp. 55–61.
17. Malkinson, A. M. (1998) Molecular comparison of human and mouse pulmonary adenocarcinomas. *Exp. Lung Res.* **24,** 541–555.
18. Herzog, C. R., Lubet, R. A., and You, M. (1997) Genetic alterations in mouse lung tumors: implications for cancer chemoprevention. *J. Cell Biochem. Suppl.* **28–29,** 49–63.
19. Thun, M. J., Lally, C. A., Flannery, J. T., Calle, E. E., Flanders, W. D., and Heath, C. W., Jr. (1997) Cigarette smoking and changes in the histopathology of lung cancer. *J. Natl. Cancer Inst.* **89,** 1580–1586.
19. Wynder, E. L. and Hoffmann, D. (1994) Smoking and lung cancer: scientific challenges and opportunities. *Cancer Res.* **54,** 5284–5295.
20. Hoffmann, D., Hoffmann, I., and Wynder, E. L. (1991) Lung cancer and the changing cigarette. *IARC. Sci. Publ.* **105,** 449–459.
21. Djordjevic, M. V., Hoffmann, D., and Hoffmann, I. (1997) Nicotine regulates smoking pattern. *Prevent. Med.* **26,** 435–440.

22. Kim, D. J., Takasuka, N., Kim, J. M., Sekine, K., Ota, T., Asamoto, M., et al. (1997) Chemoprevention by lycopene of mouse lung neoplasia after combined initiation treatment with DEN, MNU and DMH. *Cancer Lett.* **120,** 15–22.
23. Hecht, S. S. (1995) Chemoprevention by isothiocyanates. *J. Cell Biochem. Suppl.* **22,** 195–209.
24. Castonguay, A., Rioux, N., Duperron, C., and Jalbert, G. (1998) Inhibition of lung tumorigenesis by NSAIDS: a working hypothesis. *Exp. Lung Res.* **24,** 605–615.
25. Prokopczyk, B., Cox, J. E., Upadhyaya, P., Amin, S., Desai, D., Hoffmann, D., and el-Bayoumy, K. (1996) Effects of dietary 1,4-phenylenebis(methylene) selenocyanate on 4-(methylnitrosamino)-1-(3-pyridyl)-1-butanone-induced DNA adduct formation in lung and liver of A/J mice and F344 rats. *Carcinogenesis* **17,** 749–753.
26. Wattenberg, L. and Estensen, R. D. (1997) Studies of chemopreventive effects of budenoside on benzo(a)pyrene-induced neoplasia of the lung of female A/J mice. *Carcinogenesis* **18,** 2015–2017.
27. Wattenberg, L. W. and Coccia, J. B. (1991) Inhibition of 4-(methylnitrosamino)-1-(3-pyridyl)-1-butanone carcinogenesis in mice by D-limonene and citrus fruit oils. *Carcinogenesis* **12,** 115–117.
28. Lantry, L. E., Zhang, Z., Gao, F., Crist, K. A., Wang, Y., Kelloff, G. J., Lubet, R. A., and You, M. (1997) Chemopreventive effect of perillyl alcohol on 4-(methylnitrosamino)-1- (3-pyridyl)-1-butanone induced tumorigenesis in (C3H/HeJ X A/J)F1 mouse lung. *J. Cell Biochem. Suppl.* **27,** 20–25.
29. Takasaki, M., Konoshima, T., Etoh, H., Pal, S., I, Tokuda, H., and Nishino, H. (2000) Cancer chemopreventive activity of euglobal-G1 from leaves of Eucalyptus grandis. *Cancer Lett.* **155,** 61–65.
30. Cao, J., Xu, J., Chen, J., and Klaunig, J. E. (1996) Chemopreventive effects of green and black tea on pulmonary and hepatic carcinogenesis. *Fundam. Appl. Toxicol.* **29,** 244–250.
31. Stoner, G. D., Morse, M. A., and Kelloff, G. J. (1997) Perspectives in cancer chemoprevention. *Environ. Health Perspect.* **105(Suppl. 4),** 945–954.
32. De Flora, S., Bennicelli, C., and Bagnasco, M. (1999) Rationale and mechanisms of cancer chemoprevention. *Recent Results Cancer Res.* **151,** 29–44.
33. Weinstein, I. B. (2000) Disorders in cell circuitry during multistage carcinogenesis: the role of homeostasis. *Carcinogenesis* **21,** 857–864.
34. Wattenberg, L. W. (1977) Inhibition of carcinogenic effects of polycyclic hydrocarbons by benzyl isothiocyanate and related compounds. *J. Natl. Cancer Inst.* **58,** 395–398.
35. Wattenberg, L. (1985) Chemoprevention of cancer. *Cancer Res.* **45,** 1–8.
36. You, M. and Bergman, G. (1998) Preclinical and clinical models of lung cancer chemoprevention. *Hematol. Oncol. Clin. North Am.* **12,** 1037–1053.
37. Ganther, H. E. (1999) Selenium metabolism, selenoproteins and mechanisms of cancer prevention: complexities with thioredoxin reductase. *Carcinogenesis* **20,** 1657–1666.

38. Hammond, E. C. (1966) Smoking in relation to the death rates of one million men and women. *Natl. Cancer Inst. Monogr.* **19,** 127–204.
39. Wynder, E. L. and Stellman, S. D. (1977) Comparative epidemiology of tobacco-related cancers. *Cancer Res.* **37,** 4608–4622.
40. Postmus, P. E. (1998) Epidemiology of lung cancer, in *Fishman's Pulmonary Diseases and Disorders* (Fishman, A. P., Elias, J. A., Fishman, J. A., Grippi, M. A., Kaiser, L. R., and Senior, R. M., eds.), McGraw-Hill, New York, pp. 1707–1717.
41. Witschi, H. and Kennedy, A. R. (1989) Modulation of lung tumor development in mice with the soybean- derived Bowman-Birk protease inhibitor. *Carcinogenesis* **10,** 2275–2277.
42. Wattenberg, L. W. and Estensen, R. D. (1996) Chemopreventive effects of myo-inositol and dexamethasone on benzo[a]pyrene and 4-(methylnitrosoamino)-1-(3-pyridyl)-1-butanone- induced pulmonary carcinogenesis in female A/J mice. *Cancer Res.* **56,** 5132–5135.
43. Witschi, H. P. (1998). tobacco smoke as a mouse lung carcinogen. *Exp. Lung Res.* **24,** 385–394.
44. Stoner, G. D. and Shimkin, M. B. (1985) Lung tumors in strain A mice as a bioassay for carcinogenicity, in *Handbook of Carcinogen Testing* (Milman, H. and Weisburger, E. K., eds.), Noyes Publications, Park Ridge, NJ, pp. 179–214.
45. Guerin, M. R., Jenkins, R. A., and Tomkins, B. A. (1992) *The Chemistry of Environmental Tobacco Smoke: Composition and Measurement. Indoor Air Research Series.* Lewis Publishers, Boca Raton, FL.
46. Witschi, H. (2000) Successful and not so successful chemoprevention of tobacco smoke-induced lung tumors. *Exp. Lung Res.* **26,** 743–756.
47. Witschi, H. P., Espiritu, I., Peake, J. L., Wu, K., Maronpot, R. R., and Pinkerton, K. E. (1997) The carcinogenicity of environmental tobacco smoke. *Carcinogenesis* **18,** 575–586.
48. Witschi, H. P., Espiritu, I., Maronpot, R. R., Pinkerton, K. E., and Jones, A. D. (1997) The carcinogenic potential of the gas phase of environmental tobacco smoke. *Carcinogenesis* **18,** 2035–2042.
49. Witschi, H., Espiritu, I., and Uyeminami, D. (1999) Chemoprevention of tobacco smoke-induced lung tumors in A/J strain mice with dietary myo-inositol and dexamethasone. *Carcinogenesis* **20,** 1375–1378.
50. Witschi, H., Espiritu, I., Yu, M., and Willits, N. H. (1998) The effects of phenethyl isothiocyanate, N-acetylcysteine and green tea on tobacco smoke-induced lung tumors in strain A/J mice. *Carcinogenesis* **19,** 1789–1794.
51. Witschi, H., Uyeminami, D., Moran, D., and Espiritu, I. (2000) Chemoprevention of tobacco-smoke lung carcinogenesis in mice after cessation of smoke exposure. *Carcinogenesis* **21,** 977–982.
52. D'Agostini, F., Balansky, R. M., Bennicelli, C., Lubet, R. A., Kelloff, G. J., and De Flora, S. (2001) Pilot studies evaluating the lung tumor yield in cigarette smoke-exposed mice. *Int. J. Oncol.* **18,** 607–615.
53. Lu, Y. P., Lou, Y. R., Xie, J. G., Yen, P., Huang, M. T., and Conney, A. H. (1997) Inhibitory effect of black tea on the growth of established skin tumors in mice:

effects on tumor size, apoptosis, mitosis and bromodeoxyuridine incorporation into DNA. *Carcinogenesis* **18**, 2163–2169.
54. Duperron, C. and Castonguay, A. (1997) Chemopreventive efficacies of aspirin and sulindac against lung tumorigenesis in A/J mice. *Carcinogenesis* **18**, 1001–1006.
55. American Thoracic Society (1996) Cigarette smoking and health. *Am. J. Respir. Crit. Care Med.* **153**, 861–865.
56. Leuchtenberger, C. and Leuchtenberger, R. (1974) Differential response of Snell's and C57 black mice to chronic inhalation of cigarette smoke. Pulmonary carcinogenesis and vascular alterations in lung and heart. *Oncology* **29**, 122–138.
57. Wagner, B. M., Cline, M. J., Dungworth, D. L., Fischer, T. H., Gardner, D. E., Pryor, W. A., et al. (2000) A safer cigarette? A comparative study. A consensus report. *Inhal. Toxicol.* **12(Suppl. 5)**, 1–48.
58. Steimle, S. (1996) Smokeless cigarettes not harmless, researchers say. *J. Natl. Cancer Inst.* **88**, 1341–1343.
59. Estensen, R. D. and Wattenberg, L. W. (1993) Studies of chemopreventive effects of myo-inositol on benzo[a]pyrene- induced neoplasia of the lung and forestomach of female A/J mice. *Carcinogenesis* **14**, 1975–1977.
60. Wattenberg, L. W. (1999) Chemoprevention of pulmonary carcinogenesis by myo-inositol. *Anticancer Res.* **19**, 3659–3661.
61. Wattenberg, L. (1995) Chalcones, myo-inositol and other novel inhibitors of pulmonary carcinogenesis. *J. Cell Biochem. Suppl.* **22**, 162–168.
62. Wattenberg, L. W., Wiedmann, T. S., Estensen, R. D., Zimmerman, C. L., Galbraith, A. R., Steele, V. E., and Kelloff, G. J. (2000) Chemoprevention of pulmonary carcinogenesis by brief exposures to aerosolized budesonide or beclomethasone dipropionate and by the combination of aerosolized budesonide and dietary myo-inositol. *Carcinogenesis* **21**, 179–182.
63. Nagahara, N., Kitamura, H., Inoue, T., Ogawa, T., Ito, T., and Kanisawa, M. (1990) Effect of glycerol on cell kinetics and tumorigenesis in mouse lung following urethan administration. *Toxicol. Pathol.* **18**, 289–296.
64. Shamsuddin, A. M., Ullah, A., and Chakravarthy, A. K. (1989) Inositol and inositol hexaphosphate suppress cell proliferation and tumor formation in CD-1 mice. *Carcinogenesis* **10**, 1461–1463.
65. Levine, J., Barak, Y., Gonzalves, M., Szor, H., Elizur, A., Kofman, O., and Belmaker, R. H. (1995) Double-blind, controlled trial of inositol treatment of depression. *Am. J Psychiatry* **15**, 792–794.
66. Ekrami, H., Kennedy, A. R., Witschi, H., and Shen, W. C. (1993) Cationized Bowman-Birk protease inhibitor as a targeted cancer chemopreventive agent. *J. Drug Target.* **1**, 41–49.
67. Hecht, S. S., Isaacs, S., and Trushin, N. (1994) Lung tumor induction in A/J mice by the tobacco smoke carcinogens 4-(methylnitrosamino)-1-(3-pyridyl)-1-butanone and benzo[a]pyrene: a potentially useful model for evaluation of chemopreventive agents. *Carcinogenesis* **15**, 2721–2725.
68. Dahl, A. R., Grossi, I. M., Houchens, D. P., Scovell, L. J., Placke, M. E., Imondi, A. R., et al. (2000) Inhaled isotretinoin (13-cis retinoic acid) is an effective lung

cancer chemopreventive agent in A/J mice at low doses: a pilot study. *Clin. Cancer Res.* **6,** 3015–3024.
69. Smith, A. L., Hung, J., Walker, L., Rogers, T. E., Vuitch, F., Lee, E., and Gazdar, A. F. (1996) Extensive areas of aneuploidy are present in the respiratory epithelium of lung cancer patients. *Br. J Cancer* **73,** 203–209.
70. Park, I. W., Wistuba, I. I., Maitra, A., Milchgrub, S., Virmani, A. K., Minna, J. D., and Gazdar, A. F. (1999) Multiple clonal abnormalities in the bronchial epithelium of patients with lung cancer. *J. Natl. Cancer Inst.* **91,** 1863–1868.
71. Hittelman, W. N. (1999) Clones and subclones in the lung cancer field. *J. Natl. Cancer Inst.* **91,** 1796–1799.
72. De Flora, S., Astengo, M., Serra, D., and Bennicelli, C. (1986) Inhibition of urethan-induced lung tumors in mice by dietary N- acetylcysteine. *Cancer Lett.* **32,** 235–241.
73. Morse, M. A., Reinhardt, J. C., Amin, S. G., Hecht, S. S., Stoner, G. D., and Chung, F. L. (1990) Effect of dietary aromatic isothiocyanates fed subsequent to the administration of 4-(methylnitrosamino)-1-(3-pyridyl)-1-butanone on lung tumorigenicity in mice. *Cancer Lett.* **49,** 225–230.

46

Assessing the Interaction of Particulate Delivery Systems with Lung Surfactant

Timothy Scott Wiedmann

1. Introduction

The successful treatment of lung cancer, as with all cancers, is determined by the selectivity of the therapy. That is, the cells that have undergone carcinogenic changes must be eradicated, while normal cells must be preserved. With therapeutic agents that are not discriminating, significant side effects and toxicity may result. In fact, death is not rare. Because progress towards the successful treatment of lung cancer has been minimal, it is noteworthy that the lung does offer an inherent advantage (1,2). This advantage arises from its accessibility without invasive procedures. Moreover, this accessibility can increase selectivity, because therapy can be not only directed exclusively to the lung but also limited to the lung as well.

In this chapter, methods to determine the rate and extent of the solubilization of solutes in lung surfactant are reported. There are two applications of interest for such studies. The first pertains to the carcinogenesis process. Adenocarcinomas arise in the deep lung following initiation and propagation steps of cancer within the type II cells of the epithelia of the lower respiratory tract (3). We have postulated that this is a consequence of the preferential uptake of carcinogens by the type II cells in carrying out their normal physiological functions of producing, recycling, and removing lung surfactant (4). The second application follows from the first and is in the area of lung chemoprevention. Chemoprevention may be defined as the administration of a chemical entity for preventing lung cancer (5). In this regard, we have postulated that delivery of chemopreventive agents to the site of carcinogenesis will have a therapeutic advantage over those delivered to the systemic circulation. To this end, we seek compounds that are preferentially associated with lung surfactant and thereby are favorably directed to the type II cell in the lung.

The determination of the rate and extent of solubilization implies dual methodology. Prior to defining solubilization, it is necessary to define solubility. The solubility, often expressed in concentration units, is the equilibrium amount of compound dissolved in a specified volume of solvent at a given temperature and pressure *(6)*. The extent of solubilization is a measurement of the excess amount of solid dissolved in the presence of a specified amount of solubilizer. The rate of solubilization is a dynamic event involving the dissolution or dissolving of a compound introduced in the solid state. In this process, there is either formation of a solution, if completely dissolved, or formation of a suspension, if the amount of compound added given the volume of dispersion exceeds the solubility.

In the theoretical treatment of solubilization, it is essential to determine the number of phases present in the system. Thus, the question arises, does the presence of lung surfactant in the medium result in the formation of one or two phases? The answer lies in the description of the material. Lung surfactant, as it exists in the lung, is a mixture of lipids (ca. 90%) and proteins (ca. 10%) *(7)*. The major components are phospholipids, with phosphatidylcholine representing 85–90%, and phosphatidylglycerol representing 5–10%. There is also a significant amount of neutral lipid, including cholesterol, diglycerides, and triglycerides. There are four main proteins, designated A, B, C, and D, in order of their discovery *(8)*. Lung-surfactant apolipoprotein A is water-soluble but associated with lipid and accounts for 85–90% of the protein. Proteins B and C are not water-soluble, are exclusively associated with lipid, and can be readily extracted along with the lipids using organic solvents. Protein D is also water-soluble but is a minor component.

In surfactant, the overwhelming abundance of phospholipids, and in particular phosphatidylcholine, leads to the formation of lipid bilayers. As it is, there are known to be a number of structures, including multilamellar bodies, vesicles, and tubular myelin that are distinct yet are principally composed of bilayer lipids *(9)*. These structures may give rise to differences in the rate and extent of solute dissolution and solubilization, although no studies have addressed this issue to date. It is reasonable to assume that these differences would be subtle, as the fundamental structure of the aggregate remains the lipid bilayer, and the extent of solubilization is presumably determined by such. The important point is that lipid in lung surfactant is present largely in bilayer structures, and in the context of thermodynamics, should be treated as a separate phase. Thus, the answer to the defining question is that the presence of lung surfactant in an aqueous medium results in a two-phase system correctly referred to as a dispersion and not a solution. As a passing comment, the term, surfactant, is strictly correct for designation of the surface lining material of the lung (i.e., SURFace ACTive AgeNT). However, in a number of fields, surfactant is often used to refer to soaps that form micellar solutions. These are one-phase systems.

Given that surfactant is a separate phase, the extent of solubilization can be expressed as a solubility. Therefore, as the solubility of a solute in water can be rigorously defined, so also can the solubility in lung surfactant be defined as the equilibrium amount dissolved in a specified volume of surfactant. However, as surfactant is a complex mixture in an aqueous medium, the volume of material is not readily specified. A measurable quantity is the molar concentration of phospholipids, which can be obtained from a colorimetric determination of the moles of phosphorus *(10)*. The mass of lipid is problematic, since the lipids in surfactant do not have uniform molecular weights. Moreover, the presence of neutral lipids, glycerides and cholesterol, and fatty acids certainly will contribute to the solubilization, although they will not be detected in the quantification of the phospholipids. There is also the added difficulty arising in the lot-to-lot variation in commercial and isolated preparations of lung surfactant.

The albumin-equivalent protein concentration may also be determined but is of less value, since proteins would not be expected to contribute significantly to the solubilization of solutes. Protein is a minor component representing probably less than 10% of the mass. Finally, there are a number of protein assays that each yield different values of protein content *(8)*.

The rational units for solubilization in lung surfactant are mole fractions, X_s. The mole fraction is defined as the moles of solute dissolved per total moles present, or

$$X_s = N_s/N_t = N_s/(N_s + N_p)$$

where N_t is given as the sum of the moles of solute, N_s, and moles of phospholipid, N_p. A related quantity is the solubilization ratio *(4,11)*, R, defined as

$$R = N_s/N_p$$

This can be obtained from the slope of the total molar concentration of solute dissolved, C_s, plotted as a function of the total molar concentration of lipid added to solution. That is,

$$R = C_s/C_p$$

where C_p is the molar concentration of surfactant phospholipid present in solution. The mole fraction solubilization is related to the solubilization ratio as follows

$$X_s = R/(R + 1)$$

Finally, it is important to note that the total amount of drug in solution, St, has contributions from both the aqueous solubility, S_{aq}, and that solubilized in the surfactant. This is expressed as follows:

$$S_t = S_{aq} + R^* C_p$$

2. Materials

Lung surfactant may be obtained from natural sources (*see* **Subheading 3.1.**) or is commercially available as a therapeutic agent for the treatment of respiratory distress syndrome *(12)*. In the commercially available preparations, there are those that are derived from native sources (e.g., Survanta™, a bovine lung extract was obtained from Ross Laboratories) and those that are referred to as synthetic materials (e.g., Exosurf™).

Solutes used in solubility and solubilization studies must be of the highest available purity. Along this line, it is also essential that the most stable polymorphic form be used under the conditions of the study (temperature and pressure) *(6)*. Powder X-ray crystallography is an easy, but definitive, means to verify the polymorphic form used in the study. Measurement of the melting point is perhaps a more reasonable method, although more open to misinterpretation. It is prudent to make such measurements both before and after solubility/solubilization to verify the absence of solid-solid phase transitions.

Solvents used should also be of high purity, particularly in regard to organic impurities. Thus, distilled water is preferred over simply deionized water. The usual addition of sodium chloride to render the ionic strength of the solution comparable to biological fluids tends to minimize the discrepancy arising from ionic impurities. Contamination from polyvalent ions is also a concern, particularly with solutes that are ionizable. In addition, calcium is known to affect the structure of lung surfactant *(13,14)*, and presumably other polyvalent cations would also have an effect.

2.1. Lung Surfactant Isolation

2.1.1. Surgical Procedure

1. Experimental animals of choice. May include rat, rabbit, dog, cow, or pig.

2.1.2. Lavage Procedure

1. Buffered saline: 5 m*M* Tris-HCl, 100 m*M* NaCl, pH 7.4.
2. Gentamycin at a concentration of 0.75 mg/L.
3. Inert gas such as nitrogen or argon.
4. Flask with a side arm attached at the bottom.

5. Surgical clamp.
6. Flexible tubing for aspiration.

2.1.3. Processing the Lavage Fluid

1. Lung surfactant resuspension buffer: 5 mM Tris-HCl, 100 mM NaCl, pH 7.4, containing 1 mM phenylmethylsulfonylfluoride (PMFS).
2. Table top centrifuge.
3. Ultracentrifuge.

2.1.3.1. FOLCH EXTRACTION

1. 2:1 Chloroform:methanol.
2. Separatory funnel.
3. Round-bottom flask.
4. Rotary evaporator.

2.1.3.2. BUTANOL EXTRACTION

1. 1-butanol.
2. Teflon centrifuge tubes with caps.

2.1.4. Surfactant Storage

1. Cyclohexane.
2. Dry ice/acetone or dry ice/isopropanol bath.
3. Lyophilizer.

2.2. Aqueous Solubility Determination

2.2.1. Dialysis Bag Preparation

1. Cellulose ester dialysis tubing (MWCO 10,000 daltons, Spectro/Por™, Spectrum Medical Industries, Houston, TX).
2. Surgical thread.

2.2.2. Filling Dialysis Bags

1. Teflon™-lined screw cap test tube.
2. Parafilm.
3. Shaking water bath.

2.3. Solubilization of Solutes in Surfactant

2.3.1. Single concentration of surfactant

1. Dialysis tubing (MWCO 10,000 daltons, as in **Subheading 2.2.1.**).
2. Pyrex™ screw-cap test tubes with Teflon™ liners.

2.4. Dissolution of Aerosol Particles

2.4.1. Generation of Aerosol Particles

1. Ultrasonic nebulizer: piezoelectric driver or ultrasonic spray nozzle system (Sono-Tek Corp., Milton, NY) with MicroSpray™ nozzle.
2. Glass baffle with an intervening rubber gasket.

2.4.2. Drying Aerosol Droplets

1. Aluminum cylinders: 6' diameter heating ducts are most convenient and may be purchased at local hardware stores. Drying column may be obtained from TSI (Minneapolis, MN).
2. Drying agents: Charcoal for organic vapors and silica for removing water (mesh size (lines/inch) must be smaller than the mesh size of the screen) and is purchased from Sigma or other chemical supplier.
3. Screen mesh.
4. Gas tight Hamilton syringe.
5. Gas chromatograph (5830A Hewlett Packard with an 1885A terminal).
6. Erlenmeyer flasks sealed with a rubber septum.
7. Digital hygrometer (Fisher Scientific).
8. Silicone rubber heating tape.

2.4.3. Aerosol Output Measurement and Particle Size Determination

1. Microfibrile glass filters (Grade QMA filters, size 4.7 cm, Whatman International Ltd., Maidstone, UK).
2. International Clinical Centrifuge, Model CL.

2.4.4. Aerosol Particle Size Determination

1. Cascade impactor (Intox, Albuquerque, NM).
2. KaleidaGraph™ (Synergy Software, Reading, PA).

2.4.5. Capturing Aerosol Particles

1. Air sampling liquid impinger (tip is 30 mm from bottom, with an approximate 125 mL capacity) Ace Glass, Vineland, NJ.
2. 0.01% sodium dodecyl sulfate (SDS).
3. Ice bath.
4. Survanta™ used at the given concentration of 25 mg/mL or diluted.
5. Scintillation vials.
6. Tabletop centrifuge (International Equipment Co, Needham, MA).

3. Methods

3.1. Lung Surfactant Isolation

Lung surfactant has been isolated from a number of experimental animals, including rat *(15)*, rabbit *(16)*, dog *(7,17–19)*, cow *(20)*, and pig *(14)*, as well as from humans by bronchial alveolar lavage *(21)*. It requires surgical removal of the animal lung, followed by lavage, and then fluid processing by centrifugation. While the entire lung may be minced and extracted from surfactant *(22)*, lavage is considered the preferred method for isolating material most characteristic of the lung lining. Nevertheless, it is recognized that all isolation procedures have the inherent risk of altering the structure or state of aggregation of the naturally occurring lung lining material *(9)*. Therefore, isolation of lung surfactant as it occurs within the lung may not be possible.

3.1.1. Surgical Procedure

1. In the lavage procedure, sacrifice animals with minimal trauma, and open the chest cavity without puncturing the lungs.
2. The lungs are gently removed by cutting the trachea and severing the arteries and veins running between the lung and heart. It is important to minimize the contamination from blood. (*see* **Note 1**).

3.1.2. Lavage Procedure

A relatively large variety of lavage fluids has been reported in the literature *(14–21)*. In essence, they all consist of a buffered saline solution chilled to 4°C before use. The pH should be at 7.4, although 7.2 has been used without any apparent adverse effects. Gentamycin is included to minimize bacterial contamination. To minimize the oxidative damage to the lipids, it is advantageous to saturate the aqueous buffer with an inert gas such as nitrogen or argon.

1. Allow the solution to enter passively the lungs while they are held upright by the trachea (not more than 20 cm of hydrostatic pressure) (*see* **Note 2**).
2. Fill lungs from a flask with a side arm attached at the bottom to which a flexible tube has been connected. Control the flow of solution into the lung using a surgical clamp. As the solution flows into the lung, the lobes may be gently massaged.
3. After filling, the lobes should be further massaged to facilitate mixing of the lavage fluid as well as passage of the fluid into the deep regions of the lung.
4. Remove the fluid by gentle aspiration through a flexible tube. Collect the fluid in a flask held on ice. This procedure can be repeated three times, after which little additional surfactant is obtained.

3.1.3. Processing the Lavage Fluid

1. Centrifuge the resulting dispersion at low speed (150g, 10 min, 4°C) to remove cells.
2. Discard the pellet and process the supernatant by ultracentrifugation (48,000g, 30 min, 4°C).
3. The resulting pellet consists of concentrated dispersion of lung surfactant. Typically, this is resuspended in buffer and centrifuged again.
4. The second pellet will contain the lung surfactant, consisting of the lipid as well as the water-soluble and lipid-associated surfactant proteins.
5. The pellet may be centrifuged one final time in a solution appropriate for solubilization studies. The surfactant may be further processed by either a modified Folch extraction or the more common approach, a butanol extraction.

3.1.3.1. FOLCH EXTRACTION

1. Six volumes of 2:1 chloroform:methanol (1 mg phospholipid/mL) is added to the diluted surfactant pellet in a suitably sized separatory funnel. Shake the funnel and allow to stand to separate the upper aqueous layer from the lower chloroform layer.
2. Collect the lower layer into a round-bottom flask and dry into a film on a rotary evaporator at reduced pressure. During this process, the round bottom may be heated, but temperatures above 35°C should be avoided for reasons of stability.

3.1.3.2. BUTANOL EXTRACTION

1. Suspend the purified surfactant pellet in 2 mL of double-distilled water (approx 10–15 mg phospholipid/mL).
2. Add suspension rapidly to 100 mL of stirring 1-butanol and stir an additional 30 min.
3. Centrifuge the dispersion at 10,000g for 20 min and collect the supernatant containing the butanol/surfactant lipids and water insoluble proteins. Teflon centrifuge tubes with caps are particularly useful when working with organic solvents.
4. Concentrate the dispersion by rotary evaporation followed by freeze-drying.

3.1.4. Surfactant Storage

Freeze-drying or lyophilization is a very convenient means of converting the surfactant into a stable form for storage.

1. To the concentrated extracted surfactant, add four volumes high purity cyclohexane. Freeze the solution in a thin film in a round bottom or test tube by rapidly swirling in a dry ice/acetone or dry ice/isopropanol bath
2. Lyophilize under high vacuum (pressure should not exceed 50 µm of mercury). In lyophilizing surfactant, the organic solutions will be concentrated into a viscous liquid. If the solution remains frozen during drying, a porous cake of lipid/proteins is obtained that may be readily dispersed in water. If the solution melts during the drying process, either the vacuum is not sufficient or the material was frozen in a layer that was too thick.
3. When re-dispersing in water, it is critical that the dispersion not be exposed to water vapor, as it will become gummy and difficult to disperse. Therefore, the addition of water must be rapid, and thereafter the contents should be rapidly and vigorously mixed.

3.2. Aqueous Solubility Determination

One method of determining the solubility of solutes in normal saline is by dialysis *(23)*. This requires preparation of the dialysis membranes, placement of drug dispersions within the bag, and analytical determination of the solute

Interaction of Particulate Delivery Systems

concentration. The main advantage of this method is that the solid particles remain separated from the solution being assayed for drug in solution, and as such, the risk of contamination is minimized. This is particularly useful for poorly water-soluble compounds that often do not remain wetted but rather float at the surface.

3.2.1. Dialysis Bag Preparation

Tie off dialysis tubing that has been properly purified (*see* **Note 3**) on one end with surgical thread.

3.2.2. Filling Dialysis Bags

1. Add a fixed quantity of saline to the dialysis membrane along with a fixed quantity of solute for study.
2. Tie the open end of the bag and place in a teflon-lined screw-cap test tube. It is often a good precaution to wrap the caps with Parafilm to prevent leakage.
3. Submerge the test tubes containing the dialysis membranes submerged in a shaking water bath equilibrated to a temperature, usually 37°C.
4. Take aliquots periodically to determine when equilibrium is reached (*see* **Note 4**).
5. The concentration solute should be measured with a suitable, stability-indicating assay.

3.3 Solubilization of Solutes in Surfactant

Solubilization studies also require preparation of dialysis membranes, with their filling (**Subheading 3.2.**), and an analytical procedure to determine the solute concentration.

3.3.1. Single Concentration of Surfactant

The extent of solubilization of solutes can be determined in a similar manner to the aqueous solubility measurement.

1. Place a fixed, known amount of drug powder in dialysis bags (MWCO 10,000 daltons) containing a fixed volume of surfactant. In this surfactant, the lipid and protein concentration should be predetermined using the phosphorus assay *(10)* and modified Lowry's *(8)* assay, respectively.
2. Tie the bags with surgical thread and placed into Pyrex™ screw cap test tubes with Teflon™ liners containing an equivalent concentration of surfactant and a sufficient volume to suspend the dialysis tubing.
3. Oscillate the tubes with the temperature controlled, typically at 37°C, and as above, periodically assay to determine the time required for equilibration (*see* **Subheading 3.2.2.**) (*see* **Note 5**).
4. For a more thorough experiment, it is worthwhile to determine the concentration of solute dissolved as a function of surfactant concentration (*see* **Note 6**).

3.3.2. Thermodynamic Analysis

1. For thermodynamic parameters, the change in enthalpy, ΔH, and the change in entropy, ΔS, can be calculated from the slope, (slope = $-\Delta H/R$, where R is the gas law constant) and intercept of a plot of the natural logarithm of the mole fraction solubility as a function of the reciprocal of the absolute temperature *(25)*.
2. In this process, the solubility should be expressed as mole fraction of solute n, X_n, by

$$X_n = N_n/N_t$$

where N_n is the number of moles of solute n dissolved in total moles of solution, N_t. It is generally reasonable to convert the molar concentration into the mole fraction by assuming the density is unity. Analogous measurements may be made for the extent of solubilization where the extent is calculated as the mole fraction of solute dissolved in total moles of lipid and solute.

3.4. Dissolution of Aerosol Particles

Dissolution of aerosol particles requires generation of aerosol particles and their characterization, an aerosol capture method, and dissolution methodology.

3.4.1. Generation of Aerosol Particles

A convenient and easy to use apparatus for producing aerosol particles of drug is an ultrasonic nebulizer *(26,27)*. This may be a bath sonicator as is commonly used for humidifying homes or the more sophisticated ultrasonic spray nozzle system (*see* **Notes 7** and **8**). Bath sonicators typically operate at a frequency of 1–3 MHz and produce initial aerosol particles in the range of 3 µm. The ultrasonic spray nozzle system is driven by a broadband ultrasonic generator and operates at a frequency of 125 kHz. The flat-tipped probe can be equipped with a MicroSpray™ nozzle that allows for consistent atomization at low liquid flow rates. The power level can be adjusted but the dial is in arbitrary units. In carrying out experiments, the dial should be set at a consistent level.

For the bath sonicator, a flange prepared from aluminum or other suitable material (glass, stainless steel, etc.) supports an aluminum cylinder and is attached to the screw mounts of the piezoelectric crystal. The cylinder is 3.25 cm (id) and contains the solution of drug. A brass gas connector (0.5 cm) is attached to the cylinder to allow entrance of air. At the top of the aluminum cylinder, a 125 mL Erlenmeyer flask is attached by a large side arm (2.5 cm id) that extends from the lower portion of the flask. The upper conical portion of the flask is connected to the column containing charcoal. The apparatus is hermetically sealed with silicone glue minimizing the exposure to the internal solutions. Most aqueous solutions are acceptable provided the salt concentration is not too high, which can result in corrosion of the metal components. Ethanol solutions may also be used for poorly water-soluble solutes.

3.4.2. Drying Aerosol Droplets

1. For removing solvent vapor or water, construct columns of aluminum cylinders that contain an outer ring of drying agents, charcoal for organic vapors and silica for removing water *(26)*.
2. Separate the drying agent from the flow by screen mesh at the outer surface. Note that the mesh size (lines/inch) of the dying agent must be smaller than the mesh size of the screen. A vertical arrangement has the advantage of minimizing loss of particles from sedimentation. This is necessary when using the Sono-Tek system as the initial particle size is quite large and settles rapidly.
3. Test the organic solvent and water removal efficiency of the charcoal and silica column. When operating the nebulizer, determine the organic solvent vapor concentration as a function of time. Take aliquots with a gas tight Hamilton syringe at the column outlet.
4. Inject aliquots onto a gas chromatograph (GC) *(26)*. Standards are easily prepared by injecting know volumes of solvent in Erlenmeyer flasks sealed with a rubber septum. These need to be freshly prepared each day.
5. The water vapor concentration is readily determined from the relative humidity by a digital hygrometer placed in line at the column outlet. The concentration of water is calculated from the relative humidity with the ideal gas law and by assuming the saturation vapor pressure is equal to 25.934 mm Hg at 25°C *(28)*.
6. Regenerate drying agent as needed by heating the external surface of the column with heating tape while passing air through the column. Silicone rubber heating tape has the desired heat-resistant properties for this application.

3.4.3. Aerosol Output Measurement

The aerosol output and particle size distribution may be determined by the filter capture method. The output is expressed as the mass of drug aerosolized in unit time.

1. Microfibrile glass filters may be placed at the column outlet and particles collected with reduced pressure.
2. Extract the filters three times with 5 mL solvent appropriate for the solute under examination.
3. Pool the extracting solutions and centrifuge for 15 min at $500g$.
4. Determine absorbance after further dilution if necessary, and interpolate the concentration from the appropriate standard curves. Blank filters should also be extracted to correct for the background absorbance (<10 % of reading).
5. Extraction efficiencies are determined by placing know concentrations of solutions of drug on the filters, allowing the filters to air dry, and then extracting as in **Subheading 3.4.3.4.**

3.4.4. Aerosol Particle Size Determination

1. Use a cascade impactor operating at 0.5 L/min (lpm) with cut-off diameters of 5.09, 3.13, 1.93, 1.21, 0.77, 0.59, and 0.33 µm to monitor the size distribution of the aerosol. The Anderson Mark II is also commonly used, but operates at an air flow rate of 28.3 lpm.
2. The mass median aerodynamic diameter, MMAD, and geometric standard deviation, GSD, are obtained by a linear fit of the cumulative undersized collected mass plotted on a probability scale as a function of logarithm of the cutoff diameter using KaleidaGraph™ *(27)*.

3.4.5. Aerosol Particle Capture and Measurement of Dissolution Rate in Lung Surfactant

1. For the study of the dissolution of aerosol particles, an air sampling liquid impinger is filled with saturated solutions of the solute in either normal saline solution or 0.01% sodium dodecyl sulfate (SDS) and placed in an ice bath *(29)*. The SDS is added to assist in the wetting of the particles.
2. After collecting a sufficient mass of aerosol particles, initiate the dissolution process by diluting the impinger solution in a one to six (v/v) ratio with saline or solutions of SDS or Survanta™ held at 37°C.
3. Oscillate the solutions in scintillation vials on a shaking water bath at 37 ± 0.5°C.
4. Take aliquots and centrifuge in a tabletop centrifuge at high speed for 10 min to remove undissolved budesonide particles.
5. Dissolve an aliquot of the supernatant in the appropriate solvent from which the total concentration of solute in solution can be measured.
6. The concentration of dissolved solutes in the aqueous phase in samples containing Survanta™ can be determined by subjecting the samples to ultracentrifugation at 56,000g for 30 min. This causes the sedimentation of both the solute particles as well as the Survanta. Aliquots of the supernatant may be taken, and the solute measured with the appropriate assay. Thus, with these two different centrifugation steps, both the total solute and solute in the aqueous phase can be determined.

4. Notes

1. Smaller animals may be exsanguinated prior to the surgery to minimize blood contamination. With small animals, the procedure is readily carried out in a suitable animal surgical room. For large animals, the lung may be obtained from a local slaughterhouse. In the latter case, the lung may be transported on ice to the laboratory where the lavage procedure is carried out.
2. The hydrostatic pressure should be kept at this relatively modest level to minimize the baurotrauma to the lung tissues.
3. The dialysis purification procedure is provided by the manufacturer and should be closely followed.

4. In our experience *(4,11,23)*, the time typically is less than 24 h; however, there are literature reports of needing several days for equilibrium *(6)*. The actual length of time will be a function of the amount of solute added. For most solutes, there is a requirement that no decomposition of the solute under study has occurred. This in turn requires the use of a stability-indicating assay such as high-pressure liquid chromatography (HPLC) or gas chromatography (GC).

   The amount of solute added will influence the measured solubility. It appears that solids contain defects in the crystalline structure that result in slightly higher energy states. Due to the greater energy, there is also a slightly greater solubility. Slow recrystallization of the solute may reduce this phenomenon but will not eliminate it. Therefore, consistent amounts of solute should be added.

5. Generally, surfactant does not affect the transport of solutes across the membrane but greatly facilitates the wetting and thereby the dissolution rate of solutes, particularly those that have low water solubility. Thus, equilibration can be achieved more quickly in the presence of lung surfactant. When determining the solute concentration, it is essential that surfactant be included in the matrix of the controls to either account for the possible interference or determine the absence of interference. Due to the presence of unsaturated lipids in the surfactant and proteins, there is absorption of light in the ultraviolet region of the spectrum. If the lipids undergo oxidation, significant absorption can occur near 360 nm. Although this tends to be only a minor background effect, because it is a result of decomposition it often is variable *(30)*. Generally, with stability-indicating assays, this is not a critical issue.

6. To this end, experiments need to be conducted where the concentration of solute dissolved is determined at a number of surfactant concentrations. An estimated upper limit of surfactant concentration in the lung may be on the order of 50 mg lipid/mL of solution *(24)*. Moreover, it may be of interest to determine the thermodynamic parameters associated with the aqueous solubility and the extent of solubilization of solutes in surfactant. This may be accomplished by making measurements as a function of temperature. The lower temperature limit is usually the freezing point of water, and the upper limit is fixed by the chemical stability of the solutes and surfactant. For the phospholipids in the surfactant, above 40°C may be problematic since hydrolysis of the phospholipids will occur with increasing rate at elevated temperatures.

7. Ultrasonic drivers are initially charged with fluid that will undergo evaporation during the aerosol production. Thus, the concentration of solute in the bath will increase with time, which generally causes an increase in the output and a slight increase in the particle size with time of operation. In contrast, the Sono-Tek system requires an infusion pump that provides a constant output rate and particle size. A particularly desirable feature is that the output rate can be controlled by the flow rate from the infusion pump.

8. The output from the nebulizers can be directed by a glass baffle with an intervening rubber gasket. Baffles typically consist of offset concentric tubes and can be designed to entrain the aerosol particles with the carrier gas. The fitting

should be smooth with only gradual changes in diameter to prevent turbulence in the airflow. This will minimize the loss of particles due to deposition on the surface of the tubing.

Acknowledgment

The proofreading skill provided by Heather Herrington is greatly appreciated.

References

1. Wattenberg, L., Wiedmann, T. S., Estensen, R. D., Zimmerman, C. L., Steele, V. E., and Kelloff, G. J. (1998) Chemoprevention of pulmonary carcinogenesis by aerosolized budensonide in female A/J mice. *Cancer Res.* **57,** 5489–5492.
2. Wattenberg, L., Wiedmann, T. S., Estensen, R. D., Zimmerman, C. L., Steele, V. E., and Kelloff, G. J. (2000) Chemoprevention of pulmonary carcinogenesis in female A/J mice by brief exposures to aerosolized budensonide or beclomethasone dipropionate and by the combination of aerosolized budesonide and dietary myo-inositol. *Carcinogenesis* **21,** 179–182.
3. Estensen, R. D. and Wattenberg, L. W. (1993) Studies of chemopreventive effects of myo-inositol on benzo(a)pyrene-induced neoplasia of the lung and forestomach of female A/J mice. *Carcinogenesis* **14,** 1975–1977.
4. Wiedmann, T. S., Bhatia, R., and Wattenberg, L. W. (2000) Drug solubilization in lung surfactant. *J. Control. Rel.* **65,** 43–47.
5. Wattenberg, L. W. (1993) Prevention-therapy-basic science and the resolution of the cancer problem presidential address. *Cancer Res.* **53,** 5890–5896.
6. Grant, D. J. W. and Higuchi, T. (1990) Solubility behavior of organic compounds, in *Techniques of Chemistry*, vol. 21 (Saunders, Jr., W. H., ed.), John Wiley and Sons, NY, pp. 12–88.
7. King, R. J. and Clements, J. A. (1972) Surface active materials from dog lung. II. Composition and physiological correlations. *Am. J. Physiol.* **223,** 715–726.
8. Possmayer, F. (1990) The role of surfactant-associated proteins. *Am. Rev. Respir. Dis.* **142,** 749–752.
9. Stratton, C. J. (1977) The periodicity and architecture of lipid retained and extracted lung surfactant and its origin form multilamellar bodies. *Tissue Cell* **9,** 301–316.
10. Chen, Jr., P. S., Toribara, T. Y., and Warner, H. (1956) Microdetermination of phosphorus. *Anal. Chem.* **28,** 1756–1758.
11. Cai, X., Grant, D. J. W., and Wiedmann, T. S. (1997) Solubilization of drugs by bile salts and bile salt/phospholipid aggregates. *J. Pharm. Sci.* **86,** 372–377.
12. Whitsett, J. A., Ohning, B. L., Ross, G., Meuth, J., Weaver, T., Holm, B. A., et al. (1986) Hydrophobic surfactant-associated protein in whole lung surfactant and its importance for biophysical activity in lung surfactant extracts used for replacement therapy. *Pediatr. Res.* **20,** 460–467.
13. Possmayer, F., Yu, Su., and Weber, M. (1984) Calcium-protein-lipid interactions in pulmonary surfactant. *Prog. Resp. Res.* **18,** 112–120.

14. Oosterlaken-Dijksterhuis, J. A., Haagsman, H. P., van Golde, L. M. G., and Demel, R. A. (1991) Interaction of lipid vesicles with monomolecular layers containing lung surfactant proteins SP-B or SP-C. *Biochemistry* **30**, 8276–8281.
15. Claypool, W. D., Wang, D. L. Chander, A., and Fisher, A. B. (1984) An ethanol/ether soluble apoprotein from rat lung surfactant augments liposome uptake by isolated granular pneumocytes. *J. Clin. Invest.* **74**, 677–684.
16. Keough, K. M. W., Farrell, E., Cox, M., Harrell, G., and Taeusch, Jr., H. W. (1985) Physical, chemical, and physiological characteristics of isolates of pulmonary surfactant from adult rabbits. *Can. J. Physiol. Pharmacol.* **63**, 1043–1051.
17. King, R. J. and Clements, J. A. (1972) Surface active materials from dog lung. I. Method of isolation. *Am. J. Physiol.* **223**, 707–714.
18. Ryan, S. F., Hashim, S. A., Cernansky, G., Barrett, Jr., C. R., Bell, Jr., A. L. L., and Liau, D. F. (1980) Quantification of surfactant phospholipids in the dog lung. *J. Lipid Res.* **21**, 1004–1014.
19. Shiffer, K., Hawgood, S., Duzgunes, N., and Goerke, J. (1988) Interactions of the low molecular weight group surfactant-associated proteins (SP 5-18) with pulmonary surfactant lipids. *Biochemistry* **27**, 2689–2695.
20. Yu, S.-H. and Possmayer, F. (1988) Comparative studies on the biophysical activities of the low-molecular-weight hydrophobic proteins purified from bovine pulmonary surfactant. *Biochim. Biophys. Acta* **961**, 337–350.
21. Shelley, S. A., Balis, J. U., Paciga, J. E., Espinoza, C. G., and Richman, A. V. (1982) Biochemical composition of adult human lung surfactant. *Lung* **160**, 195–206.
22. Suwabe, A., Mason, R. J., Smith, D., Firestone, J. A., Browning, M. D., and Voelker, D. R. (1992) Pulmonary surfactant secretion is regulated by the physical state of extracellular phosphatidylcholine. *J. Biol. Chem.* **267**, 19884–19890.
23. Li, C-Y., Zimmerman, C., and Wiedmann, T. S. (1996) Solubilization of retinoids by bile salt/phospholipid aggregates. *Pharm. Res.* **13**, 907–913.
24. Hallman, M. (1989) Recycling of surfactant: a review of human amniotic fluid as a source of surfactant for treatment of respiratory distress syndrome. *Rev. Perinat. Med.* **6**, 197–226.
25. Grant, D. J. W. and Higuchi, T. (1990) Solubility behavior of organic compounds, in *Techniques of Chemistry*, vol. 21 (Saunders, Jr., W.H., ed.), John Wiley and Sons, NY, pp. 384–391.
26. Pham, S. and Wiedmann, T. S. (1999) A novel method for the rapid screening of poorly water soluble drugs for respiratory delivery. *Pharm. Res.* **16**, 1857–1863.
27. Wiedmann, T. S. and Ravichandran, A. (2001) Ultrasonic nebulization system for respiratory drug delivery. *Pharm. Dev. Technol.* **6**, 83–90.
28. *CRC Handbook of Chemistry and Physics.* (65th edition; West, R. C., ed.) CRC Press, 1984, New York, NY, pp. D192–D218.
29. Pham, S. and Wiedmann, T. S. (2001) Dissolution of aerosol particles of budesonide in Survanta, a model lung surfactant. *J. Pharm. Sci.* **90**, 98–104.
30. Higuchi, T., Shih, F.-M. L., Kimura, T., and Rytting, J. H. (1979) Solubility determination of barely aqueous-soluble organic solids. *J. Pharm. Sci.* **68**, 1267–1272.

47

Chemopreventive Therapeutics

Inhalation Therapies for Lung Cancer and Bronchial Premalignancy

Missak Haigentz, Jr. and Roman Perez-Soler

1. Introduction

Lung cancer is the number one cause of cancer-related death in the US. In 2001, it is estimated that approx 180,000 new patients will be diagnosed, and close to 90% of them will die of the disease *(1)*. The major reasons for such dismal prognoses are the lack of effective preventive and early therapeutic interventions that address the regional nature of the disease and the difficulty in making an early diagnosis when effective loco-regional therapies are technically possible.

2. Lung Cancer Biology

Lung cancer arises in a bronchial epithelium diffusely damaged by chronic exposure to air carcinogens in a susceptible host and is preceded by a well-defined sequence of histological lesions that are superficial and accessible by the inhalation route: squamous metaplasia, dysplasia, and carcinoma *in situ* (CIS) *(2,3)*. Bronchial metaplasia and dysplasia are commonly observed in biopsies from chronic smokers without lung cancer, suggesting that these lesions represent initial steps in the lung carcinogenesis process. The genetic alterations associated with bronchial metaplasia and dysplasia can also be found in lung cancers, further suggesting that these lesions are actually premalignant and the origin of lung cancer. The spontaneous regression of these lesions after smoking cessation in some cases suggest that they may represent a reversible stage where appropriate intervention may prevent or delay the development of the disease. However, in many cases, the genetic lesions persist after smoking cessation, thus perpetuating the risk of lung cancer occurrence (4).

The genetic basis of this sequence of morphological changes is characterized by an accumulation of interactive and synergistic genetic alterations that lead to the transformed phenotype (5). The whole respiratory tract is affected or at risk of developing the initial genetic abnormalities (field cancerization) and each successive abnormality increases the risk of acquiring another genetic abnormality (genetic instability) and progressively uncontrolled cellular proliferation (6,7). Loss of genetic material, specifically deletions in chromosomes 3p and 9p, have been consistently reported in a high proportion of lung cancers and are thought to be relatively early events (8). By the time lung cancer manifests itself clinically, there may be 10–20 accumulated genetic alterations in the lung cancer cells (6). Particularly, the loss of function of tumor-suppressor genes and the activation of dominant oncogenes play a crucial role in the pathogenesis of lung cancer. Specific gene abnormalities have been well-characterized, especially mutations of the oncogene K-ras, and loss of function by deletion, mutation, or methylation of the tumor-suppressor genes p53, p16, Rb and FHIT (6,7,9–16). Abnormal overexpression of dominant oncogenes like c-myc, epithelial growth factor receptor (EGFR), and Her-2 without genetic mutations has also been reported in a high proportion of lung cancers (7,12). Donwregulation of transforming growth factor-β (TGF-β) receptors, cadherins, and overexpression of cyclooxygenase-2 (COX-2) may also contribute to the dysregulation of cell proliferation and metastatic potential (17,18). In the near future, the diagnosis of bronchial premalignancy will be made on molecular grounds.

2. Rationale for Chemoprevention

The correction of one or more of these genetic abnormalities may possibly lead to delayed progression of existing lung cancers, thereby improving the prognosis of individuals with disease. Even more attractive, however, is the potential of reversing genetic damage done by carcinogens to the bronchial epithelium prior to the development of neoplasia. It is this concept that is actively being studied as a means for the chemoprevention of lung cancer in former tobacco smokers.

Lung cancer chemoprevention studies to date have focused on the use of oral retinoids. Although there is a body of evidence to suggest that the oral administration of 13-cis-retinoic acid results in significant suppression of new upper aerodigestive cancers, the clinical studies to date have not resulted in similar conclusions for the development of lung cancer. Additionally, the side effects of the oral retinoids are frequently debilitating and intolerable for otherwise asymptomatic individuals (19). Studies of aerosolized delivery of retinoids are in progress (20,21).

The prospect of gene therapies for lung cancer allows for targeted correction of genetic defects during the lung carcinogenesis process. Systemic or local gene therapy remains unrealistic because the available gene-transfer vectors are not effective at transforming human cells and are highly immunogenic *(22)*. Aerosolized and inhaled gene therapy, however, is a logical concept that has not been adequately explored.

2.1. Aerosolized Therapeutics: A Novel Approach to Chemoprevention

The potential advantages of aerosolized and inhaled therapies are numerous. Inhaled pharmaceuticals target the whole bronchial epithelium directly, providing high drug levels in the bronchial epithelium and bronchial fluid with minimal systemic drug exposure and, therefore, minimal systemic toxicities *(23)*. In addition, some inhaled therapeutics are cleared from the lung preferentially through the lymphatics, thus providing an opportunity for lymphatic targeting *(24)*. Inhaled anticancer therapies are therefore of potential use for the treatment of premalignant or malignant bronchial lesions that are diffuse and confined to the bronchial epithelium like dysplasia and CIS, lung malignancies that grow within the air space multifocally like bronchoalveolar carcinoma, or lung malignancies that have disseminated to the locoregional lymphatics like lymphangitic carcinomatosis or microscopic mediastinal nodal disease, which is the cause of most recurrences after surgical resection. Because of the diffuse and superficial nature of the bronchial premalignant lesions, correction of genetic and molecular defects in these lesions using inhaled agents is logical, technically feasible, and therefore worth exploring.

2.1.1. Inhalational Chemotherapeutics

Previous use of aerosolized chemotherapy in humans has been limited to 5-fluorouracil (5-FU) *(24)*. Studies with inhaled doxorubicin and paclitaxel in animals have also been reported *(25,26)*, demonstrating clinical responses in spontaneously arising primary and metastatic lung cancers without systemic toxicities. In a current clinical study in humans of inhaled doxorubicin at Memorial Sloan Kettering Cancer Center, responses to inhaled doxorubicin at 2 mg/m^2 have to date been observed in one patient with bronchoalveolar carcinoma and one patient with sarcoma *(27)*.

2.1.2. Inhalational Cytokines

Interleukin-2 (IL-2), when administered systemically, is known to have anticancer activity in patients with advanced renal cell cancer and melanoma at the expense of frequently observed significant toxicities, including the capillary leak syndrome, hypotension, and oliguria. Inhaled IL-2 has been examined as potential immunotherapy of pulmonary metastases from renal cell carcinoma, showing promising evidence of activity *(28,29)*. In a Phase I clinical trial of aerosolized IL-2 formulations, no significant toxicities were observed *(30)*. Moreover, high-dose inhaled IL-2 therapy was associated with improved stability of patient quality of life compared to low-dose systemic IL-2 therapy *(31)*. Further studies to determine the efficacy of inhaled IL-2 as an anticancer agent, either alone or in combination with systemic therapy, are warranted.

2.1.3. Inhalational Gene Therapy

Gene replacement in the bronchial epithelium via aerosolized vehicles has been most studied in the treatment of cystic fibrosis. Vehicles used in successful clinical Phase I trials of wild-type CFTR gene transfer have either been adenoviral vectors *(32–34)* or cationic liposomes *(35)*. Of the two vehicles, inhaled adenoviral vectors appear to be associated with more frequent inflammatory responses.

No trials of inhaled gene therapy for lung cancer or bronchial premalignancy have been reported to date. Our laboratory has worked on validating the concept of aerosolized gene and molecular correction of the bronchial epithelium for quite some time and is about to begin a Phase I study of aerosolized p53 gene in a cationic liposome for patients with non-small cell lung cancer (NSCLC) carrying p53 mutations.

In the remainder of this chapter, we summarize the preclinical work of our laboratory with aerosolized p53 gene therapy in mice *(36)* and the rationale behind other aerosolized approaches for the treatment and chemoprevention of lung cancer.

2.2. Aerosolized p53 Gene Delivery in a Mouse Model of Lung Cancer

Our laboratory conducted extensive formulation studies using different combinations of cationic phospholipids to carry and deliver the p53 gene. More than 20 different formulations were screened for their ability to transfect the LacZ gene. The combination of dipalmitoyl ethylphosphocholine (DPEP) and dioleoyl phosphatidylethanol amine (DOPE) at a molar ratio of 3:1 gave the best transfection efficiency results in a variety of tumor cell lines in vitro. Further optimization work led to selecting a lipid:DNA ratio of 6:1. The size of these particles is around 80 nm, as determined by a Nicomp Submicron Particle Sizer. This liposomal vehicle was used to carry and deliver wild-type p53 gene in intratracheal and aerosolized administrations.

This formulation, abbreviated as DP3, was then used to transfect the p53 gene to H358 (p53 null bronchoalveolar carcinoma) and H322 (p53 mutated

NSCLC) cells. Following a single transfection, expression of p53 protein led to cellular apoptosis in about 40% of H358 and 20% of H322 cells.

To investigate whether DP3/p53 can successfully deliver and express the gene in the lungs in vivo, p53-null mice (C57BL/6) were treated intratracheally with DP3/p53 complexes. p53 null mice treated with DP3 alone and normal wild-type p53 mice (C57BL/6) without treatment were used as controls. The dose was 8 μg DNA/48 μg DP3/day for 4 consecutive d. Animals were sacrificed on d 5 and the lungs resected and immediately frozen in liquid nitrogen until analysis. Reverse transcriptase polymerase chain reaction (RT-PCR) assays with human-p53-specific PCR primers demonstrated a strong p53 cDNA signal in the sample treated with DP3/p53.

To determine the effectiveness of DP3/p53 treatment on tumor formation in vivo, H358 and H322 lung cancer models in nude mice were developed that mimic human early lung cancer. Male nu/nu mice were inoculated with 1–2 × 10^6 cells intratracheally. Repeated experiments showed that these lung cancer cells attach to the bronchial epithelium, give rise to multiple tumors, and that the life span correlates with the number of cancer cells inoculated. In tumor growth-inhibition experiments, male nu/nu mice were inoculated with 10^6 H358 cells intratracheally. Groups of animals were treated intratracheally with pC53SN plasmid alone (containing the human p53 gene under the control of the CMV promoter), DP3 liposomes alone, or DP3/p53 complexes on d 4, 8, and 12 after H358 inoculation, respectively. The dose was 2 μg DNA/administration. Animals were sacrificed on d 74. DP3/p53 resulted in a significant tumor-growth inhibition effect with no visible tumors in the lung tissue. All three control groups (untreated; pC53SN alone; DP3 alone) showed massive replacement of lung parenchyma by tumors.

In survival experiments, male nu/nu mice were inoculated with 1 × 10^6 H358 or H322 tumor cells intratracheally. Groups of animals were administered intratracheally with pC53SN plasmid alone, DP3 liposomes alone, or DP3/p53 complexes on d 4, 8, 12, 16, and 20 after tumor inoculation, respectively. The dose was 2–8 μg DNA/administration. The average life span of H358 tumor-bearing mice treated with DP3/p53 was prolonged from 99 d to 202 d (2 μg DNA/dose; $p < 0.008$), and the average life span of H322 tumor-bearing mice treated with DP3/p53 was prolonged from 82 d to 197 d (8 μg DNA/dose; $p < 0.008$). In summary, five doses of DP3/p53 delivered intratracheally can effectively inhibit endobronchial H358 and H322 lung tumor growth and prolong the life span of tumor-bearing animals by more than twofold.

These studies have demonstrated that it is feasible to effectively treat p53 null and p53-mutant endobronchial tumors by repeated intratracheal administration of lipid/p53 complexes. These results have now also been confirmed using aerosolization. These studies have been NCI funded, and we plan to begin a

Phase I clinical trial of aerosolized liposomal p53 gene delivery in patients with lung cancer carrying p53 mutations.

3. Future Directions

Aside from the exploring the potential of gene therapy via direct wild-type gene transfection and replacement, our laboratory is actively evaluating aerosolized small molecules as regulators of endogenous bronchial epithelial cell gene expression. Specifically, we are interested in the use of the demethylating agent 5-aza-2'-deoxycytidine (decitabine), the COX-2 inhibitor celecoxib, and the EGFR inhibitor OSI-774.

3.1. Aerosolized Decitabine

Promoter hypermethylation of tumor-suppressor genes leading to inactivation is a major mechanism of lung tumorigenesis that is clearly associated with tobacco smoking and is a sound target for intervention in the prevention and treatment of smoking-related malignancies. Decitabine is a methyltransferase inhibitor that restores transcription of aberrantly methylated tumor-suppressor genes in lung cancer cell lines *(37)* and has also been shown to have chemopreventive activity against a 4-(methyl-nitrosoamino)-1-(3-pyridyl)-1-butanone (NNK)-induced mouse model of lung cancer *(38)*. Results of a pilot Phase I–II clinical study of intravenous decitabine have suggested single agent activity in advanced NSCLC *(39)*. Unfortunately, decitabine is highly toxic when administered systemically, causing severe and potentially lethal myelosuppression. However, we believe that therapeutic levels of decitabine are theoretically achievable in the bronchial epithelium by local aerosolized delivery using a much reduced dose that does not result in systemic toxicity.

3.2. Aerosolized Cyclooxygenase Inhibitors

Overexpression of COX-2, the inducible isoform of cyclooxygenase, is frequent in human lung cancer, and inhibition of its activity results in inhibited growth of human lung cancer cell lines *(40)*. Celecoxib, a selective COX-2 inhibitor, is an anti-inflammatory agent in widespread clinical use. Attention is being given to oral COX-2 inhibitors as potential chemopreventive agents against NSCLC *(41)*. Although short-term oral celecoxib use is associated with far less toxicities than those seen with nonselective COX-inhibitors, its use over months and years has not been investigated. We expect that aerosolized celecoxib use will confer protection to the bronchial epithelium against carcinogenesis and will have minimal systemic exposure and long-term toxicity.

3.3. Aerosolized EGFR Tyrosine Kinase Inhibitors

The EGFR, a tyrosine kinase receptor involved in the regulation of cellular differentiation and proliferation, is overexpressed in about 70% of human NSCLCs and is the current focus of targeted therapy of lung cancer. OSI-774 is a small molecule-selective tyrosine kinase inhibitor that has known single-agent anticancer activity in chemotherapy-refractory NSCLC patients *(42)*. It is well-tolerated, the most frequently seen side effects being diarrhea and development of a maculopapular rash. An aerosolized formulation of OSI-774 may also be utilized in the armamentarium against the development of lung cancer.

4. Summary

The lung cancer epidemic is not expected to abate in the next two decades. Smoking cessation campaigns have not been successful in reducing the prevalence of smoking to <25% of the adult population in the US. Among high school students, the prevalence of smoking is increasing (27.5% in 1991; 36.4% in 1997) *(43)*. In addition, only 1 in 8 heavy smokers develop lung cancer and about 20% of patients with lung cancer have no history of active or passive smoking, thus indicating that genetic predisposing factors and other unidentified carcinogens play a crucial role in the etiology of this disease. The bronchial epithelium is like an internal skin where lung cancer originates after chronic exposure to airborne carcinogens in a predisposed host. Although the bronchial epithelium is not readily examinable, it is easily accessible to therapeutic intervention by using inhaled therapeutics.

We hope to extend our findings using direct wild-type p53 gene replacement via intratracheal and aerosolized administrations in mice to NSCLC patients with p53 mutations. We also plan on utilizing aerosolized chemical agents to activate endogenous mechanisms of cytoprotection. If, in clinical trials, evidence of effective cytoprotection is observed in the absence of intolerable side effects, further exploration of this new strategy for the control and prevention of a neoplastic disease that accounts for a third of all cancer-related deaths will be fully justified.

References

1. Parker, S. L., Tong, T., Bolden, S., and Wingo, P. A. (1997) Cancer Statistics, 1997. *CA Cancer J. Clin.* **47,** 5–27.
2. Auerbach, O. C., Hammund, C., and Garfinkel, L. (1979) Changes in bronchial epithelium in relation to cigarette smoking 1955–1960 vs 1970–1977. *N. Engl. J. Med.* **300,** 381–386.

3. Saccomanno, G., Archer, V. E., Auerbach, O., et al. (1974) Development of carcinoma of the lung as reflected in exfoliated cells. *Cancer* **33,** 256–270.
4. Mao, L., Lee, J. S., Kurie, J. M., et al. (1997) Clonal genetic alterations in the lungs of current and former smokers. *J. Natl. Cancer Inst.* **89(12),** 857–862.
5. Chung, C. T. Y., Sundaresan, V., Haselton, P., et al. (1995) Sequencial molecular genetic changes in lung cancer development. *Oncogene* **11,** 2591–2598.
6. Minna, D. J. (1996) Tumor suppressor genes and oncogenes in lung cancer: potential clinical applications. *Adv. Oncol.* **12(1),** 3–8.
7. Kalemkerian, G. P. (1994) Biology of lung cancer. *Curr. Opin. Oncol.* **6,** 147–155.
8. Hirao, T., Nelson, H. H., Ashok, T. D. S., et al. (2001) Tobacco smoke-induced DNA damage and an early age of smoking initiation induce chromosome loss at 3p21 in lung cancer. *Cancer Res.* **61,** 612–615.
9. Sugio, K., Kishimoto, Y., Virmani, A. K., et al. (1994) K-ras mutations are a relatively late event in the pathogenesis of lung carcinomas. *Cancer Res.* **54,** 5811–5818.
10. Westra, W. H., Baas, I. G., Hruban, R. H., et al. (1996) K-ras oncogene activation in atypical hyperplasias of the human lung. *Cancer Res.* **56,** 2224–2228.
11. Mills, N. E., Fishman, C. L., Scholes, J., et al. (1995) Detection of K-ras oncogene mutations in bronchoalveolar lavage fluid for lung cancer diagnosis. *J. Natl. Cancer Inst.* **87(14),** 1056–1060.
12. Tierney, L. A., Hahn, F. F., and Lechner, J. F. (1996) p53, erbB-2 and K-ras gene alterations are rare in spontaneous and plutonium-239-induced canine lung neoplasia. *Radiat. Res.* **145,** 181–187.
13. Sozzi, G., Veronese, M. L., Negrini, M., et al. (1996) The FHIT gene at 3p14.2 is abnormal in lung cancer. *Cell* **85,** 17–26.
14. Ohta, M., Inoue, H., Cotticelli, M. G., et al. (1996) The FHIT gene, spanning the chromosome 3p14.2 fragile site and renal carcinoma-associated t(3;8) breakpoint, is abnormal in digestive tract cancers. *Cell* **84,** 587–597.
15. Kastury, K., Baffa, R., Druck, T., et al. (1996) Potential gastrointestinal tumor suppressor locus at the 3p14.2 FRA3B site identified by homozygous deletions in tumor cell lines. *Cancer Res.* **56,** 978–983.
16. Jacobson, D., Fishman, C. L., and Mills, N. E. (1995) Molecular genetic tumor markers in the early diagnosis and screening of non-small-cell lung cancer. *Ann. Oncol.* **6,** S3–8.
17. Sulzer, M. A., Leers, M. P. G., van Noord, J. A., et al.. (1998) Reduced E-cadherin expression is associated with increased lymph node metastases and unfavorable prognosis in NSCLC. *Am. J. Respir. Crit. Care Med.* **157,** 1319–1323.
18. Hida, T., Yatabe, Y., Achiva, H., et al. (1998) Increased expression of cyclooxygenase 2 occurs frequently in human lung cancers, specifically adenocarcinomas. *Cancer Res.* **58,** 3761–3764.
19. Mulshine, J. L., De Luca, L. M., Dedrick, R. L., et al. (2000) Considerations in developing successful, population-based molecular screening and prevention of lung cancer. *Cancer* **89(11),** S2465–S2467.

20. Miller, V. A., Brooks, A., Benedetti, F., et al. (1997) Inhaled retinoids for the prevention of respiratory tract cancers. *Lung Cancer* **18**, 186, (abstract A727).
21. Parthasarathy, R., Gilbert, B., and Mehta, K. (1999) Aerosol delivery of liposomal all-*trans*-retinoic acid to the lungs. *Cancer Chemother. Pharmacol.* **43**, 277–283.
22. Van Ginkel, F. W., Liu, C., Simecka, J. W., et al. (1995) Intratracheal gene delivery with adenoviral vector induces elevated systemic IgG and mucosal IgA antibodies to adenovirus and beta-galactosidase. *Human Gene Ther.* **6**, 895–903.
23. Sharma, S., White, D., Imondi, A. R., et al. (2001) Development of inhalational agents for oncologic use. *J. Clin. Oncol.* **19(6)**, 1839–1847.
24. Tatsumura, T., Koyama, S., Tsujimoto, M., et al. (1993) Further study of nebulisation chemotherapy, a new chemotherapeutic method in the treatment of lung carcinomas: fundamental and clinical. *Br. J. Cancer* **68**, 1146–1149.
25. Vail, D. M., Hershey, A. E., Kurzman, I. D., et al. (1999) Inhalation chemotherapy for primary or metastatic lung tumors: proof of principle. *AACR Proc.* **40**, 416.
26. Hershey, A. E., Kurzman, I. D., Forrest, L. J., et al. (1999) Inhalation chemotherapy for macroscopic primary or metastatic lung tumors: Proof of principle using dogs with spontaneously occurring tumors as a model. *Clin. Cancer Res.* **5**, 2653–2659.
27. Dr. Sunil Sharma, personal communication.
28. Lorenz, J., Wilhelm, K., Kessler, M., et al. (1996) Phase I trial of inhaled natural interleukin 2 for treatment of pulmonary malignancy: Toxicity, pharmacokinetics, and biological effects. *Clin. Cancer Res.* **2**, 1115–1122.
29. Enk, A. (1999) Successful therapy of lung metastases in malignant melanoma by inhalative interleukin-2 immunotherapy. *Anticancer Res.* **19**, 2002 (abstract).
30. Skubitz, K. M. and Anderson, P. M. (2000) Inhalational interleukin-2 liposomes for pulmonary metastases: a phase I clinical trial. *Anticancer Drugs* **11(7)**, 555–563.
31. Heinzer, H., Mir, T. S., Huland, E., et al. (1999) Subjective and objective prospective, long-term analysis of quality of life during inhaled interleukin-2 immunotherapy. *J. Clin. Oncol.* **17**, 3612–3620.
32. Crystal, R. G., McElvaney, N. G., Rosenfeld, M. A., et al. (1994) Administration of an adenovirus containing the human CFTR cDNA to the respiratory tract of individuals with cystic fibrosis. *Nat. Genet.* **8(1)**, 42–51.
33. Bellon, G., Michel-Calemard, L., Thouvenot, D., et al. (1997) Aerosol administration of a recombinant adenovirus expressing CFTR to cystic fibrosis patients: a phase I clinical trial. *Hum. Gene Ther.* **8(1)**, 15–25.
34. Aitken, M. L., Moss, R. B., Waltz, D. A., et al. (2001) A phase I study of aerosolized administration of tgAAVCF to cystic fibrosis subjects with mild lung disease. *Hum. Gene Ther.* **12(15)**, 1907–1916.
35. Alton, E. W., Stern, M., Farley, R., et al. (1999) Cationic lipid-mediated CFTR gene transfer to the lungs and nose of patients with cystic fibrosis: a double-blind placebo-controlled trial. *Lancet* **353(9157)**, 947–954.
36. Zou, Y., Zong, G., Ling, Y. H., et al. (1998) Effective treatment of early endobronchial cancer with regional administration of liposome-p53 complexes. *J. Natl. Cancer Inst.* **90**, 1130–1137.

37. Merlo, A., Herman, J. G., Mao, L., et al. (1995) 5'CpG island methylation is associated with transcriptional silencing of the tumor suppressor p16/CDKN2/MTS1 in human cancers. *Nat. Med.* **1(7),** 686–692.
38. Lantry, L. E., Zhang, Z., Crist, K. A., et al. (1999) 5-Aza-2'-deoxycytidine is chemopreventive in a 4-(methyl-nitrosamino)-1-(3-pyridyl)-1-butanone-induced primary mouse lung tumor model. *Carcinogenesis* **20(2),** 343–346.
39. Momparler, R. L., Bouffard, D. Y., Momparler, L. F., et al. (1997) Pilot phase I-II study on 5-aza-2'-deoxycytidine (Decitabine) in patients with metastatic lung cancer. *Anticancer Drugs* **8(4),** 358–368.
40. Hida, T., Kozaki, K., Muramatsu, H., et al. (2000) Cyclooxygenase-2 inhibitor induces apoptosis and enhances cytotoxicity of various anticancer agents in non-small cell lung cancer cell lines. *Clin. Cancer Res.* **6,** 2006–2011.
41. Dannenberg, A. J., Altorki, N. K., Zhang, F., et al. (2000) Chemoprevention of respiratory malignancies: the role of COX-2 inhibitors. *Proc. Am. Assoc. Cancer Res.* **41,** S13 (abstract).
42. Perez-Soler, R., Chachoua, A., Huberman, M., et al. (2001) A Phase II trial of the epidermal growth factor receptor (EGFR) tyrosine kinase inhibitor OSI-774, following platinum-based chemotherapy, in patients (pts) with advanced, EGFR-expressing, non-small cell lung cancer (NSCLC). *Proc. Am. Soc. Clin. Oncol.* **20,** (abstract 1235).
43. Wingo, P. A., Ries, L. A. G., Giovino, G. A., et al. (1999) Annual report to the nation on the status of cancer, 1973–1996, with a special section on lung cancer and tobacco smoking. *J. Natl. Cancer Inst.* **91,** 675–690.

Index

A
A-allele genotypes
  point estimates, 128t
ABC
  immunohistochemistry, 375
ABI PRISM 377 sequencer
  Genescan output, 253f
Abrogation multiple p53 functions
  model for sequential evolutionary consequences, 436f
ABV chemotherapy
  pulmonary KS, 104–105
Accelerated RT, 493
Acetylsalicylic acid, 744
Acetyltransferases, 122
Acquired immunodeficiency syndrome (AIDS), 79
  associated Kaposi's sarcoma clinical manifestations, 95–97
  associated KS
    Africa, 95–96
  associated pulmonary Kaposi's sarcoma, 79–107
  respiratory symptoms, 96
Acquired immunodeficiency syndrome (AIDS)-KS
  antiretroviral therapy, 85
Acrylamide gel electrophoresis
  MPE
    material, 551
    method, 556

f, figure
t, table

ACTG
  AIDS-related KS, 93–94
Activated oncogenic ras
  role, 13–15
Acute promyelocytic leukemia (APL)
  MPO, 123
AD
  MPO, 123
Adapter ligation
  SSH, 198–199
Adenocarcinoma
  immunohistochemical profile, 361t
Adenoviral gene transfer
  MPE, 547–549
Adjuvant therapy, 475–479
Adoptive immunotherapy, 697–709
  material, 698–700
  method, 700–708
Ad-p53, 532
Adriamycin, bleomycin, and vincristine (ABV) chemotherapy
  pulmonary KS, 104–105
Adult/Adolescent Spectrum of HIV Disease (ASD) project, 84
Aerosol cyclooxygenase inhibitors, 776
Aerosol decitabine, 776
Aerosol delivery, 561
Aerosol EGFR tyrosine kinase inhibitors, 777

781

Aerosol exposure to mice
  laboratory set up, 565f
Aerosol gene therapy
  PEI-p53 complexes, 607–616
    material, 609–611
    method, 611–613
Aerosol particle dissolution
  material, 759–760
  method, 764–766
Aerosol particle size analysis
  material, 562
  method, 564–565
Aerosol p53 gene delivery, 774–775
Aerosol p53 gene therapy
  material, 609–611
  method, 611–613
Aerosol therapeutics, 773–774
Aerotech-II (AT-II), 571
Africa
  AIDS-associated KS, 95–96
African-Americans
  MPO, 124
African KS, 80
  endemic
    epidemiology and clinical
      characteristics, 81t
  lymphadenopathic
    epidemiology and clinical
      characteristics, 82t
Agarose gel electrophoresis
  Southern blotting of genomic
    DNA
    material, 267
    method, 271
Age effects
  MPO, 127
A549/HL60, 315f
Ah receptors, 122

AIDS. *see* Acquired
  immunodeficiency syndrome
  (AIDS)
AIDS Clinical Trials Group (ACTG)
  AIDS-related KS, 93–94
AIDS-Malignancy Consortium
  (AMC)
  advanced KS, 104
Alkaline phosphatase, 496
Allele loss analysis, 265f
Allele-specific amplification (ASA)
  analysis, 291–303
    description, 293
    material, 295
    method, 298–301
  PCR, 326
Allelic imbalance, 251–252
Altered fractionation schema,
  492–493
Alveolar hemorrhage
  pulmonary KS, 102
Alzheimer's disease (AD)
  MPO, 123
AMC
  advanced KS, 104
Amifostine, 496–498
Amosov, N. M., 465
Amphotropic retroviral receptor, 546
Anastomosing instruments, 465
Anesthesia, 453
Aneuploid DNA histogram
  FNAB, 418f
Angiogenesis
  cellular predictive factors, 47f
Angiogenesis inhibitors, 499
Angiogenic factors, 41–42
Ankyrin repeats, 34–35
A549 NSCLC cells, 661
  caspase 3-like protease assay, 666
  Hoechst stained nuclei, 667

Anthracyclines
  pulmonary KS, 104
Anti-CAR analysis
  MPE
    material, 551
    method, 556
Antigen presenting cell (APC), 708, 711
  COX-2
    material, 727
    method, 732–733
Antigen unmasking
  immunohistochemistry, 373
Antiretroviral therapy
  AIDS-KS, 85
Antisense
  anti-PKA oligonucleotide
    material, 640–643
    method, 643–650
  bcl-2 oligonucleotide, 673
  bcl-2 transfection
    material, 674
    method, 677
  type 1 protein kinase A, 629–630
Antisense oligonucleotides (AONs), 621–632
  clinical evaluation, 622t
  definition, 621
  design, 623–624
  future directions, 630–632
  manipulation of apoptosis
    inhibitors for destruction, 655–668
    material, 657–659
    method, 660–669
  mechanisms of action, 625–626
  pharmacodynamics and toxicology, 624–625
  protein kinase C-material, 627–628
  targeting RIα subunit PKA, 637–651
  therapy toxicities, 626
  tissue distribution and cellular uptake, 624–625
  transfection
    material, 657
    method, 660
Antisense therapy, 638–639
Antisense transfected 3LL cells
  material, 725–726
  method, 728
AONs. *see* Antisense oligonucleotides (AONs)
APC. *see* Antigen presenting cell (APC)
APL
  MPO, 123
Apoptosis
  antisense Bcl-2 oligonucleotide, 671–682
  material, 673–677
  method, 677–681
  AONs
    manipulation of inhibitors, 655–668
  Bcl-2, 33, 655–656
  Bcl-xl, 655–656
  cellular predictive factors, 47f
  definition, 655
  factors, 41–42
  inhibitors, 655–668
    material, 657–659
    method, 660–669
  p53, 64–66
Aqueous solubility determination
  material, 759
  method, 762–763
Archibald, 460

Aromatic polycyclic
hydrocarbons, 740
ASA. *see* Allele-specific
amplification (ASA)
Asbestos, 6
ASD project, 84
Atherosclerosis
MPO, 123
AT-II, 571
ATP depletion, 509
Autocrine growth factor loops, 592
Autoradiographs
tumor DNA, 264, 265f
Autoradiography
RNA *in situ* hybridization
material, 168
method, 171
Avidin-biotin complex (ABC)
immunohistochemistry, 375

B

Bacillary angiomatosis *Bartonella*
organisms
KS, 101
Bacillus Calmette-Guerin (BCG),
475–477
Backbone-modified
oligonucleotides, 656
BAL, 13
BarxF1/R1, 298f
Basic fibroblast growth factor
(bFGF)
KS, 91
phosphorothioate
oligonucleotides, 656
BCG, 475–477
Bcl-2
antisense (G3139), 628–629
apoptotic cell death, 655–656

gel electrophoresis
material, 675
method, 679
gene
MA, 33
immunohistochemical dual
analysis
material, 676–677
method, 680–681
immunostained
FNAB, 414f–415f
protein expression
immunohistochemical dual
analysis, 676–677,
680–681
staining
material, 408
method, 412
Western blot analysis
material, 675–676
method, 678–680
Bcl-xl
apoptotic cell death, 655–656
Belsey, 460–461
Benign nodular KS
epidemiology and clinical
characteristics, 81t
Benzo(a)pyrene, 740–743
per cigarette, 6
Benzo(a)pyrene diol epoxide
(BPDE), 58
MPO, 122
Beta carotene, 739
B16-F10
lung metastasis
inhibition, 615f
melanoma model
material, 610
method, 612

bFGF
  KS, 91
  phosphorothioate
    oligonucleotides, 656
Biochemical analysis of
    inflammation
  PEI-DNA
    material, 564
    method, 567
Biodegradable controlled-released
    polymers, 688–689
Biological parameters
  determination by fine-needle
    aspirates
    material, 407–409
    method, 409–415
Biopsy material
  collection and processing
    material, 338–339
    method, 342
  RNA extraction
    material, 339–340
    method, 343–344
Biotin-streptavidin amplified (B-SA)
    system
  immunohistochemistry, 375
Bischloromethyl ether, 6
Bleomycin
  pulmonary KS, 104
Blocking agents, 741
Blood
  DNA extraction
    fluorescent microsatellite
      analysis, 254–255
Blotting
  material, 341
  method, 346–347
Bone marrow-derived dendritic
  isolation and propagation
    material, 713
    method, 714–715

Bowman-Birk protease inhibitor, 748
BPDE, 58
  MPO, 122
Brock, Russell, 459
Bronchial epithelial cells
  preneoplastic changes, 16–18
Bronchial lavage
  fluorescent microsatellite
    analysis, 251–259
    material, 253–254
    method, 254–257
Bronchial lavage and sputum
  DNA extraction
    fluorescent microsatellite
      analysis, 255
Bronchoalveolar lavage (BAL), 13
Bronchoscopic intratumoral
    injection, 593
Bronchoscopy, 453
Bronchus
  left lung, 460f
B-SA system
  immunohistochemistry, 375
Butanol extraction
  material, 759
  method, 762
Bystander effect, 532

C

CALGB trial, 489–492
Calu-1/HL60, 315f
cAMP-dependent PKA
  cancer drug target, 637–638
  subunits, 637
Cancer and Leukemia Group B
  (CALGB) trial, 489–492
CAP, 477
Carcinoembrionic antigen (CEA), 535
Carcinogens, 6

Carcinoma *in situ* (CIS), 507
  PDT, 515–516
Carlens, 466
Caspase-3-like protease activity
  detection
    material, 659
    method, 665–667
CAT assay
  immunohistochemistry, 569f
  PEI-DNA
    material, 563
    method, 566
    persistence, 567f
Cationic polymer, 562
CCKBR, 336
CCND1, 293–298
CD4, 537
CD4+, 106–107
  count
    KS, 92–93
    HAART, 106
Cdk. *see* Cyclin-dependent kinase (Cdk)
cDNA arrays
  tumor marker identification
    material, 241–242
    method, 244
cDNA elements
  expression profiles, 183
cDNA libraries
  subtracted high-throughput methodology
    material, 179
    method, 180–182
cDNA library analysis
  SSH
    material, 193–194
    method, 202–204
cDNA library cloning
  SSH, 201–202

cDNA library construction
  high-throughput methodology
    material, 178
    method, 179–180
cDNA microarray technology, 177, 178
cDNA preparation
  high-throughput methodology, 182
cDNA synthesis
  SSH, 197–198
cDNA target amplification
  high-throughput methodology
    material, 179
    method, 182
CD8 positive T cells, 537
CEA, 535
Cell-based methods, 432
Cell classification
  evolutionary sequences, 442–444
Cell cycle disease, 7–9, 10
Cell cycle networking, 18–19
Cell cycle specific proteins
  NSCLC, 371
Cell cycle-targeted gene therapy
  retorviral vectors
    material, 594–597
    method, 597–600
  retroviral vectors, 591–602
Cell fixation and permeabilization
  evolutionary sequences, 437
Cell lines
  maintenance and propagation
    material, 713
    method, 714
Cellular predictive factors
  drug resistance
    NSCLC, 39–50
Cetuximab, 33, 500
CF
  MPO, 123

CGH. *see* Comparative Genomic Hybridization (CGH)
Chardack, 469
CHART (Continuous Hyperfractionated Accelerated RT), 493
CHART weekend less, 493
CHART-WEL (CHART weekend less), 493
Chemoprevention
  efficiency
    against tobacco smoke, 745
  lung cancer, 739–749
    future direction, 748–749
    method, 742–748
    rationale, 772–776
    regimens, 744
  therapeutics, 771–777
    future directions, 776–777
Chemoresistance
  p53 alterations
    correlation, 65
Chemotherapeutic agents
  p53, 64–66
Chemotherapy
  sequencing, 490–492
Chest-wall resection
  extended indications, 467–473
Chlorins, 510
Cholecystokinin B receptor (CCKBR), 336
Chondroitin sulfate family
  inhibitory factor, 547
Chondroitin sulfate proteoglycans and glycosaminoglycans (CS-PG.GAG), 546
Chromosomal aberration detection
  CAD, 145–159
  FISH, 145–159
  material, 148–150
  method, 150–156

Chromosomal abnormalities, 16–18
  MA, 30–31
Chromosomal imbalances
  CGH, 227
  determination, 228f
Chromosomal instability (CIN), 251–252
Chromosome
  short arm loss, 30
Chromosome 1-22
  identification details, 232–233
Chromosome identification
  details, 232–233
  scheme, 231f
Chromosome preparations
  peripheral blood
    material, 149
    method, 150–151
Chromosome X and Y
  identification details, 233
Chromosome Y
  identification details, 233
Churchill, 460–461
Cigarettes
  benzo(a)pyrene, 6
  filter-tip, 6
  molecular damage, 5–7
  NNK, 6
  tobacco specific carcinogens
    p53, 58–59
  unfiltered, 6
CIN, 251–252
CIS, 507
  PDT, 515–516
Cisplatin, 500, 639
Clara cell secretory protein, 356
Classical cell-cycle model, 7, 8f
CMV, 83
c-Myc, 15–16

Coke ovens, 6
Coleman, 467
Comparative Genomic
    Hybridization (CGH), 194,
    209–223
  analysis
    flow chart, 210
    lung cancer, 211–216
    material, 216–218
    method, 218–229
Comparative multiplex PCR, 291–303
  description, 293
  material, 294
  method, 296–298
Computed tomography (CT), 466
  chest scan
    low-dose spiral, 4
    guided percutaneous intraluminal
      injection, 593
    pulmonary KS, 100
  spiral
    dyspnea, 100
Conservative surgery
  concept, 469–470
Continuous Hyperfractionated
    Accelerated RT, 493
Counterstaining
  biological parameters by FNA
    material, 408
    method, 412
Coupling products to microtiter
    plate
  point EXACCT
    material, 309
    method, 313
COX-2. *see* Cyclooxygenase-2
    (COX-2)
C-raf-1 antisense (ISIS 5132),
    626–627

Cryotom microdissection
  DNA extraction, 220f
CS-PG.GAG, 546
CT. *see* Computed tomography (CT)
CTL, 711
Culturing adherent lung cancer cells
  material, 675
  method, 677
Cutaneous KS
  clinical evaluation, 101–103
Cyclin-dependent kinase (Cdk), 7
  inhibitors, 8, 534
    p16, 592
Cyclin D1 gene
  allele-specific expression
    analysis, 302f
Cyclooxygenase-2 (COX-2), 776
  dependent regulation antitumor
    immunity, 723–733
    material, 725–728
    method, 728–733
  sense and antisense transfected
    3LL cells
    material, 725–726
    method, 728
Cylindrical light diffusing fibers,
    513–514
Cystic fibrosis (CF)
  MPO, 123
Cytochrome P450 family, 122
Cytokine
  from spleen
    measurement, 717
Cytokine-encapsulated microspheres
  tumor vaccination, 682–693
    material, 690–691
    method, 691–693

Cytokine gene-modified dendritic
  cells
    intratumoral therapy, 711–720
    material, 713–714
    method, 714–719
Cytokine induction
  evaluation
    material, 564
    method, 568–570
Cytokine-specific ELISA
  intratumoral therapy, 717–718
Cytomegalovirus (CMV), 83
Cytospin smears
  FNAB, 414f–415f
Cytotoxic chemotherapeutic agents
  pulmonary KS, 104–105
Cytotoxicity assay
  MAA-PEI
    material, 578
    method, 581
Cytotoxic T lymphocytes (CTL), 711
Cytoxan, adriamycin, and cisplatin
  (CAP), 477

D

DAB, 416–417
DAPI
  CGH, 224–227
Davies, Morriston, 455
DC, 711
DD
  material, 241
  method, 242–243
Defective tumor suppressor genes
  replacement, 531–534
Denaturation
  point EXACCT
    material, 310
    method, 313–314

Denaturing gradient gel
  electrophoresis (DGGE)
    p53 mutation detection, 55–56
Dendritic cells
  transduction
    material, 713–714
    method, 715–716
Dendritic cells (DC), 711
Detection systems
  immunohistochemistry, 374
Detection threshold
  calculation, 256–257
Dexamethasone, 746–748
DGGE
  p53 mutation detection, 55–56
Dialysis bag
  preparation and filling
    material, 759
    method, 762–763
Diaminobenzidine (DAB), 416–417
Differential display (DD)
  material, 241
  method, 242–243
Differential expression
  in blood samples
    material, 241
    method, 242–243
Digital image analysis
  CGH, 224–227
Dihydrofolate reductase
  MDR, 40
Dimethylene blue (DMMB) assay
  MPE
    material, 550–551
    method, 554–555
Dioleoyl phosphatidylethanol amine
  (DOPE), 774
Dipalmitoyl ethylphosphocholine
  (DPEP), 774

Direct inspection of lesions
  pulmonary KS, 102
Disease stages
  cumulative survival curves, 465f
  five-year survival rates, 462t
  localized and nonlocalized,
    463–464
Distant metastasis, 476t
DMMB assay
  MPE
    material, 550–551
    method, 554–555
DNA. *see also* cDNA
  damage, 9
DNA detection
  CGH
    material, 218
    method, 224
DNA extraction
  cryotom microdissection, 220f
  fluorescent microsatellite analysis
    material, 253–254
    method, 254–255
  Southern blotting of genomic
    DNA
    material, 266–267
    method, 269–271
DNA fragmentation
  immunohistochemical dual
    analysis
    material, 676–677
    method, 680–681
DNA isolation
  target identification tools
    material, 294
    method, 295
DNA linearization
  RNA *in situ* hybridization
    material, 165–166
    method, 168

DNA preparation
  CGH
    material, 217
    method, 220–222
  mutant enriched PCR-RFLP
    material, 327
    method, 328–329
DNA replication, 9
  MDR, 40
DNA sequencing
  LiCOR sequencer
    SSH, 202–203
DNA staining
  multiparameter analysis
    material, 440
    method, 441
DNA transfer to membrane
  Southern blotting of genomic DNA
    material, 267–268
    method, 272
DOPE, 774
Dosage regimen
  p53 gene therapy
    material, 610
    method, 611–612
Dose cloud, 494
Doxorubicin, 500, 773
  pulmonary KS, 104
DP3, 774
DPEP, 774
Drug resistance
  measured, 50

E

Early detection
  value, 239
Eastern Cooperative Oncology
  Group (ECOG), 490, 492–493
EBV, 83
ECOG, 490, 492–493

E2F, 9
EGF, 498
EGFR, 33, 500
18F-fluorodeoxyglucose (FDG)-
    PET, 499
Electron-beam radiation therapy
    pulmonary KS, 103–107
ELISA. *see* Enzyme linked
    immunosorbent assay (ELISA)
EMS1, 293–298
EMSF1/R1, 298f
En bloc mediastinal lymph node
    dissection, 463–464
En bloc resection
    superior sulcus tumor, 469
Endemic African KS
    epidemiology and clinical
        characteristics, 81t
Endonuclease digestion of PCR-
    RFLP products
    mutant enriched PCR-RFLP
        material, 328
        method, 330
Environmental carcinogens, 10–11
Enzyme linked immunosorbent
    assay (ELISA)
    MAA-PEI
        material, 577–578
        method, 580–581
    oligonucleosomal NDA
        fragmentation
        material, 676
        method, 680
Epidemic HIV-associated KS
    epidemiology and clinical
        characteristics, 82t
Epidermal growth factor (EGF), 498
Ethnic groups
    MPO polymorphism
        geographic studies, 126

Ethyol, 496–498
Etoposide
    pulmonary KS, 104
Evaluating drug resistance
    methods, 43–47
Event-triggering measurements
    evolutionary sequences, 437
Evolutionary sequences
    methodology, 437–439
Exonuclease digestion
    graphic display, 318f
Exploratory thoracotomy, 460
External beam radiation therapy
    pulmonary KS, 103–107
Extrapulmonary neuroendocrine
    tumors
    TTF-1 expression, 362t

F

FDG-PET, 101, 499
FHT (fragile histidine triad) gene, 30
Fibroblasts with adenoviral 5 vector
    transduction
    material, 713–714
    method, 715–716
Fibronectin
    TIA
        material, 550
        method, 554
Field cancerization, 16
Filter-tip cigarettes, 6
Fine-needle aspirates (FNA)
    biological parameters
        determination
        material, 407–409
        method, 409–415
FISH. *see* Fluorescence *in situ*
    hybridization (FISH)
FITC images
    CGH, 224–227

FITC-labeled oligonucleotide, 661
5-fluorouracil, 773
Five-year survival
   lung cancer, 463t
Flow cytometric DNA analysis
   biological parameters
      determination by fine-needle aspirates
         material, 408–409
         method, 413–415
   multiparameter analysis
      method, 441
Fluorescence *in situ* hybridization (FISH)
   advantages, 146–147
   application, 147–148
   CAD, 145–159
      material, 148–150
      method, 150–156
   cytogenetic assay
      material, 149
      method, 151–152
   direct labeling
      material, 150
      method, 155–156
   indirect labeling
      material, 149–150
      method, 152–154
   principles, 146
   slide denaturation, 152
      material, 149
   slide preparation
      material, 149
      method, 152
Fluorescent microsatellite analysis
   bronchial lavage, 251–259
      material, 253–254
      method, 254–257
   DNA extraction
      material, 253–254
      method, 254–255
   PCR amplification and analysis
      material, 254
      method, 255–256
Fluorodeoxyglucose (FDG)-PET, 499
Fluorouracil, 773
FNA
   biological parameters determination
      material, 407–409
      method, 409–415
Folch extraction
   material, 759
   method, 762
Forced expiratory volume in 1 s (FEV1), 518
Forced vital capacity (FVC), 518
Formalin-fixed paraffin-embedded tissues
   immunohistochemistry, 369–383
      material, 377–378
      method, 379–382
   molecular biologic substaging
      material, 377–378
      method, 379–382
   stage 1 NSCLC
      material, 377–378
      method, 379–382
Fragile histidine triad gene, 30
Fruitless thoracotomy, 466
Functional imaging, 499
FVC, 518

G

GAG
   TIA
      material, 550
      method, 554

Index

Gallium-thallium radionuclide imaging
  AIDS, 100–101
GAP, 32
Gapmer oligonucleotides, 657
Gas chromatography (GC)
  dialysis, 767
Gastrin-releasing peptide (GRP), 336
GC
  dialysis, 767
G-CSF, 105
GDP, 32
Gel electrophoresis
  bcl-2 protein
    material, 675
    method, 679
  material, 674–675
  method, 678
  target identification tools
    material, 294
    method, 298
Gel filtration analysis, 547
GEMTools Software (Incyte), 178
GEN231, 630
Gender effects
  MPO, 126–127
Gene-environmental interactions
  MPO, 128–130
Genescan output
  ABI PRISM 377 sequencer, 253f
Gene therapy, 529–538
  field, 592
Genetic aspects, 3–19
Genetic evolutionary sequences
  reconstruction, 445f
Gene-transfer efficiency
  transduction protocols and analysis
    material, 552–553
    method, 556–557

Gene walking, 672
Genomic instability, 251–252
Geno-phenotypic evolutionary sequences
  reconstruction, 431–447
Genotype images
  demonstrating LOH, 258f
Glass chip-based analysis, 178
Glucocorticoid budesonide
  synthetic, 748
Glutathione-S-transferases
  MDR, 40
Glycosaminoglycans (GAG)
  TIA
    material, 550
    method, 554
GM-CSF, 537
  KS, 91
Graham, Everts, 458–459, 458f, 461
Granulocyte-colony stimulating factor (G-CSF), 105
Granulocyte macrophage-colony stimulating factor (GM-CSF), 537
  KS, 91
Green fluorescent protein detection
  MAA-PEI
    material, 577
    method, 580
Green tea, 743
Gross tumor volume (GTV), 494, 499
Growth regulating proteins
  NSCLC, 370
GRP, 336
GTP, 32
GTPase-activating protein (GAP), 32
GTV, 494, 499
Guanosine diphosphate (GDP), 32
Guanosine triphosphate (GTP), 32

H

HAART, 79
Haight, 457–458
Hare, 469
Harvesting lungs
  histological analysis
    material, 563–564
    method, 566–568
Harvesting lymphocyte cells, 284–285
Hawaiian Tumor Registry
  MPO, 124–126
HCAs
  MPO, 122
Heat-induced denaturation
  graphic display, 318f
Heat-shock proteins
  MDR
    cellular predictive, 41
Hemoptysis
  pulmonary KS, 96
HER2/NE gene
  MA, 33
Herpes simplex virus thymidine
    kinase (HSV TK) gene, 530–531
Heterocyclic amines (HCAs)
  MPO, 122
HFTR, 492–493
HIF-1, 18–19
Highly active antiretroviral therapy
  (HAART), 79
High-pressure liquid
  chromatography (HPLC)
  dialysis, 767
High resolution computed
  tomography (HRCT), 471
High-throughput methodology
  cDNA library construction
    material, 178
    method, 179–180

cDNA preparation, 182
cDNA target amplification
  material, 179
  method, 182
identifying molecular targets
  over-expressed, 177–187
LSCC specific molecular
  markers, 183
posthybridization analysis, 183
probe synthesis, 179
reference genes, 179
subtracted cDNA libraries
  material, 179
  method, 180–182
Hilus
  left lung, 460f
Histological analysis
  harvesting lungs
    material, 563–564
    method, 566–568
Histologic classification, 473
HIV. *see* Human immunodeficiency
  virus (HIV)
Hoarseness
  pulmonary KS, 96
Homosexual men, 79, 83
Hormone response element
  MPO, 123
Hounsfield Units (HU), 471
HPLC
  dialysis, 767
HPV, 83
H-Ras. *see also* Ras protein
  antisense, 628
HRCT, 471
HRE (hormone response element)
  MPO, 123
HSA-PEI. *see* MAA-PEI
HSV TK gene, 530–531
hTERC, 163
HU, 471

Human immunodeficiency virus
   (HIV), 79
 epidemiology and clinical
   characteristics
     associated KS, 82t
     negative KS, 82t
   related KS, 96
Human papilloma virus (HPV), 83
Human serum albumin
 HSA, 576
Human tumors
 analysis, 431–447
Humoral immune responses
 MAA-PEI
   material, 577–578
   method, 580–581
Humoral mucosal immune
 MAA-PEI
   example, 586f–587f
Hutl instrument, 464
Hyaluronidase, 546
Hybridization
 CGH
   material, 218
   method, 222–224
 point EXACCT
   material, 309
   method, 313
 SCLC detection
   material, 342
   method, 347–348
 Southern blotting of genomic
   DNA
   material, 268–269
   method, 273–274
Hyperfractionated radiotherapy
 (HFTR), 492–493
Hyperfractionation, 492–493
Hypoxia inducing factor (HIF-1),
 18–19

I

IASLC
 consensus statement, 478
Iatrogenic KS, 80, 83
IF
 calculation
   schematic representation, 256f
IFN-γ
 KS, 91
Igepal CA-630 nonionic detergent,
 417
IGF. *see* Insulin-like growth factor
 (IGF)
IGFBP-3
 isolation and measurement
   material, 280
   method, 281–284
IL. *see* Interleukin (IL)
Image capture
 CGH, 224–227
Imbalance factor (IF)
 calculation
   schematic representation, 256f
Immune systems
 role, 509–512
Immunoblotting
 PKA subunit RIα, 646–647
Immunocytochemical staining
 p53 biological parameters by
   FNA
   material, 408
   method, 412
Immunocytochemistry
 biological parameters by FNA
   material, 408
   method, 410–413
Immunoenzymatic staining, 416
Immunofluorescence assay
 detection, 389–394

p53 detection
   material, 390–391
   method, 391–393
Immunofluorescence staining
   multiparameter analysis
      material, 439–440
      method, 440–441
Immunoglobulin extraction
   MPE
      material, 551
      method, 555
Immunohistochemical dual analysis
   DNA fragmentation and Bcl-2
      protein expression
         material, 676–677
         method, 680–681
Immunohistochemical staining
   SPR1 detection
      material, 398
      method, 399–400
Immunohistochemistry
   on formalin-fixed paraffin-
      embedded tissues, 369–383
      material, 377–378
      method, 379–382
   MDR, 44–47
   p53
      material, 611
      method, 613
      mutations, 59, 63
   PEI-DNA
      material, 564
      method, 568
   as tool, 372–373
Immunoprecipitation
   CAR from MPE
      material, 551
      method, 555–556

Immunostaining
   Bcl-2
      FNAB, 414f–415f
   Ki67
      FNAB, 414f–415f
   methods
      biological parameters
         determination, 411f
      paraffin sections, 361
   p53
      FNAB, 414f–415f
   TTF-1
      utilization, 355–364
Immunotherapy, 499, 535–537
Incyte, 178
Induction therapy, 478
Inhalational chemotherapeutics, 773
Inhalational cytokines, 773
Inhalational gene therapy, 774
Inhalation therapies, 771–777
Insulin-like growth factor (IGF), 279–287
Insulin-like growth factor-1 (IGF-1), 7
   isolation and measurement
      material, 280
      method, 281–284
Insulin-like growth factor-2 (IGF-2)
   isolation and measurement
      material, 280
      method, 281–284
Interassay variation
   calculation, 256–257
Interferon-γ (IFN-γ)
   KS, 91
Interleukin 1β (IL-1β)
   vs TNF-α, 570f
Interleukin (IL)
   KS, 91

Interleukin-2 (IL-2), 773–774
Interleukin-10 (IL-10), 728
  splenocyte
    material, 727
Interleukin-12 (IL-12), 728
  splenocyte production
    material, 727
    method, 731–732
International Association for the Study of Lung Cancer (IASLC) consensus statement, 478
Interphase FISH
  CAD, 145–159
Intracellular cytokine analysis
  DC, 718–719
Intrapericardial procedure, 462
Intrapleural malignancy, 545
Intrathoracic KS
  radiographic characteristics, 97–100
Intratumoral therapy
  cytokine gene-modified dendritic cells, 711–720
    material, 713–714
    method, 714–719
    histology, 719
In vitro postdrug treatment
  proliferative assay, 43–44
In vivo chemotherapy
  PKA subunit RIα
    in animals, 647–649
Ionizing radiation, 6
ISIS 2503, 628
ISIS 5132, 626–627

J

JNK4 family, 8
Johns Hopkins Oncology Center, 537

K

Kaplan-Meier curves
  stage III adenocarcinoma, 45f
Kaposi, Mortiz, 80
Kaposi's sarcoma-associated herpesvirus (KSHV), 83
  KS growth, 92
Kaposi's sarcoma (KS), 79. *see also* Pulmonary Kaposi's sarcoma
  advanced
    AMC, 104
    liposomal anthracyclines, 106
    NIH, 104
  AIDS-associated, 79–107
    clinical manifestations, 95–97
    early and progressive, 98f–99f
  bacillary angiomatosis *Bartonella* organisms, 101
  bFGF, 91
  CD4+ count, 92–93
  cutaneous
    clinical evaluation, 101–103
  epidemiology, 80–84
  epidemiology and clinical characteristics
    benign nodular, 81t
    endemic African, 81t
    epidemic HIV-associated, 82t
    HIV-negative, 82t
  following lymphangitics, 87f
  GM-CSF, 91
  histopathology, 85–88
  HIV-related, 96
  iatrogenic, 80, 83
  IFN-γ, 91
  IL, 91
  incidence
    decline, 84
  interlobular septal involvement, 90f

intrathoracic
   radiographic characteristics, 97–100
lymphadenopathic African, 80
   epidemiology and clinical characteristics, 82t
Oncostatin M, 91
parenchyma, 101
pathogenesis, 88–92
pathology, 85–88
PDGF, 91
pleural involvement, 89f
renal transplantation, 83
revised ACTG staging classification, 94t
telangiectatic variant, 88f
TNF, 91
tracheobronchial tree, 101
transplant-related, 83
   epidemiology and clinical characteristics, 80
variants
   epidemiology and clinical characteristics, 81t–82t
VEGF, 91
visceral
   liposomal anthracyclines, 106
Warthin-Starry silver staining, 101
Karyotyping
   CGH, 231
Keyhole limpet hemocyanin (KLH), 536
Ki67 immunocytochemical staining
   biological parameters by FNA
      material, 408
      method, 410–412
Ki67 immunostained
   FNAB, 414f–415f
Kip1 family, 8

Kirsten (K)-ras oncogene. *see* K-ras
KLH, 536
K-ras. *see also* Ras protein
   activated oncogenic role, 13–15
   detection, 325–331
      material, 327–328
      method, 328–330
   gene
      probes sequences, 308t
   mutant-enriched PCR-RFLP
      material, 327–328
      method, 328–330
   mutations
      NSCLC, 137
   oncogene, 13
      detection, 305–321
   point EXACCT PCR
      material, 309
      method, 311–312
   point mutations
      analysis, 316
      detection, 305–321
KS. *see* Kaposi's sarcoma (KS)

L

Lavage procedure
   material, 758–759
   method, 761
LCSG
   limited resection, 461
Leroux, 461
Levamisole immunotherapy, 477
LiCOR sequencer
   DNA sequencing
      SSH, 202–203
Li-Fraumeni cancer syndrome, 33
Ligation step
   point EXACCT
      material, 309
      method, 313

Light dosimetry, 512–513
Limited resection
  concept, 460–461
Linear and circular suturing, 465
Linearization of plasmid DNA
  RNA *in situ* hybridization
    material, 165–166
    method, 168
Linear staplers, 465
Lipofectin
  oligos, 651
Lipofectin-mediated oligonucleotide transfection and combination
    material, 640
    method, 644–645
Liposomal anthracyclines
  advanced/visceral KS, 106
Liposomal anthracyclines with BV
  pulmonary KS, 106
3LL tumor model
  COX-2
    material, 726–728
    method, 730–733
LM-6 osteosarcoma model
  material, 610
  method, 611
Lobectomy, 461
  landmark, 455
  lymph node dissection, 463–464
  techniques, 456
LOH. *see* Loss of heterozygosity (LOH)
Loss of heterozygosity (LOH), 251–253
  analysis factors, 264
  calculation, 256–257
  Genotype images, 258f
Low-dose spiral computed tomography chest scan, 4
LPHH1, 293–298, 299f

LRP, 40
L514S
  differential expression, 186f
L519S
  differential expression, 186f
L530S
  differential expression, 186f
L531S
  differential expression, 186f
L-selection^{low} cells
  isolation from TDLNs, 705–706
Luciferase assay
  MAA-PEI
    example, 582f
    injection rate, 585f
    material, 577
    method, 580
    N:P ratio, 583f
    plasmid NDA dose effect, 584f
  PEI-DNA
    material, 563
    method, 566
Lung cancer
  mortality, 5
Lung Cancer Study Group (LCSG)
  limited resection, 461
Lung KS, 95
Lung resection
  extended indications, 467–473
  first, 454
  pioneers, 455
Lung resistance-related protein (LRP), 40
Lung squamous tumor
  scatter plot of fluorescence intensities, 185f
Lung surfactant isolation
  material, 758–760
  method, 760–762

Lung tissue
  CAD, 145–159
Lymphadenopathic African KS, 80
  epidemiology and clinical
    characteristics, 82t
Lymph node
  collection and processing
    material, 338–339
    method, 342
  RNA extraction
    material, 340
    method, 344

M

MA, 251–252
MAA. *see* Macroaggregated
  albumin (MAA)
MAb, 356
Macallum, 469
Macroaggregated albumin
  (MAA), 576
Macroaggregated albumin-
  polyethylenimine conjugate
  (MAA-PEI), 575
  material, 576–578
  method, 578–581
  plasmid binding
    material, 577
    method, 579–580
  targeted delivery of expression
    plasmids
    material, 576–578
    method, 578–581
Magnetic resonance imaging (MRI),
  101, 466
Malignant pleural effusions (MPE),
  545
  viral gene therapy
    material, 549–553
    method, 553–557

Malignant transformation, 30
Manganese dismutase, 497
MAP kinase, 279–286
Mass median aerodynamic diameter
  (MMAD), 565, 571
Matrix metalloproteinase inhibitors
  (MMPIs), 499
Matrix metalloproteinase (MMP), 498
MBO
  PS-oligos
    material, 639–643
    method, 643–650
MCM proteins, 9
MDR. *see* Multidrug resistance
  (MDR)
Mediastinoscopy, 466
Membrane washing
  RNA *in situ* hybridization
    material, 168
    method, 170
Memorial Sloan Kettering Cancer
  Center, 536
Metallothonein
  MDR, 40
Metal mining, 6
Metaphase chromosome preparation
  CGH
    material, 216–217
    method, 218–219
Metastasis-related proteins
  NSCLC, 371–372
Methodological considerations
  CGH, 210
Microarray analyses
  poly A+ probes and reference
    guides, 184t
Microsatellite alterations (MA),
  251–252
Microsatellite instability (MIN),
  251–252

Microsomal epoxide hydrolase, 122
Microtiter plate coupling products point EXACCT
　material, 309
　method, 313
Microvessel density (MVD) assay, 47
MIN, 251–252
Minichromosome maintenance (MCM) proteins, 9
Mitogen-activated protein (MAP) kinase, 279–286
Mitogenic peptide hormones, 279–286
Mitomycin, vindesine, and cisplatin (MVP), 490
Mitosis (M) phase, 7
MMAD, 565, 571
MMP, 498
MMPIs, 499
Molecular abnormalities, 7–13, 54
　human solid tumors, 435
　model for determining sequence, 435f
　preferred sequence, 435f
Molecular alterations
　prognosis, 29–35
Molecular analysis
　p53 mutations, 59, 63
Molecular aspects, 3–19
Molecular-based strategies, 500
Molecular biological abnormalities, 29–35
Molecular biologic substaging, 369–383
　on formalin-fixed paraffin-embedded tissues
　　material, 377–378
　　method, 379–382
　immunohistochemistry, 369

Molecular epidemiology
　MPO, 124–130
Molecular fingerprinting, 498
Molecular markers
　in blood samples
　　material, 241–242
　　method, 242–248
　bronchial carcinogenesis, 397–402
Molecular targets over-expressed identification
　high-throughput methodology, 177–187
　material, 178–179
　method, 179–183
Monoclonal antibody (MAb), 356
Mortality
　lung cancer, 5
MPE, 545
　viral gene therapy
　　material, 549–553
　　method, 553–557
MPO. see Myeloperoxidase (MPO)
MRI, 101, 466
MRP, 40
MS
　MPO, 123
MTT assay
　PKA subunit RIα
　　material, 642
　　method, 644–645
MTT proliferation assay
　oligonucleotides
　　material, 659
　　method, 664–665
MTT proliferation assays
　A549 NSCLC cells, 665f
Multidrug resistance-associated protein (MRP), 40

Multidrug resistance (MDR)
  analysis
    comparison, 44
    detection, 46f
    family glycoprotein, 39–40
    individual factors, 47–48
    underlying causes, 39–41
Multiparameter cell-based
  measurements, 434–435
Multiparameter cell-based studies,
  432–433
Multiple radiation beams, 494
Multiple sclerosis (MS)
  MPO, 123
Multiplex assay
  PCR profile, 257t
Murine tumor models
  material, 714
Mutagen sensitivity, 279–286
  assay
    material, 281
    method, 284–285
Mutant-enriched PCR-RFLP
  detection, 325–331
  K-ras and p53 detection
    material, 327–328
    method, 328–330
Mutations
  dominant and recessive, 269
MVD assay, 47
MVP, 490
Myc, 592
  MA, 32
Myeloperoxidase (MPO), 571
  African-Americans, 124
  age effects, 127
  BPDE, 122
  carcinogenesis, 122–123
  gender effects, 126–127
  gene-environmental interactions,
    128–130
  genotypes
    lung cancer risk studies, 125t
    occupational exposures and
      lung cancer risk, 129
  HCAs, 122
  HRE, 123
  molecular basis, 122–124
  molecular epidemiology, 124–130
  PAHs, 122
  polymorphism, 123–124
    ethnic group geographic
      studies, 126
    race, 124
  promoter region polymorphism,
    121–130
  ROS, 122
Myeloperoxidase promoter region
  polymorphism
  lung cancer risk, 121–130
Myoinositol, 746–748
Myoinositol-dexamethasone
  chemopreventive effects, 747

N

N-acetylcysteine (NAC), 739
N-acetyltransferase (NAT), 122
National Cancer Institute (NCI)
  limited resection, 461
National Cancer Institute North
    American Lung Cancer Study
    Group (NCI-LGSG), 475–477
National Institute of Health (NIH)
  advanced KS, 104
  AIDS-related KS, 93–94
NCI
  limited resection, 461
NCI-LGSG, 475–477

Nd-YAG laser
  PDT, 518
  vs PDT, 519t
Necrotic debris
  removal, 514
Negative lymph nodes, 468
Neoplastic
  cellular environment, 30
  definition, 29–30
Nested PCR
  SSH, 200, 201f
Neuron restrictive silencer factor,
    335–350
  SCLC detection
    material, 338–342
    method, 342–348
Neuron restrictive silencer factor
    (NRSF or REST)
  splice variant, 336–338
  detection, 337f
Neuropeptide receptors
  SCLC detection, 335–350
    material, 338–342
    method, 342–348
Neuropeptides
  SCLC detection, 335–350
    material, 338–342
    method, 342–348
Nick translation
  CGH
    material, 217–218
    method, 222, 223f
NIH
  advanced KS, 104
  AIDS-related KS, 93–94
Nissen, 457–458
Nitrosamines, 6
NNK, 740, 776
  per cigarette, 6

Nonsmall cell lung cancer (NSCLC)
  factors predicting prognosis,
    369–372
  on formalin-fixed paraffin-
    embedded tissues
    material, 377–378
    method, 379–382
  future studies, 139–140
  immunohistochemistry, 369
  meta-analyses, 136–137
  molecular markers, 135–136
  publication bias, 137–138
  subset analyses, 138–139
  tumor markers, 135–139
Nonspecific antibody binding
  evolutionary sequences, 438–439
North American Lung Cancer Study
    Group
  limited resection, 461
Northern analysis
  COX-2
    material, 726
    method, 729
Northern blot analysis
  SSH, 203–204, 205f
N-Ras. *see* Ras protein
NRSF
  splice variant, 336–338
NSCLC. *see* Nonsmall cell lung
    cancer (NSCLC)
Nuclear apoptosis, 667
Nuclear morphology analysis
  oligonucleotides
    material, 659
    method, 667
Nuclear p53 accumulations, 57
Nucleic acid extraction
  material, 294
  method, 295–296

O

Occupationally related exposures
  lung cancer, 129
Oligonucleosomal NDA fragmentation
  ELISA-based assay
    material, 676
    method, 680
Oligonucleotide
  ligation assay
    point EXACCT, 317–318
  synthetic
    chemical modifications, 624f
Oncogene
  activation, 592
  cellular predictive factors, 47f
  inactivation, 534–535
  MA, 31–35
  pathways, 15–16
Oncostatin M
  KS, 91
Open lung biopsy
  pulmonary KS, 103
Oral candidiasis, 79
Origin recognition complex (ORC)
  proteins, 9
Osteosarcoma model
  LM-6
    material, 610
    method, 611
Overholt, 460–461

P

P14
  ARF, 14
P15
  genes
    MA, 34–35
    TGF-β, 13

P16, 10, 534
  cyclin-dependent-kinase-inhibitor, 592
  INK4A genes
    MA, 34–35
  tumor-suppressor pathway, 11–13
P18
  specific CTL elicited
    MAA-PEI-CMV-UB23, 588f
P21, 10, 11
P27, 534
P53, 7, 10, 592
  abnormalities, 433
  abrogation multiple functions
    model for sequential evolutionary consequences, 436f
  accumulations
    nuclear, 57
  Ad, 532
  aerosol gene therapy
    material, 609–611
    method, 611–613
  aerosolized gene delivery, 774–775
  alterations
    chemoresistance correlation, 65
    prognostic significance, 60t–63t
  antibodies, 66–67
  apoptosis, 64–66
  assay
    material, 611–612
    method, 613
  chemoresistance alterations
    correlation, 65
  chemotherapeutic agents, 64–66
  dosage regimen gene therapy
    material, 610
    method, 611–612
  genes
    MA, 33–34

gene therapy
  tumor inoculation, 610, 611–612
immunocytochemical staining
  material, 408
  method, 412
immunofluorescence assay
  detection
    material, 390–391
    method, 391–393
immunohistochemistry
  material, 611
  method, 613
immunostained
  FNAB, 414f–415f
molecular carcinogenesis, 9
mutant-enriched PCR-RFLP
  detection
    material, 327–328
    method, 328–330
mutation
  bronchial carcinogenesis, 57–58
  clinical implications, 53–68
  clinical prognostic studies, 59–64
  detection, 55–56, 325–331
  DGGE, 55–56
  dysplastic respiratory
    epithelium, 16
  immunohistochemistry, 59, 63
  lung cancer, 56
  molecular analysis, 59, 63
  NSCLC, 137
  sputum samples, 316
  SSCP, 55–56
nuclear accumulations, 57
PEI
  aerosol gene therapy, 607–616
protein
  detection, 389–394
  radiation, 64–66
  therapeutic response, 64–66

tobacco specific carcinogens,
  58–59
tumor inoculation gene therapy
  material, 610
  method, 611–612
wt, 611
PAbs, 356
Paclitaxel, 500, 773
PAHs, 6, 12
Pancoast, 469
PAP
  immunohistochemistry, 374–375
Particle size
  aerosol droplets, 561
Particle sizing
  equipment, 582
Particulate delivery systems
  assessing interaction, 755–758
  material, 758–760
  method, 760–766
Pathways
  activated, 30
  oncogene, 15–16
  p16
    tumor-suppressor, 11–13
  ras-MAP kinase, 14
  Rb
    molecular carcinogenesis, 9
Patient population statistics
  cellular factors, 42–43
PCD. *see* Programmed cell death
  (PCD)
PCNA, 11, 41
PCP, 79
PCR. *see* Polymerase chain reaction
  (PCR)
PDGF
  KS, 7, 91
PDT. *see* Photodynamic therapy
  (PDT)

PEI. *see* Polyethylenimine (PEI)
Penicillin
  introduction, 454
Peptide loading
  DC, 716
Percutaneous needle biopsy
  (PNB), 471
Peripheral blood
  collection and processing
    material, 338–339
    method, 342
  lymphocytes
    CAD, 145–159
  RNA extraction
    material, 339
    method, 343
Peroxidase-anti-peroxidase (PAP)
  immunohistochemistry, 374–375
PET, 466
PG
  TIA
    material, 550
    method, 554
PGE2. *see* Prostaglandin E2 (PGE2)
Phase inversion nanoencapsulation
  (PIN), 689
  microspheres characterization
    material, 690–691
    method, 691–692
  microspheres cytokine-
    encapsulated
    material, 690
    method, 691
Phenol-chloroform-extraction
  CGH, 221
  SSH, 194–195, 195f
Phorbol ester, 11
Phosphorothioate antisense bcl-2
  oligonucleotide
  synthetic, 673f

Phosphorothioate oligonucleotides
  (PS-oligos), 638–639, 656
Phosphorothioates, 656
Photochemical sensitization
  concept, 507–513
Photodynamic therapy (PDT)
  advanced endobronchial tumors,
    518–520
  clinical studies, 514–520
  integrating with radiation, 521
  in material multimodality
    approach
    role, 520–522
  mechanism of action, 507–521
  membrane injury, 508
  method and technique, 513–514
  vs Nd-YAG laser, 519t
  oxygen delivery, 508–509
  wavelength of light, 512
Photofrin, 507
Photosensitizers, 510–512
  biochemical considerations,
    510–511
  biophysical considerations,
    511–512
Phthalocyanines, 510
PIN. *see* Phase inversion
  nanoencapsulation (PIN)
PI 55 stapler, 467f
PKA. *see* Protein kinase A (PKA)
  subunit RIα
PKC
  signaling cascade, 11
Plasmid name
  content and use, 596t
Plasmid preparation
  SSH, 202
Platelet derived growth factor
  (PDGF), 7
  KS, 91

Pleural aspirates
  collection and processing
    material, 338–339
    method, 342
  RNA extraction
    material, 340
    method, 344
Pleuritic chest pain
  pulmonary KS, 96
PMN sequestration, 571
PNB, 471
Pneumocystis carinii pneumonia
  (PCP), 79
Pneumonectomy, 456–458, 461
Point-EXACCT
  detection, 305–321
Point mutations
  probes sequences, 308t
Polyclonal antibodies (PAbs), 356
Polycyclic aromatic hydrocarbons
  (PAHs), 6
  MPO, 122
Polyethylenimine (PEI), 562, 582
  DNA aerosol complexes, 561–571
    material, 562–564, 609
    method, 564–570, 611
    preparation, 609, 611
  DNA aerosol delivery
    material, 562–564
    method, 564–570
    model system, 562
  p53 complexes, 607–616
    material, 609–611
    method, 611–613
Polymerase chain reaction (PCR)
  amplification of cDNA
    material, 674
    method, 678
  ASA, 326

based mRNA analysis
  material, 674–675
  method, 675–677
comparative multiplex
  description, 293
  material, 294
  method, 296–298
fluorescent microsatellite analysis
  material, 254
  method, 255–256
FRLP analysis mutant enriched
  material, 328
  method, 330
p53 mutation detection, 55–56
product purification mutant
  enriched PCR-RFLP
  material, 328
  method, 330
pulmonary KS, 103
and *in situ* hybridization, 532
SSO, 326
Polymorphic metabolic genes
  susceptible to chemically induced
    cancer, 121
Polymorphonuclear leukocytes
  (PMN) sequestration, 571
Porfimer sodium, 507
Porphines, 510
Porphyrin-uptake studies, 508
Positron emission tomography
  (PET), 466
Positron emission tomography
  with 18-fluorodeoxyglucose
  (FDG-PET), 101, 499
Postdrug treatment proliferative
  assay, 43–44
Posthybridization analysis
  high-throughput methodology, 183
Potent radiosensitizer, 500
pRB family, 592

Pre-amplification
 mutant enriched PCR-RFLP
  material, 327–328
  method, 328–329
Preneoplastic changes
 bronchial epithelial cells, 16–18
Primary tumor, 476t
PRINS technique
 retroviral vector titration
  material, 595–597
  method, 598–600
Probe labeling
 RNA *in situ* hybridization
  material, 167
  method, 168–169
Probe synthesis
 high-throughput methodology, 179
Prognosis
 molecular alterations, 29–35
Programmed cell death (PCD). *see also* Apoptosis
 antisense Bcl-2 oligonucleotide, 671–682
  material, 673–677
  method, 677–681
 BCL-2, 33
Proliferating cell nuclear antigen (PCNA), 11, 41
Proliferation
 cellular predictive factors, 47f
Proliferative factors, 41–42
Promoter hypermethylation, 776
Promoter region polymorphism
 MPO, 121–130
Prostaglandin-endoperoxide synthase. *see* Cyclooxygenase-2
Prostaglandin E2 (PGE2), 728
 enzyme immunoassay COX-2
  material, 725–726
  method, 728

Prostaglandin H synthase. *see* Cyclooxygenase-2
Proteinase K digestion
 CGH, 221
Protein gel electrophoresis
 PKA subunit RIα, 646
Protein kinase A (PKA) subunit RIα
 MTT assay
  material, 642
  method, 644–645
 SDS-PAGE, 646
 Western blot analysis
  material, 641–642
  method, 645–647
Protein kinase C (PKC)
 signaling cascade, 11
Proteoglycan (PG)
 TIA
  material, 550
  method, 554
Proteolytic enzyme digestion
 immunohistochemistry, 373
Provitamin A, 739
PS-oligos, 638–639, 656
Pulmonary Kaposi's sarcoma
 ABV, 104–105
 AIDS
  early and progressive, 98f–99f
 AIDS-associated, 79–107
 alveolar hemorrhage, 102
 anthracyclines, 104
 bleomycin, 104
 chest radiographic features, 97t
 clinical evaluation, 101–103
 cytotoxic chemotherapeutic agents, 104–105
 direct inspection of lesions, 102
 doxorubicin, 104
 electron-beam radiation therapy, 103–107

etoposide, 104
external beam radiation therapy,
    103–107
hemoptysis, 96
hoarseness, 96
incidence, 84–85
liposomal anthracyclines with
    BV, 106
natural history, 92–95
open lung biopsy, 103
PCR, 103
pleuritic chest pain, 96
radiation therapy, 103–107
radiologic findings, 97–101
staging system, 97
stridor, 96
systemic chemotherapy, 104–105
Taxol, 106
thoracic irradiation, 104
treatment, 103–107
vinca alkaloids, 104
wheezing, 96
without mucocutaneous
    involvement, 85
Pulmonary parenchyma Kaposi's
    sarcoma, 101
Putative markers
  identification
    material, 241
    method, 242–243

Q
Quality-adjusted survival (Q-time),
    496–498

R
Racial differences, 6
Radiation
  p53, 64–66
Radiation dose escalation, 495

Radiation therapy
  pulmonary KS, 103–107
Radiation Therapy Oncology Group
    (RTOG) trial, 489–492
Radical pneumonectomy, 463–464
  concept, 461–462
Radio-labeled probe preparation
    material, 341–342
    method, 347
Radiolabeling DNA
  Southern blotting of genomic
      DNA
    material, 268
    method, 272–273
Radiotherapeutic management
  NSCLC, 489–500
    future approaches, 499
Radiotherapy (RT)
  accelerated, 493
  sequencing, 490–492
  thoracic, 489
  three-dimensional conformal,
      493–496
Radon gas, 6
Rare event imaging system
  body fluids, 423–428
    material, 424–425
    method, 425–428
  equipment picture, 427
Ras, 592
  gene
    MA, 32
    MAP kinase pathway, 14
    protein, 14 (*see also* K-ras)
  proteins
    cellular predictive factors, 47f
RATIO image
  CGH, 224–227
Rb. *see* Retinoblastoma (Rb)

Reactive oxygen species (ROS)
  MPO, 122
Real digest
  SSH, 198
Real time PCR, 240
  tumor marker identification
    material, 242
    method, 245–248
    threshold cycle, 246, 247f
Recurrence rates, 462
Reducing toxicity, 496–498
Reference genes
  high-throughput methodology, 179
Regional lymph nodes, 476t
Reinhoff, 459–460
Relative optical density, 315f
Renal transplantation
  KS, 83
Repair enzymes
  MDR, 40
Replication errors (RER), 251–252
Reporter gene assays
  harvesting lungs
    material, 563
    method, 566
RER, 251–252
Resistance
  to chemotherapy, 39
Resistance factors
  sensitivity and specificity
    analysis, 49t
Respiratory symptoms
  AIDS, 96
REST
  splice variant, 336–338
Restriction enzyme digest
  Southern blotting of genomic
    DNA
      material, 267
      method, 271

Retinoblastoma (Rb), 7
  family, 592
  gene
    MA, 34
  gene family, 592
  pathways
    molecular carcinogenesis, 9
Retorviral vectors
  cell-cycle-targeted gene therapy
    material, 594–597
    method, 597–600
Retroviral gene transfer
  MPE, 546–547
Retroviral vectors
  cell-cycle-targeted gene therapy,
    591–602
Retrovirus production, 594
Reverse transcription-PCR (RT-PCR)
  material, 674
  method, 677–678
  primer sequences, 341t
  RFLP analysis target
    identification tool
      material, 295
      method, 300
  SCLC detection, 335–350
    material, 338–342
    method, 342–348
  SPR1 detection
    material, 399
    method, 401
Reverse transcription real-time
    polymerase chain reaction
    oligonucleotides
      material, 657–658
      method, 660–662
Rhodamine, 510
Right angle Wertheim hysterectomy
    clamp, 456

RNA extraction
    SSH
        material, 192–193
        method, 194–196
    target identification tools
        material, 294
        method, 295–296
RNA *in situ* hybridization
    flow chart, 167f
    material, 165–168
    method, 168–171
    probe development, 166f
RNA quantification
    RT-PCR
        material, 340
        method, 344
ROS
    MPO, 122
RT. *see* Radiotherapy (RT)
RTOG trial, 489–492
RT-PCR. *see* Reverse transcription-PCR (RT-PCR)

S

SCLC. *see* Small cell lung cancer (SCLC)
Scope, 468f
SDS-PAGE
    Ad, 548–549
    PKA subunit RIα, 646
Sensitive assays
    lung cancer detection, 239–248
        material, 241–242
        method, 242–248
Sequence-selective oligonucleotide primers, 326
Sequence-specific conformational polymorphism, 326
Serum albumin
    phosphorothioate oligonucleotides, 656

Shushruta, 464
Signal-transduction inhibitors, 499
Silica, 6
Single cell analysis
    drawback, 444
Single-strand conformational polymorphism
    p53 mutation detection, 55–56
Single-stranded target DNA preparation
    point EXACCT
        material, 309
        method, 312–313
Sleeve lobectomy, 470
Small cell lung cancer (SCLC) detection
    neuropeptide receptors, 335–350
    neuropeptides, 335–350
    RT-PCR, 335–350
Small proline-rich protein (SPR1), 123, 397–402
Smelting, 6
Smoking. *see* Cigarettes
sNRSF splice variant
    detection, 337f
Solubilization of solutes in surfactant
    material, 759
    method, 763–764
Solution hybridization
    point EXACCT
        material, 309
        method, 313
Southern blotting
    genomic DNA, 263–276
    material, 263–269
    method, 269–273
    nylon membrane suppliers, 275

Southern blotting of genomic DNA
  agarose gel electrophoresis
    material, 267
    method, 271
  DNA extraction
    material, 266–267
    method, 269–271
  DNA transfer to membrane
    material, 267–268
    method, 272
  hybridization
    material, 268–269
    method, 273–274
  radiolabeling DNA
    material, 268
    method, 272–273
  restriction enzyme digest
    material, 267
    method, 271
South West Oncology Group
  (SWOG), 491
Specimen collection
  biological parameters
    determination by fine-needle
      aspirates
    material, 407
    method, 409–410
S-phase genes, 7, 9
Spindle tumor cells, 86f
Spiral CT scanning
  dyspnea, 100
Splenocyte IL-10 and IL-12
  production
    material, 727
    method, 731–732
SPR1 detection
  immunohistochemical staining
    material, 398
    method, 399–400
SPs, 356

SP1-transcription factor
  MPO, 123
Sputum
  cytology, 4
  DNA extraction
    fluorescent microsatellite
      analysis, 255
    K-ras point mutations, 305–321
  SSCP, 55–56, 326
  SSH. *see* Suppression subtraction
    hybridization (SSH)
Stage III adenocarcinoma
  Kaplan-Meier curves, 45f
Stage of disease
  cumulative survival curves, 465f
  five-year survival rates, 462t
  localized and nonlocalized,
    463–464
Staging classification, 473
Staining
  point EXACCT
    material, 310
    method, 314
Standard sleeve resection, 471f
Stoichiometric interaction of
  fluorochrome, 417
Streptavidin-biotin-peroxidase
  labeling
    paraffin sections, 361
Streptomyces, 546
Streptomycin
  introduction, 454
Stridor
  pulmonary KS, 96
Subtracted cDNA libraries
  high-throughput methodology
    material, 179
    method, 180–182
Subtraction-efficiency-test
  SSH, 201

Subtractive methodology, 177–178
Suicide genes, 530–531
Superior sulcus tumors, 469
Suppressing agents, 741
Suppression subtraction
  hybridization (SSH)
  expression profiling, 189–206
  material, 192–194
  method, 194–204
  flow chart, 190
  lung cancer, 191–192
Surfactant B promoter, 534
Surfactant proteins (SPs), 356
Surgical staplers, 464–466
Surgical treatment
  1880-1928, 455
  1929-1933, 456–458
  1933, 458–459
  1933-1956, 459–466
  1957-2001, 473–475
  evolution, 454f
  future, 480–481
  lung cancer, 453–481
  present, 479–480
Survivin
  cell death, 656
SWOG, 491
Synthesis phase, 7, 9
Synthetic glucocorticoid
  budesonide, 748
Synthetic oligonucleotides
  chemical modifications, 624f
Synthetic phosphorothioate
  antisense bcl-2
  oligonucleotide, 673f
Systemic chemotherapy
  pulmonary KS, 104–105
Systemic immunosuppression, 83
Systemic therapy model, 717

T

TAAs, 535, 536
Targeted delivery of expression
  plasmids, 575–588
  MAA-PEI
  material, 576–578
  method, 578–581
Target identification tools, 291–303
  material, 294–295
  method, 295–301
Targeting RIα subunit of protein
  kinase A
  antisense oligonucleotides
  (AONs), 637–651
Taxol, 639
  pulmonary KS, 106
Telomerase, 16–18
  activation
    MA, 30–31
    regulation, 164f
  analysis, 164t
  associated genes
    sequences, 165t
  catalytic component gene, 163
  RNA component, 16
  RNA gene expression
    material, 165–168
    method, 168–171
  in situ analysis, 163–175
  sequences, 165t
Telomerase reverse transcriptase
  (TERT) component, 16
Telomeres, 16–18
  analysis, 164t
Telomere-telomerase hypothesis, 16
Template DNA preparation
  point EXACCT
  material, 307–309
  method, 310–311

TERT component, 16
Tetraphenylporphine sulfate, 510
TF
  MPO, 123
TGF-β
  induces p15, 13
Therapeutic oligonucleotide
  clinical development, 626–630
Thoracic irradiation
  pulmonary KS, 104
Thoracic radiotherapy, 489
Thoracic Surgery
  organization, 454
3LL tumor model
  COX-2
    material, 726–728
    method, 730–733
Thyroid transcription factor-1
  (TTF-1) immunostaining
  adenocarcinoma, 359f
  diagnostic applications, 359–361
  expression, 356–358
  immunohistochemical detection, 357t–358t
  SCLC reacting, 360
  utilization, 355–364
    material, 361–362
    method, 362–363
Tissue fixation
  immunohistochemistry, 372
TNF
  KS, 91
TNF-α
  vs IL-1β induction, 570, 570f
TNM (Tumor, Metastasis, Nodes)
  descriptors
    categories, 473
  subsets
    stage grouping, 135, 476t, 477t

Tobacco specific carcinogens. see also Cigarettes
  p53, 58–59
Topical gene therapy
  PEI-DNA aerosol complexes, 561–571
    material, 562–564
    method, 564–570
Tracheobronchial tree
  KS, 101
Trafficking studies
  Ad-GFP transduced dendritic cells, 719
Transactivation domain, 11
Transcription factor (TF)
  MPO, 123
Transducing signal, 30
Transduction inhibition assays
  MPE
    material, 549–553
    method, 553–557
Transforming growth factor-β (TGF-β)
  induces p15, 13
Transient transfection
  retroviral vector production
    material, 594–595
    method, 597–598
Transplant-related KS, 83
  epidemiology and clinical characteristics, 80
TRAP assay, 18
TRITC images
  CGH, 224–227
TRIzol extraction
  SSH, 195–196
TTF-1. see Thyroid transcription factor-1 (TTF-1) immunostaining

Tumor, Metastasis, Nodes
  descriptors, 135, 473, 476t, 477t
Tumor-associated antigens (TAAs),
  535, 536
Tumor collective
  CGH results, 228–229
Tumor-draining lymph nodes
  adoptive immunotherapy, 698
  material, 699–700
  method, 700–708
Tumor inoculation
  p53 gene therapy
  material, 610
  method, 611–612
Tumor-invasion and metastasis-
  related proteins
  NSCLC, 371–372
Tumor markers
  NSCLC, 135–140
Tumor necrosis factor-α (TNF-α)
  vs IL-1β induction, 570, 570f
Tumor necrosis factor (TNF)
  KS, 91
Tumor-node-metastasis (TNM)
  staging system
  categories, 473
  NSCLC, 135
  subsets
    stage grouping, 476t, 477t
Tumor suppressor gene, 607
  inactivation, 592
  isolation, 264
  MA, 31–35
Tumor suppressor proteins, 534
Tumor vaccination
  cytokine-encapsulated
    microspheres, 682–693
  material, 690–691
  method, 691–693

U
Unfiltered cigarettes, 6

V
Vascular endothelial growth factor
  (VEGF), 18–19, 91, 499, 532
  KS, 91
Vascular systems
  role, 509–512
VATS, 470–472
Vector technology
  gene transfer, 529–530
Video-assisted thoracic surgery
  (VATS), 470–472
Vinblastine
  pulmonary KS, 104
Vinca alkaloids
  pulmonary KS, 104
Vincristine
  pulmonary KS, 104
Viral gene therapy, 545–557
  material, 549–553
  method, 553–557
Visualization
  lung cancer cells, 425f
Vitamin A, 739

W
Warthin-Starry silver staining
  KS, 101
Western blot analysis
  bcl-2 protein
    material, 675–676
    method, 678–680
  COX-2
    material, 726
    method, 728–730
  oligonucleotides
    material, 658–659
    method, 662–664

PKA subunit RIα
  material, 641–642
  method, 645–647
SPR1 detection
  material, 398–399
  method, 400–401

Wheezing
  pulmonary KS, 96
WHO/IASLC histologic
  classification, 474t
Wild-type CFTR gene transfer, 774
Wound closure
  method, 457

X

X-rays, 453